Topics in Differential Geometry

Topics in Differential Geometry

Peter W. Michor

Graduate Studies in Mathematics

Volume 93

American Mathematical Society
Providence, Rhode Island

2000 *Mathematics Subject Classification.* Primary 53–01.

For additional information and updates on this book, visit
www.ams.org/bookpages/gsm-93

Library of Congress Cataloging-in-Publication Data

Michor, Peter W., 1949–
Topics in differential geometry / Peter W. Michor.
p. cm. — (Graduate studies in mathematics, ISSN 1065-7339 ; v. 93)
Includes bibliographical references and index.
ISBN 978-0-8218-2003-2 (alk. paper)
1. Geometry, Differential. I. Title.

QA641.M49 2008
516.3′6—dc22 2008010629

---

10 9 8 7 6 5 4 3 2 1 13 12 11 10 09 08

To the ladies of my life,
Elli, Franziska, and Johanna

Contents

Preface

This book is an introduction to the fundamentals of differential geometry (manifolds, flows, Lie groups and their actions, invariant theory, differential forms and de Rham cohomology, bundles and connections, Riemann manifolds, isometric actions, symplectic geometry) which stresses naturality and functoriality from the beginning and is as coordinate-free as possible. The material presented in the beginning is standard — but some parts are not so easily found in text books: Among these are initial submanifolds (2.13) and the extension of the Frobenius theorem for distributions of nonconstant rank (the Stefan-Sussman theory) in (3.21) - (3.28). A quick proof of the Campbell-Baker-Hausdorff formula for Lie groups is in (4.29). Lie group actions are studied in detail: Palais' results that an infinitesimal action of a finite-dimensional Lie algebra on a manifold integrates to a local action of a Lie group and that proper actions admit slices are presented with full proofs in sections (5) and (6). The basics of invariant theory are given in section (7): The Hilbert-Nagata theorem is proved, and Schwarz's theorem on smooth invariant functions is discussed, but not proved.

In the section on vector bundles, the Lie derivative is treated for natural vector bundles, i.e., functors which associate vector bundles to manifolds and vector bundle homomorphisms to local diffeomorphisms. A formula for the Lie derivative is given in the form of a commutator, but it involves the tangent bundle of the vector bundle. So also a careful treatment of tangent bundles of vector bundles is given. Then follows a standard presentation of differential forms and de Rham cohomoloy including the theorems of de Rham and Poincaré duality. This is used to compute the cohomology of compact Lie groups, and a section on extensions of Lie algebras and Lie groups follows.

The chapter on bundles and connections starts with a thorough treatment of the Frölicher-Nijenhuis bracket via the study of all graded derivations of the algebra of differential forms. This bracket is a natural extension of the Lie bracket from vector fields to tangent bundle valued differential forms; it is one of the basic structures of differential geometry. We begin our treatment of connections in the general setting of fiber bundles (without structure group). A connection on a fiber bundle is just a projection onto the vertical bundle. Curvature and the Bianchi identity are expressed with the help of the Frölicher-Nijenhuis bracket. The parallel transport for such a general connection is not defined along the whole of the curve in the base in general — if this is the case, the connection is called complete. We show that every fiber bundle admits complete connections. For complete connections we treat holonomy groups and the holonomy Lie algebra, a subalgebra of the Lie algebra of all vector fields on the standard fiber. Then we present principal bundles and associated bundles in detail together with the most important examples. Finally we investigate principal connections by requiring equivariance under the structure group. It is remarkable how fast the usual structure equations can be derived from the basic properties of the Frölicher-Nijenhuis bracket. Induced connections are investigated thoroughly — we describe tools to recognize induced connections among general ones. If the holonomy Lie algebra of a connection on a fiber bundle with compact standard fiber turns out to be finite-dimensional, we are able to show that in fact the fiber bundle is associated to a principal bundle and the connection is an induced one. I think that the treatment of connections presented here offers some didactical advantages: The geometric content of a connection is treated first, and the additional requirement of equivariance under a structure group is seen to be additional and can be dealt with later — so the student is not required to grasp all the structures at the same time. Besides that it gives new results and new insights. This treatment is taken from [**146**].

The chapter on Riemann geometry contains a careful treatment of connections to geodesic structures to sprays to connectors and back to connections considering also the roles of the second and third tangent bundles in this. Most standard results are proved. Isometric immersions and Riemann submersions are treated in analogy to each other. A unusual feature is the Jacobi flow on the second tangent bundle. The chapter on isometric actions starts off with homogeneous Riemann manifolds and the beginnings of symmetric space theory; then Riemann G-manifolds and polar actions are treated.

The final chapter on symplectic and Poisson geometry puts some emphasis on group actions, momentum mappings and reductions.

There are some glaring omissions: The Laplace-Beltrami operator is treated only summarily, there is no spectral theory, and the structure theory of Lie algebras is not treated and used. Thus the finer theory of symmetric spaces is outside of the scope of this book.

The exposition is not always linear. Sometimes concepts treated in detail in later sections are used or pointed out earlier on when they appear in a natural way. Text cross-references to sections, subsections, theorems, numbered equations, items in a list, etc., appear in parantheses, for example, section (1), subsection (1.1), theorem (3.16), equation (3.16.3) which will be called (3) within (3.16) and its proof, property (3.22.1).

This book grew out of lectures which I have given during the last three decades on advanced differential geometry, Lie groups and their actions, Riemann geometry, and symplectic geometry. I have benefited a lot from the advise of colleagues and remarks by readers and students. In particular I want to thank Konstanze Rietsch whose write-up of my lecture course on isometric group actions was very helpful in the preparation of this book and Simon Hochgerner who helped with the last section.

Support by the Austrian FWF-projects P 4661, P 7724-PHY, P 10037-MAT, P 14195-MAT, and P 17108-N04 during the years 1983 – 2007 is acknowledged.

Kritzendorf, April 2008

CHAPTER I. Manifolds and Vector Fields

1. Differentiable Manifolds

1.1. Manifolds. A *topological manifold* is a separable metrizable space M which is locally homeomorphic to $\mathbb{R}^n$. So for any $x \in M$ there is some homeomorphism $u : U \to u(U) \subseteq \mathbb{R}^n$, where U is an open neighborhood of x in M and $u(U)$ is an open subset in $\mathbb{R}^n$. The pair (U, u) is called a *chart* on M.

One of the basic results of algebraic topology, called 'invariance of domain', conjectured by Dedekind and proved by Brouwer in 1911, says that the number n is locally constant on M; if n is constant, M is sometimes called a *pure manifold.* We will only consider pure manifolds and consequently we will omit the prefix pure.

A family $(U_\alpha, u_\alpha)_{\alpha\in A}$ of charts on M such that the U_α form a cover of M is called an *atlas.* The mappings

$$u_{\alpha\beta} := u_\alpha \circ u_\beta^{-1} : u_\beta(U_{\alpha\beta}) \to u_\alpha(U_{\alpha\beta})$$

are called the chart changings for the atlas (U_α), where we use the notation $U_{\alpha\beta} := U_\alpha \cap U_\beta$.

An atlas $(U_\alpha, u_\alpha)_{\alpha\in A}$ for a manifold M is said to be a C^k*-atlas*, if all chart changings $u_{\alpha\beta} : u_\beta(U_{\alpha\beta}) \to u_\alpha(U_{\alpha\beta})$ are differentiable of class C^k. Two C^k-atlases are called C^k*-equivalent* if their union is again a C^k-atlas for M. An equivalence class of C^k-atlases is called a C^k*-structure* on M.

From differential topology we know that if M has a C^1-structure, then it also has a C^1-equivalent C^∞-structure and even a C^1-equivalent C^ω-structure, where C^ω is shorthand for real analytic; see [**84**].

By a C^k-manifold M we mean a topological manifold together with a C^k-structure and a chart on M will be a chart belonging to some atlas of the C^k-structure.

But there are topological manifolds which do not admit differentiable structures. For example, every 4-dimensional manifold is smooth off some point, but there are such which are not smooth; see [**195**], [**62**]. There are also topological manifolds which admit several inequivalent smooth structures. The spheres from dimension 7 on have finitely many; see [**156**]. But the most surprising result is that on $\mathbb{R}^4$ there are uncountably many pairwise inequivalent (exotic) differentiable structures. This follows from the results of [**42**] and [**62**]; see [**78**] for an overview.

Note that for a Hausdorff C^∞-manifold in a more general sense the following properties are equivalent:

(1) It is paracompact.

(2) It is metrizable.

(3) It admits a Riemann metric.

(4) Each connected component is separable.

In this book a manifold will usually mean a C^∞-manifold, and smooth is used synonymously for C^∞ — it will be Hausdorff, separable, finite-dimensional, to state it precisely.

Note finally that any manifold M admits a finite atlas consisting of $\dim M + 1$ (not connected) charts. This is a consequence of topological dimension theory [**168**]; a proof for manifolds may be found in [**80**, I].

1.2. Example: Spheres. We consider the space $\mathbb{R}^{n+1}$, equipped with the standard inner product $\langle x, y\rangle = \sum x^i y^i$. The *n-sphere* S^n is then the subset $\{x \in \mathbb{R}^{n+1} : \langle x, x\rangle = 1\}$. Since $f(x) = \langle x, x\rangle$, $f : \mathbb{R}^{n+1} \to \mathbb{R}$, satisfies $df(x)y = 2\langle x, y\rangle$, it is of rank 1 off 0 and by (1.12) the sphere S^n is a submanifold of $\mathbb{R}^{n+1}$.

In order to get some feeling for the sphere, we will describe an explicit atlas for S^n, the *stereographic atlas.* Choose $a \in S^n$ ('south pole'). Let

$$U_+ := S^n \setminus \{a\}, \qquad u_+ : U_+ \to \{a\}^\perp, \qquad u_+(x) = \tfrac{x - \langle x,a\rangle a}{1 - \langle x,a\rangle},$$
$$U_- := S^n \setminus \{-a\}, \qquad u_- : U_- \to \{a\}^\perp, \qquad u_-(x) = \tfrac{x - \langle x,a\rangle a}{1 + \langle x,a\rangle}.$$

From the following drawing in the 2-plane through 0, x, and a it is easily seen that u_+ is the usual stereographic projection. We also get

$$u_+^{-1}(y) = \tfrac{|y|^2-1}{|y|^2+1}a + \tfrac{2}{|y|^2+1}y \qquad \text{for } y \in \{a\}^\perp \setminus \{0\}$$

and $(u_- \circ u_+^{-1})(y) = \frac{y}{|y|^2}$. The latter equation can directly be seen from the drawing using the intercept theorem.

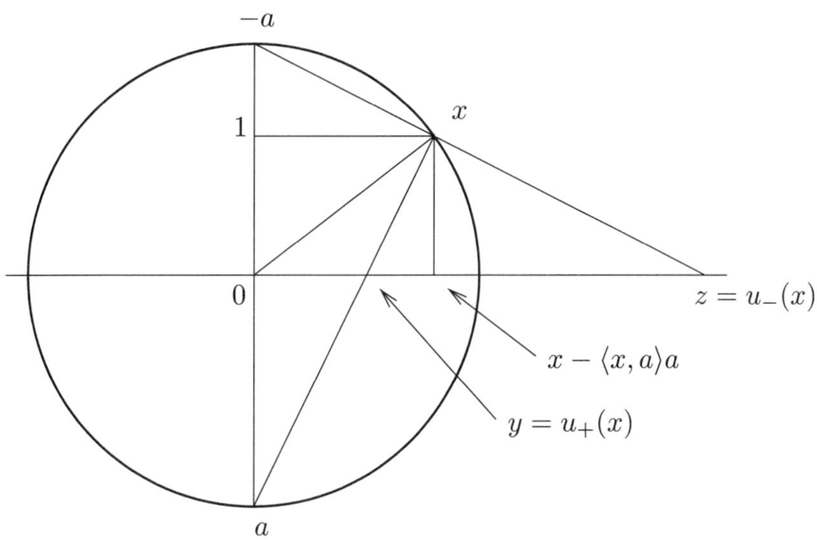

1.3. Smooth mappings. A mapping $f : M \to N$ between manifolds is said to be C^k if for each $x \in M$ and one (equivalently: any) chart (V, v) on N with $f(x) \in V$ there is a chart (U, u) on M with $x \in U$, $f(U) \subseteq V$, and $v \circ f \circ u^{-1}$ is C^k. We will denote by $C^k(M, N)$ the space of all C^k-mappings from M to N.

A C^k-mapping $f : M \to N$ is called a C^k-*diffeomorphism* if $f^{-1} : N \to M$ exists and is also C^k. Two manifolds are called *diffeomorphic* if there exists a diffeomorphism between them. From differential topology (see [**84**]) we know that if there is a C^1-diffeomorphism between M and N, then there is also a C^∞-diffeomorphism.

There are manifolds which are homeomorphic but not diffeomorphic: On $\mathbb{R}^4$ there are uncountably many pairwise nondiffeomorphic differentiable structures; on every other $\mathbb{R}^n$ the differentiable structure is unique. There are finitely many different differentiable structures on the spheres S^n for $n \geq 7$.

A mapping $f : M \to N$ between manifolds of the same dimension is called a *local diffeomorphism* if each $x \in M$ has an open neighborhood U such that $f|U : U \to f(U) \subset N$ is a diffeomorphism. Note that a local diffeomorphism need not be surjective.

1.4. Smooth functions. The set of smooth real valued functions on a manifold M will be denoted by $C^\infty(M)$, in order to distinguish it clearly from spaces of sections which will appear later. The space $C^\infty(M)$ is a real commutative algebra.

The *support* of a smooth function f is the closure of the set where it does not vanish, $\operatorname{supp}(f) = \overline{\{x \in M : f(x) \neq 0\}}$. The *zero set* of f is the set where f vanishes, $Z(f) = \{x \in M : f(x) = 0\}$.

1.5. Theorem. *Any (separable, metrizable, smooth) manifold admits smooth partitions of unity: Let $(U_\alpha)_{\alpha \in A}$ be an open cover of M.*

Then there is a family $(\varphi_\alpha)_{\alpha\in A}$ of smooth functions on M, such that:

(1) *$\varphi_\alpha(x) \geq 0$ for all $x \in M$ and all $\alpha \in A$.*

(2) $\operatorname{supp}(\varphi_\alpha) \subset U_\alpha$ *for all $\alpha \in A$.*

(3) *$(supp(\varphi_\alpha))_{\alpha\in A}$ is a locally finite family (so each $x \in M$ has an open neighborhood which meets only finitely many* $\operatorname{supp}(\varphi_\alpha)$*).*

(4) *$\sum_\alpha \varphi_\alpha = 1$ (locally this is a finite sum).*

Proof. Any (separable, metrizable) manifold is a *'Lindelöf space'*, i.e., each open cover admits a countable subcover. This can be seen as follows:

Let $\mathcal{U}$ be an open cover of M. Since M is separable, there is a countable dense subset S in M. Choose a metric on M. For each $U \in \mathcal{U}$ and each $x \in U$ there is a $y \in S$ and $n \in \mathbb{N}$ such that the ball $B_{1/n}(y)$ with respect to that metric with center y and radius $\frac{1}{n}$ contains x and is contained in U. But there are only countably many of these balls; for each of them we choose an open set $U \in \mathcal{U}$ containing it. This is then a countable subcover of $\mathcal{U}$.

Now let $(U_\alpha)_{\alpha\in A}$ be the given cover. Let us fix first α and $x \in U_\alpha$. We choose a chart (U, u) centered at x (i.e., $u(x) = 0$) and $\varepsilon > 0$ such that $\varepsilon\mathbb{D}^n \subset u(U \cap U_\alpha)$, where $\mathbb{D}^n = \{y \in \mathbb{R}^n : |y| \leq 1\}$ is the closed unit ball. Let

$$h(t) := \begin{cases} e^{-1/t} & \text{for } t > 0, \\ 0 & \text{for } t \leq 0, \end{cases}$$

a smooth function on $\mathbb{R}$. Then

$$f_{\alpha,x}(z) := \begin{cases} h(\varepsilon^2 - |u(z)|^2) & \text{for } z \in U, \\ 0 & \text{for } z \notin U \end{cases}$$

is a nonnegative smooth function on M with support in U_α which is positive at x.

We choose such a function $f_{\alpha,x}$ for each α and $x \in U_\alpha$. The interiors of the supports of these smooth functions form an open cover of M which refines

(U_α), so by the argument at the beginning of the proof there is a countable subcover with corresponding functions $f_1, f_2, \dots$. Let

$$W_n = \{x \in M : f_n(x) > 0 \text{ and } f_i(x) < \tfrac{1}{n} \text{ for } 1 \le i < n\},$$

and denote by $\overline{W}_n$ the closure. Then $(W_n)_n$ is an open cover. We claim that $(\overline{W}_n)_n$ is locally finite: Let $x \in M$. Then there is a smallest n such that $x \in W_n$. Let $V := \{y \in M : f_n(y) > \frac{1}{2} f_n(x)\}$. If $y \in V \cap \overline{W}_k$, then we have $f_n(y) > \frac{1}{2} f_n(x)$ and $f_i(y) \le \frac{1}{k}$ for $i < k$, which is possible for finitely many k only.

Consider the nonnegative smooth function

$$g_n(x) = h(f_n(x)) h(\tfrac{1}{n} - f_1(x)) \dots h(\tfrac{1}{n} - f_{n-1}(x)), \quad n \in \mathbb{N}.$$

Then obviously $\operatorname{supp}(g_n) = \overline{W}_n$. So $g := \sum_n g_n$ is smooth, since it is locally only a finite sum, and everywhere positive; thus $(g_n/g)_{n\in\mathbb{N}}$ is a smooth partition of unity on M. Since $\operatorname{supp}(g_n) = \overline{W}_n$ is contained in some $U_{\alpha(n)}$, we may put $\varphi_\alpha = \sum_{\{n:\alpha(n)=\alpha\}} \frac{g_n}{g}$ to get the required partition of unity which is subordinated to $(U_\alpha)_{\alpha\in A}$. □

1.6. Germs. Let M and N be manifolds and $x \in M$. We consider all smooth mappings $f : U_f \to N$, where U_f is some open neighborhood of x in M, and we put $f \sim_x g$ if there is some open neighborhood V of x with $f|V = g|V$. This is an equivalence relation on the set of mappings considered. The equivalence class of a mapping f is called the *germ of f at x*, sometimes denoted by $\operatorname{germ}_x f$. The set of all these germs is denoted by $C^\infty_x(M, N)$.

Note that for a germs at x of a smooth mapping only the value at x is defined. We may also consider composition of germs: $\operatorname{germ}_{f(x)} g \circ \operatorname{germ}_x f := \operatorname{germ}_x(g \circ f)$.

If $N = \mathbb{R}$, we may add and multiply germs of smooth functions, so we get the real commutative algebra $C^\infty_x(M, \mathbb{R})$ of germs of smooth functions at x. This construction works also for other types of functions like real analytic or holomorphic ones if M has a real analytic or complex structure.

Using smooth partitions of unity (1.4) it is easily seen that each germ of a smooth function has a representative which is defined on the whole of M. For germs of real analytic or holomorphic functions this is not true. So $C^\infty_x(M, \mathbb{R})$ is the quotient of the algebra $C^\infty(M)$ by the ideal of all smooth functions $f : M \to \mathbb{R}$ which vanish on some neighborhood (depending on f) of x.

1.7. The tangent space of $\mathbb{R}^n$. Let $a \in \mathbb{R}^n$. A *tangent vector* with foot point a is simply a pair (a, X) with $X \in \mathbb{R}^n$, also denoted by X_a. It induces a *derivation* $X_a : C^\infty(\mathbb{R}^n) \to \mathbb{R}$ by $X_a(f) = df(a)(X_a)$. The value depends only on the germ of f at a and we have $X_a(f \cdot g) = X_a(f) \cdot g(a) + f(a) \cdot X_a(g)$ (the derivation property).

If conversely $D : C^\infty(\mathbb{R}^n) \to \mathbb{R}$ is linear and satisfies

$$D(f \cdot g) = D(f) \cdot g(a) + f(a) \cdot D(g)$$

(a derivation at a), then D is given by the action of a tangent vector with foot point a. This can be seen as follows. For $f \in C^\infty(\mathbb{R}^n)$ we have

$$\begin{aligned} f(x) &= f(a) + \int_0^1 \tfrac{d}{dt} f(a + t(x-a))dt \\ &= f(a) + \sum_{i=1}^n \int_0^1 \tfrac{\partial f}{\partial x^i}(a + t(x-a))dt\ (x^i - a^i) \\ &= f(a) + \sum_{i=1}^n h_i(x)(x^i - a^i). \end{aligned}$$

On the constant function 1 the derivation gives $D(1) = D(1 \cdot 1) = 2D(1)$, so $D(\text{constant}) = 0$. Therefore,

$$\begin{aligned} D(f) &= D\Big(f(a) + \sum_{i=1}^n h_i(x^i - a^i)\Big) \\ &= 0 + \sum_{i=1}^n D(h_i)(a^i - a^i) + \sum_{i=1}^n h_i(a)(D(x^i) - 0) \\ &= \sum_{i=1}^n \tfrac{\partial f}{\partial x^i}(a) D(x^i), \end{aligned}$$

where x^i is the i-th coordinate function on $\mathbb{R}^n$. So we have

$$D(f) = \sum_{i=1}^n D(x^i) \tfrac{\partial}{\partial x^i}|_a(f), \qquad D = \sum_{i=1}^n D(x^i) \tfrac{\partial}{\partial x^i}|_a.$$

Thus D is induced by the tangent vector $(a, \sum_{i=1}^n D(x^i)e_i)$, where (e_i) is the standard basis of $\mathbb{R}^n$.

1.8. The tangent space of a manifold. Let M be a manifold and let $x \in M$ and $\dim M = n$. Let T_xM be the vector space of all derivations at x of $C_x^\infty(M, \mathbb{R})$, the algebra of germs of smooth functions on M at x. Using (1.5), it may easily be seen that a derivation of $C^\infty(M)$ at x factors to a derivation of $C_x^\infty(M, \mathbb{R})$.

So T_xM consists of all linear mappings $X_x : C^\infty(M) \to \mathbb{R}$ with the property $X_x(f \cdot g) = X_x(f) \cdot g(x) + f(x) \cdot X_x(g)$. The space T_xM is called the *tangent space of M at x*.

If (U, u) is a chart on M with $x \in U$, then $u^* : f \mapsto f \circ u$ induces an isomorphism of algebras $C^\infty_{u(x)}(\mathbb{R}^n, \mathbb{R}) \cong C^\infty_x(M, \mathbb{R})$, and thus also an isomorphism $T_x u : T_x M \to T_{u(x)}\mathbb{R}^n$, given by $(T_x u.X_x)(f) = X_x(f \circ u)$. So $T_x M$ is an n-dimensional vector space.

We will use the following notation: $u = (u^1, \dots, u^n)$, so u^i denotes the i-th coordinate function on U, and

$$\tfrac{\partial}{\partial u^i}|_x := (T_x u)^{-1}(\tfrac{\partial}{\partial x^i}|_{u(x)}) = (T_x u)^{-1}(u(x), e_i).$$

So $\frac{\partial}{\partial u^i}|_x \in T_x M$ is the derivation given by

$$\tfrac{\partial}{\partial u^i}|_x(f) = \frac{\partial(f \circ u^{-1})}{\partial x^i}(u(x)).$$

From (1.7) we have now

$$\begin{aligned} T_x u.X_x &= \sum_{i=1}^n (T_x u.X_x)(x^i)\tfrac{\partial}{\partial x^i}|_{u(x)} = \sum_{i=1}^n X_x(x^i \circ u)\tfrac{\partial}{\partial x^i}|_{u(x)} \\ &= \sum_{i=1}^n X_x(u^i)\tfrac{\partial}{\partial x^i}|_{u(x)}, \end{aligned}$$

$$X_x = (T_x u)^{-1}.T_x u.X_x = \sum_{i=1}^n X_x(u^i)\tfrac{\partial}{\partial u^i}|_x.$$

1.9. The tangent bundle. For a manifold M of dimension n we put $TM := \bigsqcup_{x\in M} T_x M$, the disjoint union of all tangent spaces. This is a family of vector spaces parameterized by M, with projection $\pi_M : TM \to M$ given by $\pi_M(T_x M) = x$.

For any chart (U_α, u_α) of M consider the chart $(\pi_M^{-1}(U_\alpha), Tu_\alpha)$ on TM, where $Tu_\alpha : \pi_M^{-1}(U_\alpha) \to u_\alpha(U_\alpha) \times \mathbb{R}^n$ is given by

$$Tu_\alpha.X = (u_\alpha(\pi_M(X)), T_{\pi_M(X)}u_\alpha.X).$$

Then the chart changings look as follows:

$$\begin{aligned} Tu_\beta \circ (Tu_\alpha)^{-1} : Tu_\alpha(\pi_M^{-1}(U_{\alpha\beta})) &= u_\alpha(U_{\alpha\beta}) \times \mathbb{R}^n \to \\ &\to u_\beta(U_{\alpha\beta}) \times \mathbb{R}^n = Tu_\beta(\pi_M^{-1}(U_{\alpha\beta})), \\ ((Tu_\beta \circ (Tu_\alpha)^{-1})(y, Y))(f) &= ((Tu_\alpha)^{-1}(y, Y))(f \circ u_\beta) \\ &= (y, Y)(f \circ u_\beta \circ u_\alpha^{-1}) = d(f \circ u_\beta \circ u_\alpha^{-1})(y).Y \\ &= df(u_\beta \circ u_\alpha^{-1}(y)).d(u_\beta \circ u_\alpha^{-1})(y).Y \\ &= (u_\beta \circ u_\alpha^{-1}(y), d(u_\beta \circ u_\alpha^{-1})(y).Y)(f). \end{aligned}$$

So the chart changings are smooth. We choose the topology on TM in such a way that all Tu_α become homeomorphisms. This is a Hausdorff topology, since $X, Y \in TM$ may be separated in M if $\pi(X) \neq \pi(Y)$; and they may be separated in one chart if $\pi(X) = \pi(Y)$. So TM is again a smooth manifold in a canonical way; the triple (TM, π_M, M) is called the *tangent bundle* of the manifold M.

1.10. Kinematic definition of the tangent space. Let $C_0^\infty(\mathbb{R}, M)$ denote the space of germs at 0 of smooth curves $\mathbb{R} \to M$. We put the following equivalence relation on $C_0^\infty(\mathbb{R}, M)$: the germ of c is equivalent to the germ of e if and only if $c(0) = e(0)$ and in one (equivalently: each) chart (U, u) with $c(0) = e(0) \in U$ we have $\frac{d}{dt}|_0(u \circ c)(t) = \frac{d}{dt}|_0(u \circ e)(t)$. The equivalence classes are also called velocity vectors of curves in M. We have the following diagram of mappings where $\alpha(c)(\text{germ}_{c(0)} f) = \frac{d}{dt}|_0 f(c(t))$ and $\beta : TM \to C_0^\infty(\mathbb{R}, M)$ is given by: $\beta((Tu)^{-1}(y, Y))$ is the germ at 0 of $t \mapsto u^{-1}(y + tY)$. So TM is canonically identified with the set of all possible velocity vectors of curves in M:

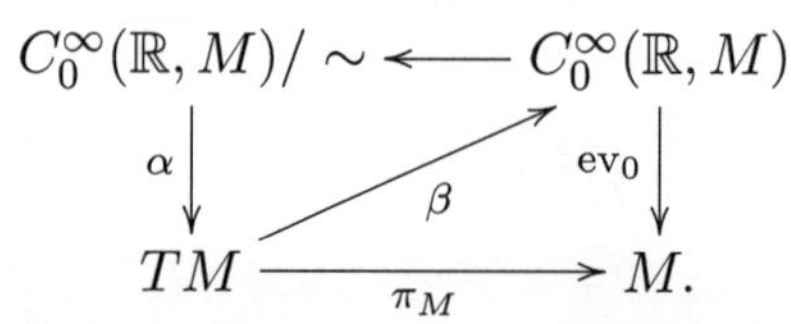

1.11. Tangent mappings. Let $f : M \to N$ be a smooth mapping between manifolds. Then f induces a linear mapping $T_x f : T_x M \to T_{f(x)} N$ for each $x \in M$ by $(T_x f.X_x)(h) = X_x(h \circ f)$ for $h \in C_{f(x)}^\infty(N, \mathbb{R})$. This mapping is well defined and linear since $f^* : C_{f(x)}^\infty(N, \mathbb{R}) \to C_x^\infty(M, \mathbb{R})$, given by $h \mapsto h \circ f$, is linear and an algebra homomorphism, and $T_x f$ is its adjoint, restricted to the subspace of derivations.

If (U, u) is a chart around x and (V, v) is one around $f(x)$, then

$$\begin{aligned}
(T_x f.\tfrac{\partial}{\partial u^i}|_x)(v^j) &= \tfrac{\partial}{\partial u^i}|_x(v^j \circ f) = \tfrac{\partial}{\partial x^i}(v^j \circ f \circ u^{-1})(u(x)),\\
T_x f.\tfrac{\partial}{\partial u^i}|_x &= \textstyle\sum_j (T_x f.\tfrac{\partial}{\partial u^i}|_x)(v^j)\tfrac{\partial}{\partial v^j}|_{f(x)} \quad \text{by (1.8)}\\
&= \textstyle\sum_j \frac{\partial(v^j \circ f \circ u^{-1})}{\partial x^i}(u(x))\tfrac{\partial}{\partial v^j}|_{f(x)}.
\end{aligned}$$

So the matrix of $T_x f : T_x M \to T_{f(x)} N$ in the bases $(\frac{\partial}{\partial u^i}|_x)$ and $(\frac{\partial}{\partial v^j}|_{f(x)})$ is just the Jacobi matrix $d(v \circ f \circ u^{-1})(u(x))$ of the mapping $v \circ f \circ u^{-1}$ at $u(x)$, so $T_{f(x)} v \circ T_x f \circ (T_x u)^{-1} = d(v \circ f \circ u^{-1})(u(x))$.

Let us denote by $Tf : TM \to TN$ the total mapping which is given by $Tf|T_x M := T_x f$. Then the composition

$$\begin{aligned}
Tv \circ Tf \circ (Tu)^{-1} &: u(U) \times \mathbb{R}^m \to v(V) \times \mathbb{R}^n,\\
(y, Y) &\mapsto ((v \circ f \circ u^{-1})(y), d(v \circ f \circ u^{-1})(y)Y),
\end{aligned}$$

is smooth; thus $Tf : TM \to TN$ is again smooth.

If $f : M \to N$ and $g : N \to P$ are smooth, then we have $T(g \circ f) = Tg \circ Tf$. This is a direct consequence of $(g \circ f)^* = f^* \circ g^*$, and it is the global version of the chain rule. Furthermore we have $T(Id_M) = Id_{TM}$.

If $f \in C^\infty(M)$, then $Tf : TM \to T\mathbb{R} = \mathbb{R} \times \mathbb{R}$. We define the *differential* of f by $df := \mathrm{pr}_2 \circ Tf : TM \to \mathbb{R}$. Let t denote the identity function on $\mathbb{R}$. Then $(Tf.X_x)(t) = X_x(t \circ f) = X_x(f)$, so we have $df(X_x) = X_x(f)$.

1.12. Submanifolds. A subset N of a manifold M is called a *submanifold* if for each $x \in N$ there is a chart (U, u) of M such that $u(U \cap N) = u(U) \cap (\mathbb{R}^k \times 0)$, where $\mathbb{R}^k \times 0 \hookrightarrow \mathbb{R}^k \times \mathbb{R}^{n-k} = \mathbb{R}^n$. Then clearly N is itself a manifold with $(U \cap N, u|(U \cap N))$ as charts, where (U, u) runs through all *submanifold charts* as above.

1.13. Let $f : \mathbb{R}^n \to \mathbb{R}^q$ be smooth. A point $x \in \mathbb{R}^q$ is called a *regular value* of f if the rank of f (more exactly: the rank of its derivative) is q at each point y of $f^{-1}(x)$. In this case, $f^{-1}(x)$ is a submanifold of $\mathbb{R}^n$ of dimension $n - q$ (or empty). This is an immediate consequence of the implicit function theorem, as follows: Let $x = 0 \in \mathbb{R}^q$. Permute the coordinates $(x^1, \dots, x^n)$ on $\mathbb{R}^n$ such that the Jacobi matrix

$$df(y) = \left(\left(\frac{\partial f^i}{\partial x^j}(y) \right)^{1 \le i \le q}_{1 \le j \le q} \middle| \left(\frac{\partial f^i}{\partial x^j}(y) \right)^{1 \le i \le q}_{q+1 \le j \le n} \right)$$

has the left hand part invertible. Then $u := (f, \mathrm{pr}_{n-q}) : \mathbb{R}^n \to \mathbb{R}^q \times \mathbb{R}^{n-q}$ has invertible differential at y, so (U, u) is a chart at any $y \in f^{-1}(0)$, and we have $f \circ u^{-1}(z^1, \dots, z^n) = (z^1, \dots, z^q)$, so $u(f^{-1}(0)) = u(U) \cap (0 \times \mathbb{R}^{n-q})$ as required.

Constant rank theorem ([41, I 10.3.1]). *Let $f : W \to \mathbb{R}^q$ be a smooth mapping, where W is an open subset of $\mathbb{R}^n$. If the derivative $df(x)$ has constant rank k for each $x \in W$, then for each $a \in W$ there are charts (U, u) of W centered at a and (V, v) of $\mathbb{R}^q$ centered at $f(a)$ such that $v \circ f \circ u^{-1} : u(U) \to v(V)$ has the following form:*

$$(x_1, \dots, x_n) \mapsto (x_1, \dots, x_k, 0, \dots, 0).$$

So $f^{-1}(b)$ is a submanifold of W of dimension $n - k$ for each $b \in f(W)$.

Proof. We will use the inverse function theorem several times. The derivative $df(a)$ has rank $k \le n, q$; without loss we may assume that the upper left $(k \times k)$-submatrix of $df(a)$ is invertible. Moreover, let $a = 0$ and $f(a) = 0$.

We consider $u : W \to \mathbb{R}^n$, $u(x^1, \dots, x^n) := (f^1(x), \dots, f^k(x), x^{k+1}, \dots, x^n)$. Then

$$du = \begin{pmatrix} \left(\frac{\partial f^i}{\partial z^j}\right)^{1 \le i \le k}_{1 \le j \le k} & \left(\frac{\partial f^i}{\partial z^j}\right)^{1 \le i \le k}_{k+1 \le j \le n} \\ 0 & \mathbb{I}_{\mathbb{R}^{n-k}} \end{pmatrix}$$

is invertible, so u is a diffeomorphism $U_1 \to U_2$ for suitable open neighborhoods of 0 in $\mathbb{R}^n$. Consider $g = f \circ u^{-1} : U_2 \to \mathbb{R}^q$. Then we have

$$\begin{aligned} g(z_1, \dots, z_n) &= (z_1, \dots, z_k, g_{k+1}(z), \dots, g_q(z)), \\ dg(z) &= \begin{pmatrix} \mathbb{I}_{\mathbb{R}^k} & 0 \\ * & (\frac{\partial g^i}{\partial z^j})^{k+1 \le i \le q}_{k+1 \le j \le n} \end{pmatrix}, \\ \operatorname{rank}(dg(z)) &= \operatorname{rank}\left(d(f \circ u^{-1})(z)\right) \\ &= \operatorname{rank}\left(df(u^{-1}(z)).du^{-1}(z)\right) = \operatorname{rank}(df(z)) = k. \end{aligned}$$

Therefore, $\dfrac{\partial g^i}{\partial z^j}(z) = 0 \qquad \text{for } k+1 \le i \le q \text{ and } k+1 \le j \le n;$

$$g^i(z^1, \dots, z^n) = g^i(z^1, \dots, z^k, 0, \dots, 0) \quad \text{for } k+1 \le i \le q.$$

Let $v : U_3 \to \mathbb{R}^q$, where $U_3 = \{y \in \mathbb{R}^q : (y^1, \dots, y^k, 0, \dots, 0) \in U_2 \subset \mathbb{R}^n\}$, be given by

$$v \begin{pmatrix} y^1 \\ \vdots \\ y^q \end{pmatrix} = \begin{pmatrix} y^1 \\ \vdots \\ y^k \\ y^{k+1} - g^{k+1}(y^1, \dots, y^k, 0, \dots, 0) \\ \vdots \\ y^q - g^q(y^1, \dots, y^k, 0, \dots, 0) \end{pmatrix} = \begin{pmatrix} y^1 \\ \vdots \\ y^k \\ y^{k+1} - g^{k+1}(\bar{y}) \\ \vdots \\ y^q - g^q(\bar{y}) \end{pmatrix},$$

where $\bar{y} = (y^1, \dots, y^q, 0, \dots, 0) \in \mathbb{R}^n$ if $q < n$ and $\bar{y} = (y^1, \dots, y^n)$ if $q \ge n$. We have $v(0) = 0$, and

$$dv = \begin{pmatrix} \mathbb{I}_{\mathbb{R}^k} & 0 \\ * & \mathbb{I}_{\mathbb{R}^{q-k}} \end{pmatrix}$$

is invertible; thus $v : V \to \mathbb{R}^q$ is a chart for a suitable neighborhood of 0. Now let $U := f^{-1}(V) \cup U_1$. Then $v \circ f \circ u^{-1} = v \circ g : \mathbb{R}^n \supseteq u(U) \to v(V) \subseteq \mathbb{R}^q$ looks as follows:

$$\begin{pmatrix} x^1 \\ \vdots \\ x^n \end{pmatrix} \xrightarrow{g} \begin{pmatrix} x^1 \\ \vdots \\ x^k \\ g^{k+1}(x) \\ \vdots \\ g^q(x) \end{pmatrix} \xrightarrow{v} \begin{pmatrix} x^1 \\ \vdots \\ x^k \\ g^{k+1}(x) - g^{k+1}(x) \\ \vdots \\ g^q(x) - g^q(x) \end{pmatrix} = \begin{pmatrix} x^1 \\ \vdots \\ x^k \\ 0 \\ \vdots \\ 0 \end{pmatrix}. \qquad \square$$

Corollary. *Let $f : M \to N$ be C^∞ with $T_x f$ of constant rank k for all $x \in M$.*

Then for each $b \in f(M)$ the set $f^{-1}(b) \subset M$ is a submanifold of M of dimension $\dim M - k$. $\square$

1.14. Products. Let M and N be smooth manifolds described by smooth atlases $(U_\alpha, u_\alpha)_{\alpha\in A}$ and $(V_\beta, v_\beta)_{\beta\in B}$, respectively. Then the family $(U_\alpha \times V_\beta, u_\alpha \times v_\beta : U_\alpha \times V_\beta \to \mathbb{R}^m \times \mathbb{R}^n)_{(\alpha,\beta)\in A\times B}$ is a smooth atlas for the cartesian product $M \times N$. Clearly the projections

$$M \xleftarrow{\mathrm{pr}_1} M \times N \xrightarrow{\mathrm{pr}_2} N$$

are also smooth. The *product* $(M \times N, \mathrm{pr}_1, \mathrm{pr}_2)$ has the following universal property:

> For any smooth manifold P and smooth mappings $f : P \to M$ and $g : P \to N$ the mapping
>
> $$(f, g) : P \to M \times N, \quad (f, g)(x) = (f(x), g(x)),$$
>
> is the unique smooth mapping with $\mathrm{pr}_1 \circ (f, g) = f$ and $\mathrm{pr}_2 \circ (f, g) = g$.

From the construction of the tangent bundle in (1.9) it is immediately clear that

$$TM \xleftarrow{T(\mathrm{pr}_1)} T(M \times N) \xrightarrow{T(\mathrm{pr}_2)} TN$$

is again a product, so that $T(M \times N) = TM \times TN$ in a canonical way.

Clearly we can form products of finitely many manifolds.

1.15. Theorem. *Let M be a connected manifold and suppose that $f : M \to M$ is smooth with $f \circ f = f$. Then the image $f(M)$ of f is a submanifold of M.*

This result can also be expressed as: 'smooth retracts' of manifolds are manifolds. If we do not suppose that M is connected, then $f(M)$ will not be a pure manifold in general; it will have different dimensions in different connected components.

Proof. We claim that there is an open neighborhood U of $f(M)$ in M such that the rank of $T_y f$ is constant for $y \in U$. Then by theorem (1.13) the result follows.

For $x \in f(M)$ we have $T_x f \circ T_x f = T_x f$; thus $\operatorname{im} T_x f = \ker(Id - T_x f)$ and $\operatorname{rank} T_x f + \operatorname{rank}(Id - T_x f) = \dim M$. Since $\operatorname{rank} T_x f$ and $\operatorname{rank}(Id - T_x f)$ cannot fall locally, $\operatorname{rank} T_x f$ is locally constant for $x \in f(M)$, and since $f(M)$ is connected, $\operatorname{rank} T_x f = r$ for all $x \in f(M)$.

But then for each $x \in f(M)$ there is an open neighborhood U_x in M with $\operatorname{rank} T_y f \geq r$ for all $y \in U_x$. On the other hand

$$\operatorname{rank} T_y f = \operatorname{rank} T_y(f \circ f) = \operatorname{rank} T_{f(y)} f \circ T_y f \leq \operatorname{rank} T_{f(y)} f = r$$

since $f(y) \in f(M)$.

So the neighborhood we need is given by $U = \bigcup_{x\in f(M)} U_x$. □

1.16. Corollary. (1) *The (separable) connected smooth manifolds are exactly the smooth retracts of connected open subsets of* $\mathbb{R}^n$*'s.*

(2) *A smooth mapping* $f : M \to N$ *is an embedding of a submanifold if and only if there is an open neighborhood* U *of* $f(M)$ *in* N *and a smooth mapping* $r : U \to M$ *with* $r \circ f = Id_M$*.*

Proof. Any manifold M may be embedded into some $\mathbb{R}^n$; see (1.19) below. Then there exists a tubular neighborhood of M in R^n (see later or [**84**, pp. 109–118]), and M is clearly a retract of such a tubular neighborhood. The converse follows from (1.15).

For the second assertion we repeat the argument for N instead of $\mathbb{R}^n$. □

1.17. Sets of Lebesque measure 0 in manifolds. An *m-cube of width $w > 0$ in $\mathbb{R}^m$* is a set of the form $C = [x_1, x_1 + w] \times \ldots \times [x_m, x_m + w]$. The measure $\mu(C)$ is then $\mu(C) = w^n$. A subset $S \subset \mathbb{R}^m$ is called a *set of (Lebesque) measure* 0 if for each $\varepsilon > 0$ these are at most countably many m-cubes C_i with $S \subset \bigcup_{i=0}^\infty C_i$ and $\sum_{i=0}^\infty \mu(C_i) < \varepsilon$. Obviously, a countable union of sets of Lebesque measure 0 is again of measure 0.

Lemma. *Let* $U \subset \mathbb{R}^m$ *be open and let* $f : U \to \mathbb{R}^m$ *be* C^1*. If* $S \subset U$ *is of measure* 0*, then also* $f(S) \subset \mathbb{R}^m$ *is of measure* 0*.*

Proof. Every point of S belongs to an open ball $B \subset U$ such that the operator norm $\|df(x)\| \le K_B$ for all $x \in B$. Then $|f(x) - f(y)| \le K_B|x - y|$ for all $x, y \in B$. So if $C \subset B$ is an m-cube of width w, then $f(C)$ is contained in an m-cube C' of width $\sqrt{m}K_B w$ and measure $\mu(C') \le m^{m/2}K_B^m \mu(C)$. Now let $S = \bigcup_{j=1}^\infty S_j$ where each S_j is a compact subset of a ball B_j as above. It suffices to show that each $f(S_j)$ is of measure 0.

For each $\varepsilon > 0$ there are m-cubes C_i in B_j with $S_j \subset \bigcup_i C_i$ and $\sum_i \mu(C_i) < \varepsilon$. As we saw above, then $f(X_j) \subset \bigcup_i C_i'$ with $\sum_i \mu(C_i') < m^{m/2}K_{B_j}^m \varepsilon$. □

Let M be a smooth (separable) manifold. A subset $S \subset M$ is called a *set of (Lebesque) measure* 0 if for each chart (U, u) of M the set $u(S \cap U)$ is of measure 0 in $\mathbb{R}^m$. By the lemma it suffices that there is some atlas whose charts have this property. Obviously, a countable union of sets of measure 0 in a manifold is again of measure 0.

An m-cube is not of measure 0. Thus a subset of $\mathbb{R}^m$ of measure 0 does not contain any m-cube; hence its interior is empty. Thus a closed set of measure 0 in a manifold is nowhere dense. More generally, let S be a subset of a manifold which is of measure 0 and σ-compact, i.e., a countable union of compact subsets. Then each of the latter is nowhere dense, so S is nowhere dense by the Baire category theorem. The complement of S is *residual*, i.e., it contains the intersection of a countable family of open dense subsets.

The Baire theorem says that a residual subset of a complete metric space is dense.

1.18. Regular values. Let $f : M \to N$ be a smooth mapping between manifolds.

(1) A point $x \in M$ is called a *singular point* of f if $T_x f$ is not surjective, and it is called a *regular point* of f if $T_x f$ is surjective.

(2) A point $y \in N$ is called a *regular value* of f if $T_x f$ is surjective for all $x \in f^{-1}(y)$. If not, y is called a *singular value.* Note that any $y \in N \setminus f(M)$ is a regular value.

Theorem ([166], [196]). *The set of all singular values of a C^k mapping $f : M \to N$ is of Lebesgue measure* 0 *in N if* $k > \max\{0, \dim(M) - \dim(N)\}$.

So any smooth mapping has regular values.

Proof. We proof this only for smooth mappings. It is sufficient to prove this locally. Thus we consider a smooth mapping $f : U \to \mathbb{R}^n$ where $U \subset \mathbb{R}^m$ is open. If $n > m$, then the result follows from lemma (1.17) above (consider the set $U \times 0 \subset \mathbb{R}^m \times \mathbb{R}^{n-m}$ of measure 0). Thus let $m \geq n$.

Let $\Sigma(f) \subset U$ denote the set of singular points of f. Let $f = (f^1, \dots, f^n)$, and let $\Sigma(f) = \Sigma_1 \cup \Sigma_2 \cup \Sigma_3$ where:

Σ_1 is the set of singular points x such that $Pf(x) = 0$ for all linear differential operators P of order $\leq \frac{m}{n}$.

Σ_2 is the set of singular points x such that $Pf(x) \neq 0$ for some differential operator P of order ≥ 2.

Σ_3 is the set of singular points x such that $\frac{\partial f^i}{x^j}(x) = 0$ for some i, j.

We first show that $f(\Sigma_1)$ has measure 0. Let $\nu = \lceil \frac{m}{n} + 1 \rceil$ be the smallest integer $> m/n$. Then each point of Σ_1 has an open neighborhood $W \subset U$ such that $|f(x) - f(y)| \leq K|x - y|^\nu$ for all $x \in \Sigma_1 \cap W$ and $y \in W$ and for some $K > 0$, by Taylor expansion. We take W to be a cube, of width w. It suffices to prove that $f(\Sigma_1 \cap W)$ has measure 0. We divide W into p^m cubes of width $\frac{w}{p}$; those which meet Si_1 will be denoted by $C_1, \dots, C_q$ for $q \leq p^m$. Each C_k is contained in a ball of radius $\frac{w}{p}\sqrt{m}$ centered at a point of $\Sigma_1 \cap W$. The set $f(C_k)$ is contained in a cube $C'_k \subset \mathbb{R}^n$ of width $2K(\frac{w}{p}\sqrt{m})^\nu$. Then

$$\sum_k \mu^n(C'_k) \leq p^m (2K)^n (\frac{w}{p}\sqrt{m})^{\nu n} = p^{m-\nu n}(2K)^n w^{\nu n} \to 0 \text{ for } p \to \infty,$$

since $m - \nu n < 0$.

Note that $\Sigma(f) = \Sigma_1$ if $n = m = 1$. So the theorem is proved in this case. We proceed by induction on m. So let $m > 1$ and assume that the theorem is true for each smooth map $P \to Q$ where $\dim(P) < m$.

We prove that $f(\Sigma_2 \setminus \Sigma_3)$ has measure 0. For each $x \in \Sigma_2 \setminus \Sigma_3$ there is a linear differential operator P such that $Pf(x) = 0$ and $\frac{\partial f^i}{\partial x^j}(x) \neq 0$ for some i, j. Let W be the set of all such points, for fixed P, i, j. It suffices to show that $f(W)$ has measure 0. By assumption, $0 \in \mathbb{R}$ is a regular value for the function $Pf^i : W \to \mathbb{R}$. Therefore W is a smooth submanifold of dimension $m-1$ in $\mathbb{R}^m$. Clearly, $\Sigma(f) \cap W$ is contained in the set of all singular points of $f|W : W \to \mathbb{R}^n$, and by induction we get that $f((\Sigma_2 \setminus \Sigma_3) \cap W) \subset f(\Sigma(f) \cap W) \subset f(\Sigma(f|W))$ has measure 0.

It remains to prove that $f(\Sigma_3)$ has measure 0. Every point of Σ_3 has an open neighborhood $W \subset U$ on which $\frac{\partial f^i}{\partial x^j} \neq 0$ for some i, j. By shrinking W if necessary and applying diffeomorphisms, we may assume that

$$\mathbb{R}^{m-1} \times \mathbb{R} \supseteq W_1 \times W_2 = W \xrightarrow{f} \mathbb{R}^{n-1} \times \mathbb{R}, \qquad (y,t) \mapsto (g(y,t),t).$$

Clearly, (y,t) is a critical point for f if and only if y is a critical point for $g(\quad,t)$. Thus $\Sigma(f) \cap W = \bigcup_{t \in W_2} (\Sigma(g(\quad,t)) \times \{t\})$. Since $\dim(W_1) = m-1$, by induction we get that $\mu^{n-1}(g(\Sigma(g(\quad,t),t))) = 0$, where μ^{n-1} is the Lebesque measure in $\mathbb{R}^{n-1}$. By Fubini's theorem we get

$$\begin{aligned} \mu^n\Big(\bigcup_{t \in W_2} (\Sigma(g(\quad,t)) \times \{t\})\Big) &= \int_{W_2} \mu^{n-1}(g(\Sigma(g(\quad,t),t)))\,dt \\ &= \int_{W_2} 0\,dt = 0. \qquad \square \end{aligned}$$

1.19. Embeddings into $\mathbb{R}^n$'s. Let M be a smooth manifold of dimension m. Then M can be embedded into $\mathbb{R}^n$ if

(1) $n = 2m + 1$ (this is due to [**228**]; see also [**84**, p. 55] or [**26**, p. 73]).
(2) $n = 2m$ (see [**228**]).
(3) Conjecture (still unproved): The minimal n is $n = 2m - \alpha(m) + 1$, where $\alpha(m)$ is the number of 1's in the dyadic expansion of m.

There exists an immersion (see section (2)) $M \to \mathbb{R}^n$ if

(4) $n = 2m$ (see [**84**]).
(5) $n = 2m - 1$ (see [**228**]).
(6) Conjecture: The minimal n is $n = 2m - \alpha(m)$. The article [**34**] claims to have proven this. The proof is believed to be incomplete.

Examples and Exercises

1.20. Discuss the following submanifolds of $\mathbb{R}^n$; in particular make drawings of them:

The unit sphere $S^{n-1} = \{x \in \mathbb{R}^n : \langle x, x\rangle = 1\} \subset \mathbb{R}^n$.

The *ellipsoid* $\{x \in \mathbb{R}^n : f(x) := \sum_{i=1}^n \frac{x_i^2}{a_i^2} = 1\}$, $a_i \neq 0$, with principal axis $a_1, \dots, a_n$.

The *hyperboloid* $\{x \in \mathbb{R}^n : f(x) := \sum_{i=1}^n \varepsilon_i \frac{x_i^2}{a_i^2} = 1\}$, $\varepsilon_i = \pm 1$, $a_i \neq 0$, with principal axis a_i and index $= \sum \varepsilon_i$.

The *saddle* $\{x \in \mathbb{R}^3 : x_3 = x_1 x_2\}$.

The *torus*: the rotation surface generated by rotation of $(y-R)^2 + z^2 = r^2$, $0 < r < R$, with center the z–axis, i.e.,

$$\{(x, y, z) : (\sqrt{x^2+y^2} - R)^2 + z^2 = r^2\}.$$

1.21. A compact surface of genus g. Let $f(x) := x(x-1)^2(x-2)^2 \dots (x-(g-1))^2(x-g)$. For small $r > 0$ the set $\{(x, y, z) : (y^2 + f(x))^2 + z^2 = r^2\}$ describes a surface of genus g (topologically a sphere with g handles) in $\mathbb{R}^3$. Visualize this:

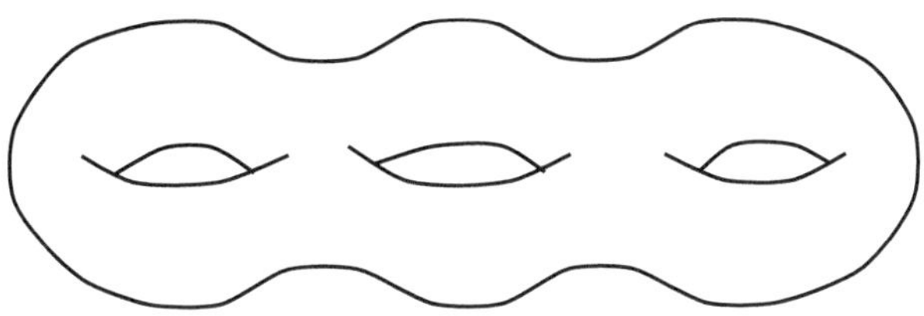

1.22. The Moebius strip. It is not the set of zeros of a regular function on an open neighborhood of $\mathbb{R}^n$. Why not? But it may be represented by the following parameterization:

$$f(r, \varphi) := \begin{pmatrix} \cos\varphi(R + r\cos(\varphi/2)) \\ \sin\varphi(R + r\cos(\varphi/2)) \\ r\sin(\varphi/2) \end{pmatrix},$$

$$(r, \varphi) \in (-1, 1) \times [0, 2\pi),$$

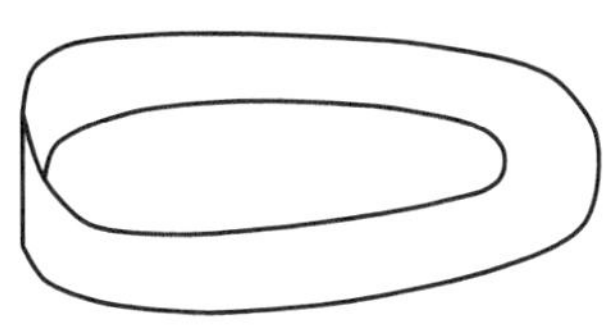

where R is quite big.

1.23. Describe an atlas for the real projective plane which consists of three charts (homogeneous coordinates) and compute the chart changings.

Then describe an atlas for the n-dimensional real projective space $P^n(\mathbb{R})$ and compute the chart changes.

1.24. Let $f : L(\mathbb{R}^n, \mathbb{R}^n) \to L(\mathbb{R}^n, \mathbb{R}^n)$ be given by $f(A) := A^\top A$. Where is f of constant rank? What is $f^{-1}(\mathbb{I}_n)$?

1.25. Let $f : L(\mathbb{R}^n, \mathbb{R}^m) \to L(\mathbb{R}^n, \mathbb{R}^n)$, $n < m$, be given by $f(A) := A^\top A$. Where is f of constant rank? What is $f^{-1}(Id_{\mathbb{R}^n})$?

1.26. Let S be a symmetric matrix, i.e., $S(x, y) := x^\top S y$ is a symmetric bilinear form on $\mathbb{R}^n$. Let $f : L(\mathbb{R}^n, \mathbb{R}^n) \to L(\mathbb{R}^n, \mathbb{R}^n)$ be given by $f(A) := A^\top S A$. Where is f of constant rank? What is $f^{-1}(S)$?

1.27. Describe $TS^2 \subset \mathbb{R}^6$.

2. Submersions and Immersions

2.1. Definition. A mapping $f : M \to N$ between manifolds is called a *submersion* at $x \in M$ if the rank of $T_x f : T_x M \to T_{f(x)} N$ equals $\dim N$. Since the rank cannot fall locally (the determinant of a submatrix of the Jacobi matrix is not 0), f is then a submersion in a whole neighborhood of x. The mapping f is said to be a *submersion* if it is a submersion at each $x \in M$.

2.2. Lemma. *If $f : M \to N$ is a submersion at $x \in M$, then for any chart (V, v) centered at $f(x)$ on N there is chart (U, u) centered at x on M such that $v \circ f \circ u^{-1}$ looks as follows:*

$$(y^1, \dots, y^n, y^{n+1}, \dots, y^m) \mapsto (y^1, \dots, y^n).$$

Proof. Use the inverse function theorem once: Apply the argument from the beginning of (1.13) to $v \circ f \circ u_1^{-1}$ for some chart (U_1, u_1) centered at the point x. □

2.3. Corollary. *Any submersion $f : M \to N$ is open: For each open $U \subset M$ the set $f(U)$ is open in N.* □

2.4. Definition. A triple (M, p, N), where $p : M \to N$ is a surjective submersion, is called a *fibered manifold*. The manifold M is called the *total space* and N is called the *base*.

A fibered manifold admits local sections: For each $x \in M$ there is an open neighborhood U of $p(x)$ in N and a smooth mapping $s : U \to M$ with $p \circ s = Id_U$ and $s(p(x)) = x$.

The existence of local sections in turn implies the following universal property:

$$\begin{array}{ccc} M & & \\ {\scriptstyle p}\downarrow & \searrow & \\ N & \xrightarrow{f} & P. \end{array}$$

If (M, p, N) is a fibered manifold and $f : N \to P$ is a mapping into some further manifold such that $f \circ p : M \to P$ is smooth, then f is smooth.

2.5. Definition. A smooth mapping $f : M \to N$ is called an *immersion at $x \in M$* if the rank of $T_x f : T_x M \to T_{f(x)} N$ equals $\dim M$. Since the rank is maximal at x and cannot fall locally, f is an immersion on a whole neighborhood of x. The mapping f is called an immersion if it is so at every $x \in M$.

2.6. Lemma. *If $f : M \to N$ is an immersion, then for any chart (U, u) centered at $x \in M$ there is a chart (V, v) centered at $f(x)$ on N such that $v \circ f \circ u^{-1}$ has the form*

$$(y^1, \ldots, y^m) \mapsto (y^1, \ldots, y^m, 0, \ldots, 0).$$

Proof. Use the inverse function theorem. □

2.7. Corollary. *If $f : M \to N$ is an immersion, then for any $x \in M$ there is an open neighborhood U of $x \in M$ such that $f(U)$ is a submanifold of N and $f|U : U \to f(U)$ is a diffeomorphism.* □

2.8. Corollary. *If an injective immersion $i : M \to N$ is a homeomorphism onto its image, then $i(M)$ is a submanifold of N.*

Proof. Use (2.7). □

2.9. Definition. If $i : M \to N$ is an injective immersion, then (M, i) is called an *immersed submanifold* of N.

A submanifold is an immersed submanifold, but the converse is wrong in general. The structure of an immersed submanifold (M, i) is in general not determined by the subset $i(M) \subset N$. All this is illustrated by the following example. Consider the curve $\gamma(t) = (\sin^3 t, \sin t.\cos t)$ in $\mathbb{R}^2$. Then $((-\pi, \pi), \gamma|(-\pi, \pi))$ and $((0, 2\pi), \gamma|(0, 2\pi))$ are two different immersed submanifolds, but the image of the embedding is in both cases just the figure eight.

2.10. Let M be a submanifold of N. Then the embedding $i : M \to N$ is an injective immersion with the following property:

(1) *For any manifold* Z *a mapping* $f : Z \to M$ *is smooth if and only if* $i \circ f : Z \to N$ *is smooth.*

There are injective immersions without property (1); see (2.9).

We want to determine all injective immersions $i : M \to N$ with property (1). To require that i is a homeomorphism onto its image is too strong as (2.11) below shows. To look for all smooth mappings $i : M \to N$ with property (2.10.1) (initial mappings in categorical terms) is too difficult as remark (2.12) below shows.

2.11. Example. We consider the 2-dimensional torus $\mathbb{T}^2 = \mathbb{R}^2/\mathbb{Z}^2$. Then the quotient mapping $\pi : \mathbb{R}^2 \to \mathbb{T}^2$ is a covering map, so locally a diffeomorphism. Let us also consider the mapping $f : \mathbb{R} \to \mathbb{R}^2$, $f(t) = (t, \alpha.t)$, where α is irrational. Then $\pi \circ f : \mathbb{R} \to \mathbb{T}^2$ is an injective immersion with dense image, and it is obviously not a homeomorphism onto its image. But $\pi \circ f$ has property (2.10.1), which follows from the fact that π is a covering map.

2.12. Remark. If $f : \mathbb{R} \to \mathbb{R}$ is a function such that f^p and f^q are smooth for some p, q which are relatively prime in $\mathbb{N}$, then f itself turns out to be smooth; see [**97**]. So the mapping $i : t \mapsto \binom{t^p}{t^q}$, $\mathbb{R} \to \mathbb{R}^2$, has property (2.10.1), but i is not an immersion at 0.

In [**98**] all germs of mappings at 0 with property (2.10.1) are characterized as in the following way: Let $g : (\mathbb{R}, 0) \to (\mathbb{R}^n, 0)$ be a germ of a C^∞-curve, $g(t) = (g_1(t), \dots, g_n(t))$. Without loss we may suppose that g is not infinitely flat at 0, so that $g_1(t) = t^r$ for $r \in \mathbb{N}$ after a suitable change of coordinates. Then g *has property* (2.10.1) *near 0 if and only if the Taylor series of* g *is not contained in any* $\mathbb{R}^n[[t^s]]$ *for* $s \geq 2$.

2.13. Definition. For an arbitrary subset A of a manifold N and $x_0 \in A$ let $C_{x_0}(A)$ denote the set of all $x \in A$ which can be joined to x_0 by a smooth curve in M lying in A.

A subset M in a manifold N is called an *initial submanifold* of dimension m if the following property is true:

(1) *For each* $x \in M$ *there exists a chart* (U, u) *centered at* x *on* N *such that* $u(C_x(U \cap M)) = u(U) \cap (\mathbb{R}^m \times 0)$.

The following three lemmas explain the name initial submanifold.

2.14. Lemma. *Let* $f : M \to N$ *be an injective immersion between manifolds with the universal property* (2.10.1). *Then* $f(M)$ *is an initial submanifold of* N.

Proof. Let $x \in M$. By (2.6) we may choose a chart (V, v) centered at $f(x)$ on N and another chart (W, w) centered at x on M such that

$$(v \circ f \circ w^{-1})(y^1, \dots, y^m) = (y^1, \dots, y^m, 0, \dots, 0).$$

Let $r > 0$ be small enough such that $\{y \in \mathbb{R}^m : |y| < 2r\} \subset w(W)$ and also $\{z \in \mathbb{R}^n : |z| < 2r\} \subset v(V)$. Put

$$U := v^{-1}(\{z \in \mathbb{R}^n : |z| < r\}) \subset N,$$
$$W_1 := w^{-1}(\{y \in \mathbb{R}^m : |y| < r\}) \subset M.$$

We claim that $(U, u = v|U)$ satisfies the condition of (2.13.1).

$$\begin{aligned} u^{-1}(u(U) \cap (\mathbb{R}^m \times 0)) &= u^{-1}(\{(y^1, \dots, y^m, 0 \dots, 0) : |y| < r\}) \\ &= f \circ w^{-1} \circ (u \circ f \circ w^{-1})^{-1}(\{(y^1, \dots, y^m, 0 \dots, 0) : |y| < r\}) \\ &= f \circ w^{-1}(\{y \in \mathbb{R}^m : |y| < r\}) = f(W_1) \subseteq C_{f(x)}(U \cap f(M)), \end{aligned}$$

since $f(W_1) \subseteq U \cap f(M)$ and $f(W_1)$ is C^∞-contractible.

Now let conversely $z \in C_{f(x)}(U \cap f(M))$. By definition there is a smooth curve $c : [0,1] \to N$ with $c(0) = f(x)$, $c(1) = z$, and $c([0,1]) \subseteq U \cap f(M)$. By property (2.10.1) the unique curve $\bar{c} : [0,1] \to M$ with $f \circ \bar{c} = c$ is smooth.

We claim that $\bar{c}([0,1]) \subseteq W_1$. If not, then there is some $t \in [0,1]$ with $\bar{c}(t) \in w^{-1}(\{y \in \mathbb{R}^m : r \le |y| < 2r\})$ since $\bar{c}$ is smooth and thus continuous. But then we have

$$\begin{aligned} (v \circ f)(\bar{c}(t)) &\in (v \circ f \circ w^{-1})(\{y \in \mathbb{R}^m : r \le |y| < 2r\}) \\ &= \{(y, 0) \in \mathbb{R}^m \times 0 : r \le |y| < 2r\} \subseteq \{z \in \mathbb{R}^n : r \le |z| < 2r\}. \end{aligned}$$

This means $(v \circ f \circ \bar{c})(t) = (v \circ c)(t) \in \{z \in \mathbb{R}^n : r \le |z| < 2r\}$, so $c(t) \notin U$, a contradiction.

So $\bar{c}([0,1]) \subseteq W_1$; thus $\bar{c}(1) = f^{-1}(z) \in W_1$ and $z \in f(W_1)$. Consequently we have $C_{f(x)}(U \cap f(M)) = f(W_1)$ and finally $f(W_1) = u^{-1}(u(U) \cap (\mathbb{R}^m \times 0))$ by the first part of the proof. □

2.15. Lemma. *Let M be an initial submanifold of a manifold N. Then there is a unique C^∞-manifold structure on M such that the injection $i : M \to N$ is an injective immersion with property* (2.10.1)*:*

(1) *For any manifold Z a mapping $f : Z \to M$ is smooth if and only if $i \circ f : Z \to N$ is smooth.*

The connected components of M are separable (but there may be uncountably many of them).

Proof. We use the sets $C_x(U_x \cap M)$ as charts for M, where $x \in M$ and (U_x, u_x) is a chart for N centered at x with the property required in (2.13.1). Then the chart changings are smooth since they are just restrictions of the chart changings on N. But the sets $C_x(U_x \cap M)$ are not open in the induced topology on M in general. So the identification topology with respect to the charts $(C_x(U_x \cap M), u_x)_{x \in M}$ yields a topology on M which is finer than the induced topology, so it is Hausdorff. Clearly $i : M \to N$ is then an injective immersion. Uniqueness of the smooth structure follows from the universal property (1) which we prove now: For $z \in Z$ we choose a chart (U, u) on N, centered at $f(z)$, such that $u(C_{f(z)}(U \cap M)) = u(U) \cap (\mathbb{R}^m \times 0)$. Then $f^{-1}(U)$ is open in Z and contains a chart (V, v) centered at z on Z with $v(V)$ a ball. Then $f(V)$ is C^∞-contractible in $U \cap M$, so $f(V) \subseteq C_{f(z)}(U \cap M)$, and $(u|C_{f(z)}(U \cap M)) \circ f \circ v^{-1} = u \circ f \circ v^{-1}$ is smooth.

Finally note that N admits a Riemann metric (22.1) which induces one on M, so each connected component of M is separable, by (1.1.4). □

2.16. Transversal mappings. Let M_1, M_2, and N be manifolds and let $f_i : M_i \to N$ be smooth mappings for $i = 1, 2$. We say that f_1 and f_2 are *transversal* at $y \in N$ if

$$\operatorname{im} T_{x_1} f_1 + \operatorname{im} T_{x_2} f_2 = T_y N \quad \text{whenever} \quad f_1(x_1) = f_2(x_2) = y.$$

Note that they are transversal at any y which is not in $f_1(M_1)$ or not in $f_2(M_2)$. The mappings f_1 and f_2 are simply said to be *transversal* if they are transversal at every $y \in N$.

If P is an initial submanifold of N with embedding $i : P \to N$, then a mapping $f : M \to N$ is said to be transversal to P if i and f are transversal.

Lemma. *In this case $f^{-1}(P)$ is an initial submanifold of M with the same codimension in M as P has in N; or $f^{-1}(P)$ is the empty set. If P is a submanifold, then also $f^{-1}(P)$ is a submanifold.*

Proof. Let $x \in f^{-1}(P)$ and let (U, u) be an initial submanifold chart for P centered at $f(x)$ on N, i.e., $u(C_{f(x)}(U \cap P)) = u(U) \cap (\mathbb{R}^p \times 0)$. Then the mapping

$$M \supseteq f^{-1}(U) \xrightarrow{f} U \xrightarrow{u} u(U) \subseteq \mathbb{R}^p \times \mathbb{R}^{n-p} \xrightarrow{\mathrm{pr}_2} \mathbb{R}^{n-p}$$

is a submersion at x since f is transversal to P. So by lemma (2.2) there is a chart (V, v) on M centered at x such that we have

$$(\mathrm{pr}_2 \circ u \circ f \circ v^{-1})(y^1, \dots, y^{n-p}, \dots, y^m) = (y^1, \dots, y^{n-p}).$$

But then $z \in C_x(f^{-1}(P) \cap V)$ if and only if $v(z) \in v(V) \cap (0 \times \mathbb{R}^{m-n+p})$, so $v(C_x(f^{-1}(P) \cap V)) = v(V) \cap (0 \times \mathbb{R}^{m-n+p})$. □

2.17. Corollary. *If $f_1 : M_1 \to N$ and $f_2 : M_2 \to N$ are smooth and transversal, then the topological pullback*

$$M_1 \underset{(f_1,N,f_2)}{\times} M_2 = M_1 \times_N M_2 := \{(x_1, x_2) \in M_1 \times M_2 : f_1(x_1) = f_2(x_2)\}$$

is a submanifold of $M_1 \times M_2$, and it has the following universal property:

> *For any smooth mappings $g_1 : P \to M_1$ and $g_2 : P \to M_2$ with $f_1 \circ g_1 = f_2 \circ g_2$ there is a unique smooth mapping $(g_1, g_2) : P \to M_1 \times_N M_2$ with $\mathrm{pr}_1 \circ (g_1, g_2) = g_1$ and $\mathrm{pr}_2 \circ (g_1, g_2) = g_2$.*

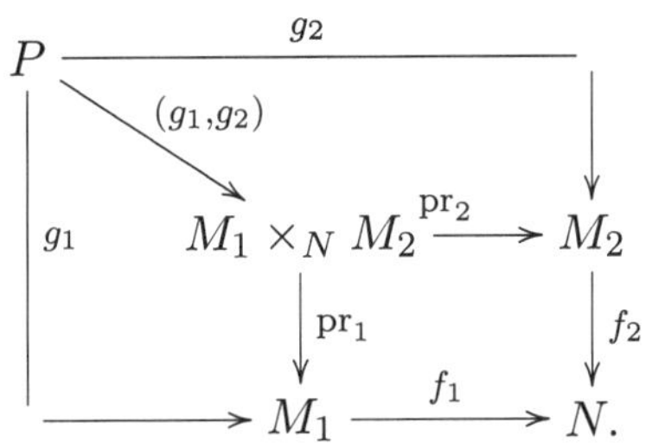

This is also called the pullback property in the category $\mathcal{M}f$ of smooth manifolds and smooth mappings. So one may say that transversal pullbacks exist in the category $\mathcal{M}f$. But there also exist pullbacks which are not transversal.

Proof. $M_1 \times_N M_2 = (f_1 \times f_2)^{-1}(\Delta)$, where $f_1 \times f_2 : M_1 \times M_2 \to N \times N$ and where Δ is the diagonal of $N \times N$, and $f_1 \times f_2$ is transversal to Δ if and only if f_1 and f_2 are transversal. □

3. Vector Fields and Flows

3.1. Definition. A *vector field* X on a manifold M is a smooth section of the tangent bundle; so $X : M \to TM$ is smooth and $\pi_M \circ X = Id_M$. A *local vector field* is a smooth section which is defined on an open subset only. We denote the set of all vector fields by $\mathfrak{X}(M)$. With pointwise addition and scalar multiplication $\mathfrak{X}(M)$ becomes a vector space.

Example. Let (U, u) be a chart on M. Then the $\frac{\partial}{\partial u^i} : U \to TM|U$, $x \mapsto \frac{\partial}{\partial u^i}|_x$, described in (1.8), are local vector fields defined on U.

Lemma. *If X is a vector field on M and (U, u) is a chart on M and $x \in U$, then we have $X(x) = \sum_{i=1}^m X(x)(u^i) \frac{\partial}{\partial u^i}|_x$. We write $X|U = \sum_{i=1}^m X(u^i) \frac{\partial}{\partial u^i}$.* □

3.2. The vector fields $(\frac{\partial}{\partial u^i})_{i=1}^m$ on U, where (U,u) is a chart on M, form a *holonomic frame field.* By a *frame field* on some open set $V \subset M$ we mean $m = \dim M$ vector fields $s_i \in \mathfrak{X}(U)$ such that $s_1(x), \dots, s_m(x)$ is a linear basis of T_xM for each $x \in V$. A frame field is said to be *holonomic* if $s_i = \frac{\partial}{\partial v^i}$ for some chart (V,v). If no such chart may be found locally, the frame field is called *anholonomic.*

With the help of partitions of unity and holonomic frame fields one may construct 'many' vector fields on M. In particular the values of a vector field can be arbitrarily preassigned on a discrete set $\{x_i\} \subset M$.

3.3. Lemma. *The space $\mathfrak{X}(M)$ of vector fields on M coincides canonically with the space of all derivations of the algebra $C^\infty(M)$ of smooth functions, i.e., those $\mathbb{R}$-linear operators $D : C^\infty(M) \to C^\infty(M)$ with*

$$D(fg) = D(f)g + fD(g).$$

Proof. Clearly each vector field $X \in \mathfrak{X}(M)$ defines a derivation (again called X; later sometimes called $\mathcal{L}_X$) of the algebra $C^\infty(M)$ by stipulating $X(f)(x) := X(x)(f) = df(X(x))$.

If conversely a derivation D of $C^\infty(M)$ is given, for any $x \in M$ we consider $D_x : C^\infty(M) \to \mathbb{R}$, $D_x(f) = D(f)(x)$. Then D_x is a derivation at x of $C^\infty(M)$ in the sense of (1.7), so $D_x = X_x$ for some $X_x \in T_xM$. In this way we get a section $X : M \to TM$. If (U,u) is a chart on M, we have $D_x = \sum_{i=1}^m X(x)(u^i)\frac{\partial}{\partial u^i}|_x$ by (1.7). Choose V open in M, $V \subset \overline{V} \subset U$, and $\varphi \in C^\infty(M,\mathbb{R})$ such that $\operatorname{supp}(\varphi) \subset U$ and $\varphi|V = 1$. Then $\varphi \cdot u^i \in C^\infty(M)$ and $(\varphi u^i)|V = u^i|V$. So $D(\varphi u^i)(x) = X(x)(\varphi u^i) = X(x)(u^i)$ and $X|V = \sum_{i=1}^m D(\varphi u^i)|V \cdot \frac{\partial}{\partial u^i}|V$ is smooth. □

3.4. The Lie bracket. By lemma (3.3) we can identify $\mathfrak{X}(M)$ with the vector space of all derivations of the algebra $C^\infty(M)$, which we will do without any notational change in the following.

If X, Y are two vector fields on M, then the mapping $f \mapsto X(Y(f)) - Y(X(f))$ is again a derivation of $C^\infty(M)$, as a simple computation shows. Thus there is a unique vector field $[X,Y] \in \mathfrak{X}(M)$ such that $[X,Y](f) = X(Y(f)) - Y(X(f))$ holds for all $f \in C^\infty(M)$.

In a local chart (U,u) on M one easily checks that for $X|U = \sum X^i\frac{\partial}{\partial u^i}$ and $Y|U = \sum Y^i\frac{\partial}{\partial u^i}$ we have

$$\Big[\sum_i X^i\tfrac{\partial}{\partial u^i}, \sum_j Y^j\tfrac{\partial}{\partial u^j}\Big] = \sum_{i,j}\big(X^i(\tfrac{\partial}{\partial u^i}Y^j) - Y^i(\tfrac{\partial}{\partial u^i}X^j)\big)\tfrac{\partial}{\partial u^j},$$

since second partial derivatives commute. The $\mathbb{R}$-bilinear mapping

$$[\quad,\quad] : \mathfrak{X}(M) \times \mathfrak{X}(M) \to \mathfrak{X}(M)$$

is called the *Lie bracket*. Note also that $\mathfrak{X}(M)$ is a module over the algebra $C^\infty(M)$ by pointwise multiplication $(f, X) \mapsto fX$.

Theorem. *The Lie bracket* $[\ \ ,\ \ ] : \mathfrak{X}(M) \times \mathfrak{X}(M) \to \mathfrak{X}(M)$ *has the following properties:*

$$\begin{aligned}
&[X,Y] = -[Y,X],\\
&[X,[Y,Z]] = [[X,Y],Z] + [Y,[X,Z]], \quad \textit{the Jacobi identity,}\\
&[fX,Y] = f[X,Y] - (Yf)X,\\
&[X,fY] = f[X,Y] + (Xf)Y.
\end{aligned}$$

The form of the Jacobi identity we have chosen says that $\operatorname{ad}(X) = [X,\ \ ]$ is a derivation for the Lie algebra $(\mathfrak{X}(M), [\ \ ,\ \ ])$. The pair $(\mathfrak{X}(M), [\ \ ,\ \ ])$ is the prototype of a *Lie algebra*. The concept of a Lie algebra is one of the most important notions of modern mathematics.

Proof. All these properties are checked easily for the commutator $[X,Y] = X \circ Y - Y \circ X$ in the space of derivations of the algebra $C^\infty(M)$. □

3.5. Integral curves. Let $c : J \to M$ be a smooth curve in a manifold M defined on an interval J. We will use the following notations: $c'(t) = \dot{c}(t) = \frac{d}{dt}c(t) := T_t c.1$. Clearly $c' : J \to TM$ is smooth. We call c' a vector field along c since we have $\pi_M \circ c' = c$:

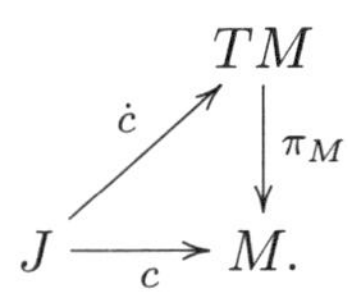

A smooth curve $c : J \to M$ will be called an *integral curve* or *flow line* of a vector field $X \in \mathfrak{X}(M)$ if $c'(t) = X(c(t))$ holds for all $t \in J$.

3.6. Lemma. *Let X be a vector field on M. Then for any $x \in M$ there is an open interval J_x containing 0 and an integral curve $c_x : J_x \to M$ for X (i.e., $c_x' = X \circ c_x$) with $c_x(0) = x$. If J_x is maximal, then c_x is unique.*

Proof. In a chart (U, u) on M with $x \in U$ the equation $c'(t) = X(c(t))$ is a system ordinary differential equations with initial condition $c(0) = x$. Since X is smooth, there is a unique local solution which even depends smoothly on the initial values, by the theorem of Picard-Lindelöf, [**41**, 10.7.4]. So on M there are always local integral curves. If $J_x = (a, b)$ and $\lim_{t \to b-} c_x(t) =: c_x(b)$ exists in M, there is a unique local solution c_1 defined in an open interval containing b with $c_1(b) = c_x(b)$. By uniqueness of the solution on the intersection of the two intervals, c_1 prolongs c_x to a larger interval. This may be repeated (also on the left hand side of J_x) as long as the limit exists. So if we suppose J_x to be maximal, J_x either equals $\mathbb{R}$ or the integral curve leaves the manifold in finite (parameter-)time in the past or future or both. □

3.7. The flow of a vector field. Let $X \in \mathfrak{X}(M)$ be a vector field. Let us write $\mathrm{Fl}^X_t(x) = \mathrm{Fl}^X(t,x) := c_x(t)$, where $c_x : J_x \to M$ is the maximally defined integral curve of X with $c_x(0) = x$, constructed in lemma (3.6).

Theorem. *For each vector field X on M, the mapping $\mathrm{Fl}^X : \mathcal{D}(X) \to M$ is smooth, where $\mathcal{D}(X) = \bigcup_{x\in M} J_x \times \{x\}$ is an open neighborhood of $0 \times M$ in $\mathbb{R} \times M$. We have*

$$\mathrm{Fl}^X(t+s,x) = \mathrm{Fl}^X(t, \mathrm{Fl}^X(s,x))$$

in the following sense. If the right hand side exists, then the left hand side exists and we have equality. If both t, $s \geq 0$ or both are ≤ 0, and if the left hand side exists, then also the right hand side exists and we have equality.

Proof. As mentioned in the proof of (3.6), $\mathrm{Fl}^X(t,x)$ is smooth in (t,x) for small t, and if it is defined for (t,x), then it is also defined for (s,y) nearby. These are local properties which follow from the theory of ordinary differential equations.

Now let us treat the equation $\mathrm{Fl}^X(t+s,x) = \mathrm{Fl}^X(t,\mathrm{Fl}^X(s,x))$. If the right hand side exists, then we consider the equation

$$\begin{cases} \frac{d}{dt}\mathrm{Fl}^X(t+s,x) = \frac{d}{du}\mathrm{Fl}^X(u,x)|_{u=t+s} = X(\mathrm{Fl}^X(t+s,x)), \\ \mathrm{Fl}^X(t+s,x)|_{t=0} = \mathrm{Fl}^X(s,x). \end{cases}$$

But the unique solution of this is $\mathrm{Fl}^X(t,\mathrm{Fl}^X(s,x))$. So the left hand side exists and equals the right hand side.

If the left hand side exists, let us suppose that $t, s \geq 0$. We put

$$c_x(u) = \begin{cases} 2\,\mathrm{Fl}^X(u,x) & \text{if } u \leq s, \\ \mathrm{Fl}^X(u-s,\mathrm{Fl}^X(s,x)) & \text{if } u \geq s. \end{cases}$$

Then we have

$$\begin{aligned} \tfrac{d}{du}c_x(u) &= \begin{cases} \frac{d}{du}\mathrm{Fl}^X(u,x) = X(\mathrm{Fl}^X(u,x)) \quad \text{for } u \leq s, \\ \frac{d}{du}\mathrm{Fl}^X(u-s,\mathrm{Fl}^X(s,x)) = X(\mathrm{Fl}^X(u-s,\mathrm{Fl}^X(s,x))) \end{cases} \\ &= X(c_x(u)) \quad \text{for } 0 \leq u \leq t+s. \end{aligned}$$

Also $c_x(0) = x$ and on the overlap both definitions coincide by the first part of the proof; thus we conclude that $c_x(u) = \mathrm{Fl}^X(u,x)$ for $0 \leq u \leq t+s$ and we have $\mathrm{Fl}^X(t,\mathrm{Fl}^X(s,x)) = c_x(t+s) = \mathrm{Fl}^X(t+s,x)$.

Now we show that $\mathcal{D}(X)$ is open and Fl^X is smooth on $\mathcal{D}(X)$. We know already that $\mathcal{D}(X)$ is a neighborhood of $0 \times M$ in $\mathbb{R} \times M$ and that Fl^X is smooth near $0 \times M$.

For $x \in M$ let J'_x be the set of all $t \in \mathbb{R}$ such that Fl^X is defined and smooth on an open neighborhood of $[0,t] \times \{x\}$ (respectively on $[t,0] \times \{x\}$ for $t < 0$) in $\mathbb{R} \times M$. We claim that $J'_x = J_x$, which finishes the proof. It suffices to show that J'_x is not empty, open and closed in J_x. It is open by construction, and not empty, since $0 \in J'_x$. If J'_x is not closed in J_x, let $t_0 \in J_x \cap (\overline{J'_x} \setminus J'_x)$ and suppose that $t_0 > 0$, say. By the local existence and smoothness Fl^X exists and is smooth near $[-\varepsilon, \varepsilon] \times \{y := \mathrm{Fl}^X(t_0, x)\}$ in $\mathbb{R} \times M$ for some $\varepsilon > 0$, and by construction Fl^X exists and is smooth near $[0, t_0 - \varepsilon] \times \{x\}$. Since $\mathrm{Fl}^X(-\varepsilon, y) = \mathrm{Fl}^X(t_0 - \varepsilon, x)$, we conclude for t near $[0, t_0 - \varepsilon]$, x' near x, and t' near $[-\varepsilon, \varepsilon]$ that $\mathrm{Fl}^X(t + t', x') = \mathrm{Fl}^X(t', \mathrm{Fl}^X(t, x'))$ exists and is smooth. So $t_0 \in J'_x$, a contradiction. □

3.8. Let $X \in \mathfrak{X}(M)$ be a vector field. Its flow Fl^X is called *global* or *complete* if its domain of definition $\mathcal{D}(X)$ equals $\mathbb{R} \times M$. Then the vector field X itself will be called a *complete vector field*. In this case Fl^X_t is also sometimes called $\exp tX$; it is a diffeomorphism of M. The *support* $\mathrm{supp}(X)$ of a vector field X is the closure of the set $\{x \in M : X(x) \neq 0\}$.

Lemma. *A vector field with compact support on M is complete.*

Proof. Let $K = \mathrm{supp}(X)$ be compact. Then the compact set $0 \times K$ has positive distance to the disjoint closed set $(\mathbb{R} \times M) \setminus \mathcal{D}(X)$ (if it is not empty), so $[-\varepsilon, \varepsilon] \times K \subset \mathcal{D}(X)$ for some $\varepsilon > 0$. If $x \notin K$, then $X(x) = 0$, so $\mathrm{Fl}^X(t, x) = x$ for all t and $\mathbb{R} \times \{x\} \subset \mathcal{D}(X)$. So we have $[-\varepsilon, \varepsilon] \times M \subset \mathcal{D}(X)$. Since $\mathrm{Fl}^X(t + \varepsilon, x) = \mathrm{Fl}^X(t, \mathrm{Fl}^X(\varepsilon, x))$ exists for $|t| \leq \varepsilon$ by theorem (3.7), we have $[-2\varepsilon, 2\varepsilon] \times M \subset \mathcal{D}(X)$ and by repeating this argument we get $\mathbb{R} \times M = \mathcal{D}(X)$. □

So on a compact manifold M each vector field is complete. If M is not compact and of dimension ≥ 2, then in general the set of complete vector fields on M is neither a vector space nor is it closed under the Lie bracket, as the following example on $\mathbb{R}^2$ shows: $X = y\frac{\partial}{\partial x}$ and $Y = \frac{x^2}{2}\frac{\partial}{\partial y}$ are complete, but neither $X + Y$ nor $[X, Y]$ is complete. In general one may embed $\mathbb{R}^2$ as a closed submanifold into M and extend the vector fields X and Y.

3.9. f-related vector fields. If $f : M \to M$ is a diffeomorphism, then for any vector field $X \in \mathfrak{X}(M)$ the mapping $Tf^{-1} \circ X \circ f$ is also a vector field, which we will denote by f^*X. We also put $f_*X := Tf \circ X \circ f^{-1} = (f^{-1})^*X$.

But if $f : M \to N$ is a smooth mapping and $Y \in \mathfrak{X}(N)$ is a vector field, there may or may not exist a vector field $X \in \mathfrak{X}(M)$ such that the following diagram commutes:

$$(1)\qquad \begin{array}{ccc} TM & \xrightarrow{Tf} & TN \\ {\scriptstyle X}\uparrow & & \uparrow{\scriptstyle Y} \\ M & \xrightarrow{f} & N. \end{array}$$

Definition. Let $f : M \to N$ be a smooth mapping. Two vector fields $X \in \mathfrak{X}(M)$ and $Y \in \mathfrak{X}(N)$ are called *f-related* if $Tf \circ X = Y \circ f$ holds, i.e., if diagram (1) commutes.

Example. If $X \in \mathfrak{X}(M)$ and $Y \in \mathfrak{X}(N)$ and if $X \times Y \in \mathfrak{X}(M \times N)$ is given by $(X \times Y)(x, y) = (X(x), Y(y))$, then we have:

(2) $X \times Y$ and X are pr_1-related.

(3) $X \times Y$ and Y are pr_2-related.

(4) X and $X \times Y$ are $\mathrm{ins}(y)$-related if and only if $Y(y) = 0$, where the mapping $\mathrm{ins}(y) : M \to M \times N$ is given by $\mathrm{ins}(y)(x) = (x, y)$.

3.10. Lemma. *Consider vector fields $X_i \in \mathfrak{X}(M)$ and $Y_i \in \mathfrak{X}(N)$ for $i = 1, 2$, and a smooth mapping $f : M \to N$. If X_i and Y_i are f-related for $i = 1, 2$, then also $\lambda_1 X_1 + \lambda_2 X_2$ and $\lambda_1 Y_1 + \lambda_2 Y_2$ are f-related, and also $[X_1, X_2]$ and $[Y_1, Y_2]$ are f-related.*

Proof. The first assertion is immediate. To prove the second, we choose $h \in C^\infty(N)$. Then by assumption we have $Tf \circ X_i = Y_i \circ f$; thus:

$$\begin{aligned}(X_i(h \circ f))(x) &= X_i(x)(h \circ f) = (T_x f . X_i(x))(h) \\ &= (Tf \circ X_i)(x)(h) = (Y_i \circ f)(x)(h) = Y_i(f(x))(h) = (Y_i(h))(f(x)),\end{aligned}$$

so $X_i(h \circ f) = (Y_i(h)) \circ f$, and we may continue:

$$\begin{aligned}[X_1, X_2](h \circ f) &= X_1(X_2(h \circ f)) - X_2(X_1(h \circ f)) \\ &= X_1(Y_2(h) \circ f) - X_2(Y_1(h) \circ f) \\ &= Y_1(Y_2(h)) \circ f - Y_2(Y_1(h)) \circ f = [Y_1, Y_2](h) \circ f.\end{aligned}$$

But this means $Tf \circ [X_1, X_2] = [Y_1, Y_2] \circ f$. □

3.11. Corollary. *If $f : M \to N$ is a local diffeomorphism (so $(T_x f)^{-1}$ makes sense for each $x \in M$), then for $Y \in \mathfrak{X}(N)$ a vector field $f^*Y \in \mathfrak{X}(M)$ is defined by $(f^*Y)(x) = (T_x f)^{-1}.Y(f(x))$. The linear mapping $f^* : \mathfrak{X}(N) \to \mathfrak{X}(M)$ is then a Lie algebra homomorphism, i.e.,*

$$f^*[Y_1, Y_2] = [f^*Y_1, f^*Y_2].$$

3.12. The Lie derivative of functions. For a vector field $X \in \mathfrak{X}(M)$ and $f \in C^\infty(M)$ we define $\mathcal{L}_X f \in C^\infty(M)$ by

$$\begin{aligned} \mathcal{L}_X f(x) &:= \tfrac{d}{dt}|_0 f(\mathrm{Fl}^X(t,x)) \quad \text{or} \\ \mathcal{L}_X f &:= \tfrac{d}{dt}|_0 (\mathrm{Fl}^X_t)^* f = \tfrac{d}{dt}|_0 (f \circ \mathrm{Fl}^X_t). \end{aligned}$$

Since $\mathrm{Fl}^X(t,x)$ is defined for small t, for any $x \in M$, the expressions above make sense.

Lemma. *We have*

$$\tfrac{d}{dt}(\mathrm{Fl}^X_t)^* f = (\mathrm{Fl}^X_t)^* X(f) = X((\mathrm{Fl}^X_t)^* f);$$

in particular for $t = 0$ *we have* $\mathcal{L}_X f = X(f) = df(X)$.

Proof. We have

$$\tfrac{d}{dt}(\mathrm{Fl}^X_t)^* f(x) = df(\tfrac{d}{dt}\mathrm{Fl}^X(t,x)) = df(X(\mathrm{Fl}^X(t,x))) = (\mathrm{Fl}^X_t)^*(Xf)(x).$$

From this we get $\mathcal{L}_X f = X(f) = df(X)$ and then in turn

$$\tfrac{d}{dt}(\mathrm{Fl}^X_t)^* f = \tfrac{d}{ds}|_0 (\mathrm{Fl}^X_t \circ \mathrm{Fl}^X_s)^* f = \tfrac{d}{ds}|_0 (\mathrm{Fl}^X_s)^* (\mathrm{Fl}^X_t)^* f = X((\mathrm{Fl}^X_t)^* f). \quad \square$$

3.13. The Lie derivative for vector fields. For $X, Y \in \mathfrak{X}(M)$ we define $\mathcal{L}_X Y \in \mathfrak{X}(M)$ by

$$\mathcal{L}_X Y := \tfrac{d}{dt}|_0 (\mathrm{Fl}^X_t)^* Y = \tfrac{d}{dt}|_0 (T(\mathrm{Fl}^X_{-t}) \circ Y \circ \mathrm{Fl}^X_t),$$

and call it the *Lie derivative* of Y along X.

Lemma. *We have*

$$\begin{aligned} \mathcal{L}_X Y &= [X, Y], \\ \tfrac{d}{dt}(\mathrm{Fl}^X_t)^* Y = (\mathrm{Fl}^X_t)^* \mathcal{L}_X Y &= (\mathrm{Fl}^X_t)^*[X,Y] = \mathcal{L}_X (\mathrm{Fl}^X_t)^* Y = [X, (\mathrm{Fl}^X_t)^* Y]. \end{aligned}$$

Proof. For $f \in C^\infty(M)$ consider the mapping $\alpha(t,s) := Y(\mathrm{Fl}^X(t,x))(f \circ \mathrm{Fl}^X_s)$, which is locally defined near 0. It satisfies

$$\begin{aligned} \alpha(t,0) &= Y(\mathrm{Fl}^X(t,x))(f), \\ \alpha(0,s) &= Y(x)(f \circ \mathrm{Fl}^X_s), \\ \tfrac{\partial}{\partial t}\alpha(0,0) &= \partial|_0 Y(\mathrm{Fl}^X(t,x))(f) = \partial|_0 (Yf)(\mathrm{Fl}^X(t,x)) = X(x)(Yf), \\ \tfrac{\partial}{\partial s}\alpha(0,0) &= \tfrac{\partial}{\partial s}|_0 Y(x)(f \circ \mathrm{Fl}^X_s) = Y(x)\tfrac{\partial}{\partial s}|_0 (f \circ \mathrm{Fl}^X_s) = Y(x)(Xf). \end{aligned}$$

But on the other hand we have

$$\begin{aligned} \tfrac{\partial}{\partial u}|_0 \alpha(u,-u) &= \tfrac{\partial}{\partial u}|_0 Y(\mathrm{Fl}^X(u,x))(f \circ \mathrm{Fl}^X_{-u}) \\ &= \tfrac{\partial}{\partial u}|_0 \left(T(\mathrm{Fl}^X_{-u}) \circ Y \circ \mathrm{Fl}^X_u\right)_x (f) = (\mathcal{L}_X Y)_x(f), \end{aligned}$$

so the first assertion follows. For the second claim we compute as follows:

$$\begin{aligned}\tfrac{\partial}{\partial t}(\mathrm{Fl}^X_t)^*Y &= \tfrac{\partial}{\partial s}|_0\left(T(\mathrm{Fl}^X_{-t})\circ T(\mathrm{Fl}^X_{-s})\circ Y\circ \mathrm{Fl}^X_s\circ \mathrm{Fl}^X_t\right)\\ &= T(\mathrm{Fl}^X_{-t})\circ \tfrac{\partial}{\partial s}|_0\left(T(\mathrm{Fl}^X_{-s})\circ Y\circ \mathrm{Fl}^X_s\right)\circ \mathrm{Fl}^X_t\\ &= T(\mathrm{Fl}^X_{-t})\circ [X,Y]\circ \mathrm{Fl}^X_t = (\mathrm{Fl}^X_t)^*[X,Y].\\ \tfrac{\partial}{\partial t}(\mathrm{Fl}^X_t)^*Y &= \tfrac{\partial}{\partial s}|_0(\mathrm{Fl}^X_s)^*(\mathrm{Fl}^X_t)^*Y = \mathcal{L}_X(\mathrm{Fl}^X_t)^*Y. \quad\square\end{aligned}$$

3.14. Lemma. *Let $X\in\mathfrak{X}(M)$ and $Y\in\mathfrak{X}(N)$ be f-related vector fields for a smooth mapping $f: M\to N$. Then we have $f\circ \mathrm{Fl}^X_t = \mathrm{Fl}^Y_t\circ f$, whenever both sides are defined. In particular, if f is a diffeomorphism, we have $\mathrm{Fl}^{f^*Y}_t = f^{-1}\circ \mathrm{Fl}^Y_t\circ f$.*

Proof. We have $\frac{d}{dt}(f\circ \mathrm{Fl}^X_t) = Tf\circ \frac{d}{dt}\mathrm{Fl}^X_t = Tf\circ X\circ \mathrm{Fl}^X_t = Y\circ f\circ Fl^X_t$ and $f(\mathrm{Fl}^X(0,x)) = f(x)$. So $t\mapsto f(\mathrm{Fl}^X(t,x))$ is an integral curve of the vector field Y on N with initial value $f(x)$, so we have $f(\mathrm{Fl}^X(t,x)) = \mathrm{Fl}^Y(t,f(x))$ or $f\circ \mathrm{Fl}^X_t = \mathrm{Fl}^Y_t\circ f$. $\square$

3.15. Corollary. *Let $X,Y\in\mathfrak{X}(M)$. Then the following assertions are equivalent:*

(1) $\mathcal{L}_XY = [X,Y] = 0$.

(2) $(\mathrm{Fl}^X_t)^*Y = Y$, *wherever defined.*

(3) $\mathrm{Fl}^X_t\circ \mathrm{Fl}^Y_s = \mathrm{Fl}^Y_s\circ \mathrm{Fl}^X_t$, *wherever defined.*

Proof. (1) $\Leftrightarrow$ (2) is immediate from lemma (3.13). To see (2) $\Leftrightarrow$ (3), we note that $\mathrm{Fl}^X_t\circ \mathrm{Fl}^Y_s = \mathrm{Fl}^Y_s\circ \mathrm{Fl}^X_t$ if and only if $\mathrm{Fl}^Y_s = \mathrm{Fl}^X_{-t}\circ \mathrm{Fl}^Y_s\circ \mathrm{Fl}^X_t = \mathrm{Fl}^{(\mathrm{Fl}^X_t)^*Y}_s$ by lemma (3.14); and this in turn is equivalent to $Y = (\mathrm{Fl}^X_t)^*Y$. $\square$

3.16. Theorem. *Let M be a manifold, let $\varphi^i: \mathbb{R}\times M\supset U_{\varphi^i}\to M$ be smooth mappings for $i=1,\dots,k$ where each U_{φ^i} is an open neighborhood of $\{0\}\times M$ in $\mathbb{R}\times M$, such that each φ^i_t is a diffeomorphism on its domain, $\varphi^i_0 = Id_M$, and $\partial|_0\varphi^i_t = X_i\in\mathfrak{X}(M)$. We put $[\varphi^i,\varphi^j]_t = [\varphi^i_t,\varphi^j_t] := (\varphi^j_t)^{-1}\circ(\varphi^i_t)^{-1}\circ\varphi^j_t\circ\varphi^i_t$. Then for each formal bracket expression P of length k we have*

$$0 = \tfrac{\partial^\ell}{\partial t^\ell}|_0 P(\varphi^1_t,\dots,\varphi^k_t)\quad \text{for } 1\le \ell < k,$$

$$P(X_1,\dots,X_k) = \tfrac{1}{k!}\tfrac{\partial^k}{\partial t^k}|_0 P(\varphi^1_t,\dots,\varphi^k_t)\in\mathfrak{X}(M)$$

in the sense explained in step 2 of the proof. In particular we have for vector fields $X,Y\in\mathfrak{X}(M)$

$$0 = \partial|_0(\mathrm{Fl}^Y_{-t}\circ \mathrm{Fl}^X_{-t}\circ \mathrm{Fl}^Y_t\circ \mathrm{Fl}^X_t),$$

$$[X,Y] = \tfrac{1}{2}\tfrac{\partial^2}{\partial t^2}|_0(\mathrm{Fl}^Y_{-t}\circ \mathrm{Fl}^X_{-t}\circ \mathrm{Fl}^Y_t\circ \mathrm{Fl}^X_t).$$

Proof. Step 1. Let $c : \mathbb{R} \to M$ be a smooth curve. If $c(0) = x \in M$, $c'(0) = 0, \dots, c^{(k-1)}(0) = 0$, then $c^{(k)}(0)$ is a well defined tangent vector in $T_x M$ which is given by the derivation $f \mapsto (f \circ c)^{(k)}(0)$ at x. Namely, we have

$$\begin{aligned}((f.g) \circ c)^{(k)}(0) &= ((f \circ c).(g \circ c))^{(k)}(0) = \sum_{j=0}^{k} \binom{k}{j} (f \circ c)^{(j)}(0)(g \circ c)^{(k-j)}(0) \\ &= (f \circ c)^{(k)}(0) g(x) + f(x)(g \circ c)^{(k)}(0),\end{aligned}$$

since all other summands vanish: $(f \circ c)^{(j)}(0) = 0$ for $1 \le j < k$.

Step 2. Let $\varphi : \mathbb{R} \times M \supset U_\varphi \to M$ be a smooth mapping where U_φ is an open neighborhood of $\{0\} \times M$ in $\mathbb{R} \times M$, such that each φ_t is a diffeomorphism on its domain and $\varphi_0 = Id_M$. We say that φ_t is a *curve of local diffeomorphisms* through Id_M.

From step 1 we see that if $\frac{\partial^j}{\partial t^j}|_0 \varphi_t = 0$ for all $1 \le j < k$, then $X := \frac{1}{k!} \frac{\partial^k}{\partial t^k}|_0 \varphi_t$ is a well defined vector field on M. We say that X is the *first nonvanishing derivative* at 0 of the curve φ_t of local diffeomorphisms. We may paraphrase this as $(\partial_t^k|_0 \varphi_t^*) f = k! \mathcal{L}_X f$.

Claim 3. Let φ_t, ψ_t be curves of local diffeomorphisms through Id_M and let $f \in C^\infty(M)$. Then we have

$$\partial_t^k|_0 (\varphi_t \circ \psi_t)^* f = \partial_t^k|_0 (\psi_t^* \circ \varphi_t^*) f = \sum_{j=0}^{k} \binom{k}{j} (\partial_t^j|_0 \psi_t^*)(\partial_t^{k-j}|_0 \varphi_t^*) f.$$

Also the multinomial version of this formula holds:

$$\partial_t^k|_0 (\varphi_t^1 \circ \dots \circ \varphi_t^\ell)^* f = \sum_{j_1 + \dots + j_\ell = k} \frac{k!}{j_1! \dots j_\ell!} (\partial_t^{j_\ell}|_0 (\varphi_t^\ell)^*) \dots (\partial_t^{j_1}|_0 (\varphi_t^1)^*) f.$$

We only show the binomial version. For a function $h(t,s)$ of two variables we have

$$\partial_t^k h(t,t) = \sum_{j=0}^{k} \binom{k}{j} \partial_t^j \partial_s^{k-j} h(t,s)|_{s=t},$$

since for $h(t,s) = f(t)g(s)$ this is just a consequence of the Leibniz rule, and linear combinations of such decomposable tensors are dense in the space of all functions of two variables in the compact C^∞-topology, so that by continuity the formula holds for all functions. In the following form it implies the claim:

$$\partial_t^k|_0 f(\varphi(t, \psi(t,x))) = \sum_{j=0}^{k} \binom{k}{j} \partial_t^j \partial_s^{k-j} f(\varphi(t, \psi(s,x)))|_{t=s=0}.$$

Claim 4. Let φ_t be a curve of local diffeomorphisms through Id_M with first nonvanishing derivative $k! X = \partial_t^k|_0 \varphi_t$. Then the inverse curve of local diffeomorphisms φ_t^{-1} has first nonvanishing derivative $-k!X = \partial_t^k|_0\varphi_t^{-1}$, for we have $\varphi_t^{-1}\circ\varphi_t = Id$, so by claim 3 we get for $1\le j\le k$

$$\begin{aligned}0 = \partial_t^j|_0(\varphi_t^{-1}\circ\varphi_t)^*f &= \sum_{i=0}^{j}\tbinom{j}{i}(\partial_t^i|_0\varphi_t^*)(\partial_t^{j-i}(\varphi_t^{-1})^*)f\\ &= \partial_t^j|_0\varphi_t^*(\varphi_0^{-1})^*f + \varphi_0^*\partial_t^j|_0(\varphi_t^{-1})^*f,\end{aligned}$$

i.e., $\partial_t^j|_0\varphi_t^*f = -\partial_t^j|_0(\varphi_t^{-1})^*f$ as required.

Claim 5. Let φ_t be a curve of local diffeomorphisms through Id_M with first nonvanishing derivative $m!X = \partial_t^m|_0\varphi_t$, and let ψ_t be a curve of local diffeomorphisms through Id_M with first nonvanishing derivative $n!Y = \partial_t^n|_0\psi_t$. Then the curve of local diffeomorphisms $[\varphi_t,\psi_t] = \psi_t^{-1}\circ\varphi_t^{-1}\circ\psi_t\circ\varphi_t$ has first nonvanishing derivative

$$(m+n)![X,Y] = \partial_t^{m+n}|_0[\varphi_t,\psi_t].$$

From this claim the theorem follows.

By the multinomial version of claim 3 we have

$$\begin{aligned}A_Nf :&= \partial_t^N|_0(\psi_t^{-1}\circ\varphi_t^{-1}\circ\psi_t\circ\varphi_t)^*f\\ &= \sum_{i+j+k+\ell=N}\frac{N!}{i!j!k!\ell!}(\partial_t^i|_0\varphi_t^*)(\partial_t^j|_0\psi_t^*)(\partial_t^k|_0(\varphi_t^{-1})^*)(\partial_t^\ell|_0(\psi_t^{-1})^*)f.\end{aligned}$$

Let us suppose that $1\le n\le m$; the case $m\le n$ is similar. If $N<n$, all summands are 0. If $N=n$, we have by claim 4

$$A_Nf = (\partial_t^n|_0\varphi_t^*)f + (\partial_t^n|_0\psi_t^*)f + (\partial_t^n|_0(\varphi_t^{-1})^*)f + (\partial_t^n|_0(\psi_t^{-1})^*)f = 0.$$

If $n<N\le m$, we have, using again claim 4:

$$\begin{aligned}A_Nf &= \sum_{j+\ell=N}\frac{N!}{j!\ell!}(\partial_t^j|_0\psi_t^*)(\partial_t^\ell|_0(\psi_t^{-1})^*)f + \delta_N^m\left((\partial_t^m|_0\varphi_t^*)f + (\partial_t^m|_0(\varphi_t^{-1})^*)f\right)\\ &= (\partial_t^N|_0(\psi_t^{-1}\circ\psi_t)^*)f + 0 = 0.\end{aligned}$$

Now we come to the difficult case $m,n<N\le m+n$.

$$\begin{aligned}A_Nf &= \partial_t^N|_0(\psi_t^{-1}\circ\varphi_t^{-1}\circ\psi_t)^*f + \tbinom{N}{m}(\partial_t^m|_0\varphi_t^*)(\partial_t^{N-m}|_0(\psi_t^{-1}\circ\varphi_t^{-1}\circ\psi_t)^*)f\\ &\quad + (\partial_t^N|_0\varphi_t^*)f, \qquad (6)\end{aligned}$$

by claim 3, since all other terms vanish; see (8) below. By claim 3 again we get:

$$\begin{aligned}&\partial_t^N|_0(\psi_t^{-1}\circ\varphi_t^{-1}\circ\psi_t)^*f\\ &\quad= \sum_{j+k+\ell=N}\frac{N!}{j!k!\ell!}(\partial_t^j|_0\psi_t^*)(\partial_t^k|_0(\varphi_t^{-1})^*)(\partial_t^\ell|_0(\psi_t^{-1})^*)f\end{aligned}$$

$$\begin{aligned}
&= \sum_{j+\ell=N} \tbinom{N}{j}(\partial_t^j|_0\psi_t^*)(\partial_t^\ell|_0(\psi_t^{-1})^*)f \\
&\quad + \tbinom{N}{m}(\partial_t^{N-m}|_0\psi_t^*)(\partial_t^m|_0(\varphi_t^{-1})^*)f \\
&\quad + \tbinom{N}{m}(\partial_t^m|_0(\varphi_t^{-1})^*)(\partial_t^{N-m}|_0(\psi_t^{-1})^*)f + \partial_t^N|_0(\varphi_t^{-1})^*f \\
&= 0 + \tbinom{N}{m}(\partial_t^{N-m}|_0\psi_t^*)m!\mathcal{L}_{-X}f + \tbinom{N}{m}m!\mathcal{L}_{-X}(\partial_t^{N-m}|_0(\psi_t^{-1})^*)f \\
&\quad + \partial_t^N|_0(\varphi_t^{-1})^*f \\
&= \delta_{m+n}^N(m+n)!(\mathcal{L}_X\mathcal{L}_Y - \mathcal{L}_Y\mathcal{L}_X)f + \partial_t^N|_0(\varphi_t^{-1})^*f
\end{aligned}$$

$$= \delta_{m+n}^N(m+n)!\mathcal{L}_{[X,Y]}f + \partial_t^N|_0(\varphi_t^{-1})^*f. \tag{7}$$

From the second expression in (7) one can also read off that

$$\partial_t^{N-m}|_0(\psi_t^{-1}\circ\varphi_t^{-1}\circ\psi_t)^*f = \partial_t^{N-m}|_0(\varphi_t^{-1})^*f. \tag{8}$$

If we put (7) and (8) into (6), we get, using claims 3 and 4 again, the final result which proves claim 5 and the theorem:

$$\begin{aligned}
A_Nf &= \delta_{m+n}^N(m+n)!\mathcal{L}_{[X,Y]}f + \partial_t^N|_0(\varphi_t^{-1})^*f \\
&\quad + \tbinom{N}{m}(\partial_t^m|_0\varphi_t^*)(\partial_t^{N-m}|_0(\varphi_t^{-1})^*)f + (\partial_t^N|_0\varphi_t^*)f \\
&= \delta_{m+n}^N(m+n)!\mathcal{L}_{[X,Y]}f + \partial_t^N|_0(\varphi_t^{-1}\circ\varphi_t)^*f \\
&= \delta_{m+n}^N(m+n)!\mathcal{L}_{[X,Y]}f + 0. \quad \square
\end{aligned}$$

3.17. Theorem. *Let $X_1, \dots, X_m$ be vector fields on M defined in a neighborhood of a point $x \in M$ such that $X_1(x), \dots, X_m(x)$ are a basis for T_xM and $[X_i, X_j] = 0$ for all i, j.*

Then there is a chart (U, u) of M centered at x such that $X_i|U = \frac{\partial}{\partial u^i}$.

Proof. For small $t = (t^1, \dots, t^m) \in \mathbb{R}^m$ we put

$$f(t^1, \dots, t^m) = (\mathrm{Fl}_{t^1}^{X_1} \circ \dots \circ \mathrm{Fl}_{t^m}^{X_m})(x).$$

By (3.15) we may interchange the order of the flows arbitrarily. Therefore

$$\tfrac{\partial}{\partial t^i} f(t^1, \dots, t^m) = \tfrac{\partial}{\partial t^i}(\mathrm{Fl}_{t^i}^{X_i} \circ \mathrm{Fl}_{t^1}^{X_1} \circ \cdots)(x) = X_i((\mathrm{Fl}_{t^1}^{x_1} \circ \cdots)(x)).$$

So T_0f is invertible, f is a local diffeomorphism, and its inverse gives a chart with the desired properties. $\square$

3.18. The theorem of Frobenius. The next three subsections will be devoted to the theorem of Frobenius for distributions of constant rank. We will give a powerful generalization for distributions of nonconstant rank below in (3.21) – (3.28).

Let M be a manifold. By a *vector subbundle* E of TM of fiber dimension k we mean a subset $E \subset TM$ such that each $E_x := E \cap T_xM$ is a linear subspace of dimension k and such that for each x im M there are k vector fields defined on an open neighborhood of M with values in E and spanning E, called a *local frame* for E. Such an E is also called a smooth *distribution* of constant rank k. See section (8) for a thorough discussion of the notion of vector bundles. The space of all vector fields with values in E will be called $\Gamma(E)$.

The vector subbundle E of TM is called *integrable* or *involutive*, if for all $X, Y \in \Gamma(E)$ we have $[X, Y] \in \Gamma(E)$.

Local version of Frobenius's theorem. *Let $E \subset TM$ be an integrable vector subbundle of fiber dimension k of TM.*

Then for each $x \in M$ there exists a chart (U, u) of M centered at x with $u(U) = V \times W \subset \mathbb{R}^k \times \mathbb{R}^{m-k}$, such that $T(u^{-1}(V \times \{y\})) = E|(u^{-1}(V \times \{y\}))$ for each $y \in W$.

Proof. Let $x \in M$. We choose a chart (U, u) of M centered at x such that there exist k vector fields $X_1, \dots, X_k \in \Gamma(E)$ which form a frame of $E|U$. Then we have $X_i = \sum_{j=1}^m f_i^j \frac{\partial}{\partial u^j}$ for $f_i^j \in C^\infty(U)$. Then $f = (f_i^j)$ is a $(k \times m)$-matrix valued smooth function on U which has rank k on U. So some $(k \times k)$-submatrix, say the top one, is invertible at x and thus we may take U so small that this top $(k \times k)$-submatrix is invertible everywhere on U. Let $g = (g_i^j)$ be the inverse of this submatrix, so that the $(k \times m)$-matrix $f.g$ is given by

$$f.g = \begin{pmatrix} \mathbb{I}_k \\ * \end{pmatrix}.$$

We put

$$\text{(1)} \qquad Y_i := \sum_{j=1}^k g_i^j X_j = \sum_{j=1}^k \sum_{l=1}^m g_i^j f_j^l \frac{\partial}{\partial u^l} = \frac{\partial}{\partial u^i} + \sum_{p \geq k+1} h_i^p \frac{\partial}{\partial u^p}.$$

We claim that $[Y_i, Y_j] = 0$ for all $1 \leq i, j \leq k$. Since E is integrable, we have $[Y_i, Y_j] = \sum_{l=1}^k c_{ij}^l Y_l$. But from (1) we conclude (using the coordinate formula in (3.4)) that $[Y_i, Y_j] = \sum_{p \geq k+1} a^p \frac{\partial}{\partial u^p}$. Again by (1) this implies that $c_{ij}^l = 0$ for all l, and the claim follows.

Now we consider an $(m-k)$-dimensional linear subspace W_1 in $\mathbb{R}^m$ which is transversal to the k vectors $T_x u.Y_i(x) \in T_0\mathbb{R}^m$ spanning $\mathbb{R}^k$, and we define $f : V \times W \to U$ by

$$f(t^1, \dots, t^k, y) := \left(\mathrm{Fl}_{t^1}^{Y_1} \circ \mathrm{Fl}_{t^2}^{Y_2} \circ \dots \circ \mathrm{Fl}_{t^k}^{Y_k}\right)(u^{-1}(y)),$$

where $t = (t^1, \dots, t^k) \in V$, a small neighborhood of 0 in $\mathbb{R}^k$, and where $y \in W$, a small neighborhood of 0 in W_1. By (3.15) we may interchange the order of the flows in the definition of f arbitrarily. Thus

$$\frac{\partial}{\partial t^i} f(t,y) = \frac{\partial}{\partial t^i}\left(\mathrm{Fl}^{Y_i}_{t^i} \circ \mathrm{Fl}^{Y_1}_{t^1} \circ \dots\right)(u^{-1}(y)) = Y_i(f(t,y)),$$
$$\frac{\partial}{\partial y^k} f(0,y) = \frac{\partial}{\partial y^k}(u^{-1})(y),$$

and so $T_0 f$ is invertible and the inverse of f on a suitable neighborhood of x gives us the required chart. □

3.19. Remark. Any charts $(U, u : U \to V \times W \subset \mathbb{R}^k \times \mathbb{R}^{m-k})$ as constructed in theorem (3.18) with V and W open balls is called a *distinguished chart* for E. The submanifolds $u^{-1}(V \times \{y\})$ are called *plaques*. Two plaques of different distinguished charts intersect in open subsets in both plaques or not at all: This follows immediately by flowing a point in the intersection into both plaques with the same construction as in the proof of (3.18). Thus an atlas of distinguished charts on M has chart change mappings which respect the submersion $\mathbb{R}^k \times \mathbb{R}^{m-k} \to \mathbb{R}^{m-k}$ (the plaque structure on M). Such an atlas (or the equivalence class of such atlases) is called the *foliation corresponding to the integrable vector subbundle $E \subset TM$.*

3.20. Global version of Frobenius's theorem. *Let $E \subsetneq TM$ be an integrable vector subbundle of TM. Then, using the restrictions of distinguished charts to plaques as charts, we get a new structure of a smooth manifold on M, which we denote by M_E. If $E \neq TM$, the topology of M_E is finer than that of M, M_E has uncountably many connected components called the leaves of the foliation, and the identity induces a bijective immersion $M_E \to M$. Each leaf L is a second countable initial submanifold of M, and it is a maximal integrable submanifold of M for E in the sense that $T_x L = E_x$ for each $x \in L$.*

Proof. Let $(U_\alpha, u_\alpha : U_\alpha \to V_\alpha \times W_\alpha \subseteq \mathbb{R}^k \times \mathbb{R}^{m-k})$ be an atlas of distinguished charts corresponding to the integrable vector subbundle $E \subset TM$, as given by theorem (3.18). Let us now use for each plaque the homeomorphisms $\mathrm{pr}_1 \circ u_\alpha|(u_\alpha^{-1}(V_\alpha \times \{y\})) : u_\alpha^{-1}(V_\alpha \times \{y\}) \to V_\alpha \subset \mathbb{R}^{m-k}$ as charts; then we describe on M a new smooth manifold structure M_E with finer topology which however has uncountably many connected components, and the identity on M induces a bijective immersion $M_E \to M$. The connected components of M_E are called the *leaves of the foliation.*

In order to check the rest of the assertions made in the theorem, let us construct the unique leaf L through an arbitrary point $x \in M$: choose a plaque containing x and take the union with any plaque meeting the first one, and keep going. Now choose $y \in L$ and a curve $c : [0,1] \to L$ with $c(0) = x$ and $c(1) = y$. Then there are finitely many distinguished charts

$(U_1, u_1), \dots, (U_n, u_n)$ and $a_1, \dots, a_n \in \mathbb{R}^{m-k}$ such that $x \in u_1^{-1}(V_1 \times \{a_1\})$, $y \in u_n^{-1}(V_n \times \{a_n\})$ and such that for each i

$$u_i^{-1}(V_i \times \{a_i\}) \cap u_{i+1}^{-1}(V_{i+1} \times \{a_{i+1}\}) \neq \emptyset. \tag{1}$$

Given u_i, u_{i+1}, and a_i, there are only countably many points a_{i+1} such that (1) holds: If not, then we get a cover of the the separable submanifold $u_i^{-1}(V_i \times \{a_i\}) \cap U_{i+1}$ by uncountably many pairwise disjoint open sets of the form given in (1), which contradicts separability.

Finally, since (each component of) M is a Lindelöf space, any distinguished atlas contains a countable subatlas. So each leaf is the union of at most countably many plaques. The rest is clear. □

3.21. Singular distributions. Let M be a manifold. Suppose that for each $x \in M$ we are given a vector subspace E_x of T_xM. The disjoint union $E = \bigsqcup_{x \in M} E_x$ is called a *(singular) distribution* on M. We do not suppose that the dimension of E_x is locally constant in x.

Let $\mathfrak{X}_{loc}(M)$ denote the set of all locally defined smooth vector fields on M, i.e., $\mathfrak{X}_{loc}(M) = \bigcup \mathfrak{X}(U)$, where U runs through all open sets in M. Furthermore let $\mathfrak{X}_E$ denote the set of all local vector fields $X \in \mathfrak{X}_{loc}(M)$ with $X(x) \in E_x$ whenever defined. We say that a subset $\mathcal{V} \subset \mathfrak{X}_E$ *spans* E if for each $x \in M$ the vector space E_x is the linear hull of the set $\{X(x) : X \in \mathcal{V}\}$. We say that E is a *smooth distribution* if $\mathfrak{X}_E$ spans E. Note that every subset $\mathcal{W} \subset \mathfrak{X}_{loc}(M)$ spans a distribution denoted by $E(\mathcal{W})$, which is obviously smooth (the linear span of the empty set is the vector space 0). From now on we will consider only smooth distributions.

An *integral manifold* of a smooth distribution E is a connected immersed submanifold (N, i) (see (2.9)) such that $T_xi(T_xN) = E_{i(x)}$ for all $x \in N$. We will see in theorem (3.25) below that any integral manifold is in fact an initial submanifold of M (see (2.13)), so that we need not specify the injective immersion i. An integral manifold of E is called *maximal* if it is not contained in any strictly larger integral manifold of E.

3.22. Lemma. *Let E be a smooth distribution on M. Then we have:*

(1) *If (N, i) is an integral manifold of E and $X \in \mathfrak{X}_E$, then i^*X makes sense and is an element of $\mathfrak{X}_{loc}(N)$, which is $i|i^{-1}(U_X)$-related to X, where $U_X \subset M$ is the open domain of X.*

(2) *If (N_j, i_j) are integral manifolds of E for $j = 1, 2$, then $i_1^{-1}(i_1(N_1) \cap i_2(N_2))$ and $i_2^{-1}(i_1(N_1) \cap i_2(N_2))$ are open subsets in N_1 and N_2, respectively; furthermore $i_2^{-1} \circ i_1$ is a diffeomorphism between them.*

(3) *If $x \in M$ is contained in some integral submanifold of E, then it is contained in a unique maximal one.*

Proof. (1) Let U_X be the open domain of $X \in \mathfrak{X}_E$. If $i(x) \in U_X$ for $x \in N$, we have $X(i(x)) \in E_{i(x)} = T_x i(T_x N)$, so $i^*X(x) := ((T_x i)^{-1} \circ X \circ i)(x)$ makes sense. The vector field i^*X is clearly defined on an open subset of N and is smooth.

(2) Let $X \in \mathfrak{X}_E$. Then $i_j^* X \in \mathfrak{X}_{loc}(N_j)$ and is i_j-related to X. So by lemma (3.14) for $j = 1, 2$ we have

$$i_j \circ \mathrm{Fl}_t^{i_j^* X} = Fl_t^X \circ i_j.$$

Now choose $x_j \in N_j$ such that $i_1(x_1) = i_2(x_2) = x_0 \in M$ and choose vector fields $X_1, \dots, X_n \in \mathfrak{X}_E$ such that $(X_1(x_0), \dots, X_n(x_0))$ is a basis of E_{x_0}. Then

$$f_j(t^1, \dots, t^n) := (\mathrm{Fl}_{t^1}^{i_j^* X_1} \circ \dots \circ \mathrm{Fl}_{t^n}^{i_j^* X_n})(x_j)$$

is a smooth local mapping $\mathbb{R}^n \to N_j$ defined near zero. Since obviously $\frac{\partial}{\partial t^k}|_0 f_j = i_j^* X_k(x_j)$ for $j = 1, 2$, we see that f_j is a diffeomorphism near 0. Finally we have

$$\begin{aligned}(i_2^{-1} \circ i_1 \circ f_1)(t^1, \dots, t^n) &= (i_2^{-1} \circ i_1 \circ \mathrm{Fl}_{t^1}^{i_1^* X_1} \circ \dots \circ \mathrm{Fl}_{t^n}^{i_1^* X_n})(x_1) \\ &= (i_2^{-1} \circ \mathrm{Fl}_{t^1}^{X_1} \circ \dots \circ \mathrm{Fl}_{t^n}^{X_n} \circ i_1)(x_1) \\ &= (\mathrm{Fl}_{t^1}^{i_2^* X_1} \circ \dots \circ \mathrm{Fl}_{t^n}^{i_2^* X_n} \circ i_2^{-1} \circ i_1)(x_1) \\ &= f_2(t^1, \dots, t^n).\end{aligned}$$

So $i_2^{-1} \circ i_1$ is a diffeomorphism, as required.

(3) Let N be the union of all integral manifolds containing x. Choose the union of all the atlases of these integral manifolds as atlas for N, which is a smooth atlas for N by (2). Note that a connected immersed submanifold of a separable manifold is automatically separable (since it carries a Riemann metric). □

3.23. Integrable singular distributions and singular foliations. A smooth singular distribution E on a manifold M is called *integrable* if each point of M is contained in some integral manifold of E. By (3.22.3) each point is then contained in a unique maximal integral manifold, so the maximal integral manifolds form a partition of M. This partition is called the *(singular) foliation* of M induced by the integrable (singular) distribution E, and each maximal integral manifold is called a *leaf* of this foliation. If $X \in \mathfrak{X}_E$, then by (3.22.1) the integral curve $t \mapsto \mathrm{Fl}^X(t, x)$ of X through $x \in M$ stays in the leaf through x.

Let us now consider an arbitrary subset $\mathcal{V} \subset \mathfrak{X}_{loc}(M)$. We say that $\mathcal{V}$ is *stable* if for all $X, Y \in \mathcal{V}$ and for all t for which it is defined the local vector field $(\mathrm{Fl}_t^X)^* Y$ is again an element of $\mathcal{V}$.

If $\mathcal{W} \subset \mathfrak{X}_{loc}(M)$ is an arbitrary subset, we call $\mathcal{S}(\mathcal{W})$ the set of all local vector fields of the form $(\mathrm{Fl}^{X_1}_{t_1} \circ \cdots \circ \mathrm{Fl}^{X_k}_{t_k})^* Y$ for $X_i, Y \in \mathcal{W}$. By lemma (3.14) the flow of this vector field is

$$\mathrm{Fl}((\mathrm{Fl}^{X_1}_{t_1} \circ \cdots \circ \mathrm{Fl}^{X_k}_{t_k})^* Y, t) = \mathrm{Fl}^{X_k}_{-t_k} \circ \cdots \circ \mathrm{Fl}^{X_1}_{-t_1} \circ \mathrm{Fl}^{Y}_{t} \circ \mathrm{Fl}^{X_1}_{t_1} \circ \cdots \circ \mathrm{Fl}^{X_k}_{t_k},$$

so $\mathcal{S}(\mathcal{W})$ is the minimal stable set of local vector fields which contains $\mathcal{W}$.

Now let F be an arbitrary distribution. A local vector field $X \in \mathfrak{X}_{\mathrm{loc}}(M)$ is called an *infinitesimal automorphism* of F if $T_x(\mathrm{Fl}^X_t)(F_x) \subset F_{\mathrm{Fl}^X(t,x)}$ whenever defined. We denote by $\mathrm{aut}(F)$ the set of all infinitesimal automorphisms of F. By arguments given just above, $\mathrm{aut}(F)$ is stable.

3.24. Lemma. *Let E be a smooth distribution on a manifold M. Then the following conditions are equivalent:*

(1) *E is integrable.*
(2) *$\mathfrak{X}_E$ is stable.*
(3) *There exists a subset $\mathcal{W} \subset \mathfrak{X}_{loc}(M)$ such that $\mathcal{S}(\mathcal{W})$ spans E.*
(4) *$aut(E) \cap \mathfrak{X}_E$ spans E.*

Proof. (1) $\Longrightarrow$ (2) Let $X \in \mathfrak{X}_E$ and let L be the leaf through $x \in M$, with $i : L \to M$ the inclusion. Then $\mathrm{Fl}^X_{-t} \circ i = i \circ \mathrm{Fl}^{i^*X}_{-t}$ by lemma (3.14), so we have

$$\begin{aligned} T_x(\mathrm{Fl}^X_{-t})(E_x) &= T(\mathrm{Fl}^X_{-t}).T_x i.T_x L = T(\mathrm{Fl}^X_{-t} \circ i).T_x L \\ &= Ti.T_x(\mathrm{Fl}^{i^*X}_{-t}).T_x L \\ &= Ti.T_{Fl^{i^*X}(-t,x)} L = E_{Fl^X(-t,x)}. \end{aligned}$$

This implies that $(\mathrm{Fl}^X_t)^* Y \in \mathfrak{X}_E$ for any $Y \in \mathfrak{X}_E$.

(2) $\Longrightarrow$ (4) In fact (2) says that $\mathfrak{X}_E \subset aut(E)$.

(4) $\Longrightarrow$ (3) We can choose $\mathcal{W} = aut(E) \cap \mathfrak{X}_E$: For $X, Y \in \mathcal{W}$ we have $(\mathrm{Fl}^X_t)^* Y \in \mathfrak{X}_E$; so $\mathcal{W} \subset \mathcal{S}(\mathcal{W}) \subset \mathfrak{X}_E$ and E is spanned by $\mathcal{W}$.

(3) $\Longrightarrow$ (1) We have to show that each point $x \in M$ is contained in some integral submanifold for the distribution E. Since $\mathcal{S}(\mathcal{W})$ spans E and is stable, we have

$$T(\mathrm{Fl}^X_t).E_x = E_{\mathrm{Fl}^X(t,x)} \tag{5}$$

for each $X \in \mathcal{S}(\mathcal{W})$. Let $\dim E_x = n$. There are $X_1, \ldots, X_n \in \mathcal{S}(\mathcal{W})$ such that $X_1(x), \ldots, X_n(x)$ is a basis of E_x, since E is smooth. As in the proof of (3.22.2) we consider the mapping

$$f(t^1, \ldots, t^n) := (\mathrm{Fl}^{X_1}_{t^1} \circ \cdots \circ \mathrm{Fl}^{X_n}_{t^n})(x),$$

defined and smooth near 0 in $\mathbb{R}^n$. Since the rank of f at 0 is n, the image under f of a small open neighborhood of 0 is a submanifold N of M. We claim that N is an integral manifold of E. The tangent space $T_{f(t^1,\dots,t^n)}N$ is linearly generated by

$$\begin{aligned}\tfrac{\partial}{\partial t^k}(\mathrm{Fl}^{X_1}_{t^1}\circ\dots\circ\mathrm{Fl}^{X_n}_{t^n})(x) &= T(\mathrm{Fl}^{X_1}_{t^1}\circ\dots\circ\mathrm{Fl}^{X_{k-1}}_{t^{k-1}})X_k((\mathrm{Fl}^{X_k}_{t^k}\circ\dots\circ\mathrm{Fl}^{X_n}_{t^n})(x))\\ &= ((\mathrm{Fl}^{X_1}_{-t^1})^*\cdots(\mathrm{Fl}^{X_{k-1}}_{-t^{k-1}})^*X_k)(f(t^1,\dots,t^n)).\end{aligned}$$

Since $\mathcal{S}(\mathcal{W})$ is stable, these vectors lie in $E_{f(t)}$. From the form of f and from (5) we see that $\dim E_{f(t)} = \dim E_x$, so these vectors even span $E_{f(t)}$ and we have $T_{f(t)}N = E_{f(t)}$ as required. □

3.25. Theorem (Local structure of singular foliations). *Let E be an integrable (singular) distribution of a manifold M. Then for each $x \in M$ there exist a chart (U,u) with $u(U) = \{y \in \mathbb{R}^m : |y^i| < \varepsilon \text{ for all } i\}$ for some $\varepsilon > 0$ and a countable subset $A \subset \mathbb{R}^{m-n}$, such that for the leaf L through x we have*

$$u(U\cap L) = \{y \in u(U) : (y^{n+1},\dots,y^m) \in A\}.$$

Each leaf is an initial submanifold.

If furthermore the distribution E has locally constant rank, this property holds for each leaf meeting U with the same n.

This chart (U,u) is called a *distinguished chart* for the (singular) distribution or the (singular) foliation. A connected component of $U\cap L$ is called a *plaque.*

Proof. Let L be the leaf through x, $\dim L = n$. Let $X_1,\dots,X_n \in \mathfrak{X}_E$ be local vector fields such that $X_1(x),\dots,X_n(x)$ is a basis of E_x. We choose a chart (V,v) centered at x on M such that the vectors

$$X_1(x),\dots,X_n(x),\ \tfrac{\partial}{\partial v^{n+1}}|_x,\dots,\tfrac{\partial}{\partial v^m}|_x$$

form a basis of T_xM. Then

$$f(t^1,\dots,t^m) = (\mathrm{Fl}^{X_1}_{t^1}\circ\dots\circ\mathrm{Fl}^{X_n}_{t^n})(v^{-1}(0,\dots,0,t^{n+1},\dots,t^m))$$

is a diffeomorphism from a neighborhood of 0 in $\mathbb{R}^m$ onto a neighborhood of x in M. Let (U,u) be the chart given by f^{-1}, suitably restricted. We have

$$y \in L \iff (\mathrm{Fl}^{X_1}_{t^1}\circ\dots\circ\mathrm{Fl}^{X_n}_{t^n})(y) \in L$$

for all y and all $t^1,\dots,t^n$ for which both expressions make sense. So we have

$$f(t^1,\dots,t^m) \in L \iff f(0,\dots,0,t^{n+1},\dots,t^m) \in L,$$

and consequently $L\cap U$ is the disjoint union of connected sets of the form $\{y \in U : (u^{n+1}(y),\dots,u^m(y)) = \text{constant}\}$. Since L is a connected immersive submanifold of M, it is second countable and only a countable set of constants can appear in the description of $u(L\cap U)$ given above. From this description it is clear that L is an initial submanifold (2.13) since $u(C_x(L\cap U)) = u(U)\cap(\mathbb{R}^n\times 0)$.

The argument given above is valid for any leaf of dimension n meeting U, so also the assertion for an integrable distribution of constant rank follows. □

3.26. Involutive singular distributions. A subset $\mathcal{V} \subset \mathfrak{X}_{loc}(M)$ is called *involutive* if $[X, Y] \in \mathcal{V}$ for all $X, Y \in \mathcal{V}$. Here $[X, Y]$ is defined on the intersection of the domains of X and Y.

A smooth distribution E on M is called *involutive* if there exists an involutive subset $\mathcal{V} \subset \mathfrak{X}_{loc}(M)$ spanning E.

For an arbitrary subset $\mathcal{W} \subset \mathfrak{X}_{loc}(M)$ let $\mathcal{L}(\mathcal{W})$ be the set consisting of all local vector fields on M which can be written as finite expressions using Lie brackets and starting from elements of $\mathcal{W}$. Clearly $\mathcal{L}(\mathcal{W})$ is the smallest involutive subset of $\mathfrak{X}_{loc}(M)$ which contains $\mathcal{W}$.

3.27. Lemma. *For each subset $\mathcal{W} \subset \mathfrak{X}_{loc}(M)$ we have*

$$E(\mathcal{W}) \subset E(\mathcal{L}(\mathcal{W})) \subset E(\mathcal{S}(\mathcal{W})).$$

In particular we have $E(\mathcal{S}(\mathcal{W})) = E(\mathcal{L}(\mathcal{S}(\mathcal{W})))$.

Proof. We will show that for $X, Y \in \mathcal{W}$ we have $[X, Y] \in \mathfrak{X}_{E(\mathcal{S}(\mathcal{W}))}$, for then by induction we get $\mathcal{L}(\mathcal{W}) \subset \mathfrak{X}_{E(\mathcal{S}(\mathcal{W}))}$ and $E(\mathcal{L}(\mathcal{W})) \subset E(\mathcal{S}(\mathcal{W}))$.

Let $x \in M$; since by (3.24) $E(\mathcal{S}(\mathcal{W}))$ is integrable, we can choose the leaf L through x, with the inclusion i. Then i^*X is i-related to X and i^*Y is i-related to Y; thus by (3.10) the local vector field $[i^*X, i^*Y] \in \mathfrak{X}_{loc}(L)$ is i-related to $[X, Y]$, and $[X, Y](x) \in E(\mathcal{S}(\mathcal{W}))_x$, as required. □

3.28. Theorem. *Let $\mathcal{V} \subset \mathfrak{X}_{loc}(M)$ be an involutive subset. Then the distribution $E(\mathcal{V})$ spanned by $\mathcal{V}$ is integrable under each of the following conditions.*

(1) *M is real analytic and $\mathcal{V}$ consists of real analytic vector fields.*

(2) *The dimension of $E(\mathcal{V})$ is constant along flow lines of vector fields in $\mathcal{V}$.*

Proof. (1) For $X, Y \in \mathcal{V}$ we have $\frac{d}{dt}(\mathrm{Fl}_t^X)^*Y = (\mathrm{Fl}_t^X)^*\mathcal{L}_XY$; consequently $\frac{d^k}{dt^k}(\mathrm{Fl}_t^X)^*Y = (\mathrm{Fl}_t^X)^*(\mathcal{L}_X)^kY$, and since everything is real analytic, we get for $x \in M$ and small t

$$(\mathrm{Fl}_t^X)^*Y(x) = \sum_{k\geq 0} \frac{t^k}{k!}\frac{d^k}{dt^k}|_0(\mathrm{Fl}_t^X)^*Y(x) = \sum_{k\geq 0}\frac{t^k}{k!}(\mathcal{L}_X)^kY(x).$$

Since $\mathcal{V}$ is involutive, all $(\mathcal{L}_X)^kY \in \mathcal{V}$. Therefore we get $(\mathrm{Fl}_t^X)^*Y(x) \in E(\mathcal{V})_x$ for small t. By the flow property of Fl^X the set of all t satisfying $(\mathrm{Fl}_t^X)^*Y(x) \in E(\mathcal{V})_x$ is open and closed, so it follows that (3.24.2) is satisfied and thus $E(\mathcal{V})$ is integrable.

(2) We choose $X_1, \dots, X_n \in \mathcal{V}$ such that $X_1(x), \dots, X_n(x)$ is a basis of $E(\mathcal{V})_x$. For any $X \in \mathcal{V}$, by hypothesis, $E(\mathcal{V})_{\mathrm{Fl}^X(t,x)}$ has also dimension n

and admits the vectors $X_1(\mathrm{Fl}^X(t,x)),\dots,X_n(\mathrm{Fl}^X(t,x))$ as basis, for small t. So there are smooth functions $f_{ij}(t)$ such that

$$[X,X_i](\mathrm{Fl}^X(t,x)) = \sum_{j=1}^n f_{ij}(t)X_j(\mathrm{Fl}^X(t,x)).$$

Therefore,

$$\begin{aligned}\tfrac{d}{dt}T(\mathrm{Fl}^X_{-t}).X_i(\mathrm{Fl}^X(t,x)) &= T(\mathrm{Fl}^X_{-t}).[X,X_i](\mathrm{Fl}^X(t,x))\\ &= \sum_{j=1}^n f_{ij}(t)T(\mathrm{Fl}^X_{-t}).X_j(\mathrm{Fl}^X(t,x)).\end{aligned}$$

So the T_xM-valued functions $g_i(t) = T(\mathrm{Fl}^X_{-t}).X_i(\mathrm{Fl}^X(t,x))$ satisfy the linear ordinary differential equation $\frac{d}{dt}g_i(t) = \sum_{j=1}^n f_{ij}(t)g_j(t)$ and have initial values in the linear subspace $E(\mathcal{V})_x$, so they have values in it for all small t. Therefore $T(\mathrm{Fl}^X_{-t})E(\mathcal{V})_{\mathrm{Fl}^X(t,x)} \subset E(\mathcal{V})_x$ for small t. Using compact time intervals and the flow property, one sees that condition (3.24.2) is satisfied and $E(\mathcal{V})$ is integrable. □

3.29. Examples. (1) The singular distribution spanned by $\mathcal{W} \subset \mathfrak{X}_{loc}(\mathbb{R}^2)$ is involutive, but not integrable, where $\mathcal{W}$ consists of all global vector fields with support in $\mathbb{R}^2 \setminus \{0\}$ and the field $\frac{\partial}{\partial x^1}$; the leaf through 0 should have dimension 1 at 0 and dimension 2 elsewhere.

(2) Let $f : \mathbb{R} \to \mathbb{R}$ be a smooth function with $f(x^1) = 0$ for $x^1 \le 0$ and $f(x^1) > 0$ for $x^1 > 0$. Then the singular distribution on $\mathbb{R}^2$ spanned by the two vector fields $X(x^1,x^2) = \frac{\partial}{\partial x^1}$ and $Y(x^1,x^2) = f(x^1)\frac{\partial}{\partial x^2}$ is involutive, but not integrable. Any leaf should pass $(0,x^2)$ tangentially to $\frac{\partial}{\partial x^1}$, should have dimension 1 for $x^1 \le 0$ and should have dimension 2 for $x^1 > 0$.

3.30. By a *time dependent vector field* on a manifold M we mean a smooth mapping $X : J \times M \to TM$ with $\pi_M \circ X = \mathrm{pr}_2$, where J is an open interval. An integral curve of X is a smooth curve $c : I \to M$ with $\dot c(t) = X(t, c(t))$ for all $t \in I$, where I is a subinterval of J.

There is an associated vector field $\bar X \in \mathfrak{X}(J \times M)$, given by $\bar X(t,x) = (\frac{\partial}{\partial t}, X(t,x)) \in T_t\mathbb{R} \times T_xM$.

By the *evolution operator* of X we mean the mapping $\Phi^X : J \times J \times M \to M$, defined in a maximal open neighborhood of $\Delta_J \times M$ (where Δ_J is the diagonal of J) and satisfying the differential equation

$$\begin{cases} \frac{d}{dt}\Phi^X(t,s,x) = X(t,\Phi^X(t,s,x))\\ \Phi^X(s,s,x) = x.\end{cases}$$

It is easily seen that $(t, \Phi^X(t,s,x)) = \mathrm{Fl}^{\bar{X}}(t-s,(s,x))$, so the maximally defined evolution operator exists and is unique, and it satisfies

$$\Phi^X_{t,s} = \Phi^X_{t,r} \circ \Phi^X_{r,s}, \qquad \text{where } \Phi^X_{t,s}(x) = \Phi(t,s,x),$$

whenever one side makes sense (with the restrictions of (3.7)).

Examples and Exercises

3.31. Compute the flow of the vector field $\xi_1(x,y) := y\frac{\partial}{\partial x}$ in $\mathbb{R}^2$. Is it a global flow? Answer the same questions for $\xi_2(x,y) := \frac{x^2}{2}\frac{\partial}{\partial y}$. Now compute $[\xi_1,\xi_2]$ and investigate its flow. This time it is not global! In fact, $Fl_t^{[\xi_1,\xi_2]}(x,y) = \left(\frac{2x}{2+xt}, \frac{y}{4}(tx+2)^2\right)$. Investigate the flow of $\xi_1+\xi_2$. It is not global either! Thus the set of complete vector fields on $\mathbb{R}^2$ is neither a vector space nor closed under the Lie bracket.

3.32. Driving a car. The phase space consists of all $(x,y,\vartheta,\varphi) \in \mathbb{R}^2 \times S^1 \times (-\pi/4, \pi/4)$, where

(x,y)	is the position of the midpoint of the rear axle,
ϑ	is the direction of the car axle,
ϕ	is the steering angle of the front wheels.

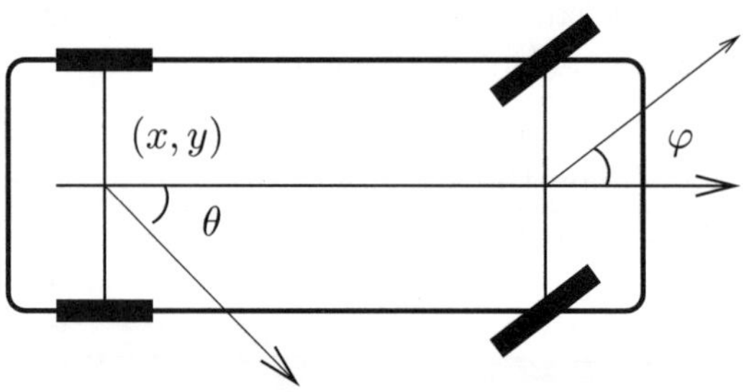

$\theta = 0$ direction

There are two 'control' vector fields:

$$\begin{aligned} \text{steer} &= \frac{\partial}{\partial \phi}, \\ \text{drive} &= \cos(\vartheta)\frac{\partial}{\partial x} + \sin(\vartheta)\frac{\partial}{\partial y} + \tan(\phi)\frac{1}{l}\frac{\partial}{\partial \vartheta} \text{ (why?)}. \end{aligned}$$

Compute [steer, drive] =: park (why?) and [drive, park], and interpret the results. Is it not convenient that the two control vector fields do not span an integrable distribution?

3.33. Describe the Lie algebra of all vector fields on S^1 in terms of Fourier expansion. This is nearly (up to a central extension) the Virasoro algebra of theoretical physics.

CHAPTER II.
Lie Groups and Group Actions

4. Lie Groups I

4.1. Definition. A *Lie group* G is a smooth manifold and a group such that the multiplication $\mu : G \times G \to G$ is smooth. We shall see in a moment that then also the inversion $\nu : G \to G$ turns out to be smooth.

We shall use the following notation:

$\mu : G \times G \to G$, multiplication, $\mu(x, y) = x.y$.

$\mu_a : G \to G$, left translation, $\mu_a(x) = a.x$.

$\mu^a : G \to G$, right translation, $\mu^a(x) = x.a$.

$\nu : G \to G$, inversion, $\nu(x) = x^{-1}$.

$e \in G$, the unit element.

Then we have

$$\mu_a \circ \mu_b = \mu_{a.b}, \qquad \mu^a \circ \mu^b = \mu^{b.a}, \qquad \mu^a \circ \mu_b = \mu_b \circ \mu^a,$$
$$\mu_a^{-1} = \mu_{a^{-1}}, \qquad (\mu^a)^{-1} = \mu^{a^{-1}}.$$

If $\varphi : G \to H$ is a smooth homomorphism between Lie groups, then we have $\varphi \circ \mu_a = \mu_{\varphi(a)} \circ \varphi$, $\varphi \circ \mu^a = \mu^{\varphi(a)} \circ \varphi$ and thus also $T\varphi.T\mu_a = T\mu_{\varphi(a)}.T\varphi$, etc. So $T_e\varphi$ is injective (surjective) if and only if $T_a\varphi$ is injective (surjective) for all $a \in G$.

4.2. Lemma. *The tangent mapping $T_{(a,b)}\mu : T_aG \times T_bG \to T_{ab}G$ of the multiplication μ is given by*

$$T_{(a,b)}\mu.(X_a, Y_b) = T_a(\mu^b).X_a + T_b(\mu_a).Y_b.$$

Proof. Let $ri_a : G \to G \times G$, $ri_a(x) = (a, x)$ be the right insertion and let $li_b : G \to G \times G$, $li_b(x) = (x, b)$ be the left insertion. Then we have

$$\begin{aligned} T_{(a,b)}\mu.(X_a, Y_b) &= T_{(a,b)}\mu.(T_a(li_b).X_a + T_b(ri_a).Y_b) \\ &= T_a(\mu \circ li_b).X_a + T_b(\mu \circ ri_a).Y_b = T_a(\mu^b).X_a + T_b(\mu_a).Y_b. \quad \square \end{aligned}$$

4.3. Corollary. *The inversion $\nu : G \to G$ is smooth and*

$$T_a\nu = -T_e(\mu^{a^{-1}}).T_a(\mu_{a^{-1}}) = -T_e(\mu_{a^{-1}}).T_a(\mu^{a^{-1}}).$$

Proof. The equation $\mu(x, \nu(x)) = e$ determines ν implicitly. The mapping ν is smooth in a neighborhood of e by the implicit function theorem since $T_e(\mu(e, \quad)) = T_e(\mu_e) = Id$. From $(\nu \circ \mu_a)(x) = x^{-1}.a^{-1} = (\mu^{a^{-1}} \circ \nu)(x)$ we may conclude that ν is everywhere smooth. Now we differentiate the equation $\mu(a, \nu(a)) = e$; this gives in turn

$$\begin{aligned} 0_e &= T_{(a,a^{-1})}\mu.(X_a, T_a\nu.X_a) = T_a(\mu^{a^{-1}}).X_a + T_{a^{-1}}(\mu_a).T_a\nu.X_a, \\ T_a\nu.X_a &= -T_e(\mu_a)^{-1}.T_a(\mu^{a^{-1}}).X_a = -T_e(\mu_{a^{-1}}).T_a(\mu^{a^{-1}}).X_a. \quad \square \end{aligned}$$

4.4. Example. The *general linear group* $GL(n, \mathbb{R})$ is the group of all invertible real $n \times n$-matrices. It is an open subset of $L(\mathbb{R}^n, \mathbb{R}^n)$, given by $\det \neq 0$ and a Lie group.

Similarly $GL(n, \mathbb{C})$, the group of invertible complex $n \times n$-matrices, is a Lie group; also $GL(n, \mathbb{H})$, the group of all invertible quaternionic $n \times n$-matrices, is a Lie group, since it is open in the real Banach algebra $L_{\mathbb{H}}(\mathbb{H}^n, \mathbb{H}^n)$ as a glance at the von Neumann series shows; but the quaternionic determinant is a more subtle instrument here.

4.5. Example. The *orthogonal group* $O(n, \mathbb{R})$ is the group of all linear isometries of $(\mathbb{R}^n, \langle \quad , \quad \rangle)$, where $\langle \quad , \quad \rangle$ is the standard positive definite inner product on $\mathbb{R}^n$. The *special orthogonal group* $SO(n, \mathbb{R}) := \{A \in O(n, \mathbb{R}) : \det A = 1\}$ is open in $O(n, \mathbb{R})$, since we have the disjoint union

$$O(n, \mathbb{R}) = SO(n, \mathbb{R}) \sqcup \begin{pmatrix} -1 & 0 \\ 0 & \mathbb{I}_{n-1} \end{pmatrix} SO(n, \mathbb{R}),$$

where $\mathbb{I}_k$ is short for the identity matrix $Id_{\mathbb{R}^k}$. We claim that $O(n, \mathbb{R})$ and $SO(n, \mathbb{R})$ are submanifolds of $L(\mathbb{R}^n, \mathbb{R}^n)$. For that we consider the mapping $f : L(\mathbb{R}^n, \mathbb{R}^n) \to L(\mathbb{R}^n, \mathbb{R}^n)$, given by $f(A) = A^\top.A$. Then $O(n, \mathbb{R}) = f^{-1}(\mathbb{I}_n)$; so $O(n, \mathbb{R})$ is closed. Since it is also bounded, $O(n, \mathbb{R})$ is compact.

We have $df(A).X = X^\top.A + A^\top.X$, so $\ker df(\mathbb{I}_n) = \{X : X^\top + X = 0\}$ is the space $\mathfrak{o}(n,\mathbb{R})$ of all skew-symmetric $n \times n$-matrices. Note that $\dim \mathfrak{o}(n,\mathbb{R}) = \frac{1}{2}(n-1)n$. If A is invertible, we get

$$\begin{aligned}\ker df(A) &= \{Y : Y^\top.A + A^\top.Y = 0\} = \{Y : A^\top.Y \in \mathfrak{o}(n,\mathbb{R})\}\\ &= (A^{-1})^\top.\mathfrak{o}(n,\mathbb{R}).\end{aligned}$$

The mapping f takes values in $L_{sym}(\mathbb{R}^n,\mathbb{R}^n)$, the space of all symmetric $n \times n$-matrices, and $\dim\ker df(A) + \dim L_{sym}(\mathbb{R}^n,\mathbb{R}^n) = \frac{1}{2}(n-1)n + \frac{1}{2}n(n+1) = n^2 = \dim L(\mathbb{R}^n,\mathbb{R}^n)$, so $f : GL(n,\mathbb{R}) \to L_{sym}(\mathbb{R}^n,\mathbb{R}^n)$ is a submersion. Since obviously $f^{-1}(\mathbb{I}_n) \subset GL(n,\mathbb{R})$, we conclude from (1.12) that $O(n,\mathbb{R})$ is a submanifold of $GL(n,\mathbb{R})$. It is also a Lie group, since the group operations are smooth as the restrictions of the ones from $GL(n,\mathbb{R})$.

4.6. Example. The *special linear group* $SL(n,\mathbb{R})$ is the group of all $n \times n$-matrices of determinant 1. The function $\det : L(\mathbb{R}^n,\mathbb{R}^n) \to \mathbb{R}$ is smooth and $d\det(A)X = \operatorname{Trace}(C(A).X)$, where $C(A)^i_j$, the cofactor of A^j_i, is the determinant of the matrix, which results from putting 1 instead of A^j_i into A and 0 in the rest of the j-th row and the i-th column of A; see (4.33). We recall Cramer's rule $C(A).A = A.C(A) = \det(A).\mathbb{I}_n$. So if $C(A) \neq 0$ (i.e., $\operatorname{rank}(A) \geq n-1$), then the linear functional $df(A)$ is nonzero. So $\det : GL(n,\mathbb{R}) \to \mathbb{R}$ is a submersion and $SL(n,\mathbb{R}) = (\det)^{-1}(1)$ is a manifold and a Lie group of dimension $n^2 - 1$. Note finally that $T_{\mathbb{I}_n}SL(n,\mathbb{R}) = \ker d\det(\mathbb{I}_n) = \{X : \operatorname{Trace}(X) = 0\}$. This space of traceless matrices is usually called $\mathfrak{sl}(n,\mathbb{R})$.

4.7. Example. The *symplectic group* $Sp(n,\mathbb{R})$ is the group of all $2n \times 2n$-matrices A such that $\omega(Ax,Ay) = \omega(x,y)$ for all $x,y \in \mathbb{R}^{2n}$, where ω is a (the standard) nondegenerate skew-symmetric bilinear form on $\mathbb{R}^{2n}$.

Such a form exists on a vector space if and only if the dimension is even, and on $\mathbb{R}^n \times (\mathbb{R}^n)^*$ the form $\omega((x,x^*),(y,y^*)) = \langle x,y^*\rangle - \langle y,x^*\rangle$ (where we use the duality pairing), in coordinates $\omega((x^i)_{i=1}^{2n},(y^j)_{j=1}^{2n}) = \sum_{i=1}^n (x^i y^{n+i} - x^{n+i}y^i)$, is such a form. Any symplectic form on $\mathbb{R}^{2n}$ looks like that after choosing a suitable basis; see (31.2) and (31.4). Let $(e_i)_{i=1}^{2n}$ be the standard basis in $\mathbb{R}^{2n}$. Then we have

$$(\omega(e_i,e_j)^i_j) = \begin{pmatrix} 0 & \mathbb{I}_n \\ -\mathbb{I}_n & 0 \end{pmatrix} =: J,$$

and the matrix J satisfies $J^\top = -J$, $J^2 = -\mathbb{I}_{2n}$, $J\binom{x}{y} = \binom{y}{-x}$ in $\mathbb{R}^n \times \mathbb{R}^n$, and $\omega(x,y) = \langle x, Jy\rangle$ in terms of the standard inner product on $\mathbb{R}^{2n}$.

For $A \in L(\mathbb{R}^{2n},\mathbb{R}^{2n})$ we have $\omega(Ax,Ay) = \langle Ax, JAy\rangle = \langle x, A^\top JAy\rangle$. Thus $A \in Sp(n,\mathbb{R})$ if and only if $A^\top JA = J$.

We consider now the mapping $f : L(\mathbb{R}^{2n}, \mathbb{R}^{2n}) \to L(\mathbb{R}^{2n}, \mathbb{R}^{2n})$ given by $f(A) = A^\top JA$. Then $f(A)^\top = (A^\top JA)^\top = -A^\top JA = -f(A)$, so f takes values in the space $\mathfrak{o}(2n, \mathbb{R})$ of skew-symmetric matrices. We have $df(A)X = X^\top JA + A^\top JX$, and therefore

$$\begin{aligned}\ker df(\mathbb{I}_{2n}) &= \{X \in L(\mathbb{R}^{2n}, \mathbb{R}^{2n}) : X^\top J + JX = 0\} \\ &= \{X : JX \text{ is symmetric}\} =: \mathfrak{sp}(n, \mathbb{R}).\end{aligned}$$

We see that $\dim \mathfrak{sp}(n, \mathbb{R}) = \frac{2n(2n+1)}{2} = \binom{2n+1}{2}$. Furthermore $\ker df(A) = \{X : X^\top JA + A^\top JX = 0\}$ and the mapping $X \mapsto A^\top JX$ is an isomorphism $\ker df(A) \to L_{sym}(\mathbb{R}^{2n}, \mathbb{R}^{2n})$ if A is invertible. Thus $\dim \ker df(A) = \binom{2n+1}{2}$ for all $A \in GL(2n, \mathbb{R})$. If $f(A) = J$, then $A^\top JA = J$, so A has rank $2n$ and is invertible, and we have $\dim \ker df(A) + \dim \mathfrak{o}(2n, \mathbb{R}) = \binom{2n+1}{2} + \frac{2n(2n-1)}{2} = 4n^2 = \dim L(\mathbb{R}^{2n}, \mathbb{R}^{2n})$. So $f : GL(2n, \mathbb{R}) \to \mathfrak{o}(2n, \mathbb{R})$ is a submersion and $f^{-1}(J) = Sp(n, \mathbb{R})$ is a manifold and a Lie group. It is the symmetry group of 'classical mechanics'.

4.8. Example. The complex general linear group $GL(n, \mathbb{C})$ of all invertible complex $n \times n$-matrices is open in $L_{\mathbb{C}}(\mathbb{C}^n, \mathbb{C}^n)$, so it is a real Lie group of real dimension $2n^2$; it is also a complex Lie group of complex dimension n^2. The complex special linear group $SL(n, \mathbb{C})$ of all matrices of determinant 1 is a submanifold of $GL(n, \mathbb{C})$ of complex codimension 1 (or real codimension 2).

The complex orthogonal group $O(n, \mathbb{C})$ is the set

$$\{A \in L(\mathbb{C}^n, \mathbb{C}^n) : g(Az, Aw) = g(z, w) \text{ for all } z, w\},$$

where $g(z, w) = \sum_{i=1}^n z^i w^i$. This is a complex Lie group of complex dimension $\frac{(n-1)n}{2}$, and it is *not* compact. Since $O(n, \mathbb{C}) = \{A : A^\top A = \mathbb{I}_n\}$, we have $1 = \det_{\mathbb{C}}(\mathbb{I}_n) = \det_{\mathbb{C}}(A^\top A) = \det_{\mathbb{C}}(A)^2$, so $\det_{\mathbb{C}}(A) = \pm 1$. Thus $SO(n, \mathbb{C}) := \{A \in O(n, \mathbb{C}) : \det_{\mathbb{C}}(A) = 1\}$ is an open subgroup of index 2 in $O(n, \mathbb{C})$.

The group $Sp(n, \mathbb{C}) = \{A \in L_{\mathbb{C}}(\mathbb{C}^{2n}, \mathbb{C}^{2n}) : A^\top JA = J\}$ is also a complex Lie group of complex dimension $n(2n + 1)$.

The groups treated here are the *classical complex Lie groups.* The groups $SL(n, \mathbb{C})$ for $n \geq 2$, $SO(n, \mathbb{C})$ for $n \geq 3$, $Sp(n, \mathbb{C})$ for $n \geq 4$, and five more exceptional groups exhaust all simple complex Lie groups up to coverings.

4.9. Example. Let $\mathbb{C}^n$ be equipped with the standard Hermitian inner product $(z, w) = \sum_{i=1}^n \bar{z}^i w^i$. The *unitary* group $U(n)$ consists of all complex $n \times n$-matrices A such that $(Az, Aw) = (z, w)$ for all z, w holds, or equivalently $U(n) = \{A : A^* A = \mathbb{I}_n\}$, where $A^* = \overline{A}^\top$.

We consider the mapping $f : L_{\mathbb{C}}(\mathbb{C}^n, \mathbb{C}^n) \to L_{\mathbb{C}}(\mathbb{C}^n, \mathbb{C}^n)$, given by $f(A) = A^*A$. Then f is smooth but not holomorphic. Its derivative is $df(A)X = X^*A + A^*X$, so $\ker df(\mathbb{I}_n) = \{X : X^* + X = 0\} =: \mathfrak{u}(n)$, the space of all skew-Hermitian matrices. We have $\dim_{\mathbb{R}} \mathfrak{u}(n) = n^2$. As above we may check that $f : GL(n, \mathbb{C}) \to L_{herm}(\mathbb{C}^n, \mathbb{C}^n)$ is a submersion, so $U(n) = f^{-1}(\mathbb{I}_n)$ is a compact real Lie group of dimension n^2.

The *special unitary* group is $SU(n) = U(n) \cap SL(n, \mathbb{C})$. For $A \in U(n)$ we have $|\det_{\mathbb{C}}(A)| = 1$; thus $\dim_{\mathbb{R}} SU(n) = n^2 - 1$.

4.10. Example. The group $Sp(n)$. Let $\mathbb{H}$ be the division algebra of quaternions. We will use the following description of quaternions: Let $(\mathbb{R}^3, \langle \quad , \quad \rangle, \Delta)$ be the oriented Euclidean space of dimension 3, where Δ is a determinant function with value 1 on a positive oriented orthonormal basis. The *vector product* on R^3 is then given by $\langle X \times Y, Z\rangle = \Delta(X, Y, Z)$. Now we let $\mathbb{H} := \mathbb{R}^3 \times \mathbb{R}$, equipped with the following product:

$$(X, s)(Y, t) := (X \times Y + sY + tX, st - \langle X, Y\rangle).$$

Now we take a positively oriented orthonormal basis of $\mathbb{R}^3$, call it (i, j, k), and identify $(0, 1)$ with 1. Then the last formula implies visibly the usual product rules for the basis $(1, i, j, k)$ of the quaternions.

The group $Sp(1) := S^3 \subset \mathbb{H} \cong \mathbb{R}^4$ is then the group of unit quaternions, obviously a Lie group.

Now let V be a right vector space over $\mathbb{H}$. Since $\mathbb{H}$ is not commutative, we have to distinguish between left and right vector spaces and we choose right ones as basic, so that matrices can multiply from the left. By choosing a basis, we get $V = \mathbb{R}^n \otimes_{\mathbb{R}} \mathbb{H} = \mathbb{H}^n$. For $u = (u^i)$, $v = (v^i) \in \mathbb{H}^n$ we put $\langle u, v\rangle := \sum_{i=1}^n \overline{u}^i v^i$. Then $\langle \quad , \quad \rangle$ is $\mathbb{R}$-bilinear and $\langle ua, vb\rangle = \overline{a}\langle u, v\rangle b$ for $a, b \in \mathbb{H}$.

An $\mathbb{R}$ linear mapping $A : V \to V$ is called $\mathbb{H}$-*linear* or *quaternionically linear* if $A(ua) = A(u)a$ holds. The space of all such mappings shall be denoted by $L_{\mathbb{H}}(V, V)$. It is real isomorphic to the space of all quaternionic $n \times n$-matrices with the usual multiplication, since for the standard basis $(e_i)_{i=1}^n$ in $V = \mathbb{H}^n$ we have $A(u) = A(\sum_i e_i u^i) = \sum_i A(e_i)u^i = \sum_{i,j} e_j A_i^j u^i$. If V is a right quaternionic vector space, then $L_{\mathbb{H}}(V, V)$ is only a real vector space — any further structure must come from a second (left) quaternionic vector space structure on V.

The group $GL(n, \mathbb{H})$ of invertible $\mathbb{H}$-linear mappings of $\mathbb{H}^n$, is a Lie group, because it is $GL(4n, \mathbb{R}) \cap L_{\mathbb{H}}(\mathbb{H}^n, \mathbb{H}^n)$ which is open in $L_{\mathbb{H}}(\mathbb{H}^n, \mathbb{H}^n)$.

A quaternionically linear mapping A is called isometric or *quaternionically unitary* if $\langle A(u), A(v)\rangle = \langle u, v\rangle$ for all $u, v \in \mathbb{H}^n$. We denote by $Sp(n)$ the group of all quaternionic isometries of $\mathbb{H}^n$, the *quaternionic unitary group*.

The reason for its name is that $Sp(n) = Sp(n, \mathbb{C}) \cap U(2n)$, since we can decompose the quaternionic Hermitian form $\langle \ , \ \rangle$ into a complex Hermitian one and a complex symplectic one. Also we have $Sp(n) \subset O(4n, \mathbb{R})$, since the real part of $\langle \ , \ \rangle$ is a positive definite real inner product. For $A \in L_{\mathbb{H}}(\mathbb{H}^n, \mathbb{H}^n)$ we put $A^* := \overline{A}^\top$. Then we have $\langle u, A(v)\rangle = \langle A^*(u), v\rangle$, so $\langle A(u), A(v)\rangle = \langle A^*A(u), v\rangle$. Thus $A \in Sp(n)$ if and only if $A^*A = Id$.

Again $f : L_{\mathbb{H}}(\mathbb{H}^n, \mathbb{H}^n) \to L_{\mathbb{H},herm}(\mathbb{H}^n, \mathbb{H}^n) = \{A : A^* = A\}$, given by $f(A) = A^*A$, is a smooth mapping with $df(A)X = X^*A + A^*X$. So we have $\ker df(Id) = \{X : X^* = -X\} =: \mathfrak{sp}(n)$, the space of quaternionic skew-Hermitian matrices. The usual proof shows that f has maximal rank on $GL(n, \mathbb{H})$, so $Sp(n) = f^{-1}(Id)$ is a compact real Lie group of dimension $2n(n-1) + 3n$.

The groups $SO(n, \mathbb{R})$ for $n \geq 3$, $SU(n)$ for $n \geq 2$, $Sp(n)$ for $n \geq 2$ and the real forms of the five exceptional complex Lie groups exhaust all simple compact Lie groups up to coverings.

4.11. Invariant vector fields and Lie algebras. Let G be a (real) Lie group. A vector field ξ on G is called *left invariant* if $\mu_a^*\xi = \xi$ for all $a \in G$, where $\mu_a^*\xi = T(\mu_{a^{-1}}) \circ \xi \circ \mu_a$ as in section (3). Since by (3.11) we have $\mu_a^*[\xi, \eta] = [\mu_a^*\xi, \mu_a^*\eta]$, the space $\mathfrak{X}_L(G)$ of all left invariant vector fields on G is closed under the Lie bracket, so it is a Lie subalgebra of $\mathfrak{X}(G)$. Any left invariant vector field ξ is uniquely determined by $\xi(e) \in T_eG$, since $\xi(a) = T_e(\mu_a).\xi(e)$. Thus the Lie algebra $\mathfrak{X}_L(G)$ of left invariant vector fields is linearly isomorphic to T_eG, and on T_eG the Lie bracket on $\mathfrak{X}_L(G)$ induces a Lie algebra structure, whose bracket is again denoted by $[\ , \ ]$. This Lie algebra will be denoted as usual by $\mathfrak{g}$, sometimes by Lie(G).

We will also give a name to the isomorphism with the space of left invariant vector fields: $L : \mathfrak{g} \to \mathfrak{X}_L(G)$, $X \mapsto L_X$, where $L_X(a) = T_e\mu_a.X$. Thus $[X, Y] = [L_X, L_Y](e)$.

A vector field η on G is called *right invariant* if $(\mu^a)^*\eta = \eta$ for all $a \in G$. If ξ is left invariant, then $\nu^*\xi$ is right invariant, since $\nu \circ \mu^a = \mu_{a^{-1}} \circ \nu$ implies that $(\mu^a)^*\nu^*\xi = (\nu \circ \mu^a)^*\xi = (\mu_{a^{-1}} \circ \nu)^*\xi = \nu^*(\mu_{a^{-1}})^*\xi = \nu^*\xi$. The right invariant vector fields form a Lie subalgebra $\mathfrak{X}_R(G)$ of $\mathfrak{X}(G)$, which is again linearly isomorphic to T_eG and induces also a Lie algebra structure on T_eG. Since $\nu^* : \mathfrak{X}_L(G) \to \mathfrak{X}_R(G)$ is an isomorphism of Lie algebras by (3.11), $T_e\nu = -Id : T_eG \to T_eG$ is an isomorphism between the two Lie algebra structures. We will denote by $R : \mathfrak{g} = T_eG \to \mathfrak{X}_R(G)$ the isomorphism discussed, which is given by $R_X(a) = T_e(\mu^a).X$.

4.12. Lemma. *If L_X is a left invariant vector field and R_Y is a right invariant one, then $[L_X, R_Y] = 0$. Thus the flows of L_X and R_Y commute.*

Proof. We consider the vector field $0 \times L_X \in \mathfrak{X}(G \times G)$, given by $(0 \times L_X)(a,b) = (0_a, L_X(b))$. Then $T_{(a,b)}\mu.(0_a, L_X(b)) = T_a\mu^b.0_a + T_b\mu_a.L_X(b) = L_X(ab)$, so $0 \times L_X$ is μ-related to L_X. Likewise $R_Y \times 0$ is μ-related to R_Y. But then $0 = [0 \times L_X, R_Y \times 0]$ is μ-related to $[L_X, R_Y]$ by (3.10). Since μ is surjective, $[L_X, R_Y] = 0$ follows. □

4.13. Lemma. *Let $\varphi : G \to H$ be a smooth homomorphism of Lie groups. Then $\varphi' := T_e\varphi : \mathfrak{g} = T_eG \to \mathfrak{h} = T_eH$ is a Lie algebra homomorphism.*

Later, in (4.21), we shall see that any continuous homomorphism between Lie groups is automatically smooth.

Proof. For $X \in \mathfrak{g}$ and $x \in G$ we have

$$\begin{aligned} T_x\varphi.L_X(x) &= T_x\varphi.T_e\mu_x.X = T_e(\varphi \circ \mu_x).X \\ &= T_e(\mu_{\varphi(x)} \circ \varphi).X = T_e(\mu_{\varphi(x)}).T_e\varphi.X = L_{\varphi'(X)}(\varphi(x)). \end{aligned}$$

So L_X is φ-related to $L_{\varphi'(X)}$. By (3.10) the field $[L_X, L_Y] = L_{[X,Y]}$ is φ-related to $[L_{\varphi'(X)}, L_{\varphi'(Y)}] = L_{[\varphi'(X),\varphi'(Y)]}$. So we have $T\varphi \circ L_{[X,Y]} = L_{[\varphi'(X),\varphi'(Y)]} \circ \varphi$. If we evaluate this at e, the result follows. □

Now we will determine the Lie algebras of all the examples given above.

4.14. For the Lie group $GL(n,\mathbb{R})$ we have $T_eGL(n,\mathbb{R}) = L(\mathbb{R}^n,\mathbb{R}^n) =: \mathfrak{gl}(n,\mathbb{R})$ and $T\,GL(n,\mathbb{R}) = GL(n,\mathbb{R}) \times L(\mathbb{R}^n,\mathbb{R}^n)$ by the affine structure of the surrounding vector space. For $A \in GL(n,\mathbb{R})$ we have $\mu_A(B) = A.B$, so μ_A extends to a linear isomorphism of $L(\mathbb{R}^n,\mathbb{R}^n)$, and for $(B,X) \in T\,GL(n,\mathbb{R})$ we get $T_B(\mu_A).(B,X) = (A.B, A.X)$. So the left invariant vector field $L_X \in \mathfrak{X}_L(GL(n,\mathbb{R}))$ is given by $L_X(A) = T_e(\mu_A).X = (A, A.X)$.

Let $f : GL(n,\mathbb{R}) \to \mathbb{R}$ be the restriction of a linear functional on $L(\mathbb{R}^n,\mathbb{R}^n)$. Then we have $L_X(f)(A) = df(A)(L_X(A)) = df(A)(A.X) = f(A.X)$, which we may write as $L_X(f) = f(\ \ .X)$. Therefore

$$\begin{aligned} L_{[X,Y]}(f) &= [L_X, L_Y](f) = L_X(L_Y(f)) - L_Y(L_X(f)) \\ &= L_X(f(\ \ .Y)) - L_Y(f(\ \ .X)) = f(\ \ .X.Y) - f(\ \ .Y.X) \\ &= f(\ \ .(XY - YX)) = L_{XY-YX}(f). \end{aligned}$$

So the Lie bracket on $\mathfrak{gl}(n,\mathbb{R}) = L(\mathbb{R}^n,\mathbb{R}^n)$ is given by $[X,Y] = XY - YX$, the usual commutator.

4.15. Example. Let V be a vector space. Then $(V,+)$ is a Lie group, $T_0V = V$ is its Lie algebra, $TV = V \times V$, left translation is $\mu_v(w) = v + w$, $T_w(\mu_v).(w,X) = (v + w, X)$. So $L_X(v) = (v, X)$, a constant vector field. Thus the Lie bracket is 0.

4.16. Example. The special linear group is $SL(n,\mathbb{R}) = \det^{-1}(1)$ and its Lie algebra is given by $T_eSL(n,\mathbb{R}) = \ker d\det(\mathbb{I}) = \{X \in L(\mathbb{R}^n,\mathbb{R}^n) : \operatorname{Trace} X = 0\} = \mathfrak{sl}(n,\mathbb{R})$ by (4.6). The injection $i : SL(n,\mathbb{R}) \to GL(n,\mathbb{R})$ is a smooth homomorphism of Lie groups, so $T_ei = i' : \mathfrak{sl}(n,\mathbb{R}) \to \mathfrak{gl}(n,\mathbb{R})$ is an injective homomorphism of Lie algebras. Thus the Lie bracket is given by $[X,Y] = XY - YX$.

The same argument gives the commutator as the Lie bracket in all other examples we have treated. We have already determined the Lie algebras as T_eG.

4.17. 1-parameter subgroups. Let G be a Lie group with Lie algebra $\mathfrak{g}$. A 1-*parameter subgroup* of G is a Lie group homomorphism $\alpha : (\mathbb{R},+) \to G$, i.e., a smooth curve α in G with $\alpha(s+t) = \alpha(s).\alpha(t)$, and hence $\alpha(0) = e$.

Lemma. *Let $\alpha : \mathbb{R} \to G$ be a smooth curve with $\alpha(0) = e$. Let $X \in \mathfrak{g}$. Then the following assertions are equivalent.*

(1) *α is a 1-parameter subgroup with $X = \partial|_0\alpha(t)$.*
(2) *$\alpha(t) = \operatorname{Fl}^{L_X}(t,e)$ for all t.*
(3) *$\alpha(t) = \operatorname{Fl}^{R_X}(t,e)$ for all t.*
(4) *$x.\alpha(t) = \operatorname{Fl}^{L_X}(t,x)$, or $\operatorname{Fl}^{L_X}_t = \mu^{\alpha(t)}$, for all t .*
(5) *$\alpha(t).x = \operatorname{Fl}^{R_X}(t,x)$, or $\operatorname{Fl}^{R_X}_t = \mu_{\alpha(t)}$, for all t.*

Proof. (1) $\Longrightarrow$ (4) We have $\frac{d}{dt}x.\alpha(t) = \frac{d}{ds}|_0 x.\alpha(t+s) = \frac{d}{ds}|_0 x.\alpha(t).\alpha(s) = \frac{d}{ds}|_0 \mu_{x.\alpha(t)}\alpha(s) = T_e(\mu_{x.\alpha(t)}).\frac{d}{ds}|_0\alpha(s) = T_e(\mu_{x.\alpha(t)}).X = L_X(x.\alpha(t))$. By uniqueness of solutions we get $x.\alpha(t) = \operatorname{Fl}^{L_X}(t,x)$.

(4) $\Longrightarrow$ (2) This is clear.

(2) $\Longrightarrow$ (1) We have

$$\begin{aligned}\tfrac{d}{ds}\alpha(t)\alpha(s) &= \tfrac{d}{ds}(\mu_{\alpha(t)}\alpha(s)) = T(\mu_{\alpha(t)})\tfrac{d}{ds}\alpha(s)\\ &= T(\mu_{\alpha(t)})L_X(\alpha(s)) = L_X(\alpha(t)\alpha(s))\end{aligned}$$

and $\alpha(t)\alpha(0) = \alpha(t)$. So we get $\alpha(t)\alpha(s) = \operatorname{Fl}^{L_X}(s,\alpha(t)) = \operatorname{Fl}^{L_X}_s \operatorname{Fl}^{L_X}_t(e) = \operatorname{Fl}^{L_X}(t+s,e) = \alpha(t+s)$.

(4) $\Longleftrightarrow$ (5) We have $\operatorname{Fl}^{\nu^*\xi}_t = \nu^{-1}\circ \operatorname{Fl}^{\xi}_t \circ\nu$ by (3.14). Therefore we have by (4.11)

$$\begin{aligned}(\operatorname{Fl}^{R_X}_t(x^{-1}))^{-1} &= (\nu\circ \operatorname{Fl}^{R_X}_t \circ\nu)(x) = \operatorname{Fl}^{\nu^*R_X}_t(x)\\ &= \operatorname{Fl}^{L_X}_{-t}(x) = x.\alpha(-t).\end{aligned}$$

So $\operatorname{Fl}^{R_X}_t(x^{-1}) = \alpha(t).x^{-1}$, and $\operatorname{Fl}^{R_X}_t(y) = \alpha(t).y$.

(5) $\Longrightarrow$ (3) $\Longrightarrow$ (1) can be shown in a similar way. □

An immediate consequence of the foregoing lemma is that left invariant and right invariant vector fields on a Lie group are always complete, so they have global flows, because a locally defined 1-parameter group can always be extended to a globally defined one by multiplying it up: $\alpha(nt) = \alpha(t)^n$.

4.18. Definition. The *exponential mapping* $\exp : \mathfrak{g} \to G$ of a Lie group is defined by

$$\exp X = \mathrm{Fl}^{L_X}(1, e) = \mathrm{Fl}^{R_X}(1, e) = \alpha_X(1),$$

where α_X is the 1-parameter subgroup of G with $\dot{\alpha}_X(0) = X$.

Theorem.

(1) $\exp : \mathfrak{g} \to G$ *is smooth.*

(2) $\exp(tX) = \mathrm{Fl}^{L_X}(t, e)$.

(3) $\mathrm{Fl}^{L_X}(t, x) = x.\exp(tX)$.

(4) $\mathrm{Fl}^{R_X}(t, x) = \exp(tX).x$.

(5) $\exp(0) = e$ *and* $T_0 \exp = Id : T_0\mathfrak{g} = \mathfrak{g} \to T_eG = \mathfrak{g}$; *thus* exp *is a diffeomorphism from a neighborhood of* 0 *in* $\mathfrak{g}$ *onto a neighborhood of* e *in* G.

Proof. (1) Let $0 \times L \in \mathfrak{X}(\mathfrak{g} \times G)$ be given by $(0 \times L)(X, x) = (0_X, L_X(x))$. Then $\mathrm{pr}_2 \mathrm{Fl}^{0\times L}(t, (X, e)) = \alpha_X(t)$ is smooth in (t, X).

(2) $\exp(tX) = \mathrm{Fl}^{t.L_X}(1, e) = \mathrm{Fl}^{L_X}(t, e) = \alpha_X(t)$.

(3) and (4) follow from lemma (4.17).

(5) $T_0 \exp .X = \frac{d}{dt}|_0 \exp(0 + t.X) = \frac{d}{dt}|_0 \mathrm{Fl}^{L_X}(t, e) = X$. □

4.19. Remark. If G is connected and $U \subset \mathfrak{g}$ is open with $0 \in U$, then the group generated by $\exp(U)$ equals G.

Namely, this group is a subgroup of G containing some open neighborhood of e, so it is open. The complement in G is also open (as union of the other cosets), so this subgroup is open and closed. Since G is connected, it coincides with G.

If G is not connected, then the subgroup generated by $\exp(U)$ is the connected component G_0 of e in G, an open connected normal subgroup.

4.20. Remark. Let $\varphi : G \to H$ be a smooth homomorphism of Lie groups. Then the diagram

$$\begin{CD} \mathfrak{g} @>{\varphi'}>> \mathfrak{h} \\ @V{\exp^G}VV @VV{\exp^H}V \\ G @>{\varphi}>> H \end{CD}$$

commutes, since $t \mapsto \varphi(\exp^G(tX))$ is a 1-parameter subgroup of H which satisfies $\frac{d}{dt}|_0 \varphi(\exp^G tX) = \varphi'(X)$, so $\varphi(\exp^G tX) = \exp^H(t\varphi'(X))$.

If G is connected and $\varphi, \psi : G \to H$ are homomorphisms of Lie groups with $\varphi' = \psi' : \mathfrak{g} \to \mathfrak{h}$, then $\varphi = \psi$. Namely, $\varphi = \psi$ on the subgroup generated by $\exp^G \mathfrak{g}$ which equals G by (4.19).

4.21. Theorem. *A continuous homomorphism $\varphi : G \to H$ between Lie groups is smooth. In particular a topological group can carry at most one compatible Lie group structure.*

Proof. Let first $\varphi = \alpha : (\mathbb{R}, +) \to G$ be a continuous 1-parameter subgroup. Then $\alpha(-\varepsilon, \varepsilon) \subset \exp(U)$, where U is an open ball with center 0 in $\mathfrak{g}$ such that $\exp \restriction 2U$ is a diffeomorphism, for some $\varepsilon > 0$. Put

$$\beta := (\exp \restriction 2U)^{-1} \circ \alpha : (-\varepsilon, \varepsilon) \to \mathfrak{g}.$$

Then for $|t| < \frac{\varepsilon}{2}$ we have

$$\exp(2\beta(t)) = \exp(\beta(t))^2 = \alpha(t)^2 = \alpha(2t) = \exp(\beta(2t)),$$

so $2\beta(t) = \beta(2t)$; thus $\beta(\frac{s}{2}) = \frac{1}{2}\beta(s)$ for $|s| < \varepsilon$. Applying exp we have $\alpha(\frac{s}{2}) = \exp(\beta(\frac{s}{2})) = \exp(\frac{1}{2}\beta(s))$ for all $|s| < \varepsilon$ and by recursion we get $\alpha(\frac{s}{2^n}) = \exp(\frac{1}{2^n}\beta(s))$ for $n \in \mathbb{N}$ and in turn

$$\alpha(\tfrac{k}{2^n}s) = \alpha(\tfrac{s}{2^n})^k = \exp(\tfrac{1}{2^n}\beta(s))^k = \exp(\tfrac{k}{2^n}\beta(s))$$

for $k \in \mathbb{Z}$. Since the $\frac{k}{2^n}$ for $k \in \mathbb{Z}$ and $n \in \mathbb{N}$ are dense in $\mathbb{R}$ and since α is continuous, we get $\alpha(ts) = \exp(t\beta(s))$ for all $t \in \mathbb{R}$. So α is smooth.

Now let $\varphi : G \to H$ be a continuous homomorphism. Let $X_1, \dots, X_n$ be a linear basis of $\mathfrak{g}$. We define $\psi : \mathbb{R}^n \to G$ by

$$\psi(t^1, \dots, t^n) = \exp(t^1 X_1) \cdots \exp(t^n X_n).$$

Then $T_0\psi$ is invertible, so ψ is a diffeomorphism near 0. Sometimes ψ^{-1} is called a coordinate system of the second kind. The curve $t \mapsto \varphi(\exp^G tX_i)$ is a continuous 1-parameter subgroup of H, so it is smooth by the first part of the proof.

We have $(\varphi \circ \psi)(t^1, \dots, t^n) = (\varphi \exp(t^1 X_1)) \cdots (\varphi \exp(t^n X_n))$, so $\varphi \circ \psi$ is smooth. Thus φ is smooth near $e \in G$ and so everywhere on G. □

4.22. Theorem. *Let G and H be Lie groups (G separable is essential here), and let $\varphi : G \to H$ be a continuous bijective homomorphism. Then φ is a diffeomorphism.*

Proof. Our first aim is to show that φ is a homeomorphism. Let V be an open e-neighborhood in G, and let K be a compact e-neighborhood in G such that $K.K^{-1} \subset V$. Since G is separable, there is a sequence $(a_i)_{i \in \mathbb{N}}$ in G such that $G = \bigcup_{i=1}^{\infty} a_i.K$. Since H is locally compact, it is a Baire space (i.e., V_i

open and dense for $i \in \mathbb{N}$ implies $\bigcap V_i$ dense). The set $\varphi(a_i)\varphi(K)$ is compact, thus closed. Since $H = \bigcup_i \varphi(a_i).\varphi(K)$, there is some i such that $\varphi(a_i)\varphi(K)$ has nonempty interior, so $\varphi(K)$ has nonempty interior. Choose $b \in G$ such that $\varphi(b)$ is an interior point of $\varphi(K)$ in H. Then $e_H = \varphi(b)\varphi(b^{-1})$ is an interior point of $\varphi(K)\varphi(K^{-1}) \subset \varphi(V)$. So if U is open in G and $a \in U$, then e_H is an interior point of $\varphi(a^{-1}U)$, so $\varphi(a)$ is in the interior of $\varphi(U)$. Thus $\varphi(U)$ is open in H, and φ is a homeomorphism.

Now by (4.21) φ and φ^{-1} are smooth. □

4.23. Examples. We first describe the exponential mapping of the general linear group $GL(n,\mathbb{R})$. Let $X \in \mathfrak{gl}(n,\mathbb{R}) = L(\mathbb{R}^n,\mathbb{R}^n)$; then the left invariant vector field is given by $L_X(A) = (A, A.X) \in GL(n,\mathbb{R}) \times \mathfrak{gl}(n,\mathbb{R})$ and the 1-parameter group $\alpha_X(t) = \mathrm{Fl}^{L_X}(t, \mathbb{I}_n)$ is given by the differential equation $\frac{d}{dt}\alpha_X(t) = L_X(\alpha_X(t)) = \alpha_X(t).X$, with initial condition $\alpha_X(0) = \mathbb{I}_n$. But the unique solution of this equation is $\alpha_X(t) = e^{tX} = \sum_{k=0}^{\infty} \frac{t^k}{k!} X^k$. So

$$\exp^{GL(n,\mathbb{R})}(X) = e^X = \sum_{k=0}^{\infty} \tfrac{1}{k!}\, X^k.$$

If $n = 1$, we get the usual exponential mapping of one real variable. For all Lie subgroups of $GL(n,\mathbb{R})$, the exponential mapping is given by the same formula $\exp(X) = e^X$; this follows from (4.20).

4.24. The adjoint representation. A *representation* of a Lie group G on a finite-dimensional vector space V (real or complex) is a homomorphism $\rho : G \to GL(V)$ of Lie groups. Its derivative $\rho' : \mathfrak{g} \to \mathfrak{gl}(V) = L(V,V)$ is a Lie algebra homomorphism by (4.13).

For $a \in G$ we define $\mathrm{conj}_a : G \to G$ by $\mathrm{conj}_a(x) = axa^{-1}$. It is called the *conjugation* or the *inner automorphism* by $a \in G$. We have $\mathrm{conj}_a(xy) = \mathrm{conj}_a(x)\,\mathrm{conj}_a(y)$, $\mathrm{conj}_{ab} = \mathrm{conj}_a \circ \mathrm{conj}_b$, and conj is smooth in all variables.

Next we define for $a \in G$ the mapping $\mathrm{Ad}(a) = (\mathrm{conj}_a)' = T_e(\mathrm{conj}_a) : \mathfrak{g} \to \mathfrak{g}$. By (4.13) the linear map $\mathrm{Ad}(a)$ is a Lie algebra homomorphism, so we have $\mathrm{Ad}(a)[X,Y] = [\mathrm{Ad}(a)X, \mathrm{Ad}(a)Y]$. Furthermore $\mathrm{Ad} : G \to GL(\mathfrak{g})$ is a representation, called the *adjoint representation* of G, since

$$\begin{aligned}\mathrm{Ad}(ab) &= T_e(\mathrm{conj}_{ab}) = T_e(\mathrm{conj}_a \circ \mathrm{conj}_b)\\ &= T_e(\mathrm{conj}_a) \circ T_e(\mathrm{conj}_b) = \mathrm{Ad}(a) \circ \mathrm{Ad}(b).\end{aligned}$$

The relations $\mathrm{Ad}(a) = T_e(\mathrm{conj}_a) = T_a(\mu^{a^{-1}}).T_e(\mu_a) = T_{a^{-1}}(\mu_a).T_e(\mu^{a^{-1}})$ will be used later.

Now we define the (lower case) *adjoint representation* of the Lie algebra $\mathfrak{g}$,

$$\mathrm{ad} : \mathfrak{g} \to \mathfrak{gl}(\mathfrak{g}) = L(\mathfrak{g},\mathfrak{g}), \quad \mathrm{ad} := \mathrm{Ad}' = T_e\,\mathrm{Ad}\,.$$

Lemma.

(1) $L_X(a) = R_{\mathrm{Ad}(a)X}(a)$ *for* $X \in \mathfrak{g}$ *and* $a \in G$.

(2) $\mathrm{ad}(X)Y = [X,Y]$ *for* $X, Y \in \mathfrak{g}$.

Proof. (1) $L_X(a) = T_e(\mu_a).X = T_e(\mu^a).T_e(\mu^{a^{-1}} \circ \mu_a).X = R_{\mathrm{Ad}(a)X}(a)$.

(2) Let $X_1, \dots, X_n$ be a linear basis of $\mathfrak{g}$ and fix $X \in \mathfrak{g}$. Then $\mathrm{Ad}(x)X = \sum_{i=1}^n f_i(x).X_i$ for $f_i \in C^\infty(G,\mathbb{R})$ and we have in turn:

$$\begin{aligned}\mathrm{Ad}'(Y)X &= T_e(\mathrm{Ad}(\quad)X)Y = d(\mathrm{Ad}(\quad)X)|_e Y = d(\textstyle\sum f_i X_i)|_e Y\\ &= \textstyle\sum df_i|_e(Y)X_i = \sum L_Y(f_i)(e).X_i.\end{aligned}$$

$$L_X(x) = R_{\mathrm{Ad}(x)X}(x) = R(\textstyle\sum f_i(x)X_i)(x) = \sum f_i(x).R_{X_i}(x) \text{ by (1).}$$

$$[L_Y, L_X] = [L_Y, \textstyle\sum f_i.R_{X_i}] = 0 + \sum L_Y(f_i).R_{X_i} \text{ by (3.4) and (4.12).}$$

$$[Y,X] = [L_Y, L_X](e) = \textstyle\sum L_Y(f_i)(e).R_{X_i}(e) = \mathrm{Ad}'(Y)X = \mathrm{ad}(Y)X. \quad \square$$

4.25. Corollary. *From* (4.20) *and* (4.23) *we have*

$$\begin{aligned}\mathrm{Ad} \circ \exp^G &= \exp^{GL(\mathfrak{g})} \circ \mathrm{ad},\\ \mathrm{Ad}(\exp^G X)Y &= \sum_{k=0}^\infty \tfrac{1}{k!}\,(\mathrm{ad}\, X)^k Y = e^{\mathrm{ad}\, X} Y\\ &= Y + [X,Y] + \tfrac{1}{2!}[X,[X,Y]] + \tfrac{1}{3!}[X,[X,[X,Y]]] + \cdots\end{aligned}$$

so that also $\mathrm{ad}(X) = \partial|_0 \mathrm{Ad}(\exp(tX))$.

4.26. The right logarithmic derivative. Let M be a manifold and let $f : M \to G$ be a smooth mapping into a Lie group G with Lie algebra $\mathfrak{g}$. We define the mapping $\delta f : TM \to \mathfrak{g}$ by the formula $\delta f(\xi_x) := T_{f(x)}(\mu^{f(x)^{-1}}).T_x f.\xi_x$. Then δf is a $\mathfrak{g}$-valued 1-form on M, $\delta f \in \Omega^1(M,\mathfrak{g})$, as we will write later. We call δf the *right logarithmic derivative* of f, since for $f : \mathbb{R} \to (\mathbb{R}^+, \cdot)$ we have $\delta f(x).1 = \frac{f'(x)}{f(x)} = (\log \circ f)'(x)$.

Lemma. *Let* $f, g : M \to G$ *be smooth. Then we have*

$$\delta(f.g)(x) = \delta f(x) + \mathrm{Ad}(f(x)).\delta g(x).$$

Proof. We compute as follows:

$$\begin{aligned}\delta(f.g)(x) &= T(\mu^{g(x)^{-1}.f(x)^{-1}}).T_x(f.g)\\ &= T(\mu^{f(x)^{-1}}).T(\mu^{g(x)^{-1}}).T_{(f(x),g(x))}\mu.(T_x f, T_x g)\\ &= T(\mu^{f(x)^{-1}}).T(\mu^{g(x)^{-1}}).\Big(T(\mu^{g(x)}).T_x f + T(\mu_{f(x)}).T_x g\Big)\\ &= \delta f(x) + \mathrm{Ad}(f(x)).\delta g(x). \quad \square\end{aligned}$$

Remark. The *left logarithmic derivative* $\delta^{\text{left}} f \in \Omega^1(M, \mathfrak{g})$ of a smooth mapping $f : M \to G$ is given by $\delta^{\text{left}} f.\xi_x = T_{f(x)}(\mu_{f(x)^{-1}}).T_x f.\xi_x$. The corresponding Leibniz rule for it is uglier than that for the right logarithmic derivative:

$$\delta^{\text{left}}(fg)(x) = \delta^{\text{left}} g(x) + \operatorname{Ad}(g(x)^{-1})\delta^{\text{left}} f(x).$$

The form $\delta^{\text{left}}(Id_G) \in \Omega^1(G, \mathfrak{g})$ is also called the *Maurer-Cartan form* of the Lie group G.

4.27. Lemma. *For* $\exp : \mathfrak{g} \to G$ *and for* $g(z) := \dfrac{e^z - 1}{z}$ *we have*

$$\delta(\exp)(X) = T(\mu^{\exp(-X)}).T_X \exp = \sum_{p=0}^{\infty} \tfrac{1}{(p+1)!} \,(\operatorname{ad} X)^p = g(\operatorname{ad} X).$$

Proof. We put $M(X) = \delta(\exp)(X) : \mathfrak{g} \to \mathfrak{g}$. Then

$$\begin{aligned}
(s+t)M((s+t)X) &= (s+t)\delta(\exp)((s+t)X) \\
&= \delta(\exp((s+t)\quad))X \qquad \text{by the chain rule} \\
&= \delta(\exp(s\quad).\exp(t\quad)).X \\
&= \delta(\exp(s\quad)).X + \operatorname{Ad}(\exp(sX)).\delta(\exp(t\quad)).X \quad \text{by (4.26)} \\
&= s.\delta(\exp)(sX) + \operatorname{Ad}(\exp(sX)).t.\delta(\exp)(tX) \\
&= s.M(sX) + \operatorname{Ad}(\exp(sX)).t.M(tX).
\end{aligned}$$

Next we put

$$\begin{aligned}
N(t) &:= t.M(tX) \in L(\mathfrak{g}, \mathfrak{g}); \\
N(s+t) &= N(s) + \operatorname{Ad}(\exp(sX)).N(t).
\end{aligned}$$

We fix t, apply $\frac{d}{ds}|_0$, and get

$$\begin{aligned}
N'(t) &= N'(0) + \operatorname{ad}(X).N(t), \\
N'(0) &= M(0) + 0 = \delta(\exp)(0) = Id_{\mathfrak{g}}.
\end{aligned}$$

So we have the differential equation

$$N'(t) = Id_{\mathfrak{g}} + \operatorname{ad}(X).N(t)$$

in $L(\mathfrak{g}, \mathfrak{g})$ with initial condition $N(0) = 0$. The unique solution is

$$N(s) = \sum_{p=0}^{\infty} \tfrac{1}{(p+1)!} \operatorname{ad}(X)^p.s^{p+1}, \quad \text{and so}$$

$$\delta(\exp)(X) = M(X) = N(1) = \sum_{p=0}^{\infty} \tfrac{1}{(p+1)!} \operatorname{ad}(X)^p. \quad \square$$

4.28. Corollary. *The tangent mapping $T_X \exp$ is bijective if and only if no eigenvalue of $\operatorname{ad}(X) : \mathfrak{g} \to \mathfrak{g}$ is of the form $\sqrt{-1}\,2k\pi$ for $k \in \mathbb{Z} \setminus \{0\}$.*

Proof. The zeros of $g(z) = \frac{e^z - 1}{z}$ are $z = 2k\pi\sqrt{-1}$ for $k \in \mathbb{Z} \setminus \{0\}$. The linear mapping $T_X \exp$ is bijective if and only if no eigenvalue of

$$g(\operatorname{ad}(X)) = T(\mu^{\exp(-X)}).T_X \exp$$

is 0. But the eigenvalues of $g(\operatorname{ad}(X))$ are the images under g of the eigenvalues of $\operatorname{ad}(X)$. □

4.29. Theorem. The Baker-Campbell-Hausdorff formula. *Let G be a Lie group with Lie algebra $\mathfrak{g}$. For complex z near 1 we consider the function $f(z) := \frac{\log(z)}{z-1} = \sum_{n\geq 0} \frac{(-1)^n}{n+1}(z-1)^n$.*
Then for X, Y near 0 in $\mathfrak{g}$ we have $\exp X.\exp Y = \exp C(X,Y)$, where

$$\begin{aligned}
C(X,Y) &= Y + \int_0^1 f(e^{t.\operatorname{ad}X}.e^{\operatorname{ad}Y}).X\,dt \\
&= X + Y + \sum_{n\geq 1} \frac{(-1)^n}{n+1} \int_0^1 \Bigg(\sum_{\substack{k,\ell\geq 0\\ k+\ell\geq 1}} \frac{t^k}{k!\,\ell!} (\operatorname{ad}X)^k (\operatorname{ad}Y)^\ell \Bigg)^n X\,dt \\
&= X + Y + \sum_{n\geq 1} \frac{(-1)^n}{n+1} \sum_{\substack{k_1,\dots,k_n\geq 0\\ \ell_1,\dots,\ell_n\geq 0\\ k_i+\ell_i\geq 1}} \frac{(\operatorname{ad}X)^{k_1}(\operatorname{ad}Y)^{\ell_1}\dots(\operatorname{ad}X)^{k_n}(\operatorname{ad}Y)^{\ell_n}}{(k_1+\dots+k_n+1)k_1!\dots k_n!\ell_1!\dots\ell_n!} X \\
&= X + Y + \tfrac{1}{2}[X,Y] + \tfrac{1}{12}([X,[X,Y]] - [Y,[Y,X]]) + \cdots .
\end{aligned}$$

Proof. Let $C(X,Y) := \exp^{-1}(\exp X.\exp Y)$ for X, Y near 0 in $\mathfrak{g}$, and let $C(t) := C(tX,Y)$. Then by (4.27) we have

$$\begin{aligned}
T(\mu^{\exp(-C(t))})\tfrac{d}{dt}(\exp C(t)) &= \delta(\exp\circ C)(t).1 = \delta\exp(C(t)).\dot{C}(t) \\
&= \textstyle\sum_{k\geq 0} \frac{1}{(k+1)!}(\operatorname{ad} C(t))^k \dot{C}(t) \\
&= g(\operatorname{ad} C(t)).\dot{C}(t),
\end{aligned}$$

where $g(z) := \frac{e^z-1}{z} = \sum_{k\geq 0} \frac{z^k}{(k+1)!}$. We have $\exp C(t) = \exp(tX)\exp Y$ and

$$\exp(-C(t)) = \exp(C(t))^{-1} = \exp(-Y)\exp(-tX);$$

therefore

$$\begin{aligned}
T(\mu^{\exp(-C(t))})\tfrac{d}{dt}(\exp C(t)) &= T(\mu^{\exp(-Y)\exp(-tX)})\tfrac{d}{dt}(\exp(tX)\exp Y) \\
&= T(\mu^{\exp(-tX)})T(\mu^{\exp(-Y)})T(\mu^{\exp Y})\tfrac{d}{dt}\exp(tX) \\
&= T(\mu^{\exp(-tX)}).R_X(\exp(tX)) = X \quad \text{by (4.18.4) and (4.11).}
\end{aligned}$$

$$X = g(\operatorname{ad} C(t)).\dot{C}(t).$$

$$\begin{aligned} e^{\operatorname{ad} C(t)} &= \operatorname{Ad}(\exp C(t)) \qquad \text{by (4.25)} \\ &= \operatorname{Ad}(\exp(tX)\exp Y) = \operatorname{Ad}(\exp(tX)).\operatorname{Ad}(\exp Y) \\ &= e^{\operatorname{ad}(tX)}.e^{\operatorname{ad} Y} = e^{t.\operatorname{ad} X}.e^{\operatorname{ad} Y}. \end{aligned}$$

If X, Y, and t are small enough, we get $\operatorname{ad} C(t) = \log(e^{t.\operatorname{ad} X}.e^{\operatorname{ad} Y})$, where $\log(z) = \sum_{n\geq 1} \frac{(-1)^{n+1}}{n}(z-1)^n$; thus we have

$$X = g(\operatorname{ad} C(t)).\dot{C}(t) = g(\log(e^{t.\operatorname{ad} X}.e^{\operatorname{ad} Y})).\dot{C}(t).$$

For z near 1 we put $f(z) := \frac{\log(z)}{z-1} = \sum_{n\geq 0}\frac{(-1)^n}{n+1}(z-1)^n$. This function satisfies $g(\log(z)).f(z) = 1$. So we have

$$X = g(\log(e^{t.\operatorname{ad} X}.e^{\operatorname{ad} Y})).\dot{C}(t) = f(e^{t.\operatorname{ad} X}.e^{\operatorname{ad} Y})^{-1}.\dot{C}(t),$$

$$\begin{cases} \dot{C}(t) = f(e^{t.\operatorname{ad} X}.e^{\operatorname{ad} Y}).X, \\ C(0) = Y. \end{cases}$$

Passing to the definite integral, we get the desired formula

$$\begin{aligned} C(X,Y) &= C(1) = C(0) + \int_0^1 \dot{C}(t)\,dt \\ &= Y + \int_0^1 f(e^{t.\operatorname{ad} X}.e^{\operatorname{ad} Y}).X\,dt \\ &= X + Y + \sum_{n\geq 1}\frac{(-1)^n}{n+1}\int_0^1 \Big(\sum_{\substack{k,\ell\geq 0\\ k+\ell\geq 1}} \frac{t^k}{k!\,\ell!}\,(\operatorname{ad} X)^k(\operatorname{ad} Y)^\ell\Big)^n X\,dt \\ &= X + Y + \sum_{n\geq 1}\frac{(-1)^n}{n+1} \sum_{\substack{k_1,\dots,k_n\geq 0\\ \ell_1,\dots,\ell_n\geq 0\\ k_i+\ell_i\geq 1}} \frac{(\operatorname{ad} X)^{k_1}(\operatorname{ad} Y)^{\ell_1}\dots(\operatorname{ad} X)^{k_n}(\operatorname{ad} Y)^{\ell_n}}{(k_1+\dots+k_n+1)k_1!\dots k_n!\ell_1!\dots\ell_n!}X \\ &= X + Y + \tfrac{1}{2}[X,Y] + \tfrac{1}{12}([X,[X,Y]] - [Y,[Y,X]]) + \cdots. \qquad \square \end{aligned}$$

4.30. Remarks. (1) If G is a Lie group of differentiability class C^2, then we may define TG and the Lie bracket of vector fields. The proof above then makes sense and the theorem shows that in the chart given by $\exp^{-1}$ the multiplication $\mu : G \times G \to G$ is C^ω near e, hence everywhere. So in this case G is a real analytic Lie group. See also remark (5.7) below.

(2) Let G be a Lie groups with Lie algebra $\mathfrak{g}$. Then *Trotter's formula* holds: For $X, Y \in \mathfrak{g}$ we have, by (4.29),

$$\begin{aligned} (\exp(\tfrac{1}{n}X)\exp(\tfrac{1}{n}))^n &= \exp(n.C(\tfrac{1}{n}X, \tfrac{1}{n}Y)) \\ &= \exp(X + Y + \tfrac{1}{n}.(\text{bounded})) \xrightarrow[n\to\infty]{} \exp(X+Y). \end{aligned}$$

(3) Similarly, by (4.29),

$$
\begin{aligned}
(\exp(\tfrac{1}{n}X)\exp(\tfrac{1}{n}Y)\exp(\tfrac{-1}{n}X)\exp(\tfrac{-1}{n}Y))^{n^2} \\
&= \exp\big(n^2 C(C(\tfrac{1}{n}X,\tfrac{1}{n}Y), C(\tfrac{-1}{n}X,\tfrac{-1}{n}Y))\big) \\
&= \exp([X,Y] + \tfrac{1}{n}(\text{bounded})) \xrightarrow[n\to\infty]{} \exp([X,Y]).
\end{aligned}
$$

(4) Let P be a formal bracket expression of length k as in (3.16). On G we use $[g,h] = ghg^{-1}h^{-1}$ as commutator. We consider smooth curves $g_i : \mathbb{R} \to G$ with $g_i(0) = e$ and $g_i'(0) = X_i \in \mathfrak{g}$. Then $\varphi_i(t,h) = h.g_i(t) = \mu^{g_i(t)}(h)$ are global curves of diffeomorphisms on G with $\partial_t|_0 \varphi_i(t,h) = L_{X_i}(h)$. Evaluating (3.16) at e, we then get

$$
\begin{aligned}
0 &= \tfrac{\partial^\ell}{\partial t^\ell}|_0 P(g_t^1,\dots,g_t^k) \quad \text{for } 1 \le \ell < k, \\
P(X_1,\dots,X_k) &= \tfrac{1}{k!}\tfrac{\partial^k}{\partial t^k}|_0 P(g_t^1,\dots,g_t^k) \in \mathfrak{X}(M).
\end{aligned}
$$

A special case of this is: For $X_i \in \mathfrak{g}$ we have

$$
\begin{aligned}
0 &= \tfrac{\partial^\ell}{\partial t^\ell}|_0 P(\exp(t.X_1),\dots,\exp(t.X_k)) \quad \text{for } 1 \le \ell < k, \\
P(X_1,\dots,X_k) &= \tfrac{1}{k!}\tfrac{\partial^k}{\partial t^k}|_0 P(\exp(t.X_1),\dots,\exp(t.X_k)) \in \mathfrak{g}.
\end{aligned}
$$

4.31. Example. The group $SO(3,\mathbb{R})$. From (4.5) and (4.16) we know that the Lie algebra $\mathfrak{o}(3,\mathbb{R})$ of $SO(3,\mathbb{R})$ is the space $L_{\text{skew}}(\mathbb{R}^3,\mathbb{R}^3)$ of all linear mappings which are skew-symmetric with respect to the inner product, with the commutator as Lie bracket.

The group $Sp(1) = S^3$ of unit quaternions has as Lie algebra $T_1 S^3 = 1^\perp$, the space of imaginary quaternions, with the commutator of the quaternion multiplications as bracket. From (4.10) we see that this is $[X,Y] = 2X \times Y$.

Then we observe that the mapping

$$
\alpha : \mathfrak{sp}(1) \to \mathfrak{o}(3,\mathbb{R}) = L_{\text{skew}}(\mathbb{R}^3,\mathbb{R}^3), \qquad \alpha(X)Y = 2X \times Y,
$$

is a linear isomorphism between two 3-dimensional vector spaces and is also an isomorphism of Lie algebras because $[\alpha(X),\alpha(Y)]Z = 4(X \times (Y \times Z) - Y \times (X \times Z)) = 4(X \times (Y \times Z) + Y \times (Z \times X)) = -4(Z \times (Y \times X)) = 2(2X \times Y) \times Z = \alpha([X,Y])Z$. Since S^3 is simply connected, we may conclude from (5.4) below that $Sp(1)$ is the universal cover of $SO(3)$.

We can also see this directly as follows: Consider the mapping $\tau : S^3 \subset \mathbb{H} \to SO(3,\mathbb{R})$ which is given by $\tau(P)X = PX\bar{P}$, where $X \in \mathbb{R}^3 \times \{0\} \subset \mathbb{H}$ is an imaginary quaternion. It is clearly a homomorphism $\tau : S^3 \to GL(3,\mathbb{R})$, and since $|\tau(P)X| = |PX\bar{P}| = |X|$ and S^3 is connected, it has values in $SO(3,\mathbb{R})$. The tangent mapping of τ is computed as $(T_1\tau.X)Y = XY1 + 1Y(-X) = 2(X \times Y) = \alpha(X)Y$, so it is an isomorphism. Thus τ is a local diffeomorphism, the image of τ is an open and compact (since S^3 is compact) subgroup of $SO(3,\mathbb{R})$, so τ is surjective since $SO(3,\mathbb{R})$ is connected. The kernel of τ is the set of all $P \in S^3$ with $PX\bar{P} = X$ for all $X \in \mathbb{R}^3$, i.e., the intersection of the center of $\mathbb{H}$ with S^3, the set $\{1,-1\}$. So τ is a two sheeted covering mapping.

So the universal cover of $SO(3,\mathbb{R})$ is the group $S^3 = Sp(1) = SU(2) = Spin(3)$. Here $Spin(n)$ is just a name for the universal cover of $SO(n)$, and the isomorphism $Sp(1) = SU(2)$ is just given by the fact that the quaternions can also be described as the set of all complex matrices

$$\begin{pmatrix} a & b \\ -\bar{b} & \bar{a} \end{pmatrix} \sim a1 + bj.$$

The fundamental group $\pi_1(SO(3,\mathbb{R})) = \mathbb{Z}_2 = \mathbb{Z}/2\mathbb{Z}$.

4.32. Example. The group $SO(4,\mathbb{R})$. We consider the smooth homomorphism $\rho : S^3 \times S^3 \to SO(4,\mathbb{R})$ given by $\rho(P,Q)Z := PZ\bar{Q}$ in terms of multiplications of quaternions. The derived mapping is $\rho'(X,Y)Z = (T_{(1,1)}\rho.(X,Y))Z = XZ1 + 1Z(-Y) = XZ - ZY$, and its kernel consists of all pairs of imaginary quaternions (X,Y) with $XZ = ZY$ for all $Z \in \mathbb{H}$. If we put $Z = 1$, we get $X = Y$; then X is in the center of $\mathbb{H}$ which intersects $\mathfrak{sp}(1)$ at 0 only. So ρ' is a Lie algebra isomorphism since the dimensions are equal, and ρ is a local diffeomorphism. Its image is open and closed in $SO(4,\mathbb{R})$, so ρ is surjective, a covering mapping. The kernel of ρ is easily seen to be $\{(1,1),(-1,-1)\} \subset S^3 \times S^3$. So the universal cover of $SO(4,\mathbb{R})$ is $S^3 \times S^3 = Sp(1) \times Sp(1) = Spin(4)$, and the fundamental group $\pi_1(SO(4,\mathbb{R})) = \mathbb{Z}_2$ again.

Examples and Exercises

4.33. Let $A \in L(\mathbb{R}^n,\mathbb{R}^n)$ be an $(n \times n)$-matrix. Let $C(A)$ be the matrix of the *signed algebraic complements* of A, i.e.,

$$C(A)^i_j := \det \begin{pmatrix} A^1_1 & \dots & A^1_{i-1} & 0 & A^1_{i+1} & \dots & A^1_n \\ \vdots & & \vdots & \vdots & \vdots & & \vdots \\ A^{j-1}_1 & \dots & A^{j-1}_{i-1} & 0 & A^{j-1}_{i+1} & \dots & A^{j-1}_n \\ 0 & \dots & 0 & 1 & 0 & \dots & 0 \\ A^{j+1}_1 & \dots & A^{j+1}_{i-1} & 0 & A^{j+1}_{i+1} & \dots & A^{j+1}_n \\ \vdots & & \vdots & \vdots & \vdots & & \vdots \\ A^n_1 & \dots & A^n_{i-1} & 0 & A^n_{i+1} & \dots & A^n_n \end{pmatrix}.$$

Prove that $C(A)A = AC(A) = \det(A) \cdot \mathbb{I}_n$ (Cramer's rule)! This can be done by remembering the expansion formula for the determinant while multiplying it out.

Prove that $d(\det)(A)X = \operatorname{Trace}(C(A)X)$! There are two ways to do this. The first one is to check that the standard inner product on $L(\mathbb{R}^n, \mathbb{R}^n)$ is given by $\langle A, X\rangle = \operatorname{Trace}(A^\top X)$ and by computing the gradient of det at A. The second way uses (14.19):

$$\det(A + t\mathbb{I}_n) = t^n + t^{n-1}\operatorname{Trace}(A) + t^{n-2}c_2^n(A) + \cdots + t\,c_{n-1}^n(A) + \det(A).$$

Assume that A is invertible. Then:

$$\begin{aligned}
\det(A + tX) &= t^n \det(t^{-1}A + X) = t^n \det(A(A^{-1}X + t^{-1}\mathbb{I}_n)) \\
&= t^n \det(A)\det(A^{-1}X + t^{-1}\mathbb{I}_n) \\
&= t^n \det(A)(t^{-n} + t^{1-n}\operatorname{Trace}(A^{-1}X) + \cdots + \det(A^{-1}X)) \\
&= \det(A)(1 + t\operatorname{Trace}(A^{-1}X) + O(t^2)), \\
d\det(A)X &= \partial|_0 \det(A + tX) = \partial|_0 \det(A)(1 + t\operatorname{Trace}(A^{-1}X) + O(t^2)) \\
&= \det(A)\operatorname{Trace}(A^{-1}X) = \operatorname{Trace}(\det(A)A^{-1}X) \\
&= \operatorname{Trace}(C(A)X).
\end{aligned}$$

Since invertible matrices are dense, the formula follows by continuity. What about $\det_{\mathbb{C}} : L_{\mathbb{C}}(\mathbb{C}^n, \mathbb{C}^n) \to \mathbb{C}$?

4.34. For a matrix $A \in L(\mathbb{R}^n, \mathbb{R}^n)$ let $e^A := \sum_{k\geq 0} \frac{1}{k!}A^k$. Prove that e^A converges everywhere, that $\det(e^A) = e^{\operatorname{Trace}(A)}$, and thus $e^A \in GL(n, \mathbb{R})$ for all $A \in L(\mathbb{R}^n, \mathbb{R}^n)$.

4.35. We can insert matrices into real analytic functions in one variable:

$$f(A) := f(0)\cdot\mathbb{I}_n + \sum_{k\geq 0} \tfrac{f^{(k)}(0)}{k!}A^k, \qquad \text{if the norm } |A| \leq \rho,$$

where ρ is the radius of convergence of f at 0. Develop some theory about that (pay attention to constants): $(f\cdot g)(A) = f(A)\cdot g(A)$, $(f \circ g)(A) = f(g(A))$, $df(A)X = f'(A)X$ if $[A, X] = 0$. What about $df(A)X$ in the general case?

4.36. Quaternions. Let $\langle\ ,\ \rangle$ denote the standard inner product on oriented $\mathbb{R}^4$. Put $1 := (0,0,0,1) \in \mathbb{R}^4$ and $\mathbb{R}^3 \cong \mathbb{R}^3 \times \{0\} = 1^\perp \subset \mathbb{R}^4$. The vector product on $\mathbb{R}^3$ is then given by $\langle x \times y, z\rangle := \det(x, y, z)$. We define a multiplication on $\mathbb{R}^4$ by $(X, s)(Y, t) := (X \times Y + sY + tX, st - \langle X, Y\rangle)$. Prove that we get the *skew-field* of *quaternions* $\mathbb{H}$, and derive all properties: associativity, $|p.q| = |p|.|q|$, $p.\bar{p} = |p|^2.1$, $p^{-1} = |p|^{-2}.p$, $\overline{p.q} = \bar{q}.\bar{p}$. How many representations of the form $x = x_0 1 + x_1 i + x_2 j + x_3 k$ can we find? Show that $\mathbb{H}$ is isomorphic to the algebra of all complex (2×2)-matrices of the form

$$\begin{pmatrix} u & v \\ -\bar{v} & \bar{u}\end{pmatrix}, \qquad u, v \in \mathbb{C}.$$

4.37. The exponential mapping for self-adjoint operators. Let V be a Euclidean vector space with positive definite inner product $(\ \mid\ )$ (or a Hermitian vector space over $\mathbb{C}$). Let $S(V)$ be the vector space of all symmetric (or self-adjoint) linear operators on V. Let $S^+(V)$ be the open subset of all positive definite symmetric operators A, so that $(Av|v) > 0$ for $v \neq 0$. Then the exponential mapping $\exp : A \mapsto e^A = \sum_{k=0}^\infty \frac{1}{k!}A^k$ maps $S(V)$ into $S^+(V)$.

Lemma. *The exponential mapping* $\exp : S(V) \to S^+(V)$ *is a diffeomorphism.*

Proof. We start with a complex Hermitian vector space V. Let $\mathbb{C}^+ := \{\lambda \in \mathbb{C} : \operatorname{Re}(\lambda) > 0\}$, and let $\log : \mathbb{C}^+ \to \mathbb{C}$ be given by $\log(\lambda) = \int_{[1,\lambda]} z^{-1}\,dz$, where $[1, \lambda]$ denotes the line segment from 1 to λ.

Let $B \in S^+(V)$. Then all eigenvalues of B are real and positive. We chose a (positively oriented) circle $\gamma \subset \mathbb{C}^+$ such that all eigenvalues of B are contained in the interior of γ. We consider $\lambda \mapsto \log(\lambda)(\lambda \operatorname{Id}_V - B)^{-1}$ as a meromorphic function in $\mathbb{C}^+$ with values in the complex vector space $\mathbb{C} \otimes S(V)$, and we define

$$\log(B) := \frac{1}{2\pi\sqrt{-1}} \int_\gamma \log(\lambda)(\lambda \operatorname{Id}_V - B)^{-1}\,d\lambda, \quad B \in S^+(V).$$

We shall see that this does not depend on the choice of γ. We may use the same choice of the curve γ for all B in an open neighborhood in $S^+(V)$; thus $\log(B)$ is real analytic in B.

We claim that $\log = \exp^{-1}$. If $B \in S^+(V)$, then B has eigenvalues $\lambda_i > 0$ with eigenvectors v_i forming an orthonormal basis of V, so that $Bv_i = \lambda_i v_i$. Thus $(\lambda \operatorname{Id}_V - B)^{-1} v_i = \frac{1}{\lambda - \lambda_i} v_i$ for $\lambda \neq \lambda_i$, and

$$(\log B)v_i = \Big(\frac{1}{2\pi\sqrt{-1}} \int_\gamma \frac{\log\lambda}{\lambda - \lambda_i}\,d\lambda\Big)\, v_i = \log(\lambda_i)v_i$$

by Cauchy's integral formula. Thus $\log(B)$ does not depend on the choice of γ and $\exp(\log(B))v_i = e^{\log(\lambda_i)} v_i = \lambda_i\, v_i = Bv_i$ for all i. Thus $\exp \circ \log = \operatorname{Id}_{S^+(V)}$. Similarly one sees that $\log \circ \exp = \operatorname{Id}_{S(V)}$.

Now let V be a real Euclidean vector space. Let $V^{\mathbb{C}} = \mathbb{C} \otimes V$ be the complexified Hermitian vector space. If $B : V \to V$ is symmetric, then $j(B) := B^{\mathbb{C}} = \operatorname{Id}_{\mathbb{C}} \otimes B : V^{\mathbb{C}} \to V^{\mathbb{C}}$ is self-adjoint. Thus we have an embedding of real vector spaces $j : S(V) \to S(V^{\mathbb{C}})$. The eigenvalues of $j(B)$ are the same as the eigenvalues of B; thus j restricts to an embedding $j : S^+(V) \to S^+(V^{\mathbb{C}})$. By definition the left hand diagram below commutes and thus also the right hand one:

$$\begin{array}{ccc} S(V) & \xrightarrow{j} & S(V^{\mathbb{C}}) \\ {\scriptstyle \exp}\downarrow & & \downarrow{\scriptstyle \exp^{\mathbb{C}}} \\ S^+(V) & \xrightarrow{j} & S^+(V^{\mathbb{C}}), \end{array} \qquad \begin{array}{ccc} S(V) & \xrightarrow{j} & S(V^{\mathbb{C}}) \\ {\scriptstyle d\exp(B)}\downarrow & & \downarrow{\scriptstyle d\exp^{\mathbb{C}}(jB)} \\ S(V) & \xrightarrow{j} & S(V^{\mathbb{C}}). \end{array}$$

Thus $d\exp(B) : S(V) \to S(V)$ is injective for each B, thus a linear isomorphism, and by the inverse function theorem $\exp : S(V) \to S^+(V)$ is locally a diffeomorphism and is injective by the diagram. It is also surjective: for $B \in S^+(V)$ we have $Bv_i = \lambda_i v_i$ for an orthonormal basis v_i, where $\lambda_i > 0$. Let $A \in S(V)$ be given by $Av_i = \log(\lambda_i)\, v_i$; then $\exp(A) = B$. □

4.38. Polar decomposition. *Let (V, g) be a Euclidean real vector space (positive definite). Then we have a real analytic diffeomorphism*

$$GL(V) \cong S^+(V, g) \times O(V, g);$$

thus each $A \in GL(V)$ decomposes uniquely and real analytically as $A = B.U$ where B is g-symmetric and g-positive definite and $U \in O(V, g)$.

Proof. The decomposition $A = BU$, if it exists, must satisfy $AA^\top = BUU^\top B^\top = B^2$. By (4.37) the exponential mapping $X \mapsto e^X$ is a real analytic diffeomorphism $\exp : S(V, g) \to S^+(V, g)$ from the real vector space of g-symmetric operators in V onto the submanifold of g-symmetric positive definite operators in $GL(V)$, with inverse $B \mapsto \log(B)$. The operator $AA^\top$ is g-symmetric and positive definite. Thus we may put $B := \sqrt{AA^\top} = \exp(\frac{1}{2}\log(AA^\top)) \in S^+(V, g)$. Moreover, B commutes with $AA^\top$. Let $U := B^{-1}A$. Then $UU^\top = B^{-1}AA^\top(B^{-1})^\top = \mathrm{Id}_V$, so $U \in O(V, g)$. □

5. Lie Groups II. Lie Subgroups and Homogeneous Spaces

5.1. Definition. Let G be a Lie group. A subgroup H of G is called a *Lie subgroup* if H is itself a Lie group (so it is separable) and the inclusion $i : H \to G$ is smooth.

In this case the inclusion is even an immersion. It suffices to check that $T_e i$ is injective: If $X \in \mathfrak{h}$ is in the kernel of $T_e i$, then $i \circ \exp^H(tX) = \exp^G(t.T_e i.X) = e$. Since i is injective, $X = 0$.

From the next result it follows that $H \subset G$ is then an initial submanifold in the sense of (2.13): If H_0 is the connected component of H, then $i(H_0)$ is the Lie subgroup of G generated by $i'(\mathfrak{h}) \subset \mathfrak{g}$, which is an initial submanifold, and this is true for all components of H.

5.2. Theorem. *Let G be a Lie group with Lie algebra $\mathfrak{g}$. If $\mathfrak{h} \subset \mathfrak{g}$ is a Lie subalgebra, then there is a unique connected Lie subgroup H of G with Lie algebra $\mathfrak{h}$. Moreover, H is an initial submanifold of G.*

Proof. Put $E_x := \{T_e(\mu_x).X : X \in \mathfrak{h}\} \subset T_xG$. Then $E := \bigsqcup_{x\in G} E_x$ is a distribution of constant rank on G. It is spanned by the involutive set $\{L_X, X \in \mathfrak{h}\}$ of vector fields. So by theorem (3.20) the distribution E is integrable and the leaf H through e is an initial submanifold. It is even a subgroup, since for $x \in H$ the initial submanifold $\mu_x H$ is again a leaf (since E is left invariant) and intersects H at x, so $\mu_x(H) = H$. Thus $H.H = H$ and consequently $H^{-1} = H$. The multiplication $\mu : H \times H \to G$ is smooth by restriction and smooth as a mapping $H \times H \to H$, since H is an initial submanifold, by lemma (2.15). □

5.3. Theorem. *Let $\mathfrak{g}$ be a finite-dimensional real Lie algebra. Then there exists a connected Lie group G whose Lie algebra is $\mathfrak{g}$.*

Sketch of Proof. By the theorem of Ado (see [**96**] or [**223**, p. 237]) $\mathfrak{g}$ has a faithful (i.e., injective) representation on a finite-dimensional vector space V, i.e., $\mathfrak{g}$ can be viewed as a Lie subalgebra of $\mathfrak{gl}(V) = L(V,V)$. By theorem (5.2) above there is a Lie subgroup G of $GL(V)$ with $\mathfrak{g}$ as its Lie algebra. □

This is a rather involved proof, since the theorem of Ado needs the structure theory of Lie algebras for its proof. There are simpler proofs available, starting from a neighborhood of e in G (a neighborhood of 0 in $\mathfrak{g}$ with the Baker-Campbell-Hausdorff formula (4.29) as multiplication) and extending the Lie group structure.

5.4. Theorem. *Let G and H be Lie groups with Lie algebras $\mathfrak{g}$ and $\mathfrak{h}$, respectively. Let $f : \mathfrak{g} \to \mathfrak{h}$ be a homomorphism of Lie algebras. Then there is a Lie group homomorphism φ, locally defined near e, from G to H, such that $\varphi' = T_e\varphi = f$. If G is simply connected, then there is a globally defined homomorphism of Lie groups $\varphi : G \to H$ with this property.*

Proof. Let $\mathfrak{k} := \operatorname{graph}(f) \subset \mathfrak{g}\times\mathfrak{h}$. Then $\mathfrak{k}$ is a Lie subalgebra of $\mathfrak{g}\times\mathfrak{h}$, since f is a homomorphism of Lie algebras. The product $\mathfrak{g}\times\mathfrak{h}$ is the Lie algebra of $G\times H$, so by theorem (5.2) there is a connected Lie subgroup $K \subset G\times H$ with algebra $\mathfrak{k}$. We consider the homomorphism $g := \operatorname{pr}_1 \circ \operatorname{incl} : K \to G\times H \to G$, whose tangent mapping satisfies

$$T_e g(X, f(X)) = T_{(e,e)} \operatorname{pr}_1 .T_e \operatorname{incl}.(X, f(X)) = X;$$

so it is invertible. Thus g is a local diffeomorphism, so $g : K \to G_0$ is a covering of the connected component G_0 of e in G. If G is simply connected, g is an isomorphism. Now we consider the homomorphism $\psi := \operatorname{pr}_2 \circ \operatorname{incl} : K \to G \times H \to H$, whose tangent mapping satisfies $T_e\psi.(X, f(X)) = f(X)$.

We see that $\varphi := \psi \circ (g \restriction U)^{-1} : G \supset U \to H$ solves the problem, where U is an e-neighborhood in K such that $g \restriction U$ is a diffeomorphism. If G is simply connected, $\varphi = \psi \circ g^{-1}$ is the global solution. □

5.5. Theorem. *Let H be a closed subgroup of a Lie group G. Then H is a Lie subgroup and a submanifold of G.*

Proof. Let $\mathfrak{g}$ be the Lie algebra of G. We consider the subset $\mathfrak{h} := \{c'(0) : c \in C^\infty(\mathbb{R}, G),\ c(\mathbb{R}) \subset H, c(0) = e\}$.

Claim 1. $\mathfrak{h}$ is a linear subspace.
If $c_i'(0) \in \mathfrak{h}$ and $t_i \in \mathbb{R}$, we define $c(t) := c_1(t_1.t).c_2(t_2.t)$. Then we have $c'(0) = T_{(e,e)}\mu.(t_1.c_1'(0), t_2.c_2'(0)) = t_1.c_1'(0) + t_2.c_2'(0) \in \mathfrak{h}$.

Claim 2. $\mathfrak{h} = \{X \in \mathfrak{g} : \exp(tX) \in H \text{ for all } t \in \mathbb{R}\}$.
Clearly we have '$\supseteq$'. To check the other inclusion, let $X = c'(0) \in \mathfrak{h}$ and consider $v(t) := (\exp^G)^{-1}c(t)$ for small t. Then we have $X = c'(0) = \frac{d}{dt}|_0 \exp(v(t)) = v'(0) = \lim_{n\to\infty} n.v(\frac{1}{n})$. We put $t_n := \frac{1}{n}$ and $X_n := n.v(\frac{1}{n})$, so that $\exp(t_n.X_n) = \exp(v(\frac{1}{n})) = c(\frac{1}{n}) \in H$. By claim 3 below we then get $\exp(tX) \in H$ for all t.

Claim 3. Let $X_n \to X$ in $\mathfrak{g}$, $0 < t_n \to 0$ in $\mathbb{R}$ with $\exp(t_nX_n) \in H$. Then $\exp(tX) \in H$ for all $t \in \mathbb{R}$.
Let $t \in \mathbb{R}$ and take $m_n \in (\frac{t}{t_n} - 1, \frac{t}{t_n}] \cap \mathbb{Z}$ for large n. Then $t_n.m_n \to t$ and $m_n.t_n.X_n \to tX$, and since H is closed, we may conclude that

$$\exp(tX) = \lim_n \exp(m_n.t_n.X_n) = \lim_n \exp(t_n.X_n)^{m_n} \in H.$$

Claim 4. Let $\mathfrak{k}$ be a complementary linear subspace for $\mathfrak{h}$ in $\mathfrak{g}$. Then there is an open 0-neighborhood W in $\mathfrak{k}$ such that $\exp(W) \cap H = \{e\}$.
If not, there are $0 \neq Y_k \in \mathfrak{k}$ with $Y_k \to 0$ such that $\exp(Y_k) \in H$. Choose a norm $|\quad|$ on $\mathfrak{g}$ and let $X_n = Y_n/|Y_n|$. Passing to a subsequence, we may assume that $X_n \to X$ in $\mathfrak{k}$; then $|X| = 1$. But $\exp(|Y_n|.X_n) = \exp(Y_n) \in H$ and $0 < |Y_n| \to 0$, so by claim 3 we have $\exp(tX) \in H$ for all $t \in \mathbb{R}$. By claim 2 we get $X \in \mathfrak{h}$, a contradiction.

Claim 5. Put $\varphi : \mathfrak{h} \times \mathfrak{k} \to G$, $\varphi(X, Y) = \exp X. \exp Y$. Then there are 0-neighborhoods V in $\mathfrak{h}$, W in $\mathfrak{k}$, and an e-neighborhood U in G such that $\varphi : V \times W \to U$ is a diffeomorphism and $U \cap H = \exp(V)$.
Choose V, W, and U so small that φ becomes a diffeomorphism. By claim 4 the set W may be chosen so small that $\exp(W) \cap H = \{e\}$. By claim 2 we have $\exp(V) \subseteq H \cap U$. Let $x \in H \cap U$. Since $x \in U$, we have $x = \exp X. \exp Y$ for unique $(X, Y) \in V \times W$. Then x and $\exp X \in H$, so $\exp Y \in H \cap \exp(W) = \{e\}$; thus $Y = 0$. So $x = \exp X \in \exp(V)$.

Claim 6. H is a submanifold and a Lie subgroup.
The pair $(U, (\varphi \restriction V \times W)^{-1} =: u)$ is a submanifold chart for H centered at e

by claim 5. For $x \in H$ the pair $(\mu_x(U), u \circ \mu_{x^{-1}})$ is a submanifold chart for H centered at x. So H is a closed submanifold of G, and the multiplication is smooth since it is a restriction. □

5.6. Theorem. *Let H be a subgroup of a Lie group G which is C^∞-pathwise connected (see* (2.13)*). Then H is a connected Lie group and an initial Lie subgroup of G.*

Proof. Let us call any smooth curve $c : \mathbb{R} \to G$ with $c(0) = e$ and $c(\mathbb{R}) \subseteq H$ an H-curve in G. As in the proof of (5.5) let $\mathfrak{h}$ be the set of $c'(0)$ for all H-curves c in G. Claim 1 in the proof of (5.5) shows that $\mathfrak{h}$ is a linear subspace of $\mathfrak{g}$. For H-curves c_i in G we use (4.30.3) to see that $[c_1'(0), c_2'(0)] = \frac{1}{2}\partial_t^2|_0 g_1(t)g_2(t)g_1(t)^{-1}g_2(t)^{-1}$ is again in $\mathfrak{h}$; so $\mathfrak{h}$ is a Lie subalgebra of $\mathfrak{g}$.

Let H_1 be the connected initial Lie subgroup of G corresponding to $\mathfrak{h}$ which is the leaf through e of the foliation given by the left invariant distribution of constant rank generated by $\mathfrak{h}$; see (5.2). For any H-curve c in G we have $T(\mu_{c(t)^{-1}}).c'(t) = \partial_s|_0 c(t)^{-1}c(t+s) \in \mathfrak{h}$. Thus c is tangent to this distribution and thus lies in the leaf H_1 through e. By assumption, any point in H is connected to e with such a curve. Thus $H \subseteq H_1$.

To prove that $H_1 \subseteq H$, we choose a basis $X_1, \dots, X_k$ of $\mathfrak{h}$ and H-curves c_i in G with $c_i'(0) = X_i$. We consider the mapping $f : \mathbb{R}^k \to H$ and H_1 which is given by $f(t_1, \dots, t_k) := c_1(t_1)\dots c_k(t_k)$. Since $T_0 f$ is invertible $\mathbb{R}^k \to \mathfrak{h}$, the mapping f is a local diffeomorphism near 0 onto an open e-neighborhood in H_1. This shows that an open e-neighborhood of H_1 is in H; thus $H_1 \subset H$. □

5.7. Remarks. The following stronger results on subgroups and the relation between topological groups and Lie groups in general are available.

Any C^0-pathwise connected subgroup of a Lie group is a connected Lie subgroup, [**230**]. Theorem (5.6) is a weaker version of this, fitting the spirit of (2.13). The proof of (5.6) works also for C^1-pathwise connected subgroups, without any changes.

Let G be a separable locally compact topological group. If it has an e-neighborhood which does not contain a proper subgroup, then G is a Lie group. This is the solution of the 5-th problem of Hilbert; see [**162**, p. 107].

Any subgroup H of a Lie group G has a coarsest Lie group structure, but it might be nonseparable. To indicate a proof of this statement, consider all continuous curves $c : \mathbb{R} \to G$ with $c(\mathbb{R}) \subset H$, and equip H with the final topology with respect to them. Then apply the Yamabe theorem cited above to the component of the identity. Or consider all smooth H-curves in G (as in the proof of (5.6)) and put the final topology with respect to these on H, and apply (5.6) to the connected component.

5.8. Let $\mathfrak{g}$ be a Lie algebra. An *ideal* $\mathfrak{k}$ in $\mathfrak{g}$ is a linear subspace $\mathfrak{k}$ such that $[\mathfrak{k}, \mathfrak{g}] \subset \mathfrak{k}$. Then the quotient space $\mathfrak{g}/\mathfrak{k}$ carries a unique Lie algebra structure such that $\mathfrak{g} \to \mathfrak{g}/\mathfrak{k}$ is a Lie algebra homomorphism.

Lemma. *A connected Lie subgroup H of a connected Lie group G is a normal subgroup if and only if its Lie algebra $\mathfrak{h}$ is an ideal in $\mathfrak{g}$.*

Proof. H normal in G means $xHx^{-1} = \operatorname{conj}_x(H) \subset H$ for all $x \in G$. By remark (4.20) this is equivalent to $T_e(\operatorname{conj}_x)(\mathfrak{h}) \subset \mathfrak{h}$, i.e., $\operatorname{Ad}(x)\mathfrak{h} \subset \mathfrak{h}$, for all $x \in G$. But this in turn is equivalent to $\operatorname{ad}(X)\mathfrak{h} \subset \mathfrak{h}$ for all $X \in \mathfrak{g}$, so to the fact that $\mathfrak{h}$ is an ideal in $\mathfrak{g}$. □

5.9. Let G be a connected Lie group. If $A \subset G$ is an arbitrary subset, the *centralizer* of A in G is the closed subgroup $Z_G(A) := \{x \in G : xa = ax \text{ for all } a \in A\}$, which by (5.5) is a Lie subgroup.

The Lie algebra $\mathfrak{z}_\mathfrak{g}(A)$ of $Z_G(A)$ consists of all $X \in \mathfrak{g}$ with $a.\exp(tX).a^{-1} = \exp(tX)$ for all $a \in A$, i.e., $\mathfrak{z}_\mathfrak{g}(A) = \{X \in \mathfrak{g} : \operatorname{Ad}(a)X = X \text{ for all } a \in A\}$.

If A is itself a connected Lie subgroup of G with Lie algebra $\mathfrak{a}$, then $\mathfrak{z}_\mathfrak{g}(A) = \{X \in \mathfrak{g} : \operatorname{ad}(Y)X = 0 \text{ for all } Y \in \mathfrak{a}\}$. This set is also called the *centralizer* of $\mathfrak{a}$ in $\mathfrak{g}$. If $A = G$ is connected, then $Z_G = Z_G(G)$ is called the *center* of G and $\mathfrak{z}_\mathfrak{g}(G) = \mathfrak{z}_\mathfrak{g} = \{X \in \mathfrak{g} : [X, Y] = 0 \text{ for all } Y \in \mathfrak{g}\}$ is then the *center* of the Lie algebra $\mathfrak{g}$.

5.10. The *normalizer* of a subset A of a connected Lie group G is the subgroup

$$N_G(A) = \{x \in G : \mu_x(A) = \mu^x(A)\} = \{x \in G : \operatorname{conj}_x(A) = A\}.$$

If A is closed, then $N_G(A)$ is also closed.

If A is a connected Lie subgroup of G, then $N_G(A) = \{x \in G : \operatorname{Ad}(x)\mathfrak{a} \subset \mathfrak{a}\}$. Its Lie algebra

$$\mathfrak{n}_G(A) = \{X \in \mathfrak{g} : \operatorname{ad}(X)\mathfrak{a} \subset \mathfrak{a}\} = \mathfrak{n}_\mathfrak{g}(\mathfrak{a})$$

is then the *normalizer* or *idealizer* of $\mathfrak{a}$ in $\mathfrak{g}$.

5.11. Homogeneous spaces. Let G be a Lie group and let $H \subset G$ be a closed subgroup. By theorem (5.5), H is a Lie subgroup of G. We denote by G/H the space of all right cosets of G, i.e., $G/H = \{gH : g \in G\}$. Let $p : G \to G/H$ be the projection. We equip G/H with the quotient topology, i.e., $U \subset G/H$ is open if and only if $p^{-1}(U)$ is open in G. Since H is closed, G/H is a Hausdorff space.

The quotient space G/H is called a *homogeneous space* of G. We have a left action of G on G/H, which is induced by the left translation and is given by $\bar{\mu}_g(g_1 H) = gg_1 H$.

Theorem. *If H is a closed subgroup of G, then there exists a unique structure of a smooth manifold on G/H such that $p : G \to G/H$ is a submersion. Thus* $\dim G/H = \dim G - \dim H$.

Proof. Surjective submersions have the universal property (2.4); thus the manifold structure on G/H is unique, if it exists. Let $\mathfrak{h}$ be the Lie algebra of the Lie subgroup H. We choose a complementary linear subspace $\mathfrak{k}$ such that $\mathfrak{g} = \mathfrak{h} \oplus \mathfrak{k}$.

Claim 1. We consider the mapping $f : \mathfrak{k} \times H \to G$, given by $f(X,h) := \exp X.h$. Then there is an open 0-neighborhood W in $\mathfrak{k}$ and an open e-neighborhood U in G such that $f : W \times H \to U$ is a diffeomorphism.

By claim 5 in the proof of theorem (5.5) there are open 0-neighborhoods V in $\mathfrak{h}$, W' in $\mathfrak{k}$, and an open e-neighborhood U' in G such that $\varphi : W' \times V \to U'$ is a diffeomorphism, where $\varphi(X,Y) = \exp X.\exp Y$, and such that $U' \cap H = \exp V$. Now we choose W in $W' \subset \mathfrak{k}$ so small that $\exp(W)^{-1}.\exp(W) \subset U'$. We will check that this W satisfies claim 1.

Claim 2. $f \restriction W \times H$ is injective.

The equality $f(X_1,h_1) = f(X_2,h_2)$ means $\exp X_1.h_1 = \exp X_2.h_2$; thus $h_2h_1^{-1} = (\exp X_2)^{-1}\exp X_1 \in \exp(W)^{-1}\exp(W) \cap H \subset U' \cap H = \exp V$. So there is a unique $Y \in V$ with $h_2h_1^{-1} = \exp Y$. But then $\varphi(X_1,0) = \exp X_1 = \exp X_2.h_2.h_1^{-1} = \exp X_2.\exp Y = \varphi(X_2,Y)$. Since φ is injective, $X_1 = X_2$ and $Y = 0$, so $h_1 = h_2$.

Claim 3. $f \restriction W \times H$ is a local diffeomorphism.

The diagram

$$\begin{array}{ccc} W \times V & \xrightarrow{Id\times\exp} & W \times (U' \cap H) \\ \varphi\downarrow & & \downarrow f \\ \varphi(W \times V) & \xrightarrow{\text{incl}} & U' \end{array}$$

commutes, and $Id_W \times \exp$ and φ are diffeomorphisms. So $f \restriction W \times (U' \cap H)$ is a diffeomorphism. Since $f(X,h) = f(X,e).h$, we conclude that $f \restriction W \times H$ is everywhere a local diffeomorphism. So finally claim 1 follows, where $U = f(W \times H)$.

Now we put $g := p \circ (\exp \restriction W) : \mathfrak{k} \supset W \to G/H$. Then the following diagram commutes:

$$\begin{array}{ccc} W \times H & \xrightarrow{f} & U \\ \text{pr}_1\downarrow & & \downarrow p \\ W & \xrightarrow{g} & G/H. \end{array}$$

Claim 4. g is a homeomorphism onto $p(U) =: \bar{U} \subset G/H$.

Clearly g is continuous, and g is open, since p is open. If $g(X_1) = g(X_2)$, then $\exp X_1 = \exp X_2.h$ for some $h \in H$, so $f(X_1, e) = f(X_2, h)$. By claim 1 we get $X_1 = X_2$, so g is injective. Finally $g(W) = \bar{U}$, so claim 4 follows.

For $a \in G$ we consider $\bar{U}_a = \bar{\mu}_a(\bar{U}) = a.\bar{U}$ and the mapping

$$u_a := g^{-1} \circ \bar{\mu}_{a^{-1}} : \bar{U}_a \to W \subset \mathfrak{k}.$$

Claim 5. $(\bar{U}_a, u_a = g^{-1} \circ \bar{\mu}_{a^{-1}} : \bar{U}_a \to W)_{a \in G}$ is a smooth atlas for G/H.

Let a, $b \in G$ such that $\bar{U}_a \cap \bar{U}_b \neq \emptyset$. Then

$$\begin{aligned} u_a \circ u_b^{-1} &= g^{-1} \circ \bar{\mu}_{a^{-1}} \circ \bar{\mu}_b \circ g : u_b(\bar{U}_a \cap \bar{U}_b) \to u_a(\bar{U}_a \cap \bar{U}_b) \\ &= g^{-1} \circ \bar{\mu}_{a^{-1}b} \circ p \circ (\exp \restriction W) \\ &= g^{-1} \circ p \circ \mu_{a^{-1}b} \circ (\exp \restriction W) \\ &= \mathrm{pr}_1 \circ f^{-1} \circ \mu_{a^{-1}b} \circ (\exp \restriction W) \quad \text{is smooth.} \quad \square \end{aligned}$$

6. Transformation Groups and G-Manifolds

6.1. Group actions. A *left action* of a Lie group G on a manifold M is a smooth mapping $\ell : G \times M \to M$ such that $\ell_g \circ \ell_h = \ell_{gh}$ and $\ell_e = Id_M$, where $\ell_g(z) = \ell(g, z)$.

A *right action* of a Lie group G on a manifold M is a smooth mapping $r : M \times G \to M$ such that $r^g \circ r^h = r^{hg}$ and $r^e = Id_M$, where $r^g(z) = r(z, g)$.

A *G-space* or a *G-manifold* is a manifold M together with a right or left action of G on M.

We will describe the following notions only for a left action of G on M. They make sense also for right actions.

The *orbit* through $z \in M$ is the set $G.z = \ell(G, z) \subset M$.

The action is called:

- *Transitive* if M is one orbit, i.e., for all $z, w \in M$ there is some $g \in G$ with $g.z = w$.
- *Free* if $g_1.z = g_2.z$ for some $z \in M$ implies already $g_1 = g_2$.
- *Effective* if $\ell_g = \ell_h$ implies $g = h$, i.e., if $\ell : G \to \mathrm{Diff}(M)$ is injective where $\mathrm{Diff}(M)$ denotes the group of all diffeomorphisms of M.
- *Infinitesimally free* if $T_e(\ell^x) : \mathfrak{g} \to T_xM$ is injective for each $x \in M$.
- *Infinitesimally transitive* if $T_e(\ell^x) : \mathfrak{g} \to T_xM$ is surjective for each $x \in M$.
- *Linear* if M is a vector space and the action defines a representation.
- *Affine* if M is an affine space, and every $\ell_g : M \to M$ is an affine map.
- *Orthogonal* if (M, γ) is a Euclidean vector space and $\ell_g \in O(M, \gamma)$ for all $g \in G$.

- *Isometric* if (M, γ) is a Riemann manifold and ℓ_g is an isometry for all $g \in G$; see section (22).
- *Symplectic* if (M, ω) is a symplectic manifold and ℓ_g is a symplectomorphism for all $g \in G$; see section (31).
- *Principal fiber bundle action* if it is free and if the projection onto the orbit space $\pi : M \to M/G$ is a principal fiber bundle; see section (17).

More generally, a continuous transformation group of a topological space M is a pair (G, M) where G is a topological group and where to each element $g \in G$ there is given a homeomorphism ℓ_g of M such that $\ell : G \times M \to M$ is continuous and $\ell_g \circ \ell_h = \ell_{gh}$. The continuity is an obvious geometrical requirement, but in accordance with the general observation that group properties often force more regularity than explicitly postulated (cf. (5.7)), differentiability follows in many situations. So, if G is locally compact, M is a smooth or real analytic manifold, all ℓ_g are smooth or real analytic homeomorphisms and the action is effective, then G is a Lie group and ℓ is smooth or real analytic, respectively; see [**162**, p. 212].

6.2. Let $\ell : G \times M \to M$ be a left action. Then we have partial mappings $\ell_a : M \to M$ and $\ell^x : G \to M$, given by $\ell_a(x) = \ell^x(a) = \ell(a, x) = a.x$, where $a \in G$ and $x \in M$.

For any $X \in \mathfrak{g}$ we define the *fundamental vector field* $\zeta_X = \zeta_X^M \in \mathfrak{X}(M)$ by

$$\zeta_X(x) = T_e(\ell^x).X = T_{(e,x)}\ell.(X, 0_x).$$

Lemma. *In this situation the following assertions hold:*

(1) $\zeta : \mathfrak{g} \to \mathfrak{X}(M)$ *is a linear mapping.*

(2) $T_x(\ell_a).\zeta_X(x) = \zeta_{\mathrm{Ad}(a)X}(a.x)$.

(3) $R_X \times 0_M \in \mathfrak{X}(G \times M)$ *is ℓ-related to* $\zeta_X \in \mathfrak{X}(M)$.

(4) $[\zeta_X, \zeta_Y] = -\zeta_{[X,Y]}$.

Proof. (1) is clear.

(2) We have $\ell_a \ell^x(b) = abx = aba^{-1}ax = \ell^{ax}\,\mathrm{conj}_a(b)$, so

$$\begin{aligned} T_x(\ell_a).\zeta_X(x) &= T_x(\ell_a).T_e(\ell^x).X = T_e(\ell_a \circ \ell^x).X \\ &= T_e(\ell^{ax}).\,\mathrm{Ad}(a).X = \zeta_{\mathrm{Ad}(a)X}(ax). \end{aligned}$$

(3) We have $\ell \circ (Id \times \ell_a) = \ell \circ (\mu^a \times Id) : G \times M \to M$, so

$$\begin{aligned} \zeta_X(\ell(a, x)) &= T_{(e,ax)}\ell.(X, 0_{ax}) = T\ell.(Id \times T(\ell_a)).(X, 0_x) \\ &= T\ell.(T(\mu^a) \times Id).(X, 0_x) = T\ell.(R_X \times 0_M)(a, x). \end{aligned}$$

(4) $[R_X \times 0_M, R_Y \times 0_M] = [R_X, R_Y] \times 0_M = -R_{[X,Y]} \times 0_M$ is ℓ-related to $[\zeta_X, \zeta_Y]$ by (3) and by (3.10). On the other hand $-R_{[X,Y]} \times 0_M$ is ℓ-related to $-\zeta_{[X,Y]}$ by (3) again. Since ℓ is surjective, we get $[\zeta_X, \zeta_Y] = -\zeta_{[X,Y]}$. □

6.3. Let $r : M \times G \to M$ be a right action, so $\check{r} : G \to \operatorname{Diff}(M)$ is a group antihomomorphism. We will use the following notation: $r^a : M \to M$ and $r_x : G \to M$, given by $r_x(a) = r^a(x) = r(x, a) = x.a$.

For any $X \in \mathfrak{g}$ we define the *fundamental vector field* $\zeta_X = \zeta_X^M \in \mathfrak{X}(M)$ by $\zeta_X(x) = T_e(r_x).X = T_{(x,e)}r.(0_x, X)$.

Lemma. *In this situation the following assertions hold:*

(1) $\zeta : \mathfrak{g} \to \mathfrak{X}(M)$ *is a linear mapping.*
(2) $T_x(r^a).\zeta_X(x) = \zeta_{\operatorname{Ad}(a^{-1})X}(x.a)$.
(3) $0_M \times L_X \in \mathfrak{X}(M \times G)$ *is r-related to* $\zeta_X \in \mathfrak{X}(M)$.
(4) $[\zeta_X, \zeta_Y] = \zeta_{[X,Y]}$. □

6.4. Theorem. *Let $\ell : G \times M \to M$ be a smooth left action. For $x \in M$ let $G_x = \{a \in G : ax = x\}$ be the isotropy subgroup or fixpoint group of x in G, a closed subgroup of G. Then $\ell^x : G \to M$ factors over $p : G \to G/G_x$ to an injective immersion $i^x : G/G_x \to M$, which is G-equivariant, i.e., $\ell_a \circ i^x = i^x \circ \bar{\mu}_a$ for all $a \in G$. The image of i^x is the orbit through x.*

The fundamental vector fields span an integrable distribution on M in the sense of (3.23). *Its leaves are the connected components of the orbits, and each orbit is an initial submanifold. Thus $i^x : G/G_x \to M$ is an initial immersion.*

Proof. Clearly ℓ^x factors over p to an injective mapping $i^x : G/G_x \to M$; by the universal property of surjective submersions i^x is smooth, and obviously it is equivariant. Thus $T_{p(a)}(i^x).T_{p(e)}(\bar{\mu}_a) = T_{p(e)}(i^x \circ \bar{\mu}_a) = T_{p(e)}(\ell_a \circ i^x) = T_x(\ell_a).T_{p(e)}(i^x)$ for all $a \in G$ and it suffices to show that $T_{p(e)}(i^x)$ is injective.

Let $X \in \mathfrak{g}$ and consider its fundamental vector field $\zeta_X \in \mathfrak{X}(M)$. By (3.14) and (6.2.3) we have

$$\ell(\exp(tX), x) = \ell(\operatorname{Fl}_t^{R_X \times 0_M}(e, x)) = \operatorname{Fl}_t^{\zeta_X}(\ell(e, x)) = \operatorname{Fl}_t^{\zeta_X}(x).$$

So $\exp(tX) \in G_x$, i.e., $X \in \mathfrak{g}_x$, if and only if $\zeta_X(x) = 0_x$. In other words, $0_x = \zeta_X(x) = T_e(\ell^x).X = T_{p(e)}(i^x).T_ep.X$ if and only if $T_ep.X = 0_{p(e)}$. Thus i^x is an immersion.

Since the connected components of the orbits are integral manifolds, the fundamental vector fields span an integrable distribution in the sense of (3.23); but also the condition (3.28.2) is satisfied. So by theorem (3.25) each orbit is an initial submanifold in the sense of (2.13). By uniqueness of the manifold structure on an initial submanifold, the mapping i^x is an initial immersion. □

6.5. Theorem (**[185]**). *Let M be a smooth manifold and let $\zeta : \mathfrak{g} \to \mathfrak{X}(M)$ be a homomorphism from a finite-dimensional Lie algebra $\mathfrak{g}$ into the Lie algebra of vector fields on M. Let G be a simply connected Lie group with Lie algebra $\mathfrak{g}$.*

Then there exists a local left action $G \times M \supset U \xrightarrow{l} M$, where U is an open neighborhood of $\{e\} \times M$ in $G \times M$ whose fundamental vector field mapping equals $-\zeta$. Here U is an open neighborhood of $\{e\} \times M$ in $G \times M$ and $l(g, l(h,x)) = l(gh,x)$ whenever both sides are defined.

Suppose moreover that each element ζ_X in the image of ζ is a complete vector field. Then there exists a left action $\ell : G \times M \to M$ of the Lie group G on the manifold M whose fundamental vector field mapping equals $-\zeta$.

The domain U of the local action cannot be chosen maximal in general: When trying to do so, one glues together open pieces of covering spaces of subsets of G and gets sets spread over G but not contained in G. See [**102**] for more information and examples.

Proof. On the product manifold $G \times M$ we consider the vector subbundle $E = \{(L_X(g), \zeta_X(x)) : (g,x) \in G \times M, X \in \mathfrak{g}\} \subset TG \times TM$ where $L_X \in \mathfrak{X}(G)$ is the left invariant vector field generated by $X \in \mathfrak{g}$. We have $\dim E_{g,x} = \dim \mathfrak{g}$. The subbundle E is integrable since $[L_X \times \zeta_X, L_Y \times \zeta_Y] = [L_X, L_Y] \times [\zeta_X, \zeta_Y] = L_{[X,Y]} \times \zeta_{[X,Y]}$. Thus by theorem (3.20) (or (3.28)) the bundle E induces a foliation on $G \times M$. Let us denote the leaf through $(g,x) \in G \times M$ by $L(g,x)$. Note that by (4.18.3) for the flow we have

$$\mathrm{Fl}_t^{L_X \times \zeta_X}(g,x) = (g.\exp(tX), \mathrm{Fl}_t^{\zeta_X}(x)). \tag{1}$$

This flow line lies in the leaf $L(g,x)$ since it is tangent to it.

Thus for ζ_X a complete vector field we have $(\mu^{\exp X} \times \mathrm{Fl}_1^{\zeta_X}) : L(g,x) \to L(g,x)$. In particular,

$$L(g.\exp(X), \mathrm{Fl}_1^{\zeta_X}(x)) = L(g,x), \quad L(g.\exp(X), y) = L(g, \mathrm{Fl}_{-1}^{\zeta_X}(y)). \tag{2}$$

If ζ_X is not complete, then (2) holds only whenever both sides (of any equation) are defined.

We have $T(\mu_g \times \mathrm{Id}_M)(L_X(h), \zeta_X(x)) = (L_X(gh), \zeta_X(x))$ for $X \in \mathfrak{g}$, $g,h \in G$, and $x \in M$. Thus $(\mu_g \times \mathrm{Id}_M)$ maps leaves to leaves:

$$L(g,x) = \{(gh,y) : (h,y) \in L(e,x)\} = (\mu_g \times \mathrm{Id}_M)(L(e,x)). \tag{3}$$

We suppose now that each vector field ζ_X for $X \in \mathfrak{g}$ is complete.

(4) *Claim. Then for any leaf $L \subset G \times M$, the restriction $\mathrm{pr}_1 |L : L \to G$ is a covering map.*

For $(g,x) \in L$ we have $T_{(g,x)}(\mathrm{pr}_1)(L_X(g), \zeta_X(x)) = L_X(g)$; thus $\mathrm{pr}_1 |L$ is locally a diffeomorphism. For any $g_1 \in G$ we can find a piecewise smooth curve c in G connecting g with g_1 consisting of pieces of the form $t \mapsto g_i.\exp(tX_i)$. Starting from $(g,x) \in L$, we can fit together corresponding pieces of the form $\mathrm{Fl}_t^{L_{X_i}\times\zeta_{X_i}}$ to obtain a curve $\tilde c$ in L with $\mathrm{pr}_1 \circ \tilde c = c$ which connects (g,x) with $(g_1, x_1) \in L$ for some $x_1 \in M$. Thus $\mathrm{pr}_1 : L \to G$ is surjective. Next we consider some absolutely convex ball $B \subset \mathfrak{g}$ such that $\exp : \mathfrak{g} \supset B \to U \subset G$ is a diffeomorphism onto an open neighborhood U of e in G. We consider the inverse image $(\mathrm{pr}_1 |L)^{-1}(g.U) \subset L$ and decompose it into its connected components, $(\mathrm{pr}_1 |L)^{-1}(g.U) = \bigsqcup V_i \subset L$. Any point in $g.U$ is of the form $g.\exp(X)$ for a unique $X \in B$, and we may lift the curve $t \mapsto g.\exp(tX)$ in G to the curve $\mathrm{Fl}_t^{L_X\times\zeta_X}(g, x_i)$ in V_i. So each V_i is diffeomorphic to $g.U$ via $\mathrm{pr}_1 |V_i$, and the claim follows.

Since G is simply connected, we conclude that for each leaf L the mapping $\mathrm{pr}_1 |L : L \to G$ is a diffeomorphism. We now define the action as follows: For $g \in G$ and $x \in M$ consider the leaf $L(e,x)$ through (e,x) and put

$$\ell(g,x) = g.x = \mathrm{pr}_2((\mathrm{pr}_1 |L(e,x))^{-1}(g^{-1})) \in M. \tag{5}$$

Obviously, ℓ is smooth.

Let us now pass to the general case, where some ζ_X may be incomplete. Then claim (4) is wrong in general. We also consider the distribution $TM \supset F := \{\zeta_X(x) : X \in \mathfrak{g}, x \in M\}$ with nonconstant rank, which is involutive and which induces a foliation on M by (3.28). Let us denote by $\mathcal{O}_{\mathfrak{g}}(x)$ the leaf through x in M (the $\mathfrak{g}$-orbit). We have the following situation. Consider the following diagram, where $W_x \subset G$ is the image of the leaf $L(e,x)$ in G:

(6)

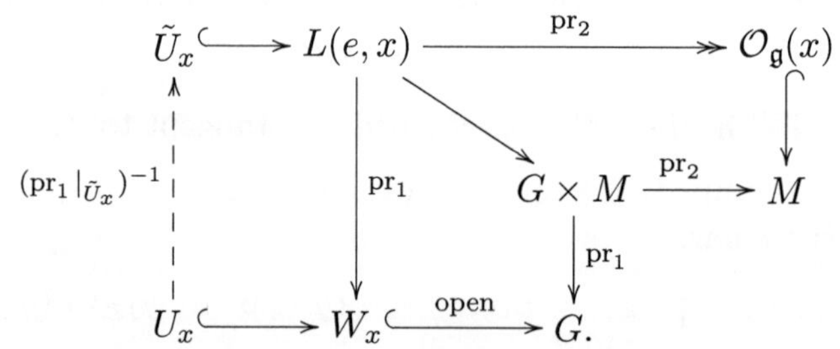

To describe U_x, we consider the vector field $\zeta \in \mathfrak{X}(\mathfrak{g}\times M)$ given by $\zeta(X,x) = (0_X, \zeta_X(x))$ with flow $\mathrm{Fl}_t^\zeta(X,x) = (X, \mathrm{Fl}_t^{\zeta_X}(x))$ defined as $\mathrm{Fl}^\zeta : \mathcal{D}(\zeta) \to M$ where the (maximal) domain of definition $\mathcal{D}(\zeta)$ is an open neighborhood of $\{0\} \times \mathfrak{g} \times M$ in $\mathbb{R} \times \mathfrak{g} \times M$ by (3.7). Let $U' = \{(X,x) \in \mathfrak{g} \times M : [-1,1] \times \{X\} \times \{x\} \subset \mathcal{D}(\zeta)\}$. Since $[-1,1]$ is compact, U' is open. Now we consider an open ball $B \subset \mathfrak{g}$ centered at 0 such that $\exp : B \to \exp(B) \subset G$ is a diffeomorphism. Then we let $U = (\exp \times \mathrm{Id}_M)(U' \cap (B \times M)) \xrightarrow{\text{open}} G \times M$ and we denote $U_x := \mathrm{pr}_1(U \cap (G \times \{x\}))$ which is open in $\exp(B)$ inside G

and which is also simply connected, since $(t, X, x) \in \mathcal{D}(\zeta) \iff (1, \frac{1}{t}X, x) \in \mathcal{D}(\zeta)$. By construction, $U_x \subset W_x$, and there is a branch $\tilde{U}_x \subset L(e, x)$ of $\text{pr}_1 : L(e, x) \to W_x$ over U_x such that $\text{pr}_1 |\tilde{U}_x : \tilde{U}_x \to U_x$ is a diffeomorphism. So all entries of diagram (6) have now been explained. We can define the local action for $(g, x) \in U$ by

$$(7) \quad G \times M \xleftarrow{\text{open}} U \xrightarrow{\ell} M, \qquad \ell(g, x) = g.x := \text{pr}_2((\text{pr}_1 |_{\tilde{U}_x})^{-1}(g^{-1})).$$

Note that (7) is the local version of (5). Again ℓ is smooth.

It remains to show that ℓ is a global or local action. Both definitions say: $g.x = y \iff (g^{-1}, y) \in L(e, x) \iff L(g^{-1}, g.x) = L(g^{-1}, y) = L(e, x)$ (for $(g, x) \in U$ in the noncomplete case). So $L(e, h.z) = L(h, z)$ determines $h.z$ uniquely; compare with (2). Applying $\mu_g \times \text{Id}_M$ and (3), we get $L(g, h.z) = L(g.h, z)$ for all $g, h \in G$ and $z \in M$. Thus $\ell : G \times M \to M$ is a (local) action, since $L(e, g.(h.x)) = L(g, h.x) = L(g.h, z) = L(e, (g.h).z)$. From the considerations in the proof of the claim (4) and from (1) and (7) it follows that for $X \in \mathfrak{g}$ we also have (for X small in the noncomplete case)

$$(8) \qquad \ell(\exp(X), x) = \exp(X).x = \text{Fl}^{\zeta_X}_{-1}(x) \in M.$$

So the fundamental vector field mapping of ℓ is $-\zeta$. □

6.6. Semidirect products of Lie groups. Let H and K be two Lie groups and let $\ell : H \times K \to K$ be a smooth left action of H in K such that each $\ell_h : K \to K$ is a group automorphism. So the associated mapping $\check{\ell} : H \to \text{Aut}(K)$ is a smooth homomorphism into the automorphism group of K. Then we can introduce the following multiplication and inversion on $K \times H$:

$$(1) \qquad (k, h)(k', h') := (k\ell_h(k'), hh'), \quad (k, h)^{-1} := (\ell_{h^{-1}}(k^{-1}), h^{-1}).$$

It is easy to see that this defines a Lie group $G = K \rtimes_\ell H$ called the *semidirect product* of H and K with respect to ℓ. If the action ℓ is clear from the context, we write $G = K \rtimes H$ only. The second projection $\text{pr}_2 : K \rtimes H \to H$ is a surjective smooth homomorphism with kernel $K \times \{e\}$, and the insertion $\text{ins}_e : H \to K \rtimes H$, $\text{ins}_e(h) = (e, h)$ is a smooth group homomorphism with $\text{pr}_2 \circ \text{ins}_e = Id_H$.

Conversely we consider an exact sequence of Lie groups and homomorphisms

$$(2) \qquad \{e\} \to K \xrightarrow{j} G \xrightarrow{p} H \to \{e\}.$$

So j is injective, p is surjective, and the kernel of p equals the image of j. We suppose furthermore that the sequence splits, so that there is a smooth homomorphism $s : H \to G$ with $p \circ s = Id_H$. Then the rule $\ell_h(k) = s(h)ks(h^{-1})$ (where we suppress j) defines a left action of H on K by automorphisms. It is easily seen that the mapping $K \rtimes_\ell H \to G$

given by $(k,h) \mapsto k.s(h)$ is an isomorphism of Lie groups with inverse $g \mapsto (g.sp(g)^{-1}, sp(g))$. Note that $g \mapsto g.sp(g)^{-1}$ is not a homomorphism of groups but only of H-modules $G \to K$. So we see that semidirect products of Lie groups correspond exactly to splitting short exact sequences.

6.7. The tangent group of a Lie group. Let G be a Lie group with Lie algebra $\mathfrak{g}$. We will use the notation from (4.1). First note that TG is also a Lie group with multiplication $T\mu$ and inversion $T\nu$, given by (see (4.2)) $T_{(a,b)}\mu.(\xi_a, \eta_b) = T_a(\mu^b).\xi_a + T_b(\mu_a).\eta_b$ and $T_a\nu.\xi_a = -T_e(\mu_{a^{-1}}).T_a(\mu^{a^{-1}}).\xi_a$.

Lemma. *Via the isomorphism given by the right trivialization $\mathfrak{g}\times G \to TG$, $(X,g) \mapsto T_e(\mu^g).X$, the group structure on TG looks as follows:*

$$(X,a).(Y,b) = (X + \operatorname{Ad}(a)Y, a.b) \quad \text{and} \quad (X,a)^{-1} = (-\operatorname{Ad}(a^{-1})X, a^{-1}).$$

So TG is isomorphic to the semidirect product $\mathfrak{g} \rtimes G$.

Proof. We compute:

$$\begin{aligned} T_{(a,b)}\mu.(T\mu^a.X, T\mu^b.Y) &= T\mu^b.T\mu^a.X + T\mu_a.T\mu^b.Y \\ &= T\mu^{ab}.X + T\mu^b.T\mu^a.T\mu^{a^{-1}}.T\mu_a.Y = T\mu^{ab}(X + \operatorname{Ad}(a)Y), \\ T_a\nu.T\mu^a.X = -T\mu^{a^{-1}}&.T\mu_{a^{-1}}.T\mu^a.X = -T\mu^{a^{-1}}.\operatorname{Ad}(a^{-1})X. \quad \square \end{aligned}$$

Remark. In the left trivialization $G \times \mathfrak{g} \to TG$, $(g,X) \mapsto T_e(\mu_g).X$, the semidirect product structure looks awkward:
$(a,X).(b,Y) = (ab, \operatorname{Ad}(b^{-1})X + Y)$ and $(a,X)^{-1} = (a^{-1}, -\operatorname{Ad}(a)X)$.

6.8. ***G*-actions and their orbit spaces.** If M is a left G-manifold, then M/G, the space of all G-orbits endowed with the quotient topology, is called the **orbit space**. We consider some examples:

The standard action of $O(n)$ on $\mathbb{R}^n$. It is orthogonal. The orbits are the concentric spheres around the fixed point 0 and 0 itself. The orbit space is $\mathbb{R}^n/O(n) \cong [0,\infty)$.

Every Lie group G acts on itself by conjugation $\operatorname{conj} : G \times G \to G$ which is defined by $(g,h) \mapsto \operatorname{conj}_g(h) := g.h.g^{-1}$ and which is a smooth left action of G on itself.

The adjoint action $\operatorname{Ad} : G \to GL(\mathfrak{g})$ of a Lie group G on its Lie algebra $\mathfrak{g}$ from (4.24). In particular, the orthogonal group acts orthogonally on $\mathfrak{o}(n)$, the Lie algebra of all skew-symmetric $n \times n$-matrices.

The $O(n)$-action on $S(n)$ treated in (7.1). Similarly, $SU(n)$ acts unitarily on the Hermitian $(n \times n)$ matrices by conjugation.

6.9. Isotropy groups. Let M be a G-manifold; then the closed subgroup $G_x = \{g \in G : g.x = x\}$ of G is called the *isotropy subgroup* of x.

The map $i^x : G/G_x \longrightarrow M$ defined by $i^x : g.G_x \mapsto g.x \in M$ is a G-equivariant initial immersion with image $G.x$, by (6.4):

$$\begin{array}{ccc} G & \xrightarrow{\ \ell^x\ } & M \\ {\scriptstyle p}\downarrow & \nearrow{\scriptstyle i^x} & \\ G/G_x. & & \end{array}$$

6.10. Lemma. *Let M be a G-manifold and $x, y \in M$; then*

(1) $G_{gx} = g.G_x.g^{-1}$,

(2) *If* $G.x \cap G.y \neq \emptyset$, *then* $G.x = G.y$,

(3) $T_x(G.x) = T_e(\ell^x).\mathfrak{g}$.

Proof. (1) $a \in G_{gx}$ means $ag.x = g.x$ or $g^{-1}ag.x = x$ and again $g^{-1}ag \in G_x$ which in turn is equivalent to $a \in g\,G_x g^{-1}$.

(2) If $z \in G.x \cap G.y$, then $z = g_1.x = g_2.y$ for some $g_1, g_2 \in G$. So $x = g_1^{-1}.g_2.y$; therefore $G.x = G.(g_1^{-1}.g_2.y) = G.y$.

(3) $X \in T_x(G.x) \Leftrightarrow X = \partial_t|_0 c(t)$ for some smooth curve $c(t) = g(t).x \in G.x$ with $g(0) = e$. So we have $X = \partial_t|_0 \ell^x(g_t) \in T_e(\ell^x).\mathfrak{g}$.

6.11. Conjugacy classes. The closed subgroups of G can be partitioned into equivalence classes by writing

$$H \sim H' \text{ if there exists } g \in G \text{ for which } H = gH'g^{-1}.$$

The equivalence class of H is denoted by (H).

Using lemma (6.10.1) we have as a first consequence: The conjugacy class of an isotropy subgroup is invariant under the action of $G : (G_x) = (G_{gx})$. Therefore we can assign to each orbit $G.x$ the conjugacy class (G_x). We will call (G_x) the *isotropy type* or the *orbit type* of the orbit through x.

If G is compact, we can define a partial ordering on the conjugacy classes simply by transferring the usual partial ordering "$\subseteq$" on the subgroups to the classes:

$$\begin{aligned} (H) \le (H') \quad :\Longleftrightarrow \quad & \exists\, K \in (H), K' \in (H') : K \subseteq K' \\ \Longleftrightarrow \quad & \exists\, g \in G : H \subseteq gH'g^{-1}. \end{aligned}$$

If G is not compact, this relation may not be reflexive. For compact G the reflexivity of this relation is a consequence of the following:

6.12. Lemma. *Let G be a compact Lie group and H a closed subgroup of G; then*

$$gHg^{-1} \subseteq H \quad \Longrightarrow \quad gHg^{-1} = H.$$

Proof. By iteration, $gHg^{-1} \subseteq H$ implies $g^n H g^{-n} \subseteq H$ for all $n \in \mathbb{N}$. Now let us study the set $A := \{g^n : n \in \mathbb{Z}_{\geq 0}\}$. We will show that g^{-1} is contained in its closure $\bar{A}$.

Suppose that e is discrete in $\bar{A}$. So there is an e-neighborhood U in G such that $U \cap \bar{A} = \{e\}$. Then $g^n U \cap \overline{\{g^k : k \geq n\}} = \{g^n\}$, so by induction $\bar{A}$ is discrete. Since G is compact, $\bar{A} = A$ is finite. Therefore $g^n = e$ for some $n > 0$, and $g^{n-1} = g^{-1} \in A$.

Suppose now that e is an accumulation point of A. Then for any neighborhood U of e there is a $g^n \in U$ where $n > 0$. This implies $g^{n-1} \in g^{-1}U \cap A$. Since the sets $g^{-1}U$ form a neighborhood basis of g^{-1}, we see that g^{-1} is an accumulation point of A as well. That is, $g^{-1} \in \bar{A}$.

Since $\operatorname{conj} : G \times G \to G$ is continuous and H is closed, we have $\operatorname{conj}(\bar{A}, H) \subseteq H$. In particular, $g^{-1}Hg \subseteq H$ which together with our premise implies that $gHg^{-1} = H$. □

6.13. Principal orbits. Let M be a G-manifold. The orbit $G.x$ is called a *principal orbit* if there is an invariant open neighborhood U of x in M and for all $y \in U$ an equivariant map

$$f : G.x \to G.y.$$

Note that f is automatically surjective: Namely, let $f(x) =: a.y$. For an arbitrary $z = g.y \in G.y$ this gives us

$$z = g.y = ga^{-1}a.y = ga^{-1}f(x) = f(ga^{-1}.x).$$

The existence of f in the above definition is equivalent to the condition: $G_x \subseteq aG_ya^{-1}$ for some $a \in G$:
If f exists, then for $g \in G_x$ we have $g.x = x$ and thus $g.f(x) = f(g.x) = f(x)$. For $f(x) =: a.y$ we get $ga.y = a.y$; thus $g \in G_{ay} = aG_ya^{-1}$ by (6.10.1).
To show the converse, we define $f : G.x \to G.y$ explicitly by $f(g.x) := ga.y$. We have to check: If $g_1.x = g_2.x$, i.e., $g := g_2^{-1}g_1 \in G_x$, then $g_1a.y = g_2a.y$ or $g \in G_{ay} = aG_ya^{-1}$. This is guaranteed by our assumption.

We call $x \in M$ a *regular point* if $G.x$ is a principal orbit. Otherwise, x is called *singular*. The subset of all regular points in M is denoted by M_{reg}, and M_{sing} denotes the subset of all singular points.

6.14. Slices. Let M be a G-manifold and $x \in M$, then a subset $S \subseteq M$ is called a *slice* at x if there is a G-invariant open neighborhood U of $G.x$ and a smooth equivariant retraction $r : U \to G.x$ such that $S = r^{-1}(x)$. Since r is equivariant, it is a submersion onto $G.x$.

6.15. Proposition. *If M is a G-manifold and $S = r^{-1}(x)$ a slice at $x \in M$, where $r : U \to G.x$ is the corresponding retraction, then:*

(1) *$x \in S$ and $G_x.S \subseteq S$.*
(2) *If $g.S \cap S \neq \emptyset$, then $g \in G_x$.*
(3) *$G.S = \{g.s : g \in G, s \in S\} = U$.*

Proof. (1) We have $x \in S$ since $S = r^{-1}(x)$ and $r(x) = x$. To show that $G_x.S \subseteq S$, take an $s \in S$ and $g \in G_x$. Then $r(g.s) = g.r(s) = g.x = x$, and therefore $g.s \in r^{-1}(x) = S$.

(2) If $g.S \cap S \neq \emptyset$, then $g.s \in S$ for some $s \in S$. So we get $x = r(g.s) = g.r(s) = g.x$; thus $g \in G_x$.

(3) We have $G.S = G.r^{-1}(x) = r^{-1}(G.x) = U$. □

6.16. Corollary. *If M is a G-manifold and S a slice at $x \in M$, then:*

(1) *S is a G_x-manifold.*
(2) *$G_s \subseteq G_x$ for all $s \in S$.*
(3) *If $G.x$ is a principal orbit and G_x compact, then $G_y = G_x$ for all $y \in S$ if the slice S at x is chosen small enough. In other words, all orbits near $G.x$ are principal as well.*
(4) *If two G_x-orbits $G_x.s_1, G_x.s_2$ in S have the same orbit type as G_x-orbits in S, then $G.s_1$ and $G.s_2$ have the same orbit type as G-orbits in M.*
(5) *$S/G_x \cong G.S/G$ is an open neighborhood of $G.x$ in the orbit space M/G.*

Proof. (1) This is clear from (6.15.1).

(2) If $g \in G_y$ then $g.y = y \in S$; thus $g \in G_x$ by (6.15.2).

(3) By (2) we have $G_y \subseteq G_x$, so G_y is compact as well. Because $G.x$ is principal it follows that for $y \in S$ close to x, G_x is conjugate to a subgroup of G_y, $G_y \subseteq G_x \subseteq g.G_y g^{-1}$. Since G_y is compact, $G_y \subseteq g.G_y g^{-1}$ implies $G_y = g.G_y g^{-1}$ by (6.12). Therefore $G_y = G_x$, and $G.y$ is also a principal orbit.

(4) For any $s \in S$ we have $(G_x)_s = G_s$, since $(G_x)_s \subseteq G_s$. Conversely, by (2), $G_s \subseteq G_x$; therefore $G_s \subseteq (G_x)_s$. So $(G_x)_{s_1} = g(G_x)_{s_2} g^{-1}$ implies $G_{s_1} = gG_{s_2}g^{-1}$ and the G-orbits have the same orbit type.

(5) The isomorphism $S/G_x \cong G.S/G$ is given by the map $G_x.s \mapsto G.s$ which is an injection by (6.15.2). Since $G.S = U$ is an open G-invariant neighborhood of $G.x$ in M by (6.15.3), we have $G.S/G$ is an open neighborhood of $G.x$ in M/G. □

6.17. Remark. The converse to (6.16.4) is generally false. If the two G-orbits $G.s_1$, $G.s_2$ are of the same type, then the isotropy groups G_{s_1} and G_{s_2} are conjugate in G. They need not be conjugate in G_x. For example, consider the following semidirect product, the compact Lie group $G := (S^1 \times S^1) \rtimes \mathbb{Z}_2$ with multiplication defined as follows. Let $\varphi_1, \varphi_2, \psi_1, \psi_2 \in S^1$ and $\alpha, \beta \in \mathbb{Z}_2$. Take on $S^1 \times S^1$ the usual multiplication by components, and as $\mathbb{Z}_2$-action:

$$\mathbb{Z}_2 \ni \bar{0} \mapsto i_0 := \mathrm{Id}_{S^1 \times S^1}, \qquad \bar{1} \mapsto (i_1 : (\varphi_1, \varphi_2) \mapsto (\varphi_2, \varphi_1)).$$

Then we consider the semidirect product structure:

$$(\varphi_1, \varphi_2, \alpha).(\psi_1, \psi_2, \beta) := \big((\varphi_1, \varphi_2).i_\alpha(\psi_1, \psi_2), \alpha + \beta\big).$$

Now we let G act on $M := V \sqcup W$ where $V = W = \mathbb{C} \times \mathbb{C}$. For any element in M we will indicate its connected component by the index $(x, y)_V$ or $(x, y)_W$. The action shall be the following:

$$(\varphi_1, \varphi_2, \bar{0}).(x, y)_V := (\varphi_1.x, \varphi_2.y)_V,$$
$$(\varphi_1, \varphi_2, \bar{1}).(x, y)_V := (\varphi_1.y, \varphi_2.x)_W.$$

The action on W is simply given by interchanging the V's and W's in the above formulae. This defines an action. Denote by H the abelian subgroup $S^1 \times S^1 \times \{\bar{0}\}$. Then H is the isotropy subgroup of $(0, 0)_V$, and V is a slice at $(0, 0)_V$. Now consider $s_1 := (0, v^1)_V$ and $s_2 := (v^2, 0)_V$, both not equal to zero. Then let

$$H_1 := G_{s_1} = S^1 \times \{1\} \times \{\bar{0}\},$$
$$H_2 := G_{s_2} = \{1\} \times S^1 \times \{\bar{0}\}.$$

The subgroups H_1 and H_2 are conjugate in G by $(1, 1, \bar{1})$. Yet they are clearly not conjugate in H since H is abelian. So $H.s_1$ and $H.s_2$ have different orbit types in H while $G.s_1$ and $G.s_2$ are of the same G-orbit type.

6.18. Proposition. *Let M be a G-manifold and S a slice at x; then there is a G-equivariant diffeomorphism of the associated bundle $G[S]$ onto $G.S$,*

$$f : G[S] := G \times_{G_x} S \to G.S$$

which maps the zero section $G \times_{G_x} \{x\}$ onto $G.x$.

See (18.7) below for more information on associated bundles.

Proof. Since $\ell(gh, h^{-1}.s) = g.s = \ell(g,s)$ for all $h \in G_x$, there is an $f : G[S] \to G.S$ such that the following diagram commutes:

$$\begin{array}{ccc} G \times S & \xrightarrow{\ \ell\ } & G.S \\ {\scriptstyle q}\downarrow & \nearrow{\scriptstyle f} & \\ G \times_{G_x} S. & & \end{array}$$

The map f is smooth because $f \circ q = \ell$ is smooth and q is a submersion. It is equivariant since ℓ and q are equivariant. Also, f maps the zero section $G \times_{G_x} \{x\}$ onto $G.x$. The map f is bijective: If $g_1.s_1 = g_2.s_2$, then $s_1 = g_1^{-1}g_2.s_2$; thus $h = g_1^{-1}g_2 \in G_x$ by (6.15.2). But then $g_1 = g_2h$ and $s_1 = h.s_2$. This is equivalent to $q(g_1, s_1) = q(g_2, s_2)$.

To see that f is a diffeomorphism, let us prove that the rank of f equals the dimension of M. First of all, note that $\operatorname{rank}(\ell_g) = \dim(g.S) = \dim S$ and $\operatorname{rank}(\ell^x) = \dim(G.x)$. Since $S = r^{-1}(x)$ and $r : G.S \to G.x$ is a submersion, it follows that $\dim(G.x) = \operatorname{codim} S$. Therefore,

$$\begin{aligned} \operatorname{rank}(f) = \operatorname{rank}(\ell) = \operatorname{rank}(\ell_g) + \operatorname{rank}(\ell^x) &= \dim S + \dim(G.x) \\ &= \dim S + \operatorname{codim} S = \dim M. \quad \square \end{aligned}$$

6.19. Remark. The converse also holds. If $\bar{f} : G \times_{G_x} S \to G.S$ is a G-equivariant diffeomorphism, then for some $\bar{g} \in G$ and $\bar{s} \in S$ whe have $\bar{f}[\bar{g}, \bar{s}] = x$. So $f[g,s] := \bar{f}[g\bar{g}, s]$ defines a G-equivariant diffeomorphism with the additional property that $x = f[e, \bar{s}]$:

$$\begin{array}{ccc} G \times_{G_x} S & \xrightarrow{\ f\ } & G.S \\ {\scriptstyle \mathrm{pr}_1}\downarrow & & \downarrow{\scriptstyle r} \\ G/G_x & \xrightarrow{\ i\ } & G.x. \end{array}$$

If we define $r := i \circ \mathrm{pr}_1 \circ f^{-1} : G.S \to G.x$, then r is again a smooth G-equivariant map, and it is a retraction onto $G.x$ since

$$x \xrightarrow{\ f^{-1}\ } [e, \bar{s}] \xrightarrow{\ \mathrm{pr}_1\ } e.G_x \xrightarrow{\ i\ } e.x.$$

Furthermore, $r^{-1}(x) = S$, making S a slice.

6.20. Proper actions. Recall that a continuous mapping between topological spaces is called *proper* if compact subsets have compact inverse images. A smooth action $\ell : G \times M \to M$ is called *proper* if it satisfies one of the following three equivalent conditions:

(1) $(\ell, \mathrm{pr}_2) : G \times M \to M \times M$, $(g,x) \mapsto (g.x, x)$, is a proper mapping

(2) $g_n.x_n \to y$ and $x_n \to x$ in M, for some $g_n \in G$ and $x_n, x, y \in M$, implies that these g_n have a convergent subsequence in G.

(3) K and L compact in M implies that $\{g \in G : g.K \cap L \neq \emptyset\}$ is compact as well.

Proof. (1) $\Rightarrow$ (2) This is a direct consequence of the definitions.

(2) $\Rightarrow$ (3) Let g_n be a sequence in $\{g \in G : g.K \cap L \neq \emptyset\}$ and $x_n \in K$ such that $g_n.x_n \in L$. Since K is compact, we can choose a convergent subsequence $x_{n_k} \to x \in K$ of x_n. Since L is compact, we can do the same for $g_{n_k}.x_{n_k}$ there. Now (2) tells us that in such a case g_n must have a convergent subsequence; therefore $\{g \in G : g.K \cap L \neq \emptyset\}$ is compact.

(3) $\Rightarrow$ (1) Let R be a compact subset of $M \times M$. Then $L := \mathrm{pr}_1(R)$ and $K := \mathrm{pr}_2(R)$ are compact, and $(\ell, \mathrm{pr}_2)^{-1}(R) \subseteq \{g \in G : g.K \cap L \neq \emptyset\} \times K$. By (3), $\{g \in G : g.K \cap L \neq \emptyset\}$ is compact. Therefore $(\ell, \mathrm{pr}_2)^{-1}(R)$ is compact, and (ℓ, pr_2) is proper. □

6.21. Remark. If G is compact, then every G-action is proper. If $\ell : G \times M \to M$ is a proper action and G is not compact, then for any noncompact $H \subseteq G$ and $x \in M$ the set $H.x$ is noncompact in M. Furthermore, all isotropy groups are compact (most easily seen from (6.20.3) by setting $K = L = \{x\}$).

6.22. Lemma. *A continuous, proper map $f : X \to Y$ between two topological spaces is closed.*

Proof (For metric spaces). Consider a closed subset $A \subseteq X$, and take a point y in the closure of $f(A)$. Let $f(a_n) \in f(A)$ converge to y ($a_n \in A$). Then the $f(a_n)$ are contained in a compact subset $K \subseteq Y$. Therefore $a_n \subseteq f^{-1}(K) \cap A$ which is now, since f is proper, a compact subset of A. Consequently, (a_n) has a convergent subsequence with limit $a \in A$, and by continuity of f, it gives a convergent subsequence of $f(a_n)$ with limit $f(a) \in f(A)$. Since $f(a_n)$ converges to y, we have $y = f(a) \in f(A)$. □

6.23. Proposition. *The orbits of a proper action $\ell : G \times M \to M$ are closed submanifolds.*

Proof. By the preceding lemma, (ℓ, pr_2) is closed. Therefore $(\ell, \mathrm{pr}_2)(G, x) = G.x \times \{x\}$, and with it $G.x$ is closed.

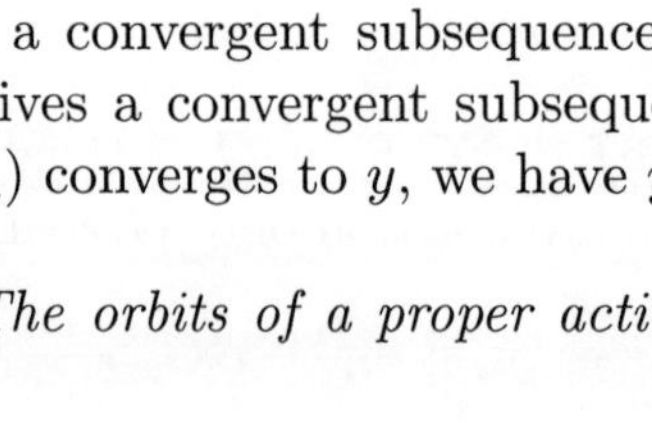

As a maximal integral manifold of the involutive distribution of (in general) nonconstant rank spanned by all fundamental vector fields, $G.x$ is an initial submanifold, and i^x is an initial immersion by (6.4). Thus $i^x : G/G_x \to G.x$ is open. □

6.24 Examples of nonproper actions. The standard action of $SL(2,\mathbb{R})$ on $\mathbb{R}^2$ has two orbits, 0 and $\mathbb{R}^2 \setminus 0$ which is not closed. By (6.23) this action is not proper.

The action of $GL(n,\mathbb{R})$ on the space $L_{\text{sym}}(\mathbb{R}^n, \mathbb{R}^n)$ of symmetric matrices given by $(g, A) \mapsto g^\top.A.g$ is not proper, since the isotropy group of the symmetric bilinear form with signature $p, n-p$ is the group $O(p, n-p, \mathbb{R})$ which is not compact for $0 < p < n$.

6.25. Lemma. *Let (M, γ) be a Riemann manifold and $\ell : G \times M \to M$ an effective isometric action (i.e., $g.x = x$ for all $x \in M \Rightarrow g = e$), such that $\ell(G) \subseteq \text{Isom}(M, \gamma)$ is closed in the compact open topology. Then ℓ is proper.*

Proof. Assume without loss that M is connected. Let $g_n \in G$ and $x_n, x, y \in M$ such that $g_n.x_n \to y$ and $x_n \to x$; then we have to show that g_n has a convergent subsequence which is the same as proving that $\{g_n : n \in \mathbb{N}\}$ is relatively compact, since $\ell(G) \subseteq \text{Isom}(M, \gamma)$ is closed.

Let us choose a compact neighborhood K of x in M. Put a metric on M (e.g., the Riemann distance function). Note first $\text{Isom}(M) \ni \varphi \mapsto \varphi|_K \in C^0(K, M)$ is an injective embedding, where we put the uniform metric on $C^0(K, M)$. Then, since the g_n act isometrically, we can find a compact neighborhood $L \subseteq M$ of y such that $\bigcup_{n=1}^\infty g_n.K$ is contained in L. So $\{g_n\}$ is bounded in $C^0(K, M)$. Furthermore, the set of all g_n is equicontinuous as subset of $\text{Isom}(M)$. Therefore, by the theorem of Ascoli-Arzela, $\{g_n : n \in \mathbb{N}\}$ is relatively compact in $\text{Isom}(M)$. □

6.26. Theorem (Existence of slices, [186]. *Let M be a G-space and let $x \in M$ be a point with compact isotropy group G_x. If for all open neighborhoods W of G_x in G there is a neighborhood V of x in M such that $\{g \in G : g.V \cap V \neq \emptyset\} \subseteq W$, then there exists a slice at x.*

Proof. Let $\tilde{\gamma}$ be any Riemann metric on M. Since G_x is compact, we can get a G_x-invariant Riemann metric on M by integrating over the Haar measure for the action of G_x; see (14.4):

$$\gamma(X, Y) := \int_{G_x} (\ell_a^* \tilde{\gamma})(X, Y) da = \int_{G_x} \tilde{\gamma}(T\ell_a X, T\ell_a Y) da.$$

We choose $\varepsilon > 0$ small enough for $T_xM \supseteq B_{0_x}(\varepsilon) \xrightarrow{\exp_x^\gamma} M$ to be a diffeomorphism onto its image and we define:

$$\tilde{S} := \exp_x^\gamma\left(T_x(G.x)^\perp \cap B_{0_x}(\varepsilon)\right) \subseteq M.$$

Then $\tilde{S}$ is a submanifold of M and the first step towards obtaining a real slice. Let us show that $\tilde{S}$ is G_x-invariant. Since G_x leaves γ unchanged and $T_x(G.x)$ is invariant under $T_x\ell_g$ (for $g \in G_x$), $T_x\ell_g$ is an isometry and leaves $T_x(G.x)^\perp \cap B_{0_x}(\varepsilon)$ invariant. Therefore:

$$\begin{array}{ccc} T_x(G.x)^\perp \cap B_{0_x}(\varepsilon) & \xrightarrow{T_x\ell_g} & T_x(G.x)^\perp \cap B_{0_x}(\varepsilon) \\ \downarrow{\scriptstyle \exp_x^\gamma} & & \downarrow{\scriptstyle \exp_x^\gamma} \\ \tilde{S} & \xrightarrow{\ell_g} & \tilde{S}. \end{array}$$

We have to shrink $\tilde{S}$ to an open subset S such that for $g \in G$ with $g.S \cap S \neq \emptyset$ we have $g \in G_x$. This property is necessary for a slice. At this point, we shall need the condition that for every open neighborhood W of G_x in G, there is a neighborhood V of x in M such that $\{g \in G : g.V \cap V \neq \emptyset\} \subseteq W$. First we must construct a W fitting our purposes. Choose an open neighborhood $U \subseteq G/G_x$ of $e.G_x$ such that there is a smooth section $\chi : U \to G$ of $\pi : G \to G/G_x$ with $\chi(e.G_x) = e$. Also, let U and possibly $\tilde{S}$ be small enough to get an embedding

$$f : U \times \tilde{S} \to M : (u, s) \mapsto \chi(u).s.$$

Our neighborhood of G_x will be $W := \pi^{-1}(U)$. Now by our assumption, there is a neighborhood V of x in M such that $\{g \in G : g.V \cap V \neq \emptyset\} \subseteq W$. Next we will prove that V can be chosen G_x-invariant. Suppose we can choose an open neighborhood $\tilde{W}$ of G_x in G such that $G_x.\tilde{W} \subseteq W$ (we will prove this below). Let V' be the neighborhood of x in M satisfying $\{g \in G : g.V' \cap V' \neq \emptyset\} \subseteq \tilde{W}$. Now $V := G_x.V'$ has the desired property, since:

$$\begin{aligned} \{g \in G : g.G_x.V' \cap G_x.V' \neq \emptyset\} &= \bigcup_{g_1, g_2 \in G_x} \{g \in G : g.g_1.V' \cap g_2.V' \neq \emptyset\} \\ &= \bigcup_{g_1, g_2 \in G_x} \{g \in G : g_2^{-1} g g_1.V' \cap V' \neq \emptyset\} \\ &= \bigcup_{g_1, g_2 \in G_x} g_2\{g \in G : g.V' \cap V' \neq \emptyset\}g_1^{-1} \\ &= G_x.\{g \in G : g.V' \cap V' \neq \emptyset\}.G_x \subseteq G_x.\tilde{W}.G_x \subseteq W.G_x = W. \end{aligned}$$

To complete the above argumentation, we have only to prove the
Claim. To any open neighborhood W of G_x in G there is an open neighborhood $\tilde{W}$ of G_x such that $G_x.\tilde{W} \subseteq W$.

The proof of this claim relies on the compactness of G_x. For all $(a,b) \in G_x \times G_x$ we choose neighborhoods $A_{a,b}$ of a and $B_{a,b}$ of b, such that $A_{a,b}.B_{a,b} \subseteq W$. This is possible by continuity, since $G_x.G_x = G_x$. $\{B_{a,b} : b \in G_x\}$ is an open cover of G_x. Then there is a finite subcover $\bigcup_{j=1}^N B_{a,b_j} := B_a \supseteq G_x$. Since $A_{a,b_j}.B_{a,b_j} \subseteq W$, we must choose $A_a := \bigcap_{j=1}^N A_{a,b_j}$, to get $A_a.B_a \subseteq W$. Now since A_a is a neighborhood of a in G_x, the A_a cover G_x again. Consider a finite subcovering $A := \bigcup_{j=1}^n A_{a_j} \supseteq G_x$, and as before define $B := \bigcap_{j=1}^n B_{a_j}$, so that $A.B \subseteq W$. In particular, this gives us $G_x.B \subseteq W$, so $\tilde{W} := B$ is an open neighborhood of G_x with the desired property.

So we have a G_x-invariant neighborhood V of x with $\{g \in G : gV \cap V \neq \emptyset\}$ contained in W. Now we define $S := \tilde{S} \cap V$ and hope for the best. The S is an open subset of $\tilde{S}$, and it is again invariant under G_x. Let us check whether we have the converse: $\{g \in G : g.S \cap S \neq \emptyset\} \subseteq G_x$. If $g.s_1 = s_2$ for some $s_1, s_2 \in S$, then $g \in W = \pi^{-1}(U)$ by the above effort. Therefore $\pi(g) \in U$. Choose $h = g^{-1}\chi(\pi(g)) \in G_x$. Then

$$f(\pi(g), h^{-1}s_1) = \chi(\pi(g))h^{-1}s_1 = g.s_1 = s_2 = f(\pi(e), s_2).$$

Since f is a diffeomorphism onto its image, we have shown that $\pi(g) = \pi(e)$, that is, $g \in G_x$.

Now, it is easy to see that $F : G \times_{G_x} S \to G.S : [g,s] \mapsto g.s$ is well defined, G-equivariant and smooth. We have the diagram

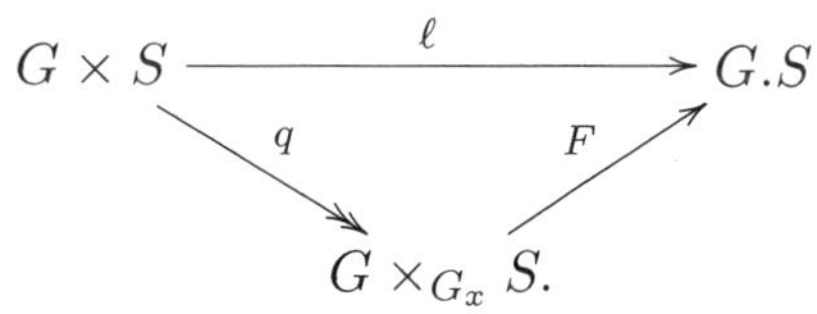

To finish the proof, we have to show that F is a diffeomorphism, according to (6.19). Firstly, F is injective because:

$$\begin{aligned} &F[g,s] = F[g',s'] \Rightarrow g.s = g'.s' \Rightarrow g^{-1}g'.s' = s \\ &\Rightarrow g^{-1}g' \in G_x \Rightarrow [g,s] = [g, g^{-1}g'.s'] = [g',s']. \end{aligned}$$

Next, we notice that $\ell(W,S) = W.S = f(U,S)$ is open in M since $f : U \times \tilde{S} \to M$ is an embedding with an open image. Consequently, $G.S = \ell(G, W.S)$ is open, since ℓ is open, and thus F is a diffeomorphism. $\square$

6.27. Theorem ([186]). *If M is a proper G-manifold, then for all $x \in M$ the conditions of the previous theorem are satisfied, so each x has slices.*

Proof. We have already shown that each isotropy group G_x is compact (6.21). Now for every neighborhood U of G_x in G, for every $x \in M$, it remains to find a neighborhood V of x in M such that

$$\{g \in G : g.V \cap V \neq \emptyset\} \subseteq U.$$

Claim. U contains an open neighborhood $\tilde{U}$ with $G_x.\tilde{U} = \tilde{U}$; so we will be able to assume $G_x.U = U$ without loss of generality.

The claim in the proof of theorem (6.26) shows the existence of a neighborhood B of G_x such that $G_x.B \subseteq U$, using only the compactness of G_x. So $\tilde{U} := G_x.B = \bigcup_{g \in G_x} g.B$ is again an open neighborhood of G_x, and it has the desired properties.

Now we can suppose $U = G_x.U$. Next, we have to construct an open neighborhood $V \subseteq M$ of x, such that $\{g \in G : g.V \cap V \neq \emptyset\} \subseteq U$. This is the same as saying $(G \setminus U).V \cap V$ should be empty. So we have to look for V in the complement of $(G \setminus U).x$.

We see that $M \setminus ((G \setminus U).x)$ is open, or rather that $(G \setminus U).x$ is closed. This is because $(G \setminus U).x \times \{x\} = (\ell, \mathrm{pr}_2)((G \setminus U) \times \{x\})$ is the image of a closed set under (ℓ, pr_2) which is a closed mapping by lemma (6.22).

Now let us choose a compact neighborhood W of x in $M \setminus ((G \setminus U).x)$. Then since G acts properly, it follows that $\{g \in G : g.W \cap W \neq \emptyset\}$ is compact; in particular $K := \{g \in G \setminus U : g.W \cap W \neq \emptyset\}$ is compact. But what we need is for $\{g \in G \setminus U : g.V \cap V \neq \emptyset\}$ to be empty. An x-neighborhood V contained in W fulfills this if $K.V \subseteq M \setminus W$. Let us find such a neighborhood.

Our choice of W guarantees $K.x \subseteq M \setminus W$. But $M \setminus W$ is open; therefore for each $k \in K$ we can choose a neighborhood Q_k of k in G and V_k of x in W, such that $Q_k.V_k \subseteq M \setminus W$. The neighborhoods Q_k cover K, and we can choose a finite subcovering $\bigcup_{j=1}^m Q_{k_j}$. Then $V := \bigcap_{j=1}^m V_{k_j}$ has the desired property: $K.V \subseteq M \setminus W$. □

6.28. Lemma. *Let M be a proper G-manifold, V a linear G-space and $f : M \to V$ smooth with compact support. Then*

$$\tilde{f} : x \mapsto \int_G g^{-1} f(g.x) d_R g$$

is a G-equivariant C^∞-map with $\tilde{f}(x) = 0$ for $x \notin G.\operatorname{supp} f$ (where d_R stands for a right Haar measure on G).

Proof. Since G acts properly, $\{g \in G : g.x \in \operatorname{supp} f\}$ is compact. Therefore the map $g \mapsto g^{-1} f(g.x)$ has compact support, and $\tilde{f}$ is well defined. To see that $\tilde{f}$ is smooth, let x_0 be in M, and let U be a compact neighborhood of x_0. Then the set $\{g \in G : g.U \cap \operatorname{supp} f \neq \emptyset\}$ is compact. Therefore,

$\tilde{f}$ restricted to U is smooth; in particular $\tilde{f}$ is smooth in x_0. Also, $\tilde{f}$ is G-equivariant, since

$$\begin{aligned}\tilde{f}(h.x) &= \int_G g^{-1} f(gh.x) d_R g = \int_G h(gh)^{-1} f(gh.x) d_R g \\ &= h. \int_G g^{-1} f(g.x) d_R g = h\tilde{f}(x).\end{aligned}$$

Furthermore, if $x \notin G.\operatorname{supp} f$, then $f(g.x) = 0$ for all $g \in G$; thus $\tilde{f}(x) = 0$. □

6.29. Corollary. *If M is a proper G-manifold, then M/G is completely regular.*

Proof. Choose $F \subseteq M/G$ closed and $\bar{x}_0 = \pi(x_0) \notin F$. Now let U be a compact neighborhood of x_0 in M fulfilling $U \cap \pi^{-1}(F) = \emptyset$, and choose $f \in C^\infty(M, [0, \infty))$ with support in U such that $f(x_0) > 0$. If we take the trivial representation of G on $\mathbb{R}$, then from lemma (6.28) it follows that

$$\tilde{f}(x) = \int_G f(g.x) d_R g$$

defines a smooth G-invariant function. Here $d_R g$ denote the right Harr measure on G; see (14.4). Moreover, $\tilde{f}(x_0) > 0$. Since $\operatorname{supp}(\tilde{f}) \subseteq G.\operatorname{supp}(f) \subseteq G.U$, we have $\operatorname{supp}(\tilde{f}) \cap \pi^{-1}(F) = \emptyset$. Since $\tilde{f} \in C^\infty(M, [0, \infty))^G$ is invariant, f factors over π to a map $\bar{f} \in C^0(M/G, [0, \infty))$, with $\bar{f}(\bar{x}_0) > 0$ and $\bar{f}|_F = 0$. □

6.30. Theorem. *If M is a proper G-manifold, then there is a G-invariant Riemann metric on M.*

Proof. By (6.27) there is a slice S_x at x for all $x \in M$. Let $\pi : M \to M/G$ be the quotient map. Notice first that M/G is Hausdorff by (6.29).

For each x choose $f_x \in C^\infty(M, [0, \infty))$ with $f_x(x > 0)$ and $\operatorname{supp}(f_x) \subseteq G.S_x$ compact; then by (6.28)

$$\bar{f}_x(y) := \int_G f_x(g.y) d_R g \in C^\infty(M, [0, \infty))^G$$

is G-invariant, positive on $G.x$, and has $\operatorname{supp}(\bar{f}_n) \subseteq G.S_n$. Moreover, $\pi(\operatorname{supp} \bar{f}_x)$ is a compact neighborhood of $\pi(x)$, so M/G is locally compact.

The interiors of the supports of the smooth functions $\bar{f}_x$ form an open cover of M. Since M is a Lindelöf-space (1.6), there is a countable subcover with corresponding functions $\bar{f}_{x_1}, \bar{f}_{x_2}, \dots$. We write $\bar{f}_n := \bar{f}_{x_n}$ and $S_N := S_{x_n}$. Let

$$W_n = \{x \in M : \bar{f}_n(x) > 0 \text{ and } \bar{f}_i(x) < \tfrac{1}{n} \text{ for } 1 \le i < n\} \subseteq G.S_n,$$

and denote by $\overline{W}_n$ the closure. Then $\{W_n\}$ is a G-invariant open cover. We claim that $\{\overline{W}_n : n \in \mathbb{N}\}$ is locally finite: Let $x \in M$. Then there is a smallest n such that $x \in W_n$. Let $V := \{y \in M : \bar{f}_n(y) > \frac{1}{2}\bar{f}_n(x)\}$. If $y \in V \cap \overline{W}_k$, then we have $\bar{f}_n(y) > \frac{1}{2}\bar{f}_n(x)$ and $\bar{f}_i(y) \le \frac{1}{k}$ for $i < k$, which is possible for finitely many k only. Let $h(t) = e^{-1/t}$ for $t > 0$ and $h(t) = 0$ for $t \le 0$. Consider the nonnegative smooth function

$$f_n(x) := h(\bar{f}_n(x))h(\tfrac{1}{n} - f_1(x)) \dots h(\tfrac{1}{n} - f_{n-1}(x))$$

for each n. Then obviously $\operatorname{supp}(f_n) = \overline{W}_n \subseteq G.S_n$.

The action of the compact group G_{x_n} on $TM|_{S_n}$ is fiber linear, so there is a G_{x_n}-invariant Riemann metric $\gamma^{(n)}$ on the vector bundle $TM|_{S_{x_n}}$ by integration over the compact group G_{x_n}. To get a Riemann metric on $TM|_{G.S_n}$ invariant under the whole group G, consider the following diagram:

$$\begin{array}{ccc}
G \times TM|_{S_{x_n}} & \xrightarrow{T_2\ell} & TM|_{G.S_{x_n}} \\
\downarrow q & \nearrow \widetilde{T_2\ell} & \\
G \times_{G_{x_n}} TM|_{S_{x_n}} & & \downarrow \\
\downarrow & & \\
G \times_{G_{x_n}} S_{x_n} & \xrightarrow{\cong} & G.S_{x_n}.
\end{array}$$

The map $T_2\ell : (g, X_s) \mapsto T_s\ell_g.X_s$ factors over q to a map $\widetilde{T_2\ell}$ which is injective, since if $T_2\ell(g, X_s) = T_2\ell(g', X_{s'})$, then on the one side $\ell(g.s) = \ell(g'.s')$ so $g^{-1}g'.s' = s$ and $g^{-1}g' \in G_x$. On the other side, $T_s\ell_g.X_s = T_{s'}\ell_{g'}.X_{s'}$. So

$$(g', X_{s'}) = \{g(g^{-1}g'), T_{s'}\ell_{g'^{-1}} T_s\ell_g.X_s\};$$

thus $q(g', X_{s'}) = q(g, X_s)$.

The Riemann metric $\gamma^{(n)}$ induces a G-invariant vector bundle metric on $G \times TM|_{S_n} \to G \times S_n$ by

$$\gamma_n((g, X_s), (g, Y_s)) := \gamma^{(n)}(X_s, Y_s).$$

It is also invariant under the right G_{x_n}-action $(g, X_s).h = (gh, T\ell_{h^{-1}}.X_s)$ and, therefore, induces a Riemann metric $\tilde{\gamma}_n$ on $G \times_{G_x} TM|_{S_n}$. This metric is again G-invariant, since the actions of G and G_x commute. Now $(\widetilde{T_2\ell})_*\tilde{\gamma}_n =: \bar{\gamma}_n$ is a G-invariant Riemann metric on $TM|_{G.S_n}$, and $\gamma := \sum_{n=1}^{\infty} f_n\bar{\gamma}_n$ is a G-invariant Riemann metric on M. □

6.31. Result ([186]). *Let G be a matrix group, that is, a Lie group with a faithful finite-dimensional representation, and let M be a proper G-space with only a finite number of orbit types. Then there is a G-equivariant embedding $f : M \to V$ into a linear G-space V.*

7. Polynomial and Smooth Invariant Theory

7.1. A motivating example. Let $S(n)$ denote the space of symmetric $n \times n$ matrices with entries in $\mathbb{R}$ and $O(n)$ the orthogonal group. Consider the action:

$$\ell : O(n) \times S(n) \to S(n), \quad (A, B) \mapsto ABA^{-1} = ABA^T.$$

If Σ is the space of all real diagonal matrices and $\mathcal{S}_n$ is the symmetric group on n letters, then we have the following:

Theorem.

(1) *This is an orthogonal $O(n)$-action on $S(n)$ for the inner product given by $\langle A, B\rangle = \operatorname{Trace}(AB^T) = \operatorname{Trace}(AB)$.*

(2) *Σ meets every $O(n)$-orbit.*

(3) *If $B \in \Sigma$, then $\ell(O(n), B) \cap \Sigma$, the intersection of the $O(n)$-orbit through B with Σ, equals the $\mathcal{S}_n$-orbit through B, where $\mathcal{S}_n$ acts on $B \in \Sigma$ by permuting the eigenvalues.*

(4) *Σ intersects each orbit orthogonally with respect to the inner product $\langle A, B\rangle = \operatorname{Trace}(AB^T) = \operatorname{Trace}(AB)$ on $S(n)$.*

(5) *$\mathbb{R}[S(n)]^{O(n)}$, the space of all $O(n)$-invariant polynomials in $S(n)$ is isomorphic to $\mathbb{R}[\Sigma]^{\mathcal{S}_n}$, the symmetric polynomials in Σ (by restriction).*

(6) *The space $C^\infty(S(n))^{O(n)}$ of $O(n)$-invariant C^∞-functions is isomorphic to $C^\infty(\Sigma)^{\mathcal{S}_n}$, the space of all symmetric C^∞-functions in Σ (again by restriction), and these again are isomorphic to the C^∞-functions in the elementary symmetric polynomials.*

(7) *The space of all $O(n)$-invariant horizontal p-forms on $S(n)$, the space of all $O(n)$-invariant p-forms ω with the property $i_X\omega = 0$ for all $X \in T_A(O(n).A)$, is isomorphic to the space of $\mathcal{S}_n$-invariant p-forms on Σ:*

$$\Omega^p_{hor}(S(n))^{O(n)} \cong \Omega^p(\Sigma)^{\mathcal{S}_n}.$$

Proof. (1) Let $A \in O(n)$ act on $H_1, H_2 \in S(n)$; then

$$\begin{aligned}\operatorname{Trace}(AH_2A^{-1}(AH_1A^{-1})^T) &= \operatorname{Trace}(AH_2A^{-1}(A^{-1})^T H_1^T A^T)\\ &= \operatorname{Trace}(AH_2A^{-1}AH_1^TA^{-1}) = \operatorname{Trace}(AH_2H_1^TA^{-1}) = \operatorname{Trace}(H_2H_1^T).\end{aligned}$$

(2) Clear from linear algebra.

(3) The transformation of a symmetric matrix into normal form is unique except for the order in which the eigenvalues appear.

(4) Take an A in Σ. For any $X \in \mathfrak{o}(n)$, that is, for any skew-symmetric X, let ζ_X denote the corresponding fundamental vector field on $S(n)$. Then we

have

$$\zeta_X(A) = \frac{d}{dt}\Big|_{t=0} \exp_e(tX) A \exp_e(tX^T)$$
$$= XA\,id + id\,AX^T = XA - AX.$$

Now the inner product with $\eta \in T_A\Sigma \cong \Sigma$ computes to

$$\langle \zeta_X(A), \eta\rangle = \mathrm{Trace}(\zeta_X(A)\eta) = \mathrm{Trace}((XA - AX)\eta)$$
$$= \mathrm{Trace}(\underbrace{XA\,\eta}_{=X\eta A}) - \mathrm{Trace}(AX\eta) = \mathrm{Trace}(X\eta A) - \mathrm{Trace}(X\eta A) = 0.$$

(5) If $p \in \mathbb{R}[S(n)]^{O(n)}$, then clearly $\tilde{p} := p|_\Sigma \in \mathbb{R}[\Sigma]^{\mathcal{S}_n}$. To construct p from $\tilde{p}$, we use the result from algebra that $\mathbb{R}[\mathbb{R}^n]^{\mathcal{S}_n}$ is just the ring of all polynomials in the elementary symmetric functions. So if we use the isomorphism

$$A := \begin{pmatrix} a_1 & 0 & \dots & 0 \\ 0 & a_2 & \dots & \\ \vdots & \vdots & \ddots & \\ 0 & 0 & \dots & a_n \end{pmatrix} \mapsto (a_1, a_2, \dots, a_n) =: a$$

to replace $\mathbb{R}^n$ by Σ, we find that each symmetric polynomial $\tilde{p}$ on Σ is of the form

$$\tilde{p}(A) = \bar{p}(\sigma_1(A), \sigma_2(A), \dots, \sigma_n(A)).$$

It can be expressed as a polynomial $\bar{p}$ in the elementary symmetric functions

$$\begin{aligned} \sigma_1 &= -x^1 - x^2 - \dots - x^n, \\ \sigma_2 &= x^1x^2 + x^1x^3 + \dots, \\ &\dots \\ \sigma_k &= (-1)^k \sum_{j_1 < \dots < j_k} x^{j_1} \dots x^{j_k}, \\ &\dots \\ \sigma_n &= (-1)^n X^1 \dots x^n. \end{aligned}$$

Let us consider the characteristic polynomial of the diagonal matrix X with eigenvalues $x^1, \dots, x^n$:

$$\begin{aligned} \prod_{i=1}^n (t - x^i) &= t^n + \sigma_1.t^{n-1} + \dots + \sigma_{n-1}.t + \sigma_n \\ &= \det(t.Id - X) \\ &= \sum_{i=0}^n (-1)^{n-i} t^i c_{n-i}(X), \qquad \text{where} \end{aligned}$$

$$c_k(Y) = \mathrm{Trace}(\bigwedge^k Y : \bigwedge^k \mathbb{R}^n \to \bigwedge^k \mathbb{R}^n)$$

is the k-th characteristic coefficient of a matrix Y; see (14.9). So the σ_i extend to $O(n)$-invariant polynomials c_i on $S(n)$. We can now extend $\tilde{p}$ to a polynomial on $S(n)$ by

$$\tilde{p}(H) := \bar{p}(c_1(H), c_2(H), \dots, c_n(H)) \qquad \text{for all } H \in S(n).$$

Therefore, $\tilde{p}$ is an $O(n)$-invariant polynomial on $S(n)$ and is unique as such due to (1).

(6) Again we have that $f \in C^\infty(S(n))^{O(n)}$ implies $\tilde{f} := f|_\Sigma \in C^\infty(\Sigma)^{\mathcal{S}_n}$. Finding an inverse map $\tilde{f} \mapsto f$ as above is possible due to the following theorem by Gerald Schwarz; see (7.13) below:

Let G be a compact Lie group with a finite-dimensional representation $G \to GL(V)$ and let $\rho_1, \rho_2, \dots, \rho_k$ be generators for the algebra $\mathbb{R}[V]^G$ of G-invariant polynomials on V. It is finitely generated as an algebra due to Hilbert; see (7.2). Then, for any smooth function $h \in C^\infty(V)^G$, there is a function $\bar{h} \in C^\infty(\mathbb{R}^k)$ such that $h(v) = \bar{h}(\rho_1(v), \dots, \rho_k(v))$.

Now we can prove the assertion as in (5) above. Again we take the symmetric polynomials $\sigma_1, \dots, \sigma_n$ as generators of $\mathbb{R}[\Sigma]^{\mathcal{S}_n}$. By Schwarz's theorem (7.13), any $\tilde{f} \in C^\infty(\Sigma)^{\mathcal{S}_n}$ can be written as a smooth function in $\sigma_1, \dots, \sigma_n$. So we have an $\bar{f} \in C^\infty(\mathbb{R}^n)$ such that

$$\tilde{f}(A) = \bar{f}(\sigma_1(A), \dots \sigma_n(A)) \qquad \text{for all } A \in \Sigma.$$

If we extend the σ_i onto $S(n)$ as in (4), we can define

$$f(H) := \bar{f}(c_1(H), c_2(H), \dots, c_n(H)) \qquad \text{for } H \in S(n).$$

Then f is again a smooth function and it is the unique $O(n)$-invariant extension of $\tilde{f}$.

(7) Consider $\sigma = (\sigma_1, \dots, \sigma_n) : \Sigma \to \mathbb{R}^n$ and put $J(x) := \det(d\sigma(x))$. For each $\alpha \in \mathcal{S}_n$ we have

$$\begin{aligned} J.dx^1 \wedge \dots \wedge dx^n &= d\sigma_1 \wedge \dots \wedge d\sigma_n \\ &= \alpha^*(d\sigma_1 \wedge \dots \wedge d\sigma_n) \\ &= (J \circ \alpha).\alpha^*(dx^1 \wedge \dots \wedge dx^n) \\ &= (J \circ \alpha).\det(\alpha).dx^1 \wedge \dots \wedge dx^n, \end{aligned}$$

$$J \circ \alpha = \det(\alpha^{-1}).J. \tag{8}$$

From this we see firstly that J is a homogeneous polynomial of degree

$$0 + 1 + \dots + (n-1) = \tfrac{n(n-1)}{2} = \binom{n}{2}.$$

The mapping σ is a local diffeomorphism on the open set $U = \Sigma \setminus J^{-1}(0)$; thus $d\sigma_1, \dots, d\sigma_n$ is a coframe on U, i.e., a basis of the cotangent bundle everywhere on U. Let (ij) be the transpositions in $\mathcal{S}_n$, and let

$$H_{(ij)} := \{x \in \Sigma : x^i - x^j = 0\}$$

be the reflection hyperplanes of the (ij). If $x \in H_{(ij)}$, then by (8) we have $J(x) = J((ij)x) = -J(x)$, so $J(X) = 0$. Thus $J|H_{(ij)} = 0$, so the polynomial J is divisible by the linear form $x^i - x^j$, for each $i < j$. By comparing degrees, we see that

(9) $$J(x) = c.\prod_{i<j}(x^i - x^j), \qquad \text{where } 0 \neq c \in \mathbb{R}.$$

By the same argument we see that:

(10) If $g \in C^\infty(\Sigma)$ satisfies $g \circ \alpha = \det(\alpha^{-1}).g$ for all $\alpha \in \mathcal{S}_n$, then $g = J.h$ for $h \in C^\infty(\Sigma)^{\mathcal{S}_n}$.

(11) **Claim.** Let $\omega \in \Omega^p(\Sigma)^{\mathcal{S}_n}$. Then we have

$$\omega = \sum_{j_1<j_2<\dots<j_p} \omega_{j_1,\dots,j_p}\, d\sigma_{j_1} \wedge \dots \wedge d\sigma_{j_p}$$

on Σ, for $\omega_{j_1,\dots,j_p} \in C^\infty(\Sigma)^{\mathcal{S}_n}$.

To prove claim (11) recall that $d\sigma_1, \dots, d\sigma_n$ is an $\mathcal{S}_n$-invariant coframe on the $\mathcal{S}_n$-invariant open set U. Thus

$$\omega|U = \sum_{j_1<j_2<\dots<j_p} \underbrace{g_{j_1,\dots,j_p}}_{\in C^\infty(U)}\, d\sigma_{j_1} \wedge \dots \wedge d\sigma_{j_p}$$

(12) $$= \sum_{j_1<j_2<\dots<j_p} \underbrace{\left(\tfrac{1}{n!}\sum_{\alpha\in\mathcal{S}_n} \alpha^* g_{j_1,\dots,j_p}\right)}_{h_{j_1,\dots,j_p}\in C^\infty(U)^{\mathcal{S}_n}} d\sigma_{j_1} \wedge \dots \wedge d\sigma_{j_p}.$$

Now choose $I = \{i_1 < \dots < i_p\} \subseteq \{1,\dots,n\}$ and let $\bar{I} = \{1,\dots,n\} \setminus I = \{i_{p+1} < \dots < i_n\}$. Then we have for a sign $\varepsilon = \pm 1$

$$\omega|U \wedge \underbrace{d\sigma_{i_{p+1}} \wedge \dots \wedge d\sigma_{i_n}}_{d\sigma^{\bar{I}}} = \varepsilon.h_I.d\sigma_1 \wedge \dots \wedge d\sigma_n$$

$$= \varepsilon.h_I.J.dx^1 \wedge \dots \wedge dx^n.$$

On the whole of Σ we have

$$\omega \wedge d\sigma^{\bar{I}} = \varepsilon.k_I.dx^1 \wedge \dots \wedge dx^n$$

for suitable $k_I \in C^\infty(\Sigma)$. By comparing the two expressions on U, we see from (8) that $k_I \circ \alpha = \det(\alpha^{-1}).k_I$ since U is dense in Σ. So from (10) we may conclude that $k_I = J.\omega_I$ for $\omega_I \in C^\infty(\Sigma)^{\mathcal{S}_n}$, but then $h_I = \omega_I|U$ and $\omega = \sum_I \omega_I\, d\sigma^I$ as asserted in claim (11).

Now we may finish the proof. By the theorem of G. Schwarz (7.13) there exist $f_I \in C^\infty(\mathbb{R}^n)$ with $\omega_I = f_I(\sigma_1, \dots, \sigma_n)$. Recall now the characteristic coefficients $c_i \in \mathbb{R}[S(n)]$ from the proof of (5) which satisfy $c_i|\Sigma = \sigma_i$. If we put now

$$\tilde{\omega} := \sum_{i_1 < \dots < i_p} f_{i_1, \dots, i_p}(c_1, \dots, c_n)\, dc_{i_1} \wedge \dots \wedge dc_{i_p} \in \Omega^p_{\text{hor}}(S(n))^{O(n)},$$

then the pullback of $\tilde{\omega}$ to Σ equals ω. □

7.2. Theorem of Hilbert and Nagata. *Let G be a Lie group with a finite-dimensional representation $G \to GL(V)$ and let one of the following conditions be fulfilled:*

(1) *G is semisimple and has only a finite number of connected components.*

(2) *V and $\langle G.f\rangle_\mathbb{R}$ are completely reducible for all $f \in \mathbb{R}[V]$; see (7.8).*

Then $\mathbb{R}[V]^G$ is finitely generated as an algebra, or equivalently, there is a finite set of polynomials $\rho_1, \dots, \rho_k \in \mathbb{R}[V]^G$, such that the map $\rho := (\rho_1, \dots, \rho_k) : V \to R^k$ induces a surjection

$$\mathbb{R}[\mathbb{R}^k] \xrightarrow{\rho^*} \mathbb{R}[V]^G.$$

Remark. The first condition is stronger than the second since for a connected, semisimple Lie group, or for one with a finite number of connected components, every finite-dimensional representation is completely reducible. To prove the theorem, we will only need to know complete reducibility for the finite-dimensional representations V and $\langle G.f\rangle_\mathbb{R}$ though as in (2).

7.3. Lemma. *Let $A = \bigoplus_{i\geq 0} A_i$ be a graded $\mathbb{R}$-algebra with $A_0 = \mathbb{R}$. If $A_+ := \bigoplus_{i>0} A_i$ is finitely generated as an A-module, then A is finitely generated as an $\mathbb{R}$-algebra.*

Proof. Let $a_1, \dots, a_n \in A_+$ be generators of A_+ as an A-module. Since they can be chosen homogeneous, we assume $a_i \in A_{d_i}$ for positive integers d_i.

Claim. The a_i generate A as an $\mathbb{R}$-algebra: $A = \mathbb{R}[a_1, \dots, a_n]$.

We will show by induction that $A_i \subseteq \mathbb{R}[a_1, \dots, a_n]$ for all i. For $i = 0$ the assertion is clearly true, since $A_0 = \mathbb{R}$. Now suppose $A_i \subseteq \mathbb{R}[a_1, \dots, a_n]$ for all $i < N$. Then we have to show that

$$A_N \subseteq \mathbb{R}[a_1, \dots, a_n]$$

as well. Take any $a \in A_N$. Then a can be expressed as

$$a = \sum_{i,j} c^i_j a_i, \qquad c^i_j \in A_j.$$

Since a is homogeneous of degree N, we can discard all $c^i_j a_i$ with total degree $j + d_i \neq N$ from the right hand side of the equation. If we set $c^i_{N-d_i} =: c^i$, we get

$$a = \sum_i c^i a_i,$$

In this equation all terms are homogeneous of degree N. In particular, any occurring a_i have degree $d_i \leq N$. Consider first the a_i of degree $d_i = N$. The corresponding c^i then automatically lie in $A_0 = \mathbb{R}$, so $c^i a_i \in \mathbb{R}[a_1, \dots, a_n]$. To handle the remaining a_i, we use the induction hypothesis. Since a_i and c^i are of degree $< N$, they are both contained in $\mathbb{R}[a_1, \dots, a_n]$. Therefore, $c^i a_i$ lies in $\mathbb{R}[a_1, \dots, a_n]$ as well. So $a = \sum c^i a_i \in \mathbb{R}[a_1, \dots, a_n]$, which completes the proof. □

Remark. If we apply this lemma for $A = \mathbb{R}[V]^G$, we see that to prove (7.2) we only have to show that $\mathbb{R}[V]^G_+$, the algebra of all invariant polynomials of strictly positive degree, is finitely generated as a module over $[V]^G$. The first step in this direction will be to prove the weaker statement:

$$B := \langle \mathbb{R}[V]^G_+ \rangle_{\mathbb{R}[V]} = \mathbb{R}[V].\mathbb{R}[V]^G_+ \text{ is finitely generated as an ideal.}$$

This is a consequence of a well known theorem by Hilbert:

7.4. Theorem (Hilbert's ideal basis theorem). *If A is a commutative Noetherian ring, then the polynomial ring $A[x]$ is Noetherian as well.*

A ring is Noetherian if every strictly ascending sequence of left ideals $I_0 \subset I_1 \subset I_2 \subset \dots$ is finite, or equivalently, if every left ideal is finitely generated. If we choose $A = \mathbb{R}$, the theorem states that $\mathbb{R}[x]$ is again Noetherian. Now consider $A = \mathbb{R}[x]$; then $\mathbb{R}[x][y] = \mathbb{R}[x, y]$ is Noetherian, and so on. By induction, we see that $\mathbb{R}[V]$ is Noetherian. Therefore, any left ideal in $\mathbb{R}[V]$, in particular B, is finitely generated.

Proof of (7.4). Take any ideal $I \subseteq A[x]$ and denote by A_i the set of leading coefficients of all i-th degree polynomials in I. Then A_i is an ideal in A, and we have a sequence of ideals

$$A_0 \subseteq A_1 \subseteq A_2 \subseteq \cdots \subseteq A.$$

Since A is Noetherian, this sequence stabilizes after a certain index r, i.e., $A_r = A_{r+1} = \cdots$. Let $\{a_{i1}, \dots, a_{in_i}\}$ be a set of generators for A_i ($i = 1, \dots, r$), and let p_{ij} be a polynomial of degree i in I with leading coefficient a_{ij}.

Claim. These polynomials generate I.

Let $\mathcal{P} = \langle p_{ij} \rangle_{A[x]} \subseteq A[x]$ be the ideal generated by the p_{ij}. Then $\mathcal{P}$ clearly contains all constants in I ($A_0 \subseteq I$). Let us show by induction that it contains all polynomials in I of degree $d > 0$ as well. Take any polynomial

p of degree d. We distinguish between two cases.
(1) Suppose $d \leq r$. Then we can find coefficients $c_1, \dots, c_{n_d} \in A$ such that

$$\tilde{p} := p - c_1 p_{d1} - c_2 p_{d2} - \dots - c_{n_d} p_{dn_d}$$

has degree $< d$.
(2) Suppose $d > r$. Then the leading coefficients of $x^{d-r} p_{r1}, \dots, x^{d-r} p_{rn_r} \in I$ generate A_d. So we can find coefficients $c_1, \dots, c_{n_r} \in A$ such that

$$\tilde{p} := p - c_1 x^{d-r} p_{r1} - c_2 x^{d-r} p_{r2} - \dots - c_{n_r} x^{d-r} p_{rn_r}$$

has degree $< d$.

In both cases we have $p \in \tilde{p} + \mathcal{P}$ and $\deg \tilde{p} < d$. Therefore by the induction hypothesis $\tilde{p}$, and with it p, lies in $\mathcal{P}$. □

To prove theorem (7.2), it remains only to show the following:

7.5. Lemma. *Let G be a Lie group acting on V such that the same conditions as in Hilbert and Nagata's theorem are satisfied. Then for $f_1, \dots, f_k \in \mathbb{R}[V]^G$:*

$$\mathbb{R}[V]^G \cap \langle f_1, \dots, f_k \rangle_{\mathbb{R}[V]} = \langle f_1, \dots, f_k \rangle_{\mathbb{R}[V]^G}$$

where the brackets denote the generated ideal (module) in the specified space.

7.6. Remark. In our case, if we take $f_i = \rho_i \in \mathbb{R}[V]^G_+$ to be the finite system of generators of B as an ideal in $\mathbb{R}[V]$, we get:

$$\mathbb{R}[V]^G_+ = \mathbb{R}[V]^G \cap B = \langle \rho_1, \dots, \rho_k \rangle_{\mathbb{R}[V]^G}.$$

That is, the ρ_i generate $\mathbb{R}[V]^G_+$ as an $\mathbb{R}[V]^G$-module. With lemma (7.3), Hilbert and Nagata's theorem (7.2) follows immediately.

7.7. Remark. The inclusion ($\supseteq$) in lemma (7.5) is trivial. If G is compact, then the opposite inclusion

$$\mathbb{R}[V]^G \cap \langle f_1, \dots, f_k \rangle_{\mathbb{R}[V]} \subseteq \langle f_1, \dots, f_k \rangle_{\mathbb{R}[V]^G}$$

is easily seen as well. Take any $f \in \mathbb{R}[V]^G \cap \langle f_1, \dots, f_k \rangle_{\mathbb{R}[V]}$. Then f can be written as

$$f = \sum p_i f_i, \qquad p_i \in \mathbb{R}[V].$$

Since G is compact, we can integrate both sides over G using the Haar measure dg; see (14.4):

$$f(x) = \int_G f(g.x)dg = \sum_i \int_G p_i(g.x) f_i(g.x) dg = \sum_i \underbrace{\Big(\int_G p_i(g.x)dg\Big)}_{=:p_i^*(x)} f_i(x).$$

The p_i^* are G-invariant polynomials; therefore f is in $\langle f_1, \dots, f_k \rangle_{\mathbb{R}[V]^G}$.

To show the lemma in its general form, we will need to find a replacement for the integral. This is done in the following central lemma.

7.8. Lemma ([169]). *Under the same conditions as theorem (7.2), for any $f \in \mathbb{R}[V]$ there exists an $f^* \in \mathbb{R}[V]^G \cap \langle G.f\rangle_{\mathbb{R}}$ such that*

$$f - f^* \in \langle Gf - Gf\rangle_{\mathbb{R}}.$$

Proof. Take $f \in \mathbb{R}[V]$. Clearly, f is contained in $M_f := \langle G.f\rangle_{\mathbb{R}}$, where f^* is supposed to lie as well. The vector space M_f is a finite-dimensional subspace of $\mathbb{R}[V]$ since it is contained in

$$M_f \subseteq \bigoplus_{i \le \deg f} \mathbb{R}[V]_i.$$

In addition we have that

$$\langle G.f - G.f\rangle_{\mathbb{R}} =: N_f \subseteq M_f$$

is an invariant subspace. So we can restrict all our considerations to the finite-dimensional G-space M_f which is completely reducible by our assumption.

If $f \in N_f$, then we can set $f^* = 0$ and we are done. Suppose $f \notin N_f$. Then the f^* we are looking for must also lie in $M_f \setminus N_f$. From the identity

$$g.f = f + \underbrace{(g.f - f)}_{\in N_f} \qquad \text{for all } g \in G$$

it follows that

$$M_f = N_f \oplus \mathbb{R}.f.$$

In particular, N_f has codimension 1 in M_f.

Since we require f^* to be G-invariant, $\mathbb{R}.f^*$ will be a 1-dimensional G-invariant subspace of M_f which is not contained in N_f. As we just saw, N_f has codimension 1 in M_f; therefore $\mathbb{R}.f^*$ will be a complementary subspace to N_f.

If we now write M_f as the direct sum

$$M_f = N_f \oplus P,$$

where P is the invariant subspace complementary to N_f guaranteed by the complete irreducibility of M_f, then P is a good place to look for f^*.

Now $P \cong M_f/N_f$ as a G-module, so let us take a look at the action of G on M_f/N_f. Every element of M_f/N_f has a representative in $\mathbb{R}.f$, so we need only consider elements of the form $\lambda f + N_f$ ($\lambda \in \mathbb{R}$). For arbitrary $g \in G$ we have:

$$g.(\lambda f + N_f) = \lambda g.f + N_f = \lambda f + \underbrace{(\lambda g.f - \lambda f)}_{\in N_f} + N_f = \lambda f + N_f.$$

So G acts trivially on M_f/N_f and therefore on P. This is good news, since now every $f' \in P$ is G-invariant and we only have to project f onto P (along N_f) to get the desired $f^* \in \mathbb{R}[V]^G \cap M_f$. □

Proof of lemma (7.5). Recall that for arbitrary $f_1, \dots, f_k$ we have to show

$$\mathbb{R}[V]^G \cap \langle f_1, \dots, f_k \rangle_{\mathbb{R}[V]} \subseteq \langle f_1, \dots, f_k \rangle_{\mathbb{R}[V]^G}.$$

We will do so by induction on k. For $k = 0$ the assertion is trivial. Suppose the lemma is valid for $k = r - 1$. Consider $f_1, \dots, f_r \in \mathbb{R}[V]^G$ and $f \in \mathbb{R}[V]^G \cap \langle f_1, \dots, f_r \rangle_{\mathbb{R}[V]}$. Then

$$f = \sum_{i=1}^{r} p_i f_i, \qquad p_i \in \mathbb{R}[V].$$

By Nagata's lemma (7.8), we can approximate p_i up to $\langle G.p_i - G.p_i \rangle_{\mathbb{R}}$ by a $p_i^* \in \mathbb{R}[V]^G$. So for some finite subset $F \subset G \times G$ we have

$$p_i = p_i^* + \sum_{s,t \in F} \lambda_{s,t}^i (s.p_i - t.p_i), \qquad \lambda_{s,t}^i \in \mathbb{R}.$$

Therefore we have

$$f - \sum_{i=1}^{r} p_i^* f_i = \sum_{i=1}^{r} \sum_{s,t \in F} \lambda_{s,t}^i (s.p_i - t.p_i) f_i \in \mathbb{R}[V]^G.$$

It remains to show that the right hand side of this lies in $\langle f_1, \dots, f_r \rangle_{\mathbb{R}[V]^G}$. Notice that by the G-invariance of f:

$$\sum_{i=1}^{r} (sp_i - tp_i) f_i = s. \sum_{i=1}^{r} p_i f_i - t. \sum_{i=1}^{r} p_i f_i = s.f - t.f = 0$$

for all $s, t \in G$. Therefore

$$\sum_{i=1}^{r-1} (s.p_i - t.p_i) f_i = (t.p_r - s.p_r) f_r.$$

Now we can use the induction hypothesis on

$$\sum_{i=1}^{r} \sum_{s,t \in F} \lambda_{s,t}^i (s.p_i - t.p_i) f_i$$
$$= \sum_{i=1}^{r-1} \sum_{s,t \in F} (\lambda_{s,t}^i - \lambda_{s,t}^r)(s.p_i - t.p_i) f_i \in \mathbb{R}[V]^G \cap \langle f_1, \dots, f_{r-1} \rangle_{\mathbb{R}[V]}$$

to complete the proof. □

7.9. Remark. With lemma (7.5), Hilbert and Nagata's theorem is proved as well. So in the setting of (7.2) we now have an exact sequence

$$0 \longrightarrow \rho^* \longrightarrow \mathbb{R}[\mathbb{R}^k] \xrightarrow{\rho^*} \mathbb{R}[V]^G \longrightarrow 0$$

where $\ker \rho^* = \{R \in \mathbb{R}[\mathbb{R}^k] : R(\rho_1, \dots, \rho_k) = 0\}$ is just the finitely generated ideal consisting of all relations between the ρ_i.
Since the action of G respects the grading of $\mathbb{R}[V] = \bigoplus_k \mathbb{R}[V]_k$, it induces an action on the space of all power series, $\mathbb{R}[[V]] = \Pi_{k=1}^{\infty} \mathbb{R}[V]_k$, and we have the following:

7.10. Theorem. *Let G be a Lie group with a finite-dimensional representation $G \to GL(V)$ satisfying the conditions of Hilbert and Nagata's theorem (7.2). Let $\rho_1, \dots, \rho_k \in \mathbb{R}[V]^G$ be generators for the algebra $\mathbb{R}[V]^G$ which exist by (7.2). Then the map*

$$\rho := (\rho_1, \dots, \rho_k) : V \to R^k$$

induces a surjection

$$\mathbb{R}[[\mathbb{R}^k]] \xrightarrow{\rho^*} \mathbb{R}[[V]]^G$$

between the spaces of formal power series.

Proof. Write the formal power series $f \in \mathbb{R}[[V]]^G$ as the sum of its homogeneous parts:

$$f(x) = f_0 + f_1(x) + f_2(x) + \dots.$$

Then to each $f_i(x) \in \mathbb{R}[V]_i^G$ there is a $g_i(y) \in \mathbb{R}[\mathbb{R}^k]$ such that $f_i(x) = g_i(\rho_1(x), \dots, \rho_k(x))$. Before we can set

$$g(y) = g_0 + g_1(y) + g_2(y) + \dots$$

to finish the proof, we have to check whether this expression is finite in each degree. This is the case, since the lowest degree λ_i that can appear in g_i goes to infinity with i:
Write explicitly $g_i = \sum_{|\alpha| \le i} A_{i,\alpha} y^\alpha$ and take an $A_{i,\alpha} \ne 0$. Then $\deg f_i = i = \alpha_1 d_1 + \dots + \alpha_k d_k$ where $d_i = \deg \rho_i$ and

$$\lambda_i = \inf\{|\alpha| : i = \sum \alpha_j d_j\} \to \infty \qquad (i \to \infty). \quad \square$$

7.11. The orbit space of a representation. If G is a Lie group acting smoothly on a manifold M, then the orbit space M/G is not generally again a smooth manifold. Yet, it still has a functional structure induced by the smooth structure on M simply by calling a function $f : M/G \to \mathbb{R}$ smooth if and only if $f \circ \pi : M \to \mathbb{R}$ is smooth (where $\pi : M \to M/G$ is the quotient map).

In the following, let G be a compact Lie group, $\ell : G \to L(V)$ a representation on $V = \mathbb{R}^n$. Let $\rho_1, \dots, \rho_k \in \mathbb{R}[V]^G$ denote a finite system of generators for the algebra $\mathbb{R}[V]^G$, and let ρ denote the polynomial mapping:

$$\rho := (\rho_1, \dots, \rho_k) : V \to \mathbb{R}^k.$$

Lemma. *Let G be a compact Lie group. Then we have*

(1) *ρ is proper* (6.20) *so ρ^{-1}(compact) is compact.*

(2) *ρ separates the orbits of G.*

(3) *There is a map $\bar{\rho} : V/G \to R^k$ such that the diagram*

$$\begin{array}{ccc} V & \xrightarrow{\ \rho\ } & \mathbb{R}^k \\ {\scriptstyle\pi}\downarrow & \nearrow{\scriptstyle\bar{\rho}} & \\ V/G & & \end{array}$$

commutes and $\bar{\rho}$ is a homeomorphism onto its image.

Proof. (1) Let $r(x) = |x|^2 = \langle x, x\rangle$, for an invariant inner product on V. Then $r \in \mathbb{R}[V]^G$. By (7.2) there is a polynomial $p \in \mathbb{R}[\mathbb{R}^k]$ such that $r(x) = p(\rho(x))$. If $(x_n) \in V$ is an unbounded sequence, then $r(x_n)$ is unbounded. Therefore $p(\rho(x_n))$ is unbounded, and, since p is a polynomial, $\rho(x_n)$ is also unbounded. For compact $K \subset \mathbb{R}^k$ then $\rho^{-1}(K)$ is closed and bounded, thus compact. So ρ is proper.

(2) Choose two different orbits $G.x \neq G.y$ $(x, y \in V)$ and consider the map:

$$f : G.x \cup G.y \to \mathbb{R}, \qquad f(v) := \begin{cases} 0 & \text{for } v \in G.x, \\ 1 & \text{for } v \in G.y. \end{cases}$$

Both orbits are compact and f is continuous. Therefore, by the Weierstrass approximation theorem, there is a polynomial $p \in \mathbb{R}[V]$ such that

$$\|p - f\|_{G.x\cup G.y} = \sup\{|p(z) - f(z)| : z \in G.x \cup G.y\} < \tfrac{1}{10}.$$

Now we can average p over the group using the Haar measure dg on G from (14.4) to get a G-invariant function

$$q(v) := \int_G p(g.v)dg.$$

Note that since the action of G is linear, q is again a polynomial. For $v \in G.x \cup G.y$, we have

$$\Big|\underbrace{\int_G f(g.v)dg}_{=f(v)} - \int_G p(g.v)dg.\Big| \le \int_G |f(g.v) - p(g.v)|\, dg \le \frac{1}{10}\underbrace{\int_G dg}_{=1}.$$

Recalling how f was defined, we get

$$\begin{aligned} |q(v)| &\le \tfrac{1}{10} \qquad \text{for } v \in G.x, \\ |1 - q(v)| &\le \tfrac{1}{10} \qquad \text{for } v \in G.y. \end{aligned}$$

Therefore $q(G.x) \neq q(G.y)$, and since q can be expressed in the Hilbert generators, we can conclude that $\rho(G.x) \neq \rho(G.y)$.
(3) Clearly, $\bar\rho$ is well defined and continuous for the quotient topology on V/G. By (2) the mapping $\bar\rho$ is injective, and by (1) it is proper, thus closed by (6.22). So it is a homeomorphism onto its image. □

7.12. Remark. (1) If $f : V \to \mathbb{R}$ is in $C^0(V)^G$, then f factors over π to a continuous map $\tilde f : V/G \to \mathbb{R}$. By (7.11.3) there is a continuous map $\bar f : \rho(V) \to R$ given by $\bar f = \tilde f \circ \bar\rho^{-1}$. It has the property $f = \bar f \circ \rho$. Since $\rho(V)$ is closed, $\bar f$ extends to a continuous function $\bar f \in C^0(\mathbb{R}^k)$ (Tietze-Urysohn). So for continuous functions we have the assertion that

$$\rho^* : C^0(\mathbb{R}^k) \to C^0(V)^G \qquad \text{is surjective.}$$

(2) The subset $\rho(V) \subset \mathbb{R}^k$ is a real semialgebraic variety, i.e., it is described by a finite number of polynomial equations and inequalities. In the complex case, the image of an algebraic variety under a polynomial map is again an algebraic variety, meaning it is described by polynomial equations only. In the real case this is already disproved by the simple polynomial map: $x \mapsto x^2$.

7.13. Result. C^∞-Invariant Theorem. *Let G be a compact Lie group, $\ell : G \to O(V)$ a finite-dimensional representation, and $\rho_1, \rho_2, \dots, \rho_k$ generators for the algebra $\mathbb{R}[V]^G$ of G-invariant polynomials on V (this space is finitely generated as an algebra by (7.2)). If*

$$\rho := (\rho_1, \dots, \rho_k) : V \to \mathbb{R}^k,$$

then

$$\rho^* : C^\infty(\mathbb{R}^k) \to C^\infty(V)^G$$

is surjective with a continuous linear section.

This theorem is due to Schwarz [**203**], who showed surjectivity. The article [**137**] extended the result to split surjective (existence of a continuous section). Later, [**18**] and [**19**] generalized this to 'semiproper real analytic mappings' ρ. For the action of $G = \{\pm 1\}$ on $\mathbb{R}^1$ the result is due to [**227**]. If $G = \mathcal{S}_n$ acting on $\mathbb{R}^n$ by the standard representation, it was shown by [**75**]. It is easy to see that $\rho^* C^\infty(\mathbb{R}^k)$ is dense in $C^\infty(V)^G$ in the compact C^∞-topology. Therefore, Schwarz's theorem is equivalent to: $\rho^* C^\infty(\mathbb{R}^k)$ is closed in $C^\infty(V)^G$. Further results in this direction were obtained by Luna who, among other things, generalized the theorem of Schwarz to reductive

Lie groups losing only the property of the Hilbert generators separating the orbits.

7.14. Result (Luna's Theorem [125]). *Consider a representation of a reductive Lie group G on $\mathbb{K}^m$ (where $\mathbb{K} = \mathbb{C}, \mathbb{R}$), and let $\sigma = (\sigma_1, \dots, \sigma_n) : \mathbb{K}^m \to \mathbb{K}^n$, where $\sigma_1, \dots, \sigma_n$ generate the algebra $\mathbb{K}[\mathbb{K}^m]^G$. Then the following assertions hold:*

(1) *If $\mathbb{K} = \mathbb{C}$, then $\sigma^* : \mathcal{O}(\mathbb{C}^n) \to \mathcal{O}(\mathbb{C}^m)^G$ is surjective.*

(2) *If $\mathbb{K} = \mathbb{R}$, then $\sigma^* : C^\omega(\mathbb{R}^n) \to C^\omega(\mathbb{R}^m)^G$ is surjective.*

(3) *If $\mathbb{K} = \mathbb{R}$, then also*

$$\sigma^* : C^\infty(\mathbb{R}^n) \longrightarrow \{f \in C^\infty(\mathbb{R}^m)^G : f \text{ constant on } \sigma^{-1}(y)\, \forall\, y \in \mathbb{R}^n\}$$

is surjective.

CHAPTER III. Differential Forms and de Rham Cohomology

8. Vector Bundles

8.1. Vector bundles. Let $p : E \to M$ be a smooth mapping between manifolds. By a *vector bundle chart* on (E, p, M) we mean a pair (U, ψ), where U is an open subset in M and where ψ is a fiber respecting diffeomorphism as in the following diagram:

$$\begin{array}{ccc} E \restriction U := p^{-1}(U) & \xrightarrow{\quad\psi\quad} & U \times V \\ & {\scriptstyle p}\searrow \quad \swarrow{\scriptstyle \mathrm{pr}_1} & \\ & U. & \end{array}$$

Here V is a fixed finite-dimensional vector space, called the *standard fiber* or the *typical fiber*, real for the moment.

Two vector bundle charts (U_1, ψ_1) and (U_2, ψ_2) are said to be *compatible* if $\psi_1 \circ \psi_2^{-1}$ is a fiber linear isomorphism, i.e.,

$$(\psi_1 \circ \psi_2^{-1})(x, v) = (x, \psi_{1,2}(x)v)$$

for some smooth mapping $\psi_{1,2} : U_{1,2} := U_1 \cap U_2 \to GL(V)$. The mapping $\psi_{1,2}$ is then unique and smooth, and it is called the *transition function* between the two vector bundle charts.

A *vector bundle atlas* $(U_\alpha, \psi_\alpha)_{\alpha \in A}$ for (E, p, M) is a set of pairwise compatible vector bundle charts (U_α, ψ_α) such that $(U_\alpha)_{\alpha \in A}$ is an open cover of

M. Two vector bundle atlases are called *equivalent* if their union is again a vector bundle atlas.

A *vector bundle* (E, p, M) consists of manifolds E (the *total space*), M (the *base*), and a smooth mapping $p : E \to M$ (the *projection*) together with an equivalence class of vector bundle atlases: So we must know at least one vector bundle atlas. The projection p turns out to be a surjective submersion.

8.2. Let us fix a vector bundle (E, p, M) for the moment. On each *fiber* $E_x := p^{-1}(x)$ (for $x \in M$) there is a unique structure of a real vector space, induced from any vector bundle chart (U_α, ψ_α) with $x \in U_\alpha$. So $0_x \in E_x$ is a special element and $0 : M \to E$, $0(x) = 0_x$, is a smooth mapping which is called the *zero section.*

A *section* u of (E, p, M) is a smooth mapping $u : M \to E$ with $p \circ u = Id_M$. The *support* of the section u is the closure of the set $\{x \in M : u(x) \neq 0_x\}$ in M. The space of all smooth sections of the bundle (E, p, M) will be denoted by either $\Gamma(E) = \Gamma(E, p, M) = \Gamma(E \to M)$. Clearly it is a vector space with fiberwise addition and scalar multiplication.

If $(U_\alpha, \psi_\alpha)_{\alpha \in A}$ is a vector bundle atlas for (E, p, M), then any smooth mapping $f_\alpha : U_\alpha \to V$ (where V is the standard fiber) defines a local section $x \mapsto \psi_\alpha^{-1}(x, f_\alpha(x))$ on U_α. If $(g_\alpha)_{\alpha \in A}$ is a partition of unity subordinated to (U_α), then a global section can be formed by $x \mapsto \sum_\alpha g_\alpha(x) \cdot \psi_\alpha^{-1}(x, f_\alpha(x))$. So a smooth vector bundle has 'many' smooth sections.

8.3. We will now give a formal description of the set of equivalence classes of vector bundles with fixed base M and fixed standard fiber V.

Let us first fix an open cover $(U_\alpha)_{\alpha \in A}$ of M. If (E, p, M) is a vector bundle which admits a vector bundle atlas (U_α, ψ_α) with the given open cover, then we have $\psi_\alpha \circ \psi_\beta^{-1}(x, v) = (x, \psi_{\alpha\beta}(x)v)$ for transition functions $\psi_{\alpha\beta} : U_{\alpha\beta} = U_\alpha \cap U_\beta \to GL(V)$, which are smooth. This family of transition functions satisfies

$$(1) \qquad \begin{cases} \psi_{\alpha\beta}(x) \cdot \psi_{\beta\gamma}(x) = \psi_{\alpha\gamma}(x) & \text{for each } x \in U_{\alpha\beta\gamma} = U_\alpha \cap U_\beta \cap U_\gamma, \\ \psi_{\alpha\alpha}(x) = e & \text{for all } x \in U_\alpha. \end{cases}$$

Condition (1) is called the *cocycle condition* and thus we call the family $(\psi_{\alpha\beta})$ the *cocycle of transition functions* for the vector bundle atlas (U_α, ψ_α).

Let us suppose now that the same vector bundle (E, p, M) is described by an equivalent vector bundle atlas $(U_\alpha, \varphi_\alpha)$ with the same open cover (U_α). Then the vector bundle charts (U_α, ψ_α) and $(U_\alpha, \varphi_\alpha)$ are compatible for each α, so

$$\varphi_\alpha \circ \psi_\alpha^{-1}(x, v) = (x, \tau_\alpha(x)v)$$

for some smooth mapping $\tau_\alpha : U_\alpha \to GL(V)$. But then we have

$$\begin{aligned}(x, \tau_\alpha(x)\psi_{\alpha\beta}(x)v) &= (\varphi_\alpha \circ \psi_\alpha^{-1})(x, \psi_{\alpha\beta}(x)v) \\ &= (\varphi_\alpha \circ \psi_\alpha^{-1} \circ \psi_\alpha \circ \psi_\beta^{-1})(x, v) = (\varphi_\alpha \circ \psi_\beta^{-1})(x, v) \\ &= (\varphi_\alpha \circ \varphi_\beta^{-1} \circ \varphi_\beta \circ \psi_\beta^{-1})(x, v) = (x, \varphi_{\alpha\beta}(x)\tau_\beta(x)v).\end{aligned}$$

So we get

$$\tau_\alpha(x)\psi_{\alpha\beta}(x) = \varphi_{\alpha\beta}(x)\tau_\beta(x) \quad \text{for all } x \in U_{\alpha\beta}. \tag{2}$$

We say that the two cocycles $(\psi_{\alpha\beta})$ and $(\varphi_{\alpha\beta})$ of transition functions over the cover (U_α) are *cohomologous*. The *cohomology classes* of cocycles $(\psi_{\alpha\beta})$ over the open cover (U_α) (where we identify cohomologous ones) form a set $\check{H}^1((U_\alpha), \underline{GL}(V))$, the first *Čech cohomology set* of the open cover (U_α) with values in the sheaf $C^\infty(\quad, GL(V)) =: \underline{GL}(V)$.

Now let $(W_i)_{i\in I}$ be an open cover of M that refines (U_α) with $W_i \subset U_{\varepsilon(i)}$, where $\varepsilon : I \to A$ is some refinement mapping; then for any cocycle $(\psi_{\alpha\beta})$ over (U_α) we define the cocycle $\varepsilon^*(\psi_{\alpha\beta}) =: (\varphi_{ij})$ by the prescription $\varphi_{ij} := \psi_{\varepsilon(i),\varepsilon(j)} \restriction W_{ij}$. The mapping ε^* respects the cohomology relations and induces therefore a mapping $\varepsilon^\sharp : \check{H}^1((U_\alpha), \underline{GL}(V)) \to \check{H}^1((W_i), \underline{GL}(V))$. One can show that the mapping ε^* depends on the choice of the refinement mapping ε only up to cohomology (use $\tau_i = \psi_{\varepsilon(i),\eta(i)} \restriction W_i$ if ε and η are two refinement mappings), so we may form the inductive limit $\varinjlim \check{H}^1(\mathcal{U}, \underline{GL}(V)) =: \check{H}^1(M, \underline{GL}(V))$ over all open covers of M directed by refinement.

Theorem. *There is a bijective correspondence between the (nonabelian if* $\dim(V) > 1$*) cohomology space* $\check{H}^1(M, \underline{GL}(V))$ *and the set of isomorphism classes of vector bundles over* M *with typical fiber* V*.*

Proof. Let $(\psi_{\alpha\beta})$ be a cocycle of transition functions $\psi_{\alpha\beta} : U_{\alpha\beta} \to GL(V)$ over some open cover (U_α) of M. We consider the disjoint union $\bigsqcup_{\alpha\in A}\{\alpha\} \times U_\alpha \times V$ and the following relation on it: $(\alpha, x, v) \sim (\beta, y, w)$ if and only if $x = y$ and $\psi_{\beta\alpha}(x)v = w$.

By the cocycle property (1) of $(\psi_{\alpha\beta})$ this is an equivalence relation. The space of all equivalence classes is denoted by $E = VB(\psi_{\alpha\beta})$ and it is equipped with the quotient topology. We put $p : E \to M$, $p[(\alpha, x, v)] = x$, and we define the vector bundle charts (U_α, ψ_α) by $\psi_\alpha[(\alpha, x, v)] = (x, v)$, $\psi_\alpha : p^{-1}(U_\alpha) =: E \restriction U_\alpha \to U_\alpha \times V$. Then the mapping $\psi_\alpha \circ \psi_\beta^{-1}(x, v) = \psi_\alpha[(\beta, x, v)] = \psi_\alpha[(\alpha, x, \psi_{\alpha\beta}(x)v)] = (x, \psi_{\alpha\beta}(x)v)$ is smooth; so E becomes a smooth manifold which is is Hausdorff: Let $u \neq v$ in E; if $p(u) \neq p(v)$, we can separate them in M and take the inverse image under p; if $p(u) = p(v)$, we can separate them in one chart. So (E, p, M) is a vector bundle.

Now suppose that we have two cocycles, $(\psi_{\alpha\beta})$ over (U_α) and (φ_{ij}) over (V_i). Then there is a common refinement (W_γ) for the two covers (U_α) and (V_i).

The construction described a moment ago gives isomorphic vector bundles if we restrict the cocycle to a finer open cover. So we may assume that $(\psi_{\alpha\beta})$ and $(\varphi_{\alpha\beta})$ are cocycles over the same open cover (U_α). If the two cocycles are cohomologous, so $\tau_\alpha \cdot \psi_{\alpha\beta} = \varphi_{\alpha\beta} \cdot \tau_\beta$ on $U_{\alpha\beta}$, then a fiber linear diffeomorphism $\tau : VB(\psi_{\alpha\beta}) \to VB(\varphi_{\alpha\beta})$ is given by $\varphi_\alpha \tau[(\alpha, x, v)] = (x, \tau_\alpha(x)v)$. By relation (2) this is well defined, so the vector bundles $VB(\psi_{\alpha\beta})$ and $VB(\varphi_{\alpha\beta})$ are isomorphic.

Most of the converse direction was already shown in the discussion before the theorem, and the argument can be easily refined to show also that isomorphic bundles give cohomologous cocycles. □

8.4. Remark. If $GL(V)$ is an abelian group (only if V is of real or complex dimension 1), then $\check{H}^1(M, \underline{GL}(V))$ is a usual cohomology group with coefficients in the sheaf $\underline{GL}(V)$ and it can be computed with the methods of algebraic topology. We will treat the two situations in a moment. If $GL(V)$ is not abelian, then the situation is rather mysterious: There is no clear definition for $\check{H}^2(M, \underline{GL}(V))$ for example. So $\check{H}^1(M, \underline{GL}(V))$ is more a notation than a mathematical concept.

A coarser relation on vector bundles (stable isomorphism) leads to the concept of topological K-theory, which can be handled much better, but is only a quotient of the real situation.

Example: Real line bundles. As an example we want to determine here the set of all *real line bundles* on a smooth manifold M. Let us first consider the following exact sequence of abelian Lie groups:

$$0 \to (\mathbb{R}, +) \xrightarrow{\exp} GL(1, \mathbb{R}) = (\mathbb{R} \setminus 0, \cdot) \xrightarrow{p} \mathbb{Z}_2 \to 0. \to 0$$

where $\mathbb{Z}_2 := \mathbb{Z}/2\mathbb{Z}$ is the two element group. This gives rise to an exact sequence of sheafs with values in abelian groups:

$$0 \to C^\infty(\quad, \mathbb{R}) \xrightarrow{\exp_*} C^\infty(\quad, GL(1, \mathbb{R})) \xrightarrow{p_*} \mathbb{Z}_2 \to 0$$

where in the end we find the constant sheaf. This induces the following long exact sequence in cohomology (the Bockstein sequence):

$$\begin{aligned} \dots \to 0 = \check{H}^1(M, C^\infty(\quad, \mathbb{R})) &\xrightarrow{\exp_*} \check{H}^1(M, C^\infty(\quad, GL(1, \mathbb{R}))) \\ &\xrightarrow{p_*} H^1(M, \mathbb{Z}_2) \xrightarrow{\delta} \check{H}^2(M, C^\infty(\quad, \mathbb{R})) = 0 \to \dots. \end{aligned}$$

Here the sheaf $C^\infty(\quad, \mathbb{R})$ has 0 cohomology in dimensions ≥ 1 since this is a fine sheaf, i.e., it admits partitions of unity; see for example [**77**]. Thus the pullback $p_* : \check{H}^1(M, C^\infty(\quad, GL(1, \mathbb{R}))) \to H^1(M, \mathbb{Z}_2)$ is an isomorphism, and by theorem (8.3) a real line bundle E over M is uniquely determined by a certain cohomology class in $H^1(M, \mathbb{Z}_2)$, namely the *first Stiefel-Whitney class* $w_1(E)$ of this line bundle.

Example: Complex line bundles. As another example we want to determine here the set of all smooth *complex line bundles* on a smooth manifold M. Again we first consider the following exact sequence of abelian Lie groups:

$$0 \to \mathbb{Z} \xrightarrow{2\pi\sqrt{-1}} (\mathbb{C}, +) \xrightarrow{\exp} GL(1, \mathbb{C}) = (\mathbb{C} \setminus 0, \cdot) \to 0.$$

This gives rise to the following exact sequence of sheafs with values in abelian groups:

$$0 \to \mathbb{Z} \to C^\infty(\quad, \mathbb{C}) \xrightarrow{\exp_*} C^\infty(\quad, GL(1, \mathbb{C})) \to 0$$

where in the beginning we find the constant sheaf. This induces the following long exact sequence in cohomology (the Bockstein sequence):

$$\begin{aligned} \dots \to 0 = \check{H}^1(M, C^\infty(\quad, \mathbb{C})) &\xrightarrow{\exp_*} \check{H}^1(M, C^\infty(\quad, GL(1, \mathbb{C}))) \\ &\xrightarrow{\delta} H^2(M, \mathbb{Z}) \xrightarrow{2\pi\sqrt{-1}} \check{H}^2(M, C^\infty(\quad, \mathbb{C})) = 0 \to \dots. \end{aligned}$$

Again the sheaf $C^\infty(\quad, \mathbb{R})$ has 0 cohomology in dimensions ≥ 1 since it is a fine sheaf. Thus $\delta : \check{H}^1(M, C^\infty(\quad, GL(1, \mathbb{C}))) \to H^2(M, \mathbb{Z})$ is an isomorphism, and by theorem (8.3) a complex smooth line bundle E over M is uniquely determined by a certain cohomology class in $H^2(M, \mathbb{Z})$, namely the *first Chern class* $c_1(E)$ of this line bundle.

8.5. Let (U_α, ψ_α) be a vector bundle atlas for a vector bundle (E, p, M). Let $(e_j)_{j=1}^k$ be a basis of the standard fiber V. We consider the section $s_j(x) := \psi_\alpha^{-1}(x, e_j)$ for $x \in U_\alpha$. Then the $s_j : U_\alpha \to E$ are local sections of E such that $(s_j(x))_{j=1}^k$ is a basis of E_x for each $x \in U_\alpha$: We say that

$$s = (s_1, \dots, s_k)$$

is a local *frame field* for E over U_α.

Now let conversely $U \subset M$ be an open set and let $s_j : U \to E$ be local sections of E such that $s = (s_1, \dots, s_k)$ is a local frame field of E over U. Then s determines a unique vector bundle chart (U, ψ) of E such that $s_j(x) = \psi^{-1}(x, e_j)$, in the following way. We define $f : U \times \mathbb{R}^k \to E \restriction U$ by $f(x, v^1, \dots, v^k) := \sum_{j=1}^k v^j s_j(x)$. Then f is smooth, invertible, and a fiber linear isomorphism, so $(U, \psi = f^{-1})$ is the vector bundle chart promised above.

8.6. Let (E,p,M) and (F,q,N) be vector bundles. A *vector bundle homomorphism* $\varphi : E \to F$ is a fiber respecting, fiber linear smooth mapping

$$\begin{array}{ccc} E & \xrightarrow{\varphi} & F \\ {\scriptstyle p}\downarrow & & \downarrow{\scriptstyle q} \\ M & \xrightarrow{\underline{\varphi}} & N. \end{array}$$

So we require that $\varphi_x : E_x \to F_{\underline{\varphi}(x)}$ is linear. We say that φ covers $\underline{\varphi}$. If φ is invertible, it is called a *vector bundle isomorphism.*

8.7. A *vector subbundle* (F,p,M) of a vector bundle (E,p,M) is a vector bundle and a vector bundle homomorphism $\tau : F \to E$, which covers Id_M, such that $\tau_x : F_x \to E_x$ is a linear embedding for each $x \in M$.

Lemma. *Let $\varphi : (E,p,M) \to (E',q,N)$ be a vector bundle homomorphism such that $\operatorname{rank}(\varphi_x : E_x \to E'_{\underline{\varphi}(x)})$ is locally constant in $x \in M$. Then $\ker\varphi$, given by $(\ker\varphi)_x = \ker(\varphi_x)$, is a vector subbundle of (E,p,M).*

Proof. This is a local question, so we may assume that both bundles are trivial: Let $E = M \times \mathbb{R}^p$ and let $F = N \times \mathbb{R}^q$; then $\varphi(x,v) = (\underline{\varphi}(x), \overline{\varphi}(x).v)$, where $\overline{\varphi} : M \to L(\mathbb{R}^p, \mathbb{R}^q)$. The matrix $\overline{\varphi}(x)$ has rank k, so by the elimination procedure we can find $p-k$ linearly independent solutions $v_i(x)$ of the equation $\overline{\varphi}(x).v = 0$. The elimination procedure (with the same lines) gives solutions $v_i(y)$ for y near x which are smooth in y, so near x we get a local frame field $v = (v_1, \dots, v_{p-k})$ for $\ker\varphi$. By (8.5), $\ker\varphi$ is then a vector subbundle. □

8.8. Constructions with vector bundles. Let $\mathcal{F}$ be a covariant functor from the category of finite-dimensional vector spaces and linear mappings into itself, such that $\mathcal{F} : L(V,W) \to L(\mathcal{F}(V), \mathcal{F}(W))$ is smooth. Then $\mathcal{F}$ will be called a *smooth functor* for shortness sake. Well known examples of smooth functors are $\mathcal{F}(V) = \bigwedge^k(V)$ (the k-th exterior power), or $\mathcal{F}(V) = \bigotimes^k V$, and the like.

If (E,p,M) is a vector bundle, described by a vector bundle atlas with cocycle of transition functions $\varphi_{\alpha\beta} : U_{\alpha\beta} \to GL(V)$, where (U_α) is an open cover of M, then we may consider the smooth functions $\mathcal{F}(\varphi_{\alpha\beta}) : x \mapsto \mathcal{F}(\varphi_{\alpha\beta}(x))$, $U_{\alpha\beta} \to GL(\mathcal{F}(V))$. Since $\mathcal{F}$ is a covariant functor, $\mathcal{F}(\varphi_{\alpha\beta})$ satisfies again the cocycle condition (8.3.1), and cohomology of cocycles (8.3.2) is respected, so there exists a unique vector bundle $(\mathcal{F}(E) := VB(\mathcal{F}(\varphi_{\alpha\beta})), p, M)$, the value at the vector bundle (E,p,M) of the canonical extension of the functor $\mathcal{F}$ to the category of vector bundles and their homomorphisms.

If $\mathcal{F}$ is a contravariant smooth functor like the duality functor $\mathcal{F}(V) = V^*$, then we have to consider the new cocycle $\mathcal{F}(\varphi_{\alpha\beta}^{-1})$ instead of $\mathcal{F}(\varphi_{\alpha\beta})$.

If $\mathcal{F}$ is a contra-covariant smooth bifunctor like $L(V,W)$, then the construction $\mathcal{F}(VB(\psi_{\alpha\beta}), VB(\varphi_{\alpha\beta})) := VB(\mathcal{F}(\psi_{\alpha\beta}^{-1}, \varphi_{\alpha\beta}))$ describes the induced canonical vector bundle construction, and similarly in other constructions.

So for vector bundles (E,p,M) and (F,q,M) we have the following vector bundles with base M: $\bigwedge^k E$, $E \oplus F$, E^*, $\bigwedge E = \bigoplus_{k\geq 0} \bigwedge^k E$, $E \otimes F$, $L(E,F) \cong E^* \otimes F$, and so on.

8.9. Pullbacks of vector bundles. Let (E,p,M) be a vector bundle and let $f : N \to M$ be smooth. Then the *pullback vector bundle* (f^*E, f^*p, N) with the same typical fiber and a vector bundle homomorphism

$$\begin{CD} f^*E @>{p^*f}>> E \\ @V{f^*p}VV @VV{p}V \\ N @>{f}>> M \end{CD}$$

is defined as follows. Let E be described by a cocycle $(\psi_{\alpha\beta})$ of transition functions over an open cover (U_α) of M, $E = VB(\psi_{\alpha\beta})$. Then $(\psi_{\alpha\beta} \circ f)$ is a cocycle of transition functions over the open cover $(f^{-1}(U_\alpha))$ of N and the bundle is given by $f^*E := VB(\psi_{\alpha\beta} \circ f)$. As a manifold we have $f^*E = N \times_{(f,M,p)} E$ in the sense of (2.17).

The vector bundle f^*E has the following universal property: For any vector bundle (F,q,P), vector bundle homomorphism $\varphi : F \to E$ and smooth $g : P \to N$ such that $f \circ g = \underline{\varphi}$, there is a unique vector bundle homomorphism $\psi : F \to f^*E$ with $\underline{\psi} = g$ and $p^*f \circ \psi = \varphi$:

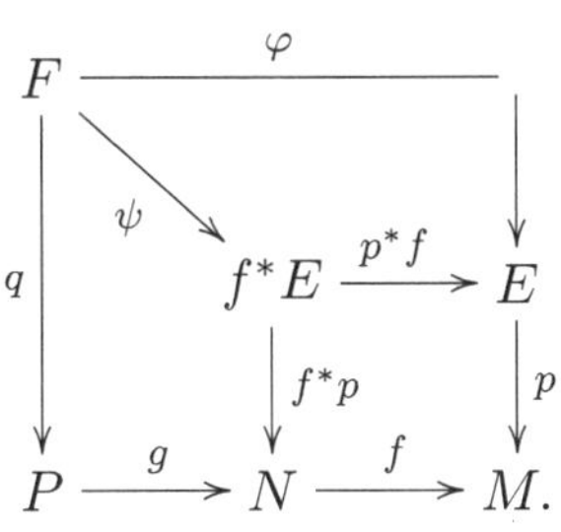

8.10. Theorem. *Any vector bundle admits a finite vector bundle atlas.*

Proof. Let (E,p,M) be the vector bundle in question, where $\dim M = m$. Let $(U_\alpha, \psi_\alpha)_{\alpha\in A}$ be a vector bundle atlas. By topological dimension theory, since M is separable, there exists a refinement of the open cover $(U_\alpha)_{\alpha\in A}$ of the form $(V_{ij})_{i=1,\dots,m+1;j\in\mathbb{N}}$, such that $V_{ij} \cap V_{ik} = \emptyset$ for $j \neq k$; see the remarks at the end of (1.1). We define the set $W_i := \bigsqcup_{j\in\mathbb{N}} V_{ij}$ (a disjoint union) and $\psi_i \restriction V_{ij} = \psi_{\alpha(i,j)}$, where $\alpha : \{1,\dots,m+1\} \times \mathbb{N} \to A$ is a refining map. Then $(W_i, \psi_i)_{i=1,\dots,m+1}$ is a finite vector bundle atlas of E. $\square$

8.11. Theorem. *For any vector bundle (E, p, M) there is a second vector bundle (F, p, M) such that $(E \oplus F, p, M)$ is a trivial vector bundle, i.e., isomorphic to $M \times \mathbb{R}^N$ for some $N \in \mathbb{N}$.*

Proof. Let $(U_i, \psi_i)_{i=1}^n$ be a finite vector bundle atlas for (E, p, M). Let (g_i) be a smooth partition of unity subordinated to the open cover (U_i). Let $\ell_i : \mathbb{R}^k \to (\mathbb{R}^k)^n = \mathbb{R}^k \times \cdots \times \mathbb{R}^k$ be the embedding on the i-th factor, where $\mathbb{R}^k$ is the typical fiber of E. Let us define $\psi : E \to M \times \mathbb{R}^{nk}$ by

$$\psi(u) = \left(p(u), \sum_{i=1}^n g_i(p(u)) \, (\ell_i \circ \operatorname{pr}_2 \circ \psi_i)(u) \right);$$

then ψ is smooth, fiber linear, and an embedding on each fiber, so E is a vector subbundle of $M \times \mathbb{R}^{nk}$ via ψ. Now we define $F_x = E_x^\perp$ in $\{x\} \times \mathbb{R}^{nk}$ with respect to the standard inner product on $\mathbb{R}^{nk}$. Then $F \to M$ is a vector bundle and $E \oplus F \cong M \times \mathbb{R}^{nk}$. □

8.12. The tangent bundle of a vector bundle. Let (E, p, M) be a vector bundle with fiber addition $+_E : E \times_M E \to E$ and fiber scalar multiplication $m_t^E : E \to E$. Then (TE, π_E, E), the tangent bundle of the manifold E, is itself a vector bundle, with fiber addition denoted by $+_{TE}$ and scalar multiplication denoted by m_t^{TE}.

If $(U_\alpha, \psi_\alpha : E \upharpoonright U_\alpha \to U_\alpha \times V)_{\alpha \in A}$ is a vector bundle atlas for E, such that (U_α, u_α) is also a manifold atlas for M, then $(E \upharpoonright U_\alpha, \psi'_\alpha)_{\alpha \in A}$ is an atlas for the manifold E, where

$$\psi'_\alpha := (u_\alpha \times Id_V) \circ \psi_\alpha : E \upharpoonright U_\alpha \to U_\alpha \times V \to u_\alpha(U_\alpha) \times V \subset \mathbb{R}^m \times V.$$

Hence the family $(T(E \upharpoonright U_\alpha), T\psi'_\alpha : T(E \upharpoonright U_\alpha) \to T(u_\alpha(U_\alpha) \times V) = u_\alpha(U_\alpha) \times V \times \mathbb{R}^m \times V)_{\alpha \in A}$ is the atlas describing the canonical vector bundle structure of (TE, π_E, E). The transition functions are in turn:

$$\begin{aligned}
(\psi_\alpha \circ \psi_\beta^{-1})(x, v) &= (x, \psi_{\alpha\beta}(x)v) \quad \text{for } x \in U_{\alpha\beta}, \\
(u_\alpha \circ u_\beta^{-1})(y) &= u_{\alpha\beta}(y) \quad \text{for } y \in u_\beta(U_{\alpha\beta}), \\
(\psi'_\alpha \circ (\psi'_\beta)^{-1})(y, v) &= (u_{\alpha\beta}(y), \psi_{\alpha\beta}(u_\beta^{-1}(y))v), \\
(T\psi'_\alpha \circ T(\psi'_\beta)^{-1})(y, v; \xi, w) &= \big(u_{\alpha\beta}(y), \psi_{\alpha\beta}(u_\beta^{-1}(y))v; d(u_{\alpha\beta})(y)\xi, \\
&\qquad (d(\psi_{\alpha\beta} \circ u_\beta^{-1})(y)\xi)v + \psi_{\alpha\beta}(u_\beta^{-1}(y))w\big).
\end{aligned}$$

So we see that for fixed (y, v) the transition functions are linear in $(\xi, w) \in \mathbb{R}^m \times V$. This describes the vector bundle structure of the tangent bundle (TE, π_E, E).

For fixed (y, ξ) the transition functions of TE are also linear in $(v, w) \in V \times V$. This gives a vector bundle structure on (TE, Tp, TM). Its fiber addition will be denoted by $T(+_E) : T(E \times_M E) = TE \times_{TM} TE \to TE$, since it is the tangent mapping of $+_E$. Likewise its scalar multiplication will be denoted by $T(m_t^E)$. One may say that the second vector bundle structure on TE, that one over TM, is the derivative of the original one on E.

The space $\{\Xi \in TE : Tp.\Xi = 0 \text{ in } TM\} = (Tp)^{-1}(0)$ is denoted by VE and is called the *vertical bundle* over E. The local form of a vertical vector Ξ is $T\psi'_\alpha.\Xi = (y, v; 0, w)$, so the transition function looks like

$$(T\psi'_\alpha \circ T(\psi'_\beta)^{-1})(y, v; 0, w) = (u_{\alpha\beta}(y), \psi_{\alpha\beta}(u_\beta^{-1}(y))v; 0, \psi_{\alpha\beta}(u_\beta^{-1}(y))w).$$

They are linear in $(v, w) \in V \times V$ for fixed y, so VE is a vector bundle over M. It coincides with $0_M^*(TE, Tp, TM)$, the pullback of the bundle $TE \to TM$ over the zero section. We have a canonical isomorphism $\mathrm{vl}_E : E \times_M E \to VE$, called the *vertical lift*, given by $\mathrm{vl}_E(u_x, v_x) := \frac{d}{dt}|_0(u_x + tv_x)$, which is fiber linear over M. The local representation of the vertical lift is $(T\psi'_\alpha \circ \mathrm{vl}_E \circ (\psi'_\alpha \times \psi'_\alpha)^{-1})((y, u), (y, v)) = (y, u; 0, v)$.

If (and only if) $\varphi : (E, p, M) \to (F, q, N)$ is a vector bundle homomorphism, then we have $\mathrm{vl}_F \circ (\varphi \times_M \varphi) = T\varphi \circ \mathrm{vl}_E : E \times_M E \to VF \subset TF$. So vl is a natural transformation between certain functors on the category of vector bundles and their homomorphisms.

The mapping $\mathrm{vpr}_E := \mathrm{pr}_2 \circ \mathrm{vl}_E^{-1} : VE \to E$ is called the *vertical projection*. Note also the relation $\mathrm{pr}_1 \circ \mathrm{vl}_E^{-1} = \pi_E \restriction VE$.

8.13. The second tangent bundle of a manifold. All of (8.12) is valid for the second tangent bundle $T^2M = TTM$ of a manifold, but here we have one more natural structure at our disposal. The *canonical flip* or *involution* $\kappa_M : T^2M \to T^2M$ is defined locally by

$$(T^2u \circ \kappa_M \circ T^2u^{-1})(x, \xi; \eta, \zeta) = (x, \eta; \xi, \zeta),$$

where (U, u) is a chart on M. Clearly this definition is invariant under changes of charts.

The flip κ_M has the following properties:

(1) $\kappa_N \circ T^2 f = T^2 f \circ \kappa_M$ for each $f \in C^\infty(M, N)$.

(2) $T(\pi_M) \circ \kappa_M = \pi_{TM}$.

(3) $\pi_{TM} \circ \kappa_M = T(\pi_M)$.

(4) $\kappa_M^{-1} = \kappa_M$.

(5) κ_M is a linear isomorphism from the bundle $(TTM, T(\pi_M), TM)$ to the bundle (TTM, π_{TM}, TM), so it interchanges the two vector bundle structures on TTM.

(6) It is the unique smooth mapping $TTM \to TTM$ which satisfies the equation $\frac{\partial}{\partial t}\frac{\partial}{\partial s}c(t, s) = \kappa_M \frac{\partial}{\partial s}\frac{\partial}{\partial t}c(t, s)$ for each $c : \mathbb{R}^2 \to M$.

All this follows from the local formula given above.

8.14. Lemma. *For vector fields $X, Y \in \mathfrak{X}(M)$ we have*

$$[X,Y] = \operatorname{vpr}_{TM} \circ (TY \circ X - \kappa_M \circ TX \circ Y),$$
$$TY \circ X - \kappa_M \circ TX \circ Y = \operatorname{vl}_{TM}(Y, [X,Y]).$$

We will give global proofs of this result later on: the first one is (8.19).

Proof. We prove this locally, so we may assume that M is open in $\mathbb{R}^m$, $X(x) = (x, \bar X(x))$, and $Y(x) = (x, \bar Y(x))$. Then by (3.4) we have

$$[X,Y](x) = (x, d\bar Y(x).\bar X(x) - d\bar X(x).\bar Y(x)),$$

and thus:

$$\begin{aligned}
(TY \circ X - \kappa_M \circ TX \circ Y)(x) &= TY.(x, \bar X(x)) - \kappa_M \circ TX.(x, \bar Y(x)) \\
&= (x, \bar Y(x); \bar X(x), d\bar Y(x).\bar X(x)) - \kappa_M(x, \bar X(x); \bar Y(x), d\bar X(x).\bar Y(x)) \\
&= (x, \bar Y(x); 0, d\bar Y(x).\bar X(x) - d\bar X(x).\bar Y(x)),
\end{aligned}$$

$$\operatorname{vpr}_{TM} \circ (TY \circ X - \kappa_M \circ TX \circ Y)(x) = (x, d\bar Y(x).\bar X(x) - d\bar X(x).\bar Y(x)). \quad \square$$

8.15. Natural vector bundles or **vector bundle functors.** Let $\mathcal{M}f_m$ denote the category of all m-dimensional smooth manifolds and local diffeomorphisms (i.e., immersions) between them. A *vector bundle functor* or *natural vector bundle* is a functor F which associates a vector bundle $(F(M), p_M, M)$ to each m-manifold M and a vector bundle homomorphism

$$\begin{array}{ccc}
F(M) & \xrightarrow{F(f)} & F(N) \\
{\scriptstyle p_M}\downarrow & & \downarrow{\scriptstyle p_N} \\
M & \xrightarrow{f} & N
\end{array}$$

to each $f : M \to N$ in $\mathcal{M}f_m$, which covers f and is fiberwise a linear isomorphism. We also require that for smooth $f : \mathbb{R} \times M \to N$ the mapping $(t,x) \mapsto F(f_t)(x)$ is smooth $\mathbb{R} \times F(M) \to F(N)$. We will say that F maps smoothly parametrized families to smoothly parametrized families. See [**108**] for more information on naturality in differential geometry.

Examples. (1) TM, the tangent bundle. This is even a functor on the category $\mathcal{M}f$ of all manifolds and all smooth mappings, not only local diffeomorphisms.

(2) T^*M, the cotangent bundle, where by (8.8) the action on morphisms is given by $(T^*f)_x := ((T_x f)^{-1})^* : T_x^*M \to T_{f(x)}^*N$. This functor is defined on $\mathcal{M}f_m$ only.

(3) $\bigwedge^k T^*M$, $\bigwedge T^*M = \bigoplus_{k\geq 0} \bigwedge^k T^*M$.

(4) $\bigotimes^k T^*M \otimes \bigotimes^\ell TM = T^*M \otimes \dots \otimes T^*M \otimes TM \otimes \dots \otimes TM$, where the action on morphisms involves Tf^{-1} in the T^*M-parts and Tf in the TM-parts.

(5) $\mathcal{F}(TM)$, where $\mathcal{F}$ is any smooth functor on the category of finite-dimensional vector spaces and linear mappings, as in (8.8).

(6) All examples discussed till now are of the following form: For a manifold of dimension m, consider the *linear frame bundle* $GL(\mathbb{R}^m, TM) = inv J^1_0(\mathbb{R}^m, M)$ (see (18.11) and (21.6)) and a representation of the structure group $\rho : GL(m,\mathbb{R}) \to GL(V)$ on some vector space V. Then the associated bundle $GL(\mathbb{R}^m, TM) \times_{GL(m,\mathbb{R})} V$ is a natural bundle. This can be generalized to frame bundles of higher order, which is described in (21.6).

8.16. Lie derivative. Let F be a vector bundle functor on $\mathcal{M}f_m$ as described in (8.15). Let M be a manifold and let $X \in \mathfrak{X}(M)$ be a vector field on M. Then the flow Fl^X_t, for fixed t, is a diffeomorphism defined on an open subset of M, which we do not specify. The mapping

$$\begin{array}{ccc} F(M) & \xrightarrow{F(\mathrm{Fl}^X_t)} & F(M) \\ {\scriptstyle p_M}\downarrow & & \downarrow{\scriptstyle p_M} \\ M & \xrightarrow{\mathrm{Fl}^X_t} & M \end{array}$$

is then a vector bundle isomorphism, defined over an open subset of M.

We consider a section $s \in \Gamma(F(M))$ of the vector bundle $(F(M), p_M, M)$ and we define for $t \in \mathbb{R}$

$$(\mathrm{Fl}^X_t)^* s := F(\mathrm{Fl}^X_{-t}) \circ s \circ \mathrm{Fl}^X_t,$$

a local section of the bundle $F(M)$. For each $x \in M$ the value $((\mathrm{Fl}^X_t)^* s)(x) \in F(M)_x$ is defined, if t is small enough (depending on x). So in the vector space $F(M)_x$ the expression $\frac{d}{dt}|_0((\mathrm{Fl}^X_t)^* s)(x)$ makes sense and therefore the section

$$\mathcal{L}_X s := \tfrac{d}{dt}|_0 (\mathrm{Fl}^X_t)^* s$$

is globally defined and is an element of $\Gamma(F(M))$. It is called the *Lie derivative* of s along X.

Lemma. *In this situation we have*

(1) $(\mathrm{Fl}^X_t)^*(\mathrm{Fl}^X_r)^* s = (\mathrm{Fl}^X_{t+r})^* s$, *wherever defined.*

(2) $\frac{d}{dt}(\mathrm{Fl}^X_t)^* s = (\mathrm{Fl}^X_t)^* \mathcal{L}_X s = \mathcal{L}_X (\mathrm{Fl}^X_t)^* s$, *so*
$[\mathcal{L}_X, (\mathrm{Fl}^X_t)^*] := \mathcal{L}_X \circ (\mathrm{Fl}^X_t)^* - (\mathrm{Fl}^X_t)^* \circ \mathcal{L}_X = 0$, *whenever defined.*

(3) $(\mathrm{Fl}^X_t)^* s = s$ *for all relevant t if and only if* $\mathcal{L}_X s = 0$.

Proof. (1) is clear. (2) is seen by the following computations:

$$\begin{aligned}
\tfrac{d}{dt}(\mathrm{Fl}_t^X)^* s &= \tfrac{d}{dr}|_0(\mathrm{Fl}_r^X)^*(\mathrm{Fl}_t^X)^* s = \mathcal{L}_X(\mathrm{Fl}_t^X)^* s,\\
\tfrac{d}{dt}((\mathrm{Fl}_t^X)^* s)(x) &= \tfrac{d}{dr}|_0((\mathrm{Fl}_t^X)^*(\mathrm{Fl}_r^X)^* s)(x)\\
&= \tfrac{d}{dr}|_0 F(\mathrm{Fl}_{-t}^X)(F(\mathrm{Fl}_{-r}^X)\circ s\circ \mathrm{Fl}_r^X)(\mathrm{Fl}_t^X(x))\\
&= F(\mathrm{Fl}_{-t}^X)\tfrac{d}{dr}|_0(F(\mathrm{Fl}_{-r}^X)\circ s\circ \mathrm{Fl}_r^X)(\mathrm{Fl}_t^X(x))\\
&= ((\mathrm{Fl}_t^X)^*\mathcal{L}_X s)(x),
\end{aligned}$$

since $F(\mathrm{Fl}_{-t}^X) : F(M)_{\mathrm{Fl}_t^X(x)} \to F(M)_x$ is linear.
(3) follows from (2). □

8.17. Let F_1, F_2 be two vector bundle functors on $\mathcal{M}f_m$. Then the (fiberwise) tensor product $(F_1 \otimes F_2)(M) := F_1(M) \otimes F_2(M)$ is again a vector bundle functor and for $s_i \in \Gamma(F_i(M))$ there is a section $s_1 \otimes s_2 \in \Gamma((F_1 \otimes F_2)(M))$, given by the pointwise tensor product.

Lemma. *In this situation, for $X \in \mathfrak{X}(M)$ we have*

$$\mathcal{L}_X(s_1 \otimes s_2) = \mathcal{L}_X s_1 \otimes s_2 + s_1 \otimes \mathcal{L}_X s_2.$$

In particular, for $f \in C^\infty(M)$ we have $\mathcal{L}_X(fs) = df(X)\, s + f\,\mathcal{L}_X s$.

Proof. Using bilinearity of the tensor product, we have

$$\begin{aligned}
\mathcal{L}_X(s_1 \otimes s_2) &= \tfrac{d}{dt}|_0(\mathrm{Fl}_t^X)^*(s_1 \otimes s_2)\\
&= \tfrac{d}{dt}|_0((\mathrm{Fl}_t^X)^* s_1 \otimes (\mathrm{Fl}_t^X)^* s_2)\\
&= \tfrac{d}{dt}|_0(\mathrm{Fl}_t^X)^* s_1 \otimes s_2 + s_1 \otimes \tfrac{d}{dt}|_0(\mathrm{Fl}_t^X)^* s_2\\
&= \mathcal{L}_X s_1 \otimes s_2 + s_1 \otimes \mathcal{L}_X s_2. \quad \square
\end{aligned}$$

8.18. Let $\varphi : F_1 \to F_2$ be a linear *natural transformation* between vector bundle functors on $\mathcal{M}f_m$. So for each $M \in \mathcal{M}f_m$ we have a vector bundle homomorphism $\varphi_M : F_1(M) \to F_2(M)$ covering the identity on M, such that $F_2(f) \circ \varphi_M = \varphi_N \circ F_1(f)$ holds for any $f : M \to N$ in $\mathcal{M}f_m$.

Example. A *tensor field* of type $\binom{p}{q}$ is a smooth section of the natural bundle $\bigotimes^q T^*M \otimes \bigotimes^p TM$. For such tensor fields, by (8.16) the Lie derivative along any vector field is defined and by (8.17) it is a derivation with respect to the tensor product. For functions and vector fields the Lie derivative was already defined in section (3). This natural bundle admits many natural transformations: Any 'contraction' like the trace $T^*M \otimes TM = L(TM, TM) \to M \times \mathbb{R}$, but applied just to one specified factor T^*M and another one of type TM, is a natural transformation. Also, any 'permutation of the same kind of factors' is a natural transformation.

Lemma. *In this situation, for $s \in \Gamma(F_1(M))$ and $X \in \mathfrak{X}(M)$ we have $\mathcal{L}_X(\varphi_M\, s) = \varphi_M(\mathcal{L}_X s)$,*

Proof. Since φ_M is fiber linear and natural, we can compute as follows:

$$\begin{aligned}\mathcal{L}_X(\varphi_M\, s)(x) &= \tfrac{d}{dt}|_0((\mathrm{Fl}^X_t)^*(\varphi_M\, s))(x) = \tfrac{d}{dt}|_0(F_2(\mathrm{Fl}^X_{-t}) \circ \varphi_M \circ s \circ \mathrm{Fl}^X_t)(x)\\ &= \varphi_M \circ \tfrac{d}{dt}|_0(F_1(\mathrm{Fl}^X_{-t}) \circ s \circ \mathrm{Fl}^X_t)(x) = (\varphi_M\, \mathcal{L}_X s)(x). \quad \square\end{aligned}$$

Thus the Lie derivative on tensor fields commutes with any kind of 'contraction' or 'permutation of the indices'.

8.19. Let F be a vector bundle functor on $\mathcal{M}f_m$ and let $X \in \mathfrak{X}(M)$ be a vector field. We consider the local vector bundle homomorphism $F(\mathrm{Fl}^X_t)$ on $F(M)$. Since $F(\mathrm{Fl}^X_t) \circ F(\mathrm{Fl}^X_s) = F(\mathrm{Fl}^X_{t+s})$ and $F(\mathrm{Fl}^X_0) = Id_{F(M)}$, we have $\frac{d}{dt}F(\mathrm{Fl}^X_t) = \frac{d}{ds}|_0 F(\mathrm{Fl}^X_s) \circ F(\mathrm{Fl}^X_t) = X^F \circ F(\mathrm{Fl}^X_t)$, so we get $F(\mathrm{Fl}^X_t) = \mathrm{Fl}^{X^F}_t$, where $X^F = \frac{d}{ds}|_0 F(\mathrm{Fl}^X_s) \in \mathfrak{X}(F(M))$ is a vector field on $F(M)$, which is called the *flow prolongation* or the *natural lift* of X to $F(M)$.

Lemma.

(1) $X^T = \kappa_M \circ TX$.

(2) $[X,Y]^F = [X^F, Y^F]$.

(3) $X^F : (F(M), p_M, M) \to (TF(M), T(p_M), TM)$ *is a vector bundle homomorphism for the $T(+)$-structure.*

(4) *For $s \in \Gamma(F(M))$ and $X \in \mathfrak{X}(M)$ we have* $\mathcal{L}_X s = \mathrm{vpr}_{F(M)} \circ (Ts \circ X - X^F \circ s)$.

(5) $\mathcal{L}_X s$ *is linear in X and s.*

Proof. (1) is an easy computation. The mapping $F(\mathrm{Fl}^X_t)$ is fiber linear and this implies (3).

(4) is seen as follows:

$$\begin{aligned}(\mathcal{L}_X s)(x) &= \tfrac{d}{dt}|_0(F(\mathrm{Fl}^X_{-t}) \circ s \circ \mathrm{Fl}^X_t)(x) \quad \text{in } F(M)_x\\ &= \mathrm{vpr}_{F(M)}(\tfrac{d}{dt}|_0(F(\mathrm{Fl}^X_{-t}) \circ s \circ \mathrm{Fl}^X_t)(x) \text{ in } VF(M))\\ &= \mathrm{vpr}_{F(M)}(-X^F \circ s \circ \mathrm{Fl}^X_0(x) + T(F(\mathrm{Fl}^X_0)) \circ Ts \circ X(x))\\ &= \mathrm{vpr}_{F(M)}(Ts \circ X - X^F \circ s)(x).\end{aligned}$$

(5) $\mathcal{L}_X s$ is homogeneous of degree 1 in X by formula (4), and it is smooth as a mapping $\mathfrak{X}(M) \to \Gamma(F(M))$, so it is linear. See [**64**] or [**113**] for the convenient calculus in infinite dimensions.

(2) Note first that F induces a smooth mapping between appropriate spaces of local diffeomorphisms which are infinite-dimensional manifolds; see [**113**].

By (3.16) we have

$$\begin{aligned} 0 &= \partial|_0(\mathrm{Fl}^Y_{-t}\circ \mathrm{Fl}^X_{-t}\circ \mathrm{Fl}^Y_t\circ \mathrm{Fl}^X_t),\\ [X,Y] &= \tfrac12 \tfrac{\partial^2}{\partial t^2}|_0(\mathrm{Fl}^Y_{-t}\circ \mathrm{Fl}^X_{-t}\circ \mathrm{Fl}^Y_t\circ \mathrm{Fl}^X_t)\\ &= \partial|_0\,\mathrm{Fl}^{[X,Y]}_t. \end{aligned}$$

Applying F to these curves of local diffeomorphisms, we get

$$\begin{aligned} 0 &= \partial|_0(\mathrm{Fl}^{Y^F}_{-t}\circ \mathrm{Fl}^{X^F}_{-t}\circ \mathrm{Fl}^{Y^F}_t\circ \mathrm{Fl}^{X^F}_t),\\ [X^F,Y^F] &= \tfrac12 \tfrac{\partial^2}{\partial t^2}|_0(\mathrm{Fl}^{Y^F}_{-t}\circ \mathrm{Fl}^{X^F}_{-t}\circ \mathrm{Fl}^{Y^F}_t\circ \mathrm{Fl}^{X^F}_t)\\ &= \tfrac12 \tfrac{\partial^2}{\partial t^2}|_0F(\mathrm{Fl}^Y_{-t}\circ \mathrm{Fl}^X_{-t}\circ \mathrm{Fl}^Y_t\circ \mathrm{Fl}^X_t)\\ &= \partial|_0F(\mathrm{Fl}^{[X,Y]}_t) = [X,Y]^F. \quad \square \end{aligned}$$

8.20. Theorem. *For any vector bundle functor F on $\mathcal{M}f_m$ and $X,Y\in \mathfrak{X}(M)$ we have*

$$[\mathcal{L}_X,\mathcal{L}_Y] := \mathcal{L}_X\circ\mathcal{L}_Y - \mathcal{L}_Y\circ\mathcal{L}_X = \mathcal{L}_{[X,Y]} : \Gamma(F(M))\to\Gamma(F(M)).$$

So $\mathcal{L}:\mathfrak{X}(M)\to \operatorname{End}\Gamma(F(M))$ is a Lie algebra homomorphism.

Proof. We need some preparation.

(1)
$$\begin{aligned} X^F\circ \mathrm{vpr}_{F(M)} &= \tfrac{d}{dt}|_0F(\mathrm{Fl}^X_t)\circ \mathrm{vpr}_{F(M)}\\ &= \tfrac{d}{dt}|_0\,\mathrm{vpr}_{F(M)}\circ TF(\mathrm{Fl}^X_t)\restriction VF(M)\\ &= T(\mathrm{vpr}_{F(M)})\circ \tfrac{d}{dt}|_0 TF(\mathrm{Fl}^X_t)\restriction VF(M)\\ &= T(\mathrm{vpr}_{F(M)})\circ\kappa_{F(M)}\circ T(\tfrac{d}{dt}|_0F(\mathrm{Fl}^X_t))\restriction VF(M)\\ &= T(\mathrm{vpr}_{F(M)})\circ\kappa_{F(M)}\circ T(X^F)\restriction VF(M). \end{aligned}$$

(2) **Sublemma.** *For any vector bundle (E,p,M) we have*

$$\mathrm{vpr}_E\circ T(\mathrm{vpr}_E)\circ\kappa_E = \mathrm{vpr}_E\circ T(\mathrm{vpr}_E) = \mathrm{vpr}_E\circ\mathrm{vpr}_{TE} : VTE\cap TVE\to E,$$

and this is linear for all three vector bundle structures on TTE.

The assertion of this sublemma is local over M, so one may assume that (E,p,M) is trivial. Then one may carefully write out the action of the three mappings on a typical element $(x,v;0,w;;0,0;0,w')\in VTE\cap TVE$ and get the result.

Now we can start the actual proof.

$$\begin{aligned} \mathcal{L}_{[X,Y]}s &= \mathrm{vpr}_{F(M)}(Ts\circ[X,Y]-[X,Y]^F\circ s) \quad \text{by (8.19)}\\ &= \mathrm{vpr}_{F(M)}\circ\big(Ts\circ\mathrm{vpr}_{TM}\circ(TY\circ X-\kappa_M\circ TX\circ Y)\\ &\qquad - \mathrm{vpr}_{TF(M)}\circ(TY^F\circ X^F-\kappa_{F(M)}\circ TX^F\circ Y^F)\circ s\big) \end{aligned}$$

$$
\begin{aligned}
&= \mathrm{vpr}_{F(M)} \circ \mathrm{vpr}_{TF(M)} \circ \big(T^2 s \circ TY \circ X - \kappa_{F(M)} \circ T^2 s \circ TX \circ Y \\
&\qquad\qquad - TY^F \circ X^F \circ s - \kappa_{F(M)} \circ TX^F \circ Y^F \circ s\big), \\
\mathcal{L}_X \mathcal{L}_Y s &= \mathcal{L}_X (\mathrm{vpr}_{F(M)} \circ (Ts \circ Y - Y^F \circ s)) \\
&= \mathrm{vpr}_{F(M)} \circ \big(T(\mathrm{vpr}_{F(M)}) \circ (T^2 s \circ TY \; T(-) \; T(Y^F) \circ Ts) \circ X \\
&\qquad\qquad - X^F \circ \mathrm{vpr}_{F(M)} \circ (Ts \circ Y - Y^F \circ s)\big) \\
&= \mathrm{vpr}_{F(M)} \circ T(\mathrm{vpr}_{F(M)}) \circ (T^2 s \circ TY \circ X \; T(-) \; T(Y^F) \circ Ts \circ X) \\
&\qquad - \mathrm{vpr}_{F(M)} \circ T(\mathrm{vpr}_{F(M)}) \circ \kappa_{F(M)} \circ T(X^F) \circ (Ts \circ Y - Y^F \circ s) \\
&= \mathrm{vpr}_{F(M)} \circ \mathrm{vpr}_{TF(M)} \circ \big(T^2 s \circ TY \circ X - T(Y^F) \circ Ts \circ X \\
&\qquad\qquad - \kappa_{F(M)} \circ T(X^F) \circ Ts \circ Y + \kappa_{F(M)} \circ T(X^F) \circ Y^F \circ s\big).
\end{aligned}
$$

Finally we have

$$
\begin{aligned}
[\mathcal{L}_X, \mathcal{L}_Y] s &= \mathcal{L}_X \mathcal{L}_Y s - \mathcal{L}_Y \mathcal{L}_X s \\
&= \mathrm{vpr}_{F(M)} \circ \mathrm{vpr}_{TF(M)} \circ \big(T^2 s \circ TY \circ X - T(Y^F) \circ Ts \circ X \\
&\qquad\qquad - \kappa_{F(M)} \circ T(X^F) \circ Ts \circ Y + \kappa_{F(M)} \circ T(X^F) \circ Y^F \circ s\big) \\
&- \mathrm{vpr}_{F(M)} \circ \mathrm{vpr}_{TF(M)} \circ \kappa_{F(M)} \circ \big(T^2 s \circ TY \circ X \; T(-) \; T(Y^F) \circ Ts \circ X \\
&\qquad T(-) \; \kappa_{F(M)} \circ T(X^F) \circ Ts \circ Y \; T(+) \; \kappa_{F(M)} \circ T(X^F) \circ Y^F \circ s\big) \\
&\qquad\qquad = \mathcal{L}_{[X,Y]} s. \qquad \square
\end{aligned}
$$

9. Differential Forms

9.1. The *cotangent bundle* of a manifold M is the vector bundle $T^*M := (TM)^*$, the (real) dual of the tangent bundle.
If (U,u) is a chart on M, then $(\frac{\partial}{\partial u^1}, \dots, \frac{\partial}{\partial u^m})$ is the associated frame field over U of TM. Since $\frac{\partial}{\partial u^i}|_x(u^j) = du^j(\frac{\partial}{\partial u^i}|_x) = \delta^j_i$, we see that $(du^1, \dots, du^m)$ is the dual frame field on T^*M over U. It is also called a *holonomic frame field.* A section of T^*M is also called a 1-*form.*

9.2. According to (8.18) a *tensor field* of type $\binom{p}{q}$ on a manifold M is a smooth section of the vector bundle

$$
\bigotimes^p TM \otimes \bigotimes^q T^*M = \overbrace{TM \otimes \dots \otimes TM}^{p \text{ times}} \otimes \overbrace{T^*M \otimes \dots \otimes T^*M}^{q \text{ times}}.
$$

The position of p (up) and q (down) can be explained as follows: If (U,u) is a chart on M, we have the holonomous frame field

$$
\left(\frac{\partial}{\partial u^{i_1}} \otimes \frac{\partial}{\partial u^{i_2}} \otimes \dots \otimes \frac{\partial}{\partial u^{i_p}} \otimes du^{j_1} \otimes \dots \otimes du^{j_q} \right)_{i \in \{1,\dots,m\}^p, j \in \{1,\dots,m\}^q}
$$

over U of this tensor bundle, and for any $\binom{p}{q}$-tensor field A we have

$$A \mid U = \sum_{i,j} A^{i_1 \dots i_p}_{j_1 \dots j_q} \frac{\partial}{\partial u^{i_1}} \otimes \dots \otimes \frac{\partial}{\partial u^{i_p}} \otimes du^{j_1} \otimes \dots \otimes du^{j_q}.$$

The coefficients have p indices up and q indices down, and they are smooth functions on U.

From a categorical point of view one should look where the indices of the frame field are, but this convention here has a long tradition.

9.3. Lemma. *Let*

$$\Phi : \mathfrak{X}(M) \times \dots \times \mathfrak{X}(M) = \mathfrak{X}(M)^k \to \Gamma(\bigotimes^l TM)$$

be a mapping which is k-linear over $C^\infty(M)$. Then Φ is given by the action of a $\binom{l}{k}$-tensor field.

Proof. For simplicity's sake we put $k = 1$, $\ell = 0$, so $\Phi : \mathfrak{X}(M) \to C^\infty(M)$ is a $C^\infty(M)$-linear mapping: $\Phi(f.X) = f.\Phi(X)$. In the general case we subject each entry to the treatment described below.

Claim 1. If $X \mid U = 0$ for some open subset $U \subset M$, then we have $\Phi(X) \mid U = 0$.

Let $x \in U$. We choose $f \in C^\infty(M)$ with $f(x) = 0$ and $f \mid M \setminus U = 1$. Then $f.X = X$, so $\Phi(X)(x) = \Phi(f.X)(x) = f(x).\Phi(X)(x) = 0$.

Claim 2. If $X(x) = 0$, then also $\Phi(X)(x) = 0$.

Let (U, u) be a chart centered at x, and let V be open with $x \in V \subset \bar{V} \subset U$. Then

$$X \mid U = \sum X^i \frac{\partial}{\partial u^i} \quad \text{and} \quad X^i(x) = 0.$$

We choose $g \in C^\infty(M)$ with $g \mid V \equiv 1$ and with support contained in U. Then $(g^2.X) \mid V = X \mid V$ and by claim 1 the restriction $\Phi(X) \mid V$ depends only on $X \mid V$; thus $g^2.X = \sum_i (g.X^i)(g.\frac{\partial}{\partial u^i})$ is a decomposition which is globally defined on M. Therefore we have

$$\begin{aligned}\Phi(X)(x) = \Phi(g^2.X)(x) &= \Phi\left(\sum_i (g.X^i)(g.\tfrac{\partial}{\partial u^i})\right)(x)\\ &= \sum (g.X^i)(x).\Phi(g.\tfrac{\partial}{\partial u^i})(x) = 0.\end{aligned}$$

So we see that for a general vector field X the value $\Phi(X)(x)$ depends only on the value $X(x)$, for each $x \in M$. So there is a linear map $\varphi_x : T_xM \to \mathbb{R}$ for each $x \in M$ with $\Phi(X)(x) = \varphi_x(X(x))$. Then $\varphi : M \to T^*M$ is smooth since $\varphi \mid V = \sum_i \Phi(g.\frac{\partial}{\partial u^i})\, du^i$ in the setting of claim 2. $\square$

9.4. Definition. A *differential form* of degree k, or a k-*form* for short, is a section of the (natural) vector bundle $\bigwedge^k T^*M$. The space of all k-forms will be denoted by $\Omega^k(M)$. It may also be viewed as the space of all skew-symmetric $\binom{0}{k}$-tensor fields, i.e., (by (9.3)) the space of all mappings

$$\varphi : \mathfrak{X}(M) \times \cdots \times \mathfrak{X}(M) = \mathfrak{X}(M)^k \to C^\infty(M),$$

which are k-linear over $C^\infty(M)$ and are skew-symmetric:

$$\varphi(X_{\sigma 1}, \dots, X_{\sigma k}) = \operatorname{sign} \sigma \cdot \varphi(X_1, \dots, X_k)$$

for each permutation $\sigma \in \mathcal{S}_k$.

We put $\Omega^0(M) := C^\infty(M)$. Then the space

$$\Omega(M) := \bigoplus_{k=0}^{\dim M} \Omega^k(M)$$

is an algebra with the following product, called the *wedge product.* For $\varphi \in \Omega^k(M)$ and $\psi \in \Omega^\ell(M)$ and for X_i in $\mathfrak{X}(M)$ (or in T_xM) we put

$$\begin{aligned} &(\varphi \wedge \psi)(X_1, \dots, X_{k+\ell}) \\ &= \tfrac{1}{k!\,\ell!} \sum_{\sigma \in \mathcal{S}_{k+\ell}} \operatorname{sign} \sigma \cdot \varphi(X_{\sigma 1}, \dots, X_{\sigma k}).\psi(X_{\sigma(k+1)}, \dots, X_{\sigma(k+\ell)}). \end{aligned}$$

This product is defined fiberwise, i.e., $(\varphi \wedge \psi)_x = \varphi_x \wedge \psi_x$ for each $x \in M$. It is also associative, i.e., $(\varphi \wedge \psi) \wedge \tau = \varphi \wedge (\psi \wedge \tau)$, and graded commutative, i.e., $\varphi \wedge \psi = (-1)^{k\ell} \psi \wedge \varphi$. There are differing conventions for the factor in the definition of the wedge product: in [**191**] the factor $\frac{1}{(k+\ell)!}$ is used. But then the insertion operator of (9.7) is no longer a graded derivation.

9.5. If $f : N \to M$ is a smooth mapping and $\varphi \in \Omega^k(M)$, then the pullback $f^*\varphi \in \Omega^k(N)$ is defined for $X_i \in T_xN$ by

$$(1) \qquad (f^*\varphi)_x(X_1, \dots, X_k) := \varphi_{f(x)}(T_xf.X_1, \dots, T_xf.X_k).$$

Then we have $f^*(\varphi \wedge \psi) = f^*\varphi \wedge f^*\psi$, so $f^* : \Omega(M) \to \Omega(N)$ is an algebra homomorphism. Moreover we have $(g \circ f)^* = f^* \circ g^* : \Omega(P) \to \Omega(N)$ if $g : M \to P$, and $(Id_M)^* = Id_{\Omega(M)}$.

So $M \mapsto \Omega(M) = \Gamma(\bigwedge T^*M)$ is a contravariant functor from the category $\mathcal{M}f$ of all manifolds and all smooth mappings into the category of real graded commutative algebras, whereas $M \mapsto \bigwedge T^*M$ is a covariant vector bundle functor defined only on $\mathcal{M}f_m$, the category of m-dimensional manifolds and local diffeomorphisms, for each m separately.

9.6. The Lie derivative of differential forms. Since $M \mapsto \bigwedge^k T^*M$ is a vector bundle functor on $\mathcal{M}f_m$, by (8.16) for $X \in \mathfrak{X}(M)$ the *Lie derivative* of a k-form φ along X is defined by

$$\mathcal{L}_X\varphi = \tfrac{d}{dt}|_0(\mathrm{Fl}_t^X)^*\varphi.$$

Lemma. *The Lie derivative has the following properties.*

(1) $\mathcal{L}_X(\varphi\wedge\psi) = \mathcal{L}_X\varphi\wedge\psi + \varphi\wedge\mathcal{L}_X\psi$, *so $\mathcal{L}_X$ is a derivation.*

(2) *For $Y_i \in \mathfrak{X}(M)$ we have*

$$(\mathcal{L}_X\varphi)(Y_1,\dots,Y_k) = X(\varphi(Y_1,\dots,Y_k)) - \sum_{i=1}^k \varphi(Y_1,\dots,[X,Y_i],\dots,Y_k).$$

(3) $[\mathcal{L}_X,\mathcal{L}_Y]\varphi = \mathcal{L}_{[X,Y]}\varphi$.

(4) $\frac{\partial}{\partial t}(\mathrm{Fl}_t^X)^*\varphi = (\mathrm{Fl}_t^X)^*\mathcal{L}_X\varphi = \mathcal{L}_X((\mathrm{Fl}_t^X)^*\varphi)$.

Proof. (1) The mapping $Alt : \bigotimes^k T^*M \to \bigwedge^k T^*M$, given by

$$(Alt A)(Y_1,\dots,Y_k) := \tfrac{1}{k!}\sum_\sigma \operatorname{sign}(\sigma)\, A(Y_{\sigma 1},\dots,Y_{\sigma k}),$$

is a linear natural transformation in the sense of (8.18) and induces an algebra homomorphism from $\bigoplus_{k\ge 0}\Gamma(\bigotimes^k T^*M)$ onto $\Omega(M)$. So (1) follows from (8.17) and (8.18).

Second, direct proof, using the definition and (9.5):

$$\begin{aligned}\mathcal{L}_X(\varphi\wedge\psi) &= \tfrac{d}{dt}|_0(\mathrm{Fl}_t^X)^*(\varphi\wedge\psi) = \tfrac{d}{dt}|_0\left((\mathrm{Fl}_t^X)^*\varphi\wedge(\mathrm{Fl}_t^X)^*\psi\right)\\ &= \tfrac{d}{dt}|_0(\mathrm{Fl}_t^X)^*\varphi\wedge(\mathrm{Fl}_0^X)^*\psi + (\mathrm{Fl}_0^X)^*\varphi\wedge\tfrac{d}{dt}|_0(\mathrm{Fl}_t^X)^*\psi\\ &= \mathcal{L}_X\varphi\wedge\psi + \varphi\wedge\mathcal{L}_X\psi.\end{aligned}$$

(2) Again by (8.17) and (8.18) we may compute as follows, where Trace is the full evaluation of the form on all vector fields:

$$\begin{aligned}X(\varphi(Y_1,\dots,Y_k)) &= \mathcal{L}_X\circ\operatorname{Trace}(\varphi\otimes Y_1\otimes\dots\otimes Y_k)\\ &= \operatorname{Trace}\circ\mathcal{L}_X(\varphi\otimes Y_1\otimes\dots\otimes Y_k)\\ &= \operatorname{Trace}\big(\mathcal{L}_X\varphi\otimes(Y_1\otimes\dots\otimes Y_k) + \varphi\otimes(\textstyle\sum_i Y_1\otimes\dots\otimes\mathcal{L}_X Y_i\otimes\dots\otimes Y_k)\big).\end{aligned}$$

Now we use $\mathcal{L}_X Y_i = [X,Y_i]$ from (3.13).

Second, independent proof:

$$\begin{aligned}X(\varphi(Y_1,\dots,Y_k)) &= \tfrac{d}{dt}|_0(\mathrm{Fl}_t^X)^*(\varphi(Y_1,\dots,Y_k))\\ &= \tfrac{d}{dt}|_0((\mathrm{Fl}_t^X)^*\varphi)((\mathrm{Fl}_t^X)^*Y_1,\dots,(\mathrm{Fl}_t^X)^*Y_k))\\ &= (\mathcal{L}_X\varphi)(Y_1,\dots,Y_k) + \sum_{i=1}^k\varphi(Y_1,\dots,\mathcal{L}_XY_i,\dots,Y_k).\end{aligned}$$

(3) is a special case of (8.20). See (9.9.7) below for another proof.

$$(4)\qquad \tfrac{\partial}{\partial t}(\mathrm{Fl}^X_t)^*\varphi = \tfrac{\partial}{\partial s}|_0\left(\bigwedge^k T(\mathrm{Fl}^X_{-t})\circ T(\mathrm{Fl}^X_{-s})^*\circ\varphi\circ\mathrm{Fl}^X_s\circ\mathrm{Fl}^X_t\right)$$

$$= \bigwedge^k T(\mathrm{Fl}^X_{-t})^*\circ\tfrac{\partial}{\partial s}|_0\left(\bigwedge^k T(\mathrm{Fl}^X_{-s})^*\circ\varphi\circ\mathrm{Fl}^X_s\right)\circ\mathrm{Fl}^X_t$$

$$= \bigwedge^k T(\mathrm{Fl}^X_{-t})^*\circ\mathcal{L}_X\varphi\circ\mathrm{Fl}^X_t = (\mathrm{Fl}^X_t)^*\mathcal{L}_X\varphi,$$

$$\tfrac{\partial}{\partial t}(\mathrm{Fl}^X_t)^*Y = \tfrac{\partial}{\partial s}|_0(\mathrm{Fl}^X_s)^*(\mathrm{Fl}^X_t)^*Y = \mathcal{L}_X(\mathrm{Fl}^X_t)^*\varphi.\quad\square$$

9.7. The insertion operator. For a vector field $X\in\mathfrak{X}(M)$ we define the *insertion operator* $i_X = i(X) : \Omega^k(M)\to\Omega^{k-1}(M)$ by

$$(i_X\varphi)(Y_1,\dots,Y_{k-1}) := \varphi(X,Y_1,\dots,Y_{k-1}).$$

Lemma.

(1) *i_X is a graded derivation of degree -1 of the graded algebra $\Omega(M)$, so we have $i_X(\varphi\wedge\psi) = i_X\varphi\wedge\psi + (-1)^{-\deg\varphi}\varphi\wedge i_X\psi$.*

(2) *$i_X\circ i_Y + i_Y\circ i_X = 0$.*

(3) *$[\mathcal{L}_X, i_Y] := \mathcal{L}_X\circ i_Y - i_Y\circ\mathcal{L}_X = i_{[X,Y]}$.*

Proof. (1) For $\varphi\in\Omega^k(M)$ and $\psi\in\Omega^\ell(M)$ we have

$$(i_{X_1}(\varphi\wedge\psi))(X_2,\dots,X_{k+\ell}) = (\varphi\wedge\psi)(X_1,\dots,X_{k+\ell})$$

$$= \tfrac{1}{k!\,\ell!}\sum_\sigma \operatorname{sign}(\sigma)\,\varphi(X_{\sigma 1},\dots,X_{\sigma k})\psi(X_{\sigma(k+1)},\dots,X_{\sigma(k+\ell)}),$$

$$(i_{X_1}\varphi\wedge\psi + (-1)^k\varphi\wedge i_{X_1}\psi)(X_2,\dots,X_{k+\ell})$$

$$= \tfrac{1}{(k-1)!\,\ell!}\sum_\sigma \operatorname{sign}(\sigma)\,\varphi(X_1,X_{\sigma 2},\dots,X_{\sigma k})\psi(X_{\sigma(k+1)},\dots,X_{\sigma(k+\ell)})$$

$$+\frac{(-1)^k}{k!\,(\ell-1)!}\sum_\sigma \operatorname{sign}(\sigma)\,\varphi(X_{\sigma 2},\dots,X_{\sigma(k+1)})\psi(X_1,X_{\sigma(k+2)},\dots).$$

Using the skew-symmetry of φ and ψ, we may distribute X_1 to each position by adding an appropriate sign. There are $k+\ell$ summands. Since

$$\tfrac{1}{(k-1)!\,\ell!} + \tfrac{1}{k!\,(\ell-1)!} = \tfrac{k+\ell}{k!\,\ell!},$$

and since we can generate each permutation in $\mathcal{S}_{k+\ell}$ in this way, the result follows.

(2) $(i_X i_Y \varphi)(Z_1, \dots, Z_{k-2}) = \varphi(Y, X, Z_1, \dots, Z_n)$

$$= -\varphi(X, Y, Z_1, \dots, Z_n) = -(i_Y i_X \varphi)(Z_1, \dots, Z_{k-2}).$$

(3) By (8.17) and (8.18) we have:

$$\begin{aligned} \mathcal{L}_X i_Y \varphi &= \mathcal{L}_X \operatorname{Trace}_1(Y \otimes \varphi) = \operatorname{Trace}_1 \mathcal{L}_X(Y \otimes \varphi) \\ &= \operatorname{Trace}_1(\mathcal{L}_X Y \otimes \varphi + Y \otimes \mathcal{L}_X \varphi) = i_{[X,Y]}\varphi + i_Y \mathcal{L}_X \varphi. \end{aligned}$$

See (9.9.6) below for another proof. □

9.8. The exterior differential. We want to construct a differential operator $\Omega^k(M) \to \Omega^{k+1}(M)$ which is natural. We will show that the simplest choice will work and (later) that it is essentially unique.

Let U be open in $\mathbb{R}^n$, and let $\varphi \in \Omega^k(U) = C^\infty(U, L^k_{alt}(\mathbb{R}^n, \mathbb{R}))$. We consider the derivative $D\varphi \in C^\infty(U, L(\mathbb{R}^n, L^k_{alt}(\mathbb{R}^n, \mathbb{R})))$, and we take its canonical image in $C^\infty(U, L^{k+1}_{alt}(\mathbb{R}^n, \mathbb{R}))$. Here we write D for the derivative in order to distinguish it from the exterior differential, which we define as

$$d\varphi := (k+1) \operatorname{Alt} D\varphi,$$

more explicitly as

$$\begin{aligned} (1) \qquad (d\varphi)_x(X_0, \dots, X_k) &= \tfrac{1}{k!} \sum_\sigma \operatorname{sign}(\sigma)\, D\varphi(x)(X_{\sigma 0})(X_{\sigma 1}, \dots, X_{\sigma k}) \\ &= \sum_{i=0}^k (-1)^i D\varphi(x)(X_i)(X_0, \dots, \widehat{X_i}, \dots, X_k), \end{aligned}$$

where the hat over a symbol means that this is to be omitted and where $X_i \in \mathbb{R}^n$.

Now we pass to an arbitrary manifold M. For a k-form $\varphi \in \Omega^k(M)$ and vector fields $X_i \in \mathfrak{X}(M)$ we try to replace $D\varphi(x)(X_i)(X_0, \dots)$ in formula (1) by Lie derivatives. We differentiate

$$\begin{aligned} &X_i(\varphi(x)(X_0, \dots)) \\ &\quad = D\varphi(x)(X_i)(X_0, \dots) + \sum_{0 \le j \le k, j \ne i} \varphi(x)(X_0, \dots, DX_j(x)X_i, \dots) \end{aligned}$$

and insert this expression into formula (1) in order to get (cf. (3.4)) our working definition

$$\begin{aligned} (2) \quad d\varphi(X_0, \dots, X_k) &:= \sum_{i=0}^k (-1)^i X_i(\varphi(X_0, \dots, \widehat{X_i}, \dots, X_k)) \\ &\quad + \sum_{i<j} (-1)^{i+j} \varphi([X_i, X_j], X_0, \dots, \widehat{X_i}, \dots, \widehat{X_j}, \dots, X_k). \end{aligned}$$

This formula gives $d\varphi$ as a $(k+1)$-linear mapping over $C^\infty(M)$, as a short computation involving (3.4) shows. It is obviously skew-symmetric, so $d\varphi$ is a $(k+1)$-form by (9.3), and the operator $d : \Omega^k(M) \to \Omega^{k+1}(M)$ is called the *exterior derivative.*

If (U, u) is a chart on M, then we have

$$\varphi \upharpoonright U = \sum_{i_1 < \dots < i_k} \varphi_{i_1,\dots,i_k} du^{i_1} \wedge \dots \wedge du^{i_k},$$

where

$$\varphi_{i_1,\dots,i_k} = \varphi(\tfrac{\partial}{\partial u^{i_1}}, \dots, \tfrac{\partial}{\partial u^{i_k}}).$$

An easy computation shows that (2) leads to

$$d\varphi \upharpoonright U = \sum_{i_1 < \dots < i_k} d\varphi_{i_1,\dots,i_k} \wedge du^{i_1} \wedge \dots \wedge du^{i_k}, \tag{3}$$

so that formulas (1) and (2) really define the same operator.

9.9. Theorem. *The exterior derivative $d : \Omega^k(M) \to \Omega^{k+1}(M)$ has the following properties:*

(1) $d(\varphi \wedge \psi) = d\varphi \wedge \psi + (-1)^{\deg \varphi} \varphi \wedge d\psi$, *so d is a graded derivation of degree 1.*
(2) $\mathcal{L}_X = i_X \circ d + d \circ i_X$ *for any vector field X.*
(3) $d^2 = d \circ d = 0$.
(4) $f^* \circ d = d \circ f^*$ *for any smooth $f : N \to M$.*
(5) $\mathcal{L}_X \circ d = d \circ \mathcal{L}_X$ *for any vector field X.*
(6) $[\mathcal{L}_X, i_Y] := \mathcal{L}_X \circ i_Y - i_Y \circ \mathcal{L}_X = i_{[X,Y]}$. *See also* (9.7.3).
(7) $[\mathcal{L}_X, \mathcal{L}_Y] = \mathcal{L}_{[X,Y]}$ *for any two vector fields X, Y.*

Remark. In terms of the graded commutator

$$[D_1, D_2] := D_1 \circ D_2 - (-1)^{\deg(D_1)\deg(D_2)} D_2 \circ D_1$$

for graded homomorphisms and graded derivations (see (16.1)) the assertions of this theorem take the following form:

(2) $\mathcal{L}_X = [i_X, d]$.
(3) $\frac{1}{2}[d, d] = 0$.
(4) $[f^*, d] = 0$.
(5) $[\mathcal{L}_X, d] = 0$.

This point of view will be developed in section (16) below. The equation (7) is a special case of (8.20).

Proof. (2) For $\varphi \in \Omega^k(M)$ and $X_i \in \mathfrak{X}(M)$ we have

$$\begin{aligned}
(\mathcal{L}_{X_0}\varphi)(X_1,\dots,X_k) &= X_0(\varphi(X_1,\dots,X_k)) \\
&\quad + \sum_{j=1}^{k}(-1)^{0+j}\varphi([X_0,X_j],X_1,\dots,\widehat{X_j},\dots,X_k) \text{ by (9.6.2)},
\end{aligned}$$

$$\begin{aligned}
(i_{X_0}d\varphi)(X_1,\dots,X_k) &= d\varphi(X_0,\dots,X_k) \\
&= \sum_{i=0}^{k}(-1)^i X_i(\varphi(X_0,\dots,\widehat{X_i},\dots,X_k)) \\
&\quad + \sum_{0\le i<j}(-1)^{i+j}\varphi([X_i,X_j],X_0,\dots,\widehat{X_i},\dots,\widehat{X_j},\dots,X_k),
\end{aligned}$$

$$\begin{aligned}
(di_{X_0}\varphi)(X_1,\dots,X_k) &= \sum_{i=1}^{k}(-1)^{i-1}X_i((i_{X_0}\varphi)(X_1,\dots,\widehat{X_i},\dots,X_k)) \\
&\quad + \sum_{1\le i<j}(-1)^{i+j-2}(i_{X_0}\varphi)([X_i,X_j],X_1,\dots,\widehat{X_i},\dots,\widehat{X_j},\dots,X_k) \\
&= -\sum_{i=1}^{k}(-1)^i X_i(\varphi(X_0,X_1,\dots,\widehat{X_i},\dots,X_k)) \\
&\quad - \sum_{1\le i<j}(-1)^{i+j}\varphi([X_i,X_j],X_0,X_1,\dots,\widehat{X_i},\dots,\widehat{X_j},\dots,X_k).
\end{aligned}$$

By summing up, the result follows.

(1) Let $\varphi \in \Omega^p(M)$ and $\psi \in \Omega^q(M)$. We prove the result by induction on $p+q$.

$p+q=0$: $d(f\cdot g) = df\cdot g + f\cdot dg$.

Suppose that (1) is true for $p+q<k$. Then for $X \in \mathfrak{X}(M)$ we have by part (2) and (9.6), (9.7) and by induction

$$\begin{aligned}
i_X\, d(\varphi\wedge\psi) &= \mathcal{L}_X(\varphi\wedge\psi) - d\, i_X(\varphi\wedge\psi) \\
&= \mathcal{L}_X\varphi\wedge\psi + \varphi\wedge\mathcal{L}_X\psi - d(i_X\varphi\wedge\psi + (-1)^p\varphi\wedge i_X\psi) \\
&= i_X d\varphi\wedge\psi + di_X\varphi\wedge\psi + \varphi\wedge i_X d\psi + \varphi\wedge di_X\psi - di_X\varphi\wedge\psi \\
&\qquad - (-1)^{p-1}i_X\varphi\wedge d\psi - (-1)^p d\varphi\wedge i_X\psi - \varphi\wedge di_X\psi \\
&= i_X(d\varphi\wedge\psi + (-1)^p\varphi\wedge d\psi).
\end{aligned}$$

Since X is arbitrary, (1) follows.

(3) By (1) the operator d is a graded derivation of degree 1, so $d^2 = \frac{1}{2}[d,d]$ is a graded derivation of degree 2; see (16.1). It is obviously local: $d^2(\varphi\wedge\psi) = d^2(\varphi)\wedge\psi + \varphi\wedge d(\psi)$. Since $\Omega(M)$ is locally generated as an algebra by $C^\infty(M)$ and $\{df : f\in C^\infty(M)\}$, it suffices to show that $d^2 f = 0$ for each

$f \in C^\infty(M)$ ($d^3 f = 0$ is a consequence). But this is easy:

$$d^2 f(X,Y) = X df(Y) - Y df(X) - df([X,Y]) = XYf - YXf - [X,Y]f = 0.$$

(4) $f^* : \Omega(M) \to \Omega(N)$ is an algebra homomorphism by (9.6), so $f^* \circ d$ and $d \circ f^*$ are both graded derivations over f^* of degree 1. So if $f^* \circ d$ and $d \circ f^*$ agree on φ and on ψ, then they also agree on $\varphi \wedge \psi$. By the same argument as in the proof of (3) above it suffices to show that they agree on g and dg for all $g \in C^\infty(M)$. We have

$$(f^* dg)_y(Y) = (dg)_{f(y)}(T_y f.Y) = (T_y f.Y)(g) = Y(g \circ f)(y) = (df^* g)_y(Y);$$

thus also $df^* dg = ddf^* g = 0$, and $f^* ddg = 0$.

(5) $d\mathcal{L}_X = d\, i_X\, d + ddi_X = di_X d + i_X dd = \mathcal{L}_X d$.

(6) We use the graded commutator alluded to in the remarks. Both $\mathcal{L}_X$ and i_Y are graded derivations; thus the graded commutator $[L_X, i_Y]$ is also a graded derivation as is $i_{[X,Y]}$. Thus it suffices to show that they agree on 0-forms $g \in C^\infty(M)$ and on exact 1-forms dg. We have

$$\begin{aligned}
[\mathcal{L}_X, i_Y]g &= \mathcal{L}_X i_Y g - i_Y \mathcal{L}_X g = \mathcal{L}_X 0 - i_Y(dg(X)) = 0 = i_{[X,Y]} g, \\
[\mathcal{L}_X, i_Y]dg &= \mathcal{L}_X i_Y dg - i_Y \mathcal{L}_X dg = \mathcal{L}_X \mathcal{L}_Y g - i_Y d\mathcal{L}_X g \\
&= (XY - YX)g = [X,Y]g = i_{[X,Y]} dg.
\end{aligned}$$

(7) By the (graded) Jacobi identity and by (6) (or lemma (9.7.3)) we have

$$\begin{aligned}
[\mathcal{L}_X, \mathcal{L}_Y] &= [\mathcal{L}_X, [i_Y, d]] = [[\mathcal{L}_X, i_Y], d] + [i_Y, [\mathcal{L}_X, d]] \\
&= [i_{[X,Y]}, d] + 0 = \mathcal{L}_{[X,Y]}. \quad \square
\end{aligned}$$

9.10. A differential form $\omega \in \Omega^k(M)$ is called *closed* if $d\omega = 0$, and it is called *exact* if $\omega = d\varphi$ for some $\varphi \in \Omega^{k-1}(M)$. Since $d^2 = 0$, any exact form is closed. The quotient space

$$H^k(M) := \frac{\ker(d : \Omega^k(M) \to \Omega^{k+1}(M))}{\operatorname{im}(d : \Omega^{k-1}(M) \to \Omega^k(M))}$$

is called the k-th de Rham cohomology space of M. As a preparation for our treatment of cohomology we finish with the

Lemma of Poincaré. *A closed differential form of degree $k \geq 1$ is locally exact. More precisely: let $\omega \in \Omega^k(M)$ with $d\omega = 0$. Then for any $x \in M$ there is an open neighborhood U of x in M and a $\varphi \in \Omega^{k-1}(U)$ with $d\varphi = \omega \restriction U$.*

Proof. Let (U, u) be a chart on M centered at x such that $u(U) = \mathbb{R}^m$. So we may just assume that $M = \mathbb{R}^m$.

We consider $\alpha : \mathbb{R} \times \mathbb{R}^m \to \mathbb{R}^m$, given by $\alpha(t,x) = \alpha_t(x) = tx$. Let $I \in \mathfrak{X}(\mathbb{R}^m)$ be the vector field $I(x) = x$; then $\alpha(e^t, x) = \mathrm{Fl}^I_t(x)$. So for $t > 0$ we have

$$\begin{aligned}\tfrac{d}{dt}\alpha_t^*\omega &= \tfrac{d}{dt}(\mathrm{Fl}^I_{\log t})^*\omega = \tfrac{1}{t}(\mathrm{Fl}^I_{\log t})^*\mathcal{L}_I\omega \\ &= \tfrac{1}{t}\alpha_t^*(i_I d\omega + d i_I\omega) = \tfrac{1}{t}d\alpha_t^* i_I\omega.\end{aligned}$$

Note that $T_x(\alpha_t) = t.Id$. Therefore

$$\begin{aligned}(\tfrac{1}{t}\alpha_t^* i_I\omega)_x(X_2,\dots,X_k) &= \tfrac{1}{t}(i_I\omega)_{tx}(tX_2,\dots,tX_k) \\ &= \tfrac{1}{t}\omega_{tx}(tx, tX_2,\dots,tX_k) = \omega_{tx}(x, tX_2,\dots,tX_k).\end{aligned}$$

So if $k \geq 1$, the $(k-1)$-form $\frac{1}{t}\alpha_t^* i_I\omega$ is defined and smooth in (t,x) for all $t \in \mathbb{R}$. Clearly $\alpha_1^*\omega = \omega$ and $\alpha_0^*\omega = 0$; thus

$$\begin{aligned}\omega &= \alpha_1^*\omega - \alpha_0^*\omega = \int_0^1 \tfrac{d}{dt}\alpha_t^*\omega dt \\ &= \int_0^1 d(\tfrac{1}{t}\alpha_t^* i_I\omega)dt = d\left(\int_0^1 \tfrac{1}{t}\alpha_t^* i_I\omega dt\right) = d\varphi. \quad \square\end{aligned}$$

10. Integration on Manifolds

10.1. Let $U \subset \mathbb{R}^n$ be an open subset, let dx denote Lebesque measure on $\mathbb{R}^n$ (which depends on the Euclidean structure), let $g : U \to g(U)$ be a diffeomorphism onto some other open subset in $\mathbb{R}^n$, and let $f : g(U) \to \mathbb{R}$ be an integrable continuous function. Then the *transformation formula for multiple integrals* reads

$$\int_{g(U)} f(y)\, dy = \int_U f(g(x))|\det dg(x)|dx.$$

This suggests that the suitable objects for integration on a manifold are sections of a 1-dimensional vector bundle whose cocycle of transition functions is given by the absolute value of the Jacobi matrix of the chart changes. They will be called *densities* below.

10.2. The volume bundle. Let M be a manifold and let (U_α, u_α) be a smooth atlas for it. The *volume bundle* $(\mathrm{Vol}(M), \pi_M, M)$ of M is the 1-dimensional vector bundle (line bundle) which is given by the following cocycle of transition functions; see (8.3):

$$\begin{gathered}\psi_{\alpha\beta} : U_{\alpha\beta} = U_\alpha \cap U_\beta \to \mathbb{R} \setminus \{0\} = GL(1,\mathbb{R}), \\ \psi_{\alpha\beta}(x) = |\det d(u_\beta \circ u_\alpha^{-1})(u_\alpha(x))| = \frac{1}{|\det d(u_\alpha \circ u_\beta^{-1})(u_\beta(x))|}.\end{gathered}$$

Lemma. Vol(M) *is a trivial line bundle over* M.

But there is no natural trivialization.

Proof. We choose a positive local section over each U_α and we glue them with a partition of unity. Since positivity is invariant under the transitions, the resulting global section μ is nowhere 0. By (8.5), μ is a global frame field and trivializes $\mathrm{Vol}(M)$. $\square$

Definition. Sections of the line bundle $\mathrm{Vol}(M)$ are called densities.

10.3. Integral of a density. Let $\mu \in \Gamma(\mathrm{Vol}(M))$ be a density with compact support on the manifold M. We define the *integral of the density* μ as follows:

> Let (U_α, u_α) be an atlas on M, and let f_α be a partition of unity with $\mathrm{supp}(f_\alpha) \subset U_\alpha$. Then we put
>
> $$\int_M \mu = \sum_\alpha \int_{U_\alpha} f_\alpha \mu := \sum_\alpha \int_{u_\alpha(U_\alpha)} f_\alpha(u_\alpha^{-1}(y)).\psi_\alpha(\mu(u_\alpha^{-1}(y)))\, dy.$$

If μ does not have compact support, we require that $\sum \int_{U_\alpha} f_\alpha\, |\mu| < \infty$. The series is then absolutely convergent.

Lemma. *$\int_M \mu$ is well defined.*

Proof. Let (V_β, v_β) be another atlas on M, and let (g_β) be a partition of unity with $\mathrm{supp}(g_\beta) \subset V_\beta$. Let (U_α, ψ_α) be the vector bundle atlas of $\mathrm{Vol}(M)$ induced by the atlas (U_α, u_α), and let (V_β, φ_β) be the one induced by (V_β, v_β). By the transformation formula of integrals for the diffeomorphisms $u_\alpha \circ v_\beta^{-1} : v_\beta(U_\alpha \cap V_\beta) \to u_\alpha(U_\alpha \cap V_\beta)$ we have:

$$\begin{aligned}
\sum_\alpha \int_{U_\alpha} f_\alpha\, \mu &= \sum_\alpha \int_{u_\alpha(U_\alpha)} (f_\alpha \circ u_\alpha^{-1})(y)\psi_\alpha(\mu(u_\alpha^{-1}(y)))\, dy \\
&= \sum_\alpha \int_{u_\alpha(U_\alpha)} \sum_\beta (g_\beta \circ u_\alpha^{-1})(y)(f_\alpha \circ u_\alpha^{-1})(y)\psi_\alpha(\mu(u_\alpha^{-1}(y)))\, dy \\
&= \sum_{\alpha\beta} \int_{u_\alpha(U_\alpha\cap V_\beta)} (g_\beta \circ u_\alpha^{-1})(y)(f_\alpha \circ u_\alpha^{-1})(y)\psi_\alpha(\mu(u_\alpha^{-1}(y)))\, dy \\
&= \sum_{\alpha\beta} \int_{v_\beta(U_\alpha\cap V_\beta)} (g_\beta \circ v_\beta^{-1})(x)(f_\alpha \circ v_\beta^{-1})(x)\cdot \\
&\qquad\qquad \cdot \psi_\alpha(\mu(v_\beta^{-1}(x)))|\det d(u_\alpha \circ v_\beta^{-1})(x)|\, dx \\
&= \sum_{\alpha\beta} \int_{v_\beta(U_\alpha\cap V_\beta)} (g_\beta \circ v_\beta^{-1})(x)(f_\alpha \circ v_\beta^{-1})(x)\varphi_\beta(\mu(v_\beta^{-1}(x)))\, dx \\
&= \sum_\beta \int_{V_\beta} g_\beta\, \mu. \quad \square
\end{aligned}$$

Remark. If $\mu \in \Gamma(\mathrm{Vol}(M))$ is an arbitrary section and $f \in C_c^\infty(M)$ is a function with compact support, then we may define the integral of f with respect to μ by $\int_M f\mu$, since $f\mu$ is a density with compact support. In this way μ defines a Radon measure on M.

For the converse we note first that (C^1 suffices) diffeomorphisms between open subsets on $\mathbb{R}^m$ map sets of Lebesque measure zero to sets of Lebesque measure zero. Thus on a manifold we have a well defined notion of sets of Lebesque measure zero — but no measure. If ν is a Radon measure on M which is absolutely continuous, i.e., the $|\nu|$-measure of a set of Lebesque measure zero is zero, then it is given by a uniquely determined measurable section of the line bundle Vol. Here a section is called measurable if in any line bundle chart it is given by a measurable function.

10.4. *p*-densities. For $0 \le p \le 1$ let $\mathrm{Vol}^p(M)$ be the line bundle defined by the cocycle of transition functions

$$\psi^p_{\alpha\beta} : U_{\alpha\beta} \to \mathbb{R} \setminus \{0\},$$
$$\psi^p_{\alpha\beta}(x) = |\det d(u_\alpha \circ u_\beta^{-1})(u_\beta(x))|^{-p}.$$

This is also a trivial line bundle. Its sections are called p-densities. Note that 1-densities are just densities and that 0-densities are functions. If μ is a p-density and ν is a q-density with $p+q \le 1$, then $\mu.\nu := \mu \otimes \nu$ is a $p+q$-density, i.e., $\mathrm{Vol}^p(M) \otimes \mathrm{Vol}^q(M) = \mathrm{Vol}^{p+q}(M)$. Thus the product of two $\frac{1}{2}$-densities with compact support can be integrated, so $\Gamma_c(\mathrm{Vol}^{1/2}(M))$ is a pre-Hilbert space in a natural way.

Distributions on M (in the sense of generalized functions) are elements of the dual space of the space $\Gamma_c(\mathrm{Vol}(M))$ of densities with compact support equipped with the inductive limit topology — so they contain functions.

10.5. Example. The density of a Riemann metric. Let g be a Riemann metric on a manifold M; see section (22) below. So g is a symmetric $\binom{0}{2}$-tensor field such that g_x is a positive definite inner product on T_xM for each $x \in M$. If (U,u) is a chart on M, then we have

$$g|U = \sum_{i,j=1}^m g^u_{ij}\, du^i \otimes du^j$$

where the functions $g^u_{ij} = g(\frac{\partial}{\partial u^i}, \frac{\partial}{\partial u^j})$ form a positive definite symmetric matrix. So $\det(g^u_{ij}) = \det((g(\frac{\partial}{\partial u^i}, \frac{\partial}{\partial u^j}))^m_{i,j=1}) > 0$. We put

$$\mathrm{vol}(g)^u := \sqrt{\det((g(\tfrac{\partial}{\partial u^i}, \tfrac{\partial}{\partial u^j}))^m_{i,j=1})}.$$

If (V, v) is another chart, we have

$$\begin{aligned}
\operatorname{vol}(g)^u &= \sqrt{\det((g(\tfrac{\partial}{\partial u^i}, \tfrac{\partial}{\partial u^j}))_{i,j=1}^m)} \\
&= \sqrt{\det((g(\sum_k \tfrac{\partial v^k}{\partial u^i}\tfrac{\partial}{\partial v^k}, \sum_\ell \tfrac{\partial v^\ell}{\partial u^j}\tfrac{\partial}{\partial v^\ell}))_{i,j=1}^m)} \\
&= \sqrt{\det((\tfrac{\partial v^k}{\partial u^i})_{k,i})^2 \det((g(\tfrac{\partial}{\partial v^\ell}, \tfrac{\partial}{\partial v^j}))_{\ell,j})} \\
&= |\det d(v \circ u^{-1})| \operatorname{vol}(g)^v,
\end{aligned}$$

so these local representatives determine a section $\operatorname{vol}(g) \in \Gamma(\operatorname{Vol}(M))$, which is called the *density or volume of the Riemann metric* g. If M is compact, then $\int_M \operatorname{vol}(g)$ is called the *volume* of the Riemann manifold (M, g).

10.6. The orientation bundle. For a manifold M with $\dim M = m$ and an atlas (U_α, u_α) for M the line bundle $\bigwedge^m T^*M$ is given by the cocycle of transition functions

$$\varphi_{\alpha\beta}(x) = \det d(u_\beta \circ u_\alpha^{-1})(u_\alpha(x)) = \bigwedge^m d(u_\beta \circ u_\alpha^{-1})(u_\alpha(x)).$$

We consider the line bundle $\operatorname{Or}(M)$ which is given by the cocycle of transition functions

$$\tau_{\alpha\beta}(x) = \operatorname{sign} \varphi_{\alpha\beta}(x) = \operatorname{sign} \det d(u_\beta \circ u_\alpha^{-1})(u_\alpha(x)).$$

Since $\tau_{\alpha\beta}(x)\varphi_{\alpha\beta}(x) = \psi_{\alpha\beta}(x)$, the cocycle of the volume bundle of (10.2), we have

$$\operatorname{Vol}(M) = \operatorname{Or}(M) \otimes \bigwedge^m T^*M,$$

$$\bigwedge^m T^*M = \operatorname{Or}(M) \otimes \operatorname{Vol}(M).$$

10.7. Definition. A manifold M is called *orientable* if the orientation bundle $\operatorname{Or}(M)$ is trivial. Obviously this is the case if and only if there exists an atlas (U_α, u_α) for the smooth structure of M such that $\det d(u_\alpha \circ u_\beta^{-1})(u_\beta(x)) > 0$ for all $x \in U_{\alpha\beta}$.

Since the transition functions of $\operatorname{Or}(M)$ take only the values $+1$ and -1, there is a well defined notion of a fiberwise absolute value on $\operatorname{Or}(M)$, given by $|s(x)| := pr_2\, \tau_\alpha(s(x))$, where (U_α, τ_α) is a vector bundle chart of $\operatorname{Or}(M)$ induced by an atlas for M. If M is orientable, there are two distinguished global frames for the orientation bundle $\operatorname{Or}(M)$, namely those with absolute value $|s(x)| = 1$.

The two normed frames s_1 and s_2 of $\operatorname{Or}(M)$ will be called the two possible *orientations* of the orientable manifold M. We call M an *oriented manifold* if one of these two normed frames of $\operatorname{Or}(M)$ is specified: We call it $\mathfrak{o}_M$.

If M is oriented, then $\operatorname{Or}(M) \cong M \times \mathbb{R}$ with the help of the orientation, so we have also

$$\bigwedge^m T^*M = \operatorname{Or}(M) \otimes \operatorname{Vol}(M) = (M \times \mathbb{R}) \otimes \operatorname{Vol}(M) = \operatorname{Vol}(M).$$

So an orientation gives us a canonical identification of m-forms and densities. Thus for any m-form $\omega \in \Omega^m(M)$ the *integral* $\int_M \omega$ is defined by the isomorphism above as the integral of the associated density; see (10.3). If (U_α, u_α) is an oriented atlas (i.e., in each induced vector bundle chart (U_α, τ_α) for $\operatorname{Or}(M)$ we have $\tau_\alpha(\mathfrak{o}_M) = 1$), then the integral of the m-form ω is given by

$$\begin{aligned}\int_M \omega = \sum_\alpha \int_{U_\alpha} f_\alpha \omega &:= \sum_\alpha \int_{U_\alpha} f_\alpha . \omega^\alpha \, du^1 \wedge \dots \wedge du^m \\ :&= \sum_\alpha \int_{u_\alpha(U_\alpha)} f_\alpha(u_\alpha^{-1}(y)) . \omega^\alpha(u_\alpha^{-1}(y)) \, dy^1 \wedge \dots \wedge dy^m,\end{aligned}$$

where the last integral has to be interpreted as an oriented integral on an open subset in $\mathbb{R}^m$.

10.8. Manifolds with boundary. A *manifold with boundary* M is a second countable metrizable topological space together with an equivalence class of smooth atlases (U_α, u_α) which consist of *charts with boundary*: So $u_\alpha : U_\alpha \to u_\alpha(U_\alpha)$ is a homeomorphism from U_α onto an open subset of a *half-space*

$$(-\infty, 0] \times \mathbb{R}^{m-1} = \{(x_1, \dots, x_m) : x_1 \leq 0\},$$

and all chart changes $u_{\alpha\beta} : u_\beta(U_\alpha \cap U_\beta) \to u_\alpha(U_\alpha \cap U_\beta)$ are smooth in the sense that they are restrictions of smooth mappings defined on open (in $\mathbb{R}^m$) neighborhoods of the respective domains. There is a more intrinsic treatment of this notion of smoothness by means of Whitney jets, [**226**], [**220**], and for the case of half-spaces and quadrants as here, [**204**].

We have $u_{\alpha\beta}(u_\beta(U_\alpha \cap U_\beta) \cap (0 \times \mathbb{R}^{m-1})) = u_\alpha(U_\alpha \cap U_\beta) \cap (0 \times \mathbb{R}^{m-1})$ since interior points (with respect to $\mathbb{R}^m$) are mapped to interior points by the inverse function theorem.

Thus the boundary of M, denoted by ∂M, is uniquely given as the set of all points $x \in M$ such that $u_\alpha(x) \in 0 \times \mathbb{R}^{m-1}$ for one (equivalently any) chart (U_α, u_α) of M. Obviously the boundary ∂M is itself a smooth manifold of dimension $m - 1$.

A simple example: The closed unit ball $B^m = \{x \in \mathbb{R}^m : |x| \leq 1\}$ is a manifold with boundary; its boundary is $\partial B^m = S^{m-1}$.

The notions of smooth functions, smooth mappings, tangent bundle (use the approach (1.9) without any change in notation) are analogous to the usual ones. If $x \in \partial M$, we may distinguish in T_xM tangent vectors pointing into the interior, pointing into the exterior, and those in $T_x(\partial M)$.

10.9. Lemma. *Let M be a manifold with boundary of dimension m. Then M is a submanifold with boundary of an m-dimensional manifold $\tilde{M}$ without boundary.*

Proof. Using partitions of unity, we construct a vector field X on M which points strictly into the interior of M. We may multiply X by a strictly positive function so that the flow Fl_t^X exists for all $0 \le t < 2\varepsilon$ for some $\varepsilon > 0$. Then $\mathrm{Fl}_\varepsilon^X : M \to M \setminus \partial M$ is a diffeomorphism onto its image which embeds M as a submanifold with boundary of $M \setminus \partial M$. □

10.10. Lemma. *Let M be an oriented manifold with boundary. Then there is a canonical induced orientation on the boundary ∂M.*

Proof. Let (U_α, u_α) be an oriented atlas for M. Then the chart changes respect the boundary,

$$u_{\alpha\beta} : u_\beta(U_{\alpha\beta} \cap \partial M) \to u_\alpha(U_{\alpha\beta} \cap \partial M).$$

Thus for $x \in u_\beta(U_{\alpha\beta} \cap \partial M)$ we have $du_{\alpha\beta}(x) : 0 \times \mathbb{R}^{m-1} \to 0 \times \mathbb{R}^{m-1}$,

$$du_{\alpha\beta}(x) = \begin{pmatrix} \lambda & 0 & \dots & 0 \\ * & & & * \end{pmatrix},$$

where $\lambda > 0$ since $du_{\alpha\beta}(x)(-e_1)$ is again pointing downwards. So

$$\det du_{\alpha\beta}(x) = \lambda \det(du_{\alpha\beta}(x)|0 \times \mathbb{R}^{m-1}) > 0;$$

consequently $\det(du_{\alpha\beta}(x)|0 \times \mathbb{R}^{m-1}) > 0$ and the restriction of the atlas (U_α, u_α) is an oriented atlas for ∂M. □

10.11. Theorem of Stokes. *Let M be an m-dimensional oriented manifold with boundary ∂M. Then for any $(m-1)$-form $\omega \in \Omega_c^{m-1}(M)$ with compact support on M we have*

$$\int_M d\omega = \int_{\partial M} i^*\omega = \int_{\partial M} \omega,$$

where $i : \partial M \to M$ is the embedding.

Proof. Clearly $d\omega$ again has compact support. Let (U_α, u_α) be an oriented smooth atlas for M and let (f_α) be a smooth partition of unity with $\mathrm{supp}(f_\alpha) \subset U_\alpha$. Then we have $\sum_\alpha f_\alpha \omega = \omega$ and $\sum_\alpha d(f_\alpha \omega) = d\omega$. Consequently

$$\int_M d\omega = \sum_\alpha \int_{U_\alpha} d(f_\alpha \omega) \quad \text{and} \quad \int_{\partial M} \omega = \sum_\alpha \int_{\partial U_\alpha} f_\alpha \omega.$$

It suffices to show that for each α we have

$$\int_{U_\alpha} d(f_\alpha \omega) = \int_{\partial U_\alpha} f_\alpha \omega.$$

For simplicity's sake we now omit the index α. The form $f\omega$ has compact support in U and we have in turn

$$f\omega = \sum_{k=1}^{m} \omega_k du^1 \wedge \cdots \wedge \widehat{du^k} \cdots \wedge du^m,$$

$$d(f\omega) = \sum_{k=1}^{m} \frac{\partial \omega_k}{\partial u^k} du^k \wedge du^1 \wedge \cdots \wedge \widehat{du^k} \cdots \wedge du^m$$

$$= \sum_{k=1}^{m} (-1)^{k-1} \frac{\partial \omega_k}{\partial u^k} du^1 \wedge \cdots \wedge du^m.$$

Since $i^* du^1 = 0$, we have $f\omega | \partial U = i^*(f\omega) = \omega_1 du^2 \wedge \cdots \wedge du^m$, where $i : \partial U \to U$ is the embedding. Finally we get

$$\begin{aligned}
\int_U d(f\omega) &= \int_U \sum_{k=1}^{m} (-1)^{k-1} \frac{\partial \omega_k}{\partial u^k} du^1 \wedge \cdots \wedge du^m \\
&= \sum_{k=1}^{m} (-1)^{k-1} \int_U \frac{\partial \omega_k}{\partial u^k} du^1 \wedge \cdots \wedge du^m \\
&= \sum_{k=1}^{m} (-1)^{k-1} \int_{u(U)} \frac{\partial \omega_k}{\partial x^k} dx^1 \wedge \cdots \wedge dx^m \\
&= \int_{\mathbb{R}^{m-1}} \left(\int_{-\infty}^{0} \frac{\partial \omega_1}{\partial x^1} dx^1 \right) dx^2 \ldots dx^m \\
&\quad + \sum_{k=2}^{m} (-1)^{k-1} \int_{(-\infty,0] \times \mathbb{R}^{m-2}} \left(\int_{-\infty}^{\infty} \frac{\partial \omega_k}{\partial x^k} dx^k \right) dx^1 \ldots \widehat{dx^k} \ldots dx^m \\
&= \int_{\mathbb{R}^{m-1}} (\omega_1(0, x^2, \ldots, x^m) - 0) dx^2 \ldots dx^m \\
&= \int_{\partial U} (\omega_1 | \partial U) du^2 \ldots du^m = \int_{\partial U} f\omega.
\end{aligned}$$

We used the fundamental theorem of calculus twice,

$$\int_{-\infty}^{0} \frac{\partial \omega_1}{\partial x^1} dx^1 = \omega_1(0, x^2, \ldots, x^m) - 0, \qquad \int_{-\infty}^{\infty} \frac{\partial \omega_k}{\partial x^k} dx^k = 0,$$

which holds since $f\omega$ has compact support in U. □

11. De Rham Cohomology

11.1. De Rham cohomology. Let M be a smooth manifold which may have boundary. We consider the graded algebra $\Omega(M) = \bigoplus_{k=0}^{\dim M} \Omega^k(M)$ of all differential forms on M. The space $Z(M) := \{\omega \in \Omega(M) : d\omega = 0\}$ of *closed forms* is a graded subalgebra of Ω, i.e., it is a subalgebra and satisfies $Z(M) = \bigoplus_{k=0}^{\dim M}(\Omega^k(M) \cap Z(M)) = \bigoplus_{k=0}^{\dim M} Z^k(M)$. The space $B(M) := \{d\varphi : \varphi \in \Omega(M)\}$ of *exact forms* is a graded ideal in $Z(M)$: $B(M) \wedge Z(M) \subset B(M)$. This follows directly from the derivation property $d(\varphi \wedge \psi) = d\varphi \wedge \psi + (-1)^{\deg \varphi}\varphi \wedge d\psi$ of the exterior derivative.

Definition. The algebra

$$H^*(M) := \frac{Z(M)}{B(M)} = \frac{\{\omega \in \Omega(M) : d\omega = 0\}}{\{d\varphi : \varphi \in \Omega(M)\}}$$

is called the *de Rham cohomology algebra* of the manifold M. It is graded by

$$H^*(M) = \bigoplus_{k=0}^{\dim M} H^k(M) = \bigoplus_{k=0}^{\dim M} \frac{\ker(d : \Omega^k(M) \to \Omega^{k+1}(M))}{\operatorname{im} d : \Omega^{k-1}(M) \to \Omega^k(M)}.$$

If $f : M \to N$ is a smooth mapping between manifolds, then $f^* : \Omega(N) \to \Omega(M)$ is a homomorphism of graded algebras by (9.5) which satisfies $d \circ f^* = f^* \circ d$ by (9.9). Thus f^* induces an algebra homomorphism which we again call $f^* : H^*(N) \to H^*(M)$.

11.2. Remark. Since $\Omega^k(M) = 0$ for $k > \dim M =: m$, we have

$$\begin{aligned}
H^m(M) &= \frac{\Omega^m(M)}{\{d\varphi : \varphi \in \Omega^{m-1}(M)\}}, \\
H^k(M) &= 0 \qquad \text{for } k > m, \\
H^0(M) &= \frac{\{f \in \Omega^0(M) = C^\infty(M) : df = 0\}}{0} \\
&= \text{the space of locally constant functions on } M \\
&= \mathbb{R}^{b_0(M)},
\end{aligned}$$

where $b_0(M)$ is the number of pathwise connected components of M. We put $b_k(M) := \dim_{\mathbb{R}} H^k(M)$ and call it the k-th *Betti number* of M. If $b_k(M) < \infty$ for all k, we put

$$f_M(t) := \sum_{k=0}^{m} b_k(M)t^k$$

and call it the *Poincaré polynomial* of M. The number

$$\chi_M := \sum_{k=0}^{m} b_k(M)(-1)^k = f_M(-1)$$

is called the *Euler-Poincaré characteristic* of M; see also (13.7) below.

11.3. Examples. We have $H^0(\mathbb{R}^m) = \mathbb{R}$ since it has only one connected component. We have $H^k(\mathbb{R}^m) = 0$ for $k > 0$ by the proof of the lemma of Poincaré (9.10).

For the 1-dimensional sphere we have $H^0(S^1) = \mathbb{R}$ since it is connected, and clearly $H^k(S^1) = 0$ for $k > 1$ by reasons of dimension. Also, we have

$$\begin{aligned} H^1(S^1) &= \frac{\{\omega \in \Omega^1(S^1) : d\omega = 0\}}{\{d\varphi : \varphi \in \Omega^0(S^1)\}} \\ &= \frac{\Omega^1(S^1)}{\{df : f \in C^\infty(S^1)\}}, \\ \Omega^1(S^1) &= \{f\, d\vartheta : f \in C^\infty(S^1)\} \\ &\cong \{f \in C^\infty(\mathbb{R}) : f \text{ is periodic with period } 2\pi\}, \end{aligned}$$

where $d\vartheta$ denotes the global coframe of T^*S^1. If $f \in C^\infty(\mathbb{R})$ is periodic with period 2π, then $f\,dt$ is exact if and only if $\int f\,dt$ is also 2π periodic, i.e., $\int_0^{2\pi} f(t)dt = 0$. So we have

$$H^1(S^1) = \frac{\{f \in C^\infty(\mathbb{R}) : f \text{ is periodic with period } 2\pi\}}{\{f \in C^\infty(\mathbb{R}) : f \text{ is periodic with period } 2\pi, \int_0^{2\pi} f\,dt = 0\}} = \mathbb{R},$$

where $f \mapsto \int_0^{2\pi} f\,dt$ factors to the isomorphism.

11.4. Lemma. *Let f, $g : M \to N$ be smooth mappings between manifolds which are C^∞-homotopic: There exists $h \in C^\infty(\mathbb{R} \times M, N)$ with $h(0,x) = f(x)$ and $h(1,x) = g(x)$. Then f and g induce the same mapping in cohomology:*

$$f^* = g^* : H(N) \to H(M).$$

Remark. f, $g \in C^\infty(M,N)$ are called homotopic if there exists a continuous mapping $h : [0,1]\times M \to N$ with $h(0,x) = f(x)$ and $h(1,x) = g(x)$. This seemingly looser relation in fact coincides with the relation of C^∞-homotopy. We sketch a proof of this statement: Let $\varphi : \mathbb{R} \to [0,1]$ be a smooth function with $\varphi((-\infty, 1/4]) = 0$ and $\varphi([3/4,\infty)) = 1$, and with φ monotone in between. Then consider $\bar{h} : \mathbb{R}\times M \to N$, given by $\bar{h}(t,x) = h(\varphi(t),x)$. Now we may approximate $\bar{h}$ by smooth functions $\tilde{h} : \mathbb{R} \times M \to N$ without changing it on $(-\infty, 1/8) \times M$ where it equals f and on $(7/8,\infty) \times M$ where it equals g. This is done chartwise by convolution with a smooth function with small support on $\mathbb{R}^m$. See [**26**] for a careful presentation of the approximation. So we will use the equivalent concept of homotopic mappings below.

Proof. For $\omega \in \Omega^k(N)$ we have $h^*\omega \in \Omega^k(\mathbb{R}\times M)$. We consider the insertion operator $\operatorname{ins}_t : M \to \mathbb{R}\times M$, given by $\operatorname{ins}_t(x) = (t,x)$. For $\varphi \in \Omega^k(\mathbb{R}\times M)$ we then have a smooth curve $t \mapsto \operatorname{ins}_t^* h^*\varphi$ in $\Omega^k(M)$ (this can be made precise with the help of the calculus in infinite dimensions of [**64**]). We define the integral operator $I_0^1 : \Omega^k(\mathbb{R}\times M) \to \Omega^k(M)$ by $I_0^1(\varphi) := \int_0^1 \operatorname{ins}_t^* \varphi\, dt$. Looking at this locally on M, one sees that it is well defined, even without infinite-dimensional calculus. Let $T := \frac{\partial}{\partial t} \in \mathfrak{X}(\mathbb{R}\times M)$ be the unit vector field in direction $\mathbb{R}$.

We have $\operatorname{ins}_{t+s} = \operatorname{Fl}_t^T \circ \operatorname{ins}_s$ for $s,t \in \mathbb{R}$, so

$$\begin{aligned}\tfrac{\partial}{\partial s} \operatorname{ins}_s^* \varphi &= \partial|_0 (\operatorname{Fl}_t^T \circ \operatorname{ins}_s)^*\varphi = \partial|_0 \operatorname{ins}_s^* (\operatorname{Fl}_t^T)^*\varphi \\ &= \operatorname{ins}_s^* \partial|_0 (\operatorname{Fl}_t^T)^*\varphi = (\operatorname{ins}_s)^* \mathcal{L}_T\varphi \qquad \text{by (9.6).}\end{aligned}$$

We have used that $(\operatorname{ins}_s)^* : \Omega^k(\mathbb{R}\times M) \to \Omega^k(M)$ is linear and continuous and so one may differentiate through it by the chain rule. This can also be checked by evaluating at $x \in M$. Then we have in turn

$$\begin{aligned} d\, I_0^1 \varphi &= d \int_0^1 \operatorname{ins}_t^* \varphi\, dt = \int_0^1 d \operatorname{ins}_t^* \varphi\, dt \\ &= \int_0^1 \operatorname{ins}_t^* d\varphi\, dt = I_0^1\, d\,\varphi \qquad \text{by (9.9.4),} \\ (\operatorname{ins}_1^* - \operatorname{ins}_0^*)\varphi &= \int_0^1 \tfrac{\partial}{\partial t} \operatorname{ins}_t^* \varphi\, dt = \int_0^1 \operatorname{ins}_t^* \mathcal{L}_T \varphi\, dt \\ &= I_0^1\, \mathcal{L}_T\varphi = I_0^1 (d\, i_T + i_T\, d)\varphi \qquad \text{by (9.9).}\end{aligned}$$

Now we define the *homotopy operator* $\bar h := I_0^1 \circ i_T \circ h^* : \Omega^k(N) \to \Omega^{k-1}(M)$. Then we get

$$\begin{aligned} g^* - f^* &= (h\circ \operatorname{ins}_1)^* - (h \circ \operatorname{ins}_0)^* = (\operatorname{ins}_1^* - \operatorname{ins}_0^*)\circ h^* \\ &= (d \circ I_0^1 \circ i_T + I_0^1 \circ i_T \circ d) \circ h^* = d \circ \bar h - \bar h \circ d,\end{aligned}$$

which implies the desired result since for $\omega \in \Omega^k(M)$ with $d\omega = 0$ we have $g^*\omega - f^*\omega = \bar h d\omega + d\bar h\omega = d\bar h\omega$. □

11.5. Lemma. *If a manifold is decomposed into a disjoint union $M = \bigsqcup_\alpha M_\alpha$ of open submanifolds, then $H^k(M) = \prod_\alpha H^k(M_\alpha)$ for all k.*

Proof. $\Omega^k(M)$ is isomorphic to $\prod_\alpha \Omega^k(M_\alpha)$ via $\varphi \mapsto (\varphi|M_\alpha)_\alpha$. This isomorphism commutes with exterior differential d and induces the result. □

11.6. The setting for the Mayer-Vietoris sequence. Let M be a smooth manifold, and let $U, V \subset M$ be open subsets such that $M = U \cup V$. We consider the following embeddings:

$$\begin{array}{ccccc} & & U \cap V & & \\ & {\scriptstyle j_U} \swarrow & & \searrow {\scriptstyle j_V} & \\ U & & & & V \\ & {\scriptstyle i_U} \searrow & & \swarrow {\scriptstyle i_V} & \\ & & M. & & \end{array}$$

Lemma. *In this situation the sequence*

$$0 \to \Omega(M) \xrightarrow{\alpha} \Omega(U) \oplus \Omega(V) \xrightarrow{\beta} \Omega(U \cap V) \to 0$$

is exact, where $\alpha(\omega) := (i_U^* \omega, i_V^* \omega)$ *and* $\beta(\varphi, \psi) = j_U^* \varphi - j_V^* \psi$. *We also have* $(d \oplus d) \circ \alpha = \alpha \circ d$ *and* $d \circ \beta = \beta \circ (d \oplus d)$.

Proof. We have to show that α is injective, $\ker \beta = \operatorname{im} \alpha$, and that β is surjective. The first two assertions are obvious and for the last one we let $\{f_U, f_V\}$ be a partition of unity with $\operatorname{supp} f_U \subset U$ and $\operatorname{supp} f_V \subset V$. For $\varphi \in \Omega(U \cap V)$ we consider $f_V \varphi \in \Omega(U \cap V)$; note that $\operatorname{supp}(f_V \varphi)$ is closed in the set $U \cap V$ which is open in U, so we may extend $f_V \varphi$ by 0 to $\varphi_U \in \Omega(U)$. Likewise we extend $-f_U \varphi$ by 0 to $\varphi_V \in \Omega(V)$. Then we have $\beta(\varphi_U, \varphi_V) = (f_U + f_V)\varphi = \varphi$. □

Now we are in the situation where we may apply the main theorem of homological algebra, (11.8). So we deviate now to develop the basics of homological algebra.

11.7. The essentials of homological algebra. A *graded differential space* (GDS) $K = (K, d)$ is a sequence

$$\cdots \to K^{n-1} \xrightarrow{d^{n-1}} K^n \xrightarrow{d^n} K^{n+1} \to \cdots$$

of abelian groups K^n and group homomorphisms $d^n : K^n \to K^{n+1}$ such that $d^{n+1} \circ d^n = 0$. In our case these are the vector spaces $K^n = \Omega^n(M)$ and the exterior derivative. The group

$$H^n(K) := \frac{\ker(d^n : K^n \to K^{n+1})}{\operatorname{im}(d^{n-1} : K^{n-1} \to K^n)}$$

is called the n-th *cohomology group* of the GDS K. We consider also the direct sum

$$H^*(K) := \bigoplus_{n=-\infty}^{\infty} H^n(K)$$

as a graded group. A homomorphism $f : K \to L$ of graded differential spaces is a sequence of homomorphisms $f^n : K^n \to L^n$ such that $d^n \circ f^n = f^{n+1} \circ d^n$. It induces a homomorphism

$$f_* = H^*(f) : H^*(K) \to H^*(L)$$

and H^* has clearly the properties of a functor from the category of graded differential spaces into the category of graded groups:

$$H^*(Id_K) = Id_{H^*(K)},$$
$$H^*(f \circ g) = H^*(f) \circ H^*(g).$$

A graded differential space (K, d) is called a graded differential algebra if $\bigoplus_n K^n$ is an associative algebra which is graded (so $K^n.K^m \subset K^{n+m}$), such that the differential d is a graded derivation: $d(x.y) = dx.y + (-1)^{\deg x} x.dy$. The cohomology group $H^*(K, d)$ of a graded differential algebra is a graded algebra; see (11.1).

By a *short exact sequence* of graded differential spaces we mean a sequence

$$0 \to K \xrightarrow{i} L \xrightarrow{p} M \to 0$$

of homomorphisms of graded differential spaces which is degreewise exact: For each n the sequence $0 \to K^n \to L^n \to M^n \to 0$ is exact.

11.8. Theorem. Mayer-Vietoris sequence. *Let*

$$0 \to K \xrightarrow{i} L \xrightarrow{p} M \to 0$$

be an exact sequence of graded differential spaces. Then there exists a graded homomorphism $\delta = (\delta^n : H^n(M) \to H^{n+1}(K))_{n\in\mathbb{Z}}$ called the 'connecting homomorphism' such that the following is an exact sequence of abelian groups:

$$\cdots \to H^{n-1}(M) \xrightarrow{\delta} H^n(K) \xrightarrow{i_*} H^n(L) \xrightarrow{p_*} H^n(M) \xrightarrow{\delta} H^{n+1}(K) \to \cdots$$

It is called the 'long exact sequence in cohomology'. Here δ is a natural transformation in the following sense: Let

$$\begin{array}{ccccccccc}
0 & \longrightarrow & K & \xrightarrow{i} & L & \xrightarrow{p} & M & \longrightarrow & 0 \\
 & & \downarrow{\scriptstyle k} & & \downarrow{\scriptstyle \ell} & & \downarrow{\scriptstyle m} & & \\
0 & \longrightarrow & K' & \xrightarrow[i']{} & L' & \xrightarrow[p']{} & M' & \longrightarrow & 0
\end{array}$$

be a commutative diagram of homomorphisms of graded differential spaces with exact lines. Then also the following diagram is commutative:

$$
\begin{array}{ccccccccc}
\cdots \longrightarrow & H^{n-1}(M) & \xrightarrow{\delta} & H^n(K) & \xrightarrow{i_*} & H^n(L) & \xrightarrow{p_*} & H^n(M) & \longrightarrow \cdots \\
 & \downarrow{\scriptstyle m_*} & & \downarrow{\scriptstyle k_*} & & \downarrow{\scriptstyle \ell_*} & & \downarrow{\scriptstyle m_*} & \\
\cdots \longrightarrow & H^{n-1}(M') & \xrightarrow[\delta']{} & H^n(K') & \xrightarrow[i'_*]{} & H^n(L') & \xrightarrow[p'_*]{} & H^n(M) & \longrightarrow \cdots
\end{array}
$$

The long exact sequence in cohomology can also be written in the following way:

$$
\begin{array}{ccc}
H^*(K) & \xrightarrow{i_*} & H^*(L) \\
 & {\scriptstyle \delta}\nwarrow \quad \swarrow{\scriptstyle p_*} & \\
 & H^*(M). &
\end{array}
$$

Definition of δ. The connecting homomorphism is defined by '$\delta = i^{-1} \circ d \circ p^{-1}$' or $\delta[p\ell] = [i^{-1} d\ell]$. This is meant as follows:

$$
\begin{array}{ccccccccc}
 & & & & L^{n-1} & \xrightarrow{p^{n-1}} & M^{n-1} & \longrightarrow & 0 \\
 & & & & \downarrow{\scriptstyle d^{n-1}} & & \downarrow{\scriptstyle d^{n-1}} & & \\
0 & \longrightarrow & K^n & \xrightarrow{i^n} & L^n & \xrightarrow{p^n} & M^n & \longrightarrow & 0 \\
 & & \downarrow{\scriptstyle d^n} & & \downarrow{\scriptstyle d^n} & & \downarrow{\scriptstyle d^n} & & \\
0 & \longrightarrow & K^{n+1} & \xrightarrow{i^{n+1}} & L^{n+1} & \xrightarrow{p^{n+1}} & M^{n+1} & \longrightarrow & 0 \\
 & & \downarrow{\scriptstyle d^{n+1}} & & \downarrow{\scriptstyle d^{n+1}} & & & & \\
0 & \longrightarrow & K^{n+2} & \xrightarrow{i^{n+2}} & L^{n+2}. & & & &
\end{array}
$$

The following argument is called a diagram chase. Let $[m] \in H^n(M)$. Then $m \in M^n$ with $dm = 0$. Since p is surjective, there is $\ell \in L^n$ with $p\ell = m$. We consider $d\ell \in L^{n+1}$ for which we have $pd\ell = dp\ell = dm = 0$, so $d\ell \in \ker p = \operatorname{im} i$; thus there is an element $k \in K^{n+1}$ with $ik = d\ell$. We have $idk = dik = dd\ell = 0$. Since i is injective, we have $dk = 0$, so $[k] \in H^{n+1}(K)$. Now we put $\delta[m] := [k]$ or $\delta[p\ell] = [i^{-1} d\ell]$.

This method of diagram chasing can be used for the whole proof of the theorem. The reader is advised to do it at least once in his life with fingers on the diagram above. For the naturality imagine two copies of the diagram lying above each other with homomorphisms going up.

11.9. Five-lemma. *Let*

$$\begin{array}{ccccccccc}
A_1 & \xrightarrow{\alpha_1} & A_2 & \xrightarrow{\alpha_2} & A_3 & \xrightarrow{\alpha_3} & A_4 & \xrightarrow{\alpha_4} & A_5 \\
\downarrow{\scriptstyle\varphi_1} & & \downarrow{\scriptstyle\varphi_2} & & \downarrow{\scriptstyle\varphi_3} & & \downarrow{\scriptstyle\varphi_4} & & \downarrow{\scriptstyle\varphi_5} \\
B_1 & \xrightarrow{\beta_1} & B_2 & \xrightarrow{\beta_2} & B_3 & \xrightarrow{\beta_3} & B_4 & \xrightarrow{\beta_4} & B_5
\end{array}$$

be a commutative diagram of abelian groups with exact lines. If φ_1, φ_2, φ_4, and φ_5 are isomorphisms, then also the middle φ_3 is an isomorphism.

Proof. Diagram chasing in this diagram leads to the result. The chase becomes simpler if one first replaces the diagram by the following equivalent one with exact lines:

$$\begin{array}{ccccccccc}
0 & \longrightarrow & A_2/\operatorname{im}\alpha_1 & \xrightarrow{\alpha_2'} & A_3 & \xrightarrow{\alpha_3'} & \ker\alpha_4 & \longrightarrow & 0 \\
& & \downarrow{\scriptstyle\varphi_2'}\cong & & \downarrow{\scriptstyle\varphi_3} & & \downarrow{\scriptstyle\varphi_4'}\cong & & \\
0 & \longrightarrow & B_2/\operatorname{im}\beta_2 & \xrightarrow{\beta_2'} & B_3 & \xrightarrow{\beta_3'} & \ker\beta_4 & \longrightarrow & 0.
\end{array}$$ □

11.10. Theorem. Mayer-Vietoris sequence. *Let U and V be open subsets in a manifold M such that $M = U \cup V$. Then there is an exact sequence*

$$\cdots \to H^k(M) \xrightarrow{\alpha_*} H^k(U) \oplus H^k(V) \xrightarrow{\beta_*} H^k(U \cap V) \xrightarrow{\delta} H^{k+1}(M) \to \cdots$$

which is natural in the triple (M, U, V) in the sense explained in (11.8). *The homomorphisms α_* and β_* are algebra homomorphisms, but δ is not.*

Proof. This follows from (11.6) and theorem (11.8). □

Since we shall need it later, we will give now a detailed description of the connecting homomorphism δ. Let $\{f_U, f_V\}$ be a partition of unity with $\operatorname{supp} f_U \subset U$ and $\operatorname{supp} f_V \subset V$. Let $\omega \in \Omega^k(U \cap V)$ with $d\omega = 0$ so that $[\omega] \in H^k(U \cap V)$. Then $(f_V.\omega, -f_U.\omega) \in \Omega^k(U) \oplus \Omega^k(V)$ is mapped to ω by β and so we have by the description of δ in (11.8)

$$\begin{aligned}
\delta[\omega] &= [\alpha^{-1} d(f_V.\omega, -f_U.\omega)] = [\alpha^{-1}(df_V \wedge \omega, -df_U \wedge \omega)] \\
&= [df_V \wedge \omega] = -[df_U \wedge \omega],
\end{aligned}$$

where we have used the following fact: $f_U + f_V = 1$ implies that on $U \cap V$ we have $df_V = -df_U$; thus $df_V \wedge \omega = -df_U \wedge \omega$ and off $U \cap V$ both are 0.

11.11. Axioms for cohomology. The de Rham cohomology is uniquely determined by the following properties which we have already verified:

(1) $H^*(\quad)$ is a contravariant functor from the category of smooth manifolds and smooth mappings into the category of $\mathbb{Z}$-graded groups and graded homomorphisms.
(2) $H^k(\text{point}) = \mathbb{R}$ for $k = 0$ and $H^k(\text{point}) = 0$ for $k \neq 0$.
(3) If f and g are C^∞-homotopic, then $H^*(f) = H^*(g)$.
(4) If $M = \bigsqcup_\alpha M_\alpha$ is a disjoint union of open subsets, then $H^*(M) = \prod_\alpha H^*(M_\alpha)$.
(5) If U and V are open in M, then there exists a connecting homomorphism $\delta : H^k(U \cap V) \to H^{k+1}(U \cup V)$ which is natural in the triple $(U \cup V, U, V)$ such that the following sequence is exact:

$$\dots \to H^k(U \cup V) \to H^k(U) \oplus H^k(V) \to H^k(U \cap V) \xrightarrow{\delta} H^{k+1}(U \cup V) \to \cdots$$

There are many other cohomology theories for topological spaces, like singular cohomology, Čech cohomology, simplicial cohomology, Alexander-Spanier cohomology, etc., which satisfy the above axioms for smooth manifolds when defined with real coefficients, so they all coincide with the de Rham cohomology on manifolds. See books on algebraic topology or sheaf theory for all this and look for the abstract theorem of de Rham in sheaf cohomology.

11.12. Example. If M is contractible (which is equivalent to the seemingly stronger concept of C^∞-contractibility; see the remark in (11.4)), then $H^0(M) = \mathbb{R}$ since M is connected, and $H^k(M) = 0$ for $k \neq 0$, because the constant mapping $c : M \to \text{ point } \to M$ onto some fixed point of M is homotopic to Id_M, so $H^*(c) = H^*(Id_M) = Id_{H^*(M)}$ by (11.4). But we have

$$\begin{array}{ccc} H^k(M) & \xrightarrow{\quad H^k(c) \quad} & H^k(M) \\ & \searrow \qquad \nearrow & \\ & H^k(\text{point}). & \end{array}$$

More generally, two manifolds M and N are called smoothly homotopy equivalent if there exist smooth mappings $f : M \to N$ and $g : N \to M$ such that $g \circ f$ is homotopic to Id_M and $f \circ g$ is homotopic to Id_N. If this is the case, both $H^*(f)$ and $H^*(g)$ are isomorphisms, since

$$H^*(g) \circ H^*(f) = Id_{H^*(M)} \quad \text{and} \quad H^*(f) \circ H^*(g) = Id_{H^*(N)}.$$

As an example consider a vector bundle (E, p, M) with zero section $0_E : M \to E$. Then $p \circ 0_E = Id_M$ whereas $0_E \circ p$ is homotopic to Id_E via $(t, u) \mapsto t.u$. Thus $H^*(E)$ is isomorphic to $H^*(M)$.

11.13. Example. The cohomology of spheres. *For $n \geq 1$ we have*

$$H^k(S^n) = \begin{cases} \mathbb{R} & \text{for } k = 0, \\ 0 & \text{for } 1 \leq k \leq n-1, \\ \mathbb{R} & \text{for } k = n, \\ 0 & \text{for } k > n, \end{cases} \qquad H^k(S^0) = \begin{cases} \mathbb{R}^2 & \text{for } k = 0, \\ 0 & \text{for } k > 0. \end{cases}$$

We may say: The cohomology of S^n has two generators as a graded vector space, one in dimension 0 and one in dimension n. The Poincaré polynomial is given by $f_{S^n}(t) = 1 + t^n$.

Proof. The assertion for S^0 is obvious, and for S^1 it was proved in (11.3) so let $n \geq 2$. Then $H^0(S^n) = \mathbb{R}$ since it is connected, so let $k > 0$. Now fix a north pole $a \in S^n$, $0 < \varepsilon < 1$, and let

$$\begin{aligned} S^n &= \{x \in \mathbb{R}^{n+1} : |x|^2 = \langle x, x \rangle = 1\}, \\ U &= \{x \in S^n : \langle x, a \rangle > -\varepsilon\}, \\ V &= \{x \in S^n : \langle x, a \rangle < \varepsilon\}, \end{aligned}$$

so U and V are overlapping northern and southern hemispheres, respectively, which are diffeomorphic to an open ball and thus smoothly contractible. Their cohomology is thus described in (11.12). Clearly $U \cup V = S^n$ and $U \cap V \cong S^{n-1} \times (-\varepsilon, \varepsilon)$ which is obviously (smoothly) homotopy equivalent to S^{n-1}. By theorem (11.10) we have the following part of the Mayer-Vietoris sequence

$$\begin{array}{ccccccc} H^k(U) \oplus H^k(V) & \longrightarrow & H^k(U \cap V) & \xrightarrow{\delta} & H^{k+1}(S^n) & \longrightarrow & H^{k+1}(U) \oplus H^{k+1}(V) \\ \| & & \| & & & & \| \\ 0 & & H^k(S^{n-1}) & & & & 0, \end{array}$$

where the vertical isomorphisms are from (11.12). Thus $H^k(S^{n-1})$ is isomorphic to $H^{k+1}(S^n)$ for $k > 0$ and $n \geq 2$.

Next we look at the initial segment of the Mayer-Vietoris sequence:

$$\begin{array}{ccccccccccc} 0 & \longrightarrow & H^0(S^n) & \longrightarrow & H^0(U \sqcup V) & \xrightarrow{\beta} & H^0(U \cap V) & \xrightarrow{\delta} & H^1(S^n) & \longrightarrow & H^1(U \sqcup V) \\ & & \| & & \| & & \| & & & & \| \\ 0 & \longrightarrow & \mathbb{R} & \xrightarrow{\alpha} & \mathbb{R}^2 & \longrightarrow & \mathbb{R} & & & & 0. \end{array}$$

From exactness we have: In the lower line α is injective, so $\dim(\ker \beta) = 1$, so β is surjective and thus $\delta = 0$. This implies that $H^1(S^n) = 0$ for $n \geq 2$. Starting from $H^k(S^1)$ for $k > 0$ the result now follows by induction on n.

By looking more closely on the initial segment of the Mayer-Vietoris sequence for $n = 1$ and taking into account the form of $\delta : H^0(S^0) \to H^1(S^1)$, we could even derive the result for S^1 without using (11.3). The reader is advised to try this. □

11.14. Example. The Stiefel manifold $V(k,n;\mathbb{R})$ of oriented orthonormal k-frames in $\mathbb{R}^n$ (see (18.5)) has the following Poincaré polynomial:

For:	$f_{V(k,n)} =$
$n = 2m,\ k = 2l+1,\ l \geq 0:$	$(1+t^{2m-1})\prod_{i=1}^{l}(1+t^{4m-4i-1})$
$n = 2m+1,\ k = 2l,\ l \geq 1:$	$\prod_{i=1}^{l}(1+t^{4m-4i+3})$
$n = 2m,\ k = 2l,\ m > l \geq 1:$	$(1+t^{2m-2l})(1+t^{2m-1})\prod_{i=1}^{l-1}(1+t^{4m-4i-1})$
$n = 2m+1, k = 2l+1,$ $m > l \geq 0:$	$(1+t^{2m-2l})\prod_{i=1}^{l-1}(1+t^{4m-4i+3})$

Since $V(n-1,n;\mathbb{R}) = SO(n;\mathbb{R})$, we get

$$f_{SO(2m;\mathbb{R})}(t) = (1+t^{2m-1})\prod_{i=1}^{m-1}(1+t^{4i-1}),$$

$$f_{SO(2m+1,\mathbb{R})}(t) = \prod_{i=1}^{m}(1+t^{4i-1}).$$

So the cohomology can be quite complicated. For a proof of these formulas using the Gysin sequence for sphere bundles, see [**80**, II].

11.15. Relative de Rham cohomology. Let $N \subset M$ be a closed submanifold and let

$$\Omega^k(M,N) := \{\omega \in \Omega^k(M) : i^*\omega = 0\},$$

where $i : N \to M$ is the embedding. Since $i^* \circ d = d \circ i^*$, we get a graded differential subalgebra $(\Omega^*(M,N),d)$ of $(\Omega^*(M),d)$. Its cohomology, denoted by $H^*(M,N)$, is called the *relative de Rham cohomology* of the *manifold pair* (M,N).

11.16. Lemma. *In the setting of* (11.15),

$$0 \to \Omega^*(M,N) \hookrightarrow \Omega^*(M) \xrightarrow{i^*} \Omega^*(N) \to 0$$

is an exact sequence of differential graded algebras. Thus by (11.8) *we have the long exact sequence in cohomology*

$$\dots \to H^k(M,N) \to H^k(M) \to H^k(N) \xrightarrow{\delta} H^{k+1}(M,N) \to \dots$$

which is natural in the manifold pair (M,N). *It is called the long exact cohomology sequence of the pair* (M,N).

Proof. We only have to show that $i^* : \Omega^*(M) \to \Omega^*(N)$ is surjective. So we have to extend each $\omega \in \Omega^k(N)$ to the whole of M. We cover N by submanifold charts of M with respect to N. These and $M \setminus N$ cover M. On each of the submanifold charts one can easily extend the restriction of ω and one can glue all these extensions by a partition of unity which is subordinated to the cover of M. □

12. Cohomology with Compact Supports and Poincaré Duality

12.1. Cohomology with compact supports. Let $\Omega_c^k(M)$ denote the space of all k-forms with compact support on the manifold M. Since $\operatorname{supp}(d\omega) \subset \operatorname{supp}(\omega)$, $\operatorname{supp}(\mathcal{L}_X\omega) \subset \operatorname{supp}(X) \cap \operatorname{supp}(\omega)$, and $\operatorname{supp}(i_X\omega) \subset \operatorname{supp}(X) \cap \operatorname{supp}(\omega)$, all formulas of section (9) are also valid in $\Omega_c^*(M) = \bigoplus_{k=0}^{\dim M} \Omega_c^k(M)$. So $\Omega_c^*(M)$ is an ideal and a differential graded subalgebra of $\Omega^*(M)$. The cohomology of $\Omega_c^*(M)$

$$H_c^k(M) := \frac{\ker(d : \Omega_c^k(M) \to \Omega_c^{k+1}(M))}{\operatorname{im} d : \Omega_c^{k-1}(M) \to \Omega_c^k(M)},$$

$$H_c^*(M) := \bigoplus_{k=0}^{\dim M} H_c^k(M)$$

is called the *de Rham cohomology algebra with compact supports* of the manifold M. It has no unit if M is not compact.

12.2. Mappings. If $f : M \to N$ is a smooth mapping between manifolds and if $\omega \in \Omega_c^k(N)$ is a form with compact support, then $f^*\omega$ is a k-form on M, in general with noncompact support. So Ω_c^* is not a functor on the category of all smooth manifolds and all smooth mappings. But if we restrict the morphisms suitably, then Ω_c^* becomes a functor. There are two ways to do this:

(1) Ω_c^* is a contravariant functor on the category of all smooth manifolds and *proper smooth mappings* (f is called proper if f^{-1}(compact set) is a compact set) by the usual pullback operation.

(2) Ω_c^* is a covariant functor on the category of all smooth manifolds and embeddings of open submanifolds: For $i : U \hookrightarrow M$ and $\omega \in \Omega_c^k(U)$ just extend ω by 0 off U to get $i_*\omega \in \Omega_c^k(M)$. Clearly $i_* \circ d = d \circ i_*$.

12.3. Remarks. (1) If a manifold M is a disjoint union, $M = \bigsqcup_\alpha M_\alpha$, then we have obviously $H_c^k(M) = \bigoplus_\alpha H_c^k(M_\alpha)$.
(2) $H_c^0(M)$ is a direct sum of copies of $\mathbb{R}$, one for each compact connected component of M.
(3) If M is compact, then $H_c^k(M) = H^k(M)$.

12.4. The Mayer-Vietoris sequence with compact supports. Let M be a smooth manifold, and let U, $V \subset M$ be open subsets such that $M = U \cup V$. We consider the following embeddings:

$$\begin{array}{ccccc} & & U \cap V & & \\ & {\scriptstyle j_U} \swarrow & & \searrow {\scriptstyle j_V} & \\ U & & & & V \\ & {\scriptstyle i_U} \searrow & & \swarrow {\scriptstyle i_V} & \\ & & M. & & \end{array}$$

Theorem. *The following sequence of graded differential algebras is exact:*

$$0 \to \Omega_c^*(U \cap V) \xrightarrow{\beta_c} \Omega_c^*(U) \oplus \Omega_c^*(V) \xrightarrow{\alpha_c} \Omega_c^*(M) \to 0,$$

where $\beta_c(\omega) := ((j_U)_*\omega, (j_V)_*\omega)$ *and* $\alpha_c(\varphi, \psi) = (i_U)_*\varphi - (i_V)_*\psi$. *So by* (11.8) *we have the long exact sequence*

$$\cdots \to H_c^{k-1}(M) \xrightarrow{\delta_c} H_c^k(U \cap V) \to H_c^k(U) \oplus H_c^k(V) \to$$
$$\to H_c^k(M) \xrightarrow{\delta_c} H_c^{k+1}(U \cap V) \to \dots$$

which is natural in the triple (M, U, V). *It is called the Mayer Vietoris sequence with compact supports.*

The connecting homomorphism $\delta_c : H_c^k(M) \to H_c^{k+1}(U \cap V)$ is given by

$$\begin{aligned} \delta_c[\varphi] &= [\beta_c^{-1}\, d\, \alpha_c^{-1}(\varphi)] = [\beta_c^{-1}\, d(f_U\varphi, -f_V\varphi)] \\ &= [df_U \wedge \varphi \restriction U \cap V] = -[df_V \wedge \varphi \restriction U \cap V]. \end{aligned}$$

Proof. The only part that is not completely obvious is that α_c is surjective. Let $\{f_U, f_V\}$ be a partition of unity with $\operatorname{supp}(f_U) \subset U$ and $\operatorname{supp}(f_V) \subset V$, and let $\varphi \in \Omega_c^k(M)$. Then $f_U\varphi \in \Omega_c^k(U)$ and $-f_V\varphi \in \Omega_c^k(V)$ satisfy $\alpha_c(f_U\varphi, -f_V\varphi) = (f_U + f_V)\varphi = \varphi$. $\square$

12.5. Proper homotopies. A smooth mapping $h : \mathbb{R} \times M \to N$ is called a *proper homotopy* if $h^{-1}(\text{compact set}) \cap ([0,1] \times M)$ is compact. A continuous homotopy $h : [0,1] \times M \to N$ is a proper homotopy if and only if it is a proper mapping.

Lemma. *Let $f, g : M \to N$ be proper and proper homotopic. Then $f^* = g^* : H_c^k(N) \to H_c^k(M)$ for all k.*

Proof. Recall the proof of lemma (11.4).
Claim. In the proof of (11.4) we have furthermore $\bar{h} : \Omega_c^k(N) \to \Omega_c^{k-1}(M)$. Let $\omega \in \Omega_c^k(N)$ and let $K_1 := \operatorname{supp}(\omega)$, a compact set in N. Then $K_2 := h^{-1}(K_1) \cap ([0,1] \times M)$ is compact in $\mathbb{R} \times M$, and finally $K_3 := pr_2(K_2)$ is compact in M. If $x \notin K_3$, then we have

$$(\bar{h}\omega)_x = ((I_0^1 \circ i_T \circ h^*)\omega)_x = \int_0^1 (\operatorname{ins}_t^*(i_T h^* \omega))_x \, dt = 0.$$

The rest of the proof is then again as in (11.4). □

12.6. Lemma.

$$H_c^k(\mathbb{R}^n) = \begin{cases} \mathbb{R} & \text{for } k = n, \\ 0 & \text{otherwise.} \end{cases}$$

Proof. We embed $\mathbb{R}^n$ into its one point compactification $\mathbb{R}^n \cup \{\infty\}$ which is diffeomorphic to S^n; see (1.2). The embedding induces the exact sequence of complexes

$$0 \to \Omega_c(\mathbb{R}^n) \to \Omega(S^n) \to \Omega(S^n)_\infty \to 0,$$

where $\Omega(S^n)_\infty$ denotes the space of germs at the point $\infty \in S^n$. For germs at a point the lemma of Poincaré (9.10) is valid, so we have $H^0(\Omega(S^n)_\infty) = \mathbb{R}$ and $H^k(\Omega(S^n)_\infty) = 0$ for $k > 0$. By theorem (11.8) there is a long exact sequence in cohomology whose beginning is:

$$\begin{array}{cccccccccccc} H_c^0(\mathbb{R}^n) & \to & H^0(S^n) & \to & H^0(\Omega(S^n)_\infty) & \xrightarrow{\delta} & H_c^1(\mathbb{R}^n) & \to & H^1(S^n) & \to & H^1(\Omega(S^n)_\infty) \\ \| & & \| & & \| & & & & & & \| \\ 0 & & \mathbb{R} & & \mathbb{R} & & & & & & 0. \end{array}$$

From this we see that $\delta = 0$ and consequently $H_c^1(\mathbb{R}^n) \cong H^1(S^n)$. Another part of this sequence for $k \geq 2$ is:

$$\begin{array}{ccccccc} H^{k-1}(\Omega(S^n)_\infty) & \xrightarrow{\delta} & H_c^k(\mathbb{R}^n) & \longrightarrow & H^k(S^n) & \longrightarrow & H^k(\Omega(S^n)_\infty) \\ \| & & & & & & \| \\ 0 & & & & & & 0. \end{array}$$

It implies $H_c^k(\mathbb{R}^n) \cong H^k(S^n)$ for all k. □

12.7. Fiber integration. Let M be a manifold, and let $\mathrm{pr}_1 : M\times\mathbb{R} \to M$. We define an operator called the *fiber integration*

$$\int_{\text{fiber}} : \Omega_c^k(M\times\mathbb{R}) \to \Omega_c^{k-1}(M)$$

as follows. Let t be the coordinate function on $\mathbb{R}$. A differential form with compact support on $M\times\mathbb{R}$ is a finite linear combination of two types of forms:

(1) $\mathrm{pr}_1^*\varphi.f(x,t)$, or for short $\varphi.f$,

(2) $\mathrm{pr}_1^*\varphi\wedge f(x,t)\,dt$, or for short $\varphi\wedge f\,dt$,

where $\varphi\in\Omega(M)$ and $f\in C_c^\infty(M\times\mathbb{R},\mathbb{R})$. We then put

(1) $\int_{\text{fiber}}\mathrm{pr}_1^*\varphi f := 0$,

(2) $\int_{\text{fiber}}\mathrm{pr}_1^*\varphi\wedge f\,dt := \varphi\int_{-\infty}^{\infty} f(\quad,t)\,dt$.

This is well defined since the only relation which we have to satisfy is $\mathrm{pr}_1^*(\varphi g)\wedge f(x,t)dt = \mathrm{pr}_1^*\varphi g(x)\wedge f(x,t)dt$.

Lemma. *We have $d\circ\int_{fiber} = \int_{fiber}\circ d$. Thus $\int_{fiber}$ induces a linear mapping in cohomology*

$$\left(\int_{fiber}\right)_* : H_c^k(M\times\mathbb{R}) \to H_c^{k-1}(M),$$

which however is not an algebra homomorphism.

Proof. In case (1) we have

$$\begin{aligned}\int_{\text{fiber}} d(\varphi.f) &= \int_{\text{fiber}} d\varphi.f + (-1)^k\int_{\text{fiber}}\varphi.d_Mf + (-1)^k\int_{\text{fiber}}\varphi.\tfrac{\partial f}{\partial t}dt\\ &= (-1)^k\varphi\int_{-\infty}^{\infty}\tfrac{\partial f}{\partial t}dt = 0 \quad\text{ since } f \text{ has compact support}\\ &= d\int_{\text{fiber}}\varphi.f.\end{aligned}$$

In case (2) we get

$$\begin{aligned}\int_{\text{fiber}} d(\varphi\wedge f dt) &= \int_{\text{fiber}} d\varphi\wedge f dt + (-1)^k\int_{\text{fiber}}\varphi\wedge d_Mf\wedge dt\\ &= d\varphi\int_{-\infty}^{\infty} f(\quad,t)dt + (-1)^k\varphi\int_{-\infty}^{\infty} d_Mf(\quad,t)dt\\ &= d\left(\varphi\int_{-\infty}^{\infty} f(\quad,t)dt\right) = d\int_{\text{fiber}}\varphi\wedge f dt. \qquad\square\end{aligned}$$

In order to find a mapping in the converse direction, we let $e = e(t)dt$ be a compactly supported 1-form on $\mathbb{R}$ with $\int_{-\infty}^{\infty} e(t)dt = 1$. We define

$e_* : \Omega_c^k(M) \to \Omega_c^{k+1}(M \times \mathbb{R})$ by $e_*(\varphi) = \varphi \wedge e$. Then

$$de_*(\varphi) = d(\varphi \wedge e) = d\varphi \wedge e + 0 = e_*(d\varphi),$$

so we have an induced mapping in cohomology $e_* : H_c^k(M) \to H_c^{k+1}(M \times \mathbb{R})$. We have $\int_{\text{fiber}} \circ e_* = Id_{\Omega_c^k(M)}$, since

$$\int_{\text{fiber}} e_*(\varphi) = \int_{\text{fiber}} \varphi \wedge e(\quad)dt = \varphi \int_{-\infty}^{\infty} e(t)dt = \varphi.$$

Next we define $K : \Omega_c^k(M \times \mathbb{R}) \to \Omega_c^{k-1}(M \times \mathbb{R})$ by

(1) $K(\varphi.f) := 0$,

(2) $K(\varphi \wedge f dt) = \varphi \int_{-\infty}^t f dt - \varphi.A(t) \int_{-\infty}^{\infty} f dt$, where $A(t) := \int_{-\infty}^t e(t)dt$.

Lemma. *Then we have*

$$Id_{\Omega_c^k(M\times\mathbb{R})} - e_* \circ \int_{\textit{fiber}} = (-1)^{k-1}(d \circ K - K \circ d). \tag{3}$$

Proof. We have to check the two cases. In case (1) we have

$$(Id - e_* \circ \int_{\text{fiber}})(\varphi.f) = \varphi.f - 0,$$

$$\begin{aligned}(d \circ K - K \circ d)(\varphi.f) &= 0 - K(d\varphi.f + (-1)^k \varphi \wedge d_1 f + (-1)^k \varphi \wedge \tfrac{\partial f}{\partial t} dt) \\ &= -(-1)^k \left(\varphi \int_{-\infty}^t \tfrac{\partial f}{\partial t} dt - \varphi.A(t) \int_{-\infty}^{\infty} \tfrac{\partial f}{\partial t} dt\right) \\ &= (-1)^{k-1} \varphi.f + 0.\end{aligned}$$

In case (2) we get

$$(Id - e_* \circ \int_{\text{fiber}})(\varphi \wedge f dt) = \varphi \wedge f dt - \varphi \int_{-\infty}^{\infty} f dt \wedge e,$$

$$\begin{aligned}(d \circ K - K \circ d)(\varphi \wedge f dt) &= d\left(\varphi \int_{-\infty}^t f dt - \varphi.A(t) \int_{-\infty}^{\infty} f dt\right) \\ &\quad - K(d\varphi \wedge f dt + (-1)^{k-1} \varphi \wedge d_1 f \wedge dt) \\ &= (-1)^{k-1} \left(\varphi \wedge f dt - \varphi \wedge e \int_{-\infty}^{\infty} f dt\right). \quad \square\end{aligned}$$

Corollary. *The induced mappings $\left(\int_{\textit{fiber}}\right)_*$ and e_* are inverse to each other and thus isomorphisms between $H_c^k(M \times \mathbb{R})$ and $H_c^{k-1}(M)$.*

Proof. This is clear from the chain homotopy (3). □

12.8. Second proof of (12.6). For $k \leq n$ we have

$$H_c^k(\mathbb{R}^n) \cong H_c^{k-1}(\mathbb{R}^{n-1}) \cong \ldots \cong H_c^0(\mathbb{R}^{n-k})$$
$$= \begin{cases} 0 & \text{for } k < n, \\ H_c^0(\mathbb{R}^0) = \mathbb{R} & \text{for } k = n. \end{cases}$$

Note that the isomorphism $H_c^n(\mathbb{R}^n) \cong \mathbb{R}$ is given by integrating the differential form with compact support with respect to the standard orientation. This is well defined since by Stokes's theorem (10.11) we have $\int_{\mathbb{R}^n} d\omega = \int_\emptyset \omega = 0$, so the integral induces a mapping $\int_* : H_c^n(\mathbb{R}^n) \to \mathbb{R}$. □

12.9. Example. We consider the open Möbius strip M in $\mathbb{R}^3$; see (1.22). Open means without boundary. Then M is contractible onto S^1; in fact M is the total space of a real line bundle over S^1. So from (11.12) we see that $H^k(M) \cong H^k(S^1) = \mathbb{R}$ for $k = 0, 1$ and $= 0$ for $k > 1$.

Now we claim that $H_c^k(M) = 0$ for all k. For that we cut the Möbius strip in two pieces which are glued at the end with one turn,

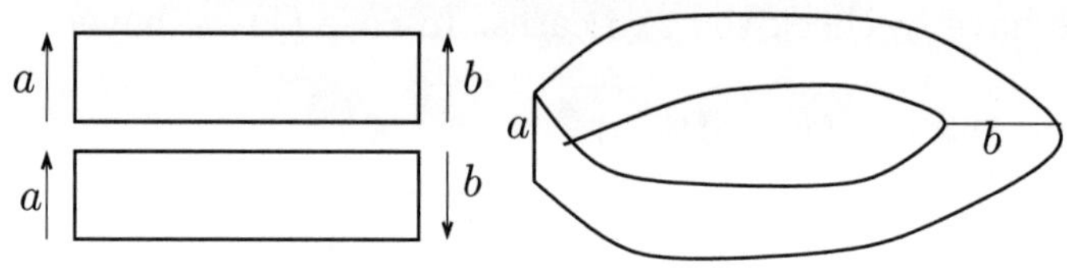

so that $M = U \cup V$ where $U \cong \mathbb{R}^2$, $V \cong \mathbb{R}^2$, and $U \cap V \cong \mathbb{R}^2 \sqcup \mathbb{R}^2$, the disjoint union. We also know that $H_c^0(M) = 0$ since M is not compact and connected. Then the Mayer-Vietoris sequence (see (12.4)) is given by

$$\begin{array}{ccccccc} H_c^1(U) \oplus H_c^1(V) & \longrightarrow & H_c^1(M) & \xrightarrow{\delta} & H_c^2(U \cap V) & \xrightarrow{\beta_c} & \\ \| & & & & \| & & \\ 0 & & & & \mathbb{R} \oplus \mathbb{R} & & \end{array}$$

$$\begin{array}{ccccccc} \xrightarrow{\beta_c} & H_c^2(U) \oplus H_c^2(V) & \longrightarrow & H_c^2(M) & \longrightarrow & H_c^3(U \cap V) \\ & \| & & & & \| \\ & \mathbb{R} \oplus \mathbb{R} & & & & 0. \end{array}$$

We shall show that the linear mapping β_c has rank 2. So we read from the sequence that $H_c^1(M) = 0$ and $H_c^2(M) = 0$. By reasons of dimension $H^k(M) = 0$ for $k > 2$.

Let φ, $\psi \in \Omega_c^2(U \cap V)$ be two forms, supported in the two connected components, respectively, with integral 1 in the orientation induced from one on U. Then $\int_U \varphi = 1$, $\int_U \psi = 1$, but for some orientation on V we have

$\int_V \varphi = 1$ and $\int_V \psi = -1$. So the matrix of the mapping β_c in these bases is $\begin{pmatrix} 1 & 1 \\ 1 & -1 \end{pmatrix}$, which has rank 2.

12.10. Mapping degree for proper mappings. Let $f : \mathbb{R}^n \to \mathbb{R}^n$ be a smooth proper mapping; then $f^* : \Omega_c^k(\mathbb{R}^n) \to \Omega_c^k(\mathbb{R}^n)$ is defined and is an algebra homomorphism. So also the induced mapping in cohomology with compact supports makes sense and by

$$\begin{array}{ccc} H_c^n(\mathbb{R}^n) & \xrightarrow{f^*} & H_c^n(\mathbb{R}^n) \\ {\scriptstyle \int_*}\downarrow{\scriptstyle \cong} & & {\scriptstyle \cong}\downarrow{\scriptstyle \int_*} \\ \mathbb{R} & \xrightarrow{\deg f} & \mathbb{R} \end{array}$$

a linear mapping $\mathbb{R} \to \mathbb{R}$, i.e., multiplication by a real number, is defined. This number $\deg f$ is called the *mapping degree* of f.

12.11. Lemma. *The mapping degree of proper mappings has the following properties:*

(1) *If $f, g : \mathbb{R}^n \to \mathbb{R}^n$ are proper, then* $\deg(f \circ g) = \deg(f).\deg(g)$.
(2) *If f and $g : \mathbb{R}^n \to \mathbb{R}^n$ are proper homotopic (see* (12.5)*), then* $\deg(f) = \deg(g)$.
(3) $\deg(Id_{\mathbb{R}^n}) = 1$.
(4) *If $f : \mathbb{R}^n \to \mathbb{R}^n$ is proper and not surjective, then* $\deg(f) = 0$.

Proof. Only statement (4) needs a proof. Since f is proper, $f(\mathbb{R}^n)$ is closed in $\mathbb{R}^n$: For K compact in $\mathbb{R}^n$ the inverse image $K_1 = f^{-1}(K)$ is compact, so $f(K_1) = f(\mathbb{R}^n) \cap K$ is compact, thus closed. By local compactness $f(\mathbb{R}^n)$ is closed.

Suppose that there exists $x \in \mathbb{R}^n \backslash f(\mathbb{R}^n)$; then there is an open neighborhood $U \subset \mathbb{R}^n \setminus f(\mathbb{R}^n)$. We choose a bump n-form α on $\mathbb{R}^n$ with support in U and $\int \alpha = 1$. Then $f^*\alpha = 0$, so $\deg(f) = 0$ since $[\alpha]$ is a generator of $H_c^n(\mathbb{R}^n)$. □

12.12. Lemma. *For a proper smooth mapping $f : \mathbb{R}^n \to \mathbb{R}^n$ the mapping degree is an integer; in fact for any regular value y of f we have*

$$\deg(f) = \sum_{x \in f^{-1}(y)} \operatorname{sign}(\det(df(x))) \in \mathbb{Z}.$$

Proof. By the Morse-Sard theorem, see (1.18), there exists a regular value y of f. If $f^{-1}(y) = \emptyset$, then f is not surjective, so $\deg(f) = 0$ by (12.11.4) and the formula holds. If $f^{-1}(y) \neq \emptyset$, then for all $x \in f^{-1}(y)$ the tangent mapping $T_x f$ is surjective, thus an isomorphism. By the inverse mapping theorem f is locally a diffeomorphism from an open neighborhood of x onto a neighborhood of y. Thus $f^{-1}(y)$ is a discrete and compact set, say $f^{-1}(y) = \{x_1, \dots, x_k\} \subset \mathbb{R}^n$.

Now we choose pairwise disjoint open neighborhoods U_i of x_i and an open neighborhood V of y such that $f : U_i \to V$ is a diffeomorphism for each i. We choose an n-form α on $\mathbb{R}^n$ with support in V and $\int \alpha = 1$. So $f^*\alpha = \sum_i (f|U_i)^*\alpha$ and moreover

$$\int_{U_i} (f|U_i)^*\alpha = \operatorname{sign}(\det(df(x_i))) \int_V \alpha = \operatorname{sign}(\det(df(x_i))),$$

$$\deg(f) = \int_{\mathbb{R}^n} f^*\alpha = \sum_i \int_{U_i} (f|U_i)^*\alpha = \sum_i^k \operatorname{sign}(\det(df(x_i))) \in \mathbb{Z}. \qquad \square$$

12.13. Example. The last result for a proper smooth mapping $f : \mathbb{R} \to \mathbb{R}$ can be interpreted as follows: Think of f as parametrizing the path of a car on an (infinite) street. A regular value of f is then a position on the street where the car never stops. Wait there and count the directions of the passes of the car: The sum is the mapping degree, the number of journeys from $-\infty$ to ∞. In dimension 1 it can be only -1, 0, or $+1$ (why?).

12.14. Poincaré duality. Let M be an oriented smooth manifold of dimension m without boundary. By Stokes's theorem (10.11), the integral operator $\int : \Omega_c^m(M) \to \mathbb{R}$ vanishes on exact forms and induces the *cohomological integral*

$$\int_* : \quad H_c^m(M) \to \mathbb{R}. \tag{1}$$

It is surjective (use a bump m-form with small support). The *Poincaré product* is the bilinear form

$$\begin{aligned} &P_M^k : H^k(M) \times H_c^{m-k}(M) \to \mathbb{R}, \\ &P_M^k([\alpha],[\beta]) = \int_* [\alpha] \wedge [\beta] = \int_M \alpha \wedge \beta. \end{aligned} \tag{2}$$

It is well defined since for β closed $d\gamma \wedge \beta = d(\gamma \wedge \beta)$, etc. If $j : U \to M$ is an orientation preserving embedding of an open submanifold, then for $[\alpha] \in H^k(M)$ and for $[\beta] \in H_c^{m-k}(U)$ we may compute as follows:

$$\begin{aligned} P_U^k(j^*[\alpha],[\beta]) &= \int_* (j^*[\alpha]) \wedge [\beta] = \int_U j^*\alpha \wedge \beta \\ &= \int_U j^*(\alpha \wedge j_*\beta) = \int_{j(U)} \alpha \wedge j_*\beta \\ &= \int_M \alpha \wedge j_*\beta = P_M^k([\alpha], j_*[\beta]). \end{aligned} \tag{3}$$

Now we define the *Poincaré duality operator*

$$
(4)\qquad \begin{aligned} D^k_M &: H^k(M) \to (H^{m-k}_c(M))^*, \\ \langle [\beta], D^k_M[\alpha]\rangle &= P^k_M([\alpha],[\beta]). \end{aligned}
$$

For example, we have

$$
D^0_{\mathbb{R}^n}(1) = \Big(\int_{\mathbb{R}^n}\Big)_* \in (H^n_c(\mathbb{R}^n))^*.
$$

Let $M = U \cup V$ with U, V open in M; then we have the two Mayer-Vietoris sequences from (11.10) and from (12.4)

$$
\cdots \to H^k(M) \xrightarrow{\alpha_*} H^k(U)\oplus H^k(V) \xrightarrow{\beta_*} H^k(U\cap V) \xrightarrow{\delta} H^{k+1}(M) \to \cdots
$$

$$
H^{m-k}_c(M) \leftarrow H^{m-k}_c(U)\oplus H^{m-k}_c(V) \leftarrow H^{m-k}_c(U\cap V) \xleftarrow{\delta_c} H^{m-(k+1)}_c(M).
$$

We take dual spaces and dual mappings in the second sequence and we replace δ in the first sequence by $(-1)^{k-1}\delta$ and get the following diagram which is commutative as we will see in a moment:

$$
(5)\qquad \begin{array}{ccc}
\vdots & & \vdots \\
\Big\downarrow{\scriptstyle (-1)^{k-2}\delta} & & \Big\downarrow{\scriptstyle \delta_c^*} \\
H^k(M) & \xrightarrow{D^k_M} & H^{m-k}_c(M)^* \\
\Big\downarrow{\scriptstyle (i_U^*,i_V^*)} & & \Big\downarrow{\scriptstyle ((i_U)_*,(i_V)_*)^*} \\
H^k(U)\oplus H^k(V) & \xrightarrow{D^k_U\oplus D^k_V} & H^{m-k}_c(U)^*\oplus H^{m-k}_c(V)^* \\
\Big\downarrow{\scriptstyle j_U^*-j_V^*} & & \Big\downarrow{\scriptstyle ((j_U)_*-(j_V)_*)^*} \\
H^k(U\cap V) & \xrightarrow{D_{U\cap V}} & H^{m-k}_c(U\cap V)^* \\
\Big\downarrow{\scriptstyle (-1)^{k-1}\delta} & & \Big\downarrow{\scriptstyle \delta_c^*} \\
H^{k+1}(M) & \xrightarrow{D^{k+1}_M} & H^{m-(k+1)}_c(M)^* \\
\Big\downarrow & & \Big\downarrow \\
\vdots & & \vdots
\end{array}
$$

12.15. Lemma. *Diagram* (12.14.5) *commutes.*

Proof. The first and the second square from the top commute by (12.14.3). So we have to check that the bottom one commutes. Let $[\alpha] \in H^k(U\cap V)$

and $[\beta] \in H_c^{m-(k+1)}(M)$, and let (f_U, f_V) be a partition of unity which is subordinated to the open cover (U, V) of M. Then we have

$$\begin{aligned}
\langle [\beta], D_M^{k+1}(-1)^{k-1}\delta[\alpha]\rangle &= P_M^{k+1}((-1)^{k-1}\delta[\alpha],[\beta]) \\
&= P_M^{k+1}((-1)^{k-1}[df_V \wedge \alpha],[\beta]) \quad \text{by (11.10)} \\
&= (-1)^{k-1}\int_M df_V \wedge \alpha \wedge \beta, \\
\langle [\beta], \delta_c^* D_{U\cap V}^k[\alpha]\rangle &= \langle \delta_c[\beta], D_{U\cap V}^k[\alpha]\rangle = P_{U\cap V}^k([\alpha], \delta_c[\beta]) \\
&= P_{U\cap V}^k([\alpha], [df_U \wedge \beta] = -[df_V \wedge \beta]) \quad \text{by (12.4)} \\
&= -\int_{U\cap V} \alpha \wedge df_V \wedge \beta = -(-1)^k \int_M df_V \wedge \alpha \wedge \beta. \quad \square
\end{aligned}$$

12.16. Theorem. Poincaré duality. *If M is an oriented manifold of dimension m without boundary, then the Poincaré duality mapping*

$$D_M^k : H^k(M) \to H_c^{m-k}(M)^*$$

is a linear isomorphism for each k.

Proof. Step 1. Let $\mathcal{O}$ be an i-base for the open sets of M, i.e., $\mathcal{O}$ is a basis containing all finite intersections of sets in $\mathcal{O}$. Let $\mathcal{O}_f$ be the set of all open sets in M which are finite unions of sets in $\mathcal{O}$. Let $\mathcal{O}_s$ be the set of all open sets in M which are at most countable disjoint unions of sets in $\mathcal{O}$. Then obviously $\mathcal{O}_f$ and $\mathcal{O}_s$ are again i-bases.

Step 2. Let $\mathcal{O}$ be an i-base for M. If $D_O : H(O) \to H_c(O)^*$ is an isomorphism for all $O \in \mathcal{O}$, then it is so also for all $O \in \mathcal{O}_f$.

Let $U \in \mathcal{O}_f$, $U = O_1 \cup \dots \cup O_k$ for $O_i \in \mathcal{O}$. We consider O_1 and $V = O_2 \cup \dots \cup O_k$. Then $O_1 \cap V = (O_1 \cap O_2) \cup \dots \cup (O_1 \cap O_k)$ is again a union of elements of $\mathcal{O}$ since it is an i-base. Now we prove the claim by induction on k. The case $k = 1$ is trivial. By induction D_{O_1}, D_V, and $D_{O_1 \cap V}$ are isomorphisms, so D_U is also an isomorphism by the five-lemma (11.9) applied to the diagram (12.14.5).

Step 3. If $\mathcal{O}$ is a basis of open sets in M such that D_O is an isomorphism for all $O \in \mathcal{O}$, then it is so also for all $O \in \mathcal{O}_s$.

If $U \in \mathcal{O}_s$, we have $U = O_1 \sqcup O_2 \sqcup \ldots = \bigsqcup_{i=1}^\infty O_i$ for $O_i \in \mathcal{O}$. But then the diagram

$$\begin{array}{ccccc}
H(U) & = & \!\!=\!\!=\!\!=\!\!=\!\!= & = & \prod_{i=1}^\infty H(O_i) \\
{\scriptstyle D_U}\downarrow & & & & \downarrow{\scriptstyle \prod D_{O_i}} \\
H_c(U)^* & = & (\bigoplus_{i=1}^\infty H_c(O_i))^* & = & \prod_{i=1}^\infty H_c(O_i)^*
\end{array}$$

commutes and implies that D_U is an isomorphism.

Step 4. If D_O is an isomorphism for each $O \in \mathcal{O}$ where $\mathcal{O}$ is an i-base for the open sets of M, then D_U is an isomorphism for each open set $U \subset M$.

Namely, $((\mathcal{O}_f)_s)_f$ contains all open sets of M. This is a consequence of the proof that each manifold admits a finite atlas. Then the result follows from steps 2 and 3.

Step 5. $D_{\mathbb{R}^m} : H(\mathbb{R}^m) \to H_c(\mathbb{R}^m)^*$ is an isomorphism.

We have

$$H^k(\mathbb{R}^m) = \begin{cases} \mathbb{R} & \text{for } k = 0, \\ 0 & \text{for } k > 0, \end{cases} \qquad H_c^k(\mathbb{R}^m) = \begin{cases} \mathbb{R} & \text{for } k = m, \\ 0 & \text{for } k \neq m. \end{cases}$$

The class $[1]$ is a generator for $H^0(\mathbb{R}^m)$, and $[\alpha]$ is a generator for $H_c^m(\mathbb{R}^m)$ where α is any m-form with compact support and $\int_M \alpha = 1$. But then $P^0_{\mathbb{R}^m}([1],[\alpha]) = \int_{\mathbb{R}^m} 1.\alpha = 1$.

Step 6. For each open subset $U \subset \mathbb{R}^m$ the mapping D_U is an isomorphism.

The set $\{\{x \in \mathbb{R}^m : a^i < x^i < b^i \text{ for all } i\} : a^i < b^i\}$ is an i-base of $\mathbb{R}^m$. Each element O in it is diffeomorphic (with orientation preserved) to $\mathbb{R}^m$, so D_O is an isomorphism by step 5. From step 4 the result follows.

Step 7. D_M is an isomorphism for each oriented manifold M.

Let $\mathcal{O}$ be the set of all open subsets of M which are diffeomorphic to an open subset of $\mathbb{R}^m$, i.e., all charts of a maximal atlas. Then $\mathcal{O}$ is an i-base for M, and D_O is an isomorphism for each $O \in \mathcal{O}$. By step 4 the operator D_U is an isomorphism for each open U in M; thus also D_M is an isomorphism. □

12.17. Corollary. *For each oriented manifold M without boundary the bilinear pairings*

$$P_M : H^*(M) \times H_c^*(M) \to \mathbb{R},$$
$$P_M^k : H^k(M) \times H_c^{m-k}(M) \to \mathbb{R}$$

are not degenerate.

12.18. Corollary. *Let $j : U \to M$ be the embedding of an open submanifold of an oriented manifold M of dimension m without boundary. Then of the following two mappings one is an isomorphism if and only if the other one is:*

$$j^* : H^k(U) \leftarrow H^k(M),$$
$$j_* : H_c^{m-k}(U) \to H_c^{m-k}(M).$$

Proof. Use (12.14.3), $P_U^k(j^*[\alpha],[\beta]) = P_M^k([\alpha], j_*[\beta])$. □

12.19. Theorem. *Let M be an oriented connected manifold of dimension m without boundary. Then the integral*

$$\int_* : H_c^m(M) \to \mathbb{R}$$

is an isomorphism. So $\ker \int_M = d(\Omega_c^{m-1}(M)) \subset \Omega_c^m(M)$.

Proof. Considering m-forms with small support shows that the integral is surjective. By Poincaré duality (12.16), $\dim_{\mathbb{R}} H_c^m(M)^* = \dim_{\mathbb{R}} H^0(M) = 1$ since M is connected. □

Definition. The uniquely defined cohomology class $\omega_M \in H_c^m(M)$ with integral $\int_M \omega_M = 1$ is called the *orientation class* of the manifold M.

12.20. Relative cohomology with compact supports. Let M be a smooth manifold and let N be a closed submanifold. Then the injection $i : N \to M$ is a proper smooth mapping. We consider the spaces

$$\Omega_c^k(M, N) := \{\omega \in \Omega_c^k(M) : \omega|N = i^*\omega = 0\}$$

whose direct sum $(\Omega_c^*(M, N), d)$ is a graded differential subalgebra of the graded differential algebra $(\Omega_c^*(M), d)$. Its cohomology, $H_c^*(M, N)$, is called the *relative de Rham cohomology with compact supports* of the *manifold pair* (M, N). The sequence of graded differential algebras

$$0 \to \Omega_c^*(M, N) \hookrightarrow \Omega_c^*(M) \xrightarrow{i^*} \Omega_c^*(N) \to 0$$

is exact. This is seen by the same proof as of (11.16) with some obvious changes. Thus by (11.8) we have the long exact sequence in cohomology

$$\dots \to H_c^k(M, N) \to H_c^k(M) \to H_c^k(N) \xrightarrow{\delta} H_c^{k+1}(M, N) \to \dots$$

which is natural in the manifold pair (M, N). It is called the *long exact cohomology sequence with compact supports of the pair* (M, N).

12.21. Now let M be an oriented smooth manifold of dimension m with boundary ∂M. Then ∂M is a closed submanifold of M. Since for $\omega \in \Omega_c^{m-1}(M, \partial M)$ we have $\int_M d\omega = \int_{\partial M} \omega = \int_{\partial M} 0 = 0$, the integral of m-forms factors as

$$\begin{array}{ccccc} \Omega_c^m(M, \partial M) & \hookrightarrow & \Omega_c^m(M) & \xrightarrow{\int_M} & \mathbb{R} \\ \downarrow & & & \nearrow_{\int_*} & \\ H_c^m(M, \partial M) & & & & \end{array}$$

to the *cohomological integral* $\int_* : H_c^m(M, \partial M) \to \mathbb{R}$.

Example. Let $I = [a, b]$ be a compact interval; then $\partial I = \{a, b\}$. We have $H^1(I) = 0$ since $f dt = d \int_a^t f(s) ds$. The long exact sequence in cohomology of the pair $(I, \partial I)$ is

$$\begin{array}{ccccccccccc} 0 \to H^0(I,\partial I) & \to & H^0(I) & \to & H^0(\partial I) & \xrightarrow{\delta} & H^1(I,\partial I) & \to & H^1(I) & \to & H^1(\partial I) \\ \| & & \| & & \| & & \int_* \downarrow \cong & & \| & & \| \\ 0 & & \mathbb{R} & & \mathbb{R}^2 & & \mathbb{R} & & 0 & & 0. \end{array}$$

The connecting homomorphism $\delta : H^0(\partial I) \to H^1(I, \partial I)$ is given by the following procedure: Let $(f(a), f(b)) \in H^0(\partial I)$, where $f \in C^\infty(I)$. Then

$$\delta(f(a), f(b)) = [df] = \int_* [df] = \int_a^b df = \int_a^b f'(t) dt = f(b) - f(a).$$

So the fundamental theorem of calculus can be interpreted as the connecting homomorphism for the long exact sequence of the relative cohomology for the pair $(I, \partial I)$.

The general situation. Let M be an oriented smooth manifold with boundary ∂M. We consider the following piece of the long exact sequence in cohomology with compact supports of the pair $(M, \partial M)$:

$$\begin{array}{ccccccc} H_c^{m-1}(M) & \longrightarrow & H_c^{m-1}(\partial M) & \xrightarrow{\delta} & H_c^m(M, \partial M) & \longrightarrow & H_c^m(M) \longrightarrow 0 \\ & & \int_* \downarrow & & \downarrow \int_* & & \\ & & \mathbb{R} & = & \mathbb{R}. & & \end{array}$$

The connecting homomorphism is given by

$$\delta[\omega | \partial M] = [d\omega]_{H_c^m(M, \partial M)}, \quad \omega \in \Omega_c^{m-1}(M),$$

so commutation of the diagram above is equivalent to the validity of Stokes's theorem.

13. De Rham Cohomology of Compact Manifolds

13.1. The oriented double cover. Let M be a manifold. We consider the orientation bundle $\operatorname{Or}(M)$ of M which we discussed in (10.6), and we consider the subset $\operatorname{or}(M) := \{v \in \operatorname{Or}(M) : |v| = 1\}$; see (10.7) for the modulus. We shall see shortly that it is a submanifold of the total space $\operatorname{Or}(M)$, that it is orientable, and that $\pi_M : \operatorname{or}(M) \to M$ is a double cover of M. The manifold $\operatorname{or}(M)$ is called the *orientable double cover* of M.

We first check that the total space $\operatorname{Or}(M)$ of the orientation bundle is orientable. Let (U_α, u_α) be an atlas for M. Then the orientation bundle is given by the cocycle of transition functions

$$\tau_{\alpha\beta}(x) = \operatorname{sign} \varphi_{\alpha\beta}(x) = \operatorname{sign} \det d(u_\beta \circ u_\alpha^{-1})(u_\alpha(x)).$$

Let (U_α, τ_α) be the induced vector bundle atlas for $\mathrm{Or}(M)$; see (8.3). We consider the mappings

$$\begin{array}{ccccc} \mathrm{Or}(M)|U_\alpha & \xrightarrow{\tau_\alpha} & U_\alpha \times \mathbb{R} & \xrightarrow{u_\alpha \times Id} & u_\alpha(U_\alpha) \times \mathbb{R} \subset \mathbb{R}^{m+1} \\ & {\scriptstyle \pi_M} \searrow & \swarrow {\scriptstyle pr_1} & & \\ & & U_\alpha & & \end{array}$$

and we use them as charts for $\mathrm{Or}(M)$. The chart changes $u_\beta(U_{\alpha\beta}) \times \mathbb{R} \to u_\alpha(U_{\alpha\beta}) \times \mathbb{R}$ are then given by

$$\begin{aligned}(y,t) &\mapsto (u_\alpha \circ u_\beta^{-1}(y), \tau_{\alpha\beta}(u_\beta^{-1}(y))t) \\ &= (u_\alpha \circ u_\beta^{-1}(y), \operatorname{sign}\det d(u_\beta \circ u_\alpha^{-1})((u_\alpha \circ u_\beta^{-1})(y))t) \\ &= (u_\alpha \circ u_\beta^{-1}(y), \operatorname{sign}\det d(u_\alpha \circ u_\beta^{-1})(y)t).\end{aligned}$$

The Jacobi matrix of this mapping is

$$\begin{pmatrix} d(u_\alpha \circ u_\beta^{-1})(y) & * \\ 0 & \operatorname{sign}\det d(u_\alpha \circ u_\beta^{-1})(y) \end{pmatrix}$$

which has positive determinant.

Now we let $Z := \{v \in \mathrm{Or}(M) : |v| \le 1\}$ which is a submanifold with boundary in $\mathrm{Or}(M)$ of the same dimension and thus orientable. Its boundary ∂Z coincides with $\mathrm{or}(M)$, which is thus orientable.

Next we consider the diffeomorphism $\varphi : \mathrm{or}(M) \to \mathrm{or}(M)$ which is induced by the multiplication with -1 in $\mathrm{Or}(M)$. We have $\varphi \circ \varphi = Id$ and $\pi_M^{-1}(x) = \{z, \varphi(z)\}$ for $z \in \mathrm{or}(M)$ and $\pi_M(z) = x$.

Suppose that the manifold M is connected. Then the oriented double cover $\mathrm{or}(M)$ has at most two connected components, since π_M is a two sheeted covering map. If $\mathrm{or}(M)$ has two components, then φ restricts to a diffeomorphism between them. The projection π_M, if restricted to one of the components, becomes invertible, so $\mathrm{Or}(M)$ admits a section which vanishes nowhere; thus M is orientable. So we see that $\mathrm{or}(M)$ is connected if and only if M is not orientable.

The pullback mapping $\varphi^* : \Omega(\mathrm{or}(M)) \to \Omega(\mathrm{or}(M))$ also satisfies $\varphi^* \circ \varphi^* = Id$. We put

$$\begin{aligned}\Omega_+(\mathrm{or}(M)) &:= \{\omega \in \Omega(\mathrm{or}(M)) : \varphi^*\omega = \omega\}, \\ \Omega_-(\mathrm{or}(M)) &:= \{\omega \in \Omega(\mathrm{or}(M)) : \varphi^*\omega = -\omega\}.\end{aligned}$$

For each $\omega \in \Omega(\mathrm{or}(M))$ we have $\omega = \frac{1}{2}(\omega + \varphi^*\omega) + \frac{1}{2}(\omega - \varphi^*\omega) \in \Omega_+(\mathrm{or}(M)) \oplus \Omega_-(\mathrm{or}(M))$, so $\Omega(\mathrm{or}(M)) = \Omega_+(\mathrm{or}(M)) \oplus \Omega_-(\mathrm{or}(M))$. Since $d \circ \varphi^* = \varphi^* \circ d$, these two subspaces are invariant under d; thus we conclude that

(1) $$H^k(\mathrm{or}(M)) = H^k(\Omega_+(\mathrm{or}(M))) \oplus H^k(\Omega_-(\mathrm{or}(M))).$$

Since $\pi_M^* : \Omega(M) \to \Omega(\mathrm{or}(M))$ is an embedding with image $\Omega_+(\mathrm{or}(M))$, we see that the induced mapping $\pi_M^* : H^k(M) \to H^k(\mathrm{or}(M))$ is also an embedding with image $H^k(\Omega_+(\mathrm{or}(M)))$.

13.2. Theorem. *For a compact manifold M we have* $\dim_{\mathbb{R}} H^*(M) < \infty$.

Proof. Step 1. If M is orientable, we have by Poincaré duality (12.16)

$$H^k(M) \xrightarrow[\cong]{D_M^k} (H_c^{m-k}(M))^* = (H^{m-k}(M))^* \xleftarrow[\cong]{(D_M^{m-k})^*} (H_c^k(M))^{**},$$

so $H^k(M)$ is finite-dimensional since otherwise $\dim(H^k(M))^* > \dim H^k(M)$.
Step 2. Let M be not orientable. Then from (13.1) we see that the oriented double cover $\mathrm{or}(M)$ of M is compact, oriented, and connected, and we have $\dim H^k(M) = \dim H^k(\Omega_+(\mathrm{or}(M))) \leq \dim H^k(\mathrm{or}(M)) < \infty$. □

13.3. Theorem. *Let M be a connected manifold of dimension m. Then*

$$H^m(M) \cong \begin{cases} \mathbb{R} & \text{if } M \text{ is compact and orientable,} \\ 0 & \text{otherwise.} \end{cases}$$

Proof. If M is compact and orientable, the integral $\int_* : H^m(M) \to \mathbb{R}$ is an isomorphism, by (12.19).

Next let M be compact but not orientable. Then the oriented double cover $\mathrm{or}(M)$ is connected, compact and oriented. Let $\omega \in \Omega^m(\mathrm{or}(M))$ be an m-form which vanishes nowhere. Then also $\varphi^*\omega$ is nowhere zero where $\varphi : \mathrm{or}(M) \to \mathrm{or}(M)$ is the covering transformation from (13.1). So $\varphi^*\omega = f\omega$ for a function $f \in C^\infty(\mathrm{or}(M))$ which vanishes nowhere. So $f > 0$ or $f < 0$. If $f > 0$, then $\alpha := \omega + \varphi^*\omega = (1+f)\omega$ is again nowhere 0 and $\varphi^*\alpha = \alpha$, so $\alpha = \pi_M^*\beta$ for an m-form β on M without zeros. So M is orientable, a contradiction. Thus $f < 0$ and φ changes the orientation.

The m-form $\gamma := \omega - \varphi^*\omega = (1-f)\omega$ has no zeros, so $\int_{\mathrm{or}(M)} \gamma > 0$ if we orient $\mathrm{or}(M)$ using ω; thus the cohomology class $[\gamma] \in H^m(\mathrm{or}(M))$ is not zero. But $\varphi^*\gamma = -\gamma$ so $\gamma \in \Omega_-(\mathrm{or}(M))$; thus $H^m(\Omega_-(\mathrm{or}(M))) \neq 0$. By the first part of the proof we have $H^m(\mathrm{or}(M)) = \mathbb{R}$ and from (13.1) we get $H^m(\mathrm{or}(M)) = H^m(\Omega_-(\mathrm{or}(M)))$, so $H^m(M) = H^m(\Omega_+(\mathrm{or}(M))) = 0$.

Finally let us suppose that M is not compact. If M is orientable, we have by Poincaré duality (12.16) and by (12.3.1) that $H^m(M) \cong H_c^0(M)^* = 0$.

If M is not orientable, then $\mathrm{or}(M)$ is connected by (13.1) and not compact, so $H^m(M) = H^m(\Omega_+(\mathrm{or}(M))) \subset H^m(\mathrm{or}(M)) = 0$. □

13.4. Corollary. *Let M be a connected manifold which is not orientable. Then* $\operatorname{or}(M)$ *is orientable and the Poincaré duality pairing of* $\operatorname{or}(M)$ *satisfies*

$$\begin{aligned}
&P^k_{\operatorname{or}(M)}(H^k_+(\operatorname{or}(M)), (H^{m-k}_c)_+(\operatorname{or}(M))) = 0,\\
&P^k_{\operatorname{or}(M)}(H^k_-(\operatorname{or}(M)), (H^{m-k}_c)_-(\operatorname{or}(M))) = 0,\\
&H^k_+(\operatorname{or}(M)) \cong (H^{m-k}_c)_-(\operatorname{or}(M))^*,\\
&H^k_-(\operatorname{or}(M)) \cong (H^{m-k}_c)_+(\operatorname{or}(M))^*.
\end{aligned}$$

Proof. From (13.1) we know that $\operatorname{or}(M)$ is connected and orientable. So $\mathbb{R} = H^0(\operatorname{or}(M)) \cong H^m_c(\operatorname{or}(M))^*$.

Now we orient $\operatorname{or}(M)$ and choose a nonnegative bump m-form ω with compact support on $\operatorname{or}(M)$ so that $\int_{\operatorname{or}(M)} \omega > 0$. From the proof of (13.3) we know that the covering transformation $\varphi : \operatorname{or}(M) \to \operatorname{or}(M)$ changes the orientation, so $\varphi^*\omega$ is negatively oriented, i.e., $\int_{\operatorname{or}(M)} \varphi^*\omega < 0$. Then $\omega - \varphi^*\omega \in \Omega^m_-(\operatorname{or}(M))$ and $\int_{\operatorname{or}(M)}(\omega - \varphi^*\omega) > 0$, so $(H^m_c)_-(\operatorname{or}(M)) = \mathbb{R}$ and $(H^m_c)_+(\operatorname{or}(M)) = 0$.

Since φ^* is an algebra homomorphism, we have

$$\begin{aligned}
&\Omega^k_+(\operatorname{or}(M)) \wedge (\Omega^{m-k}_c)_+(\operatorname{or}(M)) \subset (\Omega^m_c)_+(\operatorname{or}(M)),\\
&\Omega^k_-(\operatorname{or}(M)) \wedge (\Omega^{m-k}_c)_-(\operatorname{or}(M)) \subset (\Omega^m_c)_+(\operatorname{or}(M)).
\end{aligned}$$

From $(H^m_c)_+(\operatorname{or}(M)) = 0$ the first two results follows. The last two assertions then follow from this and $H^k(\operatorname{or}(M)) = H^k_+(\operatorname{or}(M)) \oplus H^k_-(\operatorname{or}(M))$ and the analogous decomposition of $H^k_c(\operatorname{or}(M))$. □

13.5. Theorem. *For the real projective spaces we have*

$$\begin{aligned}
&H^0(\mathbb{RP}^n) = \mathbb{R},\\
&H^k(\mathbb{RP}^n) = 0 \qquad \textit{for } 1 \le k < n,\\
&H^n(\mathbb{RP}^n) = \begin{cases} \mathbb{R} & \textit{for odd } n,\\ 0 & \textit{for even } n. \end{cases}
\end{aligned}$$

Proof. The projection $\pi : S^n \to \mathbb{RP}^n$ is a smooth covering mapping with two sheets; the covering transformation is the antipodal mapping $A : S^n \to S^n$, $x \mapsto -x$. We put $\Omega_+(S^n) = \{\omega \in \Omega(S^n) : A^*\omega = \omega\}$ and $\Omega_-(S^n) = \{\omega \in \Omega(S^n) : A^*\omega = -\omega\}$. The pullback $\pi^* : \Omega(\mathbb{RP}^n) \to \Omega(S^n)$ is an embedding onto $\Omega_+(S^n)$.

Let Δ be the determinant function on the oriented Euclidean space $\mathbb{R}^{n+1}$. We identify T_xS^n with $\{x\}^\perp$ in $\mathbb{R}^{n+1}$ and we consider the n-form $\omega_{S^n} \in$

$\Omega^n(S^n)$ which is given by $(\omega_{S^n})_x(X_1,\dots,X_n) = \Delta(x, X_1,\dots,X_n)$. Then we have

$$\begin{aligned}(A^*\omega_{S^n})_x(X_1,\dots,X_n) &= (\omega_{S^n})_{A(x)}(T_xA.X_1,\dots,T_xA.X_n)\\ &= (\omega_{S^n})_{-x}(-X_1,\dots,-X_n)\\ &= \Delta(-x,-X_1,\dots,-X_n)\\ &= (-1)^{n+1}\Delta(x,X_1,\dots,X_n)\\ &= (-1)^{n+1}(\omega_{S^n})_x(X_1,\dots,X_n).\end{aligned}$$

Since ω_{S^n} is invariant under the action of the group $SO(n+1,\mathbb{R})$, it must be the Riemann volume form, so

$$\int_{S^n}\omega_{S^n} = \operatorname{vol}(S^n) = \frac{(n+1)\pi^{\frac{n+1}{2}}}{\Gamma(\frac{n+3}{2})} = \begin{cases}\frac{2\pi^k}{(k-1)!} & \text{for } n = 2k-1,\\ \frac{2^k\pi^{k-1}}{1\cdot 3\cdot 5\dots(2k-3)} & \text{for } n = 2k-2.\end{cases}$$

Thus $[\omega_{S^n}] \in H^n(S^n)$ is a generator for the cohomology. We have $A^*\omega_{S^n} = (-1)^{n+1}\omega_{S^n}$, so

$$\omega_{S^n} \in \begin{cases}\Omega^n_+(S^n) & \text{for odd } n,\\ \Omega^n_-(S^n) & \text{for even } n.\end{cases}$$

Thus $H^n(\mathbb{RP}^n) = H^n(\Omega_+(S^n))$ equals $H^n(S^n) = \mathbb{R}$ for odd n and equals 0 for even n.

Since $\mathbb{RP}^n$ is connected, we have $H^0(\mathbb{RP}^n) = \mathbb{R}$. For $1 \le k < n$ we have $H^k(\mathbb{RP}^n) = H^k(\Omega_+(S^n)) \subset H^k(S^n) = 0$. □

13.6. Corollary. *Let M be a compact manifold. Then for all Betti numbers, we have $b_k(M) := \dim_{\mathbb{R}} H^k(M) < \infty$. If M is compact and orientable of dimension m, we have $b_k(M) = b_{m-k}(M)$.*

Proof. This follows from (13.2) and from Poincaré duality (12.16). □

13.7. Euler-Poincaré characteristic. If M is compact, then all Betti numbers are finite, so the Euler-Poincaré characteristic (see also (11.2))

$$\chi_M = \sum_{k=0}^{\dim M}(-1)^k b_k(M) = f_M(-1)$$

is defined.

Theorem. *Let M be a compact and orientable manifold of dimension m. Then we have:*

(1) *If m is odd, then $\chi_M = 0$.*
(2) *If $m = 2n$ for odd n, then $\chi_M \equiv b_n(M) \equiv 0 \mod (2)$.*
(3) *If $m = 4k$, then $\chi_M \equiv b_{2k}(M) \equiv \mathit{signature}(P^{2k}_M) \mod (2)$.*

Proof. From (13.6) we have $b_q(M) = b_{m-q}(M)$. Thus the Euler-Poincaré characteristic is $\chi_M = \sum_{q=0}^m (-1)^q b_q = \sum_{q=0}^m (-1)^q b_{m-q} = (-1)^m \chi_M$ which implies (1).
If $m = 2n$, we have $\chi_M = \sum_{q=0}^{2n}(-1)^q b_q = 2\sum_{q=0}^{n-1}(-1)^q b_q + (-1)^n b_n$, so $\chi_M \equiv b_n \mod (2)$. In general we have for a compact oriented manifold

$$P_M^q([\alpha],[\beta]) = \int_M \alpha\wedge\beta = (-1)^{q(m-q)}\int_M \beta\wedge\alpha = (-1)^{q(m-q)}P_M^{m-q}([\beta],[\alpha]).$$

For odd n and $m = 2n$ we see that P_M^n is a skew-symmetric nondegenerate bilinear form on $H^n(M)$, so b_n must be even (see (4.7) or (31.4) below) which implies (2).
(3) If $m = 4k$, then P_M^{2k} is a nondegenerate symmetric bilinear form on $H^{2k}(M)$, an inner product. By the *signature* of a nondegenerate symmetric inner product one means the number of positive eigenvalues minus the number of negative eigenvalues, so the number $\dim H^{2k}(M)_+ - \dim H^{2k}(M)_- =: a_+ - a_-$, but since $H^{2k}(M)_+ \oplus H^{2k}(M)_- = H^{2k}(M)$, we have $a_+ + a_- = b_{2k}$, so $a_+ - a_- = b_{2k} - 2a_- \equiv b_{2k} \mod (2)$. □

13.8. The mapping degree. Let M and N be smooth compact oriented manifolds, both of the same dimension m. Then for any smooth mapping $f : M \to N$ there is a real number $\deg f$, called the *degree* of f, which is given in the bottom row of the diagram

$$\begin{array}{ccc} H^m(M) & \xleftarrow{H^m(f)} & H^m(N) \\ {\scriptstyle \int_*}\downarrow{\scriptstyle \cong} & & {\scriptstyle \int_*}\downarrow{\scriptstyle \cong} \\ \mathbb{R} & \xleftarrow{\deg f} & \mathbb{R} \end{array}$$

where the vertical arrows are isomorphisms by (12.19) and where $\deg f$ is the linear mapping given by multiplication with that number. So we also have the defining relation

$$\int_M f^*\omega = \deg f \int_N \omega \qquad \text{for all } \omega \in \Omega^m(N).$$

13.9. Lemma. *The mapping degree* deg *has the following properties:*

(1) $\deg(f\circ g) = \deg f \cdot \deg g$, *and* $\deg(Id_M) = 1$.
(2) *If* $f,\, g : M \to N$ *are (smoothly) homotopic, then* $\deg f = \deg g$.
(3) *If* $\deg f \neq 0$, *then* f *is surjective.*
(4) *If* $f : M \to M$ *is a diffeomorphism, then* $\deg f = 1$ *if* f *respects the orientation and* $\deg f = -1$ *if* f *reverses the orientation.*

Proof. (1) and (2) are clear. (3) If $f(M) \neq N$, we choose a bump m-form ω on N with support in the open set $N \setminus f(M)$. Then $f^*\omega = 0$ so we have $0 = \int_M f^*\omega = \deg f \int_N \omega$. Since $\int_N \omega \neq 0$, we get $\deg f = 0$.

(4) follows either directly from the definition of the integral (10.7) or from (13.11) below. □

13.10. Examples on spheres. Let $f \in O(n+1, \mathbb{R})$ and restrict it to a mapping $f : S^n \to S^n$. Then $\deg f = \det f$. This follows from the description of the volume form on S^n given in the proof of (13.5).

Let f, $g : S^n \to S^n$ be smooth mappings. If $f(x) \neq -g(x)$ for all $x \in S^n$, then the mappings f and g are smoothly homotopic: The homotopy moves $f(x)$ along the shorter arc of the geodesic (big circle) to $g(x)$. So $\deg f = \deg g$.

If $f(x) \neq -x$ for all $x \in S^n$, then f is homotopic to Id_{S^n}, so $\deg f = 1$.

If $f(x) \neq x$ for all $x \in S^n$, then f is homotopic to $-Id_{S^n}$, so $\deg f = (-1)^{n+1}$.

The *hairy ball theorem* says that on S^n for even n each vector field vanishes somewhere. This can be seen as follows. The tangent bundle of the sphere is

$$TS^n = \{(x,y) \in \mathbb{R}^{n+1} \times \mathbb{R}^{n+1} : |x|^2 = 1, \langle x, y \rangle = 0\},$$

so a vector field without zeros is a mapping $x \mapsto (x, g(x))$ with $g(x) \perp x$; then $f(x) := g(x)/|g(x)|$ defines a smooth mapping $f : S^n \to S^n$ with $f(x) \perp x$ for all x. So $f(x) \neq x$ for all x; thus $\deg f = (-1)^{n+1} = -1$. But also $f(x) \neq -x$ for all x, so $\deg f = 1$, a contradiction.

Finally we consider the unit circle $S^1 \xrightarrow{i} \mathbb{C} = \mathbb{R}^2$. Its volume form is given by $\omega := i^*(x\,dy - y\,dx) = i^* \frac{x\,dy - y\,dx}{x^2+y^2}$; obviously we have $\int_{S^1} xdy - ydx = 2\pi$. Now let $f : S^1 \to S^1$ be smooth, $f(t) = (x(t), y(t))$ for $0 \leq t \leq 2\pi$. Then

$$\deg f = \frac{1}{2\pi} \int_{S^1} f^*(xdy - ydx)$$

is the winding number about 0 from complex analysis.

13.11. The mapping degree is an integer. Let $f : M \to N$ be a smooth mapping between compact oriented manifolds of dimension m. Let $b \in N$ be a regular value for f which exists by Sard's theorem; see (1.18). Then for each $x \in f^{-1}(b)$ the tangent mapping $T_x f$ is invertible, so f is a diffeomorphism near x. Thus $f^{-1}(b)$ is a finite set, since M is compact. We define the mapping $\varepsilon : M \to \{-1, 0, 1\}$ by

$$\varepsilon(x) = \begin{cases} 0 & \text{if } T_x f \text{ is not invertible,} \\ 1 & \text{if } T_x f \text{ is invertible and respects orientations,} \\ -1 & \text{if } T_x f \text{ is invertible and changes orientations.} \end{cases}$$

13.12. Theorem. *In the setting of* (13.11), *if* $b \in N$ *is a regular value for* f, *then*

$$\deg f = \sum_{x \in f^{-1}(b)} \varepsilon(x).$$

In particular $\deg f$ *is always an integer.*

Proof. The proof is the same as for lemma (12.12) with obvious changes. □

14. Lie Groups III. Analysis on Lie Groups

Invariant Integration on Lie Groups

14.1. Invariant differential forms on Lie groups. Let G be a real Lie group of dimension n with Lie algebra $\mathfrak{g}$. Then the tangent bundle of G is a trivial vector bundle, see (6.7), so G is orientable. Recall from section (4) the notation:

$\mu : G \times G \to G$, multiplication, $\mu(x, y) = x.y$.

$\mu_a : G \to G$, left translation, $\mu_a(x) = a.x$.

$\mu^a : G \to G$, right translation, $\mu^a(x) = x.a$.

$\nu : G \to G$, inversion, $\nu(x) = x^{-1}$.

$e \in G$, the unit element.

A differential form $\omega \in \Omega^n(G)$ is called *left invariant* if

$$\mu_x^* \omega = \omega \quad \text{for all} \quad x \in G.$$

Then ω is uniquely determined by its value

$$\omega_e \in \bigwedge^n T^*G = \bigwedge^n \mathfrak{g}^*.$$

For each determinant function Δ on $\mathfrak{g}$ there is a unique left invariant n-form L_Δ on G which is given by

$$\text{(1)} \qquad \begin{aligned} (L_\Delta)_x(X_1, \dots, X_n) &:= \Delta(T_x(\mu_{x^{-1}}).X_1, \dots, T_x(\mu_{x^{-1}}).X_n), \\ (L_\Delta)_x &= T_x(\mu_{x^{-1}})^* \Delta. \end{aligned}$$

Likewise there is a unique right invariant n-form R_Δ which is given by

$$\text{(2)} \qquad (R_\Delta)_x(X_1, \dots, X_n) := \Delta(T_x(\mu^{x^{-1}}).X_1, \dots, T_x(\mu^{x^{-1}}).X_n).$$

14.2. Lemma. *We have for all* $a \in G$

$$\text{(1)} \qquad (\mu^a)^* L_\Delta = \det(\mathrm{Ad}(a^{-1})) L_\Delta,$$

$$\text{(2)} \qquad (\mu_a)^* R_\Delta = \det(\mathrm{Ad}(a)) R_\Delta,$$

$$\text{(3)} \qquad (R_\Delta)_a = \det(\mathrm{Ad}(a)) (L_\Delta)_a.$$

Proof. We compute as follows:

$$\begin{aligned}
((\mu^a)^*L_\Delta)_x&(X_1,\dots,X_n) = (L_\Delta)_{xa}(T_x(\mu^a).X_1,\dots,T_x(\mu^a).X_n)\\
&= \Delta(T_{xa}(\mu_{(xa)^{-1}}).T_x(\mu^a).X_1,\dots,T_{xa}(\mu_{(xa)^{-1}}).T_x(\mu^a).X_n)\\
&= \Delta(T_a(\mu_{a^{-1}}).T_{xa}(\mu_{x^{-1}}).T_x(\mu^a).X_1,\dots,T_a(\mu_{a^{-1}}).T_{xa}(\mu_{x^{-1}}).T_x(\mu^a).X_n)\\
&= \Delta(T_a(\mu_{a^{-1}}).T_e(\mu^a).T_x(\mu_{x^{-1}}).X_1,\dots,T_a(\mu_{a^{-1}}).T_e(\mu^a).T_x(\mu_{x^{-1}}).X_n)\\
&= \Delta(\mathrm{Ad}(a^{-1}).T_x(\mu_{x^{-1}}).X_1,\dots,\mathrm{Ad}(a^{-1}).T_x(\mu_{x^{-1}}).X_n)\\
&= \det(\mathrm{Ad}(a^{-1}))\Delta(T_x(\mu_{x^{-1}}).X_1,\dots,T_x(\mu_{x^{-1}}).X_n)\\
&= \det(\mathrm{Ad}(a^{-1}))(L_\Delta)_x(X_1,\dots,X_n),\\
((\mu_a)^*R_\Delta)_x&(X_1,\dots,X_n) = (R_\Delta)_{ax}(T_x(\mu_a).X_1,\dots,T_x(\mu_a).X_n)\\
&= \Delta(T_{ax}(\mu^{(ax)^{-1}}).T_x(\mu_a).X_1,\dots,T_{ax}(\mu^{(ax)^{-1}}).T_x(\mu_a).X_n)\\
&= \Delta(T_a(\mu^{a^{-1}}).T_{ax}(\mu^{x^{-1}}).T_x(\mu_a).X_1,\dots,T_a(\mu^{a^{-1}}).T_{ax}(\mu^{x^{-1}}).T_x(\mu_a).X_n)\\
&= \Delta(T_a(\mu^{a^{-1}}).T_e(\mu_a).T_x(\mu^{x^{-1}}).X_1,\dots,T_a(\mu^{a^{-1}}).T_e(\mu_a).T_x(\mu^{x^{-1}}).X_n)\\
&= \Delta(\mathrm{Ad}(a).T_x(\mu^{x^{-1}}).X_1,\dots,\mathrm{Ad}(a).T_x(\mu^{x^{-1}}).X_n)\\
&= \det(\mathrm{Ad}(a))\Delta(T_x(\mu^{x^{-1}}).X_1,\dots,T_x(\mu^{x^{-1}}).X_n)\\
&= \det(\mathrm{Ad}(a))(R_\Delta)_x(X_1,\dots,X_n),\\
\det(\mathrm{Ad}(a))&(L_\Delta)_a(X_1,\dots,X_n)\\
&= \det(\mathrm{Ad}(a))\Delta(T_a(\mu_{a^{-1}}).X_1,\dots,T_a(\mu_{a^{-1}}).X_n)\\
&= \Delta(\mathrm{Ad}(a).T_a(\mu_{a^{-1}}).X_1,\dots,\mathrm{Ad}(a).T_a(\mu_{a^{-1}}).X_n)\\
&= \Delta(T_a(\mu^{a^{-1}}).T_e(\mu_a).T_a(\mu_{a^{-1}}).X_1,\dots,T_a(\mu^{a^{-1}}).T_e(\mu_a).T_a(\mu_{a^{-1}}).X_n)\\
&= \Delta(T_a(\mu^{a^{-1}}).X_1,\dots,T_a(\mu^{a^{-1}}).X_n) = (R_\Delta)_a(X_1,\dots,X_n). \quad\square
\end{aligned}$$

14.3. Corollary and Definition. *The Lie group G admits a bi-invariant (i.e., left and right invariant) n-form if and only if* $\det(\mathrm{Ad}(a)) = 1$ *for all* $a \in G$.

The Lie group G is called unimodular if $|\det(\mathrm{Ad}(a))| = 1$ *for all* $a \in G$.

Note that $\det(\mathrm{Ad}(a)) > 0$ if G is connected.

Proof. This is obvious from lemma (14.2). □

14.4. Haar measure. We orient the Lie group G by a left invariant n-form L_Δ where $n = \dim(G)$. If $f \in C_c^\infty(G,\mathbb{R})$ is a smooth function with compact support on G, then the integral $\int_G fL_\Delta$ is defined and we have

$$\int_G (\mu_a^* f)L_\Delta = \int_G \mu_a^*(fL_\Delta) = \int_G fL_\Delta,$$

because $\mu_a : G \to G$ is an orientation preserving diffeomorphism of G. Thus $f \mapsto \int_G f L_\Delta$ is a left invariant integration on G, which is also denoted by $\int_G f(x) d_L x$ and which gives rise to a left invariant measure on G, the so-called left *Haar measure*. It is unique up to a multiplicative constant, since $\dim(\bigwedge^n \mathfrak{g}^*) = 1$. In the other notation the left invariance looks like

$$\int_G f(ax) d_L x = \int_G f(x) d_L x \text{ for all } f \in C_c^\infty(G, \mathbb{R}), a \in G.$$

From lemma (14.2.1) we have

$$\int_G ((\mu^a)^* f) L_\Delta = \det(\mathrm{Ad}(a)) \int_G (\mu^a)^* (f L_\Delta) = |\det(\mathrm{Ad}(a))| \int_G f L_\Delta,$$

since the mapping μ^a is orientation preserving if and only if $\det(\mathrm{Ad}(a)) > 0$. So a left invariant Haar measure is also a right invariant one if and only if the Lie group G is unimodular.

14.5. Lemma. *Each compact Lie group is unimodular.*

Proof. The mapping $\det \circ \mathrm{Ad} : G \to GL(1, \mathbb{R})$ is a homomorphism of Lie groups, so its image is a compact subgroup of $GL(1, \mathbb{R})$. Thus $\det(\mathrm{Ad}(G))$ equals $\{1\}$ or $\{1, -1\}$. In both cases we have $|\det(\mathrm{Ad}(a))| = 1$ for all $a \in G$. □

Analysis for Mappings between Lie Groups

14.6. Definition. Let G and H be Lie groups with Lie algebras $\mathfrak{g}$ and $\mathfrak{h}$, respectively, and let $f : G \to H$ be a smooth mapping. Then we define the mapping $Df : G \to L(\mathfrak{g}, \mathfrak{h})$ by

$$Df(x) := T_{f(x)}((\mu^{f(x)})^{-1}).T_x f.T_e(\mu^x) = \delta f(x).T_e(\mu^x),$$

and we call it the *right trivialized derivative* of f.

14.7. Lemma. *The chain rule: For smooth $g : K \to G$ and $f : G \to H$ we have*

$$D(f \circ g)(x) = Df(g(x)) \circ Dg(x).$$

The product rule: For $f, h \in C^\infty(G, H)$ we have

$$D(fh)(x) = Df(x) + \mathrm{Ad}(f(x)) Dh(x).$$

Proof. We compute as follows:

$$\begin{aligned} D(f \circ g)(x) &= T(\mu^{f(g(x))^{-1}}).T_x(f \circ g).T_e(\mu^x) \\ &= T(\mu^{f(g(x))^{-1}}).T_{g(x)}(f).T_e(\mu^{g(x)}).T(\mu^{g(x)^{-1}}).T_x(g).T_e(\mu^x) \\ &= Df(g(x)).Dg(x), \end{aligned}$$

$$D(fh)(x) = T(\mu^{(f(x)h(x))^{-1}}).T_x(\mu \circ (f, h)).T_e(\mu^x)$$

$$
\begin{aligned}
&= T(\mu^{f(x)^{-1}}).T(\mu^{h(x)^{-1}}).T_{f(x),h(x)}\mu.(T_x f.T_e(\mu^x), T_x h.T_e(\mu^x)) \\
&= T(\mu^{f(x)^{-1}}).T(\mu^{h(x)^{-1}}).\Big(T(\mu^{h(x)}).T_x f.T_e(\mu^x) + T(\mu_{f(x)}).T_x h.T_e(\mu^x)\Big) \\
&= T(\mu^{f(x)^{-1}}).T_x f.T_e(\mu^x) + T(\mu^{f(x)^{-1}}).T(\mu_{f(x)}).T(\mu^{h(x)^{-1}}).T_x h.T_e(\mu^x) \\
&= Df(x) + \operatorname{Ad}(f(x)).Dh(x). \quad \square
\end{aligned}
$$

14.8. Inverse function theorem. *Let $f : G \to H$ be smooth and for some $x \in G$ let $Df(x) : \mathfrak{g} \to \mathfrak{h}$ be invertible. Then f is a diffeomorphism from a suitable neighborhood of x in G onto a neighborhood of $f(x)$ in H, and for the derivative we have $D(f^{-1})(f(x)) = (Df(x))^{-1}$.*

Proof. This follows from the usual inverse function theorem. $\square$

14.9. Lemma. *Let $f \in C^\infty(G, G)$ and let $\Delta \in \bigwedge^{\dim G} \mathfrak{g}^*$ be a determinant function on $\mathfrak{g}$. Then we have for all $x \in G$,*

$$(f^* R_\Delta)_x = \det(Df(x))(R_\Delta)_x.$$

Proof. Let $\dim G = n$. We compute as follows:

$$
\begin{aligned}
(f^* R_\Delta)_x(X_1, \dots, X_n) &= (R_\Delta)_{f(x)}(T_x f.X_1, \dots, T_x f.X_n) \\
&= \Delta(T(\mu^{f(x)^{-1}}).T_x f.X_1, \dots) \\
&= \Delta(T(\mu^{f(x)^{-1}}).T_x f.T(\mu^x).T(\mu^{x^{-1}}).X_1, \dots) \\
&= \Delta(Df(x).T(\mu^{x^{-1}}).X_1, \dots) \\
&= \det(Df(x))\Delta(T(\mu^{x^{-1}}).X_1, \dots) \\
&= \det(Df(x))(R_\Delta)_x(X_1, \dots, X_n). \quad \square
\end{aligned}
$$

14.10. Theorem. Transformation formula for multiple integrals. *Let $f : G \to G$ be a diffeomorphism, and let $\Delta \in \bigwedge^{\dim G} \mathfrak{g}^*$. Then for any $g \in C_c^\infty(G, \mathbb{R})$ we have*

$$\int_G g(f(x))|\det(Df(x))| d_R x = \int_G g(y) d_R y,$$

where $d_R x$ is the right Haar measure, given by R_Δ.

Proof. We consider the locally constant function $\varepsilon(x) = \operatorname{sign}\det(Df(x))$ which is 1 on those connected components where f respects the orientation and is -1 on the other components. Then the integral is the sum of all integrals over the connected components and we may investigate each one separately, so let us restrict attention to the component G_0 of the identity. By a right translation (which does not change the integrals) we may assume that $f(G_0) = G_0$. So finally let us assume without loss of generality that G is connected, so that ε is constant. Then by lemma (14.9) we have

$$\begin{aligned}\int_G gR_\Delta &= \varepsilon \int_G f^*(gR_\Delta) = \varepsilon \int_G f^*(g) f^*(R_\Delta)\\ &= \int_G (g\circ f)\varepsilon \det(Df) R_\Delta = \int_G (g\circ f)|\det(Df)| R_\Delta. \quad \square\end{aligned}$$

14.11. Theorem. *Let G be a compact and connected Lie group, and let $f \in C^\infty(G,G)$ and $\Delta \in \bigwedge^{\dim G} \mathfrak{g}^*$. Then we have for $g \in C^\infty(G)$,*

$$\deg f \int_G gR_\Delta = \int_G (g\circ f)\det(Df)R_\Delta, \text{ or}$$

$$\deg f \int_G g(y) d_R y = \int_G g(f(x))\det(Df(x)) d_R x.$$

Here $\deg f$*, the mapping degree of* f*, see* (13.8)*, is an integer.*

Proof. From lemma (14.9) we have $f^*R_\Delta = \det(Df)R_\Delta$. Using this and the defining relation from (13.8) for $\deg f$, we may compute as follows:

$$\begin{aligned}\deg f \int_G gR_\Delta &= \int_G f^*(gR_\Delta) = \int_G f^*(g)f^*(R_\Delta)\\ &= \int_G (g\circ f)\det(Df)R_\Delta. \quad \square\end{aligned}$$

14.12. Examples. Let G be a compact connected Lie group.

(1) If $f = \mu^a : G \to G$, then $D(\mu^a)(x) = Id_{\mathfrak{g}}$. From theorem (14.11) we get $\int_G gR_\Delta = \int_G (g\circ\mu^a)R_\Delta$, the right invariance of the right Haar measure.

(2) If $f = \mu_a : G \to G$, then $D(\mu_a)(x) = T(\mu^{(ax)^{-1}}).T_x(\mu_a).T_e(\mu^x) = \operatorname{Ad}(a)$. So the last two results give $\int_G gR_\Delta = \int_G (g\circ\mu_a)|\det \operatorname{Ad}(a)|R_\Delta$ which we already know from (14.4).

(3) If $f(x) = x^2 = \mu(x,x)$, we have

$$\begin{aligned}Df(x) &= T_{x^2}(\mu^{x^{-2}}).T_{(x,x)}\mu.(T_e(\mu^x), T_e(\mu^x))\\ &= T_x(\mu^{x^{-1}}).T_{x^2}(\mu^{x^{-1}})\,(T_x(\mu_x).T_e(\mu^x) + T_x(\mu^x).T_e(\mu^x))\\ &= \operatorname{Ad}(x) + Id_{\mathfrak{g}}.\end{aligned}$$

Let us now suppose that $\int_G R_\Delta = 1$; then we get

$$\deg((\ \ )^2) = \deg((\ \ )^2)\int_G R_\Delta = \int_G \det(Id_{\mathfrak{g}} + \operatorname{Ad}(x)) d_R x,$$

$$\int_G g(x^2)\det(Id_{\mathfrak{g}} + \operatorname{Ad}(x)) d_R x = \int_G \det(Id_{\mathfrak{g}} + \operatorname{Ad}(x)) d_R x \int_G g(x) d_R x.$$

(4) Let $f(x) = x^k$ for $k \in \mathbb{N}$, and suppose that $\int_G d_R x = 1$. Then we claim that

$$D((\ \ )^k)(x) = \sum_{i=0}^{k-1} \operatorname{Ad}(x^i).$$

This follows from induction, starting from example (3) above, since

$$\begin{aligned} D((\ \ )^k)(x) &= D(Id_G.(\ \ )^{k-1})(x) \\ &= D(Id_G)(x) + \operatorname{Ad}(x).D((\ \ )^{k-1})(x) \quad \text{by (14.7)} \\ &= Id_{\mathfrak{g}} + \operatorname{Ad}(x)(\sum_{i=0}^{k-2} \operatorname{Ad}(x^i)) = \sum_{i=0}^{k-1} \operatorname{Ad}(x^i). \end{aligned}$$

We conclude that

$$\deg(\ \ )^k = \int_G \det\left(\sum_{i=0}^{k-1} \operatorname{Ad}(x^i)\right) d_R x.$$

If G is abelian, we have $\deg(\ \ )^k = k$ since then $\operatorname{Ad}(x) = Id_{\mathfrak{g}}$.
(5) Let $f(x) = \nu(x) = x^{-1}$. Then we have $D\nu(x) = T\mu^{\nu(x)^{-1}}.T_x\nu.T_e\mu^x = -\operatorname{Ad}(x^{-1})$. Using this, we see that the result in (4) holds also for negative k if the summation is interpreted in the right way:

$$D((\ \ )^{-k})(x) = \sum_{i=-k+1}^{0} \operatorname{Ad}(x^i) = -\sum_{i=0}^{k-1} \operatorname{Ad}(x^{-i}).$$

Cohomology of Compact Connected Lie Groups

14.13. Let G be a connected Lie group with Lie algebra $\mathfrak{g}$. The de Rham cohomology of G is the cohomology of the graded differential algebra $(\Omega(G), d)$. We will investigate now what is contributed by the subcomplex of the left invariant differential forms.

Definition. A differential form $\omega \in \Omega(G)$ is called *left invariant* if $\mu_a^*\omega = \omega$ for all $a \in G$. We denote by $\Omega_L(G)$ the subspace of all left invariant forms. Clearly the mapping

$$L : \bigwedge \mathfrak{g}^* \to \Omega_L(G),$$
$$(L_\omega)_x(X_1, \dots, X_k) = \omega(T(\mu_{x^{-1}}).X_1, \dots, T(\mu_{x^{-1}}).X_k),$$

is a linear isomorphism. Since $\mu_a^* \circ d = d \circ \mu_a^*$, the space $(\Omega_L(G), d)$ is a graded differential subalgebra of $(\Omega(G), d)$.

We shall also need the representation $\widetilde{\operatorname{Ad}} : G \to GL(\bigwedge \mathfrak{g}^*)$ which is given by $\widetilde{\operatorname{Ad}}(a) = \bigwedge(\operatorname{Ad}(a^{-1})^*)$ or

$$(\widetilde{\operatorname{Ad}}(a)\omega)(X_1, \dots, X_k) = \omega(\operatorname{Ad}(a^{-1}).X_1, \dots, \operatorname{Ad}(a^{-1}).X_k).$$

14.14. Lemma. (1) *Via the isomorphism $L : \bigwedge \mathfrak{g}^* \to \Omega_L(G)$ the exterior differential d has the following form on $\bigwedge \mathfrak{g}^*$:*

$$d\omega(X_0,\dots,X_k) = \sum_{0\le i<j\le k} (-1)^{i+j}\omega([X_i,X_j],X_0,\dots,\widehat{X_i},\dots\widehat{X_j},\dots,X_k),$$

where $\omega \in \bigwedge^k \mathfrak{g}^*$ and $X_i \in \mathfrak{g}$.
(2) *For $X \in \mathfrak{g}$ we have $i(L_X)\Omega_L(G) \subset \Omega_L(G)$ and $\mathcal{L}_{L_X}\Omega_L(G) \subset \Omega_L(G)$. Thus we have induced mappings*

$$i_X : \bigwedge^k \mathfrak{g}^* \to \bigwedge^{k-1} \mathfrak{g}^*,$$
$$(i_X\omega)(X_1,\dots,X_{k-1}) = \omega(X,X_1,\dots,X_{k-1});$$
$$\mathcal{L}_X : \bigwedge^k \mathfrak{g}^* \to \bigwedge^k \mathfrak{g}^*,$$
$$(\mathcal{L}_X\omega)(X_1,\dots,X_k) = \sum_{i=1}^k (-1)^i\omega([X,X_i],X_1,\dots,\widehat{X_i},\dots X_k).$$

(3) *These mappings satisfy all the properties from section (9), in particular*

$$\mathcal{L}_X = i_X\circ d + d\circ i_X, \quad \text{see (9.9.2)},$$
$$\mathcal{L}_X\circ d = d\circ\mathcal{L}_X, \quad \text{see (9.9.5)},$$
$$[\mathcal{L}_X,\mathcal{L}_Y] = \mathcal{L}_{[X,Y]}, \quad \text{see (9.6.3)}.$$
$$[\mathcal{L}_X,i_Y] = i_{[X,Y]}, \quad \text{see (9.7.3)}.$$

(4) *The representation $\widetilde{\mathrm{Ad}} : G \to GL(\bigwedge \mathfrak{g}^*)$ has derivative $T_e\widetilde{\mathrm{Ad}}.X = \mathcal{L}_X$.*

Proof. For $\omega \in \bigwedge^k \mathfrak{g}^*$ and $X_i \in \mathfrak{g}$ the function

$$\begin{aligned}(L_\omega)_x(L_{X_1}(x),\dots,L_{X_k}(x)) &= \omega(T(\mu_{x^{-1}}).L_{X_1}(x),\dots)\\ &= \omega(T(\mu_{x^{-1}}).T(\mu_x).X_1,\dots)\\ &= \omega(X_1,\dots,X_k)\end{aligned}$$

is constant in x. This implies already that $i(L_X)\Omega_L(G) \subset \Omega_L(G)$ and the form of i_X in (2). Then by (9.8.2) we have

$$\begin{aligned}(d\omega)(X_0,\dots,X_k) &= (dL_\omega)(L_{X_0},\dots,L_{X_k})(e)\\ &= \sum_{i=0}^k (-1)^i L_{X_i}(e)(\omega(X_0,\dots\widehat{X_i},\dots X_k))\\ &\quad + \sum_{0\le i<j\le k} (-1)^{i+j}\omega([X_i,X_j],X_0,\dots,\widehat{X_i},\dots,\widehat{X_j},\dots X_k),\end{aligned}$$

from which assertion (1) follows since the first summand is 0. Similarly we have

$$
\begin{aligned}
(\mathcal{L}_X\omega)(X_1,\dots,X_k) &= (\mathcal{L}_{L_X}L_\omega)(L_{X_1},\dots,L_{X_k})(e)\\
&= L_X(e)(\omega(X_1,\dots,X_k)) + \sum_{i=1}^k(-1)^i\omega([X,X_i],X_1,\dots,\widehat{X_i},\dots X_k).
\end{aligned}
$$

Again the first summand is 0 and the second result of (2) follows.

(3) This is obvious.

(4) For X and $X_i \in \mathfrak{g}$ and for $\omega \in \bigwedge^k \mathfrak{g}^*$ we have

$$
\begin{aligned}
((T_e\widetilde{\mathrm{Ad}}.X)\omega)(X_1,\dots,X_k) &= \partial|_0(\widetilde{\mathrm{Ad}}(\exp(tX))\omega)(X_1,\dots,X_k)\\
&= \partial|_0\omega(\mathrm{Ad}(\exp(-tX)).X_1,\dots,\mathrm{Ad}(\exp(-tX)).X_k)\\
&= \sum_{i=1}^k \omega(X_1,\dots,X_{i-1},-\mathrm{ad}(X)X_i,X_{i+1},\dots X_k)\\
&= \sum_{i=1}^k(-1)^i\omega([X,X_i],X_1,\dots,\widehat{X_i},\dots X_k)\\
&= (\mathcal{L}_X\omega)(X_1,\dots,X_k). \quad\square
\end{aligned}
$$

14.15. Lemma of Maschke. *Let G be a compact Lie group, and let*

$$(0\to)V_1 \xrightarrow{i} V_2 \xrightarrow{p} V_3 \to 0$$

be an exact sequence of G-modules and module homomorphisms such that each V_i is a complete locally convex vector space, i and p are continuous, and the representation of G on each V_i consists of continuous linear mappings with $g\mapsto g.v$ continuous $G\to V_i$ for each $v\in V_i$. Then also the sequence

$$(0\to)V_1^G \xrightarrow{i} V_2^G \xrightarrow{p^G} V_3^G \to 0$$

is exact, where $V_i^G := \{v\in V_i : g.v = v \text{ for all } g\in G\}$.

Convenient vector spaces are sufficient for this lemma; see[**113**].

Proof. We prove first that p^G is surjective. Let $v_3 \in V_3^G \subset V_3$. Since $p: V_2\to V_3$ is surjective, there is a $v_2\in V_2$ with $p(v_2)=v_3$. We consider the element $\tilde v_2 := \int_G x.v_2 d_Lx$; the integral makes sense since $x\mapsto x.v_2$ is a continuous mapping $G\to V_2$, G is compact, and Riemann sums converge in the locally convex topology of V_2. We assume that $\int_G d_Lx = 1$. Then we have

$$a.\tilde v_2 = a.\int_G x.v_2 d_Lx = \int_G (ax).v_2 d_Lx = \int_G x.v_2 d_Lx = \tilde v_2$$

by the left invariance of the integral, see (14.4), where one uses continuous linear functionals to reduce to the scalar valued case. So $\tilde v_2 \in V_2^G$ and since

p is a G-homomorphism, we get

$$\begin{aligned} p^G(\tilde{v}_2) = p(\tilde{v}_2) &= p(\int_G x.v_2 d_L x) \\ &= \int_G p(x.v_2) d_L x = \int_G x.p(v_2) d_L x \\ &= \int x.v_3 d_L x = \int_G v_3 d_L x = v_3. \end{aligned}$$

So p^G is surjective.
Now we prove that the sequence is exact at V_2^G. Clearly $p^G \circ i^G = (p \circ i)|V_1^G = 0$. Suppose conversely that $v_2 \in V_2^G$ with $p^G(v_2) = p(v_2) = 0$. Then there is a $v_1 \in V_1$ with $i(v_1) = v_2$. Consider $\tilde{v}_1 := \int_G x.v_1 d_L x$. As above we see that $\tilde{v}_1 \in V_1^G$ and that $i^G(\tilde{v}_1) = v_2$. □

14.16. Theorem (Chevalley, Eilenberg). *Let G be a compact connected Lie group with Lie algebra $\mathfrak{g}$. Then we have:*

(1) $H^*(G) = H^*(\bigwedge \mathfrak{g}^*, d) =: H^*(\mathfrak{g})$.
(2) $H^*(\mathfrak{g}) = H^*(\bigwedge \mathfrak{g}^*, d) = (\bigwedge \mathfrak{g}^*)^{\mathfrak{g}} = \{\omega \in \bigwedge \mathfrak{g}^* : \mathcal{L}_X \omega = 0$ *for all* $X \in \mathfrak{g}\}$, *the space of all $\mathfrak{g}$-invariant forms on $\mathfrak{g}$.*

The algebra $H^*(\mathfrak{g}) = H(\bigwedge \mathfrak{g}^*, d)$ is called the *Chevalley cohomology of the Lie algebra* $\mathfrak{g}$. For the proof we follow [**193**].

Proof of (1). Let $Z^k(G) = \ker(d : \Omega^k(G) \to \Omega^{k+1}(G))$, and let us consider the following exact sequence of vector spaces:

$$\Omega^{k-1}(G) \xrightarrow{d} Z^k(G) \to H^k(G) \to 0. \tag{3}$$

The group G acts on $\Omega(G)$ by $a \mapsto \mu_{a^{-1}}^*$; this action commutes with d and induces thus an action of G on $Z^k(G)$ and also on $H^k(G)$. On the space $\Omega(G)$ we may consider the compact C^∞-topology (uniform convergence on the compact G, in all derivatives separately, in a fixed set of charts). In this topology d is continuous, $Z^k(G)$ is closed, and the action of G is pointwise continuous. So the assumptions of the lemma of Maschke (14.15) are satisfied and we conclude that the following sequence is also exact:

$$\Omega_L^{p-1}(G) \xrightarrow{d} Z^k(G)^G \to H^k(G)^G \to 0. \tag{4}$$

Since G is connected, for each $a \in G$ we may find a smooth curve $c : [0,1] \to G$ with $c(0) = e$ and $c(1) = a$. Then $(t,x) \mapsto \mu_{c(t)^{-1}}(x) = c(t)^{-1}x$ is a smooth homotopy between Id_G and $\mu_{a^{-1}}$, so by (11.4) the two mappings induce the same mapping in homology; we have

$$\mu_{a^{-1}}^* = Id : H^k(G) \to H^k(G) \quad \text{for each} \quad a \in G.$$

Thus $H^k(G)^G = H^k(G)$. Moreover $Z^k(G)^G = \ker(d : \Omega^k_L(G) \to \Omega^{k+1}_L(G))$, so from the exact sequence (4) we may conclude that

$$H^k(G) = H^k(G)^G = \frac{\ker(d : \Omega^k_L(G) \to \Omega^{k+1}_L(G))}{\operatorname{im}(d : \Omega^{k-1}_L(G) \to \Omega^k_L(G))} = H^k(\textstyle\bigwedge \mathfrak{g}^*, d).$$

Proof of (2). From (14.14.3) we have $\mathcal{L}_X \circ d = d \circ \mathcal{L}_X$, so by (14.14.4) we conclude that $\widetilde{\operatorname{Ad}}(a) \circ d = d \circ \widetilde{\operatorname{Ad}}(a) : \bigwedge \mathfrak{g}^* \to \bigwedge \mathfrak{g}^*$ since G is connected. Thus the sequence

$$\text{(5)} \qquad \bigwedge^{k-1} \mathfrak{g}^* \xrightarrow{d} Z^k(\mathfrak{g}^*) \to H^k(\textstyle\bigwedge \mathfrak{g}^*, d) \to 0$$

is an exact sequence of G-modules and G-homomorphisms, where $Z^k(\mathfrak{g}^*) = \ker(d : \bigwedge^k \mathfrak{g}^* \to \bigwedge^{k+1} \mathfrak{g}^*)$. All spaces are finite-dimensional, so the lemma of Maschke (14.15) is applicable and we may conclude that also the following sequence is exact:

$$\text{(6)} \qquad (\bigwedge^{k-1} \mathfrak{g}^*)^G \xrightarrow{d} Z^k(\mathfrak{g}^*)^G \to H^k(\textstyle\bigwedge \mathfrak{g}^*, d)^G \to 0.$$

The space $H^k(\bigwedge \mathfrak{g}^*, d)^G$ consists of all cohomology classes α with $\widetilde{\operatorname{Ad}}(a)\alpha = \alpha$ for all $a \in G$. Since G is connected, by (14.14.4) these are exactly the α with $\mathcal{L}_X \alpha = 0$ for all $X \in \mathfrak{g}$. For $\omega \in \bigwedge \mathfrak{g}^*$ with $d\omega = 0$ we have by (14.14.3) that $\mathcal{L}_X \omega = i_X d\omega + d i_X \omega = d i_X \omega$, so that $\mathcal{L}_X \alpha = 0$ for all $\alpha \in H^k(\bigwedge \mathfrak{g}^*, d)$. Thus we get $H^k(\bigwedge \mathfrak{g}^*, d) = H^k(\bigwedge \mathfrak{g}^*, d)^G$. Also we have $(\bigwedge \mathfrak{g}^*)^G = (\bigwedge \mathfrak{g}^*)^{\mathfrak{g}}$ so that the exact sequence (6) translates to

$$\text{(7)} \qquad H^k(\mathfrak{g}) = H^k(\textstyle\bigwedge \mathfrak{g}^*, d) = H^k((\bigwedge \mathfrak{g}^*)^{\mathfrak{g}}, d).$$

Now let $\omega \in (\bigwedge^k \mathfrak{g}^*)^{\mathfrak{g}} = \{\varphi : \mathcal{L}_X \varphi = 0 \text{ for all } X \in \mathfrak{g}\}$ and consider the inversion $\nu : G \to G$. Then we have for $\omega \in \bigwedge^k \mathfrak{g}^*$ and $X_i \in \mathfrak{g}$:

$$\begin{aligned}
&(\nu^* L_\omega)_a (T_e(\mu_a).X_1, \dots, T_e(\mu_a).X_k) \\
&= (L_\omega)_{a^{-1}} (T_a \nu.T_e(\mu_a).X_1, \dots, T_a \nu.T_e(\mu_a).X_k) \\
&= (L_\omega)_{a^{-1}} (-T(\mu^{a^{-1}}).T(\mu_{a^{-1}}).T_e(\mu_a).X_1, \dots) \\
&= (L_\omega)_{a^{-1}} (-T_e(\mu^{a^{-1}}).X_1, \dots, -T_e(\mu^{a^{-1}}).X_k) \\
&= (-1)^k \omega(T\mu_a.T\mu^{a^{-1}}.X_1, \dots, T\mu_a.T\mu^{a^{-1}}.X_k) \\
&= (-1)^k \omega(\operatorname{Ad}(a).X_1, \dots, \operatorname{Ad}(a).X_k) \\
&= (-1)^k (\widetilde{\operatorname{Ad}}(a^{-1})\omega)(X_1, \dots, X_k) \\
&= (-1)^k \omega(X_1, \dots, X_k) \qquad \text{since } \omega \in (\textstyle\bigwedge^k \mathfrak{g}^*)^{\mathfrak{g}} \\
&= (-1)^k (L_\omega)_a (T_e(\mu_a).X_1, \dots, T_e(\mu_a).X_k).
\end{aligned}$$

So for $\omega \in (\bigwedge^k \mathfrak{g}^*)^{\mathfrak{g}}$ we have $\nu^* L_\omega = (-1)^k L_\omega$ and thus also $(-1)^{k+1} L_{d\omega} = \nu^* dL_\omega = d\nu^* L_\omega = (-1)^k dL_\omega = (-1)^k L_{d\omega}$ which implies $d\omega = 0$. Hence we have $d|(\bigwedge \mathfrak{g}^*)^{\mathfrak{g}} = 0$.

From (7) we now get $H^k(\mathfrak{g}) = H^k((\bigwedge \mathfrak{g}^*)^{\mathfrak{g}}, 0) = (\bigwedge^k \mathfrak{g}^*)^{\mathfrak{g}}$ as required. □

14.17. Corollary. *Let G be a compact connected Lie group. Then its Poincaré polynomial is given by*

$$f_G(t) = \int_G \det(\mathrm{Ad}(x) + tId_{\mathfrak{g}}) d_L x.$$

Proof. Let $\dim G = n$. By (11.2) and (13.6) we have

$$f_G(t) = \sum_{k=0}^n b_k(G) t^k = \sum_{k=0}^n b_k(G) t^{n-k} = \sum_{k=0}^n \dim_{\mathbb{R}} H^k(G) t^{n-k}.$$

On the other hand we have

$$\begin{aligned}\int_G \det(\mathrm{Ad}(x) + tId_{\mathfrak{g}}) d_L x &= \int_G \det(\mathrm{Ad}(x^{-1})^* + tId_{\mathfrak{g}^*}) d_L x \\ &= \int_G \sum_{k=0}^n \mathrm{Trace}(\bigwedge^k \mathrm{Ad}(x^{-1})^*)\, t^{n-k} d_L x \quad \text{by (14.19) below} \\ &= \sum_{k=0}^n \int_G \mathrm{Trace}(\widetilde{\mathrm{Ad}}(x) | \bigwedge^k \mathfrak{g}^*) d_L x\; t^{n-k}.\end{aligned}$$

If $\rho : G \to GL(V)$ is a finite-dimensional representation of G, then the operator $\int_G \rho(x) d_L x : V \to V$ is just a projection onto V^G, the space of fixed points of the representation; see the proof of the lemma of Maschke (14.15). The trace of a projection is the dimension of the image. So

$$\begin{aligned}\int_G \mathrm{Trace}(\widetilde{\mathrm{Ad}}(a) | \bigwedge^k \mathfrak{g}^*) d_L x &= \mathrm{Trace}\left(\int_G (\widetilde{Ad}(a) | \bigwedge^k \mathfrak{g}^*) d_L x\right) \\ &= \dim(\bigwedge^k \mathfrak{g}^*)^G = \dim H^k(G). \quad \square\end{aligned}$$

14.18. Let $\mathbb{T}^n = (S^1)^n$ be the n-dimensional torus, and let $\mathfrak{t}^n$ be its Lie algebra. The Lie bracket is zero since the torus is an abelian group. From theorem (14.16) we have then that $H^*(\mathbb{T}^n) = (\bigwedge (\mathfrak{t}^n)^*)^{\mathfrak{t}^n} = \bigwedge (\mathfrak{t}^n)^*$, so the Poincaré polynomial is $f_{\mathbb{T}^n}(t) = (1+t)^n$.

14.19. Lemma. *Let V be an n-dimensional vector space and let $A : V \to V$ be a linear mapping. Then we have*

$$\det(A + tId_V) = \sum_{k=0}^{n} t^{n-k} \operatorname{Trace}(\bigwedge^k A).$$

Proof. By $\bigwedge^k A : \bigwedge^k V \to \bigwedge^k V$ we mean the mapping $v_1 \wedge \cdots \wedge v_k \mapsto Av_1 \wedge \cdots \wedge Av_k$. Let $e_1, \dots, e_n$ be a basis of V. By the definition of the determinant we have

$$\begin{aligned}\det(A + tId_V)(e_1 \wedge \cdots \wedge e_n) &= (Ae_1 + te_1) \wedge \cdots \wedge (Ae_n + te_n) \\ &= \sum_{k=0}^{n} t^{n-k} \sum_{i_1 < \cdots < i_k} e_1 \wedge \cdots \wedge Ae_{i_1} \wedge \cdots \wedge Ae_{i_k} \wedge \cdots \wedge e_n.\end{aligned}$$

The multivectors $(e_{i_1} \wedge \cdots \wedge e_{i_k})_{i_1 < \cdots < i_k}$ are a basis of $\bigwedge^k V$ and we can thus write

$$(\bigwedge^k A)(e_{i_1} \wedge \cdots \wedge e_{i_k}) = Ae_{i_1} \wedge \cdots \wedge Ae_{i_k} = \sum_{j_1 < \cdots < j_k} A^{j_1 \dots j_k}_{i_1 \dots i_k} e_{j_1} \wedge \cdots \wedge e_{j_k},$$

where $(A^{j_1 \dots j_k}_{i_1 \dots i_k})$ is the matrix of $\bigwedge^k A$ in this basis. We see that

$$e_1 \wedge \cdots \wedge Ae_{i_1} \wedge \cdots \wedge Ae_{i_k} \wedge \cdots \wedge e_n = A^{i_1 \dots i_k}_{i_1 \dots i_k} e_1 \wedge \cdots \wedge e_n.$$

Consequently we have

$$\begin{aligned}\det(A + tId_V) e_1 \wedge \cdots \wedge e_n &= \sum_{k=0}^{n} t^{n-k} \sum_{i_1 < \cdots < i_k} A^{i_1 \dots i_k}_{i_1 \dots i_k} e_1 \wedge \cdots \wedge e_n \\ &= \sum_{k=0}^{n} t^{n-k} \operatorname{Trace}(\bigwedge^{\text{k}} \text{A})\, e_1 \wedge \cdots \wedge e_n,\end{aligned}$$

which implies the result. □

15. Extensions of Lie Algebras and Lie Groups

Extension of Lie Algebras

In this section we describe first the theory of semidirect products and central extensions of Lie algebras, later the more involved theory of general extensions with noncommutative kernels. For the latter we follow the presentation from [**6**], with special emphasis on relations with the (algebraic) theory of covariant exterior derivatives, curvature and the Bianchi identity in differential geometry (see section (15.3)). The results are due to [**89**], [**163**], [**208**], and generalizations for Lie algebroids are in [**126**]. The analogous result for Lie super-algebras are available in [**7**].

15.1. Extensions. An *extension* of a Lie algebra $\mathfrak{g}$ with kernel $\mathfrak{h}$ is an exact sequence of homomorphisms of Lie algebras:

$$0 \to \mathfrak{h} \xrightarrow{i} \mathfrak{e} \xrightarrow{p} \mathfrak{g} \to 0.$$

(1) This extension is called a *semidirect product* if we can find a section $s : \mathfrak{g} \to \mathfrak{e}$ which is a Lie algebra homomorphism. Then we have a representation of the Lie algebra $\alpha : \mathfrak{g} \to L(\mathfrak{h}, \mathfrak{h})$ which is given by $\alpha_X(H) = [s(X), H]$ where we suppress the injection i. It is a representation since $\alpha_{[X,Y]}H = [s([X,Y]), H] = [[s(X), s(Y)], H] = [s(X), [s(Y), H]] - [s(Y), [s(X, H)]] = (\alpha_X\alpha_Y - \alpha_Y\alpha_X)H$. This representation takes values in the Lie algebra $\operatorname{der}(\mathfrak{h})$ of derivations of $\mathfrak{h}$, so $\alpha : \mathfrak{g} \to \operatorname{der}(\mathfrak{h})$. From the data α, s we can reconstruct the extension $\mathfrak{e}$ since on $\mathfrak{h} \times \mathfrak{g}$ we have $[H + s(X), H' + s(X')] = [H, H'] + [s(X), H'] - [s(X'), H] + [X, X'] = [H, H'] + \alpha_X(H') - \alpha_{X'}(H) + [X, X']$.
(2) The extension is called a *central extension* if $\mathfrak{h}$ or rather $i(\mathfrak{h})$ is in the center of $\mathfrak{e}$.

15.2. Describing extensions. Consider any exact sequence of homomorphisms of Lie algebras:

$$0 \to \mathfrak{h} \xrightarrow{i} \mathfrak{e} \xrightarrow{p} \mathfrak{g} \to 0.$$

Consider a linear mapping $s : \mathfrak{g} \to \mathfrak{e}$ with $p \circ s = \operatorname{Id}_{\mathfrak{g}}$. Then s induces mappings

$$\alpha : \mathfrak{g} \to \operatorname{der}(\mathfrak{h}), \qquad \alpha_X(H) = [s(X), H], \tag{1}$$

$$\rho : \textstyle\bigwedge^2 \mathfrak{g} \to \mathfrak{h}, \qquad \rho(X, Y) = [s(X), s(Y)] - s([X, Y]), \tag{2}$$

which are easily seen to satisfy

$$[\alpha_X, \alpha_Y] - \alpha_{[X,Y]} = \operatorname{ad}_{\rho(X,Y)}, \tag{3}$$

$$\sum_{\text{cyclic}\{X,Y,Z\}} \Big(\alpha_X \rho(Y, Z) - \rho([X, Y], Z)\Big) = 0. \tag{4}$$

We can completely describe the Lie algebra structure on $\mathfrak{e} = \mathfrak{h} \oplus s(\mathfrak{g})$ in terms of α and ρ :

$$\begin{aligned}
&[H_1 + s(X_1), H_2 + s(X_2)] \\
&\qquad = ([H_1, H_2] + \alpha_{X_1} H_2 - \alpha_{X_2} H_1 + \rho(X_1, X_2)) + s[X_1, X_2]
\end{aligned} \tag{5}$$

and one can check that formula (5) gives a Lie algebra structure on $\mathfrak{h} \oplus s(\mathfrak{g})$ if $\alpha : \mathfrak{g} \to \operatorname{der}(\mathfrak{h})$ and $\rho : \bigwedge^2 \mathfrak{g} \to \mathfrak{h}$ satisfy (3) and (4).

15.3. Motivation: Lie algebra extensions associated to a principal bundle. Let $\pi : P \to M = P/K$ be a principal bundle with structure group K; see section (18): P is a manifold with a free right action of a Lie group K and π is the projection on the orbit space $M = P/K$. Denote by $\mathfrak{g} = \mathfrak{X}(M)$ the Lie algebra of the vector fields on M, by $\mathfrak{e} = \mathfrak{X}(P)^K$ the Lie algebra of K-invariant vector fields on P and by $\mathfrak{h} = \mathfrak{X}_{\text{vert}}(P)^K$ the ideal of the K-invariant vertical vector fields of $\mathfrak{e}$. Geometrically, $\mathfrak{e}$ is the Lie algebra of infinitesimal automorphisms of the principal bundle P and $\mathfrak{h}$ is the ideal of infinitesimal automorphisms acting trivially on M, i.e., the Lie algebra of infinitesimal gauge transformations. We have a natural homomorphism $\pi_* : \mathfrak{e} \to \mathfrak{g}$ with the kernel $\mathfrak{h}$, i.e., $\mathfrak{e}$ is an extension of $\mathfrak{g}$ by $\mathfrak{h}$.

Note that we have additional structures of $C^\infty(M)$-modules on $\mathfrak{g}, \mathfrak{h}, \mathfrak{e}$, such that $[X, fY] = f[X,Y] + (\pi_* X)fY$, where $X, Y \in \mathfrak{e}$, $f \in C^\infty(M)$. In particular, $\mathfrak{h}$ is a Lie algebra over $C^\infty(M)$. The extension

$$0 \to \mathfrak{h} \to \mathfrak{e} \to \mathfrak{g} \to 0$$

is also an extension of $C^\infty(M)$-modules.

Assume now that the section $s : \mathfrak{g} \to \mathfrak{e}$ is a homomorphism of $C^\infty(M)$-modules. Then it can be considered as a connection in the principal bundle π, see section (19), and the $\mathfrak{h}$-valued 2-form ρ as its curvature. In this sense we interpret the constructions from section (15.1) as follows in (15.4) below. The analogy with differential geometry has also been noticed by [**117**] and [**118**].

15.4. Geometric interpretation. Note that (15.2.2) is similar to the Maurer-Cartan formula for the *curvature* on principal bundles of differential geometry (19.2.3)

$$\rho = ds + \tfrac{1}{2}[s,s]_\wedge,$$

where for an arbitrary vector space V the usual Chevalley differential, see (14.14.2), is given by

$$d : L^p_{\text{skew}}(\mathfrak{g}; V) \to L^{p+1}_{\text{skew}}(\mathfrak{g}; V),$$
$$d\varphi(X_0, \dots, X_p) = \sum_{i<j} (-1)^{i+j} \varphi([X_i, X_j], X_0, \dots, \widehat{X_i}, \dots, \widehat{X_j}, \dots, X_p)$$

and where for a vector space W and a Lie algebra $\mathfrak{f}$ the $\mathbb{N}_0$-graded Lie bracket $[\ \ ,\ \ ]_\wedge$ on $L^*_{\text{skew}}(W, \mathfrak{f})$, see (19.2), is given by

$$[\varphi, \psi]_\wedge(X_1, \dots, X_{p+q}) = \frac{1}{p!\,q!} \sum_\sigma \operatorname{sign}(\sigma)[\varphi(X_{\sigma 1}, \dots, X_{\sigma p}), \psi(X_{\sigma(p+1)}, \dots)]_{\mathfrak{f}}.$$

Similarly formula (15.2.3) reads as

$$\operatorname{ad}_\rho = d\alpha + \tfrac{1}{2}[\alpha, \alpha]_\wedge.$$

Thus we view s as a *connection* in the sense of a *horizontal lift* of vector fields on the base of a bundle and α as an *induced connection*; see (19.8). Namely, for every $\operatorname{der}(\mathfrak{h})$-module V we put

$$\alpha_\wedge : L^p_{\text{skew}}(\mathfrak{g}; V) \to L^{p+1}_{\text{skew}}(\mathfrak{g}; V),$$

$$\alpha_\wedge\varphi(X_0, \dots, X_p) = \sum_{i=0}^{p} (-1)^i \alpha_{X_i}(\varphi(X_0, \dots, \widehat{X_i}, \dots, X_p)).$$

Then we have the *covariant exterior differential* (on the sections of an associated vector bundle; see (19.12))

$$(1) \qquad \delta_\alpha : L^p_{\text{skew}}(\mathfrak{g}; V) \to L^{p+1}_{\text{skew}}(\mathfrak{g}; V), \qquad \delta_\alpha\varphi = \alpha_\wedge\varphi + d\varphi,$$

for which formula (15.2.4) looks like the *Bianchi identity*, see (19.5.6), $\delta_\alpha\rho = 0$. Moreover one can prove by direct evaluation that another well known result from differential geometry holds, namely (19.5.9), i.e.,

$$(2) \qquad \delta_\alpha\delta_\alpha(\varphi) = [\rho, \varphi]_\wedge, \quad \varphi \in L^p_{\text{skew}}(\mathfrak{g}; \mathfrak{h}).$$

If we change the linear section s to $s' = s + b$ for linear $b : \mathfrak{g} \to \mathfrak{h}$, then we get

$$(3) \qquad \alpha'_X = \alpha_X + \operatorname{ad}^{\mathfrak{h}}_{b(X)},$$

$$\begin{aligned}(4) \qquad \rho'(X, Y) &= \rho(X, Y) + \alpha_X b(Y) - \alpha_Y b(X) - b([X, Y]) + [bX, bY] \\ &= \rho(X, Y) + (\delta_\alpha b)(X, Y) + [bX, bY], \\ \rho' &= \rho + \delta_\alpha b + \tfrac{1}{2}[b, b]_\wedge.\end{aligned}$$

15.5. Theorem. *Let $\mathfrak{h}$ and $\mathfrak{g}$ be Lie algebras.*

Then isomorphism classes of extensions of $\mathfrak{g}$ over $\mathfrak{h}$, i.e., short exact sequences of Lie algebras $0 \to \mathfrak{h} \to \mathfrak{e} \to \mathfrak{g} \to 0$, modulo the equivalence described by the commutative diagram of Lie algebra homomorphisms

$$\begin{array}{ccccccccc} 0 & \longrightarrow & \mathfrak{h} & \longrightarrow & \mathfrak{e} & \longrightarrow & \mathfrak{g} & \longrightarrow & 0 \\ & & \| & & \downarrow{\scriptstyle\varphi} & & \| & & \\ 0 & \longrightarrow & \mathfrak{h} & \longrightarrow & \mathfrak{e}' & \longrightarrow & \mathfrak{g} & \longrightarrow & 0, \end{array}$$

correspond bijectively to equivalence classes of data of the following form:

(1) *a linear mapping* $\alpha : \mathfrak{g} \to \operatorname{der}(\mathfrak{h})$,

(2) *a skew-symmetric bilinear mapping* $\rho : \mathfrak{g} \times \mathfrak{g} \to \mathfrak{h}$

such that

$$(3) \qquad [\alpha_X, \alpha_Y] - \alpha_{[X,Y]} = \operatorname{ad}_{\rho(X,Y)},$$

$$(4) \qquad \sum_{\text{cyclic}} \Big(\alpha_X\rho(Y, Z) - \rho([X, Y], Z)\Big) = 0, \qquad \textit{equivalently}, \qquad \delta_\alpha\rho = 0.$$

On the vector space $\mathfrak{e} := \mathfrak{h} \oplus \mathfrak{g}$ *a Lie algebra structure is given by*

$$\begin{aligned}(5)\quad & [H_1 + X_1, H_2 + X_2]_{\mathfrak{e}} \\ & \qquad = [H_1, H_2]_{\mathfrak{h}} + \alpha_{X_1} H_2 - \alpha_{X_2} H_1 + \rho(X_1, X_2) + [X_1, X_2]_{\mathfrak{g}},\end{aligned}$$

and the associated exact sequence is

$$0 \to \mathfrak{h} \xrightarrow{i_1} \mathfrak{h} \oplus \mathfrak{g} = \mathfrak{e} \xrightarrow{\mathrm{pr}_2} \mathfrak{g} \to 0.$$

Two data (α, ρ) *and* (α', ρ') *are equivalent if there exists a linear mapping* $b : \mathfrak{g} \to \mathfrak{h}$ *such that*

$$(6)\qquad \alpha'_X = \alpha_X + \mathrm{ad}^{\mathfrak{h}}_{b(X)},$$

$$\begin{aligned}(7)\quad \rho'(X,Y) &= \rho(X,Y) + \alpha_X b(Y) - \alpha_Y b(X) - b([X,Y]) + [b(X), b(Y)], \\ \rho' &= \rho + \delta_\alpha b + \tfrac{1}{2}[b,b]_\wedge,\end{aligned}$$

the corresponding isomorphism being

$$\mathfrak{e} = \mathfrak{h} \oplus \mathfrak{g} \to \mathfrak{h} \oplus \mathfrak{g} = \mathfrak{e}', \qquad H + X \mapsto H - b(X) + X.$$

Moreover, a datum (α, ρ) *corresponds to a split extension (a semidirect product) if and only if* (α, ρ) *is equivalent to a datum of the form* $(\alpha', 0)$ *(then* α' *is a homomorphism). This is the case if and only if there exists a mapping* $b : \mathfrak{g} \to \mathfrak{h}$ *such that*

$$(8)\qquad \rho = -\delta_\alpha b - \tfrac{1}{2}[b,b]_\wedge.$$

Proof. Straightforward computations. □

15.6. Corollary ([120]). *Let* $\mathfrak{g}$ *and* $\mathfrak{h}$ *be Lie algebras such that* $\mathfrak{h}$ *has no center. Then isomorphism classes of extensions of* $\mathfrak{g}$ *over* $\mathfrak{h}$ *correspond bijectively to Lie homomorphisms*

$$\bar{\alpha} : \mathfrak{g} \to \mathrm{out}(\mathfrak{h}) = \mathrm{der}(\mathfrak{h}) / \mathrm{ad}(\mathfrak{h}).$$

Proof. If (α, ρ) is a datum, then the map $\bar{\alpha} : \mathfrak{g} \to \mathrm{der}(\mathfrak{h}) / \mathrm{ad}(\mathfrak{h})$ is a Lie algebra homomorphism by (15.5.3). Conversely, let $\bar{\alpha}$ be given. Choose a linear lift $\alpha : \mathfrak{g} \to \mathrm{der}(\mathfrak{h})$ of $\bar{\alpha}$. Since $\bar{\alpha}$ is a Lie algebra homomorphism and $\mathfrak{h}$ has no center, there is a uniquely defined skew-symmetric linear mapping $\rho : \mathfrak{g} \times \mathfrak{g} \to \mathfrak{h}$ such that $[\alpha_X, \alpha_Y] - \alpha_{[X,Y]} = \mathrm{ad}_{\rho(X,Y)}$. Condition (15.5.4) is then automatically satisfied. For later use also, we record the simple proof:

$$\begin{aligned}&\sum_{\text{cyclic } X,Y,Z} \Big[\alpha_X \rho(Y,Z) - \rho([X,Y],Z), H\Big] \\ &= \sum_{\text{cyclic } X,Y,Z} \Big(\alpha_X [\rho(Y,Z), H] - [\rho(Y,Z), \alpha_X H] - [\rho([X,Y],Z), H]\Big)\end{aligned}$$

$$= \sum_{\text{cyclic } X,Y,Z} \Big(\alpha_X[\alpha_Y,\alpha_Z] - \alpha_X\alpha_{[Y,Z]} - [\alpha_Y,\alpha_Z]\alpha_X + \alpha_{[Y,Z]}\alpha_X$$
$$- [\alpha_{[X,Y]},\alpha_Z] + \alpha_{[[X,Y]Z]}\Big)H$$
$$= \sum_{\text{cyclic } X,Y,Z} \Big([\alpha_X,[\alpha_Y,\alpha_Z]] - [\alpha_X,\alpha_{[Y,Z]}] - [\alpha_{[X,Y]},\alpha_Z] + \alpha_{[[X,Y]Z]}\Big)H = 0.$$

Thus (α, ρ) describes an extension by theorem (15.5). The rest is clear. □

15.7. Remarks. If $\mathfrak{h}$ has no center and $\bar{\alpha} : \mathfrak{g} \to \operatorname{out}(\mathfrak{h}) = \operatorname{der}(\mathfrak{h})/\operatorname{ad}(\mathfrak{h})$ is a given homomorphism, the extension corresponding to $\bar{\alpha}$ can be constructed in the following easy way: It is given by the pullback diagram

$$\begin{array}{ccccccccc}
0 & \longrightarrow & \mathfrak{h} & \longrightarrow & \operatorname{der}(\mathfrak{h}) \times_{\operatorname{out}(\mathfrak{h})} \mathfrak{g} & \xrightarrow{\operatorname{pr}_2} & \mathfrak{g} & \longrightarrow & 0 \\
 & & \| & & \downarrow{\scriptstyle \operatorname{pr}_1} & & \downarrow{\scriptstyle \bar{\alpha}} & & \\
0 & \longrightarrow & \mathfrak{h} & \longrightarrow & \operatorname{der}(\mathfrak{h}) & \xrightarrow{\pi} & \operatorname{out}(\mathfrak{h}) & \longrightarrow & 0
\end{array}$$

where $\operatorname{der}(\mathfrak{h}) \times_{\operatorname{out}(\mathfrak{h})} \mathfrak{g}$ is the Lie subalgebra

$$\operatorname{der}(\mathfrak{h}) \times_{\operatorname{out}(\mathfrak{h})} \mathfrak{g} := \{(D, X) \in \operatorname{der}(\mathfrak{h}) \times \mathfrak{g} : \pi(D) = \bar{\alpha}(X)\} \subset \operatorname{der}(\mathfrak{h}) \times \mathfrak{g}.$$

We owe this remark to E. Vinberg.

If $\mathfrak{h}$ has no center and satisfies $\operatorname{der}(\mathfrak{h}) = \mathfrak{h}$ and if $\mathfrak{h}$ is normal in a Lie algebra $\mathfrak{e}$, then $\mathfrak{e} \cong \mathfrak{h} \oplus \mathfrak{e}/\mathfrak{h}$, since $\operatorname{Out}(\mathfrak{h}) = 0$.

15.8. Theorem. *Let $\mathfrak{g}$ and $\mathfrak{h}$ be Lie algebras and let*

$$\bar{\alpha} : \mathfrak{g} \to \operatorname{out}(\mathfrak{h}) = \operatorname{der}(\mathfrak{h})/\operatorname{ad}(\mathfrak{h})$$

be a Lie algebra homomorphism. For a linear lift $\alpha : \mathfrak{g} \to \operatorname{der}(\mathfrak{h})$ of $\bar{\alpha}$ choose $\rho : \bigwedge^2 \mathfrak{g} \to \mathfrak{h}$ satisfying $([\alpha_X, \alpha_Y] - \alpha_{[X,Y]}) = \operatorname{ad}_{\rho(X,Y)}$. Then $\lambda = \lambda(\alpha, \rho) := \delta_\alpha \rho : \bigwedge^3 \mathfrak{g} \to Z(\mathfrak{h})$ is a cocycle for the cochain complex

$$\delta_{\bar{\alpha}} : L^k_{skew}(\mathfrak{g}; Z(\mathfrak{h})) \to L^{k+1}_{skew}(\mathfrak{g}; Z(\mathfrak{h})), \quad \delta_{\bar{\alpha}} \circ \delta_{\bar{\alpha}} = 0.$$

The cohomology class $[\lambda] \in H^3(\mathfrak{g}; Z(\mathfrak{h}))$ depends only on $\bar{\alpha}$ and not on the choices of α and ρ. Then the following are equivalent:

(1) *The $\delta_{\bar{\alpha}}$-cohomology class of λ vanishes: $[\lambda] = 0 \in H^3(\mathfrak{g}; Z(\mathfrak{h}))$.*
(2) *There exists an extension $0 \to \mathfrak{h} \to \mathfrak{e} \to \mathfrak{g} \to 0$ inducing the homomorphism $\bar{\alpha}$.*

If this is the case, then all extensions $0 \to \mathfrak{h} \to \mathfrak{e} \to \mathfrak{g} \to 0$ inducing the homomorphism $\bar{\alpha}$ are parameterized by $H^2(\mathfrak{g}, (Z(\mathfrak{h}), \bar{\alpha}))$, the second Chevalley cohomology space of $\mathfrak{g}$ with values in the center $Z(\mathfrak{h})$, considered as $\mathfrak{g}$-module via $\bar{\alpha}$.

Proof. Using once more the computation in the proof of corollary (15.6), we see that $\operatorname{ad}(\lambda(X,Y,Z)) = \operatorname{ad}(\delta_\alpha\rho(X,Y,Z)) = 0$ so that $\lambda(X,Y,Z) \in Z(\mathfrak{h})$. The Lie algebra $\operatorname{out}(\mathfrak{h}) = \operatorname{der}(\mathfrak{h})/\operatorname{ad}(\mathfrak{h})$ acts on the center $Z(\mathfrak{h})$; thus $Z(\mathfrak{h})$ is a $\mathfrak{g}$-module via $\bar\alpha$, and $\delta_{\bar\alpha}$ is the differential of the Chevalley cohomology. Using (15.4.2), we see that

$$\delta_{\bar\alpha}\lambda = \delta_\alpha\delta_\alpha\rho = [\rho,\rho]_\wedge = -(-1)^{2\cdot 2}[\rho,\rho]_\wedge = 0,$$

so that $[\lambda] \in H^3(\mathfrak{g}; Z(\mathfrak{h}))$.

Let us check next that the cohomology class $[\lambda]$ does not depend on the choices we made. If we are given a pair (α,ρ) as above and we take another linear lift $\alpha' : \mathfrak{g} \to \operatorname{der}(\mathfrak{h})$, then $\alpha'_X = \alpha_X + \operatorname{ad}_{b(X)}$ for some linear $b : \mathfrak{g} \to \mathfrak{h}$. We consider

$$\rho' : \bigwedge^2 \mathfrak{g} \to \mathfrak{h}, \quad \rho'(X,Y) = \rho(X,Y) + (\delta_\alpha b)(X,Y) + [b(X), b(Y)].$$

Computations involving only the definitions and the Jacobi identity show that

$$[\alpha'_X, \alpha'_Y] - \alpha'_{[X,Y]} = \operatorname{ad}_{\rho'(X,Y)}, \quad \lambda(\alpha,\rho) = \delta_\alpha\rho = \delta_{\alpha'}\rho' = \lambda(\alpha',\rho'),$$

so that even the cochain did not change. So let us consider for fixed α two linear mappings

$$\rho,\rho' : \bigwedge^2 \mathfrak{g} \to \mathfrak{h}, \quad [\alpha_X, \alpha_Y] - \alpha_{[X,Y]} = \operatorname{ad}_{\rho(X,Y)} = \operatorname{ad}_{\rho'(X,Y)}.$$

Then $\rho - \rho' =: \mu : \bigwedge^2 \mathfrak{g} \to Z(\mathfrak{h})$ and $\lambda(\alpha,\rho) - \lambda(\alpha,\rho') = \delta_\alpha\rho - \delta_\alpha\rho' = \delta_{\bar\alpha}\mu$.

If there exists an extension inducing $\bar\alpha$, then for any lift α we may find ρ as in (15.5) such that $\lambda(\alpha,\rho) = 0$. On the other hand, given a pair (α,ρ) as in (1) such that $[\lambda(\alpha,\rho)] = 0 \in H^3(\mathfrak{g}, (Z(\mathfrak{h}),\bar\alpha))$, there exists $\mu : \bigwedge^2 \mathfrak{g} \to Z(\mathfrak{h})$ such that $\delta_{\bar\alpha}\mu = \lambda$. But then

$$\operatorname{ad}_{(\rho-\mu)(X,Y)} = \operatorname{ad}_{\rho(X,Y)}, \quad \delta_\alpha(\rho - \mu) = 0,$$

so that $(\alpha, \rho - \mu)$ satisfies the conditions of (15.5) and thus defines an extension which induces $\bar\alpha$.

Finally, suppose that (1) is satisfied, and let us determine how many extensions there exist which induce $\bar\alpha$. By (15.5) we have to determine all equivalence classes of data (α,ρ) as in (15.5). We may fix the linear lift α and one mapping $\rho : \bigwedge^2 \mathfrak{g} \to \mathfrak{h}$ which satisfies (15.5.3) and (15.5.4), and we have to find all ρ' with this property. But then $\rho - \rho' = \mu : \bigwedge^2 \mathfrak{g} \to Z(\mathfrak{h})$ and

$$\delta_{\bar\alpha}\mu = \delta_\alpha\rho - \delta_\alpha\rho' = 0 - 0 = 0$$

so that μ is a 2-cocycle. We may still pass to equivalent data in the sense of (15.5) using some $b : \mathfrak{g} \to \mathfrak{h}$ which does not change α, i.e., $b : \mathfrak{g} \to Z(\mathfrak{h})$.

The corresponding ρ' is, by (15.5.7), $\rho' = \rho + \delta_\alpha b + \frac{1}{2}[b,b]_\wedge = \rho + \delta_{\bar\alpha} b$. Thus only the cohomology class of μ matters. □

15.9. Corollary. *Let $\mathfrak{g}$ and $\mathfrak{h}$ be Lie algebras such that $\mathfrak{h}$ is abelian. Then isomorphism classes of extensions of $\mathfrak{g}$ over $\mathfrak{h}$ correspond bijectively to the set of all pairs $(\alpha, [\rho])$, where $\alpha : \mathfrak{g} \to \mathfrak{gl}(\mathfrak{h}) = \operatorname{der}(\mathfrak{h})$ is a homomorphism of Lie algebras and $[\rho] \in H^2(\mathfrak{g}, \mathfrak{h})$ is a Chevalley cohomology class with coefficients in the $\mathfrak{g}$-module $\mathfrak{h}$ given by α.*

Isomorphism classes of central extensions correspond bijectively to elements $[\rho] \in H^2(\mathfrak{g}, \mathbb{R}) \otimes \mathfrak{h}$ (0 action of $\mathfrak{g}$ on $\mathfrak{h}$).

Proof. This is obvious from theorem (15.8). □

15.10. An interpretation of the class λ. Let $\mathfrak{h}$ and $\mathfrak{g}$ be Lie algebras and let a homomorphism $\bar\alpha : \mathfrak{g} \to \operatorname{der}(\mathfrak{h})/\operatorname{ad}(\mathfrak{h})$ be given. We consider the extension

$$0 \to \operatorname{ad}(\mathfrak{h}) \to \operatorname{der}(\mathfrak{h}) \to \operatorname{der}(\mathfrak{h})/\operatorname{ad}(\mathfrak{h}) \to 0$$

and the following diagram, where the bottom right hand square is a pullback (compare with remark (15.7)):

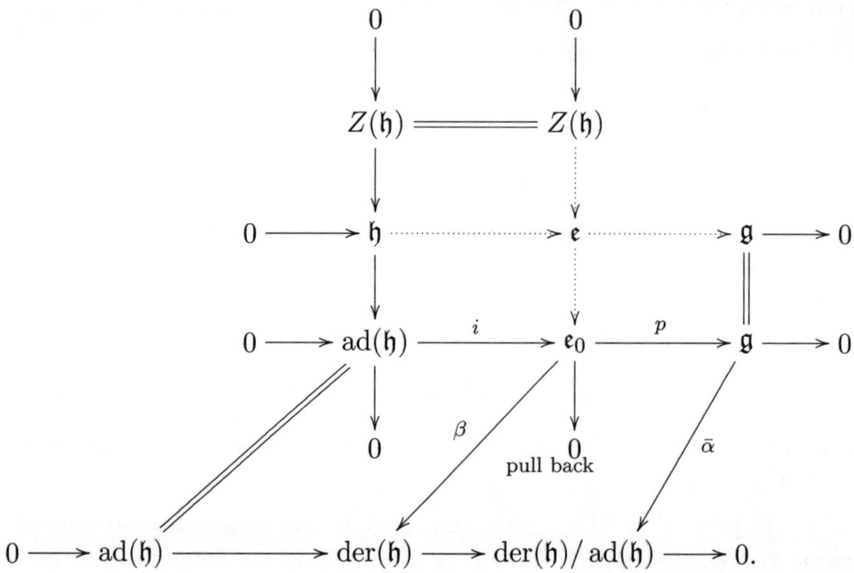

The left hand vertical column describes $\mathfrak{h}$ as a central extension of $\operatorname{ad}(\mathfrak{h})$ with abelian kernel $Z(\mathfrak{h})$ which is moreover killed under the action of $\mathfrak{g}$ via $\bar\alpha$; it is given by a cohomology class $[\nu] \in H^2(\operatorname{ad}(\mathfrak{h}); Z(\mathfrak{h}))^{\mathfrak{g}}$. In order to get an extension $\mathfrak{e}$ of $\mathfrak{g}$ with kernel $\mathfrak{h}$ as in the third row, we have to check that the cohomology class $[\nu]$ is in the image of $i^* : H^2(\mathfrak{e}_0; Z(\mathfrak{h})) \to H^2(\operatorname{ad}(\mathfrak{h}); Z(\mathfrak{h}))^{\mathfrak{g}}$. It would be interesting to express this in terms of the Hochschild-Serre exact sequence; see [**92**].

Extensions of Groups and Lie Groups

In this section we present a discussion and variants of cohomology results going back to O. Schreier [**200, 201**], R. Baer [**15**], S. Eilenberg and S. MacLane, [**56**], G. Hochschild [**88, 89**], and G. Hochschild and J.-P Serre [**91**]. A convenient source for group cohomology is [**74**]. We have to be careful when taking sections; see (15.12) for a discussion of this.

15.11. Describing extensions. Let G and N be Lie groups. An *extension of* G *over* N is an exact sequence of homomorphism of groups:

$$e \to N \xrightarrow{i} E \xrightarrow{p} G \to e.$$

Two extensions are defined to be *equivalent* if there exists a homomorphism φ fitting commutatively into the diagram

$$\begin{array}{ccccccccc} e & \longrightarrow & N & \xrightarrow{i} & E & \xrightarrow{p} & G & \longrightarrow & e \\ & & \| & & \downarrow{\scriptstyle\varphi} & & \| & & \\ e & \longrightarrow & N & \xrightarrow{i'} & E' & \xrightarrow{p'} & G & \longrightarrow & e. \end{array}$$

Note that if such a morphism φ exists, then it is an isomorphism.

For a given extension let us choose a section $s : G \to E$ of p with $s(e) = e$. We may assume that s is smooth on an open e-neighborhood U in G. Then s defines mappings

$$\alpha : G \to \operatorname{Aut}(N), \qquad \alpha_x(h) = s(x)hs(x)^{-1},$$
$$f : G \times G \to N, \qquad f(x,y) = s(x)s(y)s(xy)^{-1},$$

which are smooth near e in G and, by the definition of α and by associativity, have the following properties:

$$\alpha_x \circ \alpha_y = \operatorname{conj}_{f(x,y)} \circ \alpha_{xy},$$
$$\alpha_x(f(y,z))f(x,yz) = f(x,y)f(xy,z),$$
$$f(e,e) = f(x,e) = f(e,x) = e,$$

where $\operatorname{conj}_h(n) = hnh^{-1}$ is conjugation by H, an inner automorphism. We shall denote by $\operatorname{Int}(N) \subseteq \operatorname{Aut}(N)$ the normal subgroup of all inner automorphisms in the group of all automorphisms. If we choose another section $s' : G \to E$ which is smooth near e, then $s'(x) = b(x)s(x)$ for a mapping $b : G \to N$ which is smooth near e in G. We have

$$\alpha'_x = \operatorname{conj}_{b(x)} \circ \alpha_x,$$
$$f'(x,y) = b(x)\alpha_x(b(y))f(x,y)b(xy)^{-1}.$$

The group multiplication on E is then described in terms of α and f by

$$\begin{aligned} ms(x).ns(y) &= ms(x)ns(x)^{-1}s(x)s(y) = m\alpha_x(n)f(x,y).s(xy), \\ (ms(x))^{-1} &= \alpha_{x^{-1}}(f(x,x^{-1})^{-1}m^{-1}).s(x^{-1}) \\ &= (\alpha_x)^{-1}(m^{-1}f(x,x^{-1})^{-1})s(x^{-1}) \\ &= f(x^{-1},x)^{-1}\alpha_{x^{-1}}(m^{-1})\cdot s(x^{-1}). \end{aligned}$$

See (15.12) below for the reconstruction of the smooth structure.

15.12 Choosing sections smoothly or reconstructing the smooth manifold structure. Let

$$e \to N \xrightarrow{i} E \xrightarrow{p} G \to e$$

be an exact sequence of smooth homomorphisms of Lie groups. In particular, E is a principal fiber bundle over G with structure group N. If we are able to choose a smooth section $s : G \to E$ of p as in (15.11), then this is a trivial fiber bundle, so $E \cong N \times G$ as a smooth manifold, and we can use all constructions of (15.13)–(15.27) below to describe Lie group extensions of G over N which are topologically trivial.

Let us look at the long exact sequence in homotopy:

$$\dots \to \pi_2(G) \to \pi_1(N) \to \pi_1(E) \to \pi_1(G) \to \pi_0(N) \to \dots.$$

We always have $\pi_2(N) = 0$. So if N is connected and E is simply connected, then both N and G are simply connected. Using structure theory of Lie algebras and Lie groups, one can prove the following (see [**90**]): *If E is simply connected and N is connected, then there is a closed submanifold M of E meeting N only in $\{e\}$ transversally, such that $E \cong N \times M$. Thus there exists a global smooth section $s : G \to E$.*

For the topologically nontrivial case, we can find a global section s which is smooth only on a neighborhood U of e in G which also satisfies $U^{-1} = U$.

Lemma. *Then we can reconstruct the Lie group structure on E from the extension data (which are all smooth near e on G) and the smooth manifold structure on $N \times U \cong \tilde{U} := p^{-1}(U) \subset E$.*

Proof. Choose $e \in V \subset U$ open with $V^{-1} = V$ and $V.V \subset U$, and let $\tilde{V} := p^{-1}(V)$. In the setting of (15.11) we then have: $\alpha : U \to \operatorname{Aut}(N)$ and $f : V \times V \to N$ are smooth and the group multiplication (15.13.4) is smooth on $\tilde{V} \times \tilde{V} \to \tilde{U}$. We then use $(x.\tilde{V}, \mu_{x^{-1}} : x.\tilde{V} \to \tilde{V})_{x\in E}$ as atlas for E. The chart changes are $\mu_{y^{-1}} \circ \mu_x = \mu_{y^{-1}.x} : x^{-1}.(x.\tilde{V} \cap y.\tilde{V}) = \tilde{V} \cap (x^{-1}.y.\tilde{V}) \to (y^{-1}.x.\tilde{V} \cap V)$, so they are smooth. The resulting smooth manifold structure on E has the property that $p : E \to G$ and $i : N \to E$ are smooth, and the group structure maps μ and ν are smooth also. Moreover E is Hausdorff:

Either $p(x) = p(y)$ and then we can separate them already in one chart $x.\tilde{V} = p^{-1}(p(x).V)$, or we can separate them with open sets of the form $p^{-1}(U_1)$ and $p^{-1}(U_2)$. □

We shall use this lemma in all constructions below without mentioning it. Note that a homomorphism between Lie groups which is smooth near e is smooth everywhere.

15.13. Proposition ([200, 201]). *Let G and N be Lie groups. We consider pairs (α, f) of mappings which are smooth near e:*

$$\alpha : G \to \operatorname{Aut}(N) \quad \text{and} \quad f : G \times G \to N$$

with the properties

$$\alpha_x \circ \alpha_y = \operatorname{conj}_{f(x,y)} \circ \alpha_{xy}, \tag{1}$$

$$f(e,e) = f(x,e) = f(e,y) = e, \tag{2}$$

$$e = \alpha_x(f(y,z)) f(x,yz) f(xy,z)^{-1} f(x,y)^{-1}. \tag{3}$$

Then the following assertions hold:

(4) *Every such pair (α, f) defines a Lie group extension of G over N, given by the set $E = N \times G$, with the group structure*

$$(m,x).(n,y) = (m\alpha_x(n)f(x,y), xy),$$
$$(n,x)^{-1} = (f(x^{-1},x)^{-1}\alpha_{x^{-1}}(n^{-1}), x^{-1}).$$

Up to isomorphism, every extension of G over N can be so obtained.

(5) *Two data (α, f) and (α', f') define equivalent extensions if there exists a mapping $b : G \to N$ (smooth near e) such that*

$$\alpha'_x = \operatorname{conj}_{b(x)} \circ \alpha_x,$$
$$f'(x,y) = b(x)\alpha_x(b(y)) f(x,y) b(xy)^{-1}.$$

The induced smooth isomorphism $E \to E'$ between the extensions defined by (α, f) and (α', f') is given by $(n,x) \mapsto (nb(x)^{-1}, x)$.

(6) *A datum (α, f) describes a splitting extension (a semidirect product) if and only if it is equivalent to a datum (α', f'), where f' is constant $= e$. This is the case if and only if there exists a map $b : G \to N$ (smooth near e) with*

$$f(x,y) = b(x)\alpha_x(b(y))b(xy)^{-1}.$$

Note that for such a pair $(\alpha', f' = e)$ the map α' must be a homomorphism and thus is smooth everywhere.

Proof. (15.11) and routine calculations. □

15.14. Remarks. (1) The center $Z(N)$ of N is preserved by all automorphisms of N and pointwise fixed by all inner automorphisms, so the Lie group $\operatorname{Aut}(N)/\operatorname{Int}(N)$ acts by automorphisms on $Z(N)$. Every homomorphism $\bar{\alpha} : G \to \operatorname{Aut}(N)/\operatorname{Int}(N)$ naturally induces a homomorphism $G \to \operatorname{Aut}(Z(N))$ and therefore turns $Z(N)$ into a smooth G-module $(Z(N), \bar{\alpha})$.

Condition (15.13.1) implies that every extension of G over N induces a smooth homomorphism $\bar{\alpha} : G \to \operatorname{Aut}(N)/\operatorname{Int}(N)$, hence defines a G-module structure on $Z(N)$. Thus we have the following commutative diagram with exact rows:

$$\begin{array}{ccccccccc} e & \longrightarrow & N & \xrightarrow{i} & E & \xrightarrow{p} & G & \longrightarrow & e \\ & & \downarrow{\scriptstyle \operatorname{conj}} & & \downarrow{\scriptstyle \operatorname{conj}|_N} & & \downarrow{\scriptstyle \bar{\alpha}} & & \\ e & \longrightarrow & \operatorname{Int}(N) & \longrightarrow & \operatorname{Aut}(N) & \xrightarrow{q} & \operatorname{Aut}(N)/\operatorname{Int}(N) & \longrightarrow & e. \end{array}$$

Note that the commutativity of this diagram yields a surjective homomorphism $E \to \Gamma$, where Γ is the pullback object of the morphisms q and $\bar{\alpha}$. We shall exploit this fact later.

(2) Note that if (α, f) is the data of an extension then *every* lift $\alpha' : G \to N$ (smooth near e) of $\bar{\alpha}$ shows up in a data pair (α', f') equivalent to (α, f). This is a consequence of (15.13.5).

(3) In [**15**] and [**56**] a triplet $(N, G, \bar{\alpha})$, where N and G are groups and $\bar{\alpha}$ is a homomorphism $G \to \operatorname{Aut}(N)/\operatorname{Int}(N)$, is usually called an *abstract kernel* or *kernel* for short. The kernel $(N, G, \bar{\alpha})$ is said to be *extendible* if it can be derived from an extension of G over N.

In the following we want to characterize those smooth homomorphisms $\bar{\alpha}$ for which $(N, G, \bar{\alpha})$ is an extendible kernel.

15.15. Notation. Let us fix a smooth homomorphism of Lie groups $\bar{\alpha} : G \to \operatorname{Aut}(N)/\operatorname{Int}(N)$ and consider all pairs (α, f) consisting of a lift $\alpha : G \to \operatorname{Aut}(N)$, $x \mapsto \alpha_x$, of $\bar{\alpha}$, and of $f : G \times G \to N$ which are smooth near e and satisfy conditions (15.13.1) and (15.13.2):

$$(1) \qquad \alpha_x \circ \alpha_y = \operatorname{conj}_{f(x,y)} \circ \alpha_{xy}$$

$$(2) \qquad f(e,e) = f(x,e) = f(e,x) = e, \quad \alpha_e = \operatorname{Id}_N .$$

For the sake of brevity, we call such a pair (α, f) an *$\bar{\alpha}$-pair.* We write

$$(3) \qquad \lambda(x,y,z) = \alpha_x(f(y,z)) f(x,yz) f(xy,z)^{-1} f(x,y)^{-1}$$

for the right side of equation (15.13.3). To avoid taking inverses it will be often convenient to write (3) in the equivalent form

$$(3') \qquad \lambda(x,y,z) f(x,y) f(xy,z) = \alpha_x(f(y,z)) f(x,yz).$$

Note that the normalization condition (15.13.2) implies that λ is also normalized, i.e.,

(4) $\lambda(e,y,z) = \lambda(x,e,z) = \lambda(x,y,e) = e \qquad \text{for all } x,y,z \in G.$

Two $\bar{\alpha}$-pairs (α, f) and (α', f') are said to be *equivalent* if there exists a mapping $b : G \to N$ such that

$$\begin{aligned} \alpha'_x &= \operatorname{conj}_{b(x)} \circ \alpha_x, \\ f'(x,y) &= b(x)\alpha_x(b(y))f(x,y)b(xy)^{-1}. \end{aligned}$$

Following [**93**], the function f is traditionally called a *factor set*, and λ is called the *obstruction* of (α, f) to form an extension. We shall not use this terminology here.

15.16. Nonabelian cohomology. Let (Z, β) be a smooth G-module, i.e., an abelian Lie group with a smooth G-action. The boundary operator of group cohomology with values in (Z, β) is given by

$$\begin{aligned} &\delta_\beta : \operatorname{Map}_e(G^k, Z) \to \operatorname{Map}_e(G^{k+1}, Z), \\ &(\delta_\beta f)(x_0, x_1, \dots, x_k) = \beta_{x_0}(f(x_1, \dots, x_k)).f(x_0x_1, x_2, \dots, x_k)^{-1} \\ &\quad .f(x_0, x_1x_2, x_3, \dots, x_k) \dots f(x_0, \dots, x_{k-1}x_k)^{(-1)^k}.f(x_0, \dots, x_{k-1})^{(-1)^{k+1}} \end{aligned}$$

where Map_e denotes the space of mappings which are smooth near e. This gives rise to abelian group cohomology; here Z is abelian!

Now we discuss a nonabelian version. Inspired by condition (15.13.3) or by (15.15.3), for every map $\alpha : G \to \operatorname{Aut}(N)$ and $f : G \times G \to N$ which are smooth near e we consider

$$\begin{aligned} &\delta_\alpha f : G \times G \times G \to N, \\ &(\delta_\alpha f)(x,y,z) = \alpha_x(f(y,z))f(x,yz)f(xy,z)^{-1}f(x,y)^{-1}. \end{aligned}$$

Then δ_α looks like the nonabelian version of a coboundary — except that

(a) α is not a homomorphism, and that

(b) in comparison with the above traditional definition the order of the two middle terms of the expression for $\delta_\alpha f$ is reversed.

Likewise assertion (15.13.6) suggests to consider for $b : G \to N$ (smooth near e) the 'nonabelian coboundary'

$$\delta_\alpha b : G \times G \to N, \quad (\delta_\alpha b)(x,y) = b(x)\alpha_x(b(y))b(xy)^{-1}.$$

Also in this case the terms in the expression on the right hand side do not follow the traditional order.

A straightforward computation shows that

$$\begin{aligned}&\delta_\alpha\delta_\alpha b(x,y,z)\\ &=\alpha_x(b(y)\alpha_y(b(z))b(yz)^{-1})b(x)\alpha_x(b(yz))\alpha_{xy}(b(z))^{-1}\alpha_x(b(y))^{-1}b(x)^{-1}.\end{aligned}$$

If the image of b is central in N, then this reduces to

$$\delta_\alpha\delta_\alpha b(x,y,z)=\alpha_x\circ\alpha_y(b(z))\alpha_{xy}(b(z))^{-1}.$$

Thus we cannot expect $\delta_\alpha\delta_\alpha b=e$ in general.

15.17. Remarks. By (15.13) an $\bar\alpha$-pair (α,f) is the data of an extension if and only if the associated map $\lambda=\delta_\alpha f$ is identically $=e$.

If $\alpha':G\to\operatorname{Aut}(N)$ is another lift (smooth near e) of $\bar\alpha$, then there exists a map $f':G\times G\to N$ (smooth near e) such that (α',f') is equivalent to (α,f).

For fixed α the $\bar\alpha$-pairs (α,f) and (α,f') are equivalent if and only if there exists a map $b:G\to Z(N)$ which is smooth near e such that

$$f'(x,y)=b(x)\alpha_x(b(y))f(x,y)b(xy)^{-1},$$

that is, the maps f' and f differ only by the coboundary $\delta_{\bar\alpha}b$ with respect to cohomology with values in the G-module $(Z(N),\bar\alpha)$. Since $\alpha=\alpha'$, the equation $\alpha_x=\operatorname{conj}_{b(x)}\circ\alpha'_x$ implies $\operatorname{conj}_{b(x)}=id$, so $b(x)$ must be central.

15.18. Lemma.

(1) *For any $\bar\alpha$-pair (α,f) the associated $\lambda=\delta_\alpha f$ takes values in the center of N.*

(2) *If the pairs (α,f) and (α',f') are equivalent, then the associated maps λ and λ' coincide. In particular, if (α,f) is the data of an extension, then so is every equivalent pair (α',f').*

Proof of (1). Applying condition (15.13.1), we find

$$\begin{aligned}\operatorname{conj}_{\lambda(x,y,z)}&=\operatorname{conj}_{\alpha_x(f(y,z))}\operatorname{conj}_{f(x,yz)}\operatorname{conj}_{f(xy,z)^{-1}}\operatorname{conj}_{f(x,y)^{-1}}\\ &=\alpha_x\alpha_y\alpha_z\alpha_{yz}^{-1}\alpha_x^{-1}\alpha_x\alpha_{yz}\alpha_{xyz}^{-1}\alpha_{xyz}\alpha_z^{-1}\alpha_{xy}^{-1}\alpha_{xy}\alpha_y^{-1}\alpha_x^{-1}\\ &=\operatorname{Id}_N,\end{aligned}$$

which means that $\lambda(x,y,z)$ must lie in the center of N.

Proof of (2). Let (α',f') be equivalent to (α,f). Then there exists a map $b:G\to N$ with

(3) $\quad\alpha'_x=\operatorname{conj}_{b(x)}\circ\alpha_x,\qquad f'(x,y)=b(x)\alpha_x(b(y))f(x,y)b(xy)^{-1}.$

By definition we have

$$\lambda'(x,y,z)f'(x,y)f'(xy,z)=\alpha'_x(f'(y,z))f'(x,yz).$$

Inserting the identities (3), the left side of this equation reads

$$\begin{aligned}&\lambda'(x,y,z)\cdot f'(x,y)\cdot f'(xy,z)\\&=\lambda'(x,y,z)\cdot b(x)\alpha_x(b(y))f(x,y)b(xy)^{-1}\cdot b(xy)\alpha_{xy}(b(z))f(xy,z)b(xyz)^{-1}\\&=\lambda'(x,y,z)\cdot b(x)\alpha_x(b(y))f(x,y)\alpha_{xy}(b(z))f(xy,z)b(xyz)^{-1}.\end{aligned}$$

Since $\operatorname{conj}_{f(x,y)}\circ\alpha_{xy}=\alpha_x\circ\alpha_y$, we have $f(x,y)\alpha_{xy}(b(z))=\alpha_x\alpha_y(b(z))f(x,y)$ and therefore (using also that $\lambda'(x,y,z)$ is central in N):

$$\begin{aligned}&\lambda'(x,y,z)\cdot f'(x,y)\cdot f'(xy,z)\\&\quad=\lambda'(x,y,z)\cdot b(x)\alpha_x(b(y))\alpha_x\alpha_y(b(z))f(x,y)f(xy,z)b(xyz)^{-1}\\&\quad=\lambda'(x,y,z)\cdot b(x)\alpha_x(b(y)\alpha_y(b(z)))f(x,y)f(xy,z)b(xyz)^{-1}\\&\quad=b(x)\alpha_x(b(y)\alpha_y(b(z)))\lambda'(x,y,z)f(x,y)f(xy,z)b(xyz)^{-1}.\end{aligned}$$

Similarly, the right side can be transformed into

$$\begin{aligned}&\alpha'_x(f'(y,z))\cdot f'(x,yz)\\&=\operatorname{conj}_{b(x)}\alpha_x(b(y)\alpha_y(b(z))f(y,z)b(yz)^{-1})\cdot b(x)\alpha_x(b(yz))f(x,yz)b(xyz)^{-1}\\&=b(x)\alpha_x(b(y)\alpha_y(b(z))f(y,z)b(yz)^{-1})b(x)^{-1}b(x)\alpha_x(b(yz))f(x,yz)b(xyz)^{-1}\\&=b(x)\alpha_x(b(y)\alpha_y(b(z)))\alpha_x(f(y,z))f(x,yz)b(xyz)^{-1}.\end{aligned}$$

Canceling the term $b(x)\alpha_x(b(y)\alpha_y(b(z)))$ on the right and the term $b(xyz)^{-1}$ on the left, we see that $\lambda'(x,y,z)$ satisfies

$$\lambda'(x,y,z)f(x,y)f(xy,z)=\alpha_x(f(y,z))f(x,yz),$$

the defining equation for $\lambda(x,y,z)$. Thus $\lambda=\lambda'$. □

15.19. Lemma. *Let (α,f) be an $\bar{\alpha}$-pair and let $\lambda=\delta_\alpha f$.*

(1) *The map*

$$\lambda: G\times G\times G\to Z(N),\qquad (x,y,z)\mapsto\lambda(x,y,z),$$

is a normalized 3-cocycle with respect to $\delta_{\bar{\alpha}}$ cohomology with values in the G-module $(Z(N),\bar{\alpha})$ and is smooth near e.

(2) *The cocycles (smooth near e) in the $\delta_{\bar{\alpha}}$ cohomology class $[\lambda]$ of λ are exactly the maps $\lambda'=\delta_\alpha f'$ which are induced by an $\bar{\alpha}$-pair of the form (α,f').*

(3) *An $\bar{\alpha}$-pair (α,f') induces the same cocycle $\lambda\in[\lambda]$ as (α,f) if and only if $f=f'\cdot c$, where $c: G\times G\to Z(N)$, is a 2-cocycle with respect to $\delta_{\bar{\alpha}}$ cohomology, normalized by the condition $c(x,e)=c(e,y)=e$, and is smooth near e.*

Proof. (1) In order to show that λ is a 3-cocycle, we have to prove that, for any quadruplet (x, y, z, u) of elements in G,

$$\begin{aligned}(\delta_{\bar{\alpha}}\lambda)&(x, y, z, u)\\ &= \alpha_x(\lambda(y, z, u))\lambda(xy, z, u)^{-1}\lambda(x, yz, u)\lambda(x, y, zu)^{-1}\lambda(x, y, z) = e,\end{aligned}$$

or, equivalently, that

$$\alpha_x(\lambda(y, z, u))\lambda(x, yz, u)\lambda(x, y, z) = \lambda(xy, z, u)\lambda(x, y, zu).$$

By the definition of λ and the centrality of the λ's, we have for the right side R of this equation:

$$\begin{aligned}R &= \alpha_{xy}(f(z, u))f(xy, zu)f(xyz, u)^{-1}f(xy, z)^{-1} \cdot \lambda(x, y, zu)\\ &= \alpha_{xy}(f(z, u)) \cdot \lambda(x, y, zu) \cdot f(xy, zu)f(xyz, u)^{-1}f(xy, z)^{-1}.\end{aligned}$$

Applying the equation $\alpha_{xy} = \mathrm{conj}_{f(x,y)^{-1}} \circ \alpha_x \circ \alpha_y$, we conclude

$$Rf(xy, z)f(xyz, u) = f(x, y)^{-1}\alpha_x\alpha_y(f(z, u))f(x, y)\lambda(x, y, zu)f(xy, zu),$$

and, by the centrality of R and the λ's,

$$\begin{aligned}Rf(x, y)f(xy, z)f(xyz, u) &= \alpha_x\alpha_y(f(z, u)) \cdot \lambda(x, y, zu)f(x, y)f(xy, zu)\\ &= \alpha_x\alpha_y(f(z, u))\alpha_x(f(y, zu))f(x, yzu)f(xy, zu)^{-1}f(x, y)^{-1}f(x, y)f(xy, zu)\\ &= \alpha_x\alpha_y(f(z, u))\alpha_x(f(y, zu))f(x, yzu).\end{aligned}$$

For the left side $L = \alpha_x(\lambda(y, z, u))\lambda(x, yz, u)\lambda(x, y, z)$ we see

$$\begin{aligned}L &= \alpha_x(\lambda(y, z, u)) \cdot \lambda(x, yz, u) \cdot \lambda(x, y, z)\\ &= \alpha_x(\lambda(y, z, u)) \cdot \lambda(x, yz, u) \cdot \alpha_x(f(y, z))f(x, yz)f(xy, z)^{-1}f(x, y)^{-1}\\ &= \alpha_x(\lambda(y, z, u) \cdot f(y, z)) \cdot \lambda(x, yz, u)f(x, yz)f(xy, z)^{-1}f(x, y)^{-1}\\ &= \alpha_x(\lambda(y, z, u) \cdot f(y, z)) \cdot \alpha_x(f(yz, u))f(x, yzu)f(xyz, u)^{-1}f(x, yz)^{-1}\\ &\qquad \cdot f(x, yz)f(xy, z)^{-1}f(x, y)^{-1}\\ &= \alpha_x(\alpha_y(f(z, u))f(y, zu)f(yz, u)^{-1}f(y, z)^{-1}f(y, z)f(yz, u))\\ &\qquad \cdot f(x, yzu)f(xyz, u)^{-1}f(xy, z)^{-1}f(x, y)^{-1}\\ &= \alpha_x(\alpha_y(f(z, u))f(y, zu))f(x, yzu)f(xyz, u)^{-1}f(xy, z)^{-1}f(x, y)^{-1}.\end{aligned}$$

Thus we conclude that

$$\begin{aligned}Lf(x, y)f(xy, z)f(xyz, u) &= \alpha_x(\alpha_y(f(z, u))f(y, zu))f(x, yzu)\\ &= Rf(x, y)f(xy, z)f(xyz, u)\end{aligned}$$

and, upon cancellation, $L = R$. This finishes the proof of (1).

(2) Consider any mapping $f' : G \times G \to N$ such that (α, f') is an $\bar{\alpha}$-pair. Then

$$\mathrm{conj}_{f(x,y)} = \mathrm{conj}_{f'(x,y)} = (\alpha_{xy})^{-1}\alpha_x^{-1}\alpha_y,$$

and therefore the element $c(x,y) = f(x,y)^{-1}f'(x,y)$ lies in the center $Z(N)$ of N. Now

$$\begin{aligned}
&\lambda'(x,y,z)\lambda(x,y,z)^{-1} \\
&\quad = \alpha_x(f(y,z)c(y,z))f(x,yz)c(x,yz)c(xy,z)^{-1}f(xy,z)^{-1}c(x,y)^{-1}f(x,y)^{-1} \\
&\qquad \cdot f(x,y)f(xy,z)f(x,yz)^{-1}\alpha_x(f(y,z)^{-1}) \\
&\quad = \alpha_x(c(y,z))c(x,yz)c(xy,z)^{-1}c(x,y)^{-1} = (\delta_{\bar\alpha}c)(x,y,z)
\end{aligned}$$

so that $[\lambda] = [\lambda'] \in H^3(G,(Z(N),\bar\alpha))$.

Reading the above calculations backwards, we see that, conversely, *every* cochain λ' lying in the cohomology class of λ is induced by some pair (α, f').

(3) We have seen in the proof of (2) that the cochains $\lambda = \delta_\alpha f$, $\lambda' = \delta_\alpha f'$ induced, respectively, by the $\bar\alpha$-pairs (α, f) and (α, f') differ by the cocycle $\delta_{\bar\alpha}c$, where $c(x,y) = f(x,y)^{-1}f'(x,y)$. Thus (α, f) and (α, f') induce the same cocycle λ if and only if $\delta_{\bar\alpha}c$ vanishes. This implies that the cocycles in the cohomology class $[\lambda]$ are in 1-1 correspondence to the 2-cocycles with respect to $\delta_{\bar\alpha}$-cohomology. This finishes the proof. □

15.20. Corollary. *The cohomology class of $\lambda = \delta_\alpha f$ depends only on $\bar\alpha$, not on the particular choice of the $\bar\alpha$-pair (α, f).*

Proof. Suppose that (α', f') is another $\bar\alpha$-pair and let $\lambda' = \delta_{\alpha'}f'$. By (15.17), the pair (α', f') is equivalent to some pair (α, f''). Since by lemma (15.18.2) equivalent pairs produce the same λ, we have $\lambda' = \delta_\alpha f''$. By lemma (15.19.2), $\delta_\alpha f$ and $\delta_\alpha f''$ are in the same cohomology class. This proves the assertion. □

Notation. For given $\bar\alpha$ we henceforth write $\lambda^{\bar\alpha}$ for the cohomology class $[\lambda] \in H^3(G;(Z(N),\bar\alpha))$ (smooth near e). By the corollary above this notation is unambiguous.

15.21. Theorem ([56]). *Let G and N be Lie groups and consider a smooth homomorphism $\bar\alpha : G \to \operatorname{Aut}(N)/\operatorname{Int}(N)$. Then the following assertions hold:*

(1) *The homomorphism $\bar\alpha$ is induced by a Lie group extension if and only if the corresponding cohomology class $\lambda^{\bar\alpha} \in H^3(G, Z(N))$ vanishes.*
(2) *If $\bar\alpha$ is induced by an extension, then all extensions inducing $\bar\alpha$ are parameterized by $H^2(G, Z(N))$.*
(3) *The homomorphism $\bar\alpha$ is induced by a splitting extension if and only if it can be lifted to a (smooth) homomorphism $\alpha : G \to \operatorname{Aut}(N)$.*

Here $H^k(G, Z(N)))$ denotes the group cohomology (smooth near e) of G with values in the G-module $(Z(N),\bar\alpha)$.

Proof. (1) We know already from (15.13) that if (α, f) is the data associated with an extension, then $\lambda^{\bar{\alpha}} = e$.

Conversely, if $\lambda^{\bar{\alpha}}$ is trivial, then by (15.19) for any lift $\alpha : G \to \operatorname{Aut}(N)$ of $\bar{\alpha}$ we can find a map $f : G \times G \to N$ such that (α, f) is a pair with $\delta_\alpha f = e$; by (15.13) this pair (α, f) defines an extension inducing $\bar{\alpha}$.

(2) By (15.13) an $\bar{\alpha}$-pair (α, f) is the data of an extension if and only if $\delta_\alpha f = e$. By (15.19.3) we know that if $\delta_\alpha f = \delta_\alpha f'$, then $f = f' \cdot c$, where $c : G \times G \to Z(N)$ is a 2-cocycle. Furthermore, by (15.15.2) two such $\bar{\alpha}$-pairs are equivalent — and thus define equivalent extensions — if and only if $f = f' \cdot c$ where c is the coboundary $c = \delta_{\bar{\alpha}} b$ for $b : G \to Z(N)$. Thus we see that the extensions inducing $\bar{\alpha}$ are in 1-1 correspondence with the elements of $H^2(G; (Z(N), \bar{\alpha}))$.

(3) By (15.13.6) we know that if (α, f) is an $\bar{\alpha}$-pair inducing a splitting extension, then there exists a map $b : G \to N$ such that the map $\alpha' : G \to N$, $x \mapsto \operatorname{conj}_{b(x)} \circ \alpha$ is a homomorphism, so $\bar{\alpha}$ has a homomorphic lift. The converse is obvious. □

15.22. Corollary. *Let G and N be Lie groups, N abelian. Then isomorphism classes of Lie group extensions of G over N correspond bijectively to the set of pairs $(\alpha, [f])$ where $\alpha : G \to \operatorname{Aut}(N)$ is a smooth homomorphism and $[f] \in H^2(G; (N, \alpha))$ is an element in the second group cohomology (smooth near e) of G with values in the G-module (N, α).*

Proof. Since N is abelian, $\operatorname{Int}(N) = e$ and therefore $\bar{\alpha}$ can be considered as a homomorphism $\alpha : G \to \operatorname{Aut}(N)$. Thus we can form the semidirect product $N \rtimes_\alpha G$, so extensions inducing $\bar{\alpha}$ exist. Now theorem (15.21) applies and yields the assertion. □

15.23. Corollary ([15]). *Let G and N be Lie groups, N without center. Then isomorphism classes of Lie group extensions correspond bijectively to smooth group homomorphisms $\bar{\alpha} : G \to \operatorname{Aut}(N)/\operatorname{Int}(N)$.*

Proof. Since $Z(N) = e$, the cohomologies $H^3(G; Z(N))$ and $H^2(G; Z(N))$ obviously vanish; hence by theorem (15.21) every homomorphism $\bar{\alpha}$ induces a unique extension.

Conversely, every extension induces some $\bar{\alpha} : G \to \operatorname{Aut}(N)/\operatorname{Int}(N)$ by the construction in (15.11). □

Alternative proof of (15.23). For a given $\bar{\alpha}$ consider the group

$$\Gamma = \{(g, \varphi) \in G \times \operatorname{Aut}(N) \mid \varphi \in \bar{\alpha}(g)\}$$

which is the pullback object of the diagram

$$\begin{array}{ccc} & & G \\ & & \downarrow{\scriptstyle \bar\alpha} \\ \mathrm{Aut}(N) & \longrightarrow & \mathrm{Aut}(N)/\,\mathrm{Int}(N). \end{array}$$

By assumption, N has no center. Therefore the map $N \to \mathrm{Aut}(N), h \mapsto \mathrm{conj}_h$ is injective and hence the map $N \to \Gamma$, $h \mapsto (e, \mathrm{conj}_h)$ is a homomorphic injection. Moreover, its image N is the kernel of the quotient map $\Gamma \to G$, $(g, \varphi) \mapsto g$. Thus we have an extension

$$e \to N \longrightarrow \Gamma \longrightarrow G \to e \tag{1}$$

of G by N which induces $\bar\alpha$. Conversely, let

$$e \to N \xrightarrow{\;i\;} E \xrightarrow{\;p\;} G \to e \tag{2}$$

be an extension inducing $\bar\alpha$. Then the map $\vartheta : E \to \Gamma$, $x \mapsto (p(x), q(x))$, where $q(x)$ denotes the automorphism of N induced by conj_x, is a homomorphism. Thus (1) and (2) are equivalent extensions. □

15.24. In the general case this construction runs as follows: Define Γ and ϑ as above. Then every extension

$$e \to N \xrightarrow{\;i\;} E \xrightarrow{\;p\;} G \to e \tag{1}$$

gives rise to an extension of Γ over the center $Z(N)$ of N:

$$e \to Z(N) \xrightarrow{\;i|Z(N)\;} E \xrightarrow{\;\theta\;} \Gamma \to e, \tag{2}$$

where Γ operates on $Z(N)$ via $z \cdot (g, \varphi) = \varphi(z)$. These two extensions fit into the commutative diagram

$$\begin{array}{ccccccccc} e & \longrightarrow & Z(N) & \xrightarrow{\;i|_{Z(N)}\;} & E & \xrightarrow{\;\theta\;} & \Gamma & \longrightarrow & e \\ & & \downarrow{\scriptstyle \mathrm{incl}} & & \| & & \downarrow{\scriptstyle pr_1} & & \\ e & \longrightarrow & N & \xrightarrow{\;i\;} & E & \xrightarrow{\;p\;} & G & \longrightarrow & e. \end{array}$$

Roughly speaking, E can be regarded both as an extension of G over N and as an extension of Γ over $Z(N)$. It can be shown that if $\bar\alpha$ admits an extension, then every extension inducing $\bar\alpha$ is obtained in this way.

Note that for a given abstract kernel $(N, G, \bar\alpha)$ there is always an extension of Γ over $Z(N)$, but if $[\lambda^{\bar\alpha}] \in H^3(G, Z(N))$ is nonzero, then the inclusion $Z(N) \to E$ does not extend to an inclusion $N \to E$.

15.25. In [**56**] a pair (K, ψ), with $\psi : G \to \operatorname{Aut}(K)/\operatorname{Int}(K)$ a homomorphism (G being fixed), is called a *kernel.* As we have seen above, ψ induces a homomorphism $\psi_0 : G \to \operatorname{Aut}(Z(K))$. Consider all kernels (K, ψ), with fixed center $C = Z(K)$ and fixed restriction ψ_0. Two such kernels are said to be *similar* if they differ only by a kernel coming from a homomorphism. One of the results in [**56**] is that the similarity classes of kernels form a group under a multiplication which is defined using the amalgamated direct product of two kernels with C as amalgamating subgroup and that this group can be naturally identified with the third cohomology group $H^3(G, (C, \psi_0))$. In the following we outline the arguments.

15.26. Proposition. *Suppose that we are given an abelian group Z, a homomorphism $\alpha^0 : G \to \operatorname{Aut}(Z)$ and a normalized 3-cocycle $\lambda : G \times G \times G \to Z$. Then there exists a group N containing Z as its center and a homomorphism $\bar{\alpha} : G \to \operatorname{Aut}(N)/\operatorname{Int}(N)$ inducing both α^0 and λ.*

Proof. Let S be the free group generated by the symbols $[x, y]$ with $e \neq x \in G$ and $e \neq y \in G$. For convenience the identity e of F is identified with all symbols $[x, y]$ such that either $x = e$ or $y = e$. The group F is centerfree except in the case where G is cyclic of order two. For the moment we set aside the exceptional case.

We define N to be the direct product $F \times Z$ and, for every $g \in G$, we define a homomorphism $\alpha_g : N \to N$ by the formula

$$\alpha_g([x, y], z) = ([g, x][gx, y][g, xy]^{-1}, \lambda(g, x, y)\alpha_g^0(z)).$$

Since λ is normalized, we see that α_e is the identity. We claim that

$$\alpha_x \alpha_y = \operatorname{conj}_{([x,y],e)} \circ \alpha_{xy}. \tag{1}$$

To see this, we apply the left side of this equation to an element $([u, v], z)$:

$$\begin{aligned}
\alpha_x\alpha_y([u, v], z) &= \alpha_x\big([y, u][yu, v][y, uv]^{-1}, \lambda(y, u, v)\alpha_y^0(z)\big) \\
&= \big([x, y][xy, u][x, yu]^{-1} \cdot [x, yu][xyu, v][x, yuv]^{-1} \\
&\qquad \cdot [x, yuv][xy, uv]^{-1}[x, y]^{-1}, \\
&\qquad\qquad \lambda(x, y, u)\lambda(x, yu, v)\lambda(x, y, uv)^{-1}\alpha_x^0(\lambda(y, u, v))\alpha_{xy}^0(z)\big) \\
&= \operatorname{conj}_{([x,y],e)}\big([xy, u][xyu, v][xy, uv]^{-1}[x, y]^{-1}, \\
&\qquad\qquad \lambda(x, y, u)\lambda(x, yu, v)\lambda(x, y, uv)^{-1}\alpha_x^0(\lambda(y, u, v))\alpha_{xy}^0(z)\big).
\end{aligned}$$

Since λ is a cocycle, we have

$$\begin{aligned}
e &= \delta_{\alpha^0}\lambda(x, y, u, v) \\
&= \alpha_x^0(\lambda(y, u, v))\lambda(xy, u, v)^{-1}\lambda(x, yu, v)\lambda(x, y, uv)^{-1}\lambda(x, y, u)
\end{aligned}$$

and therefore

$$\alpha_x^0(\lambda(y,u,v))\lambda(x,yu,v)\lambda(x,y,uv)^{-1}\lambda(x,y,u) = \lambda(xy,u,v).$$

Thus we find

$$\begin{aligned}
&\alpha_x\alpha_y([u,v],z)\\
&\quad = \operatorname{conj}_{([x,y],e)}([xy,u][xyu,v][xy,uv]^{-1}[x,y]^{-1}, \lambda(xy,uv)\alpha_{xy}^0(z))\\
&\quad = \operatorname{conj}_{([x,y],e)}(\alpha_{xy}([u,v]))
\end{aligned}$$

which establishes our claim (1).

By (1) we have the equations $\alpha_{x^{-1}} \circ \alpha_x = \operatorname{conj}_{[x^{-1},x]}$ and $\alpha_x \circ \alpha_{x^{-1}} = \operatorname{conj}_{[x,x^{-1}]}$, so every homomorphism α_x, $x \in G$, is injective as well as surjective, hence an automorphism.

If we assume that G is not cyclic of order two, then $e \times Z$ is exactly the center of N and equation (1) defines a homomorphism $\bar\alpha : G \to \operatorname{Aut}(N)/\operatorname{Int}(N)$ which, by construction, induces both λ and α^0. □

15.27. An interpretation of the class λ. Let N and G be Lie groups and let a homomorphism $\alpha : G \to \operatorname{Aut}(N)/\operatorname{Int}(N)$ be given. We consider the extension

$$e \to \operatorname{Int}(N) \to \operatorname{Aut}(N) \to \operatorname{Aut}(N)/\operatorname{Int}(N) \to e$$

and the following diagram, where the bottom right hand square is a pullback (compare with the alternative proof of (15.23)):

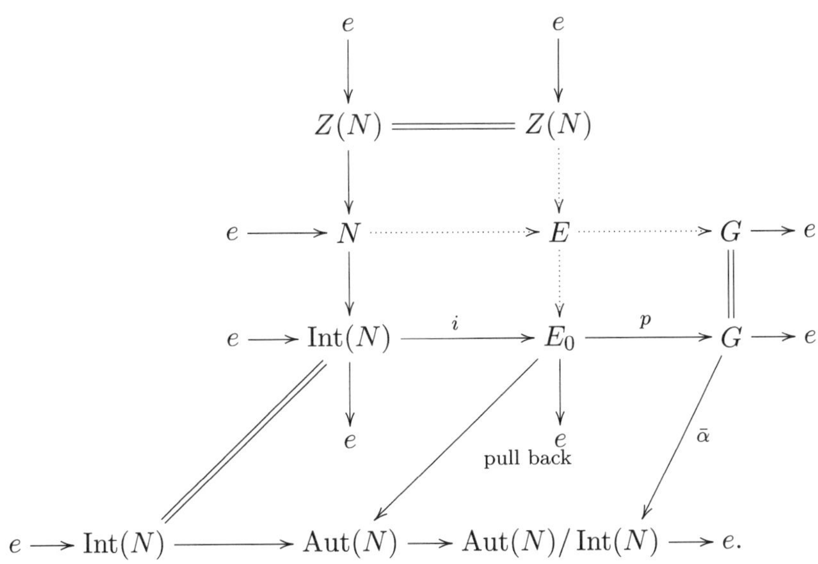

The left hand vertical column describes N as a central extension of $\operatorname{Int}(N)$ with abelian kernel $Z(N)$ which is moreover invariant under the action of G via $\bar\alpha$; it is given by a cohomology class $[\nu] \in H^2(N; Z(N))^G$. In order to get an extension E of G with kernel N as in the third row, we have to check that the cohomology class $[\nu]$ is in the image of $i^* : H^2(E_0; Z(N)) \to H^2(N; Z(N))^G$.

CHAPTER IV. Bundles and Connections

16. Derivations on the Algebra of Differential Forms

16.1. Derivations. In this section let M be a smooth manifold. We consider the graded commutative algebra

$$\Omega(M) = \bigoplus_{k=0}^{\dim M} \Omega^k(M) = \bigoplus_{k=-\infty}^{\infty} \Omega^k(M)$$

of differential forms on M, where we put $\Omega^k(M) = 0$ for $k < 0$ and $k > \dim M$. We denote by $\operatorname{Der}_k \Omega(M)$ the space of all *(graded) derivations* of degree k, i.e., all linear mappings $D : \Omega(M) \to \Omega(M)$ with $D(\Omega^\ell(M)) \subset \Omega^{k+\ell}(M)$ and $D(\varphi \wedge \psi) = D(\varphi) \wedge \psi + (-1)^{k\ell} \varphi \wedge D(\psi)$ for $\varphi \in \Omega^\ell(M)$.

Lemma. *Then the space* $\operatorname{Der} \Omega(M) = \bigoplus_k \operatorname{Der}_k \Omega(M)$ *is a graded Lie algebra with the graded commutator* $[D_1, D_2] := D_1 \circ D_2 - (-1)^{k_1 k_2} D_2 \circ D_1$ *as bracket. This means that the bracket is graded anticommutative and satisfies the graded Jacobi identity*

$$[D_1, D_2] = -(-1)^{k_1 k_2}[D_2, D_1],$$
$$[D_1, [D_2, D_3]] = [[D_1, D_2], D_3] + (-1)^{k_1 k_2}[D_2, [D_1, D_3]]$$

(so that $\operatorname{ad}(D_1) = [D_1, \quad]$ *is itself a derivation of degree* k_1*).*

Proof. Plug in the definition of the graded commutator and compute. □

In section (9) we have already met some graded derivations: For a vector field X on M the derivation i_X is of degree -1, $\mathcal{L}_X$ is of degree 0, and d is of degree 1. Note also that the important formula $\mathcal{L}_X = d\,i_X + i_X\,d$ translates to $\mathcal{L}_X = [i_X, d]$.

16.2. Algebraic derivations. A derivation $D \in \operatorname{Der}_k \Omega(M)$ is called *algebraic* if $D \mid \Omega^0(M) = 0$. Then $D(f.\omega) = f.D(\omega)$ for $f \in C^\infty(M)$, so D is of tensorial character by (9.3). So D induces a derivation $D_x \in \operatorname{Der}_k \bigwedge T_x^*M$ for each $x \in M$. It is uniquely determined by its restriction to 1-forms $D_x|T_x^*M : T_x^*M \to \bigwedge^{k+1} T^*M$ which we may view as an element $K_x \in \bigwedge^{k+1} T_x^*M \otimes T_xM$ depending smoothly on $x \in M$. To express this dependence, we write $D = i_K = i(K)$, where $K \in \Gamma(\bigwedge^{k+1} T^*M \otimes TM) =: \Omega^{k+1}(M;TM)$. Note the defining equation: $i_K(\omega) = \omega \circ K$ for $\omega \in \Omega^1(M)$. We call $\Omega(M,TM) = \bigoplus_{k=0}^{\dim M} \Omega^k(M,TM)$ the space of all *vector valued differential forms.*

Theorem. (1) *For $K \in \Omega^{k+1}(M,TM)$ the formula*

$$\begin{aligned}(i_K\omega)(X_1,\dots,X_{k+\ell}) \\ = \frac{1}{(k+1)!\,(\ell-1)!} \sum_{\sigma\in\mathcal{S}_{k+\ell}} \operatorname{sign}\sigma\,.\omega(K(X_{\sigma 1},\dots,X_{\sigma(k+1)}),X_{\sigma(k+2)},\dots)\end{aligned}$$

for $\omega \in \Omega^\ell(M)$, $X_i \in \mathfrak{X}(M)$ (or T_xM) defines an algebraic graded derivation $i_K \in \operatorname{Der}_k \Omega(M)$ and any algebraic derivation is of this form.

(2) *By $i([K,L]^\wedge) := [i_K, i_L]$ we get a bracket $[\quad,\quad]^\wedge$ on $\Omega^{*+1}(M,TM)$ which defines a graded Lie algebra structure with the grading as indicated, and for $K \in \Omega^{k+1}(M,TM)$, $L \in \Omega^{\ell+1}(M,TM)$ we have*

$$[K,L]^\wedge = i_K L - (-1)^{k\ell} i_L K$$

where $i_K(\omega \otimes X) := i_K(\omega) \otimes X$.

The bracket $[\quad,\quad]^\wedge$ is called the *algebraic bracket* or the *Nijenhuis-Richardson bracket*; see [**177**].

Proof. Since $\bigwedge T_x^*M$ is the free graded commutative algebra generated by the vector space T_x^*M, any $K \in \Omega^{k+1}(M,TM)$ extends to a graded derivation. By applying it to an exterior product of 1-forms, one can derive the formula in (1). The graded commutator of two algebraic derivations is again algebraic, so the injection $i : \Omega^{*+1}(M,TM) \to \operatorname{Der}_*(\Omega(M))$ induces a graded Lie bracket on $\Omega^{*+1}(M,TM)$ whose form can be seen by applying it to a 1-form. $\square$

16.3. Lie derivations. The exterior derivative d lies in $\operatorname{Der}_1 \Omega(M)$. In view of the formula $\mathcal{L}_X = [i_X, d] = i_X\, d + d\, i_X$ for vector fields X, we define for $K \in \Omega^k(M;TM)$ the *Lie derivative* $\mathcal{L}_K = \mathcal{L}(K) \in \operatorname{Der}_k \Omega(M)$ by $\mathcal{L}_K := [i_K, d] = i_K\, d - (-1)^{k-1} d\, i_K$.

Then the mapping $\mathcal{L} : \Omega(M, TM) \to \operatorname{Der} \Omega(M)$ is injective, since $\mathcal{L}_K f = i_K df = df \circ K$ for $f \in C^\infty(M)$.

Theorem. *For any graded derivation $D \in \operatorname{Der}_k \Omega(M)$ there are unique $K \in \Omega^k(M;TM)$ and $L \in \Omega^{k+1}(M;TM)$ such that*

$$D = \mathcal{L}_K + i_L.$$

We have $L = 0$ if and only if $[D, d] = 0$. The derivation D is algebraic if and only if $K = 0$.

Proof. Let $X_i \in \mathfrak{X}(M)$ be vector fields. Then $f \mapsto (Df)(X_1, \dots, X_k)$ is a derivation $C^\infty(M) \to C^\infty(M)$, so there exists a vector field $K(X_1, \dots, X_k) \in \mathfrak{X}(M)$ by (3.3) such that

$$(Df)(X_1, \dots, X_k) = K(X_1, \dots, X_k) f = df(K(X_1, \dots, X_k)).$$

Clearly $K(X_1, \dots, X_k)$ is $C^\infty(M)$-linear in each X_i and alternating, so K is tensorial by (9.3), $K \in \Omega^k(M;TM)$.

The defining equation for K is $Df = df \circ K = i_K df = \mathcal{L}_K f$ for $f \in C^\infty(M)$. Thus $D - \mathcal{L}_K$ is an algebraic derivation, so $D - \mathcal{L}_K = i_L$ by (16.2) for unique $L \in \Omega^{k+1}(M;TM)$.

Since we have $[d, d] = 2d^2 = 0$, by the graded Jacobi identity, we obtain $0 = [i_K, [d, d]] = [[i_K, d], d] + (-1)^{k-1}[d, [i_K, d]] = 2[\mathcal{L}_K, d]$. The mapping $K \mapsto [i_K, d] = \mathcal{L}_K$ is injective, so the last assertions follow. $\square$

16.4. Applying $i(Id_{TM})$ on a k-fold exterior product of 1-forms, we get $i(Id_{TM})\omega = k\omega$ for $\omega \in \Omega^k(M)$. Thus we have $\mathcal{L}(Id_{TM})\omega = i(Id_{TM})d\omega - d\, i(Id_{TM})\omega = (k+1)d\omega - kd\omega = d\omega$. Thus $\mathcal{L}(Id_{TM}) = d$.

16.5. Let $K \in \Omega^k(M;TM)$ and $L \in \Omega^\ell(M;TM)$. Then $[[\mathcal{L}_K, \mathcal{L}_L], d] = 0$, so we have

$$[\mathcal{L}(K), \mathcal{L}(L)] = \mathcal{L}([K, L])$$

for a uniquely defined $[K, L] \in \Omega^{k+\ell}(M;TM)$. This vector valued form $[K, L]$ is called the *Frölicher-Nijenhuis bracket* of K and L.

Theorem. *The space $\Omega(M;TM) = \bigoplus_{k=0}^{\dim M} \Omega^k(M;TM)$ with its usual grading is a graded Lie algebra for the Frölicher-Nijenhuis bracket. So we have*

$$[K, L] = -(-1)^{k\ell}[L, K],$$
$$[K_1, [K_2, K_3]] = [[K_1, K_2], K_3] + (-1)^{k_1 k_2}[K_2, [K_1, K_3]].$$

The 1-form $\mathrm{Id}_{TM} \in \Omega^1(M;TM)$ *is in the center, i.e.,* $[K, Id_{TM}] = 0$ *for all* K. *The operator* $\mathcal{L} : (\Omega(M;TM), [\quad,\quad]) \to \operatorname{Der}\Omega(M)$ *is an injective homomorphism of graded Lie algebras. For vector fields the Frölicher-Nijenhuis bracket coincides with the Lie bracket.*

Proof. $df \circ [X,Y] = \mathcal{L}([X,Y])f = [\mathcal{L}_X, \mathcal{L}_Y]f$. The rest is clear. □

16.6. Lemma. *For* $K \in \Omega^k(M;TM)$ *and* $L \in \Omega^{\ell+1}(M;TM)$ *we have*

$$[\mathcal{L}_K, i_L] = i([K,L]) - (-1)^{k\ell}\mathcal{L}(i_L K), \text{ or}$$
$$[i_L, \mathcal{L}_K] = \mathcal{L}(i_L K) - (-1)^k\, i([L,K]).$$

This generalizes (9.7.3).

Proof. For $f \in C^\infty(M)$ we have $[i_L, \mathcal{L}_K]f = i_L\, i_K\, df - 0 = i_L(df \circ K) = df \circ (i_L K) = \mathcal{L}(i_L K)f$. So $[i_L, \mathcal{L}_K] - \mathcal{L}(i_L K)$ is an algebraic derivation.

$$\begin{aligned}[][[i_L, \mathcal{L}_K], d] &= [i_L, [\mathcal{L}_K, d]] - (-1)^{k\ell}[\mathcal{L}_K, [i_L, d]] \\ &= 0 - (-1)^{k\ell}\mathcal{L}([K,L]) = (-1)^k[i([L,K]), d].\end{aligned}$$

Since $[\quad, d]$ kills the '$\mathcal{L}$'s' and is injective on the 'i's', the algebraic part of $[i_L, \mathcal{L}_K]$ is $(-1)^k\, i([L,K])$. □

16.7. Module structure. The space $\operatorname{Der}\Omega(M)$ is a graded module over the graded algebra $\Omega(M)$ with the action $(\omega \wedge D)\varphi = \omega \wedge D(\varphi)$, because $\Omega(M)$ is graded commutative.

Theorem. *Let the degree of* ω *be* q, *of* φ *be* k, *and of* ψ *be* ℓ. *Let the other degrees be as indicated. Then we have:*

(1) $$[\omega \wedge D_1, D_2] = \omega \wedge [D_1, D_2] - (-1)^{(q+k_1)k_2} D_2(\omega) \wedge D_1.$$

(2) $$i(\omega \wedge L) = \omega \wedge i(L).$$

(3) $$\omega \wedge \mathcal{L}_K = \mathcal{L}(\omega \wedge K) + (-1)^{q+k-1} i(d\omega \wedge K).$$

(4) $$[\omega \wedge L_1, L_2]^\wedge = \omega \wedge [L_1, L_2]^\wedge - (-1)^{(q+\ell_1-1)(\ell_2-1)} i(L_2)\omega \wedge L_1.$$

(5) $$\begin{aligned}[][\omega \wedge K_1, K_2] &= \omega \wedge [K_1, K_2] - (-1)^{(q+k_1)k_2}\mathcal{L}(K_2)\omega \wedge K_1 \\ &\quad + (-1)^{q+k_1} d\omega \wedge i(K_1)K_2.\end{aligned}$$

(6) $$\begin{aligned}[][\varphi \otimes X, \psi \otimes Y] &= \varphi \wedge \psi \otimes [X,Y] \\ &\quad - \Big(i_Y d\varphi \wedge \psi \otimes X - (-1)^{k\ell} i_X d\psi \wedge \varphi \otimes Y\Big) \\ &\quad - \Big(d(i_Y\varphi \wedge \psi) \otimes X - (-1)^{k\ell} d(i_X\psi \wedge \varphi) \otimes Y\Big) \\ &= \varphi \wedge \psi \otimes [X,Y] + \varphi \wedge \mathcal{L}_X\psi \otimes Y - \mathcal{L}_Y\varphi \wedge \psi \otimes X \\ &\quad + (-1)^k \left(d\varphi \wedge i_X\psi \otimes Y + i_Y\varphi \wedge d\psi \otimes X\right).\end{aligned}$$

Proof. For (1), (2), (3) write out the definitions. For (4) compute $i([\omega \wedge L_1, L_2]^\wedge)$. For (5) compute $\mathcal{L}([\omega \wedge K_1, K_2])$. For (6) use (5). $\square$

16.8. Theorem. *For $K \in \Omega^k(M;TM)$ and $\omega \in \Omega^\ell(M)$ the Lie derivative of ω along K is given by the following formula, where the X_i are vector fields on M:*

$$\begin{aligned}
&(\mathcal{L}_K\omega)(X_1,\dots,X_{k+\ell})\\
&\quad= \tfrac{1}{k!\,\ell!}\sum_\sigma \operatorname{sign}\sigma\; \mathcal{L}(K(X_{\sigma1},\dots,X_{\sigma k}))(\omega(X_{\sigma(k+1)},\dots,X_{\sigma(k+\ell)}))\\
&\qquad+ \tfrac{-1}{k!\,(\ell-1)!}\sum_\sigma \operatorname{sign}\sigma\; \omega([K(X_{\sigma1},\dots,X_{\sigma k}),X_{\sigma(k+1)}],X_{\sigma(k+2)},\dots)\\
&\qquad+ \tfrac{(-1)^{k-1}}{(k-1)!\,(\ell-1)!\,2!}\sum_\sigma \operatorname{sign}\sigma\; \omega(K([X_{\sigma1},X_{\sigma2}],X_{\sigma3},\dots),X_{\sigma(k+2)},\dots).
\end{aligned}$$

Proof. It suffices to consider $K = \varphi \otimes X$. Then by (16.7.3) we have $\mathcal{L}(\varphi \otimes X) = \varphi \wedge \mathcal{L}_X - (-1)^{k-1} d\varphi \wedge i_X$. Now use the global formulas of section (9) to expand this. $\square$

16.9. Theorem. *For $K \in \Omega^k(M;TM)$ and $L \in \Omega^\ell(M;TM)$ we have for the Frölicher-Nijenhuis bracket $[K,L]$ the following formula, where the X_i are vector fields on M:*

$$\begin{aligned}
&[K,L](X_1,\dots,X_{k+\ell})\\
&\quad= \tfrac{1}{k!\,\ell!}\sum_\sigma \operatorname{sign}\sigma\; [K(X_{\sigma1},\dots,X_{\sigma k}),L(X_{\sigma(k+1)},\dots,X_{\sigma(k+\ell)})]\\
&\qquad+ \tfrac{-1}{k!\,(\ell-1)!}\sum_\sigma \operatorname{sign}\sigma\; L([K(X_{\sigma1},\dots,X_{\sigma k}),X_{\sigma(k+1)}],X_{\sigma(k+2)},\dots)\\
&\qquad+ \tfrac{(-1)^{k\ell}}{(k-1)!\,\ell!}\sum_\sigma \operatorname{sign}\sigma\; K([L(X_{\sigma1},\dots,X_{\sigma\ell}),X_{\sigma(\ell+1)}],X_{\sigma(\ell+2)},\dots)\\
&\qquad+ \tfrac{(-1)^{k-1}}{(k-1)!\,(\ell-1)!\,2!}\sum_\sigma \operatorname{sign}\sigma\; L(K([X_{\sigma1},X_{\sigma2}],X_{\sigma3},\dots),X_{\sigma(k+2)},\dots)\\
&\qquad+ \tfrac{(-1)^{(k-1)\ell}}{(k-1)!\,(\ell-1)!\,2!}\sum_\sigma \operatorname{sign}\sigma\; K(L([X_{\sigma1},X_{\sigma2}],X_{\sigma3},\dots),X_{\sigma(\ell+2)},\dots).
\end{aligned}$$

Proof. It suffices to consider $K = \varphi \otimes X$ and $L = \psi \otimes Y$; then for $[\varphi \otimes X, \psi \otimes Y]$ we may use (16.7.6) and evaluate that at $(X_1,\dots,X_{k+\ell})$. After some combinatorial computation we get the right hand side of the above formula for $K = \varphi \otimes X$ and $L = \psi \otimes Y$. $\square$

There are more illuminating ways to prove this formula; see [**146**].

16.10. Local formulas. In a local chart (U,u) on the manifold M we put $K \mid U = \sum K^i_\alpha d^\alpha \otimes \partial_i$, $L \mid U = \sum L^j_\beta d^\beta \otimes \partial_j$, and $\omega \mid U = \sum \omega_\gamma d^\gamma$, where $\alpha = (1 \le \alpha_1 < \alpha_2 < \dots < \alpha_k \le \dim M)$ is a form index, $d^\alpha = du^{\alpha_1} \wedge \ldots \wedge du^{\alpha_k}$, $\partial_i = \frac{\partial}{\partial u^i}$ and so on.

Plugging $X_j = \partial_{i_j}$ into the global formulas (16.2), (16.8), and (16.9), we get the following local formulas:

$$\begin{aligned}
i_K\omega \mid U &= \sum K^i_{\alpha_1\dots\alpha_k}\omega_{i\alpha_{k+1}\dots\alpha_{k+\ell-1}}\, d^\alpha,\\
[K,L]^\wedge \mid U &= \sum \Big(K^i_{\alpha_1\dots\alpha_k}\, L^j_{i\alpha_{k+1}\dots\alpha_{k+\ell}}\\
&\qquad - (-1)^{(k-1)(\ell-1)} L^i_{\alpha_1\dots\alpha_\ell}\, K^j_{i\alpha_{\ell+1}\dots\alpha_{k+\ell}}\Big)\, d^\alpha \otimes \partial_j,\\
\mathcal{L}_K\omega \mid U &= \sum \Big(K^i_{\alpha_1\dots\alpha_k}\, \partial_i \omega_{\alpha_{k+1}\dots\alpha_{k+\ell}}\\
&\qquad + (-1)^k (\partial_{\alpha_1} K^i_{\alpha_2\dots\alpha_{k+1}})\, \omega_{i\alpha_{k+2}\dots\alpha_{k+\ell}}\Big) d^\alpha,\\
[K,L] \mid U &= \sum \Big(K^i_{\alpha_1\dots\alpha_k}\, \partial_i L^j_{\alpha_{k+1}\dots\alpha_{k+\ell}}\\
&\qquad - (-1)^{k\ell} L^i_{\alpha_1\dots\alpha_\ell}\, \partial_i K^j_{\alpha_{\ell+1}\dots\alpha_{k+\ell}}\\
&\qquad - k K^j_{\alpha_1\dots\alpha_{k-1} i}\, \partial_{\alpha_k} L^i_{\alpha_{k+1}\dots\alpha_{k+\ell}}\\
&\qquad + (-1)^{k\ell} \ell L^j_{\alpha_1\dots\alpha_{\ell-1} i}\, \partial_{\alpha_\ell} K^i_{\alpha_{\ell+1}\dots\alpha_{k+\ell}}\Big)\, d^\alpha \otimes \partial_j.
\end{aligned}$$

16.11. Theorem. *For $K_i \in \Omega^{k_i}(M;TM)$ and $L_i \in \Omega^{k_i+1}(M;TM)$ we have*

$$\begin{aligned}
(1)\quad [\mathcal{L}_{K_1} + i_{L_1}, \mathcal{L}_{K_2} + i_{L_2}] &= \mathcal{L}\left([K_1,K_2] + i_{L_1}K_2 - (-1)^{k_1k_2} i_{L_2}K_1 \right)\\
&\quad + i\left([L_1,L_2]^\wedge + [K_1,L_2] - (-1)^{k_1k_2}[K_2,L_1] \right).
\end{aligned}$$

Each summand of this formula looks like a semidirect product of graded Lie algebras, but the mappings

$$\begin{aligned}
i &: \Omega(M;TM) \to \operatorname{End}(\Omega(M;TM), [\quad,\quad]),\\
\operatorname{ad} &: \Omega(M;TM) \to \operatorname{End}(\Omega(M;TM), [\quad,\quad]^\wedge)
\end{aligned}$$

do not take values in the subspaces of graded derivations. We have instead for $K \in \Omega^k(M;TM)$ and $L \in \Omega^{\ell+1}(M;TM)$ the following relations:

$$\begin{aligned}
(2)\qquad i_L[K_1,K_2] &= [i_LK_1, K_2] + (-1)^{k_1\ell}[K_1, i_LK_2],\\
&\quad - \left((-1)^{k_1\ell} i([K_1,L])K_2 - (-1)^{(k_1+\ell)k_2} i([K_2,L])K_1 \right)
\end{aligned}$$

$$\begin{aligned}
(3)\qquad [K,[L_1,L_2]^\wedge] &= [[K,L_1],L_2]^\wedge + (-1)^{kk_1}[L_1,[K,L_2]]^\wedge\\
&\quad - \left((-1)^{kk_1}[i(L_1)K, L_2] - (-1)^{(k+k_1)k_2}[i(L_2)K, L_1] \right).
\end{aligned}$$

The algebraic meaning of the relations of this theorem and its consequences in group theory have been investigated in [**149**]. The corresponding product of groups is well known to algebraists under the name Zappa-Szep product.

Proof. Equation (1) is an immediate consequence of (16.6). Equations (2) and (3) follow from (1) by writing out the graded Jacobi identity or as follows: Consider $\mathcal{L}(i_L[K_1, K_2])$ and use (16.6) repeatedly to obtain $\mathcal{L}$ of the right hand side of (2). Then consider $i([K, [L_1, L_2]^\wedge])$ and use again (16.6) several times to obtain i of the right hand side of (3). □

16.12. Corollary (of (16.9)). *For K, $L \in \Omega^1(M; TM)$ we have*

$$\begin{aligned}[K, L](X, Y) = [KX, LY] - [KY, LX] - L([KX, Y] - [KY, X]) \\ - K([LX, Y] - [LY, X]) + (LK + KL)[X, Y].\end{aligned}$$

16.13. Curvature. Let $P \in \Omega^1(M; TM)$ be a fiber projection, i.e., $P \circ P = P$. This is the most general case of a (first order) *connection.* We may call $\ker P$ the *horizontal space* and $\operatorname{im} P$ the *vertical space* of the connection. If P is of constant rank, then both are vector subbundles of TM. If $\operatorname{im} P$ is some primarily fixed vector subbundle or (tangent bundle of) a foliation, P can be called a connection for it. Special cases of this will be treated extensively later on. The following result is immediate from (16.12).

Lemma. *We have*

$$[P, P] = 2R + 2\bar{R},$$

where R, $\bar{R} \in \Omega^2(M; TM)$ are given by $R(X, Y) = P[(Id - P)X, (Id - P)Y]$ and $\bar{R}(X, Y) = (Id - P)[PX, PY]$.

If P has constant rank, then R is the obstruction against integrability of the horizontal bundle $\ker P$, and $\bar{R}$ is the obstruction against integrability of the vertical bundle $\operatorname{im} P$. Thus we call R the *curvature* and $\bar{R}$ the *cocurvature* of the connection P. We will see later that for a principal fiber bundle R is just the negative of the usual curvature.

16.14. Lemma (Bianchi identity). *If $P \in \Omega^1(M; TM)$ is a connection (fiber projection) with curvature R and cocurvature $\bar{R}$, then we have*

$$\begin{aligned}&[P, R + \bar{R}] = 0,\\ &[R, P] = i_R \bar{R} + i_{\bar{R}} R.\end{aligned}$$

Proof. We have $[P, P] = 2R + 2\bar{R}$ by (16.13) and $[P, [P, P]] = 0$ by the graded Jacobi identity. So the first formula follows. We have $2R = P \circ [P, P] = i_{[P,P]}P$. By (16.11.2) we get $i_{[P,P]}[P, P] = 2[i_{[P,P]}P, P] - 0 = 4[R, P]$.

Therefore $[R,P] = \frac{1}{4}i_{[P,P]}[P,P] = i(R+\bar{R})(R+\bar{R}) = i_R\bar{R} + i_{\bar{R}}R$ since R has vertical values and kills vertical vectors, so $i_R R = 0$; likewise for $\bar{R}$. □

16.15. Naturality of the Frölicher-Nijenhuis bracket. Let $f : M \to N$ be a smooth mapping between manifolds. Two vector valued forms $K \in \Omega^k(M;TM)$ and $K' \in \Omega^k(N;TN)$ are called *f-related* or *f-dependent* if for all $X_i \in T_xM$ we have

(1) $$K'_{f(x)}(T_xf \cdot X_1, \dots, T_xf \cdot X_k) = T_xf \cdot K_x(X_1, \dots, X_k).$$

Theorem.

(2) *If K and K' as above are f-related, then $i_K \circ f^* = f^* \circ i_{K'} : \Omega(N) \to \Omega(M)$.*

(3) *If $i_K \circ f^* \mid B^1(N) = f^* \circ i_{K'} \mid B^1(N)$, then K and K' are f-related, where B^1 denotes the space of exact 1-forms.*

(4) *If K_j and K'_j are f-related for $j = 1,2$, then $i_{K_1}K_2$ and $i_{K'_1}K'_2$ are f-related, and also $[K_1,K_2]^\wedge$ and $[K'_1,K'_2]^\wedge$ are f-related.*

(5) *If K and K' are f-related, then $\mathcal{L}_K \circ f^* = f^* \circ \mathcal{L}_{K'} : \Omega(N) \to \Omega(M)$.*

(6) *If $\mathcal{L}_K \circ f^* \mid \Omega^0(N) = f^* \circ \mathcal{L}_{K'} \mid \Omega^0(N)$, then K and K' are f-related.*

(7) *If K_j and K'_j are f-related for $j = 1,2$, then their Frölicher-Nijenhuis brackets $[K_1,K_2]$ and $[K'_1,K'_2]$ are also f-related.*

Proof. (2) By (16.2) we have for $\omega \in \Omega^q(N)$ and $X_i \in T_xM$:

$$\begin{aligned}
&(i_K f^*\omega)_x(X_1,\dots,X_{q+k-1}) \\
&= \tfrac{1}{k!\,(q-1)!}\sum_\sigma \operatorname{sign}\sigma\,(f^*\omega)_x(K_x(X_{\sigma1},\dots,X_{\sigma k}),X_{\sigma(k+1)},\dots) \\
&= \tfrac{1}{k!\,(q-1)!}\sum_\sigma \operatorname{sign}\sigma\,\omega_{f(x)}(T_xf\cdot K_x(X_{\sigma1},\dots),T_xf\cdot X_{\sigma(k+1)},\dots) \\
&= \tfrac{1}{k!\,(q-1)!}\sum_\sigma \operatorname{sign}\sigma\,\omega_{f(x)}(K'_{f(x)}(T_xf\cdot X_{\sigma1},\dots),T_xf\cdot X_{\sigma(k+1)},\dots) \\
&= (f^*i_{K'}\omega)_x(X_1,\dots,X_{q+k-1}).
\end{aligned}$$

(3) follows from this computation, since the df, $f \in C^\infty(M)$, separate points.

(4) follows from the same computation for K_2 instead of ω; the result for the bracket then follows from (16.2.2).

(5) By (2) the algebra homomorphism f^* intertwines the operators i_K and $i_{K'}$, and f^* commutes with the exterior derivative d. Thus f^* intertwines the commutators $[i_K,d] = \mathcal{L}_K$ and $[i_{K'},d] = \mathcal{L}_{K'}$.

(6) For $g \in \Omega^0(N)$ we have $\mathcal{L}_K\, f^*\, g = i_K\, d\, f^*\, g = i_K\, f^*\, dg$ and $f^*\, \mathcal{L}_{K'}\, g = f^*\, i_{K'}\, dg$. By (3) the result follows.

(7) The algebra homomorphism f^* intertwines $\mathcal{L}_{K_j}$ and $\mathcal{L}_{K'_j}$, so also their graded commutators which equal $\mathcal{L}([K_1, K_2])$ and $\mathcal{L}([K'_1, K'_2])$, respectively. Now use (6). □

16.16. Let $f : M \to N$ be a local diffeomorphism. Then we can consider the pullback operator $f^* : \Omega(N; TN) \to \Omega(M; TM)$, given by

$$(1) \qquad (f^*K)_x(X_1, \dots, X_k) = (T_xf)^{-1} K_{f(x)}(T_xf \cdot X_1, \dots, T_xf \cdot X_k).$$

Note that this is a special case of the pullback operator for sections of natural vector bundles in (8.16). Clearly K and f^*K are then f-related.

Theorem. *In this situation we have:*

(2) $f^*[K, L] = [f^*K, f^*L]$.

(3) $f^* i_K L = i_{f^*K} f^*L$.

(4) $f^*[K, L]^\wedge = [f^*K, f^*L]^\wedge$.

(5) *For a vector field $X \in \mathfrak{X}(M)$ and $K \in \Omega(M; TM)$ by* (8.16) *the Lie derivative $\mathcal{L}_X K = \partial|_0 (\mathrm{Fl}^X_t)^* K$ is defined. Then we have $\mathcal{L}_X K = [X, K]$, the Frölicher-Nijenhuis bracket.*

We may say that the Frölicher-Nijenhuis bracket, $[\ \ ,\ \ ]^\wedge$, etc., are *natural bilinear mappings*.

Proof. (2) – (4) are obvious from (16.15). (5) Obviously $\mathcal{L}_X$ is $\mathbb{R}$-linear, so it suffices to check this formula for $K = \psi \otimes Y$, $\psi \in \Omega(M)$ and $Y \in \mathfrak{X}(M)$. But then

$$\begin{aligned} \mathcal{L}_X(\psi \otimes Y) &= \mathcal{L}_X\psi \otimes Y + \psi \otimes \mathcal{L}_X Y \quad \text{by (8.17)} \\ &= \mathcal{L}_X\psi \otimes Y + \psi \otimes [X, Y] \\ &= [X, \psi \otimes Y] \quad \text{by (16.7.6).} \quad \square \end{aligned}$$

16.17. Remark. At last we mention the best known application of the Frölicher-Nijenhuis bracket, which also led to its discovery. A vector valued 1-form $J \in \Omega^1(M; TM)$ with $J \circ J = -Id$ is called an *almost complex structure*; if it exists, $\dim M$ is even and J can be viewed as a fiber multiplication with $\sqrt{-1}$ on TM. By (16.12) we have

$$[J, J](X, Y) = 2([JX, JY] - [X, Y] - J[X, JY] - J[JX, Y]).$$

The vector valued form $\frac{1}{2}[J, J]$ is also called the *Nijenhuis tensor* of J. For it the following result is true:

> *A manifold M with an almost complex structure J is a complex manifold (i.e., there exists an atlas for M with holomorphic chart-change mappings) if and only if $[J, J] = 0$. See* [**172**].

17. Fiber Bundles and Connections

17.1. Definition. A *(fiber) bundle* (E, p, M, S) consists of manifolds E, M, S, and a smooth mapping $p : E \to M$; furthermore each $x \in M$ has an open neighborhood U such that $E \mid U := p^{-1}(U)$ is diffeomorphic to $U \times S$ via a fiber respecting diffeomorphism:

$$\begin{array}{ccc} E \mid U & \xrightarrow{\quad\psi\quad} & U \times S \\ & {\scriptstyle p}\searrow \quad \swarrow{\scriptstyle \mathrm{pr}_1} & \\ & U. & \end{array}$$

The manifold E is called the *total space*, M is called the *base space* or *basis*, p is a surjective submersion, called the *projection*, S is called *standard fiber*, and (U, ψ) as above is called a *fiber chart*.

A collection of fiber charts (U_α, ψ_α), such that (U_α) is an open cover of M, is called a *fiber bundle atlas*. If we fix such an atlas, then $\psi_\alpha \circ {\psi_\beta}^{-1}(x, s) = (x, \psi_{\alpha\beta}(x, s))$, where $\psi_{\alpha\beta} : (U_\alpha \cap U_\beta) \times S \to S$ is smooth and $\psi_{\alpha\beta}(x, \ )$ is a diffeomorphism of S for each $x \in U_{\alpha\beta} := U_\alpha \cap U_\beta$. We may thus consider the mappings $\psi_{\alpha\beta} : U_{\alpha\beta} \to \mathrm{Diff}(S)$ with values in the group $\mathrm{Diff}(S)$ of all diffeomorphisms of S; their differentiability is a subtle question, which will not be discussed in this book, but see [**147**]. In either form these mappings $\psi_{\alpha\beta}$ are called the *transition functions* of the bundle. They satisfy the *cocycle condition*: $\psi_{\alpha\beta}(x) \circ \psi_{\beta\gamma}(x) = \psi_{\alpha\gamma}(x)$ for $x \in U_{\alpha\beta\gamma}$ and $\psi_{\alpha\alpha}(x) = Id_S$ for $x \in U_\alpha$. Therefore the collection $(\psi_{\alpha\beta})$ is called a *cocycle of transition functions*.

Given an open cover (U_α) of a manifold M and a cocycle of transition functions $(\psi_{\alpha\beta})$, we may construct a fiber bundle (E, p, M, S) in a similar way as in (8.3).

17.2. Lemma. *Let $p : N \to M$ be a surjective submersion (a fibered manifold) which is proper, so that $p^{-1}(K)$ is compact in N for each compact $K \subset M$, and let M be connected. Then (N, p, M) is a fiber bundle.*

Proof. We have to produce a fiber chart at each $x_0 \in M$. So let (U, u) be a chart centered at x_0 on M such that $u(U) \cong \mathbb{R}^m$. For each $x \in U$ let $\xi_x(y) := (T_y u)^{-1}.u(x)$; then we have $\xi_x \in \mathfrak{X}(U)$ which depends smoothly on $x \in U$, such that $u(\mathrm{Fl}_t^{\xi_x} u^{-1}(z)) = z + t.u(x)$. Thus each ξ_x is a complete vector field on U. Since p is a submersion, with the help of a partition of unity on $p^{-1}(U)$ we may construct vector fields $\eta_x \in \mathfrak{X}(p^{-1}(U))$ which depend smoothly on $x \in U$ and are p-related to ξ_x: $Tp.\eta_x = \xi_x \circ p$. Thus $p \circ \mathrm{Fl}_t^{\eta_x} = \mathrm{Fl}_t^{\xi_x} \circ p$ by (3.14), so $\mathrm{Fl}_t^{\eta_x}$ is fiber respecting, and since p is proper and ξ_x is complete, η_x has a global flow too. Denote $p^{-1}(x_0)$ by S. Then

$\varphi : U \times S \to p^{-1}(U)$, defined by $\varphi(x,y) = \mathrm{Fl}_1^{\eta_x}(y)$, is a diffeomorphism and is fiber respecting, so (U, φ^{-1}) is a fiber chart. Since M is connected, the fibers $p^{-1}(x)$ are all diffeomorphic.

17.3. Let (E, p, M, S) be a fiber bundle; we consider the fiber linear tangent mapping $Tp : TE \to TM$ and its kernel $\ker Tp =: VE$ which is called the *vertical bundle* of E. The following is a special case of (16.13).

Definition. A *connection* on the fiber bundle (E, p, M, S) is a vector valued 1-form $\Phi \in \Omega^1(E; VE)$ with values in the vertical bundle VE such that $\Phi \circ \Phi = \Phi$ and $\mathrm{Im}\Phi = VE$; so Φ is just a projection $TE \to VE$.

Then $\ker \Phi$ is of constant rank, so $\ker \Phi$ is a vector subbundle of TE by (8.7), it is called the space of *horizontal vectors* or the *horizontal bundle* and it is denoted by $HE = \ker \Phi$. Clearly $TE = HE \oplus VE$ and $T_uE = H_uE \oplus V_uE$ for $u \in E$.

Now we consider the mapping $(Tp, \pi_E) : TE \to TM \times_M E$. Then by definition $(Tp, \pi_E)^{-1}(0_{p(u)}, u) = V_uE$, so $(Tp, \pi_E) \mid HE : HE \to TM \times_M E$ is fiber linear over E and injective, so by reason of dimensions it is a fiber linear isomorphism: Its inverse is denoted by

$$C := ((Tp, \pi_E) \mid HE)^{-1} : TM \times_M E \to HE \hookrightarrow TE.$$

So $C : TM \times_M E \to TE$ is fiber linear over E and is a right inverse for (Tp, π_E). The mapping C is called the *horizontal lift* associated to the connection Φ.

Note the formula $\Phi(\xi_u) = \xi_u - C(Tp.\xi_u, u)$ for $\xi_u \in T_uE$. So we can equally well describe a connection Φ by specifying C. Then we call Φ the *vertical projection* (no confusion with (8.12) will arise) and $\chi := \mathrm{id}_{TE} - \Phi = C \circ (Tp, \pi_E)$ will be called the *horizontal projection.*

17.4. Curvature. If $\Phi : TE \to VE$ is a connection on the bundle (E, p, M, S), then as in (16.13) the curvature R of Φ is given by

$$2R = [\Phi, \Phi] = [Id - \Phi, Id - \Phi] = [\chi, \chi] \in \Omega^2(E; VE).$$

The cocurvature $\bar{R}$ vanishes since the vertical bundle VE is integrable. We have

$$R(X, Y) = \tfrac{1}{2}[\Phi, \Phi](X, Y) = \Phi[\chi X, \chi Y],$$

so R is an obstruction against integrability of the horizontal subbundle. Note that for vector fields $\xi, \eta \in \mathfrak{X}(M)$ and their horizontal lifts $C\xi, C\eta \in \mathfrak{X}(E)$ we have

$$R(C\xi, C\eta) = [C\xi, C\eta] - C([\xi, \eta]).$$

Since the vertical bundle VE is integrable, by (16.14) we have the *Bianchi identity* $[\Phi, R] = 0$.

17.5. Pullback. Let (E, p, M, S) be a fiber bundle and consider a smooth mapping $f : N \to M$. Since p is a submersion, f and p are transversal in the sense of (2.16) and thus the pullback $N \times_{(f,M,p)} E$ exists. It will be called the *pullback* of the fiber bundle E by f and we will denote it by f^*E. The following diagram sets up some further notation for it:

$$\begin{array}{ccc} f^*E & \xrightarrow{p^*f} & E \\ {\scriptstyle f^*p}\downarrow & & \downarrow{\scriptstyle p} \\ N & \xrightarrow{f} & M. \end{array}$$

Proposition. *In the situation above we have:*

(1) *(f^*E, f^*p, N, S) is again a fiber bundle, and p^*f is a fiberwise diffeomorphism.*

(2) *If $\Phi \in \Omega^1(E; VE) \subset \Omega^1(E; TE)$ is a connection on E, then the vector valued form $f^*\Phi$, given by $(f^*\Phi)_u(X) := V_u(p^*f)^{-1}.\Phi.T_u(p^*f).X$ for $X \in T_uE$, is a connection on the bundle f^*E. The forms $f^*\Phi$ and Φ are p^*f-related in the sense of* (16.15).

(3) *The curvatures of $f^*\Phi$ and Φ are also p^*f-related.*

Proof. (1) If (U_α, ψ_α) is a fiber bundle atlas of (E, p, M, S) in the sense of (17.1), then $(f^{-1}(U_\alpha), (f^*p, \mathrm{pr}_2 \circ \psi_\alpha \circ p^*f))$ is a fiber bundle atlas for (f^*E, f^*p, N, S), by the formal universal properties of a pullback (2.17). Part (2) is obvious. Part (3) follows from (2) and (16.15.7). □

17.6. Let us suppose that a connection Φ on the bundle (E, p, M, S) has zero curvature. Then by (17.4) the horizontal bundle is integrable and gives rise to the *horizontal foliation* by (3.28.2). Each point $u \in E$ lies on a unique leaf $L(u)$ such that $T_vL(u) = H_vE$ for each $v \in L(u)$. The restriction $p \mid L(u)$ is locally a diffeomorphism, but in general it is neither surjective nor is it a covering onto its image. This is seen by devising suitable horizontal foliations on the trivial bundle $\mathrm{pr}_2 : \mathbb{R} \times S^1 \to S^1$, or $\mathrm{pr}_2\, \mathbb{R} \times \mathbb{R} \to \mathbb{R}$, like $L(0, t) = \{(\tan(s - t), s) : s \in \mathbb{R}\}$.

17.7. Local description. Let Φ be a connection on (E, p, M, S). Let us fix a fiber bundle atlas (U_α) with transition functions $(\psi_{\alpha\beta})$, and let us consider the connection $((\psi_\alpha)^{-1})^*\Phi \in \Omega^1(U_\alpha \times S; U_\alpha \times TS)$, which may be written in the form

$$\big(((\psi_\alpha)^{-1})^*\Phi\big)(\xi_x, \eta_y) =: -\Gamma^\alpha(\xi_x, y) + \eta_y \text{ for } \xi_x \in T_xU_\alpha \text{ and } \eta_y \in T_yS,$$

since it reproduces vertical vectors. The Γ^α are given by

$$(0_x, \Gamma^\alpha(\xi_x, y)) := -T(\psi_\alpha).\Phi.T(\psi_\alpha)^{-1}.(\xi_x, 0_y).$$

We consider Γ^α as an element of the space $\Omega^1(U_\alpha;\mathfrak{X}(S))$, a 1-form on U^α with values in the infinite-dimensional Lie algebra $\mathfrak{X}(S)$ of all vector fields on the standard fiber. The Γ^α are called the *Christoffel forms* of the connection Φ with respect to the bundle atlas (U_α, ψ_α).

Lemma. *The transformation law for the Christoffel forms is*

$$T_y(\psi_{\alpha\beta}(x,\quad)).\Gamma^\beta(\xi_x, y) = \Gamma^\alpha(\xi_x, \psi_{\alpha\beta}(x,y)) - T_x(\psi_{\alpha\beta}(\quad, y)).\xi_x.$$

The curvature R of Φ satisfies

$$(\psi_\alpha^{-1})^* R = d\Gamma^\alpha + [\Gamma^\alpha, \Gamma^\alpha]_{\mathfrak{X}(S)}.$$

Here $d\Gamma^\alpha$ is the exterior derivative of the 1-form $\Gamma^\alpha \in \Omega^1(U_\alpha;\mathfrak{X}(S))$ with values in the complete locally convex space $\mathfrak{X}(S)$. We will later also use the Lie derivative of it and the usual formulas apply: Consult [**113**] for calculus in infinite-dimensional spaces.

The formula for the curvature is the *Maurer-Cartan* formula which in this general setting appears only in the level of local description.

Proof. From $(\psi_\alpha \circ (\psi_\beta)^{-1})(x,y) = (x, \psi_{\alpha\beta}(x,y))$ we get that

$$T(\psi_\alpha \circ (\psi_\beta)^{-1}).(\xi_x, \eta_y) = (\xi_x, T_{(x,y)}(\psi_{\alpha\beta}).(\xi_x, \eta_y))$$

and thus:

$$\begin{aligned}
&T(\psi_\beta^{-1}).(0_x, \Gamma^\beta(\xi_x, y)) = -\Phi(T(\psi_\beta^{-1})(\xi_x, 0_y))\\
&= -\Phi(T(\psi_\alpha^{-1}).T(\psi_\alpha \circ \psi_\beta^{-1}).(\xi_x, 0_y))\\
&= -\Phi(T(\psi_\alpha^{-1})(\xi_x, T_{(x,y)}(\psi_{\alpha\beta})(\xi_x, 0_y)))\\
&= -\Phi(T(\psi_\alpha^{-1})(\xi_x, 0_{\psi_{\alpha\beta}(x,y)})) - \Phi(T(\psi_\alpha^{-1})(0_x, T_{(x,y)}\psi_{\alpha\beta}(\xi_x, 0_y))\\
&= T(\psi_\alpha^{-1}).(0_x, \Gamma^\alpha(\xi_x, \psi_{\alpha\beta}(x,y))) - T(\psi_\alpha^{-1})(0_x, T_x(\psi_{\alpha\beta}(\quad, y)).\xi_x).
\end{aligned}$$

This implies the transformation law.

For the curvature R of Φ we have by (17.4) and (17.5.3)

$$\begin{aligned}
&(\psi_\alpha^{-1})^* R\,((\xi^1, \eta^1), (\xi^2, \eta^2))\\
&\quad = (\psi_\alpha^{-1})^*\Phi\,[(Id - (\psi_\alpha^{-1})^*\Phi)(\xi^1, \eta^1), (Id - (\psi_\alpha^{-1})^*\Phi)(\xi^2, \eta^2)]\\
&\quad = (\psi_\alpha^{-1})^*\Phi[(\xi^1, \Gamma^\alpha(\xi^1)), (\xi^2, \Gamma^\alpha(\xi^2))]\\
&\quad = (\psi_\alpha^{-1})^*\Phi\left([\xi^1, \xi^2], \xi^1\Gamma^\alpha(\xi^2) - \xi^2\Gamma^\alpha(\xi^1) + [\Gamma^\alpha(\xi^1), \Gamma^\alpha(\xi^2)]\right)\\
&\quad = -\Gamma^\alpha([\xi^1, \xi^2]) + \xi^1\Gamma^\alpha(\xi^2) - \xi^2\Gamma^\alpha(\xi^1) + [\Gamma^\alpha(\xi^1), \Gamma^\alpha(\xi^2)]\\
&\quad = d\Gamma^\alpha(\xi^1, \xi^2) + [\Gamma^\alpha(\xi^1), \Gamma^\alpha(\xi^2)]_{\mathfrak{X}(S)}. \quad \square
\end{aligned}$$

17.8. Theorem (Parallel transport). *Let Φ be a connection on a bundle (E, p, M, S) and let $c : (a, b) \to M$ be a smooth curve with $0 \in (a, b)$, $c(0) = x$.*

Then there is a neighborhood U of $E_x \times \{0\}$ in $E_x \times (a, b)$ and a smooth mapping $\operatorname{Pt}_c : U \to E$ such that:

(1) *$p(\operatorname{Pt}(c, u_x, t)) = c(t)$ if defined, and $\operatorname{Pt}(c, u_x, 0) = u_x$.*
(2) *$\Phi(\frac{d}{dt} \operatorname{Pt}(c, u_x, t)) = 0$ if defined.*
(3) *Reparameterization invariance: If $f : (a', b') \to (a, b)$ is smooth with $0 \in (a', b')$, then $\operatorname{Pt}(c, u_x, f(t)) = \operatorname{Pt}(c \circ f, \operatorname{Pt}(c, u_x, f(0)), t)$ if defined.*
(4) *U is maximal for properties (1) and (2).*
(5) *In a certain sense Pt depends smoothly also on c.*

First proof. In local bundle coordinates $\Phi(\frac{d}{dt} \operatorname{Pt}(c, u_x, t)) = 0$ is an ordinary differential equation of first order, nonlinear, with initial condition $\operatorname{Pt}(c, u_x, 0) = u_x$. So there is a maximally defined local solution curve which is unique. All further properties are consequences of uniqueness.

Second proof. Consider the pullback bundle $(c^*E, c^*p, (a, b), S)$ and the pullback connection $c^*\Phi$ on it. It has zero curvature, since the horizontal bundle is 1-dimensional. By (17.6) the horizontal foliation exists and the parallel transport just follows a leaf and we may map it back to E, in detail: $\operatorname{Pt}(c, u_x, t) = p^*c((c^*p \mid L(u_x))^{-1}(t))$.

Third proof. Consider a fiber bundle atlas (U_α, ψ_α) as in (17.7). Then we have $\psi_\alpha(\operatorname{Pt}(c, \psi_\alpha^{-1}(x, y), t)) = (c(t), \gamma(y, t))$, where

$$0 = \left((\psi_\alpha^{-1})^*\Phi\right)\left(\tfrac{d}{dt}c(t), \tfrac{d}{dt}\gamma(y, t)\right) = -\Gamma^\alpha\left(\tfrac{d}{dt}c(t), \gamma(y, t)\right) + \tfrac{d}{dt}\gamma(y, t),$$

so $\gamma(y, t)$ is the integral curve (evolution line) through $y \in S$ of the time dependent vector field $\Gamma^\alpha\left(\frac{d}{dt}c(t)\right)$ on S. This vector field visibly depends smoothly on c. Clearly local solutions exist and all properties follow, even (5). For more detailed information on (5) we refer to [**142**] or [**113**]. □

17.9. A connection Φ on (E, p, M, S) is called a *complete connection* if the parallel transport Pt_c along any smooth curve $c : (a, b) \to M$ is defined on the whole of $E_{c(0)} \times (a, b)$. The third proof of theorem (17.8) shows that on a fiber bundle with compact standard fiber any connection is complete.

The following is a sufficient condition for a connection Φ to be complete:

> There exists a fiber bundle atlas (U_α, ψ_α) and complete Riemann metrics g_α on the standard fiber S such that each Christoffel form $\Gamma^\alpha \in \Omega^1(U_\alpha, \mathfrak{X}(S))$ takes values in the linear subspace of g_α-bounded vector fields on S.

This is true because in the third proof of theorem (17.8) above the time dependent vector field $\Gamma^\alpha(\frac{d}{dt}c(t))$ on S is g_α-bounded for compact time intervals. By (23.9) this vector field is complete. So by continuation the solution exists globally.

A complete connection is called an *Ehresmann connection* in [**80**, I, p. 314], where the following result is given as an exercise.

Theorem. *Each fiber bundle admits complete connections.*

Proof. Let $\dim M = m$. Let (U_α, ψ_α) be a fiber bundle atlas as in (17.1). By topological dimension theory [**168**] the open cover (U_α) of M admits a refinement such that any $m+2$ members have empty intersection; see also (1.1). Let (U_α) itself have this property. Choose a smooth partition of unity (f_α) subordinated to (U_α). Then the sets $V_\alpha := \{x : f_\alpha(x) > \frac{1}{m+2}\} \subset U_\alpha$ still form an open cover of M since $\sum f_\alpha(x) = 1$ and at most $m+1$ of the $f_\alpha(x)$ can be nonzero. By renaming, assume that each V_α is connected. Then we choose an open cover (W_α) of M such that $\overline{W_\alpha} \subset V_\alpha$.

Now let g_1 and g_2 be complete Riemann metrics on M and S, respectively (see (23.8)). For not connected Riemann manifolds complete means that each connected component is complete. Then $g_1|U_\alpha \times g_2$ is a Riemann metric on $U_\alpha \times S$ and we consider the metric $g := \sum f_\alpha \psi_\alpha^*(g_1|U_\alpha \times g_2)$ on E. Obviously $p : E \to M$ is a Riemann submersion for the metrics g and g_1: This means that $T_u p : (T_u(E_{p(u)})^\perp, g_u) \to (T_{p(u)}M, (g_1)_{p(u)})$ is an isometry for each $u \in E$. We choose now the connection $\Phi : TE \to VE$ as the orthonormal projection with respect to the Riemann metric g.

Claim. Φ is a complete connection on E.

Let $c : [0,1] \to M$ be a smooth curve. We choose a partition $0 = t_0 < t_1 < \cdots < t_k = 1$ such that $c([t_i, t_{i+1}]) \subset V_{\alpha_i}$ for suitable α_i. It suffices to show that $\operatorname{Pt}(c(t_i + \quad), u_{c(t_i)}, t)$ exists for all $0 \le t \le t_{i+1} - t_i$ and all $u_{c(t_i)}$, for all i, since then we may piece them together. So we may assume that $c : [0,1] \to V_\alpha$ for some α. Let us now assume that for $x = c(0)$ and some $y \in S$ the parallel transport $\operatorname{Pt}(c, \psi_\alpha(x,y), t)$ is defined only for $t \in [0, t')$ for some $0 < t' < 1$. By the third proof of theorem (17.8) we have

$$\operatorname{Pt}(c, \psi_\alpha^{-1}(x,y), t) = \psi_\alpha^{-1}(c(t), \gamma(t)),$$

where $\gamma : [0, t') \to S$ is the maximally defined integral curve through $y \in S$ of the time dependent vector field $\Gamma^\alpha(\frac{d}{dt}c(t), \quad)$ on S. We put $g_\alpha := (\psi_\alpha^{-1})^* g$; then

$$(g_\alpha)_{(x,y)} = (g_1)_x \times \Big(\sum_\beta f_\beta(x) \psi_{\beta\alpha}(x, \quad)^* g_2\Big)_y.$$

Since $\operatorname{pr}_1 : (V_\alpha \times S, g_\alpha) \to (V_\alpha, g_1|V_\alpha)$ is a Riemann submersion and since the connection $(\psi_\alpha^{-1})^*\Phi$ is also given by orthonormal projection onto the vertical bundle, we get

$$\begin{aligned}\infty > g_1\text{-length}_0^{t'}(c) &= g_\alpha\text{-length}(c,\gamma) = \int_0^{t'} |(c'(t), \tfrac{d}{dt}\gamma(t))|_{g_\alpha}\,dt \\ &= \int_0^{t'} \sqrt{|c'(t)|_{g_1}^2 + \textstyle\sum_\beta f_\beta(c(t))(\psi_{\alpha\beta}(c(t),-)^*g_2)(\tfrac{d}{dt}\gamma(t),\tfrac{d}{dt}\gamma(t))}\,dt \\ &\ge \int_0^{t'} \sqrt{f_\alpha(c(t))}\,|\tfrac{d}{dt}\gamma(t)|_{g_2}\,dt \ge \frac{1}{\sqrt{m+2}}\int_0^{t'} |\tfrac{d}{dt}\gamma(t)|_{g_2} dt.\end{aligned}$$

So g_2-length(γ) is finite and since the Riemann metric g_2 on S is complete, the limit $\lim_{t\to t'}\gamma(t) =: \gamma(t')$ exists in S and the integral curve γ can be continued. □

17.10. Holonomy groups and Lie algebras. Let (E,p,M,S) be a fiber bundle with a complete connection Φ, and let us assume that M is connected. We choose a fixed base point $x_0 \in M$ and we identify E_{x_0} with the standard fiber S. For each closed piecewise smooth curve $c : [0,1] \to M$ through x_0 the parallel transport $\mathrm{Pt}(c,\ \ ,1) =: \mathrm{Pt}(c,1)$ (pieced together over the smooth parts of c) is a diffeomorphism of S. All these diffeomorphisms form together the group $\mathrm{Hol}(\Phi,x_0)$, the *holonomy group* of Φ at x_0, a subgroup of the diffeomorphism group $\mathrm{Diff}(S)$. If we consider only those piecewise smooth curves which are homotopic to zero, we get a subgroup $\mathrm{Hol}_0(\Phi,x_0)$, called the *restricted holonomy group* of the connection Φ at x_0.

Now let $C : TM \times_M E \to TE$ be the horizontal lifting as in (17.3), and let R be the curvature (see (17.4)) of the connection Φ. For any $x \in M$ and $X_x \in T_xM$ the horizontal lift $C(X_x) := C(X_x,\ \ ) : E_x \to TE$ is a vector field along E_x. For X_x and $Y_x \in T_xM$ we consider $R(CX_x, CY_x) \in \mathfrak{X}(E_x)$. Now we choose any piecewise smooth curve c from x_0 to x and consider the diffeomorphism $\mathrm{Pt}(c,t) : S = E_{x_0} \to E_x$ and the pullback $\mathrm{Pt}(c,1)^*R(CX_x,CY_x) \in \mathfrak{X}(S)$. Let us denote by $\mathrm{hol}(\Phi,x_0)$ the closed linear subspace, generated by all these vector fields (for all $x \in M$, X_x, $Y_x \in T_xM$ and curves c from x_0 to x) in $\mathfrak{X}(S)$ with respect to the compact C^∞-topology, and let us call it the *holonomy Lie algebra* of Φ at x_0.

Lemma. $\mathrm{hol}(\Phi,x_0)$ *is a Lie subalgebra of* $\mathfrak{X}(S)$.

Proof. For $X \in \mathfrak{X}(M)$ we consider the local flow Fl_t^{CX} of the horizontal lift of X. It restricts to parallel transport along any of the flow lines of X in M. Then for vector fields on M the expression

$$\begin{aligned}&\tfrac{d}{dt}|_0(\mathrm{Fl}_s^{CX})^*(\mathrm{Fl}_t^{CY})^*(\mathrm{Fl}_{-s}^{CX})^*(\mathrm{Fl}_z^{CZ})^*R(CU,CV) \restriction E_{x_0} \\ &\quad= (\mathrm{Fl}_s^{CX})^*[CY, (\mathrm{Fl}_{-s}^{CX})^*(\mathrm{Fl}_z^{CZ})^*R(CU,CV)] \restriction E_{x_0} \\ &\quad= [(\mathrm{Fl}_s^{CX})^*CY, (\mathrm{Fl}_z^{CZ})^*R(CU,CV)] \restriction E_{x_0}\end{aligned}$$

is in $\operatorname{hol}(\Phi, x_0)$, since it is closed in the compact C^∞-topology and the derivative can be written as a limit. Thus

$$[(\mathrm{Fl}_s^{CX})^*[CY_1, CY_2], (\mathrm{Fl}_z^{CZ})^* R(CU, CV)] \restriction E_{x_0} \in \operatorname{hol}(\Phi, x_0)$$

by the Jacobi identity and

$$[(\mathrm{Fl}_s^{CX})^* C[Y_1, Y_2], (\mathrm{Fl}_z^{CZ})^* R(CU, CV)] \restriction E_{x_0} \in \operatorname{hol}(\Phi, x_0),$$

so also their difference

$$[(\mathrm{Fl}_s^{CX})^* R(CY_1, CY_2), (\mathrm{Fl}_z^{CZ})^* R(CU, CV)] \restriction E_{x_0}$$

is in $\operatorname{hol}(\Phi, x_0)$. □

17.11. The following theorem is a generalization of the theorem of [**173, 174**] and [**9**] on principal connections. The reader who does not know principal connections is advised to read parts of sections (18) and (19) first. We include this result here in order not to disturb the development in section (19) later.

Theorem. *Let Φ be a complete connection on the fiber bundle (E, p, M, S) and let M be connected. Suppose that for some (hence any) $x_0 \in M$ the holonomy Lie algebra* $\operatorname{hol}(\Phi, x_0)$ *is finite-dimensional and consists of complete vector fields on the fiber E_{x_0}.*

Then there is a principal bundle (P, p, M, G) with finite-dimensional structure group G, a connection ω on it and a smooth action of G on S such that the Lie algebra $\mathfrak{g}$ of G equals the holonomy Lie algebra $\operatorname{hol}(\Phi, x_0)$, *the fiber bundle E is isomorphic to the associated bundle $P[S]$, and Φ is the connection induced by ω. The structure group G equals the holonomy group* $\operatorname{Hol}(\Phi, x_0)$. *The principle bundle P and its connection ω are unique up to isomorphism.*

By a theorem of [**185**] a finite-dimensional Lie subalgebra of $\mathfrak{X}(E_{x_0})$ like $\operatorname{hol}(\Phi, x_0)$ consists of complete vector fields if and only if it is generated by complete vector fields as a Lie algebra.

Proof. Let us again identify E_{x_0} and S. Then $\mathfrak{g} := \operatorname{hol}(\Phi, x_0)$ is a finite-dimensional Lie subalgebra of $\mathfrak{X}(S)$, and since each vector field in it is complete, there is a finite-dimensional connected Lie group G_0 of diffeomorphisms of S with Lie algebra $\mathfrak{g}$, by theorem (6.5).

Claim 1. G_0 contains $\operatorname{Hol}_0(\Phi, x_0)$, the restricted holonomy group.
Let $f \in \operatorname{Hol}_0(\Phi, x_0)$; then $f = \operatorname{Pt}(c, 1)$ for a piecewise smooth closed curve c through x_0, which is nullhomotopic. Since the parallel transport is essentially invariant under reparametrization, (17.8), we can replace c by $c \circ g$, where g is smooth and flat at each corner of c. So we may assume that c itself is smooth. Since c is homotopic to zero, by approximation we may

assume that there is a smooth homotopy $H : \mathbb{R}^2 \to M$ with $H_1|[0,1] = c$ and $H_0|[0,1] = x_0$. Then $f_t := \mathrm{Pt}(H_t, 1)$ is a curve in $\mathrm{Hol}_0(\Phi, x_0)$ which is smooth as a mapping $\mathbb{R} \times S \to S$; this can be seen by using the proof of claim 2 below or as in the proof of (19.7.4). We will continue the proof of claim 1 below.

Claim 2. $(\frac{d}{dt} f_t) \circ f_t^{-1} =: Z_t$ is in $\mathfrak{g}$ for all t.
To prove claim 2, we consider the pullback bundle $H^*E \to \mathbb{R}^2$ with the induced connection $H^*\Phi$. It is sufficient to prove claim 2 there. Let $X = \frac{d}{ds}$ and $Y = \frac{d}{dt}$ be constant vector fields on $\mathbb{R}^2$, so $[X, Y] = 0$. Then $\mathrm{Pt}(c, s) = \mathrm{Fl}_s^{CX}|S$ and so on. We put

$$f_{t,s} = \mathrm{Fl}_{-s}^{CX} \circ \mathrm{Fl}_{-t}^{CY} \circ \mathrm{Fl}_s^{CX} \circ \mathrm{Fl}_t^{CY} : S \to S,$$

so $f_{t,1} = f_t$. Then we have in the vector space $\mathfrak{X}(S)$

$$\begin{aligned}
(\tfrac{d}{dt} f_{t,s}) \circ f_{t,s}^{-1} &= -(\mathrm{Fl}_s^{CX})^* CY + (\mathrm{Fl}_s^{CX})^* (\mathrm{Fl}_t^{CY})^* (\mathrm{Fl}_{-s}^{CX})^* CY, \\
(\tfrac{d}{dt} f_{t,1}) \circ f_{t,1}^{-1} &= \int_0^1 \tfrac{d}{ds} \left((\tfrac{d}{dt} f_{t,s}) \circ f_{t,s}^{-1} \right) ds \\
&= \int_0^1 \big(-(\mathrm{Fl}_s^{CX})^* [CX, CY] + (\mathrm{Fl}_s^{CX})^* [CX, (\mathrm{Fl}_t^{CY})^* (\mathrm{Fl}_{-s}^{CX})^* CY] \\
&\qquad - (\mathrm{Fl}_s^{CX})^* (\mathrm{Fl}_t^{CY})^* (\mathrm{Fl}_{-s}^{CX})^* [CX, CY] \big)\, ds.
\end{aligned}$$

Since $[X, Y] = 0$, we have $[CX, CY] = \Phi[CX, CY] = R(CX, CY)$ and $(\mathrm{Fl}_t^X)^* Y = Y$; thus

$$\begin{aligned}
(\mathrm{Fl}_t^{CX})^* CY &= C\left((\mathrm{Fl}_t^X)^* Y \right) + \Phi\left((\mathrm{Fl}_t^{CX})^* CY \right) \\
&= CY + \int_0^t \tfrac{d}{dt} \Phi (\mathrm{Fl}_t^{CX})^* CY \, dt = CY + \int_0^t \Phi (\mathrm{Fl}_t^{CX})^* [CX, CY] \, dt \\
&= CY + \int_0^t \Phi (\mathrm{Fl}_t^{CX})^* R(CX, CY) \, dt = CY + \int_0^t (\mathrm{Fl}_t^{CX})^* R(CX, CY) \, dt.
\end{aligned}$$

The flows $(\mathrm{Fl}_s^{CX})^*$ and their derivatives $\mathcal{L}_{CX} = [CX, \quad]$ do not lead out of $\mathfrak{g}$; thus all parts of the integrand above are in $\mathfrak{g}$ and so $(\frac{d}{dt} f_{t,1}) \circ f_{t,1}^{-1}$ is in $\mathfrak{g}$ for all t and claim 2 follows.

Now claim 1 can be shown as follows. There is a unique smooth curve $g(t)$ in G_0 satisfying $T_e(\mu^{g(t)}) Z_t = Z_t.g(t) = \frac{d}{dt} g(t)$ and $g(0) = e$; via the action of G_0 on S the curve $g(t)$ is a curve of diffeomorphisms on S, generated by the time dependent vector field Z_t, so $g(t) = f_t$ and $f = f_1$ is in G_0. So we get $\mathrm{Hol}_0(\Phi, x_0) \subseteq G_0$.

Claim 3. $\mathrm{Hol}_0(\Phi, x_0)$ equals G_0.
In the proof of claim 1 we have seen that $\mathrm{Hol}_0(\Phi, x_0)$ is a smoothly arcwise connected subgroup of G_0, so it is a connected Lie subgroup by the theorem (5.6). It suffices thus to show that the Lie algebra $\mathfrak{g}$ of G_0 is contained

in the Lie algebra of $\operatorname{Hol}_0(\Phi, x_0)$, and for that it is enough to show that for each ξ in a linearly spanning subset of $\mathfrak{g}$ there is a smooth mapping $f : [-1,1] \times S \to S$ such that the associated curve $\check{f}$ lies in $\operatorname{Hol}_0(\Phi, x_0)$ with $\check{f}'(0) = 0$ and $\check{f}''(0) = \xi$.

By definition we may assume $\xi = \operatorname{Pt}(c,1)^* R(CX_x, CY_x)$ for $X_x, Y_x \in T_xM$ and a smooth curve c in M from x_0 to x. We extend X_x and Y_x to vector fields X and $Y \in \mathfrak{X}(M)$ with $[X,Y] = 0$ near x. We may also suppose that $Z \in \mathfrak{X}(M)$ is a vector field which extends $c'(t)$ along $c(t)$: If c is simple, we approximate it by an embedding and can consequently extend $c'(t)$ to such a vector field. If c is not simple, we do this for each simple piece of c, and then have several vector fields Z instead of one below. So we have

$$\begin{aligned}
\xi &= (\operatorname{Fl}_1^{CZ})^* R(CX, CY) = (\operatorname{Fl}_1^{CZ})^*[CX, CY] \quad \text{since } [X,Y](x) = 0 \\
&= (\operatorname{Fl}_1^{CZ})^* \tfrac{1}{2} \tfrac{d^2}{dt^2}|_{t=0} (\operatorname{Fl}_{-t}^{CY} \circ \operatorname{Fl}_{-t}^{CX} \circ \operatorname{Fl}_t^{CY} \circ \operatorname{Fl}_t^{CX}) \quad \text{by (3.16)} \\
&= \tfrac{1}{2} \tfrac{d^2}{dt^2}|_{t=0} (\operatorname{Fl}_{-1}^{CZ} \circ \operatorname{Fl}_{-t}^{CY} \circ \operatorname{Fl}_{-t}^{CX} \circ \operatorname{Fl}_t^{CY} \circ \operatorname{Fl}_t^{CX} \circ \operatorname{Fl}_1^{CZ}),
\end{aligned}$$

where the parallel transport in the last equation first follows c from x_0 to x, then follows a small closed parallelogram near x in M (since $[X,Y] = 0$ near x) and then follows c back to x_0. This curve is clearly nullhomotopic.

Step 4. Now we make $\operatorname{Hol}(\Phi, x_0)$ into a Lie group which we call G, by taking $\operatorname{Hol}_0(\Phi, x_0) = G_0$ as its connected component of the identity. Then the quotient $\operatorname{Hol}(\Phi, x_0)/\operatorname{Hol}_0(\Phi, x_0)$ is a countable group, since the fundamental group $\pi_1(M)$ is countable (by Morse theory M is homotopy equivalent to a countable CW-complex).

Step 5. Construction of a cocycle of transition functions with values in G. Let $(U_\alpha, u_\alpha : U_\alpha \to \mathbb{R}^m)$ be a locally finite smooth atlas for M such that each $u_\alpha : U_\alpha \to \mathbb{R}^m$ is surjective. Put $x_\alpha := u_\alpha^{-1}(0)$ and choose smooth curves $c_\alpha : [0,1] \to M$ with $c_\alpha(0) = x_0$ and $c_\alpha(1) = x_\alpha$. For each $x \in U_\alpha$ let $c_\alpha^x : [0,1] \to M$ be the smooth curve $t \mapsto u_\alpha^{-1}(t.u_\alpha(x))$; then c_α^x connects x_α and x and the mapping $(x,t) \mapsto c_\alpha^x(t)$ is smooth $U_\alpha \times [0,1] \to M$. Now we define a fiber bundle atlas $(U_\alpha, \psi_\alpha : E|U_\alpha \to U_\alpha \times S)$ by $\psi_\alpha^{-1}(x,s) = \operatorname{Pt}(c_\alpha^x, 1) \operatorname{Pt}(c_\alpha, 1)\, s$. Then ψ_α is smooth since $\operatorname{Pt}(c_\alpha^x, 1) = \operatorname{Fl}_1^{CX_x}$ for a local vector field X_x depending smoothly on x. Let us investigate the transition functions:

$$\begin{aligned}
\psi_\alpha \psi_\beta^{-1}(x,s) &= \big(x, \operatorname{Pt}(c_\alpha,1)^{-1} \operatorname{Pt}(c_\alpha^x,1)^{-1} \operatorname{Pt}(c_\beta^x,1) \operatorname{Pt}(c_\beta,1)\, s\big) \\
&= \big(x, \operatorname{Pt}(c_\beta.c_\beta^x.(c_\alpha^x)^{-1}.(c_\alpha)^{-1}, 4)\, s\big) \\
&=: (x, \psi_{\alpha\beta}(x)\, s), \text{ where } \psi_{\alpha\beta} : U_{\alpha\beta} \to G.
\end{aligned}$$

Clearly $\psi_{\beta\alpha} : U_{\beta\alpha} \times S \to S$ is smooth, which implies that $\psi_{\beta\alpha} : U_{\beta\alpha} \to G$ is also smooth. $(\psi_{\alpha\beta})$ is a cocycle of transition functions and we use it to glue a principal bundle with structure group G over M which we call (P, p, M, G).

From its construction it is clear that the associated bundle $P[S] = P \times_G S$ equals (E, p, M, S).

Step 6. Lifting the connection Φ to P.

For this we have to compute the Christoffel symbols of Φ with respect to the atlas of step 5. To do this directly is quite difficult since we have to differentiate the parallel transport with respect to the curve. Fortunately there is another way. Let $c : [0,1] \to U_\alpha$ be a smooth curve. Then we have

$$\begin{aligned}
&\psi_\alpha(\operatorname{Pt}(c,t)\psi_\alpha^{-1}(c(0),s)) \\
&\quad = \Big(c(t), \operatorname{Pt}((c_\alpha)^{-1},1)\operatorname{Pt}((c_\alpha^{c(0)})^{-1},1)\operatorname{Pt}(c,t)\operatorname{Pt}(c_\alpha^{c(0)},1)\operatorname{Pt}(c_\alpha,1)s\Big) \\
&\quad = (c(t), \gamma(t).s),
\end{aligned}$$

where γ is a smooth curve in the holonomy group G. Let $\Gamma^\alpha \in \Omega^1(U_\alpha, \mathfrak{X}(S))$ be the Christoffel symbol of the connection Φ with respect to the chart (U_α, ψ_α). From the third proof of theorem (17.8) we have

$$\psi_\alpha(\operatorname{Pt}(c,t)\psi_\alpha^{-1}(c(0),s)) = (c(t), \bar\gamma(t,s)),$$

where $\bar\gamma(t,s)$ is the integral curve through s of the time dependent vector field $\Gamma^\alpha(\frac{d}{dt}c(t))$ on S. But then we get

$$\begin{aligned}
\Gamma^\alpha(\tfrac{d}{dt}c(t))(\bar\gamma(t,s)) &= \tfrac{d}{dt}\bar\gamma(t,s) = \tfrac{d}{dt}(\gamma(t).s) = (\tfrac{d}{dt}\gamma(t)).s, \\
\Gamma^\alpha(\tfrac{d}{dt}c(t)) &= (\tfrac{d}{dt}\gamma(t)) \circ \gamma(t)^{-1} \in \mathfrak{g}.
\end{aligned}$$

So Γ^α takes values in the Lie subalgebra of fundamental vector fields for the action of G on S. By theorem (19.9) below the connection Φ is thus induced by a principal connection ω on P. Since by (19.8) the principal connection ω has the 'same' holonomy group as Φ and since this is also the structure group of P, the principal connection ω is irreducible; see (19.7). □

18. Principal Fiber Bundles and G-Bundles

18.1. Definition. Let G be a Lie group and let (E, p, M, S) be a fiber bundle as in (17.1). A *G-bundle structure* on the fiber bundle consists of the following data:

(1) a left action $\ell : G \times S \to S$ of the Lie group on the standard fiber,

(2) a fiber bundle atlas (U_α, ψ_α) whose transition functions $(\psi_{\alpha\beta})$ act on S via the G-action: There is a family of smooth mappings $(\varphi_{\alpha\beta} : U_{\alpha\beta} \to G)$ which satisfies the cocycle condition $\varphi_{\alpha\beta}(x)\varphi_{\beta\gamma}(x) = \varphi_{\alpha\gamma}(x)$ for $x \in U_{\alpha\beta\gamma}$ and $\varphi_{\alpha\alpha}(x) = e$, the unit in the group, such that $\psi_{\alpha\beta}(x,s) = \ell(\varphi_{\alpha\beta}(x), s) = \varphi_{\alpha\beta}(x).s$.

A fiber bundle with a G-bundle structure is called a *G-bundle.* A fiber bundle atlas as in (2) is called a *G-atlas* and the family $(\varphi_{\alpha\beta})$ is also called a cocycle of transition functions, but now for the G-bundle.

To be more precise, two G-atlases are said to be equivalent (to describe the same G-bundle) if their union is also a G-atlas. This translates as follows to the two cocycles of transition functions, where we assume that the two coverings of M are the same (by passing to the common refinement, if necessary): $(\varphi_{\alpha\beta})$ and $(\varphi'_{\alpha\beta})$ are called *cohomologous* if there is a family $(\tau_\alpha : U_\alpha \to G)$ such that $\varphi_{\alpha\beta}(x) = \tau_\alpha(x)^{-1}.\varphi'_{\alpha\beta}(x).\tau_\beta(x)$ holds for all $x \in U_{\alpha\beta}$; compare with (8.3).

In (2) one should specify only an equivalence class of G-bundle structures or only a cohomology class of cocycles of G-valued transition functions. The proof of (8.3) now shows that from any open cover (U_α) of M, some cocycle of transition functions $(\varphi_{\alpha\beta} : U_{\alpha\beta} \to G)$ for it, and a left G-action on a manifold S, we may construct a G-bundle, which depends only on the cohomology class of the cocycle. By some abuse of notation we write (E, p, M, S, G) for a fiber bundle with specified G-bundle structure.

Examples. The tangent bundle of a manifold M is a fiber bundle with structure group $GL(m)$. More generally, a vector bundle (E, p, M, V) as in (8.1) is a fiber bundle with standard fiber the vector space V and with $GL(V)$-structure.

18.2. Definition. A *principal (fiber) bundle* (P, p, M, G) is a G-bundle with typical fiber a Lie group G, where the left action of G on G is just the left translation.

So by (18.1) we are given a bundle atlas $(U_\alpha, \varphi_\alpha : P|U_\alpha \to U_\alpha \times G)$ such that we have $\varphi_\alpha\varphi_\beta^{-1}(x, a) = (x, \varphi_{\alpha\beta}(x).a)$ for the cocycle of transition functions $(\varphi_{\alpha\beta} : U_{\alpha\beta} \to G)$. This is now called a *principal bundle atlas.* Clearly the principal bundle is uniquely specified by the cohomology class of its cocycle of transition functions.

Each principal bundle admits a unique right action $r : P \times G \to P$, called the *principal right action,* given by $\varphi_\alpha(r(\varphi_\alpha^{-1}(x, a), g)) = (x, ag)$. Since left and right translation on G commute, this is well defined. As in (6.1) we write $r(u, g) = u.g$ when the meaning is clear. The principal right action is visibly free and for any $u_x \in P_x$ the partial mapping $r_{u_x} = r(u_x, \quad) : G \to P_x$ is a diffeomorphism onto the fiber through u_x, whose inverse is denoted by $\tau_{u_x} : P_x \to G$. These inverses together give a smooth mapping $\tau : P \times_M P \to G$, whose local expression is $\tau(\varphi_\alpha^{-1}(x, a), \varphi_\alpha^{-1}(x, b)) = a^{-1}.b$. This mapping is also uniquely determined by the implicit equation $r(u_x, \tau(u_x, v_x)) = v_x$; thus we also have $\tau(u_x.g, u'_x.g') = g^{-1}.\tau(u_x, u'_x).g'$ and $\tau(u_x, u_x) = e$.

When considering principal bundles, the reader should think of frame bundles as the foremost examples for this book. They will be treated in (18.11) below.

18.3. Lemma. *Let $p : P \to M$ be a surjective submersion (a fibered manifold), and let G be a Lie group which acts freely on P such that the orbits of the action are exactly the fibers $p^{-1}(x)$ of p. Then (P, p, M, G) is a principal fiber bundle.*

Proof. Let the action be a right one by using the group inversion if necessary. Let $s_\alpha : U_\alpha \to P$ be local sections (right inverses) for $p : P \to M$ such that (U_α) is an open cover of M. Let $\varphi_\alpha^{-1} : U_\alpha \times G \to P|U_\alpha$ be given by $\varphi_\alpha^{-1}(x,a) = s_\alpha(x).a$, which is obviously injective with invertible tangent mapping, so its inverse $\varphi_\alpha : P|U_\alpha \to U_\alpha \times G$ is a fiber respecting diffeomorphism. So $(U_\alpha, \varphi_\alpha)$ is already a fiber bundle atlas. Let $\tau : P \times_M P \to G$ be given by the implicit equation $r(u_x, \tau(u_x, u'_x)) = u'_x$, where r is the right G-action. The mapping τ is smooth by the implicit function theorem and clearly we have

$$\tau(u_x, u'_x.g) = \tau(u_x, u'_x).g \quad \text{and} \quad \varphi_\alpha(u_x) = (x, \tau(s_\alpha(x), u_x)).$$

Thus we have

$$\begin{aligned}\varphi_\alpha\varphi_\beta^{-1}(x,g) &= \varphi_\alpha(s_\beta(x).g) = (x, \tau(s_\alpha(x), s_\beta(x).g))\\ &= (x, \tau(s_\alpha(x), s_\beta(x)).g)\end{aligned}$$

and $(U_\alpha, \varphi_\alpha)$ is a principal bundle atlas. □

18.4. Remarks. In the proof of lemma (18.3) we have seen that a principal bundle atlas of a principal fiber bundle (P, p, M, G) is already determined if we specify a family of smooth sections of P whose domains of definition cover the base M.

Lemma (18.3) can serve as an equivalent definition for a principal bundle. But this is true only if an implicit function theorem is available, so in topology or in infinite-dimensional differential geometry one should stick to our original definition.

From lemma (18.3) itself it follows that the pullback f^*P over a smooth mapping $f : M' \to M$ is again a principal fiber bundle.

18.5. Homogeneous spaces. Let G be a Lie group with Lie algebra $\mathfrak{g}$. Let K be a closed subgroup of G; then by theorem (5.5), K is a closed Lie subgroup whose Lie algebra will be denoted by $\mathfrak{k}$. By theorem (5.11) there is a unique structure of a smooth manifold on the quotient space G/K such that the projection $p : G \to G/K$ is a submersion, so by the implicit function theorem p admits local sections.

Theorem. *$(G, p, G/K, K)$ is a principal fiber bundle.*

Proof. The group multiplication of G restricts to a free right action $\mu : G \times K \to G$ whose orbits are exactly the fibers of p. By lemma (18.3) the result follows. □

For the convenience of the reader we discuss now the best known homogeneous spaces.

The group $SO(n)$ acts transitively on $S^{n-1} \subset \mathbb{R}^n$. The isotropy group of the 'north pole' $(1, 0, \dots, 0)$ is the subgroup

$$\begin{pmatrix} 1 & 0 \\ 0 & SO(n-1) \end{pmatrix}$$

which we identify with $SO(n-1)$. So

$$S^{n-1} = SO(n)/SO(n-1)$$

and we get a principal fiber bundle

$$(SO(n), p, S^{n-1}, SO(n-1)).$$

Likewise the follwing are principal fiber bundles:

$$(O(n), p, S^{n-1}, O(n-1)),$$
$$(SU(n), p, S^{2n-1}, SU(n-1)),$$
$$(U(n), p, S^{2n-1}, U(n-1)),$$
$$(Sp(n), p, S^{4n-1}, Sp(n-1)).$$

The *Grassmann manifold* $G(k, n; \mathbb{R})$ is the space of all k-planes containing 0 in $\mathbb{R}^n$. The group $O(n)$ acts transitively on it and the isotropy group of the k-plane $\mathbb{R}^k \times \{0\}$ is the subgroup

$$\begin{pmatrix} O(k) & 0 \\ 0 & O(n-k) \end{pmatrix};$$

therefore

$$G(k, n; \mathbb{R}) = O(n)/O(k) \times O(n-k)$$

is a compact manifold and we get the principal fiber bundle

$$(O(n), p, G(k, n; \mathbb{R}), O(k) \times O(n-k)).$$

Likewise the follwing are principal fiber bundles:

$$(SO(n), p, G(k, n; \mathbb{R}), S(O(k) \times O(n-k))),$$
$$(SO(n), p, \tilde{G}(k, n; \mathbb{R}), SO(k) \times SO(n-k)),$$
$$(U(n), p, G(k, n; \mathbb{C}), U(k) \times U(n-k)),$$
$$(Sp(n), p, G(k, n; \mathbb{H}), Sp(k) \times Sp(n-k)).$$

The *Stiefel manifold* $V(k,n;\mathbb{R})$ is the space of all orthonormal k-frames in $\mathbb{R}^n$. Clearly the group $O(n)$ acts transitively on $V(k,n;\mathbb{R})$ and the isotropy subgroup of $(e_1,\dots,e_k)$ is $\mathbb{I}_k \times O(n-k)$, so

$$V(k,n;\mathbb{R}) = O(n)/O(n-k)$$

is a compact manifold, and

$$(O(n), p, V(k,n;\mathbb{R}), O(n-k))$$

is a principal fiber bundle. But $O(k)$ also acts from the right on $V(k,n;\mathbb{R})$; its orbits are exactly the fibers of the projection $p : V(k,n;\mathbb{R}) \to G(k,n;\mathbb{R})$. So by lemma (18.3) we get a principal fiber bundle

$$(V(k,n,\mathbb{R}), p, G(k,n;\mathbb{R}), O(k)).$$

Indeed we have the following diagram where all arrows are projections of principal fiber bundles and where the respective structure groups are written on the arrows:

(1)
$$\begin{array}{ccc} O(n) & \xrightarrow{O(n-k)} & V(k,n;\mathbb{R}) \\ {\scriptstyle O(k)}\downarrow & & \downarrow{\scriptstyle O(k)} \\ V(n-k,n;\mathbb{R}) & \xrightarrow{O(n-k)} & G(k,n;\mathbb{R}). \end{array}$$

The Stiefel manifold $V(k,n;\mathbb{R})$ is also diffeomorphic to the space $\{A \in L(\mathbb{R}^k,\mathbb{R}^n) : A^\top.A = \mathbb{I}_k\}$, i.e., the space of all linear isometries $\mathbb{R}^k \to \mathbb{R}^n$. There are furthermore complex and quaternionic versions of Stiefel manifolds and flag manifolds.

18.6. Homomorphisms. Let $\chi : (P,p,M,G) \to (P',p',M',G)$ be a *principal fiber bundle homomorphism*, i.e., a smooth G-equivariant mapping $\chi : P \to P'$. Then obviously the diagram

(1)
$$\begin{array}{ccc} P & \xrightarrow{\chi} & P' \\ {\scriptstyle p}\downarrow & & \downarrow{\scriptstyle p'} \\ M & \xrightarrow{\underline{\chi}} & M' \end{array}$$

commutes for a uniquely determined smooth mapping $\underline{\chi} : M \to M'$. For each $x \in M$ the mapping $\chi_x := \chi|P_x : P_x \to P'_{\underline{\chi}(x)}$ is G-equivariant and therefore a diffeomorphism, so diagram (1) is a pullback diagram.

But the most general notion of a homomorphism of principal bundles is the following. Let $\Phi : G \to G'$ be a homomorphism of Lie groups. A mapping $\chi : (P,p,M,G) \to (P',p',M',G')$ is called a *homomorphism over Φ of principal bundles* if $\chi : P \to P'$ is smooth and $\chi(u.g) = \chi(u).\Phi(g)$ holds in general. Then χ is fiber respecting, so diagram (1) again makes sense, but it is no longer a pullback diagram in general.

If χ covers the identity on the base, it is called a *reduction of the structure group* G' *to* G for the principal bundle (P', p', M', G') — the name comes from the case when Φ is the embedding of a subgroup.

By the universal property of the pullback any general homomorphism χ of principal fiber bundles over a group homomorphism can be written as the composition of a reduction of structure groups and a pullback homomorphism as follows, where we also indicate the structure groups:

(2)
$$\begin{array}{ccccc} (P,G) & \longrightarrow & (\bar{\chi}^*P', G') & \longrightarrow & (P', G') \\ & {\scriptstyle p}\searrow & \downarrow & & \downarrow{\scriptstyle p'} \\ & & M & \xrightarrow{\bar{\chi}} & M'. \end{array}$$

18.7. Associated bundles. Let (P, p, M, G) be a principal bundle and let $\ell : G \times S \to S$ be a left action of the structure group G on a manifold S. We consider the right action $R : (P \times S) \times G \to P \times S$, given by $R((u, s), g) = (u.g, g^{-1}.s)$.

Theorem. *In this situation we have:*

(1) *The space* $P \times_G S$ *of orbits of the action* R *carries a unique smooth manifold structure such that the quotient map* $q : P \times S \to P \times_G S$ *is a submersion.*

(2) $(P \times_G S, \bar{p}, M, S, G)$ *is a* G*-bundle in a canonical way, where* $\bar{p} : P \times_G S \to M$ *is given as in the following diagram, where* $q_u : \{u\} \times S \to (P \times_G S)_{p(u)}$ *is a diffeomorphism for each* $u \in P$*:*

(a)
$$\begin{array}{ccc} P \times S & \xrightarrow{q} & P \times_G S \\ \downarrow{\scriptstyle \mathrm{pr}_1} & & \downarrow{\scriptstyle \bar{p}} \\ P & \xrightarrow{p} & M. \end{array}$$

(3) $(P \times S, q, P \times_G S, G)$ *is a principal fiber bundle with principal action* R.

(4) *If* $(U_\alpha, \varphi_\alpha : P|U_\alpha \to U_\alpha \times G)$ *is a principal bundle atlas with cocycle of transition functions* $(\varphi_{\alpha\beta} : U_{\alpha\beta} \to G)$*, then together with the left action* $\ell : G \times S \to S$ *this cocycle is also one for the* G*-bundle* $(P \times_G S, \bar{p}, M, S, G)$.

Notation. $(P \times_G S, \bar{p}, M, S, G)$ is called the *associated bundle* for the action $\ell : G \times S \to S$. We will also denote it by $P[S, \ell]$ or simply $P[S]$ and we will write p for $\bar{p}$ if no confusion is possible. We also define the smooth mapping $\tau = \tau^S : P \times_M P[S, \ell] \to S$ by $\tau(u_x, v_x) := q_{u_x}^{-1}(v_x)$ which satisfies $\tau(u, q(u, s)) = s$, $q(u_x, \tau(u_x, v_x)) = v_x$, and $\tau(u_x.g, v_x) = g^{-1}.\tau(u_x, v_x)$. In the special situation where $S = G$ and the action is left translation, so that $P[G] = P$, this mapping coincides with τ considered in (18.2).

Proof. In the setting of diagram (a) in (2) the mapping $p \circ \mathrm{pr}_1$ is constant on the R-orbits, so $\bar{p}$ exists as a mapping. Let $(U_\alpha, \varphi_\alpha : P|U_\alpha \to U_\alpha \times G)$ be a principal bundle atlas with transition functions $(\varphi_{\alpha\beta} : U_{\alpha\beta} \to G)$. We define $\psi_\alpha^{-1} : U_\alpha \times S \to \bar{p}^{-1}(U_\alpha) \subset P \times_G S$ by $\psi_\alpha^{-1}(x, s) = q(\varphi_\alpha^{-1}(x, e), s)$, which is fiber respecting. For each point in $\bar{p}^{-1}(x) \subset P \times_G S$ there is exactly one $s \in S$ such that the orbit corresponding to this point passes through $(\varphi_\alpha^{-1}(x, e), s)$, namely $s = \tau^G(u_x, \varphi_\alpha^{-1}(x, e))^{-1}.s'$ if (u_x, s') is the orbit, since the principal right action is free. Thus $\psi_\alpha^{-1}(x, \quad) : S \to \bar{p}^{-1}(x)$ is bijective. Furthermore

$$\begin{aligned}
\psi_\beta^{-1}(x, s) &= q(\varphi_\beta^{-1}(x, e), s) \\
&= q(\varphi_\alpha^{-1}(x, \varphi_{\alpha\beta}(x).e), s) = q(\varphi_\alpha^{-1}(x, e).\varphi_{\alpha\beta}(x), s) \\
&= q(\varphi_\alpha^{-1}(x, e), \varphi_{\alpha\beta}(x).s) = \psi_\alpha^{-1}(x, \varphi_{\alpha\beta}(x).s),
\end{aligned}$$

so $\psi_\alpha \psi_\beta^{-1}(x, s) = (x, \varphi_{\alpha\beta}(x).s)$. So (U_α, ψ_α) is a G-atlas for $P \times_G S$ and makes it into a smooth manifold and a G-bundle. The defining equation for ψ_α shows that q is smooth and a submersion and consequently the smooth structure on $P \times_G S$ is uniquely defined, and $\bar{p}$ is smooth by the universal properties of a submersion.

By the definition of ψ_α the diagram

(5)
$$\begin{array}{ccc}
p^{-1}(U_\alpha) \times S & \xrightarrow{\varphi_\alpha \times Id} & U_\alpha \times G \times S \\
{\scriptstyle q}\big\downarrow & & \big\downarrow{\scriptstyle Id \times \ell} \\
\bar{p}^{-1}(U_\alpha) & \xrightarrow{\quad\psi_\alpha\quad} & U_\alpha \times S
\end{array}$$

commutes; since its lines are diffeomorphisms, we conclude that $q_u : \{u\} \times S \to \bar{p}^{-1}(p(u))$ is a diffeomorphism. So (1), (2), and (4) are checked.

(3) follows directly from lemma (18.3). We give below an explicit chart construction. We rewrite the last diagram in the following form:

(6)
$$\begin{array}{ccccc}
p^{-1}(U_\alpha) \times S & \xrightarrow{=} & q^{-1}(V_\alpha) & \xrightarrow{\lambda_\alpha} & V_\alpha \times G \\
 & & {\scriptstyle q}\big\downarrow & & \big\downarrow{\scriptstyle \mathrm{pr}_1} \\
 & & \bar{p}^{-1}(U_\alpha) & \xrightarrow{\quad=\quad} & V_\alpha.
\end{array}$$

Here $V_\alpha := \bar{p}^{-1}(U_\alpha) \subset P \times_G S$, and the diffeomorphism λ_α is given by the expression $\lambda_\alpha^{-1}(\psi_\alpha^{-1}(x,s),g) := (\varphi_\alpha^{-1}(x,g), g^{-1}.s)$. Then we have

$$\begin{aligned}\lambda_\beta^{-1}(\psi_\alpha^{-1}(x,s),g) &= \lambda_\beta^{-1}(\psi_\beta^{-1}(x,\varphi_{\beta\alpha}(x).s),g)\\ &= (\varphi_\beta^{-1}(x,g), g^{-1}.\varphi_{\beta\alpha}(x).s)\\ &= (\varphi_\alpha^{-1}(x,\varphi_{\alpha\beta}(x).g), g^{-1}.\varphi_{\alpha\beta}(x)^{-1}.s)\\ &= \lambda_\alpha^{-1}(\psi_\alpha^{-1}(x,s),\varphi_{\alpha\beta}(x).g),\end{aligned}$$

so $\lambda_\alpha\lambda_\beta^{-1}(\psi_\alpha^{-1}(x,s),g) = (\psi_\alpha^{-1}(x,s),\varphi_{\alpha\beta}(x).g)$ and $(P\times S, q, P\times_G S, G)$ is a principal bundle with structure group G and the same cocycle $(\varphi_{\alpha\beta})$ we started with. □

18.8. Corollary. *Let (E,p,M,S,G) be a G-bundle, specified by a cocycle of transition functions $(\varphi_{\alpha\beta})$ with values in G and a left action ℓ of G on S. Then from the cocycle of transition functions we may glue a unique principal bundle (P,p,M,G) such that $E = P[S,\ell]$.* □

This is the usual way a differential geometer thinks of an associated bundle. He is given a bundle E and a principal bundle P, and the G-bundle structure then is described with the help of the mappings τ and q.

18.9. Equivariant mappings and associated bundles.
(1) Let (P,p,M,G) be a principal fiber bundle and consider two left actions of G, $\ell : G\times S\to S$ and $\ell' : G\times S'\to S'$. Let furthermore $f : S\to S'$ be a G-equivariant smooth mapping, so $f(g.s) = g.f(s)$ or $f\circ\ell_g = \ell'_g\circ f$. Then $Id_P\times f : P\times S\to P\times S'$ is equivariant for the actions $R : (P\times S)\times G\to P\times S$ and $R' : (P\times S')\times G\to P\times S'$ and is thus a homomorphism of principal bundles, so there is an induced mapping

$$(2)\qquad \begin{array}{ccc} P\times S & \xrightarrow{Id\times f} & P\times S' \\ {\scriptstyle q}\downarrow & & \downarrow{\scriptstyle q'} \\ P\times_G S & \xrightarrow{Id\times_G f} & P\times_G S', \end{array}$$

which is fiber respecting over M, and a homomorphism of G-bundles in the sense of the definition (18.10) below.

(3) Let $\chi : (P,p,M,G)\to(P',p',M',G)$ be a principal fiber bundle homomorphism as in (18.6). Furthermore we consider a smooth left action $\ell : G\times S\to S$. Then $\chi\times Id_S : P\times S\to P'\times S$ is G-equivariant and induces a mapping $\chi\times_G Id_S : P\times_G S\to P'\times_G S$, which is fiber respecting over M, fiberwise a diffeomorphism, and again a homomorphism of G-bundles in the sense of definition (18.10) below.

(4) We consider the situation of (1) and (2) at the same time. Given two associated bundles $P[S,\ell]$ and $P'[S',\ell']$, let

$$\chi : (P,p,M,G) \to (P',p',M',G)$$

be a principal fiber bundle homomorphism and let $f : S \to S'$ be a G-equivariant mapping. Then $\chi \times f : P \times S \to P' \times S'$ is clearly G-equivariant and therefore induces a mapping

$$\chi \times_G f : P[S,\ell] \to P'[S',\ell']$$

which again is a homomorphism of G-bundles.
(5) Let S be a point. Then $P[S] = P \times_G S = M$. Furthermore let $y \in S'$ be a fixed point of the action $\ell' : G \times S' \to S'$; then the inclusion $i : \{y\} \hookrightarrow S'$ is G-equivariant. Thus $Id_P \times i$ induces

$$Id_P \times_G i : M = P[\{y\}] \to P[S'],$$

which is a global section of the associated bundle $P[S']$.
If the action of G on S is trivial, so $g.s = s$ for all $s \in S$, then the associated bundle is trivial: $P[S] = M \times S$. For a trivial principal fiber bundle any associated bundle is trivial.

18.10. Definition. In the situation of (18.9), a smooth fiber respecting mapping $\gamma : P[S,\ell] \to P'[S',\ell']$ covering a smooth mapping $\bar\gamma : M \to M'$ of the bases is called a *homomorphism of G-bundles* if the following conditions are satisfied: P is isomorphic to the pullback $\bar\gamma^* P'$, and the local representations of γ in pullback-related fiber bundle atlases belonging to the two G-bundles are fiberwise G-equivariant.

Let us describe this in more detail now. Let $(U'_\alpha, \psi'_\alpha)$ be a G-atlas for $P'[S',\ell']$ with cocycle of transition functions $(\varphi'_{\alpha\beta})$, belonging to the principal fiber bundle atlas $(U'_\alpha, \varphi'_\alpha)$ of (P',p',M',G). Then the pullback-related principal fiber bundle atlas $(U_\alpha = \bar\gamma^{-1}(U'_\alpha), \varphi_\alpha)$ for $P = \bar\gamma^* P'$ as described in the proof of (17.5) has the cocycle of transition functions $(\varphi_{\alpha\beta} = \varphi'_{\alpha\beta} \circ \bar\gamma)$; it induces the G-atlas (U_α, ψ_α) for $P[S,\ell]$. Then $(\psi'_\alpha \circ \gamma \circ \psi_\alpha^{-1})(x,s) = (\bar\gamma(x), \gamma_\alpha(x,s))$ and $\gamma_\alpha(x, \quad) : S \to S'$ is required to be G-equivariant for all α and all $x \in U_\alpha$.

Lemma. *Let $\gamma : P[S,\ell] \to P'[S',\ell']$ be a homomorphism of G-bundles as in (18.9). Then there is a homomorphism*

$$\chi : (P,p,M,G) \to (P',p',M',G)$$

of principal bundles and a G-equivariant mapping $f : S \to S'$ such that

$$\gamma = \chi \times_G f : P[S,\ell] \to P'[S',\ell'].$$

Proof. The homomorphism $\chi : (P, p, M, G) \to (P', p', M', G)$ of principal fiber bundles is already determined by the requirement that $P = \bar{\gamma}^* P'$, and we have $\bar{\gamma} = \bar{\chi}$. The G-equivariant mapping $f : S \to S'$ can be read off the following diagram:

$$(1) \qquad \begin{array}{ccc} P \times_M P[S] & \xrightarrow{\tau^S} & S \\ \downarrow{\scriptstyle \chi \times_M \gamma} & & \downarrow{\scriptstyle f} \\ P' \times_{M'} P'[S'] & \xrightarrow{\tau^{S'}} & S', \end{array}$$

which by the assumptions is seen to be well defined in the right column. □

So a homomorphism of G-bundles is described by the whole triple $(\chi : P \to P', f : S \to S'$ (G-equivariant)$, \gamma : P[S] \to P'[S'])$, such that diagram (1) commutes.

18.11. Associated vector bundles. Let (P, p, M, G) be a principal fiber bundle, and consider a representation $\rho : G \to GL(V)$ of G on a finite-dimensional vector space V. Then $P[V, \rho]$ is an associated fiber bundle with structure group G, but also with structure group $GL(V)$, for in the canonically associated fiber bundle atlas the transition functions also have values in $GL(V)$. So by section (8), $P[V, \rho]$ is a vector bundle.

Now let $\mathcal{F}$ be a covariant smooth functor from the category of finite-dimensional vector spaces and linear mappings into itself, as considered in section (8.8). Then clearly $\mathcal{F} \circ \rho : G \to GL(V) \to GL(\mathcal{F}(V))$ is another representation of G and the associated bundle $P[\mathcal{F}(V), \mathcal{F} \circ \rho]$ coincides with the vector bundle $\mathcal{F}(P[V, \rho])$ constructed with the method of (8.8), but now it has an extra G-bundle structure. For contravariant functors $\mathcal{F}$ we have to consider the representation $\mathcal{F} \circ \rho \circ \nu$, where $\nu(g) = g^{-1}$. A similar choice works for bifunctors. In particular the bifunctor $L(V, W)$ may be applied to two different representations of two structure groups of two principal bundles over the same base M to construct a vector bundle

$$L(P[V, \rho], P'[V', \rho']) = (P \times_M P')[L(V, V'), L \circ ((\rho \circ \nu) \times \rho')].$$

If (E, p, M) is a vector bundle with n-dimensional fibers, we may consider the open subset $GL(\mathbb{R}^n, E) \subset L(M \times \mathbb{R}^n, E)$, a fiber bundle over the base M, whose fiber over $x \in M$ is the space $GL(\mathbb{R}^n, E_x)$ of all invertible linear mappings. Composition from the right by elements of $GL(n)$ gives a free right action on $GL(\mathbb{R}^n, E)$ whose orbits are exactly the fibers, so by lemma (18.3) we have a principal fiber bundle $(GL(\mathbb{R}^n, E), p, M, GL(n))$. The associated bundle $GL(\mathbb{R}^n, E)[\mathbb{R}^n]$ for the banal representation of $GL(n)$ on $\mathbb{R}^n$ is isomorphic to the vector bundle (E, p, M) we started with, for the evaluation mapping $ev : GL(\mathbb{R}^n, E) \times \mathbb{R}^n \to E$ is invariant under the right action R of $GL(n)$, and locally in the image there are smooth sections to it, so it factors to a fiber linear diffeomorphism

$$GL(\mathbb{R}^n, E)[\mathbb{R}^n] = GL(\mathbb{R}^n, E) \times_{GL(n)} \mathbb{R}^n \to E.$$

The principal bundle $GL(\mathbb{R}^n, E)$ is called the *linear frame bundle* of E. Note that local sections of $GL(\mathbb{R}^n, E)$ are exactly the local frame fields of the vector bundle E as discussed in (8.5).

To illustrate the notion of reduction of structure groups, we consider now a vector bundle $(E, p, M, \mathbb{R}^n)$ equipped with a *Riemann metric* g, that is, a section $g \in C^\infty(S^2E^*)$ such that g_x is a positive definite inner product on E_x for each $x \in M$. Any vector bundle admits Riemann metrics: local existence is clear and we may glue with the help of a partition of unity on M, since the positive definite sections form an open convex subset. Now let

$$s' = (s'_1, \dots, s'_n) \in C^\infty(GL(\mathbb{R}^n, E)|U)$$

be a local frame field of the bundle E over $U \subset M$. Now we may apply the Gram-Schmidt orthonormalization procedure to the basis $(s_1(x), \dots, s_n(x))$ of E_x for each $x \in U$. Since this procedure is smooth (even real analytic), we obtain a frame field $s = (s_1, \dots, s_n)$ of E over U which is orthonormal with respect to g. We call it an *orthonormal frame field.* Now let (U_α) be an open cover of M with orthonormal frame fields $s^\alpha = (s^\alpha_1, \dots, s^\alpha_n)$, where s^α is defined on U_α. We consider the vector bundle charts

$$(U_\alpha, \psi_\alpha : E|U_\alpha \to U_\alpha \times \mathbb{R}^n)$$

given by the orthonormal frame fields:

$$\psi_\alpha^{-1}(x, v^1, \dots, v^n) = \sum s^\alpha_i(x).v^i =: s^\alpha(x).v.$$

For $x \in U_{\alpha\beta}$ we have $s^\alpha_i(x) = \sum s^\beta_j(x).g_{\beta\alpha}{}^j_i(x)$ for C^∞-functions $g_{\alpha\beta}{}^j_i : U_{\alpha\beta} \to \mathbb{R}$. Since $s^\alpha(x)$ and $s^\beta(x)$ are both orthonormal bases of E_x, the matrix $g_{\alpha\beta}(x) = (g_{\alpha\beta}{}^j_i(x))$ is an element of $O(n, \mathbb{R})$. We write $s^\alpha = s^\beta.g_{\beta\alpha}$ for short. Then we have

$$\psi_\beta^{-1}(x, v) = s^\beta(x).v = s^\alpha(x).g_{\alpha\beta}(x).v = \psi_\alpha^{-1}(x, g_{\alpha\beta}(x).v)$$

and consequently $\psi_\alpha\psi_\beta^{-1}(x, v) = (x, g_{\alpha\beta}(x).v)$. So the $(g_{\alpha\beta} : U_{\alpha\beta} \to O(n, \mathbb{R}))$ are the cocycle of transition functions for the vector bundle atlas (U_α, ψ_α). So we have constructed an $O(n, \mathbb{R})$-structure on E. The corresponding principal fiber bundle will be denoted by $O(\mathbb{R}^n, (E, g))$; it is usually called the *orthonormal frame bundle* of E. It is derived from the linear frame bundle $GL(\mathbb{R}^n, E)$ by reduction of the structure group from $GL(n)$ to $O(n)$. The phenomenon discussed here plays a prominent role in the theory of *classifying spaces.*

18.12. Sections of associated bundles. Let (P, p, M, G) be a principal fiber bundle and $\ell : G \times S \to S$ a left action. Let $C^\infty(P, S)^G$ denote the space of all smooth mappings $f : P \to S$ which are G-equivariant in the sense that $f(u.g) = g^{-1}.f(u)$ holds for $g \in G$ and $u \in P$.

Theorem. *The sections of the associated bundle $P[S, \ell]$ correspond exactly to the G-equivariant mappings $P \to S$; we have a bijection*

$$C^\infty(P, S)^G \cong \Gamma(P[S]).$$

This result follows from (18.9) and (18.10). Since it is very important, we include a direct proof.

Proof. If $f \in C^\infty(P, S)^G$, we construct $s_f \in \Gamma(P[S])$ in the following way: The mapping graph$(f) = (Id, f) : P \to P \times S$ is G-equivariant, since $(Id, f)(u.g) = (u.g, f(u.g)) = (u.g, g^{-1}.f(u)) = ((Id, f)(u)).g$. So it induces a smooth section $s_f \in \Gamma(P[S])$ as seen from (18.9) and the diagram:

(1)
$$\begin{array}{ccccc} P \times \{\text{point}\} & \xrightarrow{\cong} & P & \xrightarrow{(Id,f)} & P \times S \\ & & {\scriptstyle p}\downarrow & & \downarrow{\scriptstyle q} \\ & & M & \xrightarrow{s_f} & P[S]. \end{array}$$

For $s \in \Gamma(P[S])$ we define $f_s \in C^\infty(P, S)^G$ by

$$f_s := \tau^S \circ (Id_P \times_M s) : P = P \times_M M \to P \times_M P[S] \to S.$$

This is G-equivariant since we have by (18.7):

$$f_s(u_x.g) = \tau^S(u_x.g, s(x)) = g^{-1}.\tau^S(u_x, s(x)) = g^{-1}.f_s(u_x).$$

These constructions are inverse to each other since we have

$$f_{s(f)}(u) = \tau^S(u, s_f(p(u))) = \tau^S(u, q(u, f(u))) = f(u),$$
$$s_{f(s)}(p(u)) = q(u, f_s(u)) = q(u, \tau^S(u, s(p(u)))) = s(p(u)). \quad \square$$

18.13. Induced representations. Let K be a closed subgroup of a Lie group G. Let $\rho : K \to GL(V)$ be a representation in a vector space V, which we assume to be finite-dimensional to begin with. Then we consider the principal fiber bundle $(G, p, G/K, K)$ and the associated vector bundle $(G[V], p, G/K)$. The smooth (or even continuous) sections of $G[V]$ correspond exactly to the K-equivariant mappings $f : G \to V$, those satisfying $f(gk) = \rho(k^{-1})f(g)$, by lemma (18.12). Each $g \in G$ acts as a principal bundle homomorphism by left translation:

$$\begin{array}{ccc} G & \xrightarrow{\mu_g} & G \\ {\scriptstyle p}\downarrow & & \downarrow{\scriptstyle p} \\ G/K & \xrightarrow{\bar{\mu}_g} & G/K. \end{array}$$

So by (18.9) we have an induced isomorphism of vector bundles

$$\begin{array}{ccc} G \times V & \xrightarrow{\mu_g \times \mathrm{Id}_V} & G \times V \\ {\scriptstyle q}\downarrow & & \downarrow{\scriptstyle q} \\ G[V] & \xrightarrow{\mu_g \times_K V} & G[V] \\ {\scriptstyle p}\downarrow & & \downarrow{\scriptstyle p} \\ G/K & \xrightarrow{\bar{\mu}_g} & G/K \end{array}$$

which gives rise to the representation $\widetilde{\operatorname{ind}}_K^G \rho$ of G in the space $\Gamma(G[V])$, defined by

$$(\widetilde{\operatorname{ind}}_K^G \rho)(g)(s) := (\mu_g \times_K V) \circ s \circ \bar{\mu}_{g^{-1}} = (\mu_g \times_K V)_*(s).$$

Now let us assume that the original representation ρ is unitary, $\rho : K \to U(V)$ for a complex vector space V with inner product $\langle \quad , \quad \rangle_V$. Then $v \mapsto \|v\|^2 = \langle v, v\rangle$ is an invariant symmetric homogeneous polynomial $V \to \mathbb{R}$ of degree 2, so it is equivariant where K acts trivially on $\mathbb{R}$. By (18.9) again we get an induced mapping $G[V] \to G[\mathbb{R}] = G/K \times \mathbb{R}$, which we can polarize to a smooth fiberwise Hermitian form $\langle \quad , \quad \rangle_{G[V]}$ on the vector bundle $G[V]$. We may also express this by

$$\begin{aligned} \langle v_x, w_x \rangle_{G[V]} &= \langle \tau^V(u_x, v_x), \tau^V(u_x, w_x)\rangle_V \\ &= \langle k^{-1}\tau^V(u_x, v_x), k^{-1}\tau^V(u_x, w_x)\rangle_V \\ &= \langle \tau^V(u_x.k, v_x), \tau^V(u_x.k, w_x)\rangle_V \end{aligned}$$

for some $u_x \in G_x$, using the mapping $\tau^V : G \times_{G/M} G[V] \to V$ from (18.7); it does not depend on the choice of u_x. Still another way to describe the fiberwise Hermitian form is

$$\begin{array}{ccc} (G \times V) \times_{G/K} (G \times V) & & \\ \downarrow & \searrow{\scriptstyle f} & \\ G[V] \times_{G/K} G[V] & \xrightarrow[\langle\ ,\ \rangle_{G[V]}]{} & \mathbb{C} \\ \downarrow & & \\ G/K; & & \end{array}$$

here $f((g_1, v_1), (g_2, v_2)) := \langle v_1, \rho(\tau^K(g_1, g_2))v_2\rangle_V$ where we use the mapping $\tau^K : G \times_{G/K} G \to K$ given by $\tau^K(g_1, g_2) = g_1^{-1}g_2$ from (18.2). From this last description it is also clear that each $g \in G$ acts as an isometric vector bundle homomorphism.

Now we consider the natural line bundle $\mathrm{Vol}^{1/2}(G/K)$ of all $\frac{1}{2}$-densities on the manifold G/K from (10.4). Then for $\frac{1}{2}$-densities $\mu_i \in \Gamma(\mathrm{Vol}^{1/2}(G/M))$ and any diffeomorphism $f : G/K \to G/K$ the pushforward $f_*\mu_i$ is defined and for those with compact support we have

$$\int_{G/K} (f_*\mu_1.f_*\mu_2) = \int_{G/K} f_*(\mu_1.\mu_2) = \int_{G/K} \mu_1.\mu_2.$$

The Hermitian inner product on $G[V]$ now defines a fiberwise Hermitian mapping

$$(G[V]\otimes \mathrm{Vol}^{1/2}(G/K)) \times_{G/K} (G[V]\otimes \mathrm{Vol}^{1/2}(G/K)) \xrightarrow{\langle\ ,\ \rangle_{G[V]}} \mathrm{Vol}^1(G/L)$$

and on the space $C^\infty_c(G[V] \otimes \mathrm{Vol}^{1/2}(G/K))$ of all smooth sections with compact support we have the following Hermitian inner product:

$$\langle \sigma_1, \sigma_2\rangle := \int_{G/K} \langle \sigma_1, \sigma_2\rangle_{G[V]}.$$

For a decomposable section $\sigma_i = s_i \otimes \alpha_i$ (where $s_i \in \Gamma(G[V])$ and where $\alpha_i \in C^\infty_c(\mathrm{Vol}^{1/2}(G/K)))$ we may consider (using (18.12)) the equivariant lifts $f_{s_i} : G \to V$, their invariant inner product $\langle f_{s_1}, f_{s_2}\rangle_V : G \to \mathbb{C}$, and its factorization to $\langle f_{s_1}, f_{s_2}\rangle_{\bar{V}} : G/K \to \mathbb{C}$. Then

$$\langle \sigma_1, \sigma_2\rangle := \int_{G/K} \langle f_{s_1}, f_{s_2}\rangle_{\bar{V}}\, \alpha_1\alpha_2.$$

Obviously the resulting action of the group G on $\Gamma(G[V] \otimes \mathrm{Vol}^{1/2}(G/K))$ is unitary with respect to the Hermitian inner product, and it can be extended to the Hilbert space completion of this space of sections. The resulting unitary representation is called the *induced representation* and is denoted by $\mathrm{ind}_K^G\, \rho$.

If the original unitary representation $\rho : K \to U(V)$ is in an infinite-dimensional Hilbert space V, one can first restrict the representation ρ to the subspace of smooth vectors, on which it is differentiable, and repeat the above construction with some modifications. See [**150**] for more details on this infinite-dimensional construction.

18.14. Theorem. *Consider a principal fiber bundle (P, p, M, G) and a closed subgroup K of G. Then the reductions of structure group from G to K correspond bijectively to the global sections of the associated bundle $P[G/K, \bar\lambda]$ in a canonical way, where $\bar\lambda : G \times G/K \to G/K$ is the left action on the homogeneous space from* (5.11).

Proof. By (18.12) the section $s \in \Gamma(P[G/K])$ corresponds to an equivariant mapping $f_s \in C^\infty(P, G/K)^G$, which is a surjective submersion since the action $\bar\lambda : G \times G/K \to G/K$ is transitive. Thus $P_s := f_s^{-1}(\bar e)$ is a submanifold of P which is stable under the right action of K on P. Furthermore the K-orbits are exactly the fibers of the mapping $p : P_s \to M$, so by lemma (18.3) we get a principal fiber bundle (P_s, p, M, K). The embedding $P_s \hookrightarrow P$ is then a reduction of structure groups as required.

If conversely we have a principal fiber bundle (P', p', M, K) and a reduction of structure groups $\chi : P' \to P$, then χ is an embedding covering the identity of M and is K-equivariant, so we may view P' as a fiber subbundle of P which is stable under the right action of K. Now we consider the mapping $\tau : P \times_M P \to G$ from (18.2) and restrict it to $P \times_M P'$. Since we have $\tau(u_x, v_x.k) = \tau(u_x, v_x).k$ for $k \in K$, this restriction induces $f : P \to G/K$ by

$$\begin{array}{ccc} P \times_M P' & \xrightarrow{\ \tau\ } & G \\ \downarrow & & \downarrow p \\ P = P \times_M P'/K & \xrightarrow{\ f\ } & G/K, \end{array}$$

since $P'/K = M$; and from $\tau(u_x.g, v_x) = g^{-1}.\tau(u_x, v_x)$ it follows that f is G-equivariant as required. Finally $f^{-1}(\bar e) = \{u \in P : \tau(u, P'_{p(u)}) \subseteq K\} = P'$, so the two constructions are inverse to each other. $\square$

18.15. The bundle of gauges. If (P, p, M, G) is a principal fiber bundle, we denote by $\operatorname{Aut}(P)$ the group of all G-equivariant diffeomorphisms $\chi : P \to P$. Then $p \circ \chi = \bar\chi \circ p$ for a unique diffeomorphism $\bar\chi$ of M, so there is a group homomorphism from $\operatorname{Aut}(P)$ into the group $\operatorname{Diff}(M)$ of all diffeomorphisms of M. The kernel of this homomorphism is called $\operatorname{Gau}(P)$, the group of *gauge transformations*. So $\operatorname{Gau}(P)$ is the space of all $\chi : P \to P$ which satisfy $p \circ \chi = p$ and $\chi(u.g) = \chi(u).g$. A vector field $\xi \in \mathfrak{X}(P)$ is an *infinitesimal gauge transformation* if its flow Fl^ξ_t consists of gauge transformations, i.e., if ξ is vertical and G-invariant, $(r^g)^*\xi = \xi$.

Theorem. *The group $Gau(P)$ of gauge transformations is equal to the space*

$$Gau(P) \cong C^\infty(P, (G, \operatorname{conj}))^G \cong \Gamma(P[G, \operatorname{conj}]).$$

The Lie algebra $\mathfrak{X}_{vert}(P)^G$ of infinitesimal gauge transformations is equal to the space

$$\mathfrak{X}_{vert}(P)^G \cong C^\infty(P, (\mathfrak{g}, \operatorname{Ad}))^G \cong \Gamma(P[\mathfrak{g}, \operatorname{Ad}]).$$

Proof. We use again the mapping $\tau : P \times_M P \to G$ from (18.2). For $\chi \in \operatorname{Gau}(P)$ we define $f_\chi \in C^\infty(P, (G, \operatorname{conj}))^G$ by $f_\chi := \tau \circ (Id, \chi)$. Then

$f_\chi(u.g) = \tau(u.g, \chi(u.g)) = g^{-1}.\tau(u, \chi(u)).g = \operatorname{conj}_{g^{-1}} f_\chi(u)$, so f_χ is indeed G-equivariant.

If conversely $f \in C^\infty(P, (G, \operatorname{conj}))^G$ is given, we define $\chi_f : P \to P$ by $\chi_f(u) := u.f(u)$. It is easy to check that χ_f is indeed in $\operatorname{Gau}(P)$ and that the two constructions are inverse to each other, namely

$$\begin{aligned}
&\chi_f(ug) = ugf(ug) = ugg^{-1}f(u)g = \chi_f(u)g,\\
&f_{\chi_f}(u) = \tau^G(u, \chi_f(u)) = \tau^G(u, u.f(u)) = \tau^G(u,u)f(u) = f(u),\\
&\chi_{f_\chi}(u) = uf_\chi(u) = u\tau^G(u, \chi(u)) = \chi(u).
\end{aligned}$$

The isomorphism $C^\infty(P, (G, \operatorname{conj}))^G \cong \Gamma(P[G, \operatorname{conj}])$ is a special case of theorem (18.12).

A vertical vector field $\xi \in \mathfrak{X}_{\text{vert}}(P) = \Gamma(VP)$ is given uniquely by a mapping $f_\xi : P \to \mathfrak{g}$ via $\xi(u) = T_e(r_u).f_\xi(u)$, and it is G-equivariant if and only if

$$\begin{aligned}
T_e(r_u).f_\xi(u) = \xi(u) &= ((r^g)^*\xi)(u) = T(r^{g^{-1}}).\xi(u.g)\\
&= T(r^{g^{-1}}).T_e(r_{u.g}).f_\xi(u.g) = T_e(r^{g^{-1}} \circ r_{u.g}).f_\xi(u.g)\\
&= T_e(r_u \circ \operatorname{conj}_g).f_\xi(u.g) = T_e(r_u).\operatorname{Ad}_g .f_\xi(u.g).
\end{aligned}$$

The isomorphism $C^\infty(P, (\mathfrak{g}, \operatorname{Ad}))^G \cong \Gamma(P[\mathfrak{g}, \operatorname{Ad}])$ is again a special case of theorem (18.12). □

18.16. The tangent bundles of homogeneous spaces. Let G be a Lie group and K a closed subgroup, with Lie algebras $\mathfrak{g}$ and $\mathfrak{k}$, respectively. We recall the mapping $\operatorname{Ad}_G : G \to \operatorname{Aut}_{\text{Lie}}(\mathfrak{g})$ from (4.24) and put $\operatorname{Ad}_{G,K} := \operatorname{Ad}_G|K : K \to \operatorname{Aut}_{\text{Lie}}(\mathfrak{g})$. For $X \in \mathfrak{k}$ and $k \in K$ we have $\operatorname{Ad}_{G,K}(k)X = \operatorname{Ad}_G(k)X = \operatorname{Ad}_K(k)X \in \mathfrak{k}$, so $\mathfrak{k}$ is an invariant subspace for the representation $\operatorname{Ad}_{G,K}$ of K in $\mathfrak{g}$, and we have the factor representation $\operatorname{Ad}^\perp : K \to GL(\mathfrak{g}/\mathfrak{k})$. Then

$$0 \to \mathfrak{k} \to \mathfrak{g} \to \mathfrak{g}/\mathfrak{k} \to 0 \tag{1}$$

is short exact and K-equivariant.

Now we consider the principal fiber bundle $(G, p, G/K, K)$ and the associated vector bundles $G[\mathfrak{g}/\mathfrak{k}, \operatorname{Ad}^\perp]$ and $G[\mathfrak{k}, \operatorname{Ad}_K]$.

Theorem. *In these circumstances we have*

$$T(G/K) = G[\mathfrak{g}/\mathfrak{k}, \operatorname{Ad}^\perp] = (G \times_K \mathfrak{g}/\mathfrak{k}, p, G/K, \mathfrak{g}/\mathfrak{k}).$$

The left action $g \mapsto T(\bar{\mu}_g)$ of G on $T(G/K)$ corresponds to the canonical left action of G on $G \times_K \mathfrak{g}/\mathfrak{k}$. Furthermore $G[\mathfrak{g}/\mathfrak{k}, \operatorname{Ad}^\perp] \oplus G[\mathfrak{k}, \operatorname{Ad}_K]$ is a trivial vector bundle.

Proof. For $p : G \to G/K$ we consider the tangent mapping $T_e p : \mathfrak{g} \to T_{\bar{e}}(G/K)$ which is linear and surjective and induces a linear isomorphism

$\overline{T_e p} : \mathfrak{g}/\mathfrak{k} \to T_{\bar{e}}(G/K)$. For $k \in K$ we have $p \circ \operatorname{conj}_k = p \circ \mu_k \circ \mu^{k^{-1}} = \bar{\mu}_k \circ p$ and consequently $T_e p \circ \operatorname{Ad}_{G,K}(k) = T_e p \circ T_e(\operatorname{conj}_k) = T_{\bar{e}}\bar{\mu}_k \circ T_e p$. Thus the isomorphism $\overline{T_e p} : \mathfrak{g}/\mathfrak{k} \to T_{\bar{e}}(G/K)$ is K-equivariant for the representations $\operatorname{Ad}^\perp$ and $T_{\bar{e}}\bar{\lambda} : k \mapsto T_{\bar{e}}\bar{\mu}_k$, where, for the moment, we use the notation $\bar{\lambda} : G \times G/K \to G/K$ for the left action.

Let us now consider the associated vector bundle

$$G[T_{\bar{e}}(G/K), T_{\bar{e}}\bar{\lambda}] = (G \times_K T_{\bar{e}}(G/K), p, G/K, T_{\bar{e}}(G/K)),$$

which is isomorphic to the vector bundle $G[\mathfrak{g}/\mathfrak{k}, \operatorname{Ad}^\perp]$, since the representation spaces are isomorphic. The mapping $T_2\bar{\lambda} : G \times T_{\bar{e}}(G/K) \to T(G/K)$ (where T_2 is the second partial tangent functor) is K-invariant, since

$$T_2\bar{\lambda}((g, X)k) = T_2\bar{\lambda}(gk, T_{\bar{e}}\bar{\mu}_{k^{-1}}.X) = T\bar{\mu}_{gk}.T\bar{\mu}_{k^{-1}}.X = T\bar{\mu}_g.X.$$

Therefore it induces a mapping ψ as in the following diagram:

(2)

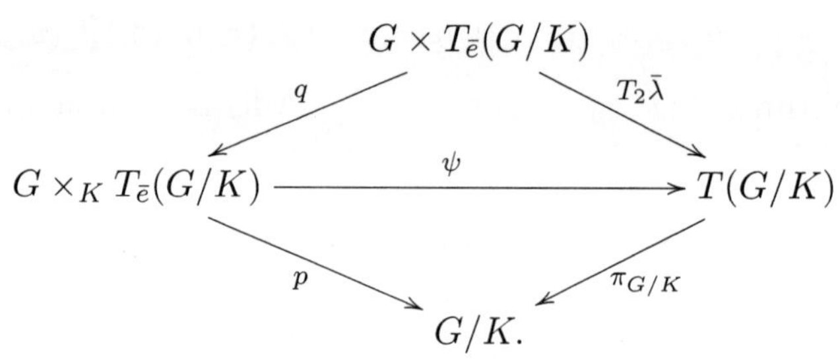

This mapping ψ is an isomorphism of vector bundles.

It remains to show the last assertion. The short exact sequence (1) induces a sequence of vector bundles over G/K:

$$G/K \times 0 \to G[\mathfrak{k}, \operatorname{Ad}_K] \to G[\mathfrak{g}, \operatorname{Ad}_{G,K}] \to G[\mathfrak{g}/\mathfrak{k}, \operatorname{Ad}^\perp] \to G/K \times 0.$$

This sequence splits fiberwise thus also locally over G/K, so we get

$$G[\mathfrak{g}/\mathfrak{k}, \operatorname{Ad}^\perp] \oplus G[\mathfrak{k}, \operatorname{Ad}_K] \cong G[\mathfrak{g}, \operatorname{Ad}_{G,K}].$$

We have to show that $G[\mathfrak{g}, \operatorname{Ad}_{G,K}]$ is a trivial vector bundle. Let $\varphi : G \times \mathfrak{g} \to G \times \mathfrak{g}$ be given by $\varphi(g, X) = (g, \operatorname{Ad}_G(g)X)$. Then for $k \in K$ we have

$$\begin{aligned} \varphi((g, X).k) &= \varphi(gk, \operatorname{Ad}_{G,K}(k^{-1})X) \\ &= (gk, \operatorname{Ad}_G(g.k.k^{-1})X) = (gk, \operatorname{Ad}_G(g)X). \end{aligned}$$

So φ is K-equivariant for the 'joint' K-action to the 'on the left' K-action and therefore induces a mapping $\bar{\varphi}$ as in the diagram:

(3)
$$\begin{array}{ccc} G \times \mathfrak{g} & \xrightarrow{\varphi} & G \times \mathfrak{g} \\ {\scriptstyle q}\downarrow & & \downarrow \\ G \times_K \mathfrak{g} & \xrightarrow{\bar{\varphi}} & G/K \times \mathfrak{g} \\ & {\scriptstyle p}\searrow \quad \swarrow{\scriptstyle \mathrm{pr}_1} & \\ & G/K. & \end{array}$$

The map $\bar{\varphi}$ is a vector bundle isomorphism. □

18.17. Tangent bundles of Grassmann manifolds. From (18.5) we know that $(V(k,n) = O(n)/O(n-k), p, G(k,n), O(k))$ is a principal fiber bundle. Using the standard representation of $O(k)$, we consider the associated vector bundle $(E_k := V(k,n)[\mathbb{R}^k], p, G(k,n))$. Recall from (18.5) the description of $V(k,n)$ as the space of all linear isometries $\mathbb{R}^k \to \mathbb{R}^n$; we get from it the evaluation mapping $ev : V(k,n) \times \mathbb{R}^k \to \mathbb{R}^n$. The mapping (p, ev) in the diagram

(1)
$$\begin{array}{ccccc} & & V(k,n) \times \mathbb{R}^k & & \\ & & {\scriptstyle q}\downarrow & \searrow^{(p,\mathrm{ev})} & \\ E_k & = & V(k,n) \times_{O(k)} \mathbb{R}^k & \xrightarrow[\psi]{} & G(k,n) \times \mathbb{R}^n \end{array}$$

is $O(k)$-invariant for the action R and factors therefore to an embedding of vector bundles $\psi : E_k \to G(k,n) \times \mathbb{R}^n$. So the fiber $(E_k)_W$ over the k-plane W in $\mathbb{R}^n$ is just the linear subspace W. Note finally that the fiberwise orthogonal complement ${E_k}^\perp$ of E_k in the trivial vector bundle $G(k,n) \times \mathbb{R}^n$ with its standard Riemann metric is isomorphic to the universal vector bundle E_{n-k} over $G(n-k,n)$, where the isomorphism covers the diffeomorphism $G(k,n) \to G(n-k,n)$ given also by the orthogonal complement mapping.

Corollary. *The tangent bundle of the Grassmann manifold is*

$$TG(k,n) \cong L(E_k, {E_k}^\perp).$$

Proof. We have $G(k,n) = O(n)/(O(k) \times O(n-k))$, so by theorem (18.16) we get

$$TG(k,n) = O(n) \underset{O(k)\times O(n-k)}{\times} (\mathfrak{so}(n)/(\mathfrak{so}(k) \times \mathfrak{so}(n-k))).$$

On the other hand we have $V(k,n) = O(n)/O(n-k)$ and the right action of $O(k)$ commutes with the right action of $O(n-k)$ on $O(n)$; therefore

$$V(k,n)[\mathbb{R}^k] = (O(n)/O(n-k)) \underset{O(k)}{\times} \mathbb{R}^k = O(n) \underset{O(k)\times O(n-k)}{\times} \mathbb{R}^k,$$

where $O(n-k)$ acts trivially on $\mathbb{R}^k$. We have

$$\begin{aligned} L(E_k, E_k{}^\perp) &= L\left(O(n) \underset{O(k)\times O(n-k)}{\times} \mathbb{R}^k, O(n) \underset{O(k)\times O(n-k)}{\times} \mathbb{R}^{n-k}\right) \\ &= O(n) \underset{O(k)\times O(n-k)}{\times} L(\mathbb{R}^k, \mathbb{R}^{n-k}), \end{aligned}$$

where $O(k) \times O(n-k)$ acts on $L(\mathbb{R}^k, \mathbb{R}^{n-k})$ by $(A,B)(C) = B.C.A^{-1}$. Finally, we have an $(O(k)\times O(n-k))$-equivariant linear isomorphism as follows:

$$L(\mathbb{R}^k, \mathbb{R}^{n-k}) \to \mathfrak{so}(n)/(\mathfrak{so}(k) \times \mathfrak{so}(n-k)),$$

$$\begin{aligned} \mathfrak{so}(n)/(\mathfrak{so}(k)\times\mathfrak{so}(n-k)) &= \big(\text{skew}\big) \Big/ \begin{pmatrix} \text{skew} & 0 \\ 0 & \text{skew} \end{pmatrix} \\ &= \left\{ \begin{pmatrix} 0 & -A^\top \\ A & 0 \end{pmatrix} : \quad A \in L(\mathbb{R}^k, \mathbb{R}^{n-k}) \right\}. \quad \square \end{aligned}$$

18.18. Tangent bundles and vertical bundles. Let (E,p,M,S) be a fiber bundle. The vector subbundle $VE = \{\xi \in TE : Tp.\xi = 0\}$ of TE is called the *vertical bundle* and is denoted by (VE, π_E, E).

Theorem. *Let (P,p,M,G) be a principal fiber bundle with principal right action $r : P\times G \to P$. Let $\ell : G\times S \to S$ be a left action. Then the following assertions hold:*

(1) *(TP, Tp, TM, TG) is again a principal fiber bundle with principal right action $Tr : TP \times TG \to TP$, where the structure group TG is the tangent group of G; see* (6.7).

(2) *The vertical bundle $(VP, \pi, P, \mathfrak{g})$ of the principal bundle is trivial as a vector bundle over P: $VP \cong P \times \mathfrak{g}$.*

(3) *The vertical bundle of the principal bundle as bundle over M is again a principal bundle: $(VP, p\circ\pi, M, TG)$.*

(4) *The tangent bundle of the associated bundle $P[S,\ell]$ is given by $T(P[S,\ell]) = TP[TS, T\ell]$.*

(5) *The vertical bundle of the associated bundle $P[S,\ell]$ is given by $V(P[S,\ell]) = P[TS, T_2\ell] = P\times_G TS$.*

Proof. Let $(U_\alpha, \varphi_\alpha : P|U_\alpha \to U_\alpha \times G)$ be a principal fiber bundle atlas with cocycle of transition functions $(\varphi_{\alpha\beta} : U_{\alpha\beta} \to G)$. Since T is a functor which respects products, $(TU_\alpha, T\varphi_\alpha : TP|TU_\alpha \to TU_\alpha \times TG)$ is again a principal fiber bundle atlas with cocycle of transition functions $(T\varphi_{\alpha\beta} : TU_{\alpha\beta} \to TG)$, describing the principal fiber bundle (TP, Tp, TM, TG). The assertion about the principal action is obvious. So (1) follows. For completeness sake we include here the transition formula for this atlas in the right trivialization of TG:

$$T(\varphi_\alpha \circ \varphi_\beta^{-1})(\xi_x, T_e(\mu^g).X) = (\xi_x, T_e(\mu^{\varphi_{\alpha\beta}(x).g}).(\delta^r\varphi_{\alpha\beta}(\xi_x) + \mathrm{Ad}(\varphi_{\alpha\beta}(x))X)),$$

where $\delta\varphi_{\alpha\beta} \in \Omega^1(U_{\alpha\beta}; \mathfrak{g})$ is the right logarithmic derivative of $\varphi_{\alpha\beta}$; see (4.26).

(2) The mapping $(u, X) \mapsto T_e(r_u).X = T_{(u,e)}r.(0_u, X)$ is a vector bundle isomorphism $P \times \mathfrak{g} \to VP$ over P.

(3) Obviously $Tr : TP \times TG \to TP$ is a free right action which acts transitively on the fibers of $Tp : TP \to TM$. Since $VP = (Tp)^{-1}(0_M)$, the bundle $VP \to M$ is isomorphic to $TP|0_M$ and Tr restricts to a free right action, which is transitive on the fibers, so by lemma (18.3) the result follows.

(4) The transition functions of the fiber bundle $P[S, \ell]$ are given by the expression $\ell \circ (\varphi_{\alpha\beta} \times Id_S) : U_{\alpha\beta} \times S \to G \times S \to S$. Then the transition functions of $T(P[S, \ell])$ are $T(\ell \circ (\varphi_{\alpha\beta} \times Id_S)) = T\ell \circ (T\varphi_{\alpha\beta} \times Id_{TS}) : TU_{\alpha\beta} \times TS \to TG \times TS \to TS$, from which the result follows.

(5) Vertical vectors in $T(P[S, \ell])$ have local representations $(0_x, \eta_s) \in TU_{\alpha\beta} \times TS$. Under the transition functions of $T(P[S, \ell])$ they transform as $T(\ell \circ (\varphi_{\alpha\beta} \times Id_S)).(0_x, \eta_s) = T\ell.(0_{\varphi_{\alpha\beta}(x)}, \eta_s) = T(\ell_{\varphi_{\alpha\beta}(x)}).\eta_s = T_2\ell.(\varphi_{\alpha\beta}(x), \eta_s)$ and this implies the result □

19. Principal and Induced Connections

19.1. Principal connections. Let (P, p, M, G) be a principal fiber bundle. Recall from (17.3) that a (general) connection on P is a fiber projection $\Phi : TP \to VP$, viewed as a 1-form in $\Omega^1(P, TP)$. Such a connection Φ is called a *principal connection* if it is G-equivariant for the principal right action $r : P \times G \to P$, so that $T(r^g).\Phi = \Phi.T(r^g)$ and Φ is r^g-related to itself, or $(r^g)^*\Phi = \Phi$ in the sense of (16.16), for all $g \in G$. By theorem (16.15.6) the curvature $R = \frac{1}{2}.[\Phi, \Phi]$ is then also r^g-related to itself for all $g \in G$.

Recall from (18.18.2) that the vertical bundle of P is trivialized as a vector bundle over P by the principal action. So

(1) $$\omega(X_u) := T_e(r_u)^{-1}.\Phi(X_u) \in \mathfrak{g}$$

and in this way we get a $\mathfrak{g}$-valued 1-form $\omega \in \Omega^1(P, \mathfrak{g})$, which is called the *(Lie algebra valued) connection form* of the connection Φ. Recall from

(6.3) the fundamental vector field mapping $\zeta : \mathfrak{g} \to \mathfrak{X}(P)$ for the principal right action given by $\zeta_X(u) = T_e(r_u)X$ which satisfies $T_u(r^g)\zeta_X(u) = \zeta_{\mathrm{Ad}(g^{-1})X}(u.g)$. The defining equation for ω can be written also as $\Phi(X_u) = \zeta_{\omega(X_u)}(u)$.

Lemma. *If $\Phi \in \Omega^1(P, VP)$ is a principal connection on the principal fiber bundle (P, p, M, G), then the connection form has the following two properties:*

(2) *ω reproduces the generators of fundamental vector fields:*

$$\omega(\zeta_X(u)) = X \quad \textit{for all} \quad X \in \mathfrak{g}.$$

(3) *ω is G-equivariant, i.e.,*

$$((r^g)^*\omega)(X_u) = \omega(T_u(r^g).X_u) = \mathrm{Ad}(g^{-1}).\omega(X_u)$$

for all $g \in G$ and $X_u \in T_uP$. Consequently we have for the Lie derivative $\mathcal{L}_{\zeta_X}\omega = -\operatorname{ad}(X).\omega$.

Conversely a 1-form $\omega \in \Omega^1(P, \mathfrak{g})$ satisfying (2) defines a connection Φ on P by $\Phi(X_u) = T_e(r_u).\omega(X_u)$, which is a principal connection if and only if (3) is satisfied.

Proof. (2) $T_e(r_u).\omega(\zeta_X(u)) = \Phi(\zeta_X(u)) = \zeta_X(u) = T_e(r_u).X$. Since $T_e(r_u) : \mathfrak{g} \to V_uP$ is an isomorphism, the result follows.

(3) Both directions follow from

$$\begin{aligned} T_e(r_{ug}).\omega(T_u(r^g).X_u) &= \zeta_{\omega(T_u(r^g).X_u)}(ug) = \Phi(T_u(r^g).X_u), \\ T_e(r_{ug}).\operatorname{Ad}(g^{-1}).\omega(X_u) &= \zeta_{\mathrm{Ad}(g^{-1}).\omega(X_u)}(ug) = T_u(r^g).\zeta_{\omega(X_u)}(u) \\ &= T_u(r^g).\Phi(X_u). \quad \square \end{aligned}$$

19.2. Curvature. Let Φ be a principal connection on the principal fiber bundle (P, p, M, G) with connection form $\omega \in \Omega^1(P, \mathfrak{g})$. We already noted in (19.1) that the curvature $R = \frac{1}{2}[\Phi, \Phi]$ is then also G-equivariant, $(r^g)^*R = R$ for all $g \in G$. Since R has vertical values, we may again define a $\mathfrak{g}$-valued 2-form

$$\Omega \in \Omega^2(P, \mathfrak{g}), \qquad \Omega(X_u, Y_u) := -T_e(r_u)^{-1}.R(X_u, Y_u),$$

which is called the *(Lie algebra valued) curvature form* of the connection. We also have

$$R(X_u, Y_u) = -\zeta_{\Omega(X_u, Y_u)}(u).$$

We take the negative sign to get the usual curvature form as in [**107**, I].

We equip the space $\Omega(P, \mathfrak{g})$ of all $\mathfrak{g}$-valued forms on P in a canonical way with the structure of a graded Lie algebra by

$$[\Psi,\Theta]_\wedge(X_1,\dots,X_{p+q})$$
$$= \frac{1}{p!\,q!}\sum_\sigma \operatorname{sign}\sigma\,[\Psi(X_{\sigma 1},\dots,X_{\sigma p}),\Theta(X_{\sigma(p+1)},\dots,X_{\sigma(p+q)})]_{\mathfrak g}$$

or equivalently by

$$[\psi\otimes X,\vartheta\otimes Y]_\wedge := \psi\wedge\vartheta\otimes[X,Y]_{\mathfrak g}.$$

From the latter description it is clear that

$$d[\Psi,\Theta]_\wedge = [d\Psi,\Theta]_\wedge + (-1)^{\deg\Psi}[\Psi,d\Theta]_\wedge.$$

In particular for $\omega\in\Omega^1(P,\mathfrak g)$ we have

$$[\omega,\omega]_\wedge(X,Y) = 2[\omega(X),\omega(Y)]_{\mathfrak g}.$$

Theorem. *The curvature form Ω of a principal connection with connection form ω has the following properties:*

(1) *Ω is horizontal, i.e., it kills vertical vectors.*
(2) *Ω is G-equivariant in the following sense: $(r^g)^*\Omega = \operatorname{Ad}(g^{-1}).\Omega$. Consequently $\mathcal L_{\zeta_X}\Omega = -\operatorname{ad}(X).\Omega$.*
(3) *The Maurer-Cartan formula holds: $\Omega = d\omega + \frac12[\omega,\omega]_\wedge$.*

Proof. (1) is true for R by (17.4). For (2) we compute as follows:

$$\begin{aligned}
T_e(r_{ug}).((r^g)^*\Omega)(X_u,Y_u) &= T_e(r_{ug}).\Omega(T_u(r^g).X_u,T_u(r^g).Y_u)\\
&= -R_{ug}(T_u(r^g).X_u,T_u(r^g).Y_u) = -T_u(r^g).((r^g)^*R)(X_u,Y_u)\\
&= -T_u(r^g).R(X_u,Y_u) = T_u(r^g).\zeta_{\Omega(X_u,Y_u)}(u)\\
&= \zeta_{\operatorname{Ad}(g^{-1}).\Omega(X_u,Y_u)}(ug) = T_e(r_{ug}).\operatorname{Ad}(g^{-1}).\Omega(X_u,Y_u), \quad \text{by (6.3).}
\end{aligned}$$

(3) For $X\in\mathfrak g$ we have $i_{\zeta_X}R=0$ by (1), and using (19.1.2), we get

$$\begin{aligned}
i_{\zeta_X}(d\omega+\frac12[\omega,\omega]_\wedge) &= i_{\zeta_X}d\omega + \frac12[i_{\zeta_X}\omega,\omega]_\wedge - \frac12[\omega,i_{\zeta_X}\omega]_\wedge\\
&= \mathcal L_{\zeta_X}\omega + [X,\omega]_\wedge = -\operatorname{ad}(X)\omega + \operatorname{ad}(X)\omega = 0.
\end{aligned}$$

So the formula holds for vertical vectors, and for horizontal vector fields $\xi,\eta\in\Gamma(H(P))$ we have

$$R(\xi,\eta) = \Phi[\xi-\Phi\xi,\eta-\Phi\eta] = \Phi[\xi,\eta] = \zeta_{\omega([\xi,\eta])},$$
$$(d\omega+\frac12[\omega,\omega])(\xi,\eta) = \xi\omega(\eta)-\eta\omega(\xi)-\omega([\xi,\eta])+0 = -\omega([\xi,\eta]). \qquad \square$$

19.3. Lemma. *Any principal fiber bundle (P,p,M,G) (with paracompact basis) admits principal connections.*

Proof. Let $(U_\alpha, \varphi_\alpha : P|U_\alpha \to U_\alpha \times G)_\alpha$ be a principal fiber bundle atlas. Let us define $\gamma_\alpha(T\varphi_\alpha^{-1}(\xi_x, T_e\mu_g.X)) := X$ for $\xi_x \in T_xU_\alpha$ and $X \in \mathfrak{g}$. Using lemma (6.3), we get

$$\begin{aligned}((r^h)^*\gamma_\alpha)(T\varphi_\alpha^{-1}(\xi_x, T_e\mu_g.X)) &= \gamma_\alpha(Tr^h.T\varphi_\alpha^{-1}(\xi_x, T_e\mu_g.X))\\ &= \gamma_\alpha(T\varphi_\alpha^{-1}(\xi_x, T\mu^h.T_e\mu_g.X))\\ &= \gamma_\alpha(T\varphi_\alpha^{-1}(\xi_x, T_e\mu_{gh}.\operatorname{Ad}(h^{-1}).X)) = \operatorname{Ad}(h^{-1}).X,\end{aligned}$$

so that $\gamma_\alpha \in \Omega^1(P|U_\alpha, \mathfrak{g})$ satisfies the requirements of lemma (19.1) and thus is a principal connection on $P|U_\alpha$. Now let (f_α) be a smooth partition of unity on M which is subordinated to the open cover (U_α), and let $\omega := \sum_\alpha (f_\alpha \circ p)\gamma_\alpha$. Since both requirements of lemma (19.1) are invariant under convex linear combinations, ω is a principal connection on P. □

19.4. Local descriptions of principal connections. We consider a principal fiber bundle (P, p, M, G) with some principal fiber bundle atlas $(U_\alpha, \varphi_\alpha : P|U_\alpha \to U_\alpha \times G)$ and corresponding cocycle $(\varphi_{\alpha\beta} : U_{\alpha\beta} \to G)$ of transition functions. We consider the sections $s_\alpha \in \Gamma(P|U_\alpha)$ which are given by $\varphi_\alpha(s_\alpha(x)) = (x, e)$ and satisfy $s_\alpha.\varphi_{\alpha\beta} = s_\beta$, since we have in turn:

$$\begin{aligned}&\varphi_\alpha(s_\beta(x)) = \varphi_\alpha\varphi_\beta^{-1}(x, e) = (x, \varphi_{\alpha\beta}(x)),\\ &\qquad s_\beta(x) = \varphi_\alpha^{-1}(x, e.\varphi_{\alpha\beta}(e)) = \varphi_\alpha^{-1}(x, e)\varphi_{\alpha\beta}(x) = s_\alpha(x)\varphi_{\alpha\beta}(x).\end{aligned}$$

(1) Let $\kappa^l \in \Omega^1(G, \mathfrak{g})$ be the left logarithmic derivative of the identity, also called the left Maurer-Cartan form, i.e., $\kappa^l(\eta_g) := T_g(\mu_{g^{-1}}).\eta_g$. We will use the forms $\kappa^l_{\alpha\beta} := {\varphi_{\alpha\beta}}^*\kappa^l \in \Omega^1(U_{\alpha\beta}, \mathfrak{g})$.

Let $\Phi = \zeta \circ \omega \in \Omega^1(P, VP)$ be a principal connection with connection form $\omega \in \Omega^1(P, \mathfrak{g})$. We may associate the following local data to the connection:

(2) $\omega_\alpha := {s_\alpha}^*\omega \in \Omega^1(U_\alpha, \mathfrak{g})$, the *physicists' version* or *Cartan moving frame version* of the connection,

(3) the Christoffel forms $\Gamma^\alpha \in \Omega^1(U_\alpha, \mathfrak{X}(G))$ from (17.7), which are given by $(0_x, \Gamma^\alpha(\xi_x, g)) = -T(\varphi_\alpha).\Phi.T(\varphi_\alpha)^{-1}(\xi_x, 0_g)$,

(4) $\gamma_\alpha := (\varphi_\alpha^{-1})^*\omega \in \Omega^1(U_\alpha \times G, \mathfrak{g})$, the local expressions of ω.

Lemma. *These local data have the following properties and are related by the following formulas.*

(5) *The forms $\omega_\alpha \in \Omega^1(U_\alpha, \mathfrak{g})$ satisfy the transition formulas*

$$\omega_\alpha = \operatorname{Ad}(\varphi_{\beta\alpha}^{-1})\omega_\beta + \kappa^l_{\beta\alpha},$$

and any set of forms like that with this transition behavior determines a unique principal connection.

(6) *We have* $\gamma_\alpha(\xi_x, T\mu_g.X) = \gamma_\alpha(\xi_x, 0_g) + X = \operatorname{Ad}(g^{-1})\omega_\alpha(\xi_x) + X$.

(7) *We have* $\Gamma^\alpha(\xi_x) = -R_{\omega_\alpha(\xi_x)}$, *a right invariant vector field, since*

$$\begin{aligned}\Gamma^\alpha(\xi_x, g) &= -T_e(\mu_g).\gamma_\alpha(\xi_x, 0_g)\\ &= -T_e(\mu_g).\operatorname{Ad}(g^{-1})\omega_\alpha(\xi_x) = -T(\mu^g)\omega_\alpha(\xi_x).\end{aligned}$$

Proof. (7) From the definition of the Christoffel forms we have

$$\begin{aligned}(0_x, \Gamma^\alpha(\xi_x, g)) &= -T(\varphi_\alpha).\Phi.T(\varphi_\alpha)^{-1}(\xi_x, 0_g)\\ &= -T(\varphi_\alpha).T_e(r_{\varphi_\alpha^{-1}(x,g)}).\omega.T(\varphi_\alpha)^{-1}(\xi_x, 0_g) \quad \text{by (19.1.1)}\\ &= -T_e(\varphi_\alpha \circ r_{\varphi_\alpha^{-1}(x,g)})\omega.T(\varphi_\alpha)^{-1}(\xi_x, 0_g)\\ &= -(0_x, T_e(\mu_g)\omega.T(\varphi_\alpha)^{-1}(\xi_x, 0_g))\\ &= -(0_x, T_e(\mu_g)\gamma_\alpha(\xi_x, 0_g)), \quad \text{by (4)},\end{aligned}$$

where we also used $\varphi_\alpha(r_{\varphi_\alpha^{-1}(x,g)}h) = \varphi_\alpha(\varphi_\alpha^{-1}(x,g)h) = \varphi_\alpha(\varphi_\alpha^{-1}(x,gh)) = (x, gh)$. This is the first part of (7). The second part follows from (6).

(6)
$$\begin{aligned}\gamma_\alpha(\xi_x, T\mu_g.X) &= \gamma_\alpha(\xi_x, 0_g) + \gamma_\alpha(0_x, T\mu_g.X)\\ &= \gamma_\alpha(\xi_x, 0_g) + \omega(T(\varphi_\alpha)^{-1}(0_x, T\mu_g.X))\\ &= \gamma_\alpha(\xi_x, 0_g) + \omega(\zeta_X(\varphi_\alpha^{-1}(x, g)))\\ &= \gamma_\alpha(\xi_x, 0_g) + X.\end{aligned}$$

So the first part of (6) holds. The second part is seen from

$$\begin{aligned}\gamma_\alpha(\xi_x, 0_g) &= \gamma_\alpha(\xi_x, T_e(\mu^g)0_e) = (\omega \circ T(\varphi_\alpha)^{-1} \circ T(Id_X \times \mu^g))(\xi_x, 0_e)\\ &= (\omega \circ T(r^g \circ \varphi_\alpha^{-1}))(\xi_x, 0_e) = \operatorname{Ad}(g^{-1})\omega(T(\varphi_\alpha^{-1})(\xi_x, 0_e))\\ &= \operatorname{Ad}(g^{-1})(s_\alpha{}^*\omega)(\xi_x) = \operatorname{Ad}(g^{-1})\omega_\alpha(\xi_x).\end{aligned}$$

(5) Via (7) the transition formulas for the ω_α are easily seen to be equivalent to the transition formulas for the Christoffel forms in lemma (17.7). A direct proof goes as follows: We have $s_\alpha(x) = s_\beta(x)\varphi_{\beta\alpha}(x) = r(s_\beta(x), \varphi_{\beta\alpha}(x))$ and thus

$$\begin{aligned}\omega_\alpha(\xi_x) &= \omega(T_x(s_\alpha).\xi_x)\\ &= (\omega \circ T_{(s_\beta(x),\varphi_{\beta\alpha}(x))}r)((T_x s_\beta.\xi_x, 0_{\varphi_{\beta\alpha}(x)}) + (0_{s_\beta}(x), T_x\varphi_{\beta\alpha}.\xi_x))\\ &= \omega(T(r^{\varphi_{\beta\alpha}(x)}).T_x(s_\beta).\xi_x) + \omega(T_{\varphi_{\beta\alpha}(x)}(r_{s_\beta(x)}).T_x(\varphi_{\beta\alpha}).\xi_x)\\ &= \operatorname{Ad}(\varphi_{\beta\alpha}(x)^{-1})\omega(T_x(s_\beta).\xi_x)\\ &\qquad + \omega(T_{\varphi_{\beta\alpha}(x)}(r_{s_\beta(x)}).T(\mu_{\varphi_{\beta\alpha}(x)} \circ \mu_{\varphi_{\beta\alpha}(x)^{-1}})T_x(\varphi_{\beta\alpha}).\xi_x)\\ &= \operatorname{Ad}(\varphi_{\beta\alpha}(x)^{-1})\omega_\beta(\xi_x)\\ &\qquad + \omega(T_e(r_{s_\beta(x)\varphi_{\beta\alpha}(x)}).\kappa^l_{\beta\alpha}.\xi_x)\\ &= \operatorname{Ad}(\varphi_{\beta\alpha}(x)^{-1})\omega_\beta(\xi_x) + \kappa^l_{\beta\alpha}(\xi_x). \quad \square\end{aligned}$$

19.5. The covariant derivative. Let (P, p, M, G) be a principal fiber bundle with principal connection $\Phi = \zeta \circ \omega$. We consider the horizontal projection $\chi = Id_{TP} - \Phi : TP \to HP$, cf. (17.3), which satisfies

$$\chi \circ \chi = \chi, \quad \operatorname{im} \chi = HP, \quad \ker \chi = VP, \quad \chi \circ T(r^g) = T(r^g) \circ \chi$$

for all $g \in G$.

If W is a finite-dimensional vector space, we consider the mapping $\chi^* : \Omega(P, W) \to \Omega(P, W)$ which is given by

$$(\chi^* \varphi)_u(X_1, \dots, X_k) = \varphi_u(\chi(X_1), \dots, \chi(X_k)).$$

The mapping χ^* is a projection onto the subspace of *horizontal differential forms*, i.e., the space $\Omega_{hor}(P, W) := \{\psi \in \Omega(P, W) : i_X \psi = 0 \text{ for } X \in VP\}$. The notion of horizontal form is independent of the choice of a connection.

The projection χ^* has the following properties: $\chi^*(\varphi \wedge \psi) = \chi^* \varphi \wedge \chi^* \psi$ if one of the two forms has values in $\mathbb{R}$; $\chi^* \circ \chi^* = \chi^*$; $\chi^* \circ (r^g)^* = (r^g)^* \circ \chi^*$ for all $g \in G$; $\chi^* \omega = 0$; and $\chi^* \circ \mathcal{L}(\zeta_X) = \mathcal{L}(\zeta_X) \circ \chi^*$. They follow easily from the corresponding properties of χ; the last property uses that $\mathrm{Fl}_t^{\zeta(X)} = r^{\exp tX}$.

We define the *covariant exterior derivative* $d_\omega : \Omega^k(P, W) \to \Omega^{k+1}(P, W)$ by prescribing $d_\omega := \chi^* \circ d$.

Theorem. *The covariant exterior derivative d_ω has the following properties.*

(1) $d_\omega(\varphi \wedge \psi) = d_\omega(\varphi) \wedge \chi^* \psi + (-1)^{\deg \varphi} \chi^* \varphi \wedge d_\omega(\psi)$ *if φ or ψ is real valued.*

(2) $\mathcal{L}(\zeta_X) \circ d_\omega = d_\omega \circ \mathcal{L}(\zeta_X)$ *for each $X \in \mathfrak{g}$.*

(3) $(r^g)^* \circ d_\omega = d_\omega \circ (r^g)^*$ *for each $g \in G$.*

(4) $d_\omega \circ p^* = d \circ p^* = p^* \circ d : \Omega(M, W) \to \Omega_{hor}(P, W)$.

(5) $d_\omega \omega = \Omega$, *the curvature form.*

(6) $d_\omega \Omega = 0$, *the Bianchi identity.*

(7) $d_\omega \circ \chi^* - d_\omega = \chi^* \circ i(R)$, *where R is the curvature.*

(8) $d_\omega \circ d_\omega = \chi^* \circ i(R) \circ d$.

(9) *Let $\Omega_{hor}(P, \mathfrak{g})^G$ be the algebra of all horizontal G-equivariant $\mathfrak{g}$-valued forms, i.e., $(r^g)^* \psi = \mathrm{Ad}(g^{-1}) \psi$. Then for any $\psi \in \Omega_{hor}(P, \mathfrak{g})^G$ we have $d_\omega \psi = d\psi + [\omega, \psi]_\wedge$.*

(10) *The mapping $\psi \mapsto \zeta_\psi$, where $\zeta_\psi(X_1, \dots, X_k)(u) = \zeta_{\psi(X_1, \dots, X_k)(u)}(u)$, is an isomorphism between $\Omega_{hor}(P, \mathfrak{g})^G$ and the algebra $\Omega_{hor}(P, VP)^G$ of all horizontal G-equivariant forms with values in the vertical bundle VP. Then we have $\zeta_{d_\omega \psi} = -[\Phi, \zeta_\psi]$.*

Proof. Parts (1) through (4) follow from the properties of χ^*.

(5) We have

$$\begin{aligned}(d_\omega\omega)(\xi,\eta) &= (\chi^*d\omega)(\xi,\eta) = d\omega(\chi\xi,\chi\eta)\\ &= (\chi\xi)\omega(\chi\eta) - (\chi\eta)\omega(\chi\xi) - \omega([\chi\xi,\chi\eta])\\ &= -\omega([\chi\xi,\chi\eta]),\\ -\zeta(\Omega(\xi,\eta)) &= R(\xi,\eta) = \Phi[\chi\xi,\chi\eta] = \zeta_{\omega([\chi\xi,\chi\eta])}.\end{aligned}$$

(6) Using (19.2), we have

$$\begin{aligned}d_\omega\Omega &= d_\omega(d\omega + \tfrac{1}{2}[\omega,\omega]_\wedge)\\ &= \chi^*dd\omega + \tfrac{1}{2}\chi^*d[\omega,\omega]_\wedge\\ &= \tfrac{1}{2}\chi^*([d\omega,\omega]_\wedge - [\omega,d\omega]_\wedge) = \chi^*[d\omega,\omega]_\wedge\\ &= [\chi^*d\omega,\chi^*\omega]_\wedge = 0, \text{ since } \chi^*\omega = 0.\end{aligned}$$

(7) For $\varphi \in \Omega(P,W)$ we have

$$\begin{aligned}(d_\omega\chi^*\varphi)(X_0,\dots,X_k) &= (d\chi^*\varphi)(\chi(X_0),\dots,\chi(X_k))\\ &= \sum_{0\le i\le k}(-1)^i\chi(X_i)((\chi^*\varphi)(\chi(X_0),\dots,\widehat{\chi(X_i)},\dots,\chi(X_k)))\\ &\quad+\sum_{i<j}(-1)^{i+j}(\chi^*\varphi)([\chi(X_i),\chi(X_j)],\chi(X_0),\dots\\ &\qquad\qquad\dots,\widehat{\chi(X_i)},\dots,\widehat{\chi(X_j)},\dots)\\ &= \sum_{0\le i\le k}(-1)^i\chi(X_i)(\varphi(\chi(X_0),\dots,\widehat{\chi(X_i)},\dots,\chi(X_k)))\\ &\quad+\sum_{i<j}(-1)^{i+j}\varphi([\chi(X_i),\chi(X_j)] - \Phi[\chi(X_i),\chi(X_j)],\chi(X_0),\dots\\ &\qquad\qquad\dots,\widehat{\chi(X_i)},\dots,\widehat{\chi(X_j)},\dots)\\ &= (d\varphi)(\chi(X_0),\dots,\chi(X_k)) + (i_R\varphi)(\chi(X_0),\dots,\chi(X_k))\\ &= (d_\omega + \chi^*i_R)(\varphi)(X_0,\dots,X_k).\end{aligned}$$

(8) $d_\omega d_\omega = \chi^*d\chi^*d = (\chi^*i_R + \chi^*d)d = \chi^*i_Rd$ holds by (7).

(9) If we insert one vertical vector field, say ζ_X for $X \in \mathfrak{g}$, into $d_\omega\psi$, we get 0 by definition. For the right hand side we use $i_{\zeta_X}\psi = 0$ and $\mathcal{L}_{\zeta_X}\psi = \partial|_0(\mathrm{Fl}_t^{\zeta_X})^*\psi = \partial|_0(r^{\exp tX}) * \psi = \partial|_0\operatorname{Ad}(\exp(-tX))\psi = -\operatorname{ad}(X)\psi$ to get

$$\begin{aligned}i_{\zeta_X}(d\psi + [\omega,\psi]_\wedge) &= i_{\zeta_X}d\psi + di_{\zeta_X}\psi + [i_{\zeta_X}\omega,\psi] - [\omega,i_{\zeta_X}\psi]\\ &= \mathcal{L}_{\zeta_X}\psi + [X,\psi] = -\operatorname{ad}(X)\psi + [X,\psi] = 0.\end{aligned}$$

Now let all vector fields ξ_i be horizontal; then we get

$$\begin{aligned}(d_\omega\psi)(\xi_0,\dots,\xi_k) &= (\chi^*d\psi)(\xi_0,\dots,\xi_k) = d\psi(\xi_0,\dots,\xi_k),\\ (d\psi + [\omega,\psi]_\wedge)(\xi_0,\dots,\xi_k) &= d\psi(\xi_0,\dots,\xi_k).\end{aligned}$$

So the first formula holds.

(10) We proceed in a similar manner. Let Ψ be in the space $\Omega^\ell_{\text{hor}}(P, VP)^G$ of all horizontal G-equivariant forms with vertical values. Then for each $X \in \mathfrak{g}$ we have $i_{\zeta_X}\Psi = 0$; furthermore the G-equivariance $(r^g)^*\Psi = \Psi$ implies that $\mathcal{L}_{\zeta_X}\Psi = [\zeta_X, \Psi] = 0$ by (16.16.5). Using formula (16.11.2), we have

$$\begin{aligned} i_{\zeta_X}[\Phi, \Psi] &= [i_{\zeta_X}\Phi, \Psi] - [\Phi, i_{\zeta_X}\Psi] + i([\Phi, \zeta_X])\Psi + i([\Psi, \zeta_X])\Phi \\ &= [\zeta_X, \Psi] - 0 + 0 + 0 = 0. \end{aligned}$$

Now let all vector fields ξ_i again be horizontal; then from the huge formula (16.9) for the Frölicher-Nijenhuis bracket only the following terms in the third and fifth lines survive:

$$\begin{aligned} [\Phi, \Psi](\xi_1, \dots, \xi_{\ell+1}) &= \tfrac{(-1)^\ell}{\ell!} \sum_\sigma \operatorname{sign}\sigma \; \Phi([\Psi(\xi_{\sigma 1}, \dots, \xi_{\sigma \ell}), \xi_{\sigma(\ell+1)}]) \\ &\quad + \tfrac{1}{(\ell-1)!\,2!} \sum_\sigma \operatorname{sign}\sigma \; \Phi(\Psi([\xi_{\sigma 1}, \xi_{\sigma 2}], \xi_{\sigma 3}, \dots, \xi_{\sigma(\ell+1)})). \end{aligned}$$

For $f : P \to \mathfrak{g}$ and horizontal ξ we have $\Phi[\xi, \zeta_f] = \zeta_{\xi(f)} = \zeta_{df(\xi)}$: It is $C^\infty(P)$-linear in ξ; or imagine it in local coordinates. So the last expression becomes

$$-\zeta(d_\omega\psi(\xi_0, \dots, \xi_k)) = -\zeta(d\psi(\xi_0, \dots, \xi_k)) = -\zeta((d\psi + [\omega, \psi]_\wedge)(\xi_0, \dots, \xi_k))$$

as required. □

19.6. Theorem. *Let (P, p, M, G) be a principal fiber bundle with principal connection ω. Then the parallel transport for the principal connection is globally defined and G-equivariant.*

In detail: For each smooth curve $c : \mathbb{R} \to M$ there is a smooth mapping $\operatorname{Pt}_c : \mathbb{R} \times P_{c(0)} \to P$ such that the following hold:

(1) $\operatorname{Pt}(c, t, u) \in P_{c(t)}$, $\operatorname{Pt}(c, 0) = Id_{P_{c(0)}}$, *and* $\omega(\frac{d}{dt}\operatorname{Pt}(c, t, u)) = 0$.
(2) $\operatorname{Pt}(c, t) : P_{c(0)} \to P_{c(t)}$ *is G-equivariant, i.e.,* $\operatorname{Pt}(c, t, u.g) = \operatorname{Pt}(c, t, u).g$ *holds for all $g \in G$ and $u \in P$. Moreover we have* $\operatorname{Pt}(c, t)^*(\zeta_X | P_{c(t)}) = \zeta_X | P_{c(0)}$ *for all $X \in \mathfrak{g}$.*
(3) *For any smooth function $f : \mathbb{R} \to \mathbb{R}$ we have* $\operatorname{Pt}(c, f(t), u) = \operatorname{Pt}(c \circ f, t, \operatorname{Pt}(c, f(0), u))$.

Proof. By (19.4) the Christoffel forms $\Gamma^\alpha \in \Omega^1(U_\alpha, \mathfrak{X}(G))$ of the connection ω with respect to a principal fiber bundle atlas $(U_\alpha, \varphi_\alpha)$ are given by $\Gamma^\alpha(\xi_x) = R_{\omega_\alpha(\xi_x)}$, so they take values in the Lie subalgebra $\mathfrak{X}_R(G)$ of all right invariant vector fields on G, which are bounded with respect to any right invariant Riemann metric on G. Each right invariant metric on a Lie group is complete. So the connection is complete by proposition (23.9).

Properties (1) and (3) follow from theorem (17.8), and (2) is seen as follows: We have $\omega(\frac{d}{dt}\operatorname{Pt}(c,t,u).g) = \operatorname{Ad}(g^{-1})\omega(\frac{d}{dt}\operatorname{Pt}(c,t,u)) = 0$, and this implies $\operatorname{Pt}(c,t,u).g = \operatorname{Pt}(c,t,u.g)$. For the second assertion we compute for $u \in P_{c(0)}$:

$$\begin{aligned}
\operatorname{Pt}(c,t)^*(\zeta_X|P_{c(t)})(u) &= T\operatorname{Pt}(c,t)^{-1}\zeta_X(\operatorname{Pt}(c,t,u))\\
&= T\operatorname{Pt}(c,t)^{-1}\tfrac{d}{ds}|_0\operatorname{Pt}(c,t,u).\exp(sX)\\
&= T\operatorname{Pt}(c,t)^{-1}\tfrac{d}{ds}|_0\operatorname{Pt}(c,t,u.\exp(sX))\\
&= \tfrac{d}{ds}|_0\operatorname{Pt}(c,t)^{-1}\operatorname{Pt}(c,t,u.\exp(sX))\\
&= \tfrac{d}{ds}|_0 u.\exp(sX) = \zeta_X(u). \qquad \square
\end{aligned}$$

19.7. Holonomy groups. Let (P,p,M,G) be a principal fiber bundle with principal connection $\Phi = \zeta \circ \omega$. We assume that M is connected and we fix $x_0 \in M$.

In (17.10) we defined the holonomy group $\operatorname{Hol}(\Phi,x_0) \subset \operatorname{Diff}(P_{x_0})$ as the group of all $\operatorname{Pt}(c,1) : P_{x_0} \to P_{x_0}$ for c any piecewise smooth closed loop through x_0. (Reparameterizing c by a function which is flat at each corner of c, we may assume that any c is smooth.) If we consider only those curves c which are nullhomotopic, we obtain the restricted holonomy group $\operatorname{Hol}_0(\Phi,x_0)$, a normal subgroup.

Now let us fix $u_0 \in P_{x_0}$. The elements $\tau(u_0,\operatorname{Pt}(c,1,u_0)) \in G$ (for c all piecewise smooth closed loops through x_0) form a subgroup of the structure group G which is isomorphic to $\operatorname{Hol}(\Phi,x_0)$; we denote it by $\operatorname{Hol}(\omega,u_0)$ and we call it again the *holonomy group* of the connection. Considering only nullhomotopic curves, we get the *restricted holonomy group* $\operatorname{Hol}_0(\omega,u_0)$, a normal subgroup of $\operatorname{Hol}(\omega,u_0)$.

Theorem. *Let (P,p,M,G) be a principal fiber bundle with principal connection $\Phi = \zeta\circ\omega$. We assume that M is connected and we fix $x_0 \in M$ and $u_0 \in P_{x_0}$.*

(1) *We have an isomorphism* $\operatorname{Hol}(\omega,u_0) \to \operatorname{Hol}(\Phi,x_0)$ *given by* $g \mapsto (u \mapsto f_g(u) = u_0.g.\tau(u_0,u))$ *with inverse* $g_f := \tau(u_0,f(u_0)) \leftarrow f$.

(2) *We have* $\operatorname{Hol}(\omega,u_0.g) = \operatorname{conj}(g^{-1})\operatorname{Hol}(\omega,u_0)$ *and* $\operatorname{Hol}_0(\omega,u_0.g) = \operatorname{conj}(g^{-1})\operatorname{Hol}_0(\omega,u_0)$.

(3) *For any curve c with $c(0)=x_0$ we have* $\operatorname{Hol}(\omega,\operatorname{Pt}(c,t,u_0)) = \operatorname{Hol}(\omega,u_0)$ *and* $\operatorname{Hol}_0(\omega,\operatorname{Pt}(c,t,u_0)) = \operatorname{Hol}_0(\omega,u_0)$.

(4) *The restricted holonomy group* $\operatorname{Hol}_0(\omega,u_0)$ *is a connected Lie subgroup of G. The quotient group* $\operatorname{Hol}(\omega,u_0)/\operatorname{Hol}_0(\omega,u_0)$ *is at most countable, so* $\operatorname{Hol}(\omega,u_0)$ *is also a Lie subgroup of G.*

(5) *The Lie algebra* $\mathrm{hol}(\omega, u_0) \subset \mathfrak{g}$ *of* $\mathrm{Hol}(\omega, u_0)$ *is generated by* $\{\Omega(X_u, Y_u) : X_u, Y_u \in T_uP, u = \mathrm{Pt}(c, 1, u_0), c : [0,1] \to M, c(0) = x_0\}$ *as a vector space. It is isomorphic to the Lie algebra* $\mathrm{hol}(\Phi, x_0)$ *we considered in* (17.10).

(6) *For* $u_0 \in P_{x_0}$ *let* $P(\omega, u_0)$ *be the set of all* $\mathrm{Pt}(c, t, u_0)$ *for* c *any (piecewise) smooth curve in* M *with* $c(0) = x_0$ *and for* $t \in \mathbb{R}$. *Then* $P(\omega, u_0)$ *is a fiber subbundle of* P *which is invariant under the right action of* $\mathrm{Hol}(\omega, u_0)$*; so it is itself a principal fiber bundle over* M *with structure group* $\mathrm{Hol}(\omega, u_0)$ *and we have a reduction of structure group; see* (18.6) *and* (18.14). *The pullback of* ω *to* $P(\omega, u_0)$ *is also a principal connection form* $i^*\omega \in \Omega^1(P(\omega, u_0); \mathrm{hol}(\omega, u_0))$.

(7) P *is foliated by the leaves* $P(\omega, u)$, $u \in P_{x_0}$.

(8) *If the curvature* $\Omega = 0$, *then* $\mathrm{Hol}_0(\omega, u_0) = \{e\}$ *and each* $P(\omega, u)$ *is a covering of* M. *The leaves* $P(\omega, u)$ *are all isomorphic and are associated to the universal covering of* M, *which is a principal fiber bundle with structure group the fundamental group* $\pi_1(M)$.

In view of assertion (6) a principal connection ω is called *irreducible* if $\mathrm{Hol}(\omega, u_0)$ equals the structure group G for some (equivalently: any) $u_0 \in P_{x_0}$.

Proof. (1) follows from the definition of $\mathrm{Hol}(\omega, u_0)$.

(2) This follows from the properties of the mapping τ from (18.2) and from the G-equivariance of the parallel transport:

$$\tau(u_0.g, \mathrm{Pt}(c, 1, u_0.g)) = \tau(u_0, \mathrm{Pt}(c, 1, u_0).g) = g^{-1}.\tau(u_0, \mathrm{Pt}(c, 1, u_0)).g.$$

So via the diffeomorphism $\tau(u_0, \quad) : P_{x_0} \to G$ the action of the holonomy group $\mathrm{Hol}(\Phi, u_0)$ on P_{x_0} is conjugate to the left translation of $\mathrm{Hol}(\omega, u_0)$ on G.

(3) By reparameterizing the curve c, we may assume that $t = 1$, and we put $\mathrm{Pt}(c, 1, u_0) =: u_1$. Then by definition, for an element $g \in G$ we have $g \in \mathrm{Hol}(\omega, u_1)$ if and only if $g = \tau(u_1, \mathrm{Pt}(e, 1, u_1))$ for some closed smooth loop e through $x_1 := c(1) = p(u_1)$, i.e.,

$$\begin{aligned}
\mathrm{Pt}(c,1)(u_0.g) &= \mathrm{Pt}(c,1)(r^g(u_0)) = r^g(\mathrm{Pt}(c,1)(u_0)) = u_1 g \\
&= \mathrm{Pt}(e,1)(\mathrm{Pt}(c,1)(u_0)) \\
u_0.g &= \mathrm{Pt}(c,1)^{-1}\,\mathrm{Pt}(e,1)\,\mathrm{Pt}(c,1)(u_0) = \mathrm{Pt}(c.e.c^{-1}, 3)(u_0),
\end{aligned}$$

where $c.e.c^{-1}$ is the curve traveling along $c(t)$ for $0 \le t \le 1$, along $e(t-1)$ for $1 \le t \le 2$, and along $c(3-t)$ for $2 \le t \le 3$. This is equivalent to $g \in \mathrm{Hol}(\omega, u_0)$. Furthermore e is nullhomotopic if and only if $c.e.c^{-1}$ is nullhomotopic, so we also have $\mathrm{Hol}_0(\omega, u_1) = \mathrm{Hol}_0(\omega, u_0)$.

(4) Let $c : [0,1] \to M$ be a nullhomotopic curve through x_0 and let $h : \mathbb{R}^2 \to M$ be a smooth homotopy with $h_1|[0,1] = c$ and $h(0,s) = h(t,0) = h(t,1) = x_0$. We consider the pullback bundle

$$\begin{array}{ccc} h^*P & \xrightarrow{p^*h} & P \\ {\scriptstyle h^*p}\downarrow & & \downarrow{\scriptstyle p} \\ \mathbb{R}^2 & \xrightarrow{h} & M. \end{array}$$

Then for the parallel transport Pt^Φ on P and for the parallel transport $\operatorname{Pt}^{h^*\Phi}$ of the pulled back connection we have

$$\operatorname{Pt}^\Phi(h_t, 1, u_0) = (p^*h)\operatorname{Pt}^{h^*\Phi}((t, \quad), 1, u_0) = (p^*h)\operatorname{Fl}_1^{C^{h^*\Phi}\partial_s}(t, u_0).$$

So $t \mapsto \tau(u_0, \operatorname{Pt}^\Phi(h_t, 1, u_0))$ is a smooth curve in the Lie group G starting from e, so $\operatorname{Hol}_0(\omega, u_0)$ is a smoothly arcwise connected subgroup of G. By theorem (5.6) the subgroup $\operatorname{Hol}_0(\omega, u_0)$ is a Lie subgroup of G. The quotient group $\operatorname{Hol}(\omega, u_0)/\operatorname{Hol}_0(\omega, u_0)$ is a countable group, since by Morse theory M is homotopy equivalent to a countable CW-complex, so the fundamental group $\pi_1(M)$ is countably generated, thus countable.

(5) Note first that for $g \in G$ and $X \in \mathfrak{X}(M)$ we have for the horizontal lift $(r^g)^*CX = CX$, since $(r^g)^*\Phi = \Phi$ implies $T_u(r^g).H_uP = H_{u.g}P$ and thus

$$\begin{aligned} T_u(r^g).C(X,u) &= T_u(r^g).(T_up|H_uP)^{-1}(X(p(u))) \\ &= (T_{u.g}p|H_{u.g}P)^{-1}(X(p(u))) = C(X, u.g). \end{aligned}$$

The vector space $\operatorname{hol}(\omega) \subset \mathfrak{g}$ is normalized by the subgroup $\operatorname{Hol}(\omega, u_0) \subseteq G$ since for $g = \tau(u_0, \operatorname{Pt}(c, 1, u_0))$ (where c is a loop at x_0) and for $u = \operatorname{Pt}(c_1, 1, u_0)$ (where $c_1(0) = x_0$) we have

$$\begin{aligned} \operatorname{Ad}(g^{-1})\Omega(C(X,u), C(Y,u)) &= \Omega(T_u(r^g).C(X,u), T_u(r^g).C(Y,u)) \\ &= \Omega(C(X,u.g), C(Y,u.g)) \in \operatorname{hol}(\omega), \\ u.g = \operatorname{Pt}(c_1, 1, u_0).g &= \operatorname{Pt}(c_1, 1, u_0.g) = \operatorname{Pt}(c_1, 1, \operatorname{Pt}(c, 1, u_0)) \\ &= \operatorname{Pt}(c_1.c, 2, u_0). \end{aligned}$$

We consider now the mapping

$$\begin{aligned} &\xi^{u_0} : \operatorname{hol}(\omega) \to \mathfrak{X}(P_{x_0}), \\ &\xi^{u_0}_X(u) = \zeta_{\operatorname{Ad}(\tau(u_0,u)^{-1})X}(u). \end{aligned}$$

It turns out that $\xi^{u_0}_X$ is related to the right invariant vector field R_X on G under the diffeomorphism $\tau(u_0, \quad) = (r_{u_0})^{-1} : P_{x_0} \to G$, since we have

$$\begin{aligned} T_g(r_{u_0}).R_X(g) &= T_g(r_{u_0}).T_e(\mu^g).X = T_{u_0}(r^g).T_e(r_{u_0}).X \\ &= T_{u_0}(r^g)\zeta_X(u_0) = \zeta_{\operatorname{Ad}(g^{-1})X}(u_0.g) = \xi^{u_0}_X(u_0.g). \end{aligned}$$

Thus ξ^{u_0} is the restriction to $\mathrm{hol}(\omega) \subseteq \mathfrak{g}$ of a Lie algebra antihomomorphism $\mathfrak{g} \to \mathfrak{X}(P_{x_0})$, and each vector field $\xi^{u_0}_X$ on P_{x_0} is complete. The dependence of ξ^{u_0} on u_0 is explained by

$$\begin{aligned}\xi^{u_0 g}_X(u) &= \zeta_{\mathrm{Ad}(\tau(u_0 g,u)^{-1})X}(u) = \zeta_{\mathrm{Ad}(\tau(u_0,u)^{-1})\,\mathrm{Ad}(g)X}(u)\\ &= \xi^{u_0}_{\mathrm{Ad}(g)X}(u).\end{aligned}$$

Recall now that the holonomy Lie algebra $\mathrm{hol}(\Phi, x_0)$ is the closed linear span of all vector fields of the form $\mathrm{Pt}(c,1)^*R(CX,CY)$, where $X, Y \in T_xM$ and c is a curve from x_0 to x. Then we have for $u = \mathrm{Pt}(c,1,u_0)$

$$\begin{aligned}R(C(X,u),C(Y,u)) &= \zeta_{\Omega(C(X,u),C(Y,u))}(u),\\ R(CX,CY)(ug) &= T(r^g)R(CX,CY)(u) = T(r^g)\zeta_{\Omega(C(X,u),C(Y,u))}(u)\\ &= \zeta_{\mathrm{Ad}(g^{-1})\Omega(C(X,u),C(Y,u))}(ug) = \xi^u_{\Omega(C(X,u),C(Y,u))}(ug),\\ (\mathrm{Pt}(c,1)^*R(CX,CY))(u_0.g) &=\\ &= T(\mathrm{Pt}(c,1)^{-1})\zeta_{\mathrm{Ad}(g^{-1})\Omega(C(X,u),C(Y,u))}(\mathrm{Pt}(c,1,u_0.g))\\ &= (\mathrm{Pt}(c,1)^*\zeta_{\mathrm{Ad}(g^{-1})\Omega(C(X,u),C(Y,u))})(u_0.g)\\ &= \zeta_{\mathrm{Ad}(g^{-1})\Omega(C(X,u),C(Y,u))}(u_0.g) \qquad \text{by (19.6.2)}\\ &= \xi^{u_0}_{\Omega(C(X,u),C(Y,u))}(u_0.g).\end{aligned}$$

So $\xi^{u_0} : \mathrm{hol}(\omega) \to \mathrm{hol}(\Phi, x_0)$ is a linear isomorphic. Since $\mathrm{hol}(\Phi, x_0)$ is a Lie subalgebra of $\mathfrak{X}(P_{x_0})$ by (17.10) and $\xi^{u_0} : \mathfrak{g} \to \mathfrak{X}(P_{x_0})$ is a Lie algebra antihomomorphism, $\mathrm{hol}(\omega)$ is a Lie subalgebra of $\mathfrak{g}$. Moreover $\mathrm{hol}(\Phi, x_0)$ consists of complete vector fields and we may apply theorem (17.11) (only claim 3) which tells us that the Lie algebra of the Lie group $\mathrm{Hol}(\Phi, x_0)$ is $\mathrm{hol}(\Phi, x_0)$. The diffeomorphism $\tau(u_0,\quad) : P_{x_0} \to G$ intertwines the actions and the infinitesimal actions in the right way.

(6) We define the vector subbundle $E \subset TP$ by $E_u := H_uP + T_e(r_u).\mathrm{hol}(\omega)$. From the proof of (4) it follows that $\xi^{u_0}_X$ are sections of E for each $X \in \mathrm{hol}(\omega)$; thus E is a vector bundle. Any vector field $\eta \in \mathfrak{X}(P)$ with values in E is a linear combination with coefficients in $C^\infty(P)$ of horizontal vector fields CX for $X \in \mathfrak{X}(M)$ and of ζ_Z for $Z \in \mathrm{hol}(\omega)$. Their Lie brackets are in turn

$$\begin{aligned}[CX,CY](u) &= C[X,Y](u) + R(CX,CY)(u)\\ &= C[X,Y](u) + \zeta_{\Omega(C(X,u),C(Y,u))}(u) \in \Gamma(E),\\ [\zeta_Z, CX] &= \mathcal{L}_{\zeta_Z}CX = \tfrac{d}{dt}|_0(\mathrm{Fl}^{\zeta_Z}_t)^*CX = 0,\end{aligned}$$

since $(r^g)^*CX = CX$; see step (5) above. So E is an integrable subbundle and induces a foliation by (3.28.2). Let $L(u_0)$ be the leaf of the foliation through u_0. Since for a curve c in M the parallel transport $\mathrm{Pt}(c,t,u_0)$ is tangent to the leaf, we have $P(\omega,u_0) \subseteq L(u_0)$. By definition the holonomy group $\mathrm{Hol}(\Phi,x_0)$ acts transitively and freely on $P(\omega,u_0) \cap P_{x_0}$, and by (5) the restricted holonomy group $\mathrm{Hol}_0(\Phi, x_0)$ acts transitively on each connected component of $L(u_0) \cap P_{x_0}$, since the vertical part of E is spanned by the generating vector fields of this action. This is true for any fiber since we may conjugate the holonomy groups by a suitable parallel transport to each fiber. Thus $P(\omega, u_0) = L(u_0)$ and by lemma (18.3) the fiber subbundle $P(\omega, x_0)$ is a principal fiber bundle with structure group $\mathrm{Hol}(\omega, u_0)$. Since all horizontal spaces $H_u P$ with $u \in P(\omega, x_0)$ are tangential to $P(\omega, x_0)$, the connection Φ restricts to a principal connection on $P(\omega, x_0)$ and we obtain the reduction we looked for of the structure group.

(7) This is obvious from the proof of (6).

(8) If the curvature Ω is everywhere 0, the holonomy Lie algebra is zero, so $P(\omega, u)$ is a principal fiber bundle with discrete structure group; thus $p|P(\omega, u) : P(\omega, u) \to M$ is a local diffeomorphism, since $T_u P(\omega, u) = H_u P$ and Tp is invertible on it. By the right action of the structure group we may translate each local section of p to any point of the fiber, so p is a covering map. Parallel transport defines a group homomorphism $\varphi : \pi_1(M, x_0) \to \mathrm{Hol}(\Phi, u_0) \cong \mathrm{Hol}(\omega, u_0)$ (see the proof of (4)). Let $\tilde{M}$ be the universal covering space of M; then from topology one knows that $\tilde{M} \to M$ is a principal fiber bundle with discrete structure group $\pi_1(M, x_0)$. Let $\pi_1(M)$ act on $\mathrm{Hol}(\omega, u_0)$ by left translation via φ; then the mapping $f : \tilde{M} \times \mathrm{Hol}(\omega, u_0) \to P(\omega, u_0)$ which is given by $f([c], g) = \mathrm{Pt}(c, 1, u_0).g$ is $\pi_1(M)$-invariant and thus factors to a mapping

$$\tilde{M} \times_{\pi_1(M)} \mathrm{Hol}(\omega, u_0) = \tilde{M}[\mathrm{Hol}(\omega, u_0)] \to P(\omega, u_0)$$

which is an isomorphism of $\mathrm{Hol}(\omega, u_0)$-bundles since the upper mapping admits local sections by the curve lifting property of the universal cover. □

19.8. Inducing principal connections on associated bundles. We consider a principal bundle (P, p, M, G) with principal right action $r : P \times G \to P$ and let $\ell : G \times S \to S$ be a left action of the structure group G on some manifold S. Then we consider the associated bundle $P[S] = P[S, \ell] = P \times_G S$, constructed in (18.7). Recall from (18.18) that its tangent and vertical bundle are given by $T(P[S, \ell]) = TP[TS, T\ell] = TP \times_{TG} TS$ and $V(P[S, \ell]) = P[TS, T_2\ell] = P \times_G TS$.

Let $\Phi = \zeta \circ \omega \in \Omega^1(P, TP)$ be a principal connection on the principal bundle P. We construct the *induced connection* $\bar{\Phi} \in \Omega^1(P[S], T(P[S]))$ by factorizing as in the following diagram:

$$\begin{array}{ccccc}
TP \times TS & \xrightarrow{\Phi \times Id} & TP \times TS & \xrightarrow{=} & T(P \times S) \\
\Big\downarrow{\scriptstyle Tq = q'} & & \Big\downarrow{\scriptstyle q'} & & \Big\downarrow{\scriptstyle Tq} \\
TP \times_{TG} TS & \xrightarrow{\bar{\Phi}} & TP \times_{TG} TS & \xrightarrow{=} & T(P \times_G S).
\end{array}$$

Let us first check that the top mapping $\Phi \times Id$ is TG-equivariant. For $g \in G$ and $X \in \mathfrak{g}$ the inverse of $T_e(\mu_g)X$ in the Lie group TG is denoted by $(T_e(\mu_g)X)^{-1}$; see lemma (6.7). Furthermore by (6.3) we have

$$\begin{aligned} Tr(\xi_u, T_e(\mu_g)X) &= T_u(r^g)\xi_u + Tr((0_P \times L_X)(u,g)) \\ &= T_u(r^g)\xi_u + T_g(r_u)(T_e(\mu_g)X) \\ &= T_u(r^g)\xi_u + \zeta_X(ug). \end{aligned}$$

We may compute

$$\begin{aligned} &(\Phi \times Id)\big(Tr(\xi_u, T_e(\mu_g)X), T\ell((T_e(\mu_g)X)^{-1}, \eta_s)\big) \\ &\quad = \big(\Phi(T_u(r^g)\xi_u + \zeta_X(ug)), T\ell((T_e(\mu_g)X)^{-1}, \eta_s)\big) \\ &\quad = \big(\Phi(T_u(r^g)\xi_u) + \Phi(\zeta_X(ug)), T\ell((T_e(\mu_g)X)^{-1}, \eta_s)\big) \\ &\quad = \big((T_u(r^g)\Phi\xi_u) + \zeta_X(ug), T\ell((T_e(\mu_g)X)^{-1}, \eta_s)\big) \\ &\quad = \big(Tr(\Phi(\xi_u), T_e(\mu_g)X), T\ell((T_e(\mu_g)X)^{-1}, \eta_s)\big). \end{aligned}$$

So the mapping $\Phi \times Id$ factors to $\bar\Phi$ as indicated in the diagram, and we have $\bar\Phi \circ \bar\Phi = \bar\Phi$ from $(\Phi \times Id) \circ (\Phi \times Id) = \Phi \times Id$. The mapping $\bar\Phi$ is fiberwise linear, since $\Phi \times Id$ and $q' = Tq$ are. The image of $\bar\Phi$ is

$$\begin{aligned} q'(VP \times TS) &= q'(\ker(Tp : TP \times TS \to TM)) \\ &= \ker(Tp : TP \times_{TG} TS \to TM) = V(P[S,\ell]). \end{aligned}$$

Thus $\bar\Phi$ is a connection on the associated bundle $P[S]$. We call it the *induced connection.*

From the diagram it also follows that the vector valued forms

$$\begin{aligned} &\Phi \times Id \in \Omega^1(P \times S, TP \times TS) \\ \text{and} \quad &\bar\Phi \in \Omega^1(P[S], T(P[S])) \end{aligned}$$

are $(q : P \times S \to P[S])$-related. So by (16.15) we have for the curvatures

$$\begin{aligned} R_{\Phi\times Id} &= \tfrac{1}{2}[\Phi \times Id, \Phi \times Id] = \tfrac{1}{2}[\Phi, \Phi] \times 0 \\ &= R_\Phi \times 0, \\ R_{\bar\Phi} &= \tfrac{1}{2}[\bar\Phi, \bar\Phi] \end{aligned}$$

that they are also q-related, i.e., $Tq \circ (R_\Phi \times 0) = R_{\bar\Phi} \circ (Tq \times_M Tq)$.

By uniqueness of the solutions of the defining differential equation we also get that

$$\mathrm{Pt}_{\bar\Phi}(c, t, q(u,s)) = q(\mathrm{Pt}_\Phi(c, t, u), s).$$

19.9. Recognizing induced connections. We consider again a principal fiber bundle (P, p, M, G) and a left action $\ell : G \times S \to S$. Suppose that we have a connection $\Psi \in \Omega^1(P[S], T(P[S]))$ on the associated bundle $P[S] = P[S, \ell]$. Then the following question arises: When is the connection Ψ induced from a principal connection on P? If this is the case, we say that Ψ is compatible with the G-structure on $P[S]$. The answer is given in the following

Theorem. *Let Ψ be a (general) connection on the associated bundle $P[S]$. Let us suppose that the action ℓ is infinitesimally effective, i.e., the fundamental vector field mapping $\zeta : \mathfrak{g} \to \mathfrak{X}(S)$ is injective.*

Then the connection Ψ is induced from a principal connection ω on P if and only if the following condition is satisfied:

- *In some (equivalently: any) fiber bundle atlas (U_α, ψ_α) of $P[S]$ belonging to the G-structure of the associated bundle the Christoffel forms $\Gamma^\alpha \in \Omega^1(U_\alpha, \mathfrak{X}(S))$ have values in the Lie subalgebra $\mathfrak{X}_{fund}(S)$ of fundamental vector fields for the action ℓ.*

Proof. Let $(U_\alpha, \varphi_\alpha : P|U_\alpha \to U_\alpha \times G)$ be a principal fiber bundle atlas for P. Then by the proof of theorem (18.7) the induced fiber bundle atlas

$$(U_\alpha, \psi_\alpha : P[S]|U_\alpha \to U_\alpha \times S)$$

is given by

(1) $$\psi_\alpha^{-1}(x, s) = q(\varphi_\alpha^{-1}(x, e), s),$$

(2) $$(\psi_\alpha \circ q)(\varphi_\alpha^{-1}(x, g), s) = (x, g.s).$$

Let $\Phi = \zeta \circ \omega$ be a principal connection on P and let $\bar{\Phi}$ be the induced connection on the associated bundle $P[S]$. By (17.7) its Christoffel symbols are given by

$$\begin{aligned}
(0_x, \Gamma^\alpha_{\bar\Phi}(\xi_x, s)) &= -(T(\psi_\alpha) \circ \bar\Phi \circ T(\psi_\alpha^{-1}))(\xi_x, 0_s) \\
&= -(T(\psi_\alpha) \circ \bar\Phi \circ Tq \circ (T(\varphi_\alpha^{-1}) \times Id))(\xi_x, 0_e, 0_s) && \text{by (1)} \\
&= -(T(\psi_\alpha) \circ Tq \circ (\Phi \times Id))(T(\varphi_\alpha^{-1})(\xi_x, 0_e), 0_s) && \text{by (19.8)} \\
&= -(T(\psi_\alpha) \circ Tq)(\Phi(T(\varphi_\alpha^{-1})(\xi_x, 0_e)), 0_s) \\
&= (T(\psi_\alpha) \circ Tq)(T(\varphi_\alpha^{-1})(0_x, \Gamma^\alpha_\Phi(\xi_x, e)), 0_s) && \text{by (19.4.3)} \\
&= -T(\psi_\alpha \circ q \circ (\varphi_\alpha^{-1} \times Id))(0_x, \omega_\alpha(\xi_x), 0_s) && \text{by (19.4.7)} \\
&= -T_e(\ell^s)\omega_\alpha(\xi_x) && \text{by (2)} \\
&= -\zeta_{\omega_\alpha(\xi_x)}(s).
\end{aligned}$$

So the condition is necessary.

Now let us conversely suppose that a connection Ψ on $P[S]$ is given such that the Christoffel forms Γ^α_Ψ with respect to a fiber bundle atlas of the G-structure have values in $\mathfrak{X}_{fund}(S)$. Then unique $\mathfrak{g}$-valued forms $\omega_\alpha \in \Omega^1(U_\alpha, \mathfrak{g})$ are given by the equation

$$\Gamma^\alpha_\Psi(\xi_x) = -\zeta(\omega_\alpha(\xi_x)),$$

since the action is infinitesimally effective. From the transition formulas (17.7) for the Γ^α_Ψ follow the transition formulas (19.4.5) for the ω^α, so that they give a unique principal connection on P, which by the first part of the proof induces the given connection Ψ on $P[S]$. $\square$

19.10. Inducing principal connections on associated vector bundles. Let (P, p, M, G) be a principal fiber bundle and let $\rho : G \to GL(W)$ be a representation of the structure group G on a finite-dimensional vector space W. We consider the associated vector bundle $(E := P[W, \rho], p, M, W)$, which was treated in some detail in (18.11):

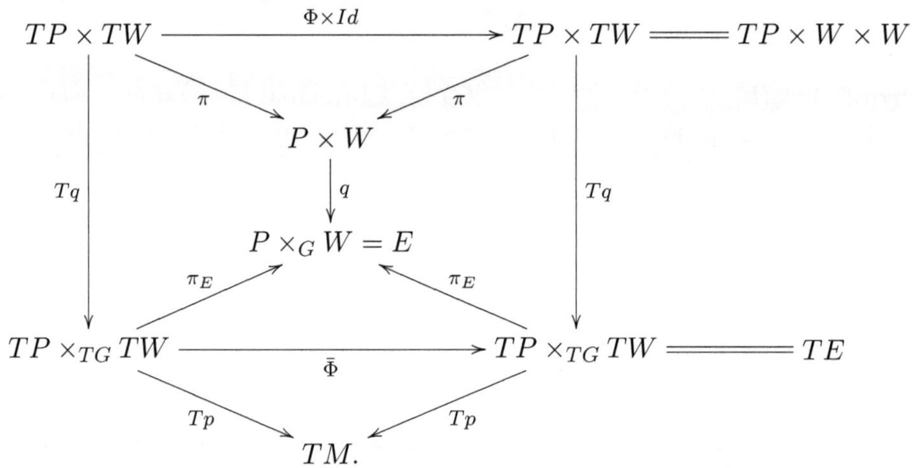

Recall from (8.12) that $T(E) = TP \times_{TG} TW$ has two vector bundle structures with the projections

$$\pi_E : T(E) = TP \times_{TG} TW \to P \times_G W = E,$$
$$Tp \circ \mathrm{pr}_1 : T(E) = TP \times_{TG} TW \to TM.$$

Now let $\Phi = \zeta \circ \omega \in \Omega^1(P, TP)$ be a principal connection on P. We consider the induced connection $\bar\Phi \in \Omega^1(E, T(E))$ from (19.8).

A look at the diagram above shows that the induced connection is linear in both vector bundle structures. We say that it is a *linear connection* on the associated bundle.

Recall now from (8.12) the vertical lift $\mathrm{vl}_E : E \times_M E \to VE$, which is an isomorphism, pr_1-π_E-fiberwise linear and also pr_2-Tp-fiberwise linear.

Now we define the *connector* K of the linear connection $\bar{\Phi}$ by

$$K := \mathrm{pr}_2 \circ (\mathrm{vl}_E)^{-1} \circ \bar{\Phi} : TE \to VE \to E \times_M E \to E.$$

Lemma. *The connector $K : TE \to E$ is a vector bundle homomorphism for both vector bundle structures on TE and satisfies*

$$K \circ \mathrm{vl}_E = \mathrm{pr}_2 : E \times_M E \to TE \to E.$$

So K is π_E-p-fiberwise linear and Tp-p-fiberwise linear.

Proof. This follows from the fiberwise linearity of the components of K and from its definition. □

19.11. Linear connections. If (E, p, M) is a vector bundle, a connection $\Psi \in \Omega^1(E, TE)$ such that $\Psi : TE \to VE \to TE$ is also Tp-Tp-fiberwise linear is called a *linear connection.* An easy check with (19.9) or a direct construction shows that Ψ is then induced from a unique principal connection on the linear frame bundle $GL(\mathbb{R}^n, E)$ of E (where n is the fiber dimension of E).

Equivalently, a linear connection may be specified by a connector

$$K : TE \to E$$

with the three properties of lemma (19.10). For then

$$HE := \{\xi_u : K(\xi_u) = 0_{p(u)}\}$$

is a complement to VE in TE which is Tp-fiberwise linearly chosen.

19.12. Covariant derivative on vector bundles. Let (E, p, M) be a vector bundle with a linear connection, given by a connector $K : TE \to E$ with the properties in lemma (19.10).

For any manifold N, smooth mapping $s : N \to E$, and vector field $X \in \mathfrak{X}(N)$ we define the *covariant derivative* of s along X by

(1) $$\nabla_X s := K \circ Ts \circ X : N \to TN \to TE \to E.$$

If $f : N \to M$ is a fixed smooth mapping, let us denote by $C^\infty_f(N, E)$ the vector space of all smooth mappings $s : N \to E$ with $p \circ s = f$ — they are called sections of E along f. From the universal property of the pullback it follows that the vector space $C^\infty_f(N, E)$ is canonically linearly isomorphic to the space $\Gamma(f^*E)$ of sections of the pullback bundle. Then the covariant derivative may be viewed as a bilinear mapping

(2) $$\nabla : \mathfrak{X}(N) \times C^\infty_f(N, E) \to C^\infty_f(N, E).$$

In particular for $f = Id_M$ we have

$$\nabla : \mathfrak{X}(M) \times \Gamma(E) \to \Gamma(E).$$

Lemma. *This covariant derivative has the following properties:*

(3) $\nabla_X s$ *is* $C^\infty(N)$*-linear in* $X \in \mathfrak{X}(N)$*. So for a tangent vector* $X_x \in T_xN$ *the mapping* $\nabla_{X_x} : C^\infty_f(N,E) \to E_{f(x)}$ *makes sense and we have* $(\nabla_X s)(x) = \nabla_{X(x)} s$.

(4) $\nabla_X s$ *is* $\mathbb{R}$*-linear in* $s \in C^\infty_f(N,E)$.

(5) $\nabla_X(h.s) = dh(X).s + h.\nabla_X s$ *for* $h \in C^\infty(N)$*, the derivation property of* ∇_X.

(6) *For any manifold* Q *and smooth mapping* $g : Q \to N$ *and* $Y_y \in T_yQ$ *we have* $\nabla_{Tg.Y_y} s = \nabla_{Y_y}(s \circ g)$*. If* $Y \in \mathfrak{X}(Q)$ *and* $X \in \mathfrak{X}(N)$ *are* g*-related, then we have* $\nabla_Y(s \circ g) = (\nabla_X s) \circ g$.

Proof. All these properties follow easily from the definition (1). □

Property (6) is not well understood in some differential geometric literature. For vector fields $X, Y \in \mathfrak{X}(M)$ and a section $s \in \Gamma(E)$ an easy computation shows that

$$\begin{aligned} R^E(X,Y)s :&= \nabla_X\nabla_Y s - \nabla_Y\nabla_X s - \nabla_{[X,Y]}s \\ &= ([\nabla_X, \nabla_Y] - \nabla_{[X,Y]})s \end{aligned}$$

is $C^\infty(M)$-linear in X, Y, and s. By the method of (9.3) it follows that R^E is a 2-form on M with values in the vector bundle $L(E,E)$, i.e., $R^E \in \Omega^2(M, L(E,E))$. It is called the *curvature* of the covariant derivative. See (19.16) below for the relation to the principal curvature if E is an associated bundle.

For $f : N \to M$, vector fields $X, Y \in \mathfrak{X}(N)$ and a section $s \in C^\infty_f(N,E)$ along f one may prove that

$$\nabla_X\nabla_Y s - \nabla_Y\nabla_X s - \nabla_{[X,Y]}s = (f^*R^E)(X,Y)s := R^E(Tf.X, Tf.Y)s.$$

19.13. Covariant exterior derivative. Let (E,p,M) be a vector bundle with a linear connection, given by a connector $K : TE \to E$.

For a smooth mapping $f : N \to M$ let $\Omega(N, f^*E)$ be the vector space of all forms on N with values in the vector bundle f^*E. We can also view them as forms on N with values along f in E, but we do not introduce an extra notation for this.

The graded space $\Omega(N, f^*E)$ is a graded $\Omega(N)$-module via

$$\begin{aligned} &(\varphi \wedge \Phi)(X_1, \dots, X_{p+q}) \\ &\qquad = \tfrac{1}{p!\,q!} \sum_\sigma \operatorname{sign}(\sigma)\, \varphi(X_{\sigma 1}, \dots, X_{\sigma p})\Phi(X_{\sigma(p+1)}, \dots, X_{\sigma(p+q)}). \end{aligned}$$

The graded module homomorphisms $H : \Omega(N, f^*E) \to \Omega(N, f^*E)$ (so that $H(\varphi \wedge \Phi) = (-1)^{\deg H.\deg \varphi}\varphi \wedge H(\Phi))$ are easily seen to coincide with the mappings $\mu(A)$ for $A \in \Omega^p(N, f^*L(E,E))$, which are given by

$$\begin{aligned}(\mu(A)\Phi)&(X_1, \dots, X_{p+q})\\ &= \tfrac{1}{p!\,q!}\sum_\sigma \operatorname{sign}(\sigma)\, A(X_{\sigma 1}, \dots, X_{\sigma p})(\Phi(X_{\sigma(p+1)}, \dots, X_{\sigma(p+q)})).\end{aligned}$$

The *covariant exterior derivative* $d_\nabla : \Omega^p(N, f^*E) \to \Omega^{p+1}(N, f^*E)$ is defined by (where the X_i are vector fields on N)

$$\begin{aligned}(d_\nabla\Phi)(X_0, \dots, X_p) &= \sum_{i=0}^{p}(-1)^i\nabla_{X_i}\Phi(X_0, \dots, \widehat{X_i}, \dots, X_p)\\ &+ \sum_{0\le i<j\le p}(-1)^{i+j}\Phi([X_i, X_j], X_0, \dots, \widehat{X_i}, \dots, \widehat{X_j}, \dots, X_p).\end{aligned}$$

Lemma. *The covariant exterior derivative is well defined and has the following properties.*

(1) *For $s \in \Gamma(f^*E) = \Omega^0(N, f^*E)$ we have $(d_\nabla s)(X) = \nabla_X s$.*
(2) *$d_\nabla(\varphi \wedge \Phi) = d\varphi \wedge \Phi + (-1)^{\deg\varphi}\varphi \wedge d_\nabla\Phi$.*
(3) *For smooth $g : Q \to N$ and $\Phi \in \Omega(N, f^*E)$ we have $d_\nabla(g^*\Phi) = g^*(d_\nabla\Phi)$.*
(4) *$d_\nabla d_\nabla\Phi = \mu(f^*R^E)\Phi$.*

Proof. It suffices to investigate decomposable forms $\Phi = \varphi \otimes s$ where $\varphi \in \Omega^p(N)$ and $s \in \Gamma(f^*E)$. Then from the definition we have

$$d_\nabla(\varphi \otimes s) = d\varphi \otimes s + (-1)^p\varphi \wedge d_\nabla s.$$

Since $d_\nabla s \in \Omega^1(N, f^*E)$ by (19.12.3), the mapping d_∇ is well defined. This formula also implies (2) immediately. Part (3) follows from (19.12.6). Part (4) is checked as follows:

$$\begin{aligned}d_\nabla d_\nabla(\varphi \otimes s) &= d_\nabla(d\varphi \otimes s + (-1)^p\varphi \wedge d_\nabla s) \text{ by (2)}\\ &= 0 + (-1)^{2p}\varphi \wedge d_\nabla d_\nabla s\\ &= \varphi \wedge \mu(f^*R^E)s \text{ by the definition of } R^E\\ &= \mu(f^*R^E)(\varphi \otimes s). \quad \square\end{aligned}$$

19.14. Let (P, p, M, G) be a principal fiber bundle and let $\rho : G \to GL(W)$ be a representation of the structure group G on a finite-dimensional vector space W.

Theorem. *There is a canonical isomorphism from the space of $P[W,\rho]$-valued differential forms on M onto the space of horizontal G-equivariant W-valued differential forms on P:*

$$\begin{aligned} q^\sharp : \Omega(M, P[W,\rho]) \to \Omega_{hor}(P, W)^G = \{&\varphi \in \Omega(P, W) : i_X\varphi = 0 \\ &\text{for all } X \in VP, (r^g)^*\varphi = \rho(g^{-1}) \circ \varphi \text{ for all } g \in G\}. \end{aligned}$$

In particular for $W = \mathbb{R}$ with trivial representation we see that

$$p^* : \Omega(M) \to \Omega_{hor}(P)^G = \{\varphi \in \Omega_{hor}(P) : (r^g)^*\varphi = \varphi\}$$

is also an isomorphism. The isomorphism

$$q^\sharp : \Omega^0(M, P[W]) = \Gamma(P[W]) \to \Omega^0_{hor}(P, W)^G = C^\infty(P, W)^G$$

is a special case of the one from (18.12).

Proof. Recall the smooth mapping $\tau^G : P \times_M P \to G$ from (18.2) with

$$\begin{aligned} r(u_x, \tau^G(u_x, v_x)) &= v_x, \\ \tau^G(u_x.g, u'_x.g') &= g^{-1}.\tau^G(u_x, u'_x).g', \\ \tau^G(u_x, u_x) &= e. \end{aligned}$$

Let $\varphi \in \Omega^k_{hor}(P, W)^G$, $X_1, \dots, X_k \in T_uP$, and $X'_1, \dots, X'_k \in T_{u'}P$ such that $T_up.X_i = T_{u'}p.X'_i$ for each i. For $g = \tau^G(u, u')$, so that $ug = u'$, we then have:

$$\begin{aligned} q(u, \varphi_u(X_1, \dots, X_k)) &= q(ug, \rho(g^{-1})\varphi_u(X_1, \dots, X_k)) \\ &= q(u', ((r^g)^*\varphi)_u(X_1, \dots, X_k)) \\ &= q(u', \varphi_{ug}(T_u(r^g).X_1, \dots, T_u(r^g).X_k)) \\ &= q(u', \varphi_{u'}(X'_1, \dots, X'_k)), \text{ since } T_u(r^g)X_i - X'_i \in V_{u'}P. \end{aligned}$$

By this a vector bundle valued form $\Phi \in \Omega^k(M, P[W])$ is uniquely determined.

For the converse recall the smooth mapping $\tau^W : P \times_M P[W,\rho] \to W$ from (18.7), which satisfies

$$\begin{aligned} \tau^W(u, q(u, w)) &= w, \\ q(u_x, \tau^W(u_x, v_x)) &= v_x, \\ \tau^W(u_xg, v_x) &= \rho(g^{-1})\tau^W(u_x, v_x). \end{aligned}$$

For $\Phi \in \Omega^k(M, P[W])$ we define $q^\sharp\Phi \in \Omega^k(P, W)$ as follows. For $X_i \in T_uP$ we put

$$(q^\sharp\Phi)_u(X_1, \dots, X_k) := \tau^W(u, \Phi_{p(u)}(T_up.X_1, \dots, T_up.X_k)).$$

Then $q^\sharp\Phi$ is smooth and horizontal. For $g \in G$ we have

$$\begin{aligned}((r^g)^*(q^\sharp\Phi))_u(X_1, \dots, X_k) &= (q^\sharp\Phi)_{ug}(T_u(r^g).X_1, \dots, T_u(r^g).X_k)\\ &= \tau^W(ug, \Phi_{p(ug)}(T_{ug}p.T_u(r^g).X_1, \dots, T_{ug}p.T_u(r^g).X_k))\\ &= \rho(g^{-1})\tau^W(u, \Phi_{p(u)}(T_up.X_1, \dots, T_up.X_k))\\ &= \rho(g^{-1})(q^\sharp\Phi)_u(X_1, \dots, X_k).\end{aligned}$$

Clearly the two constructions are inverse to each other. □

19.15. Let (P, p, M, G) be a principal fiber bundle with a principal connection $\Phi = \zeta \circ \omega$, and let $\rho : G \to GL(W)$ be a representation of the structure group G on a finite-dimensional vector space W. We consider the associated vector bundle $(E := P[W, \rho], p, M, W)$, the induced connection $\bar{\Phi}$ on it and the corresponding covariant derivative.

Theorem. *The covariant exterior derivative d_ω from* (19.5) *on P and the covariant exterior derivative for $P[W]$-valued forms on M are connected by the mapping $q^\sharp$ from* (19.14)*, as follows:*

$$q^\sharp \circ d_\nabla = d_\omega \circ q^\sharp : \Omega(M, P[W]) \to \Omega_{hor}(P, W)^G.$$

Proof. Let us consider first $f \in \Omega^0_{hor}(P, W)^G = C^\infty(P, W)^G$; then $f = q^\sharp s$ for $s \in \Gamma(P[W])$ and we have $f(u) = \tau^W(u, s(p(u)))$ and $s(p(u)) = q(u, f(u))$ by (19.14) and (18.12). Therefore we have $Ts.Tp.X_u = Tq(X_u, Tf.X_u)$, where $Tf.X_u = (f(u), df(X_u)) \in TW = W \times W$. If $\chi : TP \to HP$ is the horizontal projection as in (19.5), we have $Ts.Tp.X_u = Ts.Tp.\chi.X_u = Tq(\chi.X_u, Tf.\chi.X_u)$. So we get

$$\begin{aligned}(q^\sharp d_\nabla s)(X_u) &= \tau^W(u, (d_\nabla s)(Tp.X_u))\\ &= \tau^W(u, \nabla_{Tp.X_u} s) && \text{by (19.13.1)}\\ &= \tau^W(u, K.Ts.Tp.X_u) && \text{by (19.12.1)}\\ &= \tau^W\big(u, K.Tq(\chi.X_u, Tf.\chi.X_u)\big) && \text{from above}\\ &= \tau^W\big(u, \mathrm{pr}_2 . \mathrm{vl}^{-1}_{P[W]} .\bar{\Phi}.Tq(\chi.X_u, Tf.\chi.X_u)\big) && \text{by (19.10)}\\ &= \tau^W\big(u, \mathrm{pr}_2 . \mathrm{vl}^{-1}_{P[W]} .Tq.(\Phi \times Id)(\chi.X_u, Tf.\chi.X_u)\big) && \text{by (19.8)}\\ &= \tau^W\big(u, \mathrm{pr}_2 . \mathrm{vl}^{-1}_{P[W]} .Tq(0_u, Tf.\chi.X_u)\big) && \text{since } \Phi.\chi = 0\\ &= \tau^W\big(u, q. \mathrm{pr}_2 . \mathrm{vl}^{-1}_{P\times W} .(0_u, Tf.\chi.X_u)\big) && \text{since } q \text{ is fiber linear}\\ &= \tau^W(u, q(u, df.\chi.X_u)) = (\chi^* df)(X_u) = (d_\omega q^\sharp s)(X_u).\end{aligned}$$

Now we turn to the general case. It suffices to check the formula for a decomposable $P[W]$-valued form $\Psi = \psi \otimes s \in \Omega^k(M, P[W])$, where $\psi \in \Omega^k(M)$ and $s \in \Gamma(P[W])$. Then we have

$$\begin{aligned}
d_\omega q^\sharp(\psi \otimes s) &= d_\omega(p^*\psi \cdot q^\sharp s) \\
&= d_\omega(p^*\psi) \cdot q^\sharp s + (-1)^k \chi^* p^*\psi \wedge d_\omega q^\sharp s && \text{by (19.5.1)} \\
&= \chi^* p^* d\psi \cdot q^\sharp s + (-1)^k p^*\psi \wedge q^\sharp d_\nabla s && \text{from above and (19.5.4)} \\
&= p^* d\psi \cdot q^\sharp s + (-1)^k p^*\psi \wedge q^\sharp d_\nabla s \\
&= q^\sharp (d\psi \otimes s + (-1)^k \psi \wedge d_\nabla s) \\
&= q^\sharp d_\nabla(\psi \otimes s). \quad \square
\end{aligned}$$

19.16. Corollary. *In the situation of theorem* (19.15), *the curvature* $R^{P[W]} \in \Omega^2(M, L(P[W], P[W]))$ *is related to the Lie algebra valued curvature form* $\Omega \in \Omega^2_{hor}(P, \mathfrak{g})$ *by*

$$q^\sharp_{L(P[W],P[W])} R^{P[W]} = \rho' \circ \Omega,$$

where $\rho' = T_e\rho : \mathfrak{g} \to L(W, W)$ *is the derivative of the representation* ρ.

Proof. We use the notation of the proof of theorem (19.15). By this theorem we have for $X, Y \in T_uP$

$$\begin{aligned}
(d_\omega d_\omega q^\sharp_{P[W]} s)_u(X, Y) &= (q^\sharp d_\nabla d_\nabla s)_u(X, Y) \\
&= (q^\sharp R^{P[W]} s)_u(X, Y) \\
&= \tau^W(u, R^{P[W]}(T_u p.X, T_u p.Y) s(p(u))) \\
&= (q^\sharp_{L(P[W],P[W])} R^{P[W]})_u(X, Y)(q^\sharp_{P[W]} s)(u).
\end{aligned}$$

On the other hand we have by theorem (19.5.8)

$$\begin{aligned}
(d_\omega d_\omega q^\sharp s)_u(X, Y) &= (\chi^* i_R d q^\sharp s)_u(X, Y) \\
&= (dq^\sharp s)_u(R(X, Y)) && \text{since } R \text{ is horizontal} \\
&= (dq^\sharp s)(-\zeta_{\Omega(X,Y)}(u)) && \text{by (19.2)} \\
&= \partial|_0 (q^\sharp s)(\mathrm{Fl}^{\zeta_{\Omega(X,Y)}}_{-t}(u)) \\
&= \partial|_0 \tau^W(u.\exp(-t\Omega(X, Y)), s(p(u.\exp(-t\Omega(X, Y))))) \\
&= \partial|_0 \tau^W(u.\exp(-t\Omega(X, Y)), s(p(u))) \\
&= \partial|_0 \rho(\exp t\Omega(X, Y)) \tau^W(u, s(p(u))) && \text{by (18.7)} \\
&= \rho'(\Omega(X, Y))(q^\sharp s)(u). \quad \square
\end{aligned}$$

20. Characteristic Classes

20.1. Invariants of Lie algebras. Let G be a Lie group with Lie algebra $\mathfrak{g}$; let $\bigotimes \mathfrak{g}^*$ be the tensor algebra over the dual space $\mathfrak{g}^*$, the graded space of all multilinear real (or complex) functionals on $\mathfrak{g}$. Let $S(\mathfrak{g}^*)$ be the symmetric algebra over $\mathfrak{g}^*$ which corresponds to the algebra of polynomial functions on $\mathfrak{g}$. The adjoint representation $\operatorname{Ad} : G \to L(\mathfrak{g}, \mathfrak{g})$ induces representations $\operatorname{Ad}^* : G \to L(\bigotimes \mathfrak{g}^*, \bigotimes \mathfrak{g}^*)$ and also $\operatorname{Ad}^* : G \to L(S(\mathfrak{g}^*), S(\mathfrak{g}^*))$, which are both given by $\operatorname{Ad}^*(g)f = f \circ (\operatorname{Ad}(g^{-1}) \otimes \cdots \otimes \operatorname{Ad}(g^{-1}))$. A tensor $f \in \bigotimes \mathfrak{g}^*$ (or a polynomial $f \in S(\mathfrak{g}^*)$) is called an *invariant of the Lie algebra* if $\operatorname{Ad}^*(g)f = f$ for all $g \in G$. If the Lie group G is connected, f is an invariant if and only if $\mathcal{L}_X f = 0$ for all $X \in \mathfrak{g}$, where $\mathcal{L}_X$ is the restriction of the Lie derivative to left invariant tensor fields on G, which coincides with the unique extension of $\operatorname{ad}(X)^* : \mathfrak{g}^* \to \mathfrak{g}^*$ to a derivation on $\bigotimes \mathfrak{g}^*$ or $S(\mathfrak{g}^*)$, respectively. Compare this with the proof of (14.16.2). Obviously the space of all invariants is a graded subalgebra of $\bigotimes \mathfrak{g}^*$ or $S(\mathfrak{g}^*)$, respectively. The usual notation for the algebra of invariant polynomials is

$$I(G) := \bigoplus_{k\geq 0} I^k(G) = S(\mathfrak{g}^*)^G = \bigoplus_{k\geq 0} S^k(\mathfrak{g}^*)^G.$$

20.2. The Chern-Weil forms. Let (P, p, M, G) be a principal fiber bundle with principal connection $\Phi = \zeta \circ \omega$ and curvature $R = \zeta \circ \Omega$. For $\psi_i \in \Omega^{p_i}(P, \mathfrak{g})$ and $f \in S^k(\mathfrak{g}^*) \subset \bigotimes^k \mathfrak{g}^*$ we have the differential forms

$$\begin{aligned} \psi_1 \otimes_\wedge \cdots \otimes_\wedge \psi_k &\in \Omega^{p_1+\cdots+p_k}(P, \mathfrak{g} \otimes \cdots \otimes \mathfrak{g}), \\ f \circ (\psi_1 \otimes_\wedge \cdots \otimes_\wedge \psi_k) &\in \Omega^{p_1+\cdots+p_k}(P). \end{aligned}$$

The exterior derivative of the latter one is clearly given by

$$\begin{aligned} d(f \circ (\psi_1 \otimes_\wedge \cdots \otimes_\wedge \psi_k)) &= f \circ d(\psi_1 \otimes_\wedge \cdots \otimes_\wedge \psi_k) \\ &= f \circ \left(\textstyle\sum_{i=1}^k (-1)^{p_1+\cdots+p_{i-1}} \psi_1 \otimes_\wedge \cdots \otimes_\wedge d\psi_i \otimes_\wedge \cdots \otimes_\wedge \psi_k\right). \end{aligned}$$

Let us now consider an invariant polynomial $f \in I^k(G)$ and the curvature form $\Omega \in \Omega^2_{\text{hor}}(P, \mathfrak{g})^G$. Then the $2k$-form $f \circ (\Omega \otimes_\wedge \cdots \otimes_\wedge \Omega)$ is horizontal since Ω is horizontal by (19.2.2). It is also G-invariant since by (19.2.2) we have

$$\begin{aligned} (r^g)^*(f \circ (\Omega \otimes_\wedge \cdots \otimes_\wedge \Omega)) &= f \circ ((r^g)^*\Omega \otimes_\wedge \cdots \otimes_\wedge (r^g)^*\Omega) \\ &= f \circ (\operatorname{Ad}(g^{-1})\Omega \otimes_\wedge \cdots \otimes_\wedge \operatorname{Ad}(g^{-1})\Omega) \\ &= f \circ (\Omega \otimes_\wedge \cdots \otimes_\wedge \Omega). \end{aligned}$$

So by theorem (19.14) there is a uniquely defined $2k$-form $\operatorname{cw}(f, P, \omega) \in \Omega^{2k}(M)$ with $p^* \operatorname{cw}(f, P, \omega) = f \circ (\Omega \otimes_\wedge \cdots \otimes_\wedge \Omega)$, which we will call the *Chern-Weil form* of f.

If $h : N \to M$ is a smooth mapping, then for the pullback bundle h^*P the Chern-Weil form is given by $\operatorname{cw}(f, h^*P, h^*\omega) = h^* \operatorname{cw}(f, P, \omega)$, which is easily seen by applying p^*.

20.3. Theorem (Chern-Weil homomorphism). *In the setting of* (20.2) *we have:*

(1) *For $f \in I^k(G)$ the Chern-Weil form $\operatorname{cw}(f, P, \omega)$ is a closed differential form: $d \operatorname{cw}(f, P, \omega) = 0$. So there is a well defined cohomology class $\operatorname{Cw}(f, P) = [\operatorname{cw}(f, P, \omega)] \in H^{2k}(M)$, called the characteristic class of the invariant polynomial f.*

(2) *The characteristic class $\operatorname{Cw}(f, P)$ does not depend on the choice of the principal connection ω.*

(3) *The mapping $\operatorname{Cw}_P : I^*(G) \to H^{2*}(M)$ is a homomorphism of commutative algebras, and it is called the Chern-Weil homomorphism.*

(4) *If $h : N \to M$ is a smooth mapping, then the Chern-Weil homomorphism for the pullback bundle h^*P is given by*

$$\operatorname{Cw}_{h^*P} = h^* \circ \operatorname{Cw}_P : I^*(G) \to H^{2*}(N).$$

Proof. (1) Since $f \in I^k(G)$ is invariant, we have for any $X \in \mathfrak{g}$

$$\begin{aligned}
0 &= \tfrac{d}{dt}|_0 \operatorname{Ad}(\exp(tX_0))^* f(X_1, \dots, X_k) \\
&= \tfrac{d}{dt}|_0 f(\operatorname{Ad}(\exp(tX_0))X_1, \dots, \operatorname{Ad}(\exp(tX_0))X_k) \\
&= \textstyle\sum_{i=1}^k f(X_1, \dots, [X_0, X_i], \dots, X_k) \\
&= \textstyle\sum_{i=1}^k f([X_0, X_i], X_1, \dots, \widehat{X_i} \dots, X_k).
\end{aligned}$$

This implies that

$$\begin{aligned}
d(f \circ (\Omega \otimes_\wedge \dots \otimes_\wedge \Omega)) &= f \circ \left(\textstyle\sum_{i=1}^k \Omega \otimes_\wedge \dots \otimes_\wedge d\Omega \otimes_\wedge \dots \otimes_\wedge \Omega \right) \\
&= k\, f \circ (d\Omega \otimes_\wedge \dots \otimes_\wedge \Omega) + k\, f \circ ([\omega, \Omega]_\wedge \otimes_\wedge \dots \otimes_\wedge \Omega) \\
&= k\, f \circ (d_\omega \Omega \otimes_\wedge \Omega \otimes_\wedge \dots \otimes_\wedge \Omega) = 0 \quad \text{by (19.5.6)}, \\
p^* d \operatorname{cw}(f, P, \omega) &= d\, p^* \operatorname{cw}(f, P, \omega) \\
&= d\, (f \circ (\Omega \otimes_\wedge \dots \otimes_\wedge \Omega)) = 0,
\end{aligned}$$

and thus $d \operatorname{cw}(f, P, \omega) = 0$ since p^* is injective.

(2) Let $\omega_0, \omega_1 \in \Omega^1(P, \mathfrak{g})^G$ be two principal connections. Then we consider the principal bundle $(P \times \mathbb{R}, p \times Id, M \times \mathbb{R}, G)$ and the principal connection $\tilde{\omega} = (1 - t)\omega_0 + t\omega_1 = (1 - t)(\operatorname{pr}_1)^*\omega_0 + t(\operatorname{pr}_1)^*\omega_1$ on it, where t is the coordinate function on $\mathbb{R}$. Let $\tilde{\Omega}$ be the curvature form of $\tilde{\omega}$. Let $\operatorname{ins}_s : P \to P \times \mathbb{R}$ be the embedding at level s, $\operatorname{ins}_s(u) = (u, s)$. Then we have in turn

by (19.2.3) for $s = 0, 1$

$$\begin{aligned}\omega_s &= (\mathrm{ins}_s)^*\tilde{\omega},\\ \Omega_s &= d\omega_s + \tfrac{1}{2}[\omega_s, \omega_s]_\wedge\\ &= d(\mathrm{ins}_s)^*\tilde{\omega} + \tfrac{1}{2}[(\mathrm{ins}_s)^*\tilde{\omega}, (\mathrm{ins}_s)^*\tilde{\omega}]_\wedge\\ &= (\mathrm{ins}_s)^*(d\tilde{\omega} + \tfrac{1}{2}[\tilde{\omega}, \tilde{\omega}]_\wedge)\\ &= (\mathrm{ins}_s)^*\tilde{\Omega}.\end{aligned}$$

So we get for $s = 0, 1$

$$\begin{aligned}p^*(\mathrm{ins}_s)^* \operatorname{cw}(f, P \times \mathbb{R}, \tilde{\omega}) &= (\mathrm{ins}_s)^*(p \times Id_{\mathbb{R}})^* \operatorname{cw}(f, P \times \mathbb{R}, \tilde{\omega})\\ &= (\mathrm{ins}_s)^*(f \circ (\tilde{\Omega} \otimes_\wedge \cdots \otimes_\wedge \tilde{\Omega}))\\ &= f \circ ((\mathrm{ins}_s)^*\tilde{\Omega} \otimes_\wedge \cdots \otimes_\wedge (\mathrm{ins}_s)^*\tilde{\Omega})\\ &= f \circ (\Omega_s \otimes_\wedge \cdots \otimes_\wedge \Omega_s)\\ &= p^* \operatorname{cw}(f, P, \omega_s).\end{aligned}$$

Since p^* is injective, we get $(\mathrm{ins}_s)^* \operatorname{cw}(f, P\times\mathbb{R}, \tilde{\omega}) = \operatorname{cw}(f, P, \omega_s)$ for $s = 0, 1$, and since ins_0 and ins_1 are smoothly homotopic, the cohomology classes coincide.

(3) and (4) are obvious. □

20.4. Local description of characteristic classes. Let (P, p, M, G) be a principal fiber bundle with a principal connection $\omega \in \Omega^1(P, \mathfrak{g})^G$. Let $s_\alpha \in \Gamma(P|U_\alpha)$ be a collection of local smooth sections of the bundle such that (U_α) is an open cover of M. Recall (from the proof of (18.3) for example) that then $\varphi_\alpha = (p, \tau^G(s_\alpha \circ p, \quad)) : P|U_\alpha \to U_\alpha \times G$ is a principal fiber bundle atlas with transition functions $\varphi_{\alpha\beta}(x) = \tau^G(s_\alpha(x), s_\beta(x))$.

Then we consider the physicists' version from (19.4) of the connection ω which is described by the forms $\omega_\alpha := s_\alpha^*\omega \in \Omega^1(U_\alpha, \mathfrak{g})$. They transform according to $\omega_\alpha = \mathrm{Ad}(\varphi_{\beta\alpha}^{-1})\omega_\beta + \Theta_{\beta\alpha}$, where $\Theta_{\beta\alpha} = \varphi_{\beta\alpha}^{-1}d\varphi_{\alpha\beta}$ if G is a matrix group; see lemma (19.4). This affine transformation law is due to the fact that ω is not horizontal. Let $\Omega = d\omega + \frac{1}{2}[\omega, \omega]_\wedge \in \Omega^2_{\mathrm{hor}}(P, \mathfrak{g})^G$ be the curvature of ω; then we consider again the local forms of the curvature:

$$\begin{aligned}\Omega_\alpha :&= s_\alpha^*\Omega = s_\alpha^*(d\omega + \tfrac{1}{2}[\omega, \omega]_\wedge)\\ &= d(s_\alpha^*\omega) + \tfrac{1}{2}[s_\alpha^*\omega, s_\alpha^*\omega]_\wedge\\ &= d\omega_\alpha + \tfrac{1}{2}[\omega_\alpha, \omega_\alpha]_\wedge.\end{aligned}$$

Recall from theorem (19.14) that we have an isomorphism

$$q^\sharp : \Omega(M, P[\mathfrak{g}, \mathrm{Ad}]) \to \Omega_{\mathrm{hor}}(P, \mathfrak{g})^G.$$

Then $\Omega_\alpha = s_\alpha^*\Omega$ is the local frame expression of $(q^\sharp)^{-1}(\Omega)$ for the induced chart $P[\mathfrak{g}]|U_\alpha \to U_\alpha \times \mathfrak{g}$; thus we have the the simple transformation formula $\Omega_\alpha = \operatorname{Ad}(\varphi_{\alpha\beta})\Omega_\beta$.

If now $f \in I^k(G)$ is an invariant of G, for the Chern-Weil form $\operatorname{cw}(f, P, \omega)$ we have

$$\begin{aligned} \operatorname{cw}(f,P,\omega)|U_\alpha :&= s_\alpha^*(p^* \operatorname{cw}(f,P,\omega)) = s_\alpha^*(f \circ (\Omega \otimes_\wedge \cdots \otimes_\wedge \Omega)) \\ &= f \circ (s_\alpha^*\Omega \otimes_\wedge \cdots \otimes_\wedge s_\alpha^*\Omega) \\ &= f \circ (\Omega_\alpha \otimes_\wedge \cdots \otimes_\wedge \Omega_\alpha), \end{aligned}$$

where $\Omega_\alpha \otimes_\wedge \cdots \otimes_\wedge \Omega_\alpha \in \Omega^{2k}(U_\alpha, \mathfrak{g} \otimes \cdots \otimes \mathfrak{g})$.

20.5. Characteristic classes for vector bundles. For a real vector bundle $(E, p, M, \mathbb{R}^n)$ the characteristic classes are by definition the characteristic classes of the linear frame bundle $(GL(\mathbb{R}^n, E), p, M, GL(n, \mathbb{R}))$. We write $\operatorname{Cw}(f, E) := \operatorname{Cw}(f, GL(\mathbb{R}^n, E))$ for short and likewise for complex vector bundles.

Let (P, p, M, G) be a principal bundle and let $\rho : G \to GL(V)$ be a representation in a finite-dimensional vector space. If ω is a principal connection form on P with curvature form Ω, then for the induced covariant derivative ∇ on the associated vector bundle $P[V]$ and its curvature $R^{P[V]}$ we have $q^\sharp R^{P[V]} = \rho' \circ \Omega$ by corollary (19.16). So if the representation ρ is infinitesimally effective, i.e., if $\rho' : \mathfrak{g} \to L(V, V)$ is injective, then we see that actually $R^{P[V]} \in \Omega^2(M, P[\mathfrak{g}])$. If $f \in I^k(G)$ is an invariant, then we have the induced mapping

$$\begin{array}{ccc} P \times (\bigotimes^k \mathfrak{g}) & \xrightarrow{Id_P \times f} & P \times \mathbb{R} \\ \Big\downarrow{\scriptstyle q} & & \Big\downarrow{\scriptstyle q} \\ P[\bigotimes^k \mathfrak{g}] & \xrightarrow{P[f]} & M \times \mathbb{R}. \end{array}$$

So the Chern-Weil form can also be written as (omitting $P[(\rho')^{-1}]$)

$$\operatorname{cw}(f, P, \omega) = P[f] \circ (R^{P[V]} \otimes_\wedge \cdots \otimes_\wedge R^{P[V]}).$$

Sometimes we will make use of this expression.

All characteristic classes for a trivial vector bundle are zero, since the frame bundle is then trivial and admits a principal connection with curvature 0.

We will determine the classical bases for the algebra of invariants for the matrix groups $GL(n, \mathbb{R})$, $GL(n, \mathbb{C})$, $O(n, \mathbb{R})$, $SO(n, \mathbb{R})$, $U(n)$, and we will discuss the resulting characteristic classes for vector bundles.

20.6. The characteristic coefficients. . For a matrix $A \in \mathfrak{gl}(n, \mathbb{R}) = L(\mathbb{R}^n, \mathbb{R}^n)$ we consider the characteristic coefficients $c_k^n(A)$ which are given by the implicit equation

$$\det(A + t\mathbb{I}) = \sum_{k=0}^{n} c_k^n(A).t^{n-k}. \tag{1}$$

From lemma (14.19) we have

$$c_k^n(A) = \operatorname{Trace}(\bigwedge^k A : \bigwedge^k \mathbb{R}^n \to \bigwedge^k \mathbb{R}^n).$$

The characteristic coefficient c_k^n is a homogeneous invariant polynomial of degree k, since we have

$$\det(\operatorname{Ad}(g)A + t\mathbb{I}) = \det(gAg^{-1} + t\mathbb{I}) = \det(g(A + t\mathbb{I})g^{-1}) = \det(A + t\mathbb{I}).$$

Lemma. *We have*

$$c_k^{n+m}\left(\begin{pmatrix} A & 0 \\ 0 & B \end{pmatrix}\right) = \sum_{j=0}^{k} c_j^n(A) c_{k-j}^m(B).$$

Proof. We have

$$\begin{aligned}
\det\left(\begin{pmatrix} A & 0 \\ 0 & B \end{pmatrix} + t\mathbb{I}_{n+m}\right) &= \det(A + t\mathbb{I}_n)\det(B + t\mathbb{I}_m) \\
&= \left(\sum_{k=0}^{n} c_k^n(A) t^{n-k}\right)\left(\sum_{j=0}^{m} c_j^m(A) t^{m-l}\right) \\
&= \sum_{k=0}^{n+m}\left(\sum_{j=0}^{k} c_j^n(A) c_{k-j}^m(B)\right) t^{n+m-k}. \quad \square
\end{aligned}$$

20.7. Pontryagin classes. Let (E, p, M) be a real vector bundle. Then the *Pontryagin classes* are given by

$$\begin{aligned}
p_k(E) &:= \left(\frac{-1}{2\pi\sqrt{-1}}\right)^{2k} \operatorname{Cw}(c_{2k}^{\dim E}, E) \in H^{4k}(M; \mathbb{R}), \\
p_0(E) &:= 1 \in H^0(M; \mathbb{R}).
\end{aligned}$$

The factor $\frac{-1}{2\pi\sqrt{-1}}$ makes this class to be an integer class (in $H^{4k}(M, \mathbb{Z})$) and makes several integral formulas (like the Gauß-Bonnet-Chern formula) more beautiful. In principle one should always replace the curvature Ω by $\frac{-1}{2\pi\sqrt{-1}}\Omega$. The inhomogeneous cohomology class

$$p(E) := \sum_{k \geq 0} p_k(E) \in H^{4*}(M, \mathbb{R})$$

is called the *total Pontryagin class.*

Theorem. *For the Pontryagin classes we have:*

(1) *If E_1 and E_2 are two real vector bundles over a manifold M, then for the fiberwise direct sum we have*

$$p(E_1 \oplus E_2) = p(E_1) \wedge p(E_2) \in H^{4*}(M, \mathbb{R}).$$

(2) *For the pullback of a vector bundle along $f : N \to M$ we have*

$$p(f^*E) = f^*p(E).$$

(3) *For a real vector bundle and an invariant $f \in I^k(GL(n,\mathbb{R}))$ for odd k we have $\mathrm{Cw}(f,E) = 0$. Thus the Pontryagin classes exist only in dimension $0, 4, 8, 12, \dots$.*

Proof. (1) If $\omega^i \in \Omega^1(GL(\mathbb{R}^{n_i}, E_i), \mathfrak{gl}(n_i))^{GL(n_i)}$ are principal connection forms for the frame bundles of the two vector bundles, then for local frames of the two bundles $s^i_\alpha \in \Gamma(GL(\mathbb{R}^{n_i}, E_i|U_\alpha)$, the forms

$$\omega_\alpha := \begin{pmatrix} \omega^1_\alpha & 0 \\ 0 & \omega^2_\alpha \end{pmatrix} \in \Omega^1(U_\alpha, \mathfrak{gl}(n_1 + n_2))$$

are exactly the local expressions of the direct sum connection, and from lemma (20.6) we see that $p_k(E_1 \oplus E_2) = \sum_{j=0}^k p_j(E_1)p_{k-j}(E_2)$ holds, which implies the desired result.

(2) This follows from (20.3.4).

(3) Choose a fiber Riemann metric g on E, consider the corresponding orthonormal frame bundle $(O(\mathbb{R}^n, E), p, M, O(n,\mathbb{R}))$, and choose a principal connection ω for it. Then the local expression with respect to local orthonormal frame fields s_α are skew-symmetric matrices of 1-forms. So the local curvature forms are also skew-symmetric. As we will show shortly, there exists a matrix $C \in O(n,\mathbb{R})$ such that $CAC^{-1} = A^\top = -A$ for any real skew-symmetric matrix; thus $C\Omega_\alpha C^{-1} = -\Omega_\alpha$. But then

$$\begin{aligned} f \circ (\Omega_\alpha \otimes_\wedge \cdots \otimes_\wedge \Omega_\alpha) &= f \circ (g_\alpha \Omega_\alpha g_\alpha^{-1} \otimes_\wedge \cdots \otimes_\wedge g_\alpha \Omega_\alpha g_\alpha^{-1}) \\ &= f \circ ((-\Omega_\alpha) \otimes_\wedge \cdots \otimes_\wedge (-\Omega_\alpha)) \\ &= (-1)^k f \circ (\Omega_\alpha \otimes_\wedge \cdots \otimes_\wedge \Omega_\alpha). \end{aligned}$$

This implies that $\mathrm{Cw}(f,E) = 0$ if k is odd.

Claim. There exists a matrix $C \in O(n,\mathbb{R})$ such that $CAC^{-1} = A^\top$ for each real matrix with 0's on the main diagonal.

Note first that

$$\begin{pmatrix} 0 & 1 \\ 1 & 0 \end{pmatrix} \begin{pmatrix} a & b \\ c & d \end{pmatrix} \begin{pmatrix} 0 & 1 \\ 1 & 0 \end{pmatrix} = \begin{pmatrix} d & b \\ c & a \end{pmatrix}.$$

Let E_{ij} be the matrix which has 1 in the position (i,j) in the i-th row and j-th column. Then the (ij)-transposition matrix $P_{ij} = \mathbb{I}_n - E_{ii} - E_{jj} + E_{ij} + E_{ji}$ acts by conjugation on an arbitrary matrix A by exchanging the pair A_{ij} and A_{ji} and also exchanging the pair A_{ii} and A_{jj} on the main diagonal. So the product $C = \prod_{i<j} P_{ij}$ has the required effect on a matrix with zeros on the main diagonal.

By the way, $\operatorname{Ad}(C)$ acts on the main diagonal via the longest element in the permutation group, with respect to the canonical system of positive roots in $\mathfrak{sl}(n)$:

$$\begin{pmatrix} 1 & 2 & \dots & n-1 & n \\ n & n-1 & \dots & 2 & 1 \end{pmatrix}. \quad \square$$

20.8. Remarks. (1) If two vector bundles E and F are *stably equivalent*, i.e., $E \oplus (M \times \mathbb{R}^m) \cong F \oplus (M \times \mathbb{R}^n)$ for some m and n, then $p(E) = p(F)$. This follows from (20.7.1) and (20.7.2).

(2) If for a vector bundle E for some k the bundle $\overbrace{E \oplus \dots \oplus E}^{k} \oplus (M \times \mathbb{R}^l)$ is trivial, then $p(E) = 1$ since $p(E)^k = 1$.

(3) Let (E, p, M) be a vector bundle over a compact oriented manifold M. For $j_i \in \mathbb{N}_0$ we put

$$\lambda_{j_1,\dots,j_r}(E) := \int_M p_1(E)^{j_1} \dots p_r(E)^{j_r} \in \mathbb{R},$$

where the integral is set to be 0 on each degree which is not equal to $\dim M$. Then these *Pontryagin numbers* are indeed integers; see [**157**]. For example we have

$$\lambda_{j_1,\dots,j_r}(T(\mathbb{C}P^n)) = \binom{2n+1}{j_1} \dots \binom{2n+1}{j_r}.$$

20.9. The trace coefficients. For a matrix $A \in \mathfrak{gl}(n,\mathbb{R}) = L(\mathbb{R}^n, \mathbb{R}^n)$ the *trace coefficients* are given by

$$\operatorname{tr}_k^n(A) := \operatorname{Trace}(A^k) = \operatorname{Trace}(\overbrace{A \circ \dots \circ A}^{k}).$$

Obviously tr_k^n is an invariant polynomial, homogeneous of degree k. To a direct sum of two matrices $A \in \mathfrak{gl}(n)$ and $B \in \mathfrak{gl}(m)$ it reacts clearly by

$$\operatorname{tr}_k^{n+m}\begin{pmatrix} A & 0 \\ 0 & B \end{pmatrix} = \operatorname{Trace}\begin{pmatrix} A^k & 0 \\ 0 & B^k \end{pmatrix} = \operatorname{tr}_k^n(A) + \operatorname{tr}_k^m(B).$$

The tensor product (sometimes also called the Kronecker product) of A and B is given by $A \otimes B = (A^i_j B^k_l)_{(i,k),(j,l) \in n \times m}$ in terms of the canonical bases.

Since we have $\operatorname{Trace}(A \otimes B) = \sum_{i,k} A^i_i B^k_k = \operatorname{Trace}(A) \operatorname{Trace}(B)$, we also get

$$\begin{aligned}\operatorname{tr}^{nm}_k(A \otimes B) &= \operatorname{Trace}((A \otimes B)^k) = \operatorname{Trace}(A^k \otimes B^k) = \operatorname{Trace}(A^k) \operatorname{Trace}(B^k)\\ &= \operatorname{tr}^n_k(A) \operatorname{tr}^m_k(B).\end{aligned}$$

Lemma. *The trace coefficients and the characteristic coefficients are connected by the following recursive equation:*

$$c^n_k(A) = \tfrac{1}{k} \sum_{j=0}^{k-1} (-1)^{k-j-1} c^n_j(A) \operatorname{tr}^n_{k-j}(A).$$

Proof. For a matrix $A \in \mathfrak{gl}(n)$ let us denote by $C(A)$ the matrix of the signed algebraic complements of A (also called the classical adjoint) as in (4.33). Then Cramer's rule reads

$$A.C(A) = C(A).A = \det(A).\mathbb{I}, \tag{1}$$

and the derivative of the determinant is given by (4.33):

$$d\det(A)X = \operatorname{Trace}(C(A)X). \tag{2}$$

Note that $C(A)$ is a homogeneous matrix valued polynomial of degree $n-1$ in A. We define now matrix valued polynomials $a_k(A)$ by

$$C(A + t\mathbb{I}) = \sum_{k=0}^{n-1} a_k(A) t^{n-k-1}. \tag{3}$$

We claim that for $A \in \mathfrak{gl}(n)$ and $k = 0, 1, \dots, n-1$ we have

$$a_k(A) = \sum_{j=0}^{k} (-1)^j c^n_{k-j}(A) A^j. \tag{4}$$

We prove this in the following way: From (1) we have

$$(A + t\mathbb{I})C(A + t\mathbb{I}) = \det(A + t\mathbb{I})\mathbb{I},$$

and we insert (3) and (20.6.1) to get in turn

$$(A + t\mathbb{I}) \sum_{k=0}^{n-1} a_k(A) t^{n-k-1} = \sum_{j=0}^{n} c^n_j(A) t^{n-j} \mathbb{I},$$

$$\sum_{k=0}^{n-1} A.a_k(A) t^{n-k-1} + \sum_{k=0}^{n-1} a_k(A) t^{n-k} = \sum_{j=0}^{n} c^n_j(A) t^{n-j} \mathbb{I}.$$

We put $a_{-1}(A) := 0 =: a_n(A)$ and compare coefficients of t^{n-k} in the last equation to get the recursion formula

$$A.a_{k-1}(A) + a_k(A) = c^n_k(A).\mathbb{I}$$

which immediately leads to the desired formula (4), even for $k = 0, 1, \dots, n$. If we start this computation with the two factors in (1) reversed, we get $A.a_k(A) = a_k(A).A$. Note that (4) for $k = n$ is exactly the *Caley-Hamilton equation*

$$0 = a_n(A) = \sum_{j=0}^{n} c^n_{n-j}(A)A^j.$$

We claim that

$$\text{(5)} \qquad \operatorname{Trace}(a_k(A)) = (n-k)c^n_k(A).$$

We use (2) for the proof:

$$\begin{aligned}
\partial|_0(\det(A + t\mathbb{I})) &= d\det(A+t\mathbb{I})\partial|_0(A+t\mathbb{I}) = \operatorname{Trace}(C(A+t\mathbb{I})\mathbb{I}) \\
&= \operatorname{Trace}\left(\sum_{k=0}^{n-1} a_k(A)t^{n-k-1}\right) = \sum_{k=0}^{n-1}\operatorname{Trace}(a_k(A))t^{n-k-1}, \\
\partial|_0(\det(A + t\mathbb{I})) &= \partial|_0\left(\sum_{k=0}^{n} c^n_k(A)t^{n-k}\right) \\
&= \sum_{k=0}^{n}(n-k)c^n_k(A)t^{n-k-1}.
\end{aligned}$$

Comparing coefficients leads to the result (5).

Now we may prove the lemma itself by the following computation:

$$\begin{aligned}
(n-k)c^n_k(A) &= \operatorname{Trace}(a_k(A)) \qquad \text{by (5)} \\
&= \operatorname{Trace}\left(\sum_{j=0}^{k}(-1)^j c^n_{k-j}(A)A^j\right) \qquad \text{by (4)} \\
&= \sum_{j=0}^{k}(-1)^j c^n_{k-j}(A)\operatorname{Trace}(A^j) \\
&= n\,c^n_k(A) + \sum_{j=1}^{k}(-1)^j c^n_{k-j}(A)\operatorname{tr}^n_j(A), \\
c^n_k(A) &= -\tfrac{1}{k}\sum_{j=1}^{k}(-1)^j c^n_{k-j}(A)\operatorname{tr}^n_j(A) \\
&= \tfrac{1}{k}\sum_{j=0}^{k-1}(-1)^{k-j-1}c^n_j(A)\operatorname{tr}^n_{k-j}(A). \qquad \square
\end{aligned}$$

20.10. The trace classes. Let (E, p, M) be a real vector bundle. Then the trace classes are given by

$$\operatorname{tr}_k(E) := \left(\frac{-1}{2\pi\sqrt{-1}}\right)^{2k} \operatorname{Cw}(\operatorname{tr}_{2k}^{\dim E}, E) \in H^{4k}(M, \mathbb{R}). \tag{1}$$

Between the trace classes and the Pontryagin classes there are the following relations for $k \geq 1$

$$p_k(E) = \frac{-1}{2k} \sum_{j=0}^{k-1} p_j(E) \wedge \operatorname{tr}_{k-j}(E), \tag{2}$$

which follows directly from lemma (20.9) above.

The inhomogeneous cohomology class

$$\operatorname{tr}(E) = \sum_{k=0}^{\infty} \frac{1}{(2k)!} \operatorname{tr}_k(E) = \operatorname{Cw}(\operatorname{Trace} \circ \exp, E) \tag{3}$$

is called the *Pontryagin character* of E. In the second expression we use the smooth invariant function $\operatorname{Trace} \circ \exp : \mathfrak{gl}(n) \to \mathbb{R}$ which is given by

$$\operatorname{Trace}(\exp(A)) = \operatorname{Trace}\left(\sum_{k \geq 0} \frac{A^k}{k!}\right) = \sum_{k \geq 0} \frac{1}{k!} \operatorname{Trace}(A^k).$$

Of course one should first take the Taylor series at 0 of $\operatorname{Trace} \circ \exp$ and then take the Chern-Weil class of each homogeneous part separately.

Theorem. *Let (E_i, p, M) be vector bundles over the same base manifold M. Then we have:*

(4) $\operatorname{tr}(E_1 \oplus E_2) = \operatorname{tr}(E_1) + \operatorname{tr}(E_2)$.

(5) $\operatorname{tr}(E_1 \otimes E_2) = \operatorname{tr}(E_1) \wedge \operatorname{tr}(E_2)$.

(6) $\operatorname{tr}(g^* E) = g^* \operatorname{tr}(E)$ *for any smooth mapping $g : N \to M$.*

Clearly stably equivalent vector bundles have equal Pontryagin characters. Statements (4) and (5) say that one may view the Pontryagin character as a ring homomorphism from the real K-theory into cohomology,

$$\operatorname{tr} : K_{\mathbb{R}}(M) \to H^{4*}(M; \mathbb{R}).$$

Statement (6) says that it is even a natural transformation.

Proof. (4) This can be proved in the same way as (20.7.1), but we indicate another method which will be used also in the proof of (5) below. Covariant derivatives for E_1 and E_2 induce a covariant derivative on $E_1 \oplus E_2$ by

$\nabla_X^{E_1 \oplus E_2}(s_1, s_2) = (\nabla_X^{E_1} s_1, \nabla_X^{E_2}, s_2)$. For the curvature operators we clearly have

$$R^{E_1 \oplus E_2} = R^{E_1} \oplus R^{E_2} = \begin{pmatrix} R^{E_1} & 0 \\ 0 & R^{E_2} \end{pmatrix}.$$

So the result follows from (20.9) with the help of (20.5).

(5) We have an induced covariant derivative on $E_1 \otimes E_2$ given by $\nabla_X^{E_1 \otimes E_2} s_1 \otimes s_2 = (\nabla_X^{E_1} s_1) \otimes s_2 + s_1 \otimes (\nabla_X^{E_2} s_2)$. Then for the curvatures we get obviously $R^{E_1 \otimes E_2}(X, Y) = R^{E_1}(X, Y) \otimes Id_{E_2} + Id_{E_1} \otimes R^{E_2}(X, Y)$. The two summands of the last expression commute, so we get

$$(R^{E_1} \otimes Id_{E_2} + Id_{E_1} \otimes R^{E_2})^{\circ_\wedge, k} = \sum_{j=0}^{k} \binom{k}{j} (R^{E_1})^{\circ_\wedge, j} \otimes_\wedge (R^{E_2})^{\circ_\wedge, k-j},$$

where the product involved is given as in

$$(R^E \circ_\wedge R^E)(X_1, \dots, X_4) = \tfrac{1}{2!2!} \sum_\sigma \operatorname{sign}(\sigma) R^E(X_{\sigma 1}, X_{\sigma 2}) \circ R^E(X_{\sigma 3}, X_{\sigma 4}),$$

which makes $(\Omega(M, L(E, E)), \circ_\wedge)$ into a graded associative algebra. The next computation takes place in a commutative subalgebra of it:

$$\begin{aligned} \operatorname{tr}(E_1 \otimes E_2) &= [\operatorname{Trace} \exp(R^{E_1} \otimes Id_{E_2} + Id_{E_1} \otimes R^{E_2})]_{H(M)} \\ &= [\operatorname{Trace}(\exp(R^{E_1}) \otimes_\wedge \exp(R^{E_2}))]_{H(M)} \\ &= [\operatorname{Trace}(\exp(R^{E_1})) \wedge \operatorname{Trace}(\exp(R^{E_2}))]_{H(M)} \\ &= \operatorname{tr}(E_1) \wedge \operatorname{tr}(E_2). \end{aligned}$$

(6) This is a general fact. □

20.11. The Pfaffian. Let (V, g) be a real Euclidian vector space of dimension n, with a positive definite inner product g. Then for each p we have an induced inner product on $\bigwedge^p V$, see also (25.11), which is given by

$$\langle x_1 \wedge \dots \wedge x_p, y_1 \wedge \dots \wedge y_p \rangle_g = \det(g(x_i, y_j)_{i,j}).$$

The inner product g, when viewed as a linear isomorphism $g : V \to V^*$, induces an isomorphism $\beta : \bigwedge^2 V \to L_{g,\text{skew}}(V, V)$ which is given on decomposable forms by $\beta(x \wedge y)(z) = g(x, z)y - g(y, z)x$. We also have

$$\beta^{-1}(A) = A \circ g^{-1} \in L_{\text{skew}}(V^*, V) = \{B \in L(V^*, V) : B^\top = -B\} \cong \bigwedge^2 V,$$
$$\text{where} \quad B^\top : V^* \xrightarrow{B^*} V^{**} \xrightarrow{\cong} V.$$

Now we assume that V is of even dimension n and is oriented. Then there is a unique element $e \in \bigwedge^n V$ which is positive and normed: $\langle e, e \rangle_g = 1$. We define the *Pfaffian* of a skew-symmetric matrix A by:

$$\operatorname{Pf}^g(A) := \frac{1}{n!}\langle e, \overbrace{\beta^{-1}(A)\wedge\dots\wedge\beta^{-1}(A)}^{n/2}\rangle_g, \qquad A\in\mathfrak{so}(n,\mathbb{R}).$$

This is a homogeneous polynomial of degree $n/2$ on $\mathfrak{so}(n,\mathbb{R})$. Its polarization is the $n/2$-linear symmetric functional

$$\operatorname{Pf}^g(A_1,\dots,A_{n/2}) = \frac{1}{n!}\langle e, \beta^{-1}(A_1)\wedge\dots\wedge\beta^{-1}(A_{n/2})\rangle_g.$$

Lemma. *For an even-dimensional oriented Euclidean vector space (V,g) and skew-symmetric A we have:*

(1) *For $B\in L(V,V)$ we have $\operatorname{Pf}^g(B.A.B^\top) = \det(B)\operatorname{Pf}(A)$ where $B^\top$ is the transpose with respect to g.*
(2) *If $U\in O(V,g)$, then $\operatorname{Pf}^g(U.A.U^{-1}) = \det(U)\operatorname{Pf}^g(A)$, so Pf^g is invariant under the adjoint action of $SO(V,g)$.*
(3) *If $X\in L_{g,skew}(V,V)=\mathfrak{o}(V,g)$, then we have*

$$\sum_{i=1}^{n/2}\operatorname{Pf}^g(A_1,\dots,[X,A_i],\dots,A_{n/2}) = 0.$$

(4) $\operatorname{Pf}(rA) = r^{n/2}\operatorname{Pf}(A)$ *for $r\in\mathbb{R}$ and thus also* $\operatorname{Pf}(A^\top) = (-1)^{n/2}\operatorname{Pf}(A)$.
(5) $\operatorname{Pf}(A)^2 = \det(A)$.
(6) *We have*

$$\operatorname{Pf}(A) = \frac{1}{2^{n/2}(n/2)!}\sum \sigma\in\mathcal{S}_n\operatorname{sign}(\sigma)\prod_{i=1}^{n/2}A_{\sigma(2i-1),\sigma(2i)}.$$

(7) *(**[190]**, **[57]**)*

$$\operatorname{Pf}(A) = \sum_{j=1}^{n}A_{i,j}(-1)^{i+j}\operatorname{sign}(i-j)\operatorname{Pf}(A(ij,ij))$$

where $A(ij,ij)$ is the matrix obtained from A by deleting the lines and columns numbered i and j.

Proof. (1) The transposed $B^\top$ is given by $g(Bx,z) = g(x,B^\top z)$. So $\beta(Bx\wedge By) = B.\beta(x\wedge y).B^\top$ and thus $\beta^{-1}(B.A.B^\top) = \bigwedge^2 B\beta^{-1}(A)$. Then we have:

$$\begin{aligned}\operatorname{Pf}^g(B.A.B^\top) &= \frac{1}{n!}\langle e, \bigwedge^n(B)(\beta^{-1}(A)\wedge\dots\wedge\beta^{-1}(A))\rangle_g\\ &= \frac{1}{n!}\det(B)\langle\bigwedge^n(U)e, \bigwedge^n(U)(\beta^{-1}(A)\wedge\dots\wedge\beta^{-1}(A))\rangle_g\end{aligned}$$

$$= \frac{1}{n!} \det(B) \langle e, \beta^{-1}(A) \wedge \cdots \wedge \beta^{-1}(A) \rangle_g$$
$$= \det(B) \operatorname{Pf}^g(A).$$

(2) We have $U \in O(V, g)$ if and only if $U^\top = U^{-1}$. So this follows from (1).

(3) This follows from (2) by differentiation; see the beginning of the proof of (20.3).

(4) is obvious. The rest is left as an exercise. □

20.12. The Pfaffian class. Let (E, p, M, V) be a vector bundle which is fiber oriented and of even fiber dimension. If we choose a fiberwise Riemann metric on E, we in fact reduce the linear frame bundle of E to the oriented orthonormal one, $SO(\mathbb{R}^n, E)$. On the Lie algebra $\mathfrak{o}(n, \mathbb{R})$ of the structure group $SO(n, \mathbb{R})$ the Pfaffian form Pf of the standard inner product is an invariant, $\operatorname{Pf} \in I^{n/2}(SO(n, \mathbb{R}))$. We define the *Pfaffian class* of the oriented bundle E by

$$\operatorname{Pf}(E) := \left(\frac{-1}{2\pi\sqrt{-1}} \right)^{n/2} \operatorname{Cw}(\operatorname{Pf}, SO(\mathbb{R}^n, E)) \in H^n(M).$$

It does not depend on the choice of the Riemann metric on E, since for any two fiberwise Riemann metrics g_1 and g_2 on E there is an isometric vector bundle isomorphism $f : (E, g_1) \to (E, g_2)$ covering the identity of M, which pulls an $SO(n)$-connection for (E, g_2) to an $SO(n)$-connection for (E, g_1). So the two Pfaffian classes coincide since then $\operatorname{Pf}^1 \circ (f^*\Omega_2 \otimes_\wedge \cdots \otimes_\wedge f^*\Omega_2) = \operatorname{Pf}^2 \circ (\Omega_2 \otimes_\wedge \cdots \otimes_\wedge \Omega_2)$.

Theorem. *The Pfaffian class of oriented even-dimensional vector bundles has the following properties:*

(1) $\operatorname{Pf}(E)^2 = (-1)^{n/2} p_{n/2}(E)$ *where n is the fiber dimension of E.*

(2) $\operatorname{Pf}(E_1 \oplus E_2) = \operatorname{Pf}(E_1) \wedge \operatorname{Pf}(E_2)$.

(3) $\operatorname{Pf}(g^*E) = g^* \operatorname{Pf}(E)$ *for smooth $g : N \to M$.*

Proof. This is left as an exercise for the reader. □

20.13. Chern classes. Let (E, p, M) be a complex vector bundle over the smooth manifold M. So the structure group is $GL(n, \mathbb{C})$ where n is the fiber dimension. Recall now the explanation of the characteristic coefficients c_k^n in (20.6) and insert complex numbers everywhere. Then we get the characteristic coefficients $c_k^n \in I^k(GL(n, \mathbb{C}))$, which are just the extensions of the real ones to the complexification.

We define then the *Chern classes* by

$$c_k(E) := \left(\frac{-1}{2\pi\sqrt{-1}}\right)^k \operatorname{Cw}(c_k^{\dim E}, E) \in H^{2k}(M;\mathbb{R}). \tag{1}$$

The *total Chern class* is again the inhomogeneous cohomology class

$$c(E) := \sum_{k=0}^{\dim_{\mathbb{C}} E} c_k(E) \in H^{2*}(M;\mathbb{R}). \tag{2}$$

It has the following properties:

$$c(\bar{E}) = (-1)^{\dim_{\mathbb{C}} E} c(E), \tag{3}$$
$$c(E_1 \oplus E_2) = c(E_1) \wedge c(E_2), \tag{4}$$
$$c(g^*E) = g^*c(E) \quad \text{for smooth } g : N \to M. \tag{5}$$

One can show (see [**157**]) that (3), (4), (5), and the following normalization determine the total Chern class already completely: The total Chern class of the canonical complex line bundle over S^2 (the square root of the tangent bundle with respect to the tensor product) is $1 + \omega_{S^2}$, where ω_{S^2} is the canonical volume form on S^2 with total volume 1.

Lemma. *Then Chern classes are real cohomology classes.*

Proof. We choose a Hermitian metric on the complex vector bundle E, i.e., we reduce the structure group from $GL(n,\mathbb{C})$ to $U(n)$. Then the curvature Ω of a $U(n)$-principal connection has values in the Lie algebra $\mathfrak{u}(n)$ of skew-Hermitian matrices A with $A^* = -A$. But then we have $c_k^n(-\sqrt{-1}A) \in \mathbb{R}$ since $\overline{\det_{\mathbb{C}}(-\sqrt{-1}A + t\mathbb{I})} = \det_{\mathbb{C}}(\overline{-\sqrt{-1}A + t\mathbb{I}}) = \det_{\mathbb{C}}(-\sqrt{-1}A + t\mathbb{I})$. □

20.14. The Chern character. The *trace classes of a complex vector bundle* are given by

$$\operatorname{tr}_k(E) := \left(\frac{-1}{2\pi\sqrt{-1}}\right)^k \operatorname{Cw}(\operatorname{tr}_k^{\dim E}, E) \in H^{2k}(M,\mathbb{R}). \tag{1}$$

They are also real cohomology classes, and we have $\operatorname{tr}_0(E) = \dim_{\mathbb{C}} E$, the fiber dimension of E, and $\operatorname{tr}_1(E) = c_1(E)$. In general we have the following recursive relation between the Chern classes and the trace classes:

$$c_k(E) = \frac{-1}{k} \sum_{j=0}^{k-1} c_j(E) \wedge \operatorname{tr}_{k-j}(E), \tag{2}$$

which follows directly from lemma (20.9). The inhomogeneous cohomology class

$$\operatorname{ch}(E) := \sum_{k\geq 0} \frac{1}{k!} \operatorname{tr}_k(E) \in H^{2*}(M,\mathbb{R}) \tag{3}$$

is called the *Chern character* of the complex vector bundle E. With the same methods as for the Pontryagin character one can show that the Chern character satisfies the following properties:

$$\operatorname{ch}(E_1 \oplus E_2) = \operatorname{ch}(E_1) + \operatorname{ch}(E_2), \tag{4}$$

$$\operatorname{ch}(E_1 \otimes E_2) = \operatorname{ch}(E_1) \wedge \operatorname{ch}(E_2), \tag{5}$$

$$\operatorname{ch}(g^* E) = g^* \operatorname{ch}(E). \tag{6}$$

From these it clearly follows that the Chern character can be viewed as a ring homomorphism from complex K-theory into even cohomology,

$$\operatorname{ch} : K_{\mathbb{C}}(M) \to H^{2*}(M, \mathbb{R}),$$

which is natural.

Finally we remark that the Pfaffian class of the underlying real vector bundle of a complex vector bundle E of complex fiber dimension n coincides with the Chern class $c_n(E)$. But there is a new class, the Todd class; see below.

20.15. The Todd class. On the vector space $\mathfrak{gl}(n, \mathbb{C})$ of all complex $(n \times n)$-matrices we consider the smooth function

$$f(A) := \det{}_{\mathbb{C}} \left(\sum_{k=0}^{\infty} \frac{(-1)^k}{(k+1)!} A^k \right). \tag{1}$$

It is the unique smooth function which satisfies the functional equation

$$\det(A).f(A) = \det(\mathbb{I} - \exp(-A)).$$

Clearly f is invariant under $\operatorname{Ad}(GL(n, \mathbb{C}))$ and $f(0) = 1$, so we may consider the invariant smooth function, defined near 0, $\operatorname{Td} : \mathfrak{gl}(n, \mathbb{C}) \supset U \to \mathbb{C}$, which is given by $\operatorname{Td}(A) = 1/f(A)$. It is uniquely defined by the functional equations

$$\det(A) = \operatorname{Td}(A) \det(\mathbb{I} - \exp(-A)),$$

$$\det(\tfrac{1}{2}A) \det(\exp(\tfrac{1}{2}A)) = \operatorname{Td}(A) \det(\sinh(\tfrac{1}{2}A)).$$

The *Todd class* of a complex vector bundle is then given by

$$\begin{aligned} \operatorname{Td}(E) &= \left[GL(\mathbb{C}^n, E)[\operatorname{Td}] \left(\sum_{k\geq 0} \left(\frac{-1}{2\pi\sqrt{-1}} R^E \right)^{\otimes_\wedge, k} \right) \right]_{H^{2*}(M,\mathbb{R})} \\ &= \operatorname{Cw}(\operatorname{Td}, E). \end{aligned} \tag{2}$$

The Todd class is a real cohomology class since for $A \in \mathfrak{u}(n)$ we have $\operatorname{Td}(-A) = \operatorname{Td}(A^*) = \overline{\operatorname{Td}(A)}$. Since $\operatorname{Td}(0) = 1$, the Todd class $\operatorname{Td}(E)$ is an invertible element of $H^{2*}(M, \mathbb{R})$.

20.16. The Atiyah-Singer index formula (roughly). Let E_i be complex vector bundles over a compact manifold M, and let $D : \Gamma(E_1) \to \Gamma(E_2)$ be an elliptic pseudodifferential operator of order p. Then for appropriate Sobolev completions D prolongs to a bounded Fredholm operator between Hilbert spaces $D : \mathcal{H}^{d+p}(E_1) \to \mathcal{H}^d(E_2)$. Its *index* $\operatorname{index}(D)$ is defined as the dimension of the kernel minus the dimension of the cokernel, which does not depend on d if it is high enough. The Atiyah-Singer index formula says that

$$\operatorname{index}(D) = (-1)^{\dim M} \int_{TM} \operatorname{ch}(\sigma(D)) \operatorname{Td}(TM \otimes \mathbb{C}),$$

where $\sigma(D)$ is a virtual vector bundle (with compact support) on $TM \setminus 0$, a formal difference of two vector bundles, the so-called symbol bundle of D.

See [**21**] for a somewhat informal introduction, [**207**] for a very short introduction, [**73**] for an analytical treatment using the heat kernel method, [**116**] for a recent treatment and the papers by Atiyah and Singer for the real thing.

Special cases are the Gauß-Bonnet-Chern formula and the Riemann-Roch-Hirzebruch formula.

21. Jets

Jet spaces or jet bundles consist of the invariant expressions of Taylor developments up to a certain order of smooth mappings between manifolds. Their invention goes back to Ehresmann [**53**]. We could have treated them from the beginning and could have mixed them into every chapter, but it is also fine to have all results collected in one place.

21.1. Contact. Recall that smooth functions $f, g : \mathbb{R} \to \mathbb{R}$ are said to have *contact of order* k at 0 if all their values and all derivatives up to order k coincide.

Lemma. *Let $f, g : M \to N$ be smooth mappings between smooth manifolds and let $x \in M$. Then the following conditions are equivalent.*

(1) *For each smooth curve $c : \mathbb{R} \to M$ with $c(0) = x$ and for each smooth function $h \in C^\infty(M)$ the two functions $h \circ f \circ c$ and $h \circ g \circ c$ have contact of order k at 0.*

(2) *For each chart (U, u) of M centered at x and each chart (V, v) of N with $f(x) \in V$ the two mappings $v \circ f \circ u^{-1}$ and $v \circ g \circ u^{-1}$, defined near 0 in $\mathbb{R}^m$, with values in $\mathbb{R}^n$, have the same Taylor development up to order k at 0.*

(3) *For some charts (U,u) of M and (V,v) of N with $x \in U$ and $f(x) \in V$ we have*

$$\frac{\partial^{|\alpha|}}{\partial u^\alpha}\bigg|_x (v \circ f) = \frac{\partial^{|\alpha|}}{\partial u^\alpha}\bigg|_x (v \circ g)$$

for all multiindices $\alpha \in \mathbb{N}_0^m$ with $0 \le |\alpha| \le k$.

(4) *$T_x^k f = T_x^k g$, where T^k is the k-th iterated tangent bundle functor.*

Proof. This is an easy exercise in analysis.

21.2. Definition. If the equivalent conditions of lemma (21.1) are satisfied, we say that f and g have the same *k-jet* at x and we write $j^k f(x)$ or $j^k_x f$ for the resulting equivalence class and call it the *k-jet at x* of f; x is called the *source* of the k-jet, and $f(x)$ is its *target.*

The space of all k-jets of smooth mappings from M to N is denoted by $J^k(M,N)$. We have the *source mapping* $\alpha : J^k(M,N) \to M$ and the *target mapping* $\beta : J^k(M,N) \to N$, given by $\alpha(j^k f(x)) = x$ and $\beta(j^k f(x)) = f(x)$. We will also write $J^k_x(M,N) := \alpha^{-1}(x)$, $J^k(M,N)_y := \beta^{-1}(y)$, and $J^k_x(M,N)_y := J^k_x(M,N) \cap J^k(M,N)_y$ for the spaces of jets with source x, target y, and both, respectively. For $l < k$ we have a canonical surjective mapping $\pi^k_l : J^k(M,N) \to J^l(M,N)$, given by $\pi^k_l(j^k f(x)) := j^l f(x)$. This mapping respects the fibers of α and β and $\pi^k_0 = (\alpha,\beta) : J^k(M,N) \to M \times N$.

21.3. Jets on vector spaces. Now we look at the case $M = \mathbb{R}^m$ and $N = \mathbb{R}^n$.

Let $f : \mathbb{R}^m \to \mathbb{R}^n$ be a smooth mapping. Then by (21.1.3) the k-jet $j^k f(x)$ of f at x has a canonical representative, namely the Taylor polynomial of order k of f at x:

$$\begin{aligned} f(x+y) &= f(x) + df(x).y + \frac{1}{2!}d^2 f(x)y^2 + \cdots + \frac{1}{k!}d^k f(x).y^k + o(|y|^k) \\ &=: f(x) + \operatorname{Tay}^k_x f(y) + o(|y|^k). \end{aligned}$$

Here y^k is short for $(y,y,\dots,y)$, k-times. The 'Taylor polynomial without constant'

$$\operatorname{Tay}^k_x f : y \mapsto \operatorname{Tay}^k_x(y) := df(x).y + \frac{1}{2!}d^2 f(x).y^2 + \cdots + \frac{1}{k!}d^k f(x).y^k$$

is an element of the linear space

$$P^k(m,n) := \bigoplus_{j=1}^k L^j_{sym}(\mathbb{R}^m, \mathbb{R}^n),$$

where $L^j_{sym}(\mathbb{R}^m, \mathbb{R}^n)$ is the vector space of all j-linear symmetric mappings $\mathbb{R}^m \times \cdots \times \mathbb{R}^m \to \mathbb{R}^n$, where we silently use the total polarization of polynomials. Conversely each polynomial $p \in P^k(m,n)$ defines a k-jet $j^k_0(y \mapsto z + p(x+y))$ with arbitrary source x and target z. So we get canonical identifications $J^k_x(\mathbb{R}^m, \mathbb{R}^n)_z \cong P^k(m,n)$ and

$$J^k(\mathbb{R}^m, \mathbb{R}^n) \cong \mathbb{R}^m \times \mathbb{R}^n \times P^k(m,n).$$

If $U \subset \mathbb{R}^m$ and $V \subset \mathbb{R}^n$ are open subsets, then clearly $J^k(U,V) \cong U \times V \times P^k(m,n)$ in the same canonical way.

For later uses we consider now the *truncated composition*

$$\bullet : P^k(m,n) \times P^k(p,m) \to P^k(p,n),$$

where $p \bullet q$ is just the polynomial $p \circ q$ without all terms of order $> k$. Obviously it is a polynomial, thus a real analytic mapping. Now let $U \subset \mathbb{R}^m$, $V \subset \mathbb{R}^n$, and $W \subset \mathbb{R}^p$ be open subsets and consider the fibered product

$$\begin{aligned} J^k(U,V) \times_U J^k(W,U) &= \{(\sigma,\tau) \in J^k(U,V) \times J^k(W,U) : \alpha(\sigma) = \beta(\tau)\} \\ &= U \times V \times W \times P^k(m,n) \times P^k(p,m). \end{aligned}$$

Then the mapping

$$\begin{gathered} \gamma : J^k(U,V) \times_U J^k(W,U) \to J^k(W,V), \\ \gamma(\sigma,\tau) = \gamma((\alpha(\sigma),\beta(\sigma),\bar\sigma),(\alpha(\tau),\beta(\tau),\bar\tau)) = (\alpha(\tau),\beta(\sigma),\bar\sigma \bullet \bar\tau) \end{gathered}$$

is a real analytic mapping, called the *fibered composition of jets*.

Let U, $U' \subset \mathbb{R}^m$ and $V \subset \mathbb{R}^n$ be open subsets and let $g : U' \to U$ be a smooth diffeomorphism. We define a mapping $J^k(g,V) : J^k(U,V) \to J^k(U',V)$ by $J^k(g,V)(j^k f(x)) = j^k(f \circ g)(g^{-1}(x))$. Using the canonical representation of jets from above, we get $J^k(g,V)(\sigma) = \gamma(\sigma, j^k g(g^{-1}(x)))$ or $J^k(g,V)(x,y,\bar\sigma) = (g^{-1}(x), y, \bar\sigma \bullet \operatorname{Tay}^k_{g^{-1}(x)} g)$. If g is a C^p diffeomorphism, then $J^k(g,V)$ is a C^{p-k} diffeomorphism. If $g' : U'' \to U'$ is another diffeomorphism, then clearly $J^k(g',V) \circ J^k(g,V) = J^k(g \circ g', V)$ and $J^k(\quad, V)$ is a contravariant functor acting on diffeomorphisms between open subsets of $\mathbb{R}^m$. Since the truncated composition $\bar\sigma \mapsto \bar\sigma \bullet \operatorname{Tay}^k_{g^{-1}(x)} g$ is linear, the mapping $J^k_x(g,\mathbb{R}^n) := J^k(g,\mathbb{R}^n) | J^k_x(U,\mathbb{R}^n) : J^k_x(U,\mathbb{R}^n) \to J^k_{g^{-1}(x)}(U',\mathbb{R}^n)$ is also linear.

If more generally $g : M' \to M$ is a diffeomorphism between manifolds, the same formula as above defines a bijective mapping $J^k(g,N) : J^k(M,N) \to J^k(M',N)$ and clearly $J^k(\quad, N)$ is a contravariant functor defined on the category of manifolds and diffeomorphisms.

Now let $U \subset \mathbb{R}^m$, $V \subset \mathbb{R}^n$, and $W \subset \mathbb{R}^p$ be open subsets and let $h : V \to W$ be a smooth mapping. Then we define $J^k(U,h) : J^k(U,V) \to J^k(U,W)$ by

$J^k(U,h)(j^k f(x)) = j^k(h \circ f)(x)$ or equivalently by

$$J^k(U,h)(x,y,\bar\sigma) = (x, h(y), \mathrm{Tay}^k_y h \bullet \bar\sigma).$$

If h is C^p, then $J^k(U,h)$ is C^{p-k}. Clearly $J^k(U,\quad)$ is a covariant functor acting on smooth mappings between open subsets of finite-dimensional vector spaces. The mapping $J^k_x(U,h)_y : J^k_x(U,V)_y \to J^k(U,W)_{h(y)}$ is linear if and only if the mapping $\bar\sigma \mapsto \mathrm{Tay}^k_y h \bullet \bar\sigma$ is linear, i.e., if h is affine or if $k=1$.

If $h : N \to N'$ is a smooth mapping between manifolds, we have by the same procedure a mapping $J^k(M,h) : J^k(M,N) \to J^k(M,N')$ and $J^k(M,\quad)$ turns out to be a functor on the category of manifolds and smooth mappings.

21.4. The differential group G^k_m. The k-jets at 0 of diffeomorphisms of $\mathbb{R}^m$ which map 0 to 0 form a group under truncated composition, which will be denoted by $GL^k(m,\mathbb{R})$ or G^k_m for short, and will be called the *differential group of order* k. Clearly an arbitrary 0-respecting k-jet $\sigma \in P^k(m,m)$ is in G^k_m if and only if its linear part is invertible; thus

$$G^k_m = GL^k(m,\mathbb{R}) = GL(m) \oplus \bigoplus_{j=2}^k L^j_{\text{sym}}(\mathbb{R}^m,\mathbb{R}^m) =: GL(m) \times P^k_2(m),$$

where we put $P^k_2(m) = \bigoplus_{j=2}^k L^j_{\text{sym}}(\mathbb{R}^m,\mathbb{R}^m)$ for the space of all polynomial mappings without constant and linear term of degree $\le k$. Since the truncated composition is a polynomial mapping, G^k_m is a Lie group, and the mapping $\pi^k_l : G^k_m \to G^l_m$ is a homomorphism of Lie groups, so $\ker(\pi^k_l) = \bigoplus_{j=l+1}^k L^j_{\text{sym}}(\mathbb{R}^m,\mathbb{R}^m) =: P^k_{l+1}(m)$ is a normal subgroup for all l. The exact sequence of groups

$$\{e\} \to P^k_{l+1}(m) \to G^k_m \to G^l_m \to \{e\}$$

splits if and only if $l=1$; only then do we have a semidirect product.

21.5. Theorem. *For smooth manifolds M and N we have:*

(1) *$J^k(M,N)$ is a smooth manifold (it is of class C^{r-k} if M and N are of class C^r); a canonical atlas is given by all charts $(J^k(U,V), J^k(u^{-1},v))$, where (U,u) is a chart on M and (V,v) is a chart on N.*

(2) *$(J^k(M,N),(\alpha,\beta), M\times N, P^k(m,n), G^k_m \times G^k_n)$ is a fiber bundle with structure group, where $m = \dim M$, $n = \dim N$, and where $(\gamma,\chi) \in G^k_m \times G^k_n$ acts on $\sigma \in P^k(m,n)$ by $(\gamma,\chi).\sigma = \chi \bullet \sigma \bullet \gamma^{-1}$.*

(3) *If $f : M \to N$ is a smooth mapping, then $j^k f : M \to J^k(M,N)$ is also smooth (it is C^{r-k} if f is C^r), sometimes called the k-jet extension of f. We have $\alpha \circ j^k f = Id_M$ and $\beta \circ j^k f = f$.*

(4) *If $g : M' \to M$ is a (C^r-)diffeomorphism, then also the induced mapping $J^k(g,N) : J^k(M,N) \to J^k(M',N)$ is a (C^{r-k}-)diffeomorphism.*

(5) *If* $h : N \to N'$ *is a* (C^r-)*mapping, then*

$$J^k(M, h) : J^k(M, N) \to J^k(M, N')$$

is a (C^{r-k}-)*mapping. We get a covariant functor* $J^k(M, \quad)$ *from the category of smooth manifolds and smooth mappings into itself which maps each of the following classes of mappings into itself: immersions, embeddings, closed embeddings, submersions, surjective submersions, fiber bundle projections. Furthermore* $J^k(\quad,\quad)$ *is a contra-covariant bifunctor.*

(6) *The projections* $\pi^k_l : J^k(M,N) \to J^l(M,N)$ *are smooth and natural, i.e., they commute with the mappings from* (4) *and* (5).

(7) $(J^k(M,N), \pi^k_l, J^l(M,N), P^k_{l+1}(m,n))$ *are fiber bundles for all* l. *The bundle* $(J^k(M,N), \pi^k_{k-1}, J^{k-1}(M,N), L^k_{sym}(\mathbb{R}^m, \mathbb{R}^n))$ *is an affine bundle. The first jet space* $J^1(M,N)$ *is a vector bundle, and it is isomorphic to the bundle* $(L(TM,TN), (\pi_M, \pi_N), M \times N)$. *Moreover we have* $J^1_0(\mathbb{R}, N) = TN$ *and* $J^1(M, \mathbb{R})_0 = T^*M$.

Proof. We use (21.3) heavily. Let (U_γ, u_γ) be an atlas of M and let $(V_\varepsilon, v_\varepsilon)$ be an atlas of N. Then

$$J^k(u_\gamma^{-1}, v_\varepsilon) : (\alpha, \beta)^{-1}(U_\gamma \times V_\varepsilon) \to J^k(u_\gamma(U_\gamma), v_\varepsilon(V_\varepsilon))$$

is a bijective mapping and the chart change looks like

$$J^k(u_\gamma^{-1}, v_\varepsilon) \circ J^k(u_\delta^{-1}, v_\nu)^{-1} = J^k(u_\delta \circ u_\gamma^{-1}, v_\varepsilon \circ v_\nu^{-1})$$

by the functorial properties of $J^k(\quad,\quad)$. The space $J^k(M,N)$ is Hausdorff in the identification topology, since it is a fiber bundle and the usual argument for gluing fiber bundles applies. So (1) follows.

Now we make this manifold atlas into a fiber bundle by using as charts

$$\Big(U_\gamma \times V_\varepsilon, \psi_{(\gamma,\varepsilon)} : J^k(M,N)|U_\gamma \times V_\varepsilon \to U_\gamma \times V_\varepsilon \times P^k(m,n)\Big),$$
$$\psi_{(\gamma,\varepsilon)}(\sigma) = \big(\alpha(\sigma), \beta(\sigma), J^k_{\alpha(\sigma)}(u_\gamma^{-1}, v_\varepsilon)_{\beta(\sigma)}\big).$$

We then get as transition functions

$$\begin{aligned}\psi_{(\gamma,\varepsilon)}\psi_{(\delta,\nu)}(x, y, \bar\sigma) &= (x, y, J^k_{u_\delta(x)}(u_\delta \circ u_\gamma^{-1}, v_\varepsilon \circ v_\nu^{-1})(\bar\sigma)) \\ &= \big(x, y, \mathrm{Tay}^k_{v_\nu(y)}(v_\varepsilon \circ v_\nu^{-1}) \bullet \bar\sigma \bullet \mathrm{Tay}^k_{u_\gamma(x)}(u_\delta \circ u_\gamma^{-1})\big),\end{aligned}$$

and (2) follows.

(3), (4), and (6) are obvious from (21.3), mainly by the functorial properties of $J^k(\quad,\quad)$.

(5) It is clear from (21.3) that $J^k(M,h)$ is a smooth mapping. The rest follows by looking at special chart representations of h and the induced chart representations for $J^k(M,h)$.

It remains to show (7) and here we concentrate on the affine bundle. Let $a_1 + a \in GL(n) \times P_2^k(n,n)$, $\sigma + \sigma_k \in P^{k-1}(m,n) \oplus L^k_{\text{sym}}(\mathbb{R}^m, \mathbb{R}^n)$, and $b_1 + b \in GL(m) \times P_2^k(m,m)$; then the only term of degree k containing σ_k in $(a+a_k) \bullet (\sigma + \sigma_k) \bullet (b + b_k)$ is $a_1 \circ \sigma_k \circ b_1^k$, which depends linearly on σ_k. To this the degree k components of compositions of the lower order terms of σ with the higher order terms of a and b are added, and these may be quite arbitrary. So an affine bundle results.

We have $J^1(M,N) = L(TM,TN)$ since both bundles have the same transition functions. Finally we have $J^1_0(\mathbb{R}, N) = L(T_0\mathbb{R}, TN) = TN$, and $J^1(M,\mathbb{R})_0 = L(TM, T_0\mathbb{R}) = T^*M$. □

21.6. Frame bundles and natural bundles. Let M be a manifold of dimension m. We consider the jet bundle $J^1_0(\mathbb{R}^m, M) = L(T_0\mathbb{R}^m, TM)$ and the open subset $inv J^1_0(\mathbb{R}^m, M)$ of all invertible jets. This is visibly equal to the linear frame bundle of TM as treated in (18.11).

Note that a mapping $f : \mathbb{R}^m \to M$ is locally invertible near 0 if and only if $j^1 f(0)$ is invertible. A jet σ will be called *invertible* if its order 1 part $\pi^k_1(\sigma) \in J^1_0(\mathbb{R}^m, M)$ is invertible. Let us now consider the open subset $inv J^k_0(\mathbb{R}^m, M) \subset J^k_0(\mathbb{R}^m, M)$ of all invertible jets and let us denote it by P^kM. Then by (18.2) we have a principal fiber bundle (P^kM, π_M, M, G^k_m) which is called the *k-th order frame bundle* of the manifold M. Its principal right action r can be described in several ways: by the fiber composition of jets:

$$r = \gamma : inv J^k_0(\mathbb{R}^m, \mathbb{R}^m) \times inv J^k_0(\mathbb{R}^m, M) = G^k_m \times P^kM \to P^kM$$

or by the functorial property of the jet bundle:

$$r^{j^k g(0)} = inv J^k_0(g, M)$$

for a local diffeomorphism $g : \mathbb{R}^m, 0 \to \mathbb{R}^m, 0$.

If $h : M \to M'$ is a local diffeomorphism, the induced mapping $J^k_0(\mathbb{R}^m, h)$ maps the open subset P^kM into P^kM'. By the second description of the principal right action this induced mapping is a homomorphism of principal fiber bundles which we will denote by $P^k(h) : P^kM \to P^kM'$. Thus P^k becomes a covariant functor from the category $\mathcal{M}f_m$ of m-dimensional manifolds and local diffeomorphisms into the category of all principal fiber bundles with structure group G^k_m over m-dimensional manifolds and homomorphisms of principal fiber bundles covering local diffeomorphisms.

If we are given any smooth left action $\ell : G^k_m \times S \to S$ on some manifold S, the associated bundle construction from theorem (18.7) gives us a fiber bundle $P^kM[S,\ell] = P^kM \times_{G^k_m} S$ over M for each m-dimensional manifold M; by (18.9.3) this describes a functor $P^k(\quad)[S,\ell]$ from the category $\mathcal{M}f_m$ into the category of all fiber bundles over m-dimensional manifolds with standard fiber S and G^k_m-structure, and homomorphisms of fiber bundles covering local diffeomorphisms. These bundles are also called *natural bundles* or *geometric objects*.

21.7. Theorem. *If (E, p, M, S) is a fiber bundle, let us denote by $J^k(E) \to M$ the space of all k-jets of sections of E. Then we have:*

(1) *$J^k(E)$ is a closed submanifold of $J^k(M, E)$.*

(2) *The first jet bundle $J^1(E) \to M \times E$ is an affine subbundle of the vector bundle $J^1(M, E) = L(TM, TE)$; in fact we have $J^1(E) = \{\sigma \in L(TM, TE) : Tp \circ \sigma = Id_{TM}\}$.*

(3) *$(J^k(E), \pi^k_{k-1}, J^{k-1}(E))$ is an affine bundle.*

(4) *If (E, p, M) is a vector bundle, then $(J^k(E), \alpha, M)$ is also a vector bundle. If $\phi : E \to E'$ is a homomorphism of vector bundles covering the identity, then $J^k(\varphi)$ is of the same kind.*

Proof. (1) By (21.5.5) the mapping $J^k(M, p)$ is a submersion; thus $J^k(E) = J^k(M, p)^{-1}(j^k(Id_M))$ is a submanifold. Part (2) is clear. Parts (3) and (4) are seen by looking at appropriate canonical charts. □

CHAPTER V. Riemann Manifolds

22. Pseudo-Riemann Metrics and Covariant Derivatives

22.1. Riemann metrics. Let M be a smooth manifold of dimension m. A *Riemann metric* g on M is a symmetric $\binom{0}{2}$-tensor field such that $g_x : T_xM \times T_xM \to \mathbb{R}$ is a positive definite inner product for each $x \in M$. A *pseudo-Riemann metric* g on M is a symmetric $\binom{0}{2}$-tensor field such that g_x is nondegenerate, i.e., $\check{g}_x : TxM \to T_x^*M$ is bijective for each $x \in M$. If (U, u) is a chart on M, then we have

$$g|U = \sum_{i,j=0}^{m} g(\tfrac{\partial}{\partial u^i}, \tfrac{\partial}{\partial u^j})\, du^i \otimes du^j =: \sum_{i,j} g_{ij} du^i \otimes du^j.$$

Here $(g_{ij}(x))$ is a symmetric invertible $(m \times m)$-matrix for each $x \in M$, positive definite in the case of a Riemann metric; thus $(g_{ij}) : U \to \mathrm{Mat}_{\mathrm{sym}}(m \times m)$. In the case of a pseudo-Riemann metric, the matrix (g_{ij}) has p positive eigenvalues and q negative ones; (p, q) is called the *signature of the metric* and $q = m - p$ is called the *index of the metric*; both are locally constant on M and we shall always assume that it is constant on M.

Lemma. *One each manifold M there exist many Riemann metrics. But there need not exist a pseudo-Riemann metric of some given signature.*

Proof. Let (U_α, u_α) be an atlas on M with a subordinated partition of unity (f_α). Choose smooth mappings (g_{ij}^α) from U_α to the convex cone of all positive definite symmetric $(m \times m)$-matrices for each α and put $g = \sum_\alpha f_\alpha \sum_{ij} g_{ij}^\alpha\, du_\alpha^i \otimes d_\alpha^j$.

For example, on any even-dimensional sphere S^{2n} there does not exist a pseudo-Riemann metric g of signature $(1, 2n-1)$: Otherwise there would exist a line subbundle $L \subset TS^2$ with $g(v,v) > 0$ for $0 \neq v \in L$. But since the Euler characteristic $\chi(S^{2n}) = 2$, such a line subbundle of the tangent bundle cannot exist; see [**80**, I, p. 399]. □

22.2. Length and energy of a curve. Let $c : [a,b] \to M$ be a smooth curve. In the Riemann case the *length* of the curve c is then given by

$$L_a^b(c) := \int_a^b g(c'(t), c'(t))^{1/2} dt = \int_a^b |c'(t)|_g \, dt.$$

In both cases the *energy* of the curve c is given by

$$E_a^b(c) := \tfrac{1}{2} \int_a^b g(c'(t), c'(t)) dt.$$

In the Riemann case we have by the Cauchy-Schwarz inequality

$$L_a^b(c)^2 = \Big(\int_a^b |c'|_g.1 \, dt \Big)^2 \leq \int_a^b |c'|_g^2 \, dt.(b-a) = 2(b-a) E_a^b(c).$$

For piecewise smooth curves the length and the energy are defined by taking it for the smooth pieces and then by summing up over all the pieces. In the pseudo-Riemann case for the length one has to distinguish different classes of curves according to the sign of $g(c'(t), c'(t))$ (the sign then should be assumed constant) and by taking an appropriate sign before taking the root. These leads to the concept of 'time-like' curves (with speed less than the speed of light) and 'space-like' curves (travelling faster than light).

The length is invariant under reparameterizations of the curve:

$$\begin{aligned} L_a^b(c \circ f) &= \int_a^b g((c \circ f)'(t), (c \circ f)'(t))^{1/2} dt \\ &= \int_a^b g(f'(t) c'(f(t)), f'(t) c'(f(t)))^{1/2} dt \\ &= \int_a^b g(c'(f(t)), c'(f(t)))^{1/2} |f'(t)| dt \\ &= \int_a^b g(c'(t), c'(t))^{1/2} dt = L_a^b(c). \end{aligned}$$

The energy is not invariant under reparameterizations.

22.3. Theorem (First variational formula). *Let g be a pseudo-Riemann metric on an open subset $U \subseteq \mathbb{R}^m$. Let $\gamma : [a,b] \times (-\varepsilon, \varepsilon) \to U$ be a smooth variation of the curve $c = \gamma(\quad, 0) : [a,b] \to U$. Let $r(t) = \frac{\partial}{\partial s}|_0 \gamma(t,s) = T_{(t,0)}\gamma.(0,1) \in T_{c(t)}U$ be the variational vector field along c.*

Then we have:

$$
\begin{aligned}
\tfrac{\partial}{\partial s}|_0(E_a^b(\gamma(\quad,s))) = \int_a^b \Big(&-g(c(t))(c''(t),r(t)) \\
&- dg(c(t))(c'(t))(c'(t),r(t)) \\
&+ \tfrac12 dg(c(t))(r(t))(c'(t),c'(t))\Big)\,dt \\
&+ g(c(b))(c'(b),r(b)) - g(c(a))(c'(a),r(a)).
\end{aligned}
$$

Proof. We have the Taylor expansion $\gamma(t,s) = \gamma(t,0) + s\,\gamma_s(t,0) + O(s^2) = c(t) + sr(t) + O(s^2)$ where the remainder $O(s^2) = s^2R(s,t)$ is smooth and uniformly bounded in t. We plug this into the energy and take also the Taylor expansion of g as follows:

$$
\begin{aligned}
E_a^b(\gamma(\quad,s)) &= \tfrac12 \int_a^b g(\gamma(t,s))\big(\gamma_t(t,s),\gamma_t(t,s)\big)\,dt \\
&= \tfrac12 \int_a^b g(c(t)+sr(t)+O(s^2))\Big(c'(t)+sr'(t)+O(s^2), \\
&\qquad\qquad c'(t)+sr'(t)+O(s^2)\Big)\,dt \\
&= \tfrac12 \int_a^b \Big(g(c(t)) + sg'(c(t))(r(t)) + O(s^2)\Big)\Big(\dots,\dots\Big)\,dt \\
&= \tfrac12 \int_a^b \Big(g(c(t))(c'(t),c'(t)) + 2sg(c(t))(c'(t),r'(t)) \\
&\qquad + sg'(c(t))(r(t))(c'(t),c'(t))\Big)\,dt + O(s^2) \\
&= E_a^b(c) + s\int_a^b g(c(t))(c'(t),r'(t))\,dt \\
&\qquad + \tfrac12 s\int_a^b g'(c(t))(r(t))(c'(t),c'(t))\,dt + O(s^2).
\end{aligned}
$$

Thus for the derivative we get, using partial integration:

$$
\begin{aligned}
\tfrac{\partial}{\partial s}|_0 E_a^b(\gamma(\quad,s)) &= \lim_{s\to0}\tfrac1s\Big(E_a^b(\gamma(\quad,s)) - E_a^b(\gamma(\quad,0))\Big) \\
&= \tfrac12 \int_a^b g'(c(t))(r(t))(c'(t),c'(t))\,dt + \int_a^b g(c(t))(c'(t),r'(t))\,dt \\
&= \tfrac12 \int_a^b g'(c(t))(r(t))(c'(t),c'(t))\,dt + g(c(t))(c'(t),r(t))|_{t=a}^{t=b} \\
&\quad - \int_a^b \Big(g'(c(t))(c'(t))(c'(t),r(t)) + g(c(t))(c''(t),r(t))\Big)\,dt
\end{aligned}
$$

$$= \int_a^b \Big(-g(c(t))(c''(t), r(t)) - g'(c(t))(c'(t))(c'(t), r(t)) \\ + \tfrac{1}{2} g'(c(t))(r(t))(c'(t), c'(t))\Big)\, dt \\ + g(c(b))(c'(b), r(b)) - g(c(a))(c'(a), r(a)). \quad \square$$

22.4. Christoffel symbols and geodesics. On a pseudo-Riemann manifold (M, g), by theorem (22.3), we have $\frac{\partial}{\partial s}|_0 E_a^b(\gamma(\quad, s)) = 0$ for all variations γ of the curve c with fixed end points ($r(a) = r(b) = 0$) in a chart (U, u) if and only if the integral in theorem (22.3) vanishes. This is the case if and only if we have in $u(U) \subset \mathbb{R}^m$:

$$\begin{aligned} g(c(t))(c''(t),\quad) &= \tfrac{1}{2} g'(c(t))(\quad)(c'(t), c'(t)) \\ &\quad - \tfrac{1}{2} g'(c(t))(c'(t))(c'(t),\quad) \\ &\quad - \tfrac{1}{2} g'(c(t))(c'(t))(\quad, c'(t)). \end{aligned}$$

For $x \in u(U)$ and $X, Y, Z \in \mathbb{R}^m$ we consider the polarized version of the last equation:

$$\begin{aligned} g(x)(\Gamma_x(X, Y), Z) =& \tfrac{1}{2} g'(x)(Z)(X, Y) \\ & - \tfrac{1}{2} g'(x)(X)(Y, Z) \\ & - \tfrac{1}{2} g'(x)(Y)(Z, X) \end{aligned} \tag{1}$$

which is a well defined smooth mapping

$$\Gamma : u(U) \to L^2_{\text{sym}}(\mathbb{R}^m; \mathbb{R}^m).$$

Back on $U \subset M$ we have in coordinates

$$\begin{aligned} \Gamma_x(X, Y) &= \Gamma_x\Big(\sum_i X^i \tfrac{\partial}{\partial u^i}, \sum_j Y^j \tfrac{\partial}{\partial u^j}\Big) = \sum_{i,j} \Gamma_x\Big(\tfrac{\partial}{\partial u^i}, \tfrac{\partial}{\partial u^j}\Big) X^i Y^j \\ &=: \sum_{i,j} \Gamma_{ij}(x) X^i Y^j =: \sum_{i,j,k} \Gamma_{ij}^k(x) X^i Y^j \tfrac{\partial}{\partial u^k}, \end{aligned}$$

where the $\Gamma^k_{ij} : U \to \mathbb{R}$ are smooth functions, which are called the *Christoffel symbols* in the chart (U, u). Attention: Most of the literature uses the negative of the Christoffel symbols.

Lemma. *If $g|U = \sum_{i,j} g_{ij} du^i \otimes du^j$ and if $(g_{ij})^{-1} = (g^{ij})$ denotes the inverse matrix, then we have*

$$\Gamma^k_{ij} = \frac{1}{2} \sum_l g^{kl} \Big(\frac{\partial g_{ij}}{\partial u^l} - \frac{\partial g_{lj}}{\partial u^i} - \frac{\partial g_{il}}{\partial u^j} \Big). \tag{2}$$

Proof. We have

$$\begin{aligned}\sum_k \Gamma^k_{ij} g_{kl} &= \sum_k \Gamma^k_{ij} g(\tfrac{\partial}{\partial u^k}, \tfrac{\partial}{\partial u^l}) = g\Big(\sum_k \Gamma^k_{ij} \tfrac{\partial}{\partial u^k}, \tfrac{\partial}{\partial u^l}\Big) = g(\Gamma(\tfrac{\partial}{\partial u^i}, \tfrac{\partial}{\partial u^j}), \tfrac{\partial}{\partial u^l}) \\ &= \tfrac{1}{2} g'(\tfrac{\partial}{\partial u^l})(\tfrac{\partial}{\partial u^i}, \tfrac{\partial}{\partial u^j}) - \tfrac{1}{2} g'(\tfrac{\partial}{\partial u^i})(\tfrac{\partial}{\partial u^j}, \tfrac{\partial}{\partial u^l}) - \tfrac{1}{2} g'(\tfrac{\partial}{\partial u^j})(\tfrac{\partial}{\partial u^l}, \tfrac{\partial}{\partial u^i}) \\ &= \frac{1}{2}\frac{\partial g_{ij}}{\partial u^l} - \frac{1}{2}\frac{\partial g_{lj}}{\partial u^i} - \frac{1}{2}\frac{\partial g_{il}}{\partial u^j}. \qquad \square\end{aligned}$$

Let $c : [a,b] \to M$ be a smooth curve in the pseudo-Riemann manifold (M, g). The curve c is called a *geodesic* on M if in each chart (U, u) for the Christoffel symbols of this chart we have

$$c''(t) = \Gamma_{c(t)}(c'(t), c'(t)). \tag{3}$$

The reason for this name is: If the energy E^b_a of (each piece of) the curve is minimal under all variations with fixed end points, then by (22.3) the integral

$$\int_a^b g_{c(t)}(c''(t) - \Gamma_{c(t)}(c'(t), c'(t)), r(t))\, dt = 0$$

for each vector field r along c with $r(a) = r(b) = 0$. This implies (3). Thus (local) infima of the energy functional E^b_a are geodesics, and more generally any curve on which the energy functional E^b_a has vanishing derivative (with respect to local variations with constant ends) is called a geodesic.

Finally we should compute how the Christoffel symbols react to a chart change. Since this is easily done and since we will see soon that the Christoffel symbols indeed are coordinate expressions of an entity which belongs to the second tangent bundle TTM, we leave this exercise to the interested reader.

22.5. Covariant derivatives. Let (M, g) be a pseudo-Riemann manifold. A covariant derivative on M is a mapping $\nabla : \mathfrak{X}(M) \times \mathfrak{X}(M) \to \mathfrak{X}(M)$, denoted by $(X, Y) \mapsto \nabla_X Y$, which satisfies the following conditions:

(1) $\nabla_X Y$ is $C^\infty(N)$-linear in $X \in \mathfrak{X}(M)$, i.e., $\nabla_{f_1 X_1 + f_2 X_2} Y = f_1 \nabla_{X_1} Y + f_2 \nabla_{X_2} Y$. So for a tangent vector $X_x \in T_x M$ the mapping $\nabla_{X_x} : \mathfrak{X}(M) \to T_x M$ makes sense and we have $(\nabla_X s)(x) = \nabla_{X(x)} s$.

(2) $\nabla_X Y$ is $\mathbb{R}$-linear in $Y \in \mathfrak{X}(M)$.

(3) $\nabla_X (f.Y) = df(X).Y + f.\nabla_X Y$ for $f \in C^\infty(M)$, the derivation property of ∇_X.

The covariant derivative ∇ is called *symmetric* or *torsion-free* if moreover the following holds:

(4) $\nabla_X Y - \nabla_Y X = [X, Y]$.

The covariant derivative ∇ is called *compatible with the pseudo-Riemann metric* if we have:

(5) $X(g(Y,Z)) = g(\nabla_X Y, Z) + g(Y, \nabla_X Z)$ for all $X, Y, Z \in \mathfrak{X}(M)$.

Compare with (19.12) where we treat the covariant derivative on vector bundles.

Theorem. *On any pseudo-Riemann manifold (M,g) there exists a unique torsion-free covariant derivative $\nabla = \nabla^g$ which is compatible with the metric g. In a chart (U,u) we have*

$$\nabla_{\frac{\partial}{\partial u^i}} \frac{\partial}{\partial u^j} = -\sum_k \Gamma^k_{ij} \frac{\partial}{\partial u^k}, \tag{6}$$

where the Γ^k_{ij} are the Christoffel symbols from (22.4).

This unique covariant derivative is called the *Levi-Civita covariant derivative.*

Proof. We write the cyclic permutations of property (5) equipped with the signs $+,+,-$:

$$\begin{aligned} X(g(Y,Z)) &= g(\nabla_X Y, Z) + g(Y, \nabla_X Z),\\ Y(g(Z,X)) &= g(\nabla_Y Z, X) + g(Z, \nabla_Y X),\\ -Z(g(X,Y)) &= -g(\nabla_Z X, Y) - g(X, \nabla_Z Y). \end{aligned}$$

We add these three equations and use the torsion-free property (4) to get

$$\begin{aligned} &X(g(Y,Z)) + Y(g(Z,X)) - Z(g(X,Y))\\ &= g(\nabla_X Y + \nabla_Y X, Z) + g(\nabla_X Z - \nabla_Z X, Y) + g(\nabla_Y Z - \nabla_Z Y, X)\\ &= g(2\nabla_X Y - [X,Y], Z) - g([Z,X],Y) + g([Y,Z],X), \end{aligned}$$

which we rewrite as an implicit defining equation for $\nabla_X Y$:

$$\begin{aligned} 2g(\nabla_X Y, Z) &= X(g(Y,Z)) + Y(g(Z,X)) - Z(g(X,Y))\\ &\quad - g(X,[Y,Z]) + g(Y,[Z,X]) + g(Z,[X,Y]). \end{aligned} \tag{7}$$

This by (7) uniquely determined bilinear mapping $(X,Y) \mapsto \nabla_X Y$ indeed satisfies (1)–(5), which is tedious but easy to check. The final assertion of the theorem follows by using (7) once more:

$$\begin{aligned} 2g(\nabla_{\frac{\partial}{\partial u^i}} \tfrac{\partial}{\partial u^j}, \tfrac{\partial}{\partial u^l}) &= \tfrac{\partial}{\partial u^i}(g(\tfrac{\partial}{\partial u^j}, \tfrac{\partial}{\partial u^l})) + \tfrac{\partial}{\partial u^j}(g(\tfrac{\partial}{\partial u^l}, \tfrac{\partial}{\partial u^i})) - \tfrac{\partial}{\partial u^l}(g(\tfrac{\partial}{\partial u^i}, \tfrac{\partial}{\partial u^j}))\\ &= -2\sum_k \Gamma^k_{ij} g_{kl}, \qquad \text{by (22.4.2).} \quad \square \end{aligned}$$

22.6. Geodesic structures and sprays. By (22.5.6) and (22.4.3) we see that a smooth curve $c : (a, b) \to (M, g)$ is a geodesic in a pseudo-Riemann manifold if $\nabla_{\partial_t} c' = 0$, in a sense which we will make precise later in (22.9.6) when we discuss how we can apply ∇ to vector fields which are only defined along curves or mappings. In each chart (U, u) this is an ordinary differential equation

$$c''(t) = \Gamma_{c(t)}(c'(t), c'(t)),$$

$$\frac{d^2}{dt^2} c^k(t) = \sum_{i,j} \Gamma^k_{ij}(c(t)) \frac{d}{dt} c^i(t) \frac{d}{dt} c^j(t), \qquad c = (c^1, \dots, c^m),$$

which is of second order, linear in the second derivative, quadratic in the first derivative, and in general completely nonlinear in $c(t)$ itself. By the theorem of Picard-Lindelöf for ordinary differential equations there exists a unique solution for each given initial condition $c(t_0), c'(t_0)$, depending smoothly on the initial conditions. Thus we may piece together the local solutions and get a geodesic structure in the following sense: A *geodesic structure on a manifold* M is a smooth mapping $\operatorname{geo} : TM \times \mathbb{R} \supset U \to M$, where U is an open neighborhood of $TM \times \{0\}$ in $TM \times \mathbb{R}$, which satisfies the following:

(1) $\operatorname{geo}(X_x)(0) = x$ and $\partial|_0 \operatorname{geo}(X_x)(t) = X_x$.
(2) $\operatorname{geo}(t.X_x)(s) = \operatorname{geo}(X_x)(t.s)$.
(3) $\operatorname{geo}(\operatorname{geo}(X_x)'(s))(t) = \operatorname{geo}(X_x)(t + s)$.
(4) $U \cap (X_x \times \mathbb{R}) = \{X_x\} \times$ interval .

One could also require that U be maximal with respect to all these properties. But we shall not elaborate on this since we will reduce everything to the geodesic vector field shortly.

If we are given a geodesic structure $\operatorname{geo} : U \to M$ as above, then the mapping

$$(X, t) \mapsto \operatorname{geo}(X)'(t) = \tfrac{\partial}{\partial t} \operatorname{geo}(X)(t) \in TM$$

is the flow for the vector field $S \in \mathfrak{X}(TM)$ which is given by

$$S(X) = \partial|_0 \tfrac{\partial}{\partial t} \operatorname{geo}(X)(t) \in T^2M,$$

since we have

$$\begin{aligned}
\tfrac{\partial}{\partial t}\tfrac{\partial}{\partial t} \operatorname{geo}(X)(t) &= \tfrac{\partial}{\partial s}|_0 \tfrac{\partial}{\partial s} \operatorname{geo}(X)(t + s) \\
&= \tfrac{\partial}{\partial s}|_0 \tfrac{\partial}{\partial s} \operatorname{geo}(\tfrac{\partial}{\partial t} \operatorname{geo}(X)(t))(s) \quad \text{by (3)} \\
&= S(\tfrac{\partial}{\partial t} \operatorname{geo}(X)(t)), \\
\operatorname{geo}(X)'(0) &= X.
\end{aligned}$$

The smooth vector field $S \in \mathfrak{X}(TM)$ is called the *geodesic spray* of the geodesic structure.

Recall now the chart structure on the second tangent bundle T^2M and the canonical flip mapping $\kappa_M : T^2M \to T^2M$ from (8.12) and (8.13). Let (U,u) be a chart on M and let $c_{(x,y)}(t) = u(\text{geo}(Tu^{-1}(x,y))(t)) \in U$. Then we have

$$\begin{aligned}
Tu(\text{geo}(Tu^{-1}(x,y))'(t)) &= (c_{(x,y)}(t), c'_{(x,y)}(t)),\\
T^2u(\text{geo}(Tu^{-1}(x,y))''(t)) &= \big(c_{(x,y)}(t), c'_{(x,y)}(t); c'_{(x,y)}(t), c''_{(x,y)}(t)\big),\\
T^2u.S(Tu^{-1}(x,y)) &= T^2u(\text{geo}(Tu^{-1}(x,y))''(0))\\
&= \big(c_{(x,y)}(0), c'_{(x,y)}(0); c'_{(x,y)}(0), c''_{(x,y)}(0)\big)\\
&= \big(x, y; y, \bar S(x,y)\big).
\end{aligned} \tag{5}$$

Property (2) of the geodesic structure implies in turn

$$\begin{aligned}
c_{(x,ty)}(s) &= c_{(x,y)}(ts),\\
c'_{(x,ty)}(s) &= t.c'_{(x,y)}(ts),\\
c''_{(x,ty)}(0) &= t^2.c''_{(x,y)}(0),\\
\bar S(x,ty) &= t^2\bar S(x,y),
\end{aligned}$$

so that $\bar S(x,\quad) : \mathbb{R}^m \to \mathbb{R}^m$ is homogenous of degree 2. By polarizing or taking the second derivative with respect to y, we get

$$\begin{aligned}
\bar S(x,y) &= \Gamma_x(y,y), \qquad \text{for} \quad \Gamma : u(U) \to L^2_{\text{sym}}(\mathbb{R}^m;\mathbb{R}^m),\\
\Gamma_x(y,z) &= \tfrac12(\bar S(x,y+z) - \bar S(x,y) - \bar S(x,z)).
\end{aligned}$$

If the geodesic structure is induced by a pseudo-Riemann metric on M, then we have seen that

$$c''_{(x,y)}(t) = \Gamma_{c_{(x,y)}(t)}(c'_{(x,y)}(t), c'_{(x,y)}(t))$$

for the Christoffel symbols in the chart (U,u). Thus the geodesic spray is given in terms of the Christoffel symbols by

$$T^2u(S(Tu^{-1}(x,y))) = (x,y;y,\Gamma_x(y,y)). \tag{6}$$

22.7. The geodesic exponential mapping. Let M be a smooth manifold and let $S \in \mathfrak{X}(TM)$ be a vector field with the following properties:

(1) $\pi_{TM} \circ S = \text{Id}_{TM}$; S is a vector field.

(2) $T(\pi_M) \circ S = \text{Id}_{TM}$; S is a 'differential equation of second order'.

(3) Let $m_t^M : TM \to TM$ and $m_t^{TM} : T^2M \to T^2M$ be the scalar multiplications. Then $S \circ m_t^M = T(m_t^M).m_t^{TM}.S$.

A vector field with these properties is called a *spray*.

Theorem. *Given a spray $S \in \mathfrak{X}(TM)$ on a manifold M, we can write* $\operatorname{geo}(X)(t) := \pi_M(\operatorname{Fl}^S_t(X))$. *Then this is a geodesic structure on M in the sense of* (22.6).

If we put $\exp(X) := \pi_M(\operatorname{Fl}^S_1(X)) = \operatorname{geo}(X)(1)$, *then*

$$\exp : TM \supset V \to M$$

is a smooth mapping, defined on an open neighborhood V of the zero section in TM, which is called the exponential mapping of the spray S and which has the following properties:

(4) $T_{0_x}(\exp|T_xM) = \operatorname{Id}_{T_xM}$ *(via $T_{0_x}(T_xM) = T_xM$). Thus by the inverse function theorem* $\exp_x := \exp|T_xM : V_x \to W_x$ *is a diffeomorphism from an open neighborhood V_x of 0_x in TM onto an open neighborhood W_x of x in M. The chart $(W_x, \exp_x^{-1})$ is called a Riemann normal coordinate system at x.*

(5) $\operatorname{geo}(X)(t) = \exp(t.X)$.

(6) *The mapping*

$$(\pi_M, \exp) : TM \supset \tilde{V} \to M \times M$$

is a diffeomorphism from an open neighborhood $\tilde{V}$ of the zero section in TM onto an open neighborhood of the diagonal in $M \times M$.

Proof. By properties (1) and (2) the local expression of the spray S is given by $(x,y) \mapsto (x,y;y,\bar{S}(x,y))$, as in (22.6.5). By (3) we have

$$(x,ty;ty,\bar{S}(x,ty)) = T(m^M_t).m^{TM}_t.(x,y;y,\bar{S}(x,y)) = (x,ty;ty,t^2\bar{S}(x,y)),$$

so that $\bar{S}(x,ty) = t^2\bar{S}(x,y)$ as in (22.6).

(7) We have $\operatorname{Fl}^S_t(s.X) = s.\operatorname{Fl}^S_{s.t}(X)$ if one side exists, by uniqueness of solutions of differential equations:

$$\begin{aligned} \tfrac{\partial}{\partial t} s.\operatorname{Fl}^S_{s.t}(X) &= \tfrac{\partial}{\partial t} m^M_s \operatorname{Fl}^S_{s.t}(X) = T(m^M_s)\tfrac{\partial}{\partial t}\operatorname{Fl}^S_{s.t}(X) \\ &= T(m^M_s).m^{TM}_s S(\operatorname{Fl}^S_{s.t}(X)) \overset{(3)}{=} S(s.\operatorname{Fl}^S_{s.t}(X)), \\ s.\operatorname{Fl}^S_{s.0}(X) &= s.X, \qquad \text{thus} \qquad s.\operatorname{Fl}^S_{s.t}(X) = \operatorname{Fl}^S_t(s.X). \end{aligned}$$

We check that $\operatorname{geo} = \pi_M \circ \operatorname{Fl}^S$ is a geodesic structure, i.e., (22.6.1)–(22.6.4) hold:

$$\begin{aligned} \operatorname{geo}(X_x)(0) &= \pi_M(\operatorname{Fl}^S_0(X_x)) = \pi_M(X_x) = x, \\ \partial|_0 \operatorname{geo}(X_x)(t) &= \partial|_0 \pi_M(\operatorname{Fl}^S_t(X_x)) = T(\pi_M)\partial|_0 \operatorname{Fl}^S_t(X_x) \\ &= T(\pi_M)S(X_x) \overset{(2)}{=} X_x, \\ \operatorname{geo}(s.X_x)(t) &= \pi_M(\operatorname{Fl}^S_t(s.X_x)) = \pi_M(s.\operatorname{Fl}^S_{s.t}(X_x)), \quad \text{see above}, \\ &= \operatorname{geo}(X_x)(s.t), \end{aligned}$$

$$\begin{aligned}
\operatorname{geo}(\tfrac{\partial}{\partial s}\operatorname{geo}(X_x)(s))(t) &= \pi_M(\operatorname{Fl}^S_t(\tfrac{\partial}{\partial s}\pi_M \operatorname{Fl}^S_s(X_x)))\\
&= \pi_M(\operatorname{Fl}^S_t(T(\pi_M)S(\operatorname{Fl}^S_s(X_x))))\\
&= \pi_M(\operatorname{Fl}^S_t(\operatorname{Fl}^S_s(X_x))) \quad \text{by (2)}\\
&= \pi_M(\operatorname{Fl}^S_{t+s}(X_x)) = \operatorname{geo}(X_x)(t+s).
\end{aligned}$$

Let us investigate the exponential mapping. For $\varepsilon > 0$ let X_x be so small that $(\frac{1}{\varepsilon}X_x, \varepsilon)$ is in the domain of definition of the flow Fl^S. Then

$$\begin{aligned}
\exp_x(X_x) &= \pi_M(\operatorname{Fl}^S_1(X_x)) = \pi_M(\operatorname{Fl}^S_1(\varepsilon.\tfrac{1}{\varepsilon}.X_x))\\
&= \pi_M(\varepsilon.\operatorname{Fl}^S_\varepsilon(\tfrac{1}{\varepsilon}.X_x)) = \pi_M(\operatorname{Fl}^S_\varepsilon(\tfrac{1}{\varepsilon}.X_x)), \quad \text{by (7)}.
\end{aligned}$$

We check the properties of the exponential mapping. The tangent mapping satifies:

$$\begin{aligned}
(4)\qquad T_{0_x}(\exp_x).X_x &= \partial|_0 \exp_x(0_x + t.X_x)\\
&= \partial|_0\pi_M(\operatorname{Fl}^S_1(t.X_x))\\
&= \partial|_0\pi_M(t.\operatorname{Fl}^S_t(X_x))\\
&= \partial|_0\pi_M(\operatorname{Fl}^S_t(X_x)), \quad \text{by (7)}\\
&= T(\pi_M)\partial|_0(\operatorname{Fl}^S_t(X_x)) = T(\pi_M)(S(X_x)) = X_x.
\end{aligned}$$

Moreover we have:

$$\begin{aligned}
(5)\qquad \exp_x(t.X_x) &= \pi_M(\operatorname{Fl}^S_1(t.X_x))\\
&= \pi_M(t.\operatorname{Fl}^S_t(X_x))\\
&= \pi_M(\operatorname{Fl}^S_t(X_x)) = \operatorname{geo}(X_x)(t).
\end{aligned}$$

(6) By (4) we have $T_{0_x}(\pi_M, \exp) = \left(\begin{smallmatrix} \mathbb{I} & 0 \\ * & \mathbb{I}\end{smallmatrix}\right)$; thus $(\pi_M, \exp)$ is a local diffeomorphism. Again by (4) the mapping $(\pi_M, \exp)$ is injective on a small neighborhood of the zero section. □

22.8. Linear connections and connectors. Let M be a smooth manifold. A smooth mapping $C : TM \times_M TM \to T^2M$ is called a *linear connection* or *horizontal lift* on M if it has the following properties:

(1) $(T(\pi_M), \pi_{TM}) \circ C = \operatorname{Id}_{TM\times_M TM}$.

(2) $C(\ \ , X_x) : T_xM \to T_{X_x}(TM)$ is linear; this is the first vector bundle structure on T^2M treated in (8.13).

(3) $C(X_x,\ \ ) : T_xM \to T(\pi_M)^{-1}(X_x)$ is linear; this is the second vector bundle structure on T^2M treated in (8.13).

The connection C is called *symmetric* or *torsion-free* if moreover the following property holds:

(4) $\kappa_M \circ C = C \circ \text{flip} : TM \times_M TM \to T^2M$, where $\kappa_M : T^2M \to T^2M$ is the canonical flip mapping treated in (8.13).

From the properties (1)–(3) it follows that for a chart (U_α, u_α) on M the mapping C is given by

(5) $T^2(u_\alpha) \circ C \circ (T(u_\alpha)^{-1} \times_M T(u_\alpha)^{-1})((x,y),(x,z)) = (x,z;y,\Gamma^\alpha_x(y,z))$,

where the *Christoffel symbol* $\Gamma^\alpha_x(y,z) \in \mathbb{R}^m$ $(m = \dim(M))$ is smooth in $x \in u_\alpha(U_\alpha)$ and is bilinear in $(y,z) \in \mathbb{R}^m \times \mathbb{R}^m$. For the sake of completeness let us also note the transformation rule of the Christoffel symbols which follows now directly from the chart change of the second tangent bundle in (8.12) and (8.13). The chart change on M

$$u_{\alpha\beta} = u_\alpha \circ u_\beta^{-1} : u_\beta(U_\alpha \cap U_\beta) \to u_\alpha(U_\alpha \cap U_\beta)$$

induces the following tranformation of the Christoffel symbols:

(6) $$\Gamma^\alpha_{u_{\alpha\beta}(x)}(d(u_{\alpha\beta})(x)y, d(u_{\alpha\beta})(x)z) = d(u_{\alpha\beta})(x)\Gamma^\beta_x(y,z) + d^2(u_{\alpha\beta})(x)(y,z).$$

We have seen in (22.6.6) that a spray S on a manifold determines symmetric Christoffel symbols and thus a symmetric connection C. If the spray S is induced by a pseudo-Riemann metric g on M, then the Christoffel symbols are the same as we found by determining the singular curves of the energy in (22.4). The promised geometric description of the Christoffel symbols is (5), which also explains their transformation behavior under chart changes: They belong to the vertical part of the second tangent bundle.

Consider now a linear connection $C : TM \times_M TM \to T^2M$. For $\xi \in T^2M$ we have

$$\xi - C(T(\pi_M).\xi, \pi_{TM}(\xi)) \in V(TM) = T(\pi_M)^{-1}(0)$$

which is an element of the vertical bundle, since

$$T(\pi_M)(\xi - C(T(\pi_M).\xi, \pi_{TM}(\xi))) = T(\pi_M).\xi - T(\pi_M).\xi = 0$$

by (1). Thus we may define the *connector* $K : T^2M \to TM$ by

(7) $$K(\xi) = \text{vpr}_{TM}(\xi - C(T(\pi_M).\xi, \pi_{TM}(\xi))), \quad \text{where } \xi \in T^2M,$$

where the vertical projection vpr_{TM} was defined in (8.12). In coordinates induced by a chart on M we have

(8) $$K(x,y;a,b) = \text{vpr}(x,y;0,b-\Gamma_x(a,y)) = (x, b - \Gamma_x(a,y)).$$

Obviously the connector K has the following three properties:

(9) We have
$$K \circ \text{vl}_{TM} = \text{pr}_2 : TM \times_M TM \to TM$$
where $\text{vl}_{TM}(X_x, Y_x) = \partial|_0(X_x + tY_x)$ is the vertical lift introduced in (8.12).

(10) The mapping $K : TTM \to TM$ is linear for the (first) vector bundle structure on $\pi_{TM} : TTM \to TM$.

(11) The mapping $K : TTM \to TM$ is linear for the (second) vector bundle structure on $T(\pi_M) : TTM \to TM$.

A connector, defined as a mapping satisfying (9)–(11), is equivalent to a connection, since one can reconstruct it (which is most easily checked in a chart) by

$$C(\quad, X_x) = (T(\pi_M)|\ker(K : T_{X_x}(TM) \to T_x M))^{-1}.$$

The connecter K is associated to a symmetric connection if and only if $K \circ \kappa_M = K$. The connector treated here is a special case of the one in (19.11).

22.9. Covariant derivatives, revisited. We describe here the passage from a linear connection $C : TM \times_M TM \to T^2M$ and its associated connector $K : T^2M \to TM$ to the covariant derivative. In the more general setting of vector bundles this is treated in (19.12). Namely, for any manifold N, a smooth mapping $s : N \to TM$ (a vector field along $f := \pi_M \circ s$) and a vector field $X \in \mathfrak{X}(N)$ we define

(1) $$\nabla_X s := K \circ Ts \circ X : N \to TN \to T^2M \to TM$$

which is again a vector field along f:

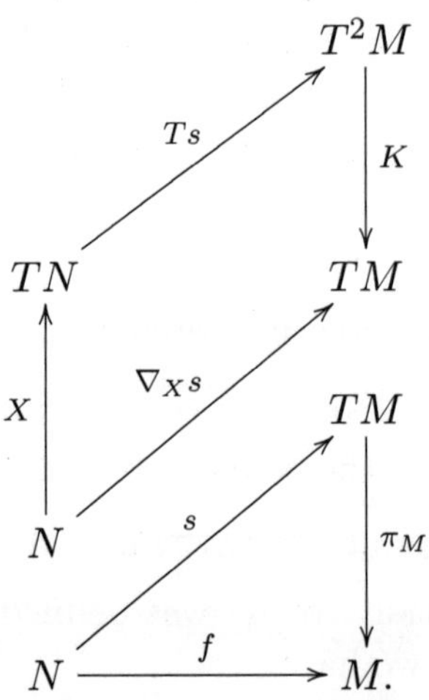

If $f : N \to M$ is a fixed smooth mapping, let us denote by $C^\infty_f(N, TM) \cong \Gamma(f^*TM)$ the vector space of all smooth mappings $s : N \to TM$ with $\pi_M \circ s = f$ – vector fields along f. Then the covariant derivative may be viewed as a bilinear mapping

(2) $$\nabla : \mathfrak{X}(N) \times C^\infty_f(N, TM) \to C^\infty_f(N, TM).$$

In particular for $f = Id_M$ we have $\nabla : \mathfrak{X}(M) \times \mathfrak{X}(M) \to \mathfrak{X}(M)$ as in (22.5).

Lemma. *This covariant derivative has the following properties:*

(3) $\nabla_X s$ *is* $C^\infty(N)$*-linear in* $X \in \mathfrak{X}(N)$*. So for a tangent vector* $X_x \in T_x N$ *the mapping* $\nabla_{X_x} : C^\infty_f(N, TM) \to T_{f(x)}M$ *makes sense and we have* $(\nabla_X s)(x) = \nabla_{X(x)} s$.

(4) $\nabla_X s$ *is* $\mathbb{R}$*-linear in* $s \in C^\infty_f(N, TM)$.

(5) $\nabla_X(h.s) = dh(X).s + h.\nabla_X s$ *for* $h \in C^\infty(N)$*; this is the derivation property of* ∇_X.

(6) *For any manifold* Q *and smooth mapping* $g : Q \to N$ *and* $Z_y \in T_y Q$ *we have* $\nabla_{Tg.Z_y} s = \nabla_{Z_y}(s \circ g)$*. If* $Z \in \mathfrak{X}(Q)$ *and* $X \in \mathfrak{X}(N)$ *are* g*-related, then we have* $\nabla_Z(s \circ g) = (\nabla_X s) \circ g$,

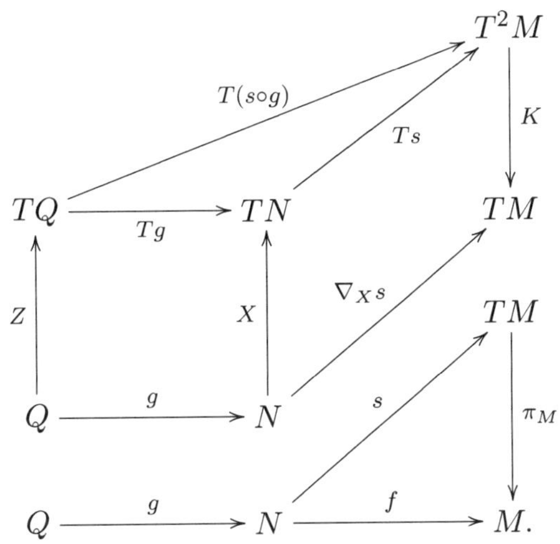

(7) *In charts on* N *and* M*, for* $s(x) = (\bar f(x), \bar s(x))$ *and* $X(x) = (x, \bar X(x))$ *we have* $(\nabla_X s)(x) = (\bar f(x), d\bar s(x).\bar X(x) - \Gamma_{\bar f(x)}(\bar s(x), d\bar f(x)\bar X(x)))$.

(8) *The connection is symmetric if and only if* $\nabla_X Y - \nabla_Y X = [X, Y]$.

Proof. All these properties follow easily from the definition (1). □

Remark. Property (6) is not well understood in some differential geometric literature. It is the reason why in the beginning of (22.6) we wrote $\nabla_{\partial_t} c' = 0$ for the geodesic equation and not $\nabla_{c'} c' = 0$ which one finds in the literature.

22.10. Torsion. Let ∇ be a general covariant derivative on a manifold M. Then the *torsion* is given by

$$\text{(1)} \qquad \operatorname{Tor}(X, Y) := \nabla_X Y - \nabla_Y X - [X, Y], \quad X, Y \in \mathfrak{X}(M).$$

It is skew-symmetric and $C^\infty(M)$-linear in $X, Y \in \mathfrak{X}(M)$ and is thus a 2-form with values in TM: $\operatorname{Tor} \in \Omega^2(M; TM) = \Gamma(\bigwedge^2 T^*M \otimes TM)$, since we have

$$\begin{aligned}\operatorname{Tor}(f.X,Y) &= \nabla_{f.X}Y - \nabla_Y(f.X) - [f.X,Y]\\ &= f.\nabla_X Y - Y(f).X - f.\nabla_Y(X) - f.[X,Y] + Y(f).X\\ &= f.\operatorname{Tor}(X,Y).\end{aligned}$$

Locally on a chart (U,u) we have

$$\begin{aligned}(2)\qquad \operatorname{Tor}|U &= \sum_{i,j}\operatorname{Tor}\Big(\tfrac{\partial}{\partial u^i},\tfrac{\partial}{\partial u^j}\Big)\otimes du^i\otimes du^j\\ &= \sum_{i,j}\Big(\nabla_{\frac{\partial}{\partial u^i}}\tfrac{\partial}{\partial u^j} - \nabla_{\frac{\partial}{\partial u^j}}\tfrac{\partial}{\partial u^i} - [\tfrac{\partial}{\partial u^i},\tfrac{\partial}{\partial u^j}]\Big)\otimes du^i\otimes du^j\\ &= \sum_{i,j}(-\Gamma^k_{ij}+\Gamma^k_{ji})du^i\otimes du^j\otimes\tfrac{\partial}{\partial u^k}\\ &= -\sum_{i,j}\Gamma^k_{ij}du^i\wedge du^j\otimes\tfrac{\partial}{\partial u^k} = -2\sum_{i<j}\Gamma^k_{ij}du^i\wedge du^j\otimes\tfrac{\partial}{\partial u^k}.\end{aligned}$$

We may add an arbitrary form $T\in\Omega^2(M;TM)$ to a given covariant derivative and we get a new covariant derivative with the same spray and geodesic structure, since the symmetrization of the Christoffel symbols stays the same.

Lemma. *Let $K: TTM\to M$ be the connector of the covariant derivative ∇, and let $X,Y\in\mathfrak{X}(M)$. Then the torsion is given by*

$$(3)\qquad \operatorname{Tor}(X,Y) = (K\circ\kappa_M - K)\circ TX\circ Y.$$

If moreover $f:N\to M$ is smooth and $U,V\in\mathfrak{X}(N)$, then we get also

$$\begin{aligned}(4)\qquad \operatorname{Tor}(Tf.U,Tf.V) &= \nabla_U(Tf\circ V) - \nabla_V(Tf\circ U) - Tf\circ[U,V]\\ &= (K\circ\kappa_M - K)\circ TTf\circ TU\circ V.\end{aligned}$$

Proof. By (22.9.1), (8.14) (or (8.19)), and (22.8.9) we have

$$\begin{aligned}\operatorname{Tor}(X,Y) &= \nabla_X Y - \nabla_Y X - [X,Y]\\ &= K\circ TY\circ X - K\circ TX\circ Y - K\circ \mathrm{vl}_{TM}\circ(Y,[X,Y]),\\ K\circ\mathrm{vl}_{TM}\circ(Y,[X,Y]) &= K\circ(TY\circ X - \kappa_M\circ TX\circ Y)\\ &= K\circ TY\circ X - K\circ\kappa_M\circ TX\circ Y.\end{aligned}$$

Similarly we get

$$\begin{aligned}K\circ\mathrm{vl}_{TM}\circ(Tf\circ V, Tf\circ[U,V]) &= K\circ TTf\circ\mathrm{vl}_{TN}\circ(V,[U,V])\\ &= K\circ TTf\circ(TV\circ U - \kappa_N\circ TU\circ V)\\ &= K\circ TTf\circ TV\circ U - K\circ\kappa_M\circ TTf\circ TU\circ V,\end{aligned}$$

and also

$$\begin{aligned}
\nabla_U(Tf\circ V) - \nabla_V(Tf\circ U) - Tf\circ[X,Y] \\
&= K\circ TTf\circ TV\circ U - K\circ TTf\circ TU\circ V \\
&\qquad - K\circ \mathrm{vl}_{TM}\circ(Tf\circ V, Tf\circ[U,V]) \\
&= (K\circ\kappa_M - K)\circ TTf\circ TU\circ V.
\end{aligned}$$

The rest will be proved locally, so let us assume now that M is open in $\mathbb{R}^m$ and $U(x) = (x, \bar U(x))$, etc. Then by (22.8.8) we have

$$\begin{aligned}
&(TTf\circ TU\circ V)(x) \\
&= TTf(x,\bar U(x);\bar V(x), d\bar U(x)\bar V(x)) \\
&= \big(f(x), df(x).\bar U(x); df(x).\bar V(x), d^2f(x)(\bar V(x),\bar U(x)) + df(x).d\bar U(x).\bar V(x)\big)
\end{aligned}$$

and also

$$\begin{aligned}
&((K\circ\kappa_M - K)\circ TTf\circ TU\circ V)(x) \\
&\quad= \big(f(x), d^2f(x)(\bar V(x),\bar U(x)) + df(x).d\bar U(x).\bar V(x) \\
&\qquad\qquad - \Gamma_{f(x)}(df(x).\bar U(x), df(x).\bar V(x))\big) \\
&\quad\quad - \big(f(x), d^2f(x)(\bar V(x),\bar U(x)) + df(x).d\bar U(x).\bar V(x) \\
&\qquad\qquad - \Gamma_{f(x)}(df(x).\bar V(x), df(x).\bar U(x))\big) \\
&\quad= \big(f(x), -\Gamma_{f(x)}(df(x).\bar U(x), df(x).\bar V(x)) \\
&\qquad\qquad + \Gamma_{f(x)}(df(x).\bar V(x), df(x).\bar U(x))\big) \\
&\quad= \mathrm{Tor}(Tf\circ U, Tf\circ V)(x). \quad \square
\end{aligned}$$

22.11. The space of all covariant derivatives. If ∇^0 and ∇^1 are two covariant derivatives on a manifold M, then $\nabla^1_X Y - \nabla^0_X Y$ turns out to be $C^\infty(M)$-linear in $X, Y \in \mathfrak{X}(M)$ and is thus a $\binom{1}{2}$-tensor field on M; see (22.10). Conversely, one may add an arbitrary $\binom{1}{2}$-tensor field A to a given covariant derivative and get a new covariant derivative. Thus the space of all covariant derivatives is an affine space with modeling vector space $\Gamma(T^*M\otimes T^*M\otimes TM)$.

22.12. The covariant derivative of tensor fields. Let ∇ be covariant derivative on a manifold M, and let $X \in \mathfrak{X}(M)$. Then the ∇_X can be extended uniquely to an operator ∇_X on the space of all tensor fields on M with the following properties:

(1) For $f \in C^\infty(M)$ we have $\nabla_X f = X(f) = df(X)$.

(2) ∇_X respects the spaces of $\binom{p}{q}$-tenor fields.

(3) $\nabla_X(A \otimes B) = (\nabla_X A) \otimes B + A \otimes (\nabla_X B)$, a derivation with respect to the tensor product.

(4) ∇_X commutes with any kind of contraction C (i.e., any trace; see (8.18)): So for $\omega \in \Omega^1(M)$ and $Y \in \mathfrak{X}(M)$ we have

$$\nabla_X(\omega(Y)) = (\nabla_X \omega)(Y) + \omega(\nabla_X Y).$$

The correct way to understand this is to use the concepts of (19.9)–(19.12): Recognize the linear connection as induced from a principal connection on the linear frame bundle $GL(\mathbb{R}^m, TM)$ and induce it then to all vector bundles associated to the representations of the structure group $GL(m, \mathbb{R})$ in all tensor spaces. Contractions are then equivariant mappings and thus intertwine the induced covariant derivatives, which is most clearly seen from (19.15).

Nevertheless, we discuss here the traditional proof, since it helps in actual computations. For $\omega \in \Omega^1(M)$ and $Y \in \mathfrak{X}(M)$ and the total contraction C we have

$$\begin{aligned}
\nabla_X(\omega(Y)) &= \nabla_X(C(\omega \otimes Y)) \\
&= C(\nabla_X \omega \otimes Y + \omega \otimes \nabla_X Y) \\
&= (\nabla_X \omega)(Y) + \omega(\nabla_X Y), \\
(\nabla_X \omega)(Y) &= \nabla_X(\omega(Y)) - \omega(\nabla_X Y),
\end{aligned}$$

which is easily seen (as in (22.10)) to be $C^\infty(M)$-linear in Y. Thus $\nabla_X \omega$ is again a 1-form.

For a $\binom{p}{q}$-tensor field A we choose $X_i \in \mathfrak{X}(M)$ and $\omega^j \in \Omega^1(M)$ and arrive (similarly using again the total contraction) at

$$\begin{aligned}
(\nabla_X A)(X_1, \dots, X_q, \omega^1, \dots, \omega^p) &= X(A(X_1, \dots, X_q, \omega^1, \dots, \omega^p)) \\
- A(\nabla_X X_1, \dots, X_q, \omega^1, \dots, \omega^p) &- \dots - A(X_1, \dots, \nabla_X X_q, \omega^1, \dots, \omega^p) \\
- A(X_1, \dots, X_q, \nabla_X \omega^1, \dots, \omega^p) &- \dots - A(X_1, \dots, X_q, \omega^1, \dots, \nabla_X \omega^p).
\end{aligned}$$

This expression is again $C^\infty(M)$-linear in each entry X_i or ω^j and defines thus the $\binom{p}{q}$-tensor field $\nabla_X A$. Obviously ∇_X is a derivation with respect to the tensor product of fields and commutes with all contractions.

For the sake of completeness we also list the local expression

$$\begin{aligned}
\nabla_{\frac{\partial}{\partial u^i}} du^j &= \sum_k \Big(\nabla_{\frac{\partial}{\partial u^i}} du^j\Big)\big(\tfrac{\partial}{\partial u^k}\big) du^k \\
&= \sum_k \Big(\tfrac{\partial}{\partial u^i} \delta^k_j - du^j\big(\nabla_{\frac{\partial}{\partial u^i}} \tfrac{\partial}{\partial u^k}\big)\Big) du^k = \sum_k \Gamma^j_{ik} du^k
\end{aligned}$$

from which one can easily derive the expression for an arbitrary tensor field:

$$\begin{aligned}
\nabla_{\frac{\partial}{\partial u^i}} A &= \sum \Big(\nabla_{\frac{\partial}{\partial u^i}} A\Big)\big(\tfrac{\partial}{\partial u^{i_1}}, \dots, \tfrac{\partial}{\partial u^{i_q}}, du^{j_1}, \dots, du^{j_p}\big) du^{i_1} \otimes \dots \otimes \tfrac{\partial}{\partial u^{j_p}} \\
&= \sum \Big(\tfrac{\partial}{\partial u^i}\Big(A\Big(\tfrac{\partial}{\partial u^{i_1}}, \dots, du^{j_p}\Big)\Big) - A\Big(\nabla_{\frac{\partial}{\partial u^i}} \tfrac{\partial}{\partial u^{i_1}}, \dots, du^{j_p}\Big) \\
&\qquad - \dots - A\Big(\tfrac{\partial}{\partial u^{i_1}}, \dots, \nabla_{\frac{\partial}{\partial u^i}} du^{j_p}\Big)\Big) du^{i_1} \otimes \dots \otimes \tfrac{\partial}{\partial u^{j_q}} \\
&= \sum \Big(\tfrac{\partial}{\partial u^i} A^{j_1,\dots,j_p}_{i_1,\dots,i_q} + A^{j_1,\dots,j_p}_{k,i_2,\dots,i_q} \Gamma^k_{i,i_1} + \dots + A^{j_1,\dots,j_p}_{i_1,\dots,i_{q-1},k} \Gamma^k_{i,i_q} \\
&\qquad - A^{k,j_2,\dots,j_p}_{i_1,\dots,i_q} \Gamma^{j_1}_{i,k} - \dots - A^{j_1,\dots,j_{p-1},k}_{i_1,\dots,i_q} \Gamma^{j_p}_{i,k}\Big) du^{i_1} \otimes \dots \otimes \tfrac{\partial}{\partial u^{j_q}}.
\end{aligned}$$

23. Geometry of Geodesics

23.1. Geodesics. On a pseudo-Riemann manifold (M, g) we have a geodesic structure which is described by the flow of the geodesic spray on TM. The geodesic with initial value $X_x \in T_xM$ is denoted by $t \mapsto \exp(t.X_x)$ in terms of the pseudo-Riemann exponential mapping exp and $\exp_x = \exp|T_xM$. We recall the properties of the geodesics which we will use.

(1) $\exp_x : T_xM \supset U_x \to M$ is defined on a maximal 'radial' open zero neighborhood U_x in T_xM. Here radial means that for $X_x \in V_x$ we also have $[0,1].X_x \subset V_x$. This follows from the flow properties since $\exp_x = \pi_M(\mathrm{Fl}^S_1 |T_xM)$ by (22.7).

(2) $T_{0_x}(\exp|T_xM) = \mathrm{Id}_{T_xM}$; thus $\partial|_0 \exp_x(t.X_x) = X_x$. See (22.7.4).

(3) $\exp(s.(\frac{\partial}{\partial t}\exp(t.X))) = \exp((t+s)X)$. See (22.6.3).

(4) $t \mapsto g(\frac{\partial}{\partial t}\exp(t.X), \frac{\partial}{\partial t}\exp(t.X))$ is constant in t: for $c(t) = \exp(t.X)$ we have $\partial_t g(c', c') = 2g(\nabla_{\partial_t} c', c') = 0$. Thus in the Riemann case the length $|\frac{\partial}{\partial t}\exp(t.X)|_g = \sqrt{g(\frac{\partial}{\partial t}\exp(t.X), \frac{\partial}{\partial t}\exp(t.X))}$ is also constant.

If for a geodesic c the (by (4)) constant $|c'(t)|_g$ is 1, we say that c is *parameterized by arc-length.*

23.2. Lemma (Gauß). *Let (M, g) be a Riemann manifold. For $x \in M$ let $\varepsilon > 0$ be so small that $\exp_x : D_x(\varepsilon) := \{X \in T_xM : |X|_g < \varepsilon\} \to M$ is a diffeomorphism on its image. Then in $\exp_x(D_x(\varepsilon))$ the geodesic rays starting from x are all orthogonal to the 'geodesic spheres' $\{\exp_x(X) : |X|_g = k\} = \exp_x(k.S(T_xM, g))$ for $k < \varepsilon$.*

On pseudo-Riemann manifolds this result holds too, with the following adaptation: Since the unit spheres in (T_xM, g_x) are hyperboloids, they are not small and may not lie in the domain of definition of the geodesic exponential mapping; the result only holds in this domain.

Proof. $\exp_x(k.S(T_xM,g))$ is a submanifold of M since $\exp_x$ is a diffeomorphism on $D_x(\varepsilon)$. Let $s\mapsto v(s)$ be a smooth curve in $kS(T_xM,g)\subset T_xM$, and let $\gamma(t,s):=\exp_x(t.v(s))$. Then γ is a variation of the geodesic $\gamma(t,0)=\exp_x(t.v(0))=:c(t)$. In the energy of the geodesic $t\mapsto\gamma(t,s)$ the integrand is constant by (23.1.4):

$$\begin{aligned}E_0^1(\gamma(\quad,s)) &= \tfrac12\int_0^1 g(\tfrac{\partial}{\partial t}\gamma(t,s),\tfrac{\partial}{\partial t}\gamma(t,s))\,dt\\ &= \tfrac12 g(\partial|_0\gamma(t,s),\partial|_0\gamma(t,s))\,dt\\ &= \tfrac12 k^2.\end{aligned}$$

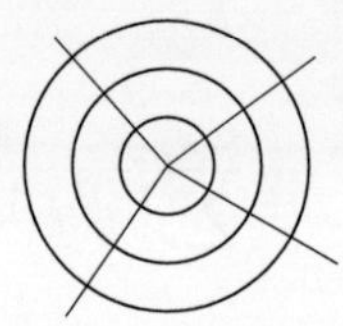

Comparing this with the first variational formula (22.3), i.e.,

$$\tfrac{\partial}{\partial s}|_0(E_0^1(\gamma(\quad,s)))=\int_0^1 0\,dt+g(c(1))(c'(1),\tfrac{\partial}{\partial s}|_0\gamma(1,s))-g(c(0))(c'(0),0),$$

we get $0=g(c(1))(c'(1),\frac{\partial}{\partial s}|_0\gamma(1,s))$, where $\frac{\partial}{\partial s}|_0\gamma(1,s)$ is an arbitrary tangent vector of $\exp_x(kS(T_xM,g))$. □

23.3. Corollary. *Let (M,g) be a Riemann manifold, $x\in M$, and $\varepsilon>0$ be such that $\exp_x: D_x(\varepsilon):=\{X\in T_xM:|X|_g<\varepsilon\}\to M$ is a diffeomorphism on its image. Let $c:[a,b]\to\exp_x(D_x(\varepsilon))\setminus\{x\}$ be a piecewise smooth curve, so that $c(t)=\exp_x(u(t).v(t))$ where $0<u(t)<\varepsilon$ and $|v(t)|_{g_x}=1$.*

Then for the length we have $L_a^b(c)\geq|u(b)-u(a)|$ with equality if and only if u is monotone and v is constant, so that c is a radial geodesic, reparameterized by u.

On pseudo-Riemann manifolds this result holds only for in the domain of definition of the geodesic exponential mapping and only for curves with positive velocity vectors (time-like curves).

Proof. We may assume that c is smooth by treating each smooth piece of c separately. Let $\alpha(u,t):=\exp_x(u.v(t))$. Then

$$\begin{aligned}c(t)&=\alpha(u(t),t),\\ \tfrac{\partial}{\partial t}c(t)&=\tfrac{\partial\alpha}{\partial u}(u(t),t).u'(t)+\tfrac{\partial\alpha}{\partial t}(u(t),t),\\ |\tfrac{\partial\alpha}{\partial u}|_{g_x}&=|v(t)|_{g_x}=1,\\ 0&=g(\tfrac{\partial\alpha}{\partial u},\tfrac{\partial\alpha}{\partial t}),\qquad\text{by lemma (23.2).}\end{aligned}$$

Putting this together, we get

$$\begin{aligned}|c'|_g^2=g(c',c')&=g(\tfrac{\partial\alpha}{\partial u}.u'+\tfrac{\partial\alpha}{\partial t},\tfrac{\partial\alpha}{\partial u}.u'+\tfrac{\partial\alpha}{\partial t})\\ &=|u'|^2|\tfrac{\partial\alpha}{\partial u}|_g^2+|\tfrac{\partial\alpha}{\partial t}|_g^2=|u'|^2+|\tfrac{\partial\alpha}{\partial t}|_g^2\geq|u'|^2\end{aligned}$$

with equality if and only if $|\frac{\partial\alpha}{\partial t}|_g = 0$; thus $\frac{\partial\alpha}{\partial t} = 0$ and $v(t) = $ constant. So finally:

$$L_a^b(c) = \int_a^b |c'(t)|_g\,dt \ge \int_a^b |u'(t)|\,dt \ge \left|\int_a^b u'(t)\,dt\right| = |u(b) - u(a)|$$

with equality if and only if u is monotone and v is constant. □

23.4. Corollary. *Let (M, g) be a Riemann manifold. Let $\varepsilon : M \to \mathbb{R}_{>0}$ be a continuous function such that for $\tilde V = \{X_x \in T_xM : |X_x| < \varepsilon(x)$ for all $x \in M\}$ the mapping $(\pi_m, \exp) : TM \supseteq \tilde V \to W \subseteq M \times M$ is a diffeomorphism from the open neighborhood $\tilde V$ of the zero section in TM onto an open neighborhood W of the diagonal in $M \times M$, as shown in (22.7.6).*

Then for each $(x, y) \in W$ there exists a unique geodesic c in M which connects x and y and has minimal length: For each piecewise smooth curve γ from x to y we have $L(\gamma) \ge L(c)$ with equality if and only if γ is a reparameterization of c.

Proof. The set $\tilde V \cap T_xM = D_x(\varepsilon(x))$ satisfies the condition of corollary (23.3). For $X_x = \exp_x^{-1}(y) = ((\pi_M, \exp)|\tilde V)^{-1}(x, y)$ the geodesic $t \mapsto c(t) = \exp_x(t.X_x)$ leads from x to y. Let $\delta > 0$ be small. Then c contains a segment which connects the geodesic spheres $\exp_x(\delta.S(T_xM, g))$ and $\exp_x(|X_x|_{g_x}.S(T_xM, g))$. By corollary (23.3) the length of this segment is $\ge |X_x|_g - \delta$ with equality if and only if this segment is radial, thus a reparameterization of c. Since this holds for all $\delta > 0$, the result follows. □

23.5. The geodesic distance. On a Riemann manifold (M, g) there is a natural topological metric defined by

$$\begin{aligned}&\operatorname{dist}^g(x, y)\\ &\quad := \inf\left\{L_0^1(c) : c : [0,1] \to M \text{ piecewise smooth, } c(0) = x, c(1) = y\right\},\end{aligned}$$

which we call the *geodesic distance* (since 'metric' is heavily used). We either assume that M is connected or we take the distance of points in different connected components as ∞.

Lemma. *On a Riemann manifold (M, g) the geodesic distance is a topological metric which generates the topology of M. For $\varepsilon_x > 0$ small enough the open ball*

$$B_x(\varepsilon_x) = \{y \in M : \operatorname{dist}^g(x, y) < \varepsilon_x\}$$

has the property that any two points in it can be connected by a geodesic of minimal length.

Proof. This follows by (23.3) and (23.4). The triangle inequality is easy to check since we admit piecewise smooth curves. □

23.6. Theorem (Hopf, Rinov). *For a Riemann manifold (M, g) the following assertions are equivalent:*

(1) $(M, \operatorname{dist}^g)$ *is a complete metrical space (Cauchy sequences converge).*

(2) *Each closed subset of M which is bounded for the geodesic distance is compact.*

(3) *Any geodesic is maximally definable on the whole of $\mathbb{R}$.*

(4) $\exp : TM \to M$ *is defined on the whole of TM.*

(5) *There exists a point x such that $\exp_x : T_xM \to M$ is defined on the whole of T_xM, in each connected component of M.*

If these equivalent conditions are satisfied, then (M, g) is called a complete Riemann manifold. In this case we even have:

(6) *On a complete connected Riemann manifold any two points can be connected by a geodesic of minimal length.*

Condition (6) does not imply the other conditions: Consider an open convex in $\mathbb{R}^m$.

Proof. (2) $\implies$ (1) is obvious.

(1) $\implies$ (3) Let c be a maximally defined geodesic, parametrized by arc-length. If c is defined on the interval (a, b) and if $b < \infty$, say, then by the definition of the distance (23.5) the sequence $c(b - \frac{1}{n})$ is a Cauchy sequence; thus by (1), $\lim_{n\to\infty} c(b - \frac{1}{n}) =: c(b)$ exists in M. For m, n large enough $(c(b - \frac{1}{n}), c(b - \frac{1}{m})) \in W$ where W is the open neighborhood of the diagonal in $M \times M$ from (23.4); thus the segment of c between $c(b - \frac{1}{n})$ and $c(b - \frac{1}{m})$ is of minimal length: $\operatorname{dist}^g(c(b - \frac{1}{n}), c(b - \frac{1}{m})) = |\frac{1}{n} - \frac{1}{m}|$. By continuity $\operatorname{dist}^g(c(b - \frac{1}{n}), c(b)) = |\frac{1}{n}|$. Now let us apply corollary (23.3) with center $c(b)$: In $\exp_{c(b)}(D_{c(b)}(\varepsilon))$ the curve $t \mapsto c(b + t)$ is a piecewise smooth curve of minimal length; thus by (23.3) a radial geodesic. Thus $\lim_{t\to b} c'(t) =: c'(b)$ exists and $t \mapsto \exp_{c(b)}((t - b)c'(b))$ equals $c(t)$ for $t < b$ and prolongs the geodesic c for $t \geq b$.

(3) $\implies$ (4) is obvious.

(4) $\implies$ (5) is obvious.

(5) $\implies$ (6) for special points, in each connected component separately. In detail: Let x, y be in one connected component of M where x is the special point with $\exp_x : T_xM \to M$ defined on the whole of T_xM. We shall prove that x can be connected to y by a geodesic of minimal length.

Let $\operatorname{dist}^g(x, y) = r > 0$. Consider the compact set $S := \exp_x(\delta.S(T_xM, g)) \subset \exp_x(T_xM)$ for $0 < \delta < r$ so small that $\exp_x$ is a diffeomorphism on $\{X \in T_xM : |X|_g < 2\delta\}$. There exists a unit vector $X_x \in S(T_xM, g_x)$ such that $z = \exp_x(\delta X_x)$ has the property that $\operatorname{dist}^g(z, y) = \min\{\operatorname{dist}^g(s, y) : s \in S\}$.

(7) **Claim.** The curve $c(t) = \exp_x(t.X_x)$ satisfies the condition

$$(*) \qquad \operatorname{dist}^g(c(t), y) = r - t$$

for all $0 \leq t \leq r$. It will take some effort to prove this claim.

Since any piecewise smooth curve from x to y hits S (its initial segment does so in the diffeomorphic preimage in T_xM), we have

$$\begin{aligned} r = \operatorname{dist}^g(x,y) &= \inf_{s\in S}(\operatorname{dist}^g(x,s) + \operatorname{dist}^g(s,y)) = \inf_{s\in S}(\delta + \operatorname{dist}^g(s,y)) \\ &= \delta + \min_{s\in S} \operatorname{dist}^g(s,y) = \delta + \operatorname{dist}^g(z,y), \\ \operatorname{dist}^g(z,y) &= r - \delta; \qquad \text{thus } (*) \text{ holds for } t = \delta. \end{aligned}$$

(8) **Claim.** If (*) holds for $t \in [\delta, r]$, then it also holds for all t' with $\delta \leq t' \leq t$, since we have:

$$\begin{aligned} &\operatorname{dist}^g(c(t'), y) \leq \operatorname{dist}^g(c(t'), c(t)) + \operatorname{dist}^g(c(t), y) \leq t - t' + r - t = r - t', \\ &r = \operatorname{dist}^g(x,y) \leq \operatorname{dist}^g(x, c(t')) + \operatorname{dist}^g(c(t'), y), \\ &\operatorname{dist}^g(c(t'), y) \geq r - \operatorname{dist}^g(x, c(t')) \geq r - t' \qquad \Longrightarrow \text{ claim (8).} \end{aligned}$$

Now let $t_0 = \sup\{t \in [\delta, r] : (*) \text{ holds for } t\}$. By continuity (*) is then also valid for t_0. Assume for contradiction that $t_0 < r$.

Let S' be the geodesic sphere with (small) radius δ' centered at $c(t_0)$, and let $z' \in S'$ be a point with minimal distance to y:

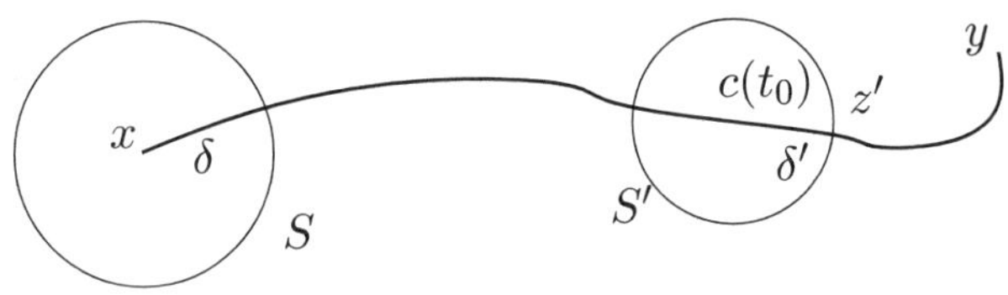

As above we see that

$$\begin{aligned} r - t_0 \overset{(*)}{=} \operatorname{dist}^g(c(t_0), y) &= \inf_{s'\in S'}(\operatorname{dist}^g(c(t_0), s') + \operatorname{dist}^g(s', y)) \\ &= \delta' + \operatorname{dist}^g(z', y), \\ (**) \qquad \operatorname{dist}^g(z', y) &= (r - t_0) - \delta', \\ \operatorname{dist}^g(x, z') &= \operatorname{dist}^g(x,y) - \operatorname{dist}^g(z', y) \\ &= r - (r - t_0) + \delta' = t_0 + \delta'. \end{aligned}$$

We consider now the piecewise smooth curve $\bar{c}$ which initially follows c from x to $c(t_0)$ and then the minimal geodesic from $c(t_0)$ to z', parameterized by arc-length. We just checked that the curve $\bar{c}$ has minimal length $t_0 + \delta'$. Thus each piece of $\bar{c}$ has also minimal length, in particular the piece between $\bar{c}(t_1)$ and $\bar{c}(t_2)$, where $t_1 < t_0 < t_2$. Since we may choose these two points near to each other, $\bar{c}$ is a minimal geodesic between them by (23.4). Thus $\bar{c}$

equals c, $z' = c(t_0 + \delta)$, $\operatorname{dist}^g(c(t_0+\delta'), y) = \operatorname{dist}^g(z', y) = r - (\delta' + t_0)$ by (**), and (*) holds for $t_0 + \delta'$ also, which contradicts the maximality of t_0 for the validity of (*). Thus the assumption $t_0 < r$ is wrong and claim (7) follows.

Finally, by claim (7) we have $\operatorname{dist}^g(c(r), y) = r - r = 0$; thus $c(t) = \exp_x(t.X_x)$ is a geodesic from x to y of length $r = \operatorname{dist}^g(x, y)$, thus of minimal length, so (6) for the special points follows.

(4) $\implies$ (6), by the foregoing proof, since then any point is special.

(5) $\implies$ (2) Let $A \subset M$ be closed and bounded for the geodesic distance. Suppose that A has diameter $r < \infty$. Then A is completely contained in one connected component of M, by (23.5). Let x be the special point in this connected component with $\exp_x$ defined on the whole of T_xM. Take $y \in A$. By (6) for the special point x (which follows from (5)), there exists a geodesic from x to y of minimal length $\operatorname{dist}^g(x, y) =: s < \infty$, and each point z of A can be connected to x by a geodesic of minimal length

$$\operatorname{dist}^g(x, z) \le \operatorname{dist}^g(x, y) + \operatorname{dist}^g(y, z) \le r + s.$$

Thus the compact set (as continuous image of a compact ball) $\exp_x\{X_x \in T_xM : |X_x|_g \le r + s\}$ contains A. Since A is closed, it is compact too. □

23.7. Conformal metrics. Two Riemann metrics g_1 and g_2 on a manifold M are called *conformal* if there exists a smooth nowhere vanishing function f with $g_2 = f^2.g_1$. Then g_1 and g_2 have the same angles, but not the same lengths. A local diffeomorphism $\varphi : (M_1, g_1) \to (M_2, g_2)$ is called *conformal* if $\varphi^* g_2$ is conformal to g_1.

As an example, which also explains the name, we mention that any holomorphic mapping with nonvanishing derivative between open domains in $\mathbb{C}$ is conformal for the Euclidean inner product. This is clear from the polar decomposition $\varphi'(z) = |\varphi'(z)|e^{i\arg(\varphi'(z))}$ of the derivative.

As another, not unrelated, example we note that *the stereographic projection from* (1.2) *is a conformal mapping:*

$$u_+ : (S^n \setminus \{a\}, g^{S^n}) \to \{a\}^\perp \to (\mathbb{R}^n, \langle\ ,\ \rangle), \qquad u_+(x) = \tfrac{x - \langle x,a\rangle a}{1 - \langle x,a\rangle}.$$

To see this, take $X \in T_xS^n \subset T_x\mathbb{R}^{n+1}$, so that $\langle X, x\rangle = 0$. Then we get:

$$\begin{aligned}
du_+(x)X &= \tfrac{(1-\langle x,a\rangle)(X - \langle X,a\rangle a) + \langle X,a\rangle(x - \langle x,a\rangle a)}{(1-\langle x,a\rangle)^2} \\
&= \tfrac{1}{(1-\langle x,a\rangle)^2}\big((1 - \langle x,a\rangle)X + \langle X,a\rangle x - \langle x,a\rangle a\big), \\
\langle du_+(x)X, du_+(x)Y\rangle &= \tfrac{1}{(1-\langle x,a\rangle)^2}\langle X, Y\rangle = \tfrac{1}{(1-\langle x,a\rangle)^2}\,(g^{S^n})_x(X, Y).
\end{aligned}$$

23.8. Theorem (Nomizu-Ozeki, Morrow). *Let (M, g) be a connected Riemann manifold. Then we have:*

(1) *There exist complete Riemann metrics on M which are conformal to g and are equal to g on any given compact subset of M.*

(2) *There also exist Riemann metrics on M such that M has finite diameter which are conformal to g and are equal to g on any given compact subset of M. If M is not compact, then by* (23.6.2) *a Riemann metric for which M has finite diameter is not complete.*

Thus the sets of all complete Riemann metrics and of all Riemann metrics with bounded diameter are both dense in the compact C^∞-topology on the space of all Riemann metrics.

Proof of (1). For $x \in M$ let

$$r(x) := \sup\{r : B_x(r) = \{y \in M : \operatorname{dist}^g(x,y) \le r\} \text{ is compact in } M\}.$$

If $r(x) = \infty$ for one x, then g is a complete metric by (23.6.2). Since $\exp_x$ is a diffeomorphism near 0_x, $r(x) > 0$ for all x. We assume that $r(x) < \infty$ for all x.

Claim. $|r(x) - r(y)| \le \operatorname{dist}^g(x,y)$; thus $r : M \to \mathbb{R}$ is continuous, since: For small $\varepsilon > 0$ the set $B_x(r(x) - \varepsilon)$ is compact, $\operatorname{dist}^g(z,x) \le \operatorname{dist}^g(z,y) + \operatorname{dist}^g(y,x)$ implies that $B_y(r(x) - \varepsilon - \operatorname{dist}^g(x,y)) \subseteq B_x(r(x) - \varepsilon)$ is compact, and thus $r(y) \ge r(x) - \operatorname{dist}^g(x,y) - \varepsilon$ and $r(x) - r(y) \le \operatorname{dist}^g(x,y)$. Now interchange x and y.

By a partition of unity argument we now construct a smooth function $f \in C^\infty(M, \mathbb{R}_{>0})$ with $f(x) > \frac{1}{r(x)}$. Consider the Riemann metric $\bar g = f^2 g$.

Claim. $\bar B_x(\frac14) := \{y \in M : \operatorname{dist}^{\bar g}(x,y) \le \frac14\} \subset B_x(\frac12 r(x))$; thus it is compact.
Suppose $y \notin B_x(\frac12 r(x))$. For any piecewise smooth curve c from x to y we have

$$L^g(c) = \int_0^1 |c'(t)|_g\, dt > \frac{r(x)}{2},$$

$$L^{\bar g}(c) = \int f(c(t)).|c'(t)|_g\, dt = f(c(t_0)) \int_0^1 |c'(t)|_g\, dt > \frac{L^g(c)}{r(c(t_0))},$$

for some $t_0 \in [0,1]$, by the mean value theorem of integral calculus. Moreover,

$$\begin{aligned} |r(c(t_0)) - r(x)| &\le \operatorname{dist}^g(c(t_0), x) \le L^g(c) =: L, \\ r(c(t_0)) &\le r(x) + L, \\ L^{\bar g}(c) &\ge \frac{L}{r(x) + L} \ge \frac{L}{3L} = \frac13, \end{aligned}$$

so $y \notin \bar{B}_x(\frac{1}{4})$ either.

Claim. $(M, \bar{g})$ is a complete Riemann manifold.

Let $X \in T_xM$ with $|X|_{\bar{g}} = 1$. Then $\exp^{\bar{g}}(t.X)$ is defined for $|t| \le \frac{1}{5} < \frac{1}{4}$. But also $\exp^{\bar{g}}(s.\frac{\partial}{\partial t}|_{t=\pm 1/5}\exp^{\bar{g}}(t.X))$ is defined for $|s| < \frac{1}{4}$ which equals $\exp^{\bar{g}}((\pm\frac{1}{5}+s)X)$, and so on. Thus $\exp^{\bar{g}}(t.X)$ is defined for all $t \in \mathbb{R}$, and by (23.6.4) the metric $\bar{g}$ is complete.

Claim. We may choose f in such a way that $f = 1$ on a neighborhood of any given compact set $K \subset M$.

Let $C = \max\{\frac{1}{r(x)} : x \in K\} + 1$. By a partition of unity argument we construct a smooth function f with $f = 1$ on a neighborhood of K and $Cf(x) > \frac{1}{r(x)}$ for all x. By the arguments above, $C^2 f^2 g$ is then a complete metric; thus so is $f^2 g$.

Proof of (2). Let g be a complete Riemann metric on M. We choose $x \in M$, a smooth function h with $h(y) > \operatorname{dist}^g(x, y)$, and we consider the Riemann metric $\tilde{g}_y = e^{-2h(y)} g_y$. By (23.6.6) for any $y \in M$ there exists a minimal g-geodesic c from x to y, parameterized by arc-length. Then $h(c(s)) > \operatorname{dist}^g(x, c(s)) = s$ for all $s \le \operatorname{dist}^g(x, y) =: L$. But then

$$L^{\tilde{g}}(c) = \int_0^L e^{-h(c(s))}|c'(s)|_g\,ds < \int_0^L e^{-s}1\,ds < \int_0^\infty e^{-s}ds = 1,$$

so that M has diameter 1 for the Riemann metric $\tilde{g}$. We may also obtain that $\tilde{g} = g$ on a compact set as above. □

23.9. Proposition. *Let (M, g) be a complete Riemann manifold. Let $X \in \mathfrak{X}(M)$ be a vector field which is bounded with respect to g, $|X|_g \le C$. Then X is a complete vector field; it admits a global flow.*

Proof. The flow of X is given by the differential equation $\frac{\partial}{\partial t}\operatorname{Fl}^X_t(x) = X(\operatorname{Fl}^X_t(x))$ with initial value $\operatorname{Fl}^X_0(x) = x$. Suppose that $c(t) = \operatorname{Fl}^X_t(x)$ is defined on (a, b) and that $b < \infty$, say. Then

$$\begin{aligned}\operatorname{dist}^g(c(b-1/n), c(b-1/m)) &\le L_{b-1/n}^{b-1/m}(c) = \int_{b-1/n}^{b-1/m} |c'(t)|_g\,dt \\ &= \int_{b-1/n}^{b-1/m} |X(c(t))|_g\,dt \le \int_{b-1/n}^{b-1/m} C\,dt = C.(\tfrac{1}{m} - \tfrac{1}{n}) \to 0,\end{aligned}$$

so that $c(b-1/n)$ is a Cauchy sequence in the complete metrical space M and the limit $c(b) = \lim_{n\to\infty} c(b-1/n)$ exists. But then we may continue the flow beyond b by $\operatorname{Fl}^X_s(\operatorname{Fl}^X_b(x)) = \operatorname{Fl}^X_{b+s}$. □

23.10. Problem (Unsolved until May 2, 2008, to the author's knowledge). *Let X be a complete vector field on a manifold M. Does there exist a complete Riemann metric g on M such that the vector field X is bounded with respect to g?*

The only inroad towards this problem is the following:

Proposition ([**76**]). *Let X be a complete vector field on a connected manifold M. Then there exists a complete Riemann metric g on the manifold $M \times \mathbb{R}$ such that the vector field $X \times \partial_t \in \mathfrak{X}(M \times \mathbb{R})$ is bounded with respect to g.*

Proof. Since $\mathrm{Fl}_t^{X\times\partial_t}(x,s) = (\mathrm{Fl}_t^X(x), s+t)$, the vector field $X \times \partial_t$ is also complete. It is nowhere 0.

Choose a smooth proper function f_1 on M; for example, if a smooth function f_1 satisfies $f_1(x) > \mathrm{dist}^{\bar g}(x_0, x)$ for a complete Riemann metric $\bar g$ on M, then f_1 is proper by (23.6.2).

For a Riemann metric $\bar g$ on M we consider the Riemann metric $\tilde g$ on the product $M \times \mathbb{R}$ which equals g_x on $T_xM \cong T_xM \times 0_t = T_{(x,t)}(M \times \{t\})$ and satisfies

$$|X \times \partial_t|_{\tilde g} = 1 \quad \text{and} \quad \tilde g_{(x,t)}((X\times\partial_t)(x,t), T_{(x,t)}(M\times\{t\})) = 0.$$

We will also use the fiberwise $\tilde g$-orthogonal projections

$$\begin{aligned} &\mathrm{pr}_M : T(M\times\mathbb{R}) \to TM \times 0 \quad \text{and} \\ &\mathrm{pr}_X : T(M\times\mathbb{R}) \to \mathbb{R}.(X\times\partial_t) \cong \mathbb{R}. \end{aligned}$$

The smooth function $f_2(x,s) = f_1(\mathrm{Fl}_{-s}^X(x)) + s$ satisfies the following and is thus still proper:

$$\begin{aligned} (\mathcal{L}_{X\times\partial_t} f_2)(x,s) &= \partial|_0 f_2(\mathrm{Fl}_t^{X\times\partial_t}(x,s)) \\ &= \partial|_0 f_2(\mathrm{Fl}_t^X(x), s+t) \\ &= \partial|_0 \left(f_1(\mathrm{Fl}_{-s-t}^X(\mathrm{Fl}_t^X(x))) + s + t\right) \\ &= \partial|_0 f_1(\mathrm{Fl}_{-s}^X(x)) + 1 = 1. \end{aligned}$$

By a partition of unity argument we construct another smooth function $f_3 : M \times \mathbb{R} \to \mathbb{R}$ which satisfies

$$f_3(x,s)^2 > \max\left\{|Y(f_2)|^2 : Y \in T_{(x,s)}(M\times\{s\}), |Y|_{\tilde g} = 1\right\}.$$

Finally we define a Riemann metric g on $M \times \mathbb{R}$ by

$$g_{(x,t)}(Y,Z) = f_3(x,t)^2\, \tilde g_{(x,t)}(\mathrm{pr}_M(Y), \mathrm{pr}_M(Z)) + \mathrm{pr}_X(Y)\cdot \mathrm{pr}_X(Z)$$

for $Y, Z \in T_{(x,t)}(M \times \mathbb{R})$, which satisfies $|X \times \partial_t|_g = 1$.

Claim. g is a complete Riemann metric on $M \times \mathbb{R}$.
Let c be a piecewise smooth curve parameterized by g-arc-length. Then

$$
\begin{aligned}
|c'|_g = 1, &\quad \text{thus also} \quad |\operatorname{pr}_M(c')|_g \le 1, \quad |\operatorname{pr}_X(c')| \le 1,\\
\tfrac{\partial}{\partial t} f_2(c(t)) &= df_2(c'(t))\\
&= (\operatorname{pr}_M(c'(t)))(f_2) + \operatorname{pr}_X(c'(t))(f_2),\\
|\tfrac{\partial}{\partial t} f_2(c(t))| &\le \left|\frac{\operatorname{pr}_M(c'(t))}{|\operatorname{pr}_M(c'(t))|_g}(f_2)\right| + \left|\frac{\operatorname{pr}_X(c'(t))}{|\operatorname{pr}_X(c'(t))|_g}(f_2)\right|\\
&= \left|\frac{1}{f_3(c(t))}\frac{\operatorname{pr}_M(c'(t))}{|\operatorname{pr}_M(c'(t))|_{\bar g}}(f_2)\right| + |\mathcal{L}_{X\times\partial_t} f_2| < 2
\end{aligned}
$$

by the definition of g and the properties of f_3 and f_2. Thus

$$|f_2(c(t)) - f_2(c(0))| \le \int_0^t |\tfrac{\partial}{\partial t} f_2(c(t))|\, dt \le 2t.$$

Since this holds for every such c, we conclude that

$$|f_2(x) - f_2(y)| \le 2\operatorname{dist}^g(x,y)$$

and thus each closed and dist^g-bounded set is contained in some

$$\{y \in M \times \mathbb{R} : \operatorname{dist}^g(x,y) \le R\} \subset f_2^{-1}([f_2(x) - \tfrac{R}{2}, f_2(x) + \tfrac{R}{2}])$$

which is compact since f_2 is proper. So $(M \times \mathbb{R}, g)$ is a complete Riemann manifold by (23.6.2). □

24. Parallel Transport and Curvature

24.1. Parallel transport. Let (M, ∇) be a manifold with a covariant derivative, as treated in (22.7). The pair (M, ∇) is also sometimes called an *affine manifold.*

A vector field $Y : N \to TM$ along a smooth mapping $f = \pi_M \circ Y : N \to M$ is called *parallel* if $\nabla_X Y = 0$ for any vector field $X \in \mathfrak{X}(N)$.

If $Y : \mathbb{R} \to TM$ is a vector field along a given curve $c = \pi_M \circ Y : \mathbb{R} \to M$, then

$$\nabla_{\partial_t} Y = K \circ TY \circ \partial_t = 0$$

takes the following form in a local chart, by (22.7.7):

$$
\begin{aligned}
K \circ TY \circ \partial_t &= K(\bar c(t), \bar Y(t); \bar c'(t), \bar Y'(t))\\
&= (\bar c(t), \bar Y'(t) - \Gamma_{\bar c(t)}(\bar Y(t), \bar c'(t))).
\end{aligned}
$$

This is a linear ordinary differential equation of first order for $\bar Y$ (since $\bar c$ is given). Thus for every initial value $Y(t_0)$ for $t_0 \in \mathbb{R}$ the parallel vector field Y along c is uniquely determined for the whole parameter space $\mathbb{R}$.

We formalize this by defining the *parallel transport* along the curve $c : \mathbb{R} \to M$ as

$$\operatorname{Pt}(c,t) : T_{c(0)}M \to T_{c(t)}M, \quad \operatorname{Pt}(c,t).Y(0) = Y(t),$$

where Y is any parallel vector field along c. Note that we treat this notion for principal bundles in (19.6) and for general fiber bundles in (17.8). This is a special case here.

Theorem. *On an affine manifold (M, ∇) the parallel transport has the following properties:*

(1) $\operatorname{Pt}(c,t) : T_{c(0)}M \to T_{c(t)}M$ *is a linear isomorphism for each $t \in \mathbb{R}$ and each curve $c : \mathbb{R} \to M$.*

(2) *For smooth $f : \mathbb{R} \to \mathbb{R}$ we have* $\operatorname{Pt}(c, f(t)) = \operatorname{Pt}(c \circ f, t) \operatorname{Pt}(c, f(0))$*; the reparameterization invariance.*

(3) $\operatorname{Pt}(c,t)^{-1} = \operatorname{Pt}(c(\ \ +t), -t)$.

(4) *If the covariant derivative is compatible with a pseudo-Riemann metric g on M, then $\operatorname{Pt}(c,t)$ is isometric, i.e.,*

$$g_{c(t)}(\operatorname{Pt}(c,t)X, \operatorname{Pt}(c,t)Y) = g_{c(0)}(X,Y).$$

Proof. (1) follows from the linearity of the differential equation.

(2) See also (17.8). Let X be parallel along c, $\nabla_{\partial_t} X = 0$ or $X(t) = \operatorname{Pt}(c,t)X(0)$. Then we have by (22.7.6)

$$\nabla_{\partial_t}(X \circ f) = \nabla_{T_t f.\partial_t} X = \nabla_{f'(t)\partial_t} X = f'(t)\nabla_{\partial_t} X = 0;$$

thus $X \circ f$ is also a parallel vector field along $c \circ f$, with initial value $X(f(0)) = \operatorname{Pt}(c, f(0))X(0)$. So

$$\operatorname{Pt}(c, f(t))X(0) = X(f(t)) = \operatorname{Pt}(c \circ f, t) \operatorname{Pt}(c, f(0))X(0).$$

(3) follows from (2).

(4) Let X and Y be parallel vector fields along c, i.e., $\nabla_{\partial_t} X = 0$, etc. Then

$$\partial_t g(X(t), Y(t)) = g(\nabla_{\partial_t} X(t), Y(t)) + g(X(t), \nabla_{\partial_t} Y(t)) = 0;$$

thus $g(X(t), Y(t))$ is constant in t. □

24.2. Flows and parallel transports. Let $X \in \mathfrak{X}(M)$ be a vector field on an affine manifold (M, ∇). Let $C : TM \times_M TM \to T^2M$ be the linear connection for the covariant derivative ∇; see (22.7). The *horizontal lift of the vector field* X is then given by $C(X,\ \ ) \in \mathfrak{X}(TM)$ which is π_M-related to X: $T(\pi_M) \circ C(X,\ \ ) = X \circ \pi_M$. A flow line $\operatorname{Fl}_t^{C(X,\ \ )}(Y_x)$ is then a smooth curve in TM whose tangent vector is everywhere horizontal, so the curve is parallel, and $\pi_M(\operatorname{Fl}_t^{C(X,\ \ )}(Y_x)) = \operatorname{Fl}_t^X(x)$ by (3.14). Thus

$$\operatorname{Pt}(\operatorname{Fl}^X, t) = \operatorname{Fl}_t^{C(X,\ \ )}. \tag{1}$$

Proposition. *For vector fields $X, Y \in \mathfrak{X}(M)$ we have:*

$$\begin{aligned}\nabla_X Y &= \partial|_0 (\mathrm{Fl}^{C(X,\ \ )}_{-t} \circ Y \circ \mathrm{Fl}^X_t) \\ &= \partial|_0 \operatorname{Pt}(\mathrm{Fl}^X, -t) \circ Y \circ \mathrm{Fl}^X_t \qquad (2) \\ &=: \partial|_0 \operatorname{Pt}(\mathrm{Fl}^X, t)^* Y,\end{aligned}$$

and more generally,

$$\begin{aligned}\tfrac{\partial}{\partial t} \operatorname{Pt}(\mathrm{Fl}^X, -t) \circ Y \circ \mathrm{Fl}^X_t &= \tfrac{\partial}{\partial t} \operatorname{Pt}(\mathrm{Fl}^X, t)^* Y \qquad (3) \\ &= \operatorname{Pt}(\mathrm{Fl}^X, t)^* \nabla_X Y \\ &= \operatorname{Pt}(\mathrm{Fl}^X, -t) \circ \nabla_X Y \circ \mathrm{Fl}^X_t \\ &= \nabla_X (\operatorname{Pt}(\mathrm{Fl}^X, t)^* Y).\end{aligned}$$

(4) *The local vector bundle isomorphism $\operatorname{Pt}(\mathrm{Fl}^X, t)$ over Fl^X_t induces vector bundle isomorphisms $\operatorname{Pt}^{\otimes}(\mathrm{Fl}^X, t)$ on all tensor bundles $\bigotimes^p TM \otimes \bigotimes^q T^*M$ over Fl^X_t. For each tensor field A we have*

$$(2') \qquad \nabla_X A = \partial|_0 \operatorname{Pt}^{\otimes}(\mathrm{Fl}^X, -t) \circ A \circ \mathrm{Fl}^X_t = \partial|_0 \operatorname{Pt}^{\otimes}(\mathrm{Fl}^X, t)^* A.$$

$$\begin{aligned}(3') \qquad \tfrac{\partial}{\partial t} \operatorname{Pt}^{\otimes}(\mathrm{Fl}^X, t)^* A &= \operatorname{Pt}^{\otimes}(\mathrm{Fl}^X, t)^* \nabla_X A = \operatorname{Pt}(\mathrm{Fl}^X, -t) \circ \nabla_X A \circ \mathrm{Fl}^X_t \\ &= \nabla_X (\operatorname{Pt}^{\otimes}(\mathrm{Fl}^X, t)^* A).\end{aligned}$$

Proof. (2) We compute

$$\begin{aligned}&\partial|_0 \mathrm{Fl}^{C(X,\ \ )}_{-t} (Y(\mathrm{Fl}^X_t(x))) \\ &\quad = -C\left(X, \mathrm{Fl}^{C(X,\ \ )}_0 (Y(\mathrm{Fl}^X_0(x)))\right) + T(\mathrm{Fl}^{C(X,\ \ )}_0) \partial|_0 (Y(\mathrm{Fl}^X_t(x))) \\ &\quad = -C(X(x), Y(x)) + TY.X(x) \\ &\quad = TY.X(x) - C(T(\pi_M).TY.X(x), \pi_{TM}(TY.X(x))) \\ &\quad = (\mathrm{Id}_{T^2M} - (\text{horizontal projection}))TY.X(x) \\ &\quad = \mathrm{vl}(Y(x), K.TY.X(x)) = \mathrm{vl}(Y(x), (\nabla_X Y)(x)).\end{aligned}$$

The vertical lift disappears if we identify the tangent space to the fiber T_xM with the fiber.

(3) We did this several times already; see (3.13), (8.16), and (9.6):

$$\begin{aligned}\tfrac{\partial}{\partial t} \operatorname{Pt}(\mathrm{Fl}^X, t)^* Y &= \tfrac{\partial}{\partial s}|_0 \left(\operatorname{Pt}(\mathrm{Fl}^X, -t) \circ \operatorname{Pt}(\mathrm{Fl}^X, -s) \circ Y \circ \mathrm{Fl}^X_s \circ \mathrm{Fl}^X_t\right) \\ &= \operatorname{Pt}(\mathrm{Fl}^X, -t) \circ \tfrac{\partial}{\partial s}|_0 \left(\operatorname{Pt}(\mathrm{Fl}^X, -s) \circ Y \circ \mathrm{Fl}^X_s\right) \circ \mathrm{Fl}^X_t \\ &= \operatorname{Pt}(\mathrm{Fl}^X, -t) \circ (\nabla_X Y) \circ \mathrm{Fl}^X_t = \operatorname{Pt}(\mathrm{Fl}^X, t)^* \nabla_X Y, \\ \tfrac{\partial}{\partial t} \operatorname{Pt}(\mathrm{Fl}^X, t)^* Y &= \tfrac{\partial}{\partial s}|_0 \operatorname{Pt}(\mathrm{Fl}^X, s)^* \operatorname{Pt}(\mathrm{Fl}^X, t)^* Y = \nabla_X (\operatorname{Pt}(\mathrm{Fl}^X, t)^* Y).\end{aligned}$$

(4) For a tensor A with foot point $\mathrm{Fl}^X_t(x)$ let us define $\operatorname{Pt}^{\otimes}(\mathrm{Fl}^X, t)^* A$ with foot point x by

$$\begin{aligned}
&(\mathrm{Pt}^{\otimes}(\mathrm{Fl}^X,t)A)(X_1,\dots,X_q,\omega^1,\dots,\omega^p)\\
&= A(\mathrm{Pt}(\mathrm{Fl}^X,t)X_1,\dots,\mathrm{Pt}(\mathrm{Fl}^X,t)X_q,\mathrm{Pt}(\mathrm{Fl}^X,-t)^*\omega^1,\dots,\mathrm{Pt}(\mathrm{Fl}^X,-t)^*\omega^p).
\end{aligned}$$

Thus $\mathrm{Pt}^{\otimes}(\mathrm{Fl}^X,t)$ is fiberwise an algebra homomorphism of the tensor algebra which commutes with all contractions. Thus $\partial|_0 \mathrm{Pt}^{\otimes}(\mathrm{Fl}^X,t)^*$ becomes a derivation on the algebra of all tensor fields which commutes with contractions and equals ∇_X on vector fields. Thus by (22.12) it coincides with ∇_X on all tensor fields. This implies (2′).

(3′) can be proved in the same way as (3). □

24.3. Curvature. Let (M,∇) be an affine manifold. The *curvature of the covariant derivative* ∇ is given by

(1) $$\begin{aligned} R(X,Y)Z &= \nabla_X\nabla_Y Z - \nabla_Y\nabla_X Z - \nabla_{[X,Y]}Z\\ &= ([\nabla_X,\nabla_Y]-\nabla_{[X,Y]})Z, \end{aligned}$$

for $X,Y,Z\in\mathfrak{X}(M)$.

A straightforward computation shows that $R(X,Y)Z$ is $C^\infty(M)$-linear in each entry; thus R is a $\binom{1}{3}$-tensor field on M.

In a local chart (U,u) we have (where $\partial_i = \frac{\partial}{\partial u^i}$):

$$\begin{aligned}
X|_U &= \sum X^i\partial_i, \qquad Y|_U = \sum Y^j\partial_j, \qquad Z|_U=\sum Z^k\partial_k,\\
R(X,Y)(Z)|_U &= \sum X^iY^jZ^kR(\partial_i,\partial_j)(\partial_k)\\
&=: \Big(\sum R^l_{i,j,k}\,du^i\otimes du^j\otimes du^k\otimes\partial_l\Big)(X,Y,Z),\\
\sum R^l_{i,j,k}\partial_l &= R(\partial_i,\partial_j)(\partial_k)\\
&= \nabla_{\partial_i}\nabla_{\partial_j}\partial_k - \nabla_{\partial_j}\nabla_{\partial_i}\partial_k - 0\\
&= \nabla_{\partial_i}(-\sum\Gamma^m_{j,k}\partial_m) - \nabla_{\partial_j}(-\sum\Gamma^m_{i,k}\partial_m)\\
&= -\sum\partial_i\Gamma^m_{j,k}\partial_m - \sum\Gamma^m_{j,k}\nabla_{\partial_i}\partial_m + \sum\partial_j\Gamma^m_{i,k}\partial_m + \sum\Gamma^m_{i,k}\nabla_{\partial_j}\partial_m\\
&= -\sum\partial_i\Gamma^l_{j,k}\partial_l + \sum\Gamma^m_{j,k}\Gamma^l_{i,m}\partial_l + \sum\partial_j\Gamma^l_{i,k}\partial_l - \sum\Gamma^m_{i,k}\Gamma^l_{j,m}\partial_l.
\end{aligned}$$

We can collect all local formulas here, also from (22.9.7) or (22.5.6) and from (22.4.2) in the case of a Levi-Civita connection (where $X=(x,\bar X)$, etc.):

(2) $$\begin{aligned}
\nabla_{\partial_i}\partial_j &= -\sum\Gamma^l_{i,j},\\
\Gamma^k_{ij} &= \tfrac12\sum g^{kl}(\partial_l g_{ij} - \partial_i g_{lj} - \partial_j g_{il}),\\
R^l_{i,j,k} &= -\partial_i\Gamma^l_{j,k} + \partial_j\Gamma^l_{i,k} + \sum\Gamma^m_{j,k}\Gamma^l_{i,m} - \sum\Gamma^m_{i,k}\Gamma^l_{j,m},\\
\bar R(\bar X,\bar Y)\bar Z &= -d\Gamma(x)(\bar X)(\bar Y,\bar Z) + d\Gamma(x)(\bar Y)(\bar X,\bar Z)\\
&\quad + \Gamma_x(\bar X,\Gamma_x(\bar Y,\bar Z)) - \Gamma_x(\bar Y,\Gamma_x(\bar X,\bar Z)).
\end{aligned}$$

24.4. Theorem. *Let ∇ be a covariant derivative on a manifold M, with torsion* Tor*; see* (22.10)*. Then the curvature R has the following properties, where $X, Y, Z, U \in \mathfrak{X}(M)$:*

(1) $$R(X,Y)Z = -R(Y,X)Z.$$

(2) $$\sum_{cyclic} R(X,Y)Z = \sum_{cyclic} \Big((\nabla_X \operatorname{Tor})(Y,Z) + \operatorname{Tor}(\operatorname{Tor}(X,Y),Z)\Big),$$

algebraic Bianchi identity.

(3) $$\sum_{cyclic} \Big((\nabla_X R)(Y,Z) + R(\operatorname{Tor}(X,Y),Z)\Big) = 0,$$ *Bianchi identity.*

If the connection ∇ is torsion-free, we have:

(2′) $$\sum_{cyclic} R(X,Y)Z = 0,$$ *algebraic Bianchi identity.*

(3′) $$\sum_{\substack{cyclic\\ X,Y,Z}} (\nabla_X R)(Y,Z) = 0,$$ *Bianchi identity.*

If ∇ is the (torsion-free) Levi-Civita connection of a pseudo-Riemann metric g, then we have moreover:

(4) $$g(R(X,Y)Z,U) = g(R(Z,U)X,Y),$$

(5) $$g(R(X,Y)Z,U) = -g(R(X,Y)U,Z).$$

Proof. (2) The extension of ∇_X to tensor fields was treated in (22.12):

(6) $$(\nabla_X \operatorname{Tor})(Y,Z) = \nabla_X(\operatorname{Tor}(Y,Z)) - \operatorname{Tor}(\nabla_X Y, Z) - \operatorname{Tor}(Y, \nabla_X Z).$$

From the definition (22.10.1) of the torsion:

$$\begin{aligned}\operatorname{Tor}(\operatorname{Tor}(X,Y),Z) &= \operatorname{Tor}(\nabla_X Y - \nabla_Y X - [X,Y], Z)\\ &= \operatorname{Tor}(\nabla_X Y, Z) + \operatorname{Tor}(Z, \nabla_Y X) - \operatorname{Tor}([X,Y],Z).\end{aligned}$$

These combine to

$$\begin{aligned}&\sum_{\text{cyclic}} \operatorname{Tor}(\operatorname{Tor}(X,Y),Z)\\ &\qquad = \sum_{\text{cyclic}} \Big(\nabla_X(\operatorname{Tor}(Y,Z)) - (\nabla_X \operatorname{Tor})(Y,Z) - \operatorname{Tor}([X,Y],Z)\Big)\end{aligned}$$

and then

$$\begin{aligned}&\sum_{\text{cyclic}} \Big((\nabla_X \operatorname{Tor})(Y,Z) + \operatorname{Tor}(\operatorname{Tor}(X,Y),Z)\Big)\\ &\qquad = \sum_{\text{cyclic}} \Big(\nabla_X(\operatorname{Tor}(Y,Z)) - \operatorname{Tor}([X,Y],Z)\Big)\end{aligned}$$

$$
\begin{aligned}
&= \sum_{\text{cyclic}} \Big(\nabla_X \nabla_Y Z - \nabla_X \nabla_Z Y - \nabla_X [Y, Z] \\
&\qquad\qquad - \nabla_{[X,Y]} Z + \nabla_Z [X, Y] + [[X, Y], Z] \Big) \\
&= \sum_{\text{cyclic}} \Big(\nabla_X \nabla_Y Z - \nabla_X \nabla_Z Y - \nabla_{[X,Y]} Z \Big) = \sum_{\text{cyclic}} R(X, Y) Z.
\end{aligned}
$$

(3) We have

$$
\begin{aligned}
\sum_{\text{cyclic}} R(\operatorname{Tor}(X, Y), Z) &= \sum_{\text{cyclic}} R(\nabla_X Y - \nabla_Y X - [X, Y], Z) \\
&= \sum_{\text{cyclic}} \Big(R(\nabla_X Y, Z) + R(Z, \nabla_Y X) - R([X, Y], Z) \Big)
\end{aligned}
$$

and

$$
\begin{aligned}
&\sum_{\text{cyclic}} (\nabla_X R)(Y, Z) \\
&\quad = \sum_{\text{cyclic}} \Big(\nabla_X R(Y, Z) - R(\nabla_X Y, Z) - R(Y, \nabla_X Z) - R(Y, Z) \nabla_X \Big)
\end{aligned}
$$

which combines to

$$
\begin{aligned}
&\sum_{\text{cyclic}} \Big((\nabla_X R)(Y, Z) + R(\operatorname{Tor}(X, Y), Z) \Big) \\
&\quad = \sum_{\text{cyclic}} \Big(\nabla_X R(Y, Z) - R(Y, Z) \nabla_X - R([X, Y], Z) \Big) \\
&\quad = \sum_{\text{cyclic}} \Big(\nabla_X \nabla_Y \nabla_Z - \nabla_X \nabla_Z \nabla_Y - \nabla_X \nabla_{[Y,Z]} \\
&\qquad\qquad - \nabla_Y \nabla_Z \nabla_X + \nabla_Z \nabla_Y \nabla_X + \nabla_{[Y,Z]} \nabla_X \\
&\qquad\qquad - \nabla_{[X,Y]} \nabla_Z + \nabla_Z \nabla_{[X,Y]} + \nabla_{[[X,Y],Z]} \Big) = 0.
\end{aligned}
$$

(5) It suffices to prove $g(R(X, Y)Z, Z) = 0$:

$$
\begin{aligned}
0 &= \mathcal{L}_0(g(Z, Z)) \\
&= (XY - YX - [X, Y]) g(Z, Z) \\
&= 2X g(\nabla_Y Z, Z) - 2Y g(\nabla_X Z, Z) - 2g(\nabla_{[X,Y]} Z, Z) \\
&= 2g(\nabla_X \nabla_Y Z, Z) + 2g(\nabla_Y Z, \nabla_X Z) \\
&\quad - 2g(\nabla_Y \nabla_X Z, Z) - 2g(\nabla_X Z, \nabla_Y Z) - 2g(\nabla_{[X,Y]} Z, Z) \\
&= 2g((\nabla_X \nabla_Y - \nabla_Y \nabla_X - \nabla_{[X,Y]}) Z, Z) = 2g(R(X, Y)Z, Z).
\end{aligned}
$$

(4) is an algebraic consequence of (1), (2′), and (5). Take (2′) four times, cyclically permuted, with different signs:

$$\begin{aligned} g(R(X,Y)Z,U)+g(R(Y,Z)X,U)+g(R(Z,X)Y,U)&=0,\\ g(R(Y,Z)U,X)+g(R(Z,U)Y,X)+g(R(U,Y)Z,X)&=0,\\ -g(R(Z,U)X,Y)-g(R(U,X)Z,Y)-g(R(X,Z)U,Y)&=0,\\ -g(R(U,X)Y,Z)-g(R(X,Y)U,Z)-g(R(Y,U)X,Z)&=0. \end{aligned}$$

Add these:

$$2g(R(X,Y)Z,U)-2g(R(Z,U)X,Y)=0. \quad \square$$

24.5. Theorem. *Let $K : TTM \to TM$ be the connector of the covariant derivative ∇ on M. If $s : N \to TM$ is a vector field along $f := \pi_M \circ s : N \to M$, then we have for vector fields X, $Y \in \mathfrak{X}(N)$*

$$\begin{aligned} &\nabla_X\nabla_Y s-\nabla_Y\nabla_X s-\nabla_{[X,Y]}s\\ &\qquad=(K\circ TK\circ\kappa_{TM}-K\circ TK)\circ TTs\circ TX\circ Y\\ &\qquad=R\circ(Tf\circ X,Tf\circ Y)s : N\to TM, \end{aligned}$$

where $R \in \Omega^2(M; L(TM,TM))$ is the curvature.

Proof. Recall from (22.9) that $\nabla_X s = K \circ Ts \circ X$. For $A, B \in T_Z(TM)$ we have

$$\begin{aligned} \mathrm{vl}_{TM}(K(A),K(B))&=\partial_t|_0(K(A)+tK(B))=\partial_t|_0K(A+tB)\\ &=TK\circ\partial_t|_0(A+tB)=TK\circ\mathrm{vl}_{(TTM,\pi_{TM},TM)}(A,B). \end{aligned}$$

We use then (22.8.9) and some obvious commutation relations:

$$\begin{aligned} &\nabla_X\nabla_Y s-\nabla_Y\nabla_X s-\nabla_{[X,Y]}s\\ &\quad=K\circ T(K\circ Ts\circ Y)\circ X-K\circ T(K\circ Ts\circ X)\circ Y-K\circ Ts\circ[X,Y],\\ &K\circ Ts\circ[X,Y]=K\circ\mathrm{vl}_{TM}\circ(K\circ Ts\circ Y,K\circ Ts\circ[X,Y]) \quad \text{by (22.8.9)}\\ &\quad=K\circ TK\circ\mathrm{vl}_{TTM}\circ(Ts\circ Y,Ts\circ[X,Y])\\ &\quad=K\circ TK\circ TTs\circ\mathrm{vl}_{TN}\circ(Y,[X,Y])\\ &\quad=K\circ TK\circ TTs\circ(TY\circ X-\kappa_N\circ TX\circ Y) \quad \text{by (8.14)}\\ &\quad=K\circ TK\circ TTs\circ TY\circ X-K\circ TK\circ TTs\circ\kappa_N\circ TX\circ Y. \end{aligned}$$

Now we sum up and use $TTs\circ\kappa_N=\kappa_{TM}\circ TTs$ to get the first result. If in particular we choose $f = \mathrm{Id}_M$ so that X, Y, s are vector fields on M, then we get the curvature R.

To see that in the general case $(K\circ TK\circ\kappa_E-K\circ TK)\circ TTs\circ TX\circ Y$ coincides with $R(Tf\circ X, Tf\circ Y)s$, we have to write out the expression $(TTs\circ TX\circ Y)(x) \in TTTM$ in canonical charts induced from charts of

N and M. There we have $X(x) = (x, \bar X(x))$, $Y(x) = (x, \bar Y(x))$, and also $s(x) = (f(x), \bar s(x))$. So we get:

$$\begin{aligned}(TTs\circ TX\circ Y)(x) &= TTs(x,\bar X(x);\bar Y(x), d\bar X(x)\bar Y(x))\\
&= \Big(f(x),\bar s(x), df(x).\bar X(x), d\bar s(x).\bar X(x); df(x).\bar Y(x), d\bar s(x).\bar Y(x),\\
&\qquad d^2f(x)(\bar Y(x),\bar X(x)) + df(x).d\bar X(x).\bar Y(x),\\
&\qquad d^2\bar s(x)(\bar Y(x),\bar X(x)) + d\bar s(x).d\bar X(x).\bar Y(x)\Big). \qquad (1)\end{aligned}$$

Recall (22.8.7) which said $K(x,y;a,b) = (x, b-\Gamma_x(a,y))$. Differentiating this, we get

$$\begin{aligned}&TK(x,y,a,b;\xi,\eta,\alpha,\beta)\\
&\quad= \Big(x, b-\Gamma_x(a,y);\xi,\beta - d\Gamma(x)(\xi)(a,y) - \Gamma_x(\alpha,y) - \Gamma_x(a,\eta)\Big).\end{aligned}$$

Thus

$$\begin{aligned}&(K\circ TK\circ \kappa_{TM} - K\circ TK)(x,y,a,b;\xi,\eta,\alpha,\beta)\\
&\quad= (K\circ TK)(x,y,\xi,\eta;a,b,\alpha,\beta) - (K\circ TK)(x,y,a,b;\xi,\eta,\alpha,\beta)\\
&\quad= K\big(x,\eta-\Gamma_x(\xi,y);a,\beta - d\Gamma(x)(a)(\xi,y) - \Gamma_x(\alpha,y) - \Gamma_x(\xi,b)\big)\\
&\qquad - K\big(x,b-\Gamma_x(a,y);\xi,\beta - d\Gamma(x)(\xi)(a,y) - \Gamma_x(\alpha,y) - \Gamma_x(a,\eta)\big)\\
&\quad= (x, -d\Gamma(x)(a)(\xi,y)\\
&\qquad + d\Gamma(x)(\xi)(a,y) + \Gamma_x(a,\Gamma_x(\xi,y)) - \Gamma_x(\xi,\Gamma_x(a,y)))\,. \qquad (2)\end{aligned}$$

Now we insert (1) into (2) and get

$$(K\circ TK\circ\kappa_{TM} - K\circ TK)\circ TTs\circ TX\circ Y = R\circ(Tf\circ X, Tf\circ Y)s. \qquad \square$$

24.6. Curvature and integrability of the horizontal bundle. What is it that the curvature is measuring? We give several answers; one of them is the following, which is intimately related to (16.13), (17.4), (19.2).

Let $C : TM\times_M TM \to T^2M$ be the linear connection corresponding to a covariant derivative ∇. For $X\in\mathfrak{X}(M)$ we denoted by $C(X,\quad)\in\mathfrak{X}(TM)$ the horizontal lift of the vector field X.

Lemma. *In this situation we have for $X, Y\in\mathfrak{X}(M)$ and $Z\in TM$*

$$[C(X,\quad),C(Y,\quad)](Z) - C([X,Y],Z) = -\operatorname{vl}(Z, R(X,Y)Z).$$

Proof. We compute locally, in charts induced by a chart (U,u) on M. A global proof can be found in (17.4) for general fiber bundles and in (19.2) for principal fiber bundles; see also (19.16). Writing $X(x) = (x, \bar{X}(x))$, $Y(x) = (x, \bar{Y}(x))$, and $Z = (x, \bar{Z})$, we have

$$\begin{aligned}
&C(X,Z) = (x,\bar{Z};\bar{X}(x),\Gamma_x(\bar{X}(x),\bar{Z})),\\
&C(Y,Z) = (x,\bar{Z};\bar{Y}(x),\Gamma_x(\bar{Y}(x),\bar{Z})),\\
&[C(X,\quad),C(Y,\quad)](Z)\\
&= \big(x,\bar{Z};d\bar{Y}(x).\bar{X}(x),d\Gamma(x)(\bar{X}(x))(\bar{Y}(x),\bar{Z})+\Gamma_x(d\bar{Y}(x).\bar{X}(x),\bar{Z})\\
&\qquad\qquad\qquad\qquad +\Gamma_x(\bar{Y}(x),\Gamma_x(\bar{X}(x),\bar{Z}))\big)\\
&\quad -\big(x,\bar{Z};d\bar{X}(x).\bar{Y}(x),d\Gamma(x)(\bar{Y}(x))(\bar{X}(x),\bar{Z})+\Gamma_x(d\bar{X}(x).\bar{Y}(x),\bar{Z})\\
&\qquad\qquad\qquad\qquad +\Gamma_x(\bar{X}(x),\Gamma_x(\bar{Y}(x),\bar{Z}))\big)\\
&= \big(x,\bar{Z};d\bar{Y}(x).\bar{X}(x),-d\bar{X}(x).\bar{Y}(x),\\
&\qquad \Gamma_x(d\bar{Y}(x).\bar{X}(x)-d\bar{X}(x).\bar{Y}(x),\bar{Z})\\
&\qquad +d\Gamma(x)(\bar{X}(x))(\bar{Y}(x),\bar{Z})-d\Gamma(x)(\bar{Y}(x))(\bar{X}(x),\bar{Z})\\
&\qquad +\Gamma_x(\bar{Y}(x),\Gamma_x(\bar{X}(x),\bar{Z}))-\Gamma_x(\bar{X}(x),\Gamma_x(\bar{Y}(x),\bar{Z}))\big)\\
&= \big(x,\bar{Z};\overline{[X,Y]}(x),\Gamma_x(\overline{[X,Y]}(x),\bar{Z})\big)\\
&\qquad +\big(x,\bar{Z};0,d\Gamma(x)(\bar{X}(x))(\bar{Y}(x),\bar{Z})-d\Gamma(x)(\bar{Y}(x))(\bar{X}(x),\bar{Z})\\
&\qquad\qquad +\Gamma_x(\bar{Y}(x),\Gamma_x(\bar{X}(x),\bar{Z}))-\Gamma_x(\bar{X}(x),\Gamma_x(\bar{Y}(x),\bar{Z}))\big)\\
&= C([X,Y],Z)+\mathrm{vl}(Z,-R(X(x),Y(x))Z), \quad \text{by (24.3.2).} \quad \square
\end{aligned}$$

The horizontal lift mapping $C(X, \quad)$ is a section of the horizontal bundle $C(TM, \quad) \subset T(TM)$, and any section is of that form. If the curvature vanishes, then by the theorem of Frobenius (3.20) the horizontal bundle is integrable and we get the leaves of the horizontal foliation.

Lemma. *Let M be a manifold and let ∇ be a flat covariant derivative on M (with vanishing curvature). Let $H \subset TM$ be a leaf of the horizontal foliation. Then $\pi_M|_H : H \to M$ is a covering map.*

Proof. Since $T(\pi_M|_H) = T(\pi_M)|C(TM, \quad)$ is fiberwise a linear isomorphism, $\pi_M : H \to M$ is a local diffeomorphism. For $x \in M$ we use a chart $(U, u : U \to u(U) = \mathbb{R}^m)$ of M centered at x and let $X \in (\pi_M|_H)^{-1}(x)$. Consider

$$s : U \to H, \qquad s(u^{-1}(z)) = \mathrm{Pt}(u^{-1}(t \mapsto t.z), 1).X.$$

Then $\pi_M \circ s = \mathrm{Id}_U$ and $s(U) \subset H$ is diffeomorphic to U, the branch of H through X over U. Since $X \in (\pi_M|_H)^{-1}(x)$ was arbitrary, the set $(\pi_M|_H)^{-1}(U)$ is the disjoint union of open subsets which are all diffeomorphic via π_M to U. Thus $\pi_M : H \to M$ is a covering map. $\square$

24.7. Theorem. *Let (M,g) be a pseudo-Riemann manifold with vanishing curvature. Then M is locally isometric to $\mathbb{R}^m$ with the standard inner product of the same signature: For each $x \in M$ there exists a chart (U,u) centered at x such that $g|U = u^*\langle\ ,\ \rangle$.*

Proof. Choose an orthonormal basis $X_1(x), \dots, X_m(x)$ of (T_xM, g_x); this means $g_x(X_i(x), X_j(x)) = \eta_{ii}\delta_{ij}$, where $\eta = \operatorname{diag}(1,\dots,1,-1,\dots,-1)$ is the standard inner product of signature (p,q). Since the curvature R vanishes, we may consider the horizontal foliation of (24.6). Let H_i denote the horizontal leaf through $X_i(x)$ and define $X_i : U \to TM$ by $X_i = (\pi_M|_{H_i})^{-1} : U \to H_i \subset TM$, where U is a suitable (simply connected) neighborhood of x in M. Since $X_i \circ c$ is horizontal in TM for any curve c in U, we have $\nabla_X X_i = 0$ for any $X \in \mathfrak{X}(M)$ for the Levi-Civita covariant derivative of g. Vector fields X_i with this property are called *Killing fields*. Moreover $X(g(X_i,X_j)) = g(\nabla_X X_i, X_j) + g(X_i, \nabla_X X_j) = 0$; thus $g(X_i,X_j) =$ constant $= g(X_i(x), X_j(x)) = \eta_{ii}\delta_{ij}$ and $X_i, \dots, X_j$ is an orthonormal frame on U. Since ∇ has no torsion, we have

$$0 = \operatorname{Tor}(X_i,X_j) = \nabla_{X_i}X_j - \nabla_{X_j}X_i - [X_i,X_j] = [X_i,X_j].$$

By theorem (3.17) there exists a chart (U,u) on M centered at x such that $X_i = \frac{\partial}{\partial u^i}$, i.e., $Tu.X_i(x) = (u(x), e_i)$ for the standard basis e_i of $\mathbb{R}^m$. Thus Tu maps an orthonormal frame on U to an orthonormal frame on $u(U) \in \mathbb{R}^m$, and u is an isometry. □

24.8. Sectional curvature. Let (M,g) be a Riemann manifold, let $P_x \subset T_xM$ be a 2-dimensional linear subspace of T_xM, and let X_x, Y_x be an orthonormal basis of P_x. Then the number

$$k(P_x) := -g(R(X_x,Y_x)X_x,Y_x) \tag{1}$$

is called the *sectional curvature* of this subspace. That $k(P_x)$ does not depend on the choice of the orthonormal basis is shown by the following lemma.

For pseudo-Riemann manifolds one can define the sectional curvature only for those subspaces P_x on which g_x is nondegenerate. This notion is rarely used in general relativity.

Lemma.

(2) *Let $A = (A^i_j)$ be a real (2×2)-matrix and let $X_1, X_2 \in T_xM$. Then for $X'_i = A^1_i X_1 + A^2_i X_2$ we have*

$$g(R(X'_1,X'_2)X'_1,X'_2) = \det(A)^2\, g(R(X_1,X_2)X_1,X_2).$$

(3) *Let X', Y' be linearly independent in $P_x \subset T_xM$; then*

$$k(P_x) = -\frac{g(R(X',Y')X',Y')}{|X'|^2|Y'|^2 - g(X',Y')^2}.$$

Proof. (2) Since $g(R(X_i,X_j)X_k,X_l) = 0$ for $i = j$ or $k = l$, we have

$$\begin{aligned} g(R(X_1',X_2')X_1',X_2') &= \sum A_1^i A_2^j A_1^k A_2^l g(R(X_i,X_j)X_k,X_l) \\ &= g(R(X_1,X_2)X_1,X_2) \\ &\quad \cdot (A_1^1A_2^2A_1^1A_2^2 - A_1^1A_2^2A_1^2A_2^1 - A_1^2A_2^1A_1^1A_2^2 + A_1^2A_2^1A_1^2A_2^1) \\ &= g(R(X_1,X_2)X_1,X_2)(A_1^1A_2^2 - A_2^1A_1^2)^2. \quad \square \end{aligned}$$

(3) Let X, Y be an orthonormal basis of P_x, let $X' = A_1^1X + A_1^2Y$ and let $Y' = A_2^1X + A_2^2Y$. Then $\det(A)^2$ equals the area^2 of the parallelogram spanned by X' and Y' which is $|X'|^2|Y'|^2 - g(X',Y')^2$. Now use (2). $\square$

24.9. Computing the sectional curvature. *Let $g : U \to S^2(\mathbb{R}^m)$ be a pseudo-Riemann metric in an open subset of $\mathbb{R}^m$. Then for $X, Y \in T_x\mathbb{R}^m$ we have:*

$$\begin{aligned} 2R_x(X,Y,X,Y) &= 2g_x(R_x(X,Y)X,Y) \\ &= -2d^2g(x)(X,Y)(Y,X) + d^2g(x)(X,X)(Y,Y) + d^2g(x)(Y,Y)(X,X) \\ &\quad - 2g(\Gamma(Y,X),\Gamma(X,Y)) + 2g(\Gamma(X,X),\Gamma(Y,Y)). \end{aligned}$$

Proof. The Christoffels $\Gamma : U \times \mathbb{R}^m \times \mathbb{R}^m \to \mathbb{R}^m$ are given by (22.4.1):

(1) $2g_x(\Gamma_x(Y,Z),U) = dg(x)(U)(Y,Z) - dg(x)(Y)(Z,U) - dg(x)(Z)(U,Y)$,

and the curvature is given in terms of the Christoffels is (24.3.2):

$$\begin{aligned} R(X,Y)Z &= (\nabla_X\nabla_Y - \nabla_Y\nabla_X - \nabla_{[X,Y]})Z \\ \text{(2)} \qquad &= -d\Gamma(X)(Y,Z) + d\Gamma(Y)(X,Z) + \Gamma(X,\Gamma(Y,Z)) - \Gamma(Y,\Gamma(X,Z)). \end{aligned}$$

We differentiate (1) once more:

$$\begin{aligned} &2dg(x)(X)(\Gamma_x(Y,Z),U) + 2g_x(d\Gamma(x)(X)(Y,Z),U) \\ \text{(3)} \quad &= +d^2g(x)(X,U)(Y,Z) - d^2g(x)(X,Y)(Z,U) - d^2g(x)(X,Z)(U,Y). \end{aligned}$$

Let us compute the combination from (2), using (3):

$$\begin{aligned} &- 2g_x(d\Gamma(x)(X)(Y,Z),U) + 2g_x(d\Gamma(x)(Y)(X,Z),U) \\ &= 2dg(x)(X)(\Gamma_x(Y,Z),U) - 2dg(x)(Y)(\Gamma_x(X,Z),U) \\ &\quad - d^2g(x)(X,U)(Y,Z) + d^2g(x)(X,Y)(Z,U) + d^2g(x)(X,Z)(U,Y) \\ &\quad + d^2g(x)(Y,U)(X,Z) - d^2g(x)(Y,X)(Z,U) - d^2g(x)(Y,Z)(U,X) \\ &= 2dg(x)(X)(\Gamma_x(Y,Z),U) - 2dg(x)(Y)(\Gamma_x(X,Z),U) \end{aligned}$$

$$
\begin{aligned}
&- d^2g(x)(X,U)(Y,Z) + d^2g(x)(X,Z)(U,Y)\\
&+ d^2g(x)(Y,U)(X,Z) - d^2g(x)(Y,Z)(U,X).
\end{aligned}
$$

Thus we have

$$
\begin{aligned}
&2R_x(X,Y,Z,U) := 2g_x(R_x(X,Y)Z,U)\\
&= 2g\Big(-d\Gamma(X)(Y,Z) + d\Gamma(Y)(X,Z) + \Gamma(X,\Gamma(Y,Z)) - \Gamma(Y,\Gamma(X,Z)),U\Big)\\
&= 2dg(x)(X)(\Gamma_x(Y,Z),U) - 2dg(x)(Y)(\Gamma_x(X,Z),U)\\
&\quad - d^2g(x)(X,U)(Y,Z) + d^2g(x)(X,Z)(U,Y)\\
&\quad + d^2g(x)(Y,U)(X,Z) - d^2g(x)(Y,Z)(U,X)\\
&\quad + 2g(\Gamma(X,\Gamma(Y,Z)),U) - 2g(\Gamma(Y,\Gamma(X,Z)),U)
\end{aligned}
$$

and for the sectional curvature we get

$$
\begin{aligned}
(4)\quad &2R_x(X,Y,X,Y) = 2g_x(R_x(X,Y)X,Y)\\
&= 2dg(x)(X)(\Gamma_x(Y,X),Y) - 2dg(x)(Y)(\Gamma_x(X,X),Y)\\
&\quad - 2d^2g(x)(X,Y)(Y,X) + d^2g(x)(X,X)(Y,Y) + d^2g(x)(Y,Y)(X,X)\\
&\quad + 2g(\Gamma(X,\Gamma(Y,X)),Y) - 2g(\Gamma(Y,\Gamma(X,X)),Y).
\end{aligned}
$$

Let us check how skew-symmetric the Christoffels are. From (1) we get

$$
\begin{aligned}
&2g_x(\Gamma_x(Y,Z),U) + 2g_x(Z,\Gamma_x(Y,U)) = 2g_x(\Gamma_x(Y,Z),U) + 2g_x(\Gamma_x(Y,U),Z)\\
&= +dg(x)(U)(Y,Z) - dg(x)(Y)(Z,U) - dg(x)(Z)(U,Y)\\
&\quad + dg(x)(Z)(Y,U) - dg(x)(Y)(U,Z) - dg(x)(U)(Z,Y)\\
&= -2dg(x)(Y)(Z,U).
\end{aligned}
$$

Thus

$$
2dg(x)(Y)(\Gamma(X,V),U) = -2g(\Gamma(Y,\Gamma(X,V)),U) - 2g(\Gamma(X,V),\Gamma(Y,U)).
$$

Using this in (4), we get finally

$$
\begin{aligned}
(5)\quad &2R_x(X,Y,X,Y) = 2g_x(R_x(X,Y)X,Y)\\
&= -2g(\Gamma(X,\Gamma(Y,X)),Y) - 2g(\Gamma(Y,X),\Gamma(X,Y))\\
&\quad + 2g(\Gamma(Y,\Gamma(X,X)),Y) + 2g(\Gamma(X,X),\Gamma(Y,Y))\\
&\quad - 2d^2g(x)(X,Y)(Y,X) + d^2g(x)(X,X)(Y,Y) + d^2g(x)(Y,Y)(X,X)\\
&\quad + 2g(\Gamma(X,\Gamma(Y,X)),Y) - 2g(\Gamma(Y,\Gamma(X,X)),Y)\\
&= -2d^2g(x)(X,Y)(Y,X) + d^2g(x)(X,X)(Y,Y) + d^2g(x)(Y,Y)(X,X)\\
&\quad - 2g(\Gamma(Y,X),\Gamma(X,Y)) + 2g(\Gamma(X,X),\Gamma(Y,Y)). \quad \square
\end{aligned}
$$

25. Computing with Adapted Frames and Examples

25.1. Frames. We recall that a *local frame* or *frame field* s on an open subset U of a pseudo-Riemann manifold (M, g) of dimension m is an m-tuple $s_1, \dots, s_m$ of vector fields on U such that $s_1(x), \dots, s_m(x)$ is a basis of the tangent space T_xM for each $x \in U$. Note that then s is a local section of the linear frame bundle $GL(\mathbb{R}^m, TM) \to M$, a principal fiber bundle, as we treat it in (18.11). We view $s(x) = (s_1(x), \dots, s_m(x))$ as a linear isomorphism $s(x) : \mathbb{R}^m \to T_xM$. The frame field s is called an *orthonormal frame* if $s_1(x), \dots, s_m(x)$ is an orthonormal basis of (T_xM, g_x) for each $x \in U$. By this we mean that $g_x(X_i(x), X_j(x)) = \eta_{ii}\delta_{ij}$, where $\eta = \operatorname{diag}(1, \dots, 1, -1, \dots, -1)$ is the standard inner product of signature $(p, q = m - p)$.

If (U, u) is a chart on M, then $\frac{\partial}{\partial u^1}, \dots, \frac{\partial}{\partial u^m}$ is a frame field on U. Out of this we can easily build one which contains no isotropic vectors (i.e., ones with $g(X, X) = 0$) and order them in such a way that the fields with $g(X, X) > 0$ are at the beginning. Using the Gram-Schmidt orthonormalization procedure, we can change this frame field then into an orthonormal one on a possibly smaller open set U. Thus there always exist orthonormal frame fields.

If $s = (s_1, \dots, s_m)$ and $s' = (s'_1, \dots, s'_m)$ are two frame fields on $U, V \subset M$, respectively, then on $U \cap V$ we have

$$s' = s.h, \quad s'_i = \textstyle\sum_j s_j h^j_i, \quad s'_i(x) = \sum_j s_j(x) h^j_i(x),$$
$$h = (h^i_j) : U \cap V \to GL(m, \mathbb{R}).$$

25.2. Connection forms. If s is a local frame on an open subset U in a manifold M and if ∇ is a covariant derivative on M, we put

$$\text{(1)} \qquad \nabla_X s_i = \textstyle\sum_j s_j.\omega^j_i(X), \quad \nabla_X s = s.\omega(X), \quad \nabla s = s.\omega,$$
$$\omega = (\omega^j_i) \in \Omega^1(U, \mathfrak{gl}(m)), \qquad \text{the connection form of } \nabla.$$

We saw this construction in (19.4) already.

Proposition. *We have:*

(2) *If* $Y = \sum s_j u^j \in \mathfrak{X}(U)$, *then*

$$\nabla Y = \textstyle\sum_k s_k(\sum_j \omega^k_j u^j + du^k) = s.\omega.u + s.du.$$

(3) *Let* s *and* $s' = s.h$ *be two local frames on* U; *then the connection forms* $\omega, \omega' \in \Omega^1(U, \mathfrak{gl}(m))$ *are related by*

$$h.\omega' = dh + \omega.h.$$

(4) *If s is a local orthonormal frame for a Riemann metric g which is respected by ∇, then*

$$\omega_i^j = -\omega_j^i, \quad \omega = (\omega_i^j) \in \Omega^1(U, \mathfrak{so}(m)).$$

If s is a local orthonormal frame for a pseudo-Riemann metric g which is respected by ∇ and if $\eta_{ij} = g(s_i, s_j) = \operatorname{diag}(1, \dots, 1, -1, \dots, -1)$ is the standard inner product matrix of the same signature (p, q), then

$$\eta_{jj}\omega_i^j = -\eta_{ii}\omega_j^i, \quad \omega = (\omega_i^j) \in \Omega^1(U, \mathfrak{so}(p, q)).$$

Proof. We use direct computations.

(2)
$$\begin{aligned}\nabla_X Y &= \nabla_X(\textstyle\sum_j s_j u^j) = \sum_j (\nabla_X s_j) u^j + \sum_j s_j X(u^j)\\ &= \textstyle\sum_k s_k \sum_j \omega_j^k(X) u^j + \sum_k s_k du^k(X).\end{aligned}$$

(3)
$$\begin{aligned}\nabla s' &= s'.\omega' = s.h.\omega',\\ \nabla s' &= \nabla(s.h) = (\nabla s).h + s.dh = s.\omega.h + s.dh.\end{aligned}$$

(4) It suffices to prove the second assertion. We differentiate the constant $\eta_{ij} = g(s_i, s_j)$:

$$\begin{aligned}0 &= X(g(s_i, s_j)) = g(\nabla_X s_i, s_j) + g(s_i, \nabla_X s_j)\\ &= g(\textstyle\sum s_k \omega_i^k(X), s_j) + g(s_i, \sum s_k \omega_j^k(X))\\ &= \textstyle\sum g(s_k, s_j)\omega_i^k(X) + \sum g(s_i, s_k)\omega_j^k(X) = \eta_{jj}\omega_i^j(X) + \eta_{ii}\omega_j^i(X). \quad \square\end{aligned}$$

25.3. Curvature forms. Let s be a local frame on U, and let ∇ be a covariant derivative with curvature R. We write

$$R(X, Y)s = (R(X, Y)s_1, \dots, R(X, Y)s_m).$$

Then we have

(1)
$$Rs_j = \textstyle\sum s_k.(d\omega_j^k + \sum \omega_l^k \wedge \omega_j^l), \qquad Rs = s.(d\omega + \omega \wedge \omega),$$

where $\omega \wedge \omega = (\sum \omega_k^i \wedge \omega_j^k)_j^i \in \Omega^2(U, \mathfrak{gl}(m))$, since

$$\begin{aligned}R(X, Y)s &= \nabla_X \nabla_Y s - \nabla_Y \nabla_X s - \nabla_{[X,Y]} s\\ &= \nabla_X(s.\omega(Y)) - \nabla_Y(s.\omega(X)) - s.\omega([X, Y])\\ &= s.X(\omega(Y)) + s.\omega(X).\omega(Y) - s.Y(\omega(X)) - s.\omega(Y).\omega(X) - s.\omega([X, Y])\\ &= s.\big(X(\omega(Y)) - Y(\omega(X)) - \omega([X, Y]) + \omega(X).\omega(Y) - \omega(Y).\omega(X)\big)\\ &= s.(d\omega + \omega \wedge \omega)(X, Y).\end{aligned}$$

We thus get the *curvature matrix*

(2)
$$\Omega = d\omega + \omega \wedge \omega \in \Omega^2(U, \mathfrak{gl}(m)),$$

and we note its defining equation $R.s = s.\Omega$.

Proposition.

(3) *If s and $s' = s.h$ are two local frames, then the curvature matrices are related by*

$$h.\Omega' = \Omega.h.$$

(4) *The second Bianchi identity becomes*

$$d\Omega + \omega \wedge \Omega - \Omega \wedge \omega = 0.$$

(5) *If s is a local orthonormal frame for a Riemann metric g which is respected by ∇, then*

$$\Omega^j_i = -\Omega^i_j, \quad \Omega = (\Omega^j_i) \in \Omega^2(U, \mathfrak{so}(m)).$$

If s is a local orthonormal frame for a pseudo-Riemann metric g which is respected by ∇ and if $\eta_{ij} = g(s_i, s_j) = \operatorname{diag}(1, \dots, 1, -1, \dots, -1)$ is the standard inner product matrix of the same signature (p, q), then

$$\eta_{jj}\Omega^j_i = -\eta_{ii}\Omega^i_j, \quad \Omega = (\Omega^j_i) \in \Omega^2(U, \mathfrak{so}(p, q)).$$

Proof. (3) Since R is a tensor field, we have $s.h.\Omega' = s'.\Omega' = Rs' = Rs.h = s.\Omega.h$.

A second, direct proof goes as follows. By (25.2.3) we have $h.\omega' = \omega.h + dh$; thus

$$\begin{aligned}
h.\Omega' &= h.(d\omega' + \omega' \wedge \omega') \\
&= h.d(h^{-1}.\omega.h + h^{-1}.dh) + (\omega.h + dh) \wedge (h^{-1}.\omega.h + h^{-1}.dh) \\
&= h.(-h^{-1}.dh.h) \wedge \omega.h + h.h^{-1}.d\omega.h - h.h^{-1}.\omega \wedge dh \\
&\quad + h.(-h^{-1}.dh.h^{-1}) \wedge dh + h.h^{-1}.ddh \\
&\quad + \omega \wedge h.h^{-1}.\omega + \omega \wedge h.h^{-1}.dh + dh.h^{-1} \wedge \omega.h + dh.h^{-1} \wedge dh \\
&= d\omega.h + \omega \wedge \omega.h = \Omega.h.
\end{aligned}$$

(4) $d\Omega = d(d\omega + \omega\wedge\omega) = 0 + d\omega\wedge\omega - \omega\wedge d\omega = (d\omega+\omega\wedge\omega)\wedge\omega - \omega\wedge(d\omega+\omega\wedge\omega)$.

(5) We prove only the second case.

$$\begin{aligned}
\eta_{jj}\Omega^j_i &= \eta_{jj} d\omega^j_i + \textstyle\sum_k \eta_{jj}\omega^j_k \wedge \omega^k_i = -\eta_{ii} d\omega^i_j - \textstyle\sum_k \eta_{kk}\omega^k_j \wedge \omega^k_i \\
&= -\eta_{ii} d\omega^i_j + \textstyle\sum_k \eta_{ii}\omega^k_j \wedge \omega^i_k = -\eta_{ii}(d\omega^i_j + \textstyle\sum_k \omega^i_k \wedge \omega^k_j) = -\eta_{ii}\Omega^i_j. \quad \square
\end{aligned}$$

25.4. Coframes. For a local frame $s = (s_1, \dots, s_m)$ on $U \subset M$ we consider the dual coframe

$$\sigma = \begin{pmatrix} \sigma^1 \\ \vdots \\ \sigma^m \end{pmatrix}, \qquad \sigma^i \in \Omega^1(U),$$

which forms the dual basis of T_x^*M for each $x \in U$; it satisfies $\langle \sigma^i, s_j \rangle = \sigma^i(s_j) = \delta^i_j$. If $s' = s.h$ is another local frame, then its *dual coframe* is given by

(1) $$\sigma' = h^{-1}.\sigma, \quad \sigma'^i = \textstyle\sum_k (h^{-1})^i_k \sigma^k,$$

since

$$\langle \textstyle\sum_k (h^{-1})^i_k \sigma^k, s'_j \rangle = \sum_{k,l} (h^{-1})^i_k \, \langle \sigma^k, s_l \rangle \, h^l_j = \delta^i_j.$$

Let s be a local frame on U, and let ∇ be a covariant derivative. We define the *torsion form* Θ by

(2) $$\mathrm{Tor} = s.\Theta, \quad \mathrm{Tor}(X,Y) =: \textstyle\sum_j s_j \Theta^j(X,Y), \quad \Theta \in \Omega^2(U, \mathbb{R}^m).$$

Proposition.

(3) *If s and $s' = s.h$ are two local frames, then the torsion forms of a covariant derivative are related by*
$$\Theta' = h^{-1}.\Theta.$$

(4) *If s is a local frame with dual coframe σ, then for a covariant derivative with connection form $\omega \in \Omega^1(U, \mathfrak{gl}(m))$ and torsion form $\Theta \in \Omega^2(U, \mathbb{R}^m)$ we have*
$$d\sigma = -\omega \wedge \sigma + \Theta, \quad d\sigma^i = -\textstyle\sum_k \omega^i_k \wedge \sigma^k + \Theta^i.$$

(5) *The algebraic Bianchi identity for a covariant derivative takes the following form:*
$$d\Theta + \omega \wedge \Theta = \Omega \wedge \sigma, \quad d\Theta^k + \textstyle\sum_l \omega^k_l \wedge \Theta^l = \sum_l \Omega^k_l \wedge \sigma^l.$$

Proof. (3) Since Tor is a tensor field, we have $s.\Theta = \mathrm{Tor} = s'\Theta' = s.h.\Theta'$; thus $h.\Theta' = \Theta$ and $\Theta' = h^{-1}.\Theta$.

(4) For $X \in \mathfrak{X}(U)$ we have $X = \sum_i s_i.\sigma^i(X)$, for short $X = s.\sigma(X)$. Then

$$\begin{aligned}
\nabla_X Y &= \nabla_X(s.\sigma(Y)) = (\nabla_X s).\sigma(Y) + s.X(\sigma(Y)) \\
&= s.\omega(X).\sigma(Y) + s.X(\sigma(Y)), \\
s.\Theta(X,Y) &= \mathrm{Tor}(X,Y) = \nabla_X Y - \nabla_Y X - [X,Y] \\
&= s.\omega(X).\sigma(Y) + s.X(\sigma(Y)) - s.\omega(Y).\sigma(X) - s.Y(\sigma(X)) - s.\sigma([X,Y]) \\
&= s.(\omega(X).\sigma(Y) - \omega(Y).\sigma(X) + X(\sigma(Y)) - Y(\sigma(X)) - \sigma([X,Y])) \\
&= s.(\omega \wedge \sigma(X) + d\sigma)(X,Y).
\end{aligned}$$

Direct proof of (3):

$$\begin{aligned}
\Theta' &= \omega' \wedge \sigma' + d\sigma' = (h^{-1}.\omega.h + h^{-1}.dh) \wedge h^{-1}.\sigma + d(h^{-1}.\sigma) \\
&= h^{-1}.\omega \wedge \sigma + h^{-1}.dh \wedge h^{-1}.\sigma - h^{-1}.dh.h^{-1}.\sigma + h^{-1}.d\sigma \\
&= h^{-1}(\omega \wedge \sigma + d\sigma) = h^{-1}.\Theta.
\end{aligned}$$

(5) $$\begin{aligned} d\Theta &= d(\omega\wedge\sigma + d\sigma) = d\omega\wedge\sigma - \omega\wedge d\sigma + 0 \\ &= (d\omega + \omega\wedge\omega)\wedge\sigma - \omega\wedge(\omega\wedge\sigma + d\sigma) = \Omega\wedge\sigma - \omega\wedge\Theta. \quad \square \end{aligned}$$

25.5. Resumé of computing with adapted frames. *Let (M,g) be a Riemann manifold, let s be an orthonormal local frame on U with dual coframe σ, and let ∇ be the Levi-Civita covariant derivative. Then we have:*

(1) $g|_U = \sum_i \sigma^i \otimes \sigma^i$.

(2) $\nabla s = s.\omega$, $\omega^i_j = -\omega^j_i$, *so* $\omega \in \Omega^1(U, \mathfrak{so}(m))$.

(3) $d\sigma + \omega\wedge\sigma = 0$, $d\sigma^i + \sum_k \omega^i_k \wedge \sigma^k = 0$.

(4) $Rs = s.\Omega$, $\Omega = d\omega + \omega\wedge\omega \in \Omega^2(U, \mathfrak{so}(m))$, $\Omega^i_j = d\omega^i_j + \sum_k \omega^i_k \wedge \omega^k_j$,

(5) $\Omega\wedge\sigma = 0$, $\sum_k \Omega^i_k \wedge \sigma^k = 0$, *the first Bianchi identity.*

(6) $d\Omega + \omega\wedge\Omega - \Omega\wedge\omega = d\Omega + [\omega,\Omega]_\wedge = 0$, *the second Bianchi identity.*

For a pseudo-Riemann manifold (M,g) we consider standard inner product matrix $\eta_{ij} = g(s_i, s_j) = \operatorname{diag}(1,\dots,1,-1,\dots,-1)$ of the same signature (p,q). Then we have instead:

(1′) $g = \sum_i \eta_{ii}\sigma^i \otimes \sigma^i$.

(2′) $\eta_{jj}\omega^j_i = -\eta_{ii}\omega^i_j$; *thus* $\omega = (\omega^j_i) \in \Omega^1(U, \mathfrak{so}(p,q))$.

(4′) $\eta_{jj}\Omega^j_i = -\eta_{ii}\Omega^i_j$; *thus* $\Omega = (\Omega^j_i) \in \Omega^2(U, \mathfrak{so}(p,q))$.

25.6. Interpretation in terms of the orthonormal frame bundle. For a pseudo-Riemann manifold (M,g) of dimension m we consider the *orthonormal frame bundle*

$$\mathcal{O}(M) = O(\mathbb{R}^m, TM) \xrightarrow{\pi_M} M.$$

Its fiber $\mathcal{O}(M)_x$ consists of all linear isometries $(\mathbb{R}^m, \eta) \to (T_xM, g)$ where η is the standard inner product with the same signature as g. It is a principal bundle with structure group $O(p,q)$ (acting by composition from the right), and it has one further structure, the *soldering form* which encodes the fact that the associated bundle $\mathcal{O}(M) \times_{O(p,q)} \mathbb{R}^m$ is the tangent bundle. The soldering form is described as follows: Let $s = (s_1,\dots,s_m) \in \mathcal{O}(M)_x$ be an orthonormal frame of T_xM with orthonormal coframe (dual basis)

$$\sigma_s = \begin{pmatrix} \sigma^1_s \\ \vdots \\ \sigma^m_s \end{pmatrix}, \qquad \sigma^i_s \in T^*_xM.$$

The soldering form is then given as (with a slight abuse of notation):

$$\sigma \in \Omega^1(\mathcal{O}(M), \mathbb{R}^m),$$

$$\sigma_s(\Xi_s) = \sigma_s(T(\pi_m).\Xi_s) = \begin{pmatrix} \sigma_s^1(T_s(\pi_M).\Xi_s) \\ \vdots \\ \sigma_s^m(T_s(\pi_M).\Xi_s) \end{pmatrix} \in \mathbb{R}^m.$$

For $h \in O(p,q)$ we have

$$((r^h)^*\sigma)_s(\Xi_s) = \sigma_{s.h}(T(r^h).\Xi_s) = h^{-1}.\sigma_s(T(\pi_M).T(r^h)\Xi) = h^{-1}.\sigma_s(\Xi).$$

So σ is $O(p,q)$-equivariant and horizontal: It kills vertical tangent vectors. By (19.14), σ induces a differential form on M with values in the associated bundle $O(M) \times_{O(p,q)} \mathbb{R}^m$; it is a vector bundle isomorphism $TM \to O(M) \times_{O(p,q)} \mathbb{R}^m$. If s is a local orthonormal frame, i.e., a local section of $\mathcal{O}(M)$, then $s^*\sigma = \sigma_s$, the dual coframe.

For the description of the principal connection form ω on $\mathcal{O}(M)$ inducing the Levi-Civita connection ∇ on M we fix an open cover $(U_\alpha)_\alpha$ of M and local orthonormal frames $s_\alpha : U_\alpha \to \mathcal{O}(M)$. Then $\sigma_\alpha = {s_\alpha}^*\sigma$ are the dual coframes, and by (25.5) the connection form $\omega_\alpha \in \Omega^1(U_\alpha, \mathfrak{o}(p,q))$ is determined by $d\sigma_\alpha + \omega_\alpha \wedge \sigma_\alpha = 0$. The local frames s_α induce a principal fiber bundle atlas

$$(U_\alpha, \varphi_\alpha : \mathcal{O}(M)|_{U_\alpha} \to U_\alpha \times O(p,q)), \qquad \varphi_\alpha(u_x) = \big(x, (s_\alpha|_x)^{-1}.u_x\big).$$

For $X \in \mathfrak{o}(p,q)$ the fundamental vector field for the principal right action is given by $\zeta_X(u_x) = \partial_t|_0 u_x \circ \exp(tX) = \mathrm{vl}(u_x, u_x.X)$ in terms of the vertical lift vl from (8.12).

The local expression $\gamma_\alpha = (\varphi_\alpha^{-1})^*\omega \in \Omega^1(U_\alpha \times O(p,q), \mathfrak{o}(p,q))$ of the principal connection ω is given by (19.4.6), where $\xi_x \in T_xM$, $h \in O(p,q)$, and $X \in \mathfrak{o}(p,q)$. Thus we have:

$$\begin{aligned} \omega(T(\varphi_\alpha^{-1})(\xi_x, T(\mu_h).X)) &= \gamma_\alpha(\xi_x, T(\mu_h).X) \\ &= \gamma_\alpha(\xi_x, 0_h) + X = \mathrm{Ad}(h^{-1})\omega_\alpha(\xi_x) + X \\ &= h^{-1}.\omega_\alpha(\xi_x).h + X. \end{aligned}$$

25.7. Example: The sphere $S^2 \subset \mathbb{R}^3$. We consider the parameterization (leaving out one longitude):

$$f : (0, 2\pi) \times (-\pi, \pi) \to \mathbb{R}^3,$$

$$f(\varphi, \vartheta) = \begin{pmatrix} \cos\varphi\ \cos\vartheta \\ \sin\varphi\ \cos\vartheta \\ \sin\vartheta \end{pmatrix},$$

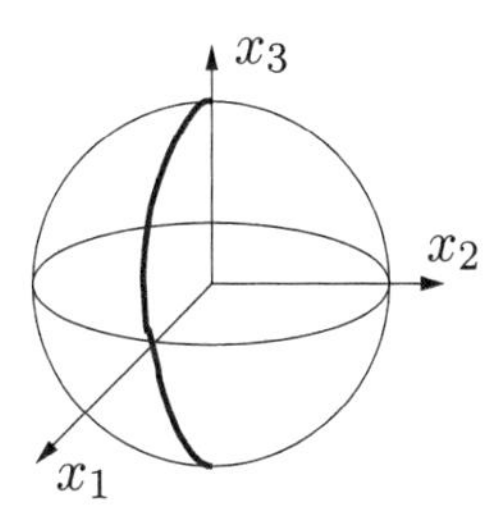

$$\begin{aligned} g &= f^*(\text{metric}) = f^*(\textstyle\sum_i dx^i \otimes dx^i) \\ &= \sum_{i=1}^{3} df^i \otimes df^i = \cos^2\vartheta\, d\varphi \otimes d\varphi + d\vartheta \otimes d\vartheta. \end{aligned}$$

From this we can read off the orthonormal coframe and then the orthonormal frame:

$$\sigma^1 = d\vartheta, \quad \sigma^2 = \cos\vartheta\, d\varphi, \qquad s_1 = \frac{\partial}{\partial\vartheta}, \quad s_2 = \frac{1}{\cos\vartheta}\frac{\partial}{\partial\varphi}.$$

We compute $d\sigma^1 = 0$ and $d\sigma^2 = -\sin\vartheta\, d\vartheta \wedge d\varphi = -\tan\vartheta\, \sigma^1 \wedge \sigma^2$. For the connection forms we have $\omega_1^1 = \omega_2^2 = 0$ by skew-symmetry. The off-diagonal terms we compute from (25.5.3), i.e., $d\sigma + \omega \wedge \sigma = 0$:

$$\begin{aligned} -d\sigma^1 &= 0 + \omega_2^1 \wedge \sigma^2 = 0 && \Rightarrow \omega_2^1 = c(\varphi, \vartheta)\sigma^2, \\ -d\sigma^2 &= \omega_1^2 \wedge \sigma^1 + 0 = \tan\vartheta\, \sigma^1 \wedge \sigma^2 && \Rightarrow \omega_2^1 = \tan\vartheta\, \sigma^2 = \sin\vartheta\, d\varphi, \\ \omega &= \begin{pmatrix} 0 & \sin\vartheta\, d\varphi \\ -\sin\vartheta\, d\varphi & 0 \end{pmatrix}. \end{aligned}$$

For the curvature forms we have again $\Omega_1^1 = \Omega_2^2 = 0$ by skew-symmetry, and then we may compute the curvature:

$$\begin{aligned} \Omega_2^1 &= d\omega_2^1 + \omega_1^1 \wedge \omega_2^1 + \omega_2^1 \wedge \omega_2^2 = d(\sin\vartheta\, d\varphi) = \cos\vartheta\, d\vartheta \wedge d\varphi = \sigma^1 \wedge \sigma^2, \\ \Omega &= \begin{pmatrix} 0 & \sigma^1 \wedge \sigma^2 \\ -\sigma^1 \wedge \sigma^2 & 0 \end{pmatrix}. \end{aligned}$$

For the sectional curvature we get

$$\begin{aligned} k(S^2) &= -g(R(s_1, s_2)s_1, s_2) = -g(\textstyle\sum_k s_k \Omega_1^k(s_1, s_2), s_2) \\ &= -g(s_2(-\sigma^1 \wedge \sigma^2)(s_1, s_2), s_2) = 1. \end{aligned}$$

25.8. Example: The Poincaré upper half-plane.
This is the set $H_+^2 = \{(x, y) \in \mathbb{R}^2 : y > 0\}$ with metric $ds^2 = \frac{1}{y^2}(dx^2 + dy^2)$ or

$$g = \frac{1}{y}dx \otimes \frac{1}{y}dx + \frac{1}{y}dy \otimes \frac{1}{y}dy,$$

which is conformal with the standard inner product.

The curvature. The orthonormal coframe and frame are then, by (25.5.1):

$$\sigma^1 = \frac{1}{y}dx, \quad \sigma^2 = \frac{1}{y}dy, \qquad s_1 = y\frac{\partial}{\partial x}, \quad s_2 = y\frac{\partial}{\partial y}.$$

We have $d\sigma^1 = d(\frac{1}{y}dx) = \frac{1}{y^2}dx \wedge dy = \sigma^1 \wedge \sigma^2$ and $d\sigma^2 = 0$. The connection forms we compute from (25.5.3), i.e, $d\sigma + \omega \wedge \sigma = 0$:

$$\begin{aligned} -d\sigma^1 &= 0 + \omega_2^1 \wedge \sigma^2 = -\sigma^1 \wedge \sigma^2, \\ -d\sigma^2 &= \omega_1^2 \wedge \sigma^1 + 0 = 0 \qquad \Rightarrow \omega_2^1 = -\sigma^1 = -y^{-1}dx, \\ \omega &= \begin{pmatrix} 0 & -\sigma^1 \\ \sigma^1 & 0 \end{pmatrix}. \end{aligned}$$

For the curvature forms we get

$$\Omega^1_2 = d\omega^1_2 + \omega^1_1 \wedge \omega^1_2 + \omega^1_2 \wedge \omega^2_2 = d(-y^{-1}dx) = -\sigma^1 \wedge \sigma^2,$$

$$\Omega = \begin{pmatrix} 0 & -\sigma^1 \wedge \sigma^2 \\ +\sigma^1 \wedge \sigma^2 & 0 \end{pmatrix}.$$

For the sectional curvature we get

$$\begin{aligned} k(H^2_+) &= -g(R(s_1,s_2)s_1,s_2) = -g(\textstyle\sum_k s_k \Omega^k_1(s_1,s_2),s_2) \\ &= -g(s_2(\sigma^1\wedge\sigma^2)(s_1,s_2),s_2) = -1. \end{aligned}$$

The geodesics. For deriving the geodesic equation let:

$$c(t) = \begin{pmatrix} x(t) \\ y(t) \end{pmatrix}, \quad c'(t) = \begin{pmatrix} x'(t) \\ y'(t) \end{pmatrix} = \frac{x'}{y} y \frac{\partial}{\partial x} + \frac{y'}{y} y \frac{\partial}{\partial y} = \frac{x'}{y} s_1 + \frac{y'}{y} s_2 =: (s \circ c).u.$$

The geodesic equation is then

$$\begin{aligned} \nabla_{\partial_t} c' &= \nabla_{\partial_t}((s\circ c).u) = s.\omega(c').u + s.du(\partial_t) \\ &= (s_1,s_2)\begin{pmatrix} 0 & \omega^1_2(c') \\ -\omega^1_2(c') & 0 \end{pmatrix}\begin{pmatrix} \frac{x'}{y} \\ \frac{y'}{y} \end{pmatrix} + (s_1,s_2)\begin{pmatrix} (\frac{x'}{y})' \\ (\frac{y'}{y})' \end{pmatrix} \\ &= \frac{x'^2}{y}\frac{\partial}{\partial y} - \frac{x'y'}{y}\frac{\partial}{\partial x} + \frac{x''y - x'y'}{y}\frac{\partial}{\partial x} + \frac{y''y - y'^2}{y}\frac{\partial}{\partial y} = 0, \end{aligned}$$

$$\begin{cases} x''y - 2x'y' = 0, \\ x'^2 + y''y - y'^2 = 0. \end{cases}$$

To see the shape of the geodesics, we first investigate $x(t) =$ constant. Then $y''y - y'^2 = 0$ has a unique solution for each initial value $y(0), y'(0)$; thus the verticals $t \mapsto \binom{\text{constant}}{y(t)}$ are geodesics. If $x'(t) = 0$ for a single t, then it is for all t since then the geodesic is already vertical. If $x'(t) \neq 0$, we claim that the geodesics are upper half-circles with center $M(t)$ on the x-axis:

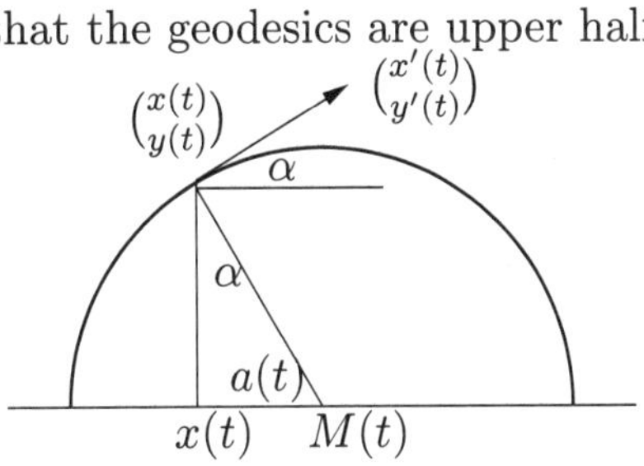

$$\frac{y'(t)}{x'(t)} = \tan\alpha(t) = \frac{a(t)}{y(t)} \quad \Rightarrow a = \frac{y'y}{x'},$$

$$M(t) = x + \frac{y'y}{x'} = \frac{x'x + y'y}{x'},$$

$$M'(t) = \left(\frac{x'x+y'y}{x'}\right)' = \dots = 0.$$

Thus $M(t) = M$, a constant. Moreover,

$$\left|\begin{pmatrix} x(t) \\ y(t) \end{pmatrix} - \begin{pmatrix} M \\ 0 \end{pmatrix}\right|^2 = (x-M)^2 + y^2 = \left(\frac{y'y}{x'}\right)^2 + y^2,$$

$$\frac{d}{dt}\left|\begin{pmatrix} x(t) \\ y(t) \end{pmatrix} - \begin{pmatrix} M \\ 0 \end{pmatrix}\right|^2 = \left(\left(\frac{y'y}{x'}\right)^2 + y^2\right)' = \dots = 0.$$

Thus the geodesics are half-circles as asserted. Note that this violates Euclid's parallel axiom: We have a non-Euclidean geometry.

Isometries and the Poincaré upper half-plane as symmetric space. The projective action of the Lie group $SL(2,\mathbb{R})$ on $\mathbb{C}P^1$, viewed in the projective chart $\mathbb{C} \ni z \mapsto [z:1]$, preserves the upper half-plane: A matrix $\left(\begin{smallmatrix} a & b \\ c & d \end{smallmatrix}\right)$ acts by $[z:1] \mapsto [az+b : cz+d] = [\frac{az+b}{cz+d} : 1]$. Moreover for $z = x+iy$ the expression

$$\begin{aligned}\frac{az+b}{cz+d} &= \frac{(az+b)(c\bar{z}+d)}{|cz+d|^2} \\ &= \frac{ac(x^2+y^2)+(ad+bc)x+db}{(cx+d)^2+(cy)^2} + i\frac{(ad-bc)y}{(cx+d)^2+(cy)^2}\end{aligned}$$

has imaginary part > 0 if and only if $y > 0$. We denote the action by $m : SL(2,\mathbb{R}) \times H^2_+ \to H^2_+$, so that $m\left(\begin{smallmatrix} a & b \\ c & d \end{smallmatrix}\right)(z) = \frac{az+b}{cz+d}$. Transformations of this form are called a *fractional linear transformations* or *Möbius transformations.*

(1) $SL(2,\mathbb{R})$ acts transitively on H^2_+, since $m\left(\begin{smallmatrix} \sqrt{y} & x/\sqrt{y} \\ 0 & 1/\sqrt{y} \end{smallmatrix}\right)(i) = x+iy$. The isotropy group fixing i is $SO(2) \subset SL(2)$, since $i = \frac{ai+b}{ci+d} = \frac{bd+ac+i}{c^2+d^2}$ if and only if $cd+ac=0$ and $c^2+d^2=1$. Thus

$$H^2_+ = SL(2,\mathbb{R})/SO(2,\mathbb{R}).$$

Any Möbius transformation by an element of $SL(2)$ *is an isometry*:

$$\begin{aligned} A &:= \left(\begin{smallmatrix} a & b \\ c & d \end{smallmatrix}\right) \in SL(2,\mathbb{R}), \\ m_A(z) - m_A(z') &= \frac{az+b}{cz+d} - \frac{az'+b}{cz'+d} = \dots = \frac{z-z'}{(cz+d)(cz'+d)}, \\ (m_A)'(z) &= \lim_{z'\to z} \frac{1}{z-z'}\frac{z-z'}{(cz+d)(cz'+d)} = \frac{1}{(cz+d)^2}, \\ m_A(z) - m_A(z') &= \sqrt{(m_A)'(z)}\sqrt{(m_A)'(z')}(z-z'), \end{aligned}$$

always for the same branch of $\sqrt{(m_A)'(z)}$. Expressing the metric in the complex variable, we then have

$$\begin{aligned} g &= \frac{1}{y^2}(dx^2+dy^2) = \frac{1}{\operatorname{Im}(z)^2}\operatorname{Re}(dz.d\bar{z}), \\ (m_A)^*g &= (m_A)^*\left(\frac{1}{\operatorname{Im}(z)^2}\operatorname{Re}(dz.d\bar{z})\right) \\ &= \frac{1}{\operatorname{Im}((m_A)(z))^2}\operatorname{Re}\big((m_A)'(z)dz.(m_A)'(\bar{z})d\bar{z}\big) \\ &= \operatorname{Im}((m_A)(z))^{-2}|cz+d|^{-4}\operatorname{Re}\big(dz.d\bar{z}\big) = \frac{1}{\operatorname{Im}(z)^2}\operatorname{Re}(dz.d\bar{z}), \end{aligned}$$

since

$$\begin{aligned}\operatorname{Im}((m_A)(z))|cz+d|^2 &= \frac{1}{2i}(m_A(z)-m_A(\bar z))|cz+d|^2\\ &= \frac{1}{2i}\frac{z-\bar z}{(cz+d)(c\bar z+d)}|cz+d|^2 = \operatorname{Im}(z).\end{aligned}$$

(2) For further use we note the Möbius transformations

$$\begin{aligned}m_1 &= m\left(\begin{smallmatrix}1 & r\\ 0 & 1\end{smallmatrix}\right) : z\mapsto z+r, \quad r\in\mathbb{R},\\ m_2 &= m\left(\begin{smallmatrix}\sqrt{r} & 0\\ 0 & 1/\sqrt{r}\end{smallmatrix}\right) : z\mapsto r.z, \quad r\in\mathbb{R}_{>0},\\ m_3 &= m\left(\begin{smallmatrix}0 & -1\\ 1 & 0\end{smallmatrix}\right) : z\mapsto \frac{-1}{z} = \frac{-\bar z}{|z|^2} = \frac{-x+iy}{x^2+y^2}.\end{aligned}$$

We can now use these three isometries to determine again the form of all geodesics in H^2_+. For this note that: *If the fixed point set $(H^2_x)^m = \{z\in H^2_+ : m(z)=z\}$ of an isometry m is a connected 1-dimensional submanifold, then this is the image of a geodesic,* since for any vector $X_z\in T_zH^2_+$ tangent to the fixed point set we have $m(\exp(tX)) = \exp(tT_zm.X) = \exp(tX)$. We first use the isometry $\psi(x,y) = (-x,y)$ which is not a Möbius transformation since it reverses the orientation. Its fixed point set is the vertical line $\{(0,y): y>0\}$ which thus is a geodesic. The image under m_1 is then the geodesic $\{(r,y): y>0\}$. The fixed point set of the isometry $\psi\circ m_3$ is the upper half of the unit circle, which thus is a geodesic. By applying m_1 and m_2, we may map it to any upper half-circle with center in the real axis.

(3) *The group $SL(2,\mathbb{R})$ acts isometrically doubly transitively on H^2_+: Any two pairs of points with the same geodesic distance can be mapped to each other by a Möbius transformation.* For $A = \left(\begin{smallmatrix}a & b\\ c & d\end{smallmatrix}\right)$ in the isotropy group $SO(2)$ of i we have $m'_A(i) = \frac{1}{(ci+d)^2}$; it double covers the unit circle in $T_i(H^2_+)$. Thus $SL(2,\mathbb{R})$ acts transitively on the set of all unit tangent vectors in H^2_+, and a shortest geodesic from z_1 to z_2 can thus be mapped by a Möbius transformation to a shortest geodesic of the same length from z'_1 to z'_2.

(4) *H^2_+ is a complete Riemann manifold, and the geodesic distance is given by*

$$\operatorname{dist}(z_1,z_2) = 2\operatorname{artanh}\left|\frac{z_1-z_2}{z_1-\bar z_2}\right|.$$

The shortest curve from iy_1 to iy_2 is obviously on the vertical line since for $z(t) = x(t)+iy(t)$ the length

$$L(c) = \int_0^1 \frac{1}{y(t)}\sqrt{x'(t)^2+y'(t)^2}\,dt$$

is minimal for $x'(t) = 0$; thus $x(t) =$ constant. By the invariance under reparameterizations of the length we have

$$\operatorname{dist}(iy_1, iy_2) = \Big|\int_{y_1}^{y_2} \frac{1}{t}\,dt\Big| = |\log y_2 - \log y_1| = |\log(\tfrac{y_2}{y_1})|.$$

From the formulas in (1) we see that the *double ratio* $|\frac{z_1-z_2}{z_1-\bar z_2}|$ is invariant under $SL(2,\mathbb{R})$ since:

$$\left|\frac{m_A(z_1)-m_A(z_2)}{m_A(z_1)-\overline{m_A(z_2)}}\right| = \left|\frac{\frac{z_1-z_2}{(cz_1+d)(cz_2+d)}}{\frac{z_1-\bar z_2}{(cz_1+d)(c\bar z_2+d)}}\right| = \left|\frac{z_1-z_2}{z_1-\bar z_2}\right|.$$

On the vertical geodesic we have

$$\left|\frac{iy_1-iy_2}{iy_1+iy_2}\right| = \left|\frac{\frac{y_1}{y_2}-1}{\frac{y_1}{y_2}+1}\right| = \left|\frac{e^{\log(\frac{y_1}{y_2})}-1}{e^{\log(\frac{y_1}{y_2})}+1}\right| = \left|\frac{e^{\frac12|\log(\frac{y_1}{y_2})|}-e^{-\frac12|\log(\frac{y_1}{y_2})|}}{e^{\frac12|\log(\frac{y_1}{y_2})|}+e^{-\frac12|\log(\frac{y_1}{y_2})|}}\right|$$
$$= \tanh(\tfrac12 \operatorname{dist}(iy_1, iy_2)).$$

Since $SL(2,\mathbb{R})$ acts isometrically doubly transitively by (3) and since both sides are invariant, the result follows.

(5) *The geodesic exponential mapping.* We have $\exp_i(ti) = e^t.i$ since by (4) we have $\operatorname{dist}(i, e^t i) = \log \frac{e^t i}{i} = t$. Now let $X \in T_i(H^2_+)$ with $|X| = 1$. In (3) we saw that there exists φ with

$$m\begin{pmatrix}\cos\varphi & -\sin\varphi\\ \sin\varphi & \cos\varphi\end{pmatrix}'(i)i = \frac{i}{(i\sin\varphi+\cos\varphi)^2} = e^{-2i\varphi}.i = X,$$
$$\varphi = \frac{\pi}{4} - \frac{\arg(X)}{2} + \pi\mathbb{Z},$$
$$\exp_i(tX) = m\begin{pmatrix}\cos\varphi & -\sin\varphi\\ \sin\varphi & \cos\varphi\end{pmatrix}(e^t i) = \frac{\cos\varphi.e^t.i - \sin\varphi}{\sin\varphi.e^t i + \cos\varphi}.$$

(6) *Hyperbolic area of a geodesic polygon.* By (10.5) the Riemann metric $g = \frac{1}{y^2}(dx^2+dy^2)$ has density $\operatorname{vol}(g) = \sqrt{\det g_{ij}}\,dx\,dy = \frac{1}{y^2}dx\,dy$; thus:

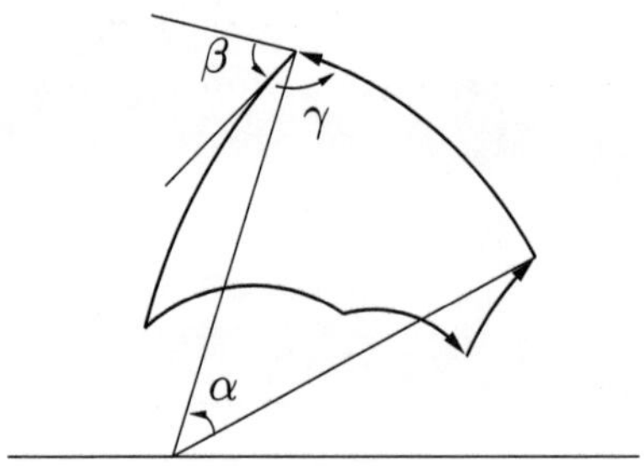

$$\operatorname{Vol}^{H^2_+}(P) = \int_P \frac{dx\wedge dy}{y^2} = \int_P d\left(\frac{dx}{y}\right)$$
$$= \int_{\partial P}\frac{dx}{y} = -\int_{\partial P} d\theta,$$

since each geodesic is part of a circle $z - a = re^{i\theta}$, $a\in\mathbb{R}$. On it we have

$$\frac{dx}{y} = \frac{d(r\cos\theta + a)}{r\sin\theta} = \frac{-r\sin\theta\,d\theta}{r\sin\theta} = -d\theta.$$

The integral is thus the total increase of the tangent angle. For a simply connected polygon the total increase of the tangent angle is 2π if we also add the exterior angles at the corners: $\int_{\partial P} d\vartheta + \sum_i \beta_i = \sum_i \alpha_i + \sum_i \beta_i = 2\pi$. We change to the inner angles $\gamma_i = \pi - \beta_i$ and get:

$$\mathrm{Vol}^{H^2_+}(P) = -\int_{\partial P} d\vartheta = -2\pi + \sum_i \beta_i = (n-2)\pi - \sum_i \gamma_i.$$

This is a particular instance of the theorem of Gauß-Bonnet.

25.9. The 3-sphere S^3. We use the parameterization of $S^3 \subset \mathbb{R}^4$ given by

$$f(\varphi, \vartheta, \tau) = \begin{pmatrix} \cos\varphi\ \cos\vartheta\ \cos\tau \\ \sin\varphi\ \cos\vartheta\ \cos\tau \\ \sin\vartheta\ \cos\tau \\ \sin\tau \end{pmatrix}, \qquad \begin{array}{c} 0 < \varphi < 2\pi, \\ -\frac{\pi}{2} < \vartheta < \frac{\pi}{2}, \\ -\frac{\pi}{2} < \tau < \frac{\pi}{2}. \end{array}$$

We write $f_1^1 = \partial_\varphi f^1$, etc. Then the induced metric is given by:

$$g_{11} = \langle f_1, f_1 \rangle = f_1^1 f_1^1 + f_1^2 f_1^2 + f_1^3 f_1^3 + f_1^4 f_1^4 = \cos^2\vartheta \cos^2\tau,$$

$$g_{12} = \langle f_1, f_2 \rangle = 0, \qquad g_{13} = 0, \qquad g_{22} = \cos^2\tau, \qquad g_{23} = 0, \qquad g_{33} = 1.$$

$$g = \cos^2\vartheta\ \cos^2\tau\ d\varphi \otimes d\varphi + \cos^2\tau\ d\vartheta \otimes d\vartheta + d\tau \otimes d\tau.$$

$$\sigma^1 = \cos\vartheta\ \cos\tau\ d\varphi, \qquad \sigma^2 = \cos\tau\ d\vartheta, \qquad \sigma^3 = d\tau.$$

$$d\sigma^1 = -\sin\vartheta\ \cos\tau\ d\vartheta \wedge d\varphi - \cos\vartheta\ \sin\tau\ d\tau \wedge d\varphi,$$

$$d\sigma^2 = -\sin\tau\ d\tau \wedge d\vartheta, \qquad d\sigma^3 = 0.$$

Now we use the first structure equation (25.5.3), i.e., $d\sigma + \omega \wedge \sigma = 0$:

$$\begin{aligned}
d\sigma^1 &= -0 - \omega_2^1 \wedge \sigma^2 - \omega_3^1 \wedge \sigma^3 = \sin\vartheta\ \cos\tau\ d\varphi \wedge d\vartheta + \cos\vartheta\ \sin\tau\ d\varphi \wedge d\tau, \\
d\sigma^2 &= -\omega_1^2 \wedge \sigma^1 - 0 - \omega_3^2 \wedge \sigma^3 = \sin\tau\ d\vartheta \wedge d\tau, \\
d\sigma^3 &= -\omega_1^3 \wedge \sigma^1 - \omega_2^3 \wedge \sigma^2 - 0 = 0.
\end{aligned}$$

$$\begin{aligned}
&-\omega_2^1 \wedge \cos\tau\ d\vartheta - \omega_3^1 \wedge d\tau = \sin\vartheta\ \cos\tau\ d\varphi \wedge d\vartheta + \cos\vartheta\ \sin\tau\ d\varphi \wedge d\tau, \\
&-\omega_1^2 \wedge \cos\vartheta\ \cos\tau\ d\varphi - \omega_3^2 \wedge d\tau = \sin\tau\ d\vartheta \wedge d\tau, \\
&-\omega_1^3 \wedge \cos\vartheta\ \cos\tau\ d\varphi - \omega_2^3 \wedge \cos\tau\ d\vartheta = 0.
\end{aligned}$$

$$\begin{cases} \omega_3^1 = -\cos\vartheta\ \sin\tau\ d\varphi, \\ \omega_3^2 = -\sin\tau\ d\vartheta, \\ \omega_2^1 = -\sin\vartheta\ d\varphi, \end{cases}$$

$$\omega = \begin{pmatrix} 0 & -\sin\vartheta\ d\varphi & -\cos\vartheta\ \sin\tau\ d\varphi \\ \sin\vartheta\ d\varphi & 0 & -\sin\tau\ d\vartheta \\ \cos\vartheta\ \sin\tau\ d\varphi & \sin\tau\ d\vartheta & 0 \end{pmatrix}.$$

From this we can compute the curvature:

$$\begin{aligned}\Omega_2^1 &= d\omega_2^1 + 0 + 0 + \omega_3^1 \wedge \omega_2^3\\ &= -\cos\vartheta\; d\vartheta \wedge d\varphi - \cos\vartheta\; \sin\tau\; d\varphi \wedge \sin\tau\; d\vartheta\\ &= \cos\vartheta\; \cos^2\tau\; d\varphi \wedge d\vartheta = \sigma^1 \wedge \sigma^2,\\ \Omega_3^1 &= d\omega_3^1 + 0 + \omega_2^1 \wedge \omega_3^2 + 0\\ &= \sin\vartheta\; \sin\tau\; d\vartheta \wedge d\varphi - \cos\vartheta\; \cos\tau\; d\tau \wedge d\varphi + \sin\vartheta\; d\varphi \wedge \sin\tau\; d\vartheta\\ &= \cos\vartheta\; \cos\tau\; d\varphi \wedge d\tau = \sigma^1 \wedge \sigma^3,\\ \Omega_3^2 &= d\omega_3^2 + \omega_1^2 \wedge \omega_3^1 + 0 + 0\\ &= -\cos\tau\; d\tau \wedge d\vartheta + 0\\ &= \cos\tau\; d\vartheta \wedge d\tau = \sigma^2 \wedge \sigma^3,\\ \Omega &= \begin{pmatrix} 0 & \sigma^1\wedge\sigma^2 & \sigma^1\wedge\sigma^3\\ -\sigma^1\wedge\sigma^2 & 0 & \sigma^2\wedge\sigma^3\\ -\sigma^1\wedge\sigma^3 & -\sigma^2\wedge\sigma^3 & 0\end{pmatrix} = \begin{pmatrix}\sigma^1\\ \sigma^2\\ \sigma^3\end{pmatrix} \wedge (\sigma^1, \sigma^2, \sigma^3).\end{aligned}$$

Another representation of the 3-sphere with radius $1/\sqrt{k}$. The induced metric is given by

$$g = \frac{1}{k}\big(\cos^2\vartheta\; \cos^2\tau\; d\varphi \otimes d\varphi + \cos^2\tau\; d\vartheta \otimes d\vartheta + d\tau \otimes d\tau\big),$$

where $0 < \varphi < 2\pi$, $-\frac{\pi}{2} < \vartheta < \frac{\pi}{2}$, and $-\frac{\pi}{2} < \tau < \frac{\pi}{2}$. Now we introduce the coordinate function r by $\cos^2\tau = k\,r^2$, more precisely by

$$r = \begin{cases} -\frac{1}{\sqrt{k}}\cos\tau & \text{for} \quad -\frac{\pi}{2} < \tau < 0,\\ \frac{1}{\sqrt{k}}\cos\tau & \text{for} \quad 0 < \tau < \frac{\pi}{2},\end{cases} \qquad 0 < |r| < \tfrac{1}{\sqrt{k}}.$$

Then $\operatorname{sign}\tau\; \cos\tau = \sqrt{k}\,r$; thus $-\operatorname{sign}\tau\; \sin\tau\; d\tau = \sqrt{k}\,dr$, and since $\sin^2\tau = 1 - \cos^2\tau = 1 - k\,r^2$, we finally get

$$(1 - k\,r^2)\,d\tau \otimes d\tau = \sin^2\tau\; d\tau \otimes d\tau = k\,dr \otimes dr.$$

Furthermore we replace ϑ by $\vartheta + \frac{\pi}{2}$. Then the metric becomes:

$$\begin{aligned} g &= \frac{1}{k}\left(\sin^2\vartheta\; k\,r^2\, d\varphi \otimes d\varphi + k\,r^2\, d\vartheta \otimes d\vartheta + \frac{k}{1-kr^2} dr \otimes dr\right)\\ &= \frac{1}{1-kr^2} dr \otimes dr + r^2\, d\vartheta \otimes d\vartheta + r^2\, \sin^2\vartheta\; d\varphi \otimes d\varphi, \qquad \text{where} \qquad (1)\end{aligned}$$

$$0 < \varphi < 2\pi, \qquad 0 < \vartheta < \pi, \qquad 0 < |r| < \frac{1}{\sqrt{k}}.$$

25.10. The Robertson-Walker metric in general relativity. This is the metric of signature $(+---)$ of the form

$$g = dt \otimes dt - R(t)^2 \left(\frac{1}{1-kr^2} dr \otimes dr + r^2\, d\vartheta \otimes d\vartheta + r^2 \sin^2\vartheta\, d\varphi \otimes d\varphi \right)$$

$$\text{for } 0 < \varphi < 2\pi, \quad 0 < \vartheta < \pi, \quad 0 < |r| < \frac{1}{\sqrt{k}},$$

$$= \rho^0 \otimes \rho^0 - \rho^1 \otimes \rho^1 - \rho^2 \otimes \rho^2 - \rho^3 \otimes \rho^3,$$

$$\rho^0 = dt, \qquad \rho^1 = \frac{R}{w}\, dr, \quad \text{where } w := \sqrt{1-kr^2},$$

$$\rho^2 = Rr\, d\vartheta, \qquad \rho^3 = Rr\, \sin\vartheta\, d\varphi.$$

The differential of the coframe is:

$$\begin{aligned}
d\rho^0 &= 0,\\
d\rho^1 &= \frac{\dot R}{w}\, dt \wedge dr = \frac{\dot R}{R}\rho^0 \wedge \rho^1,\\
d\rho^2 &= \dot R r\, dt \wedge d\vartheta + R\, dr \wedge d\vartheta = \frac{\dot R}{R}\rho^0 \wedge \rho^2 + \frac{w}{Rr}\rho^1 \wedge \rho^2,\\
d\rho^3 &= \dot R r\, \sin\vartheta\, d\vartheta \wedge d\varphi + R \sin\vartheta\, dr \wedge d\varphi + Rr\, \cos\vartheta\, d\vartheta \wedge d\varphi\\
&= \frac{\dot R}{R}\rho^0 \wedge \rho^3 + \frac{w}{Rr}\rho^1 \wedge \rho^3 + \frac{\operatorname{cotan}\vartheta}{Rr}\rho^2 \wedge \rho^3.
\end{aligned}$$

Now we use $d\rho + \omega \wedge \rho = 0$, $\omega^i_j = -\omega^j_i$ for $1 \le i, j \le 3$, $\omega^i_i = 0$, and $\omega^0_i = \omega^i_0$:

$$\begin{aligned}
d\rho^0 &= -\omega^0_1 \wedge \rho^1 - \omega^0_2 \wedge \rho^2 - \omega^0_3 \wedge \rho^3 = 0,\\
d\rho^1 &= -\omega^1_0 \wedge \rho^0 - \omega^1_2 \wedge \rho^2 - \omega^1_3 \wedge \rho^3 = \frac{\dot R}{R}\rho^0 \wedge \rho^1,\\
d\rho^2 &= -\omega^2_0 \wedge \rho^0 - \omega^2_1 \wedge \rho^1 - \omega^2_3 \wedge \rho^3 = \frac{\dot R}{R}\rho^0 \wedge \rho^2 + \frac{w}{Rr}\rho^1 \wedge \rho^2,\\
d\rho^3 &= -\omega^3_0 \wedge \rho^0 - \omega^3_1 \wedge \rho^1 - \omega^3_2 \wedge \rho^2\\
&= \frac{\dot R}{R}\rho^0 \wedge \rho^3 + \frac{w}{Rr}\rho^1 \wedge \rho^3 + \frac{\operatorname{cotan}\vartheta}{Rr}\rho^2 \wedge \rho^3.
\end{aligned}$$

This is a linear system of equations with a unique solution for the ω^i_j. Guided by (25.9) we assume that ω^0_1 is a multiple of ρ^1, etc., and we get the solutions

$$\begin{aligned}
\omega^1_0 &= \frac{\dot R}{R}\rho^1 = \frac{\dot R}{w}\, dr, & \omega^2_0 &= \frac{\dot R}{R}\rho^2 = \dot R\, r\, d\vartheta,\\
\omega^3_0 &= \frac{\dot R}{R}\rho^3 = \dot R\, r\, \sin\vartheta\, d\varphi, & \omega^2_1 &= \frac{w}{Rr}\rho^2 = w\, d\vartheta,\\
\omega^3_1 &= \frac{w}{Rr}\rho^3 = w\, \sin\vartheta\, d\varphi, & \omega^3_2 &= \frac{\operatorname{cotan}\vartheta}{Rr}\rho^3 = \cos\vartheta\, d\varphi.
\end{aligned}$$

From these we can compute the curvature 2-forms, using (25.5.5), that is, $\Omega = d\omega + \omega\wedge\omega$:

$$\Omega_0^1 = -\frac{\ddot R}{R}\rho^1\wedge\rho^0, \qquad \Omega_0^2 = -\frac{\ddot R}{R}\rho^2\wedge\rho^0,$$
$$\Omega_0^3 = -\frac{\ddot R}{R}\rho^3\wedge\rho^0, \qquad \Omega_1^2 = \frac{k+\dot R^2}{R^2}\rho^2\wedge\rho^1,$$
$$\Omega_1^3 = -\frac{-k+\dot R^2}{R^2}\rho^3\wedge\rho^1, \qquad \Omega_2^3 = \frac{k+\dot R^2}{R^2}\rho^3\wedge\rho^2.$$

25.11. The Hodge $*$-operator. Let (M,g) be an oriented pseudo-Riemann manifold of signature (p,q). Viewing $g : TM\to T^*M$, we let $g^{-1} : T^*M\to TM$ denote the dual bundle metric on T^*M. Then g^{-1} induces a symmetric nondegenerate bundle metric

$$\bigwedge^k g^{-1} : \bigwedge^k T^*M \to \bigwedge^k TM$$

on the bundle $\bigwedge^k T^*M$ of k-forms which is given by

$$g^{-1}(\varphi_1\wedge\dots\wedge\varphi_k, \psi_1\wedge\dots\wedge\psi_k) = \det(g^{-1}(\varphi_i,\psi_j)_{i,j=1}^k), \quad \varphi_i,\psi_j\in\Omega^1(M).$$

Let $\eta_{ij} = g(s_i,s_j) = \operatorname{diag}(1,\dots,1,-1,\dots,-1)$ be the standard inner product matrix of the same signature (p,q), and let $s=(s_1,\dots,s_m)$ be an orthonormal frame on $U\subseteq M$ with orthonormal coframe $\sigma = (\sigma_1,\dots,\sigma_m)$ as in (25.5) so that $g=\sum_i \eta_{ii}\sigma^i\otimes\sigma^i$; then for $\varphi^k,\psi^k\in\Omega^k(M)$ we have

$$g^{-1}(\varphi^k,\psi^k) = \sum_{\substack{i_1<\dots<i_k\\ j_1<\dots<j_k}} \varphi^k(s_{i_1},\dots,s_{i_k})\psi^k(s_{j_1},\dots,s_{j_k})\eta^{i_1j_1}\dots\eta^{i_kj_k}.$$

Note that $g^{-1}(\sigma^1\wedge\dots\wedge\sigma^m,\sigma^1\wedge\dots\wedge\sigma^m) = (-1)^q$.

If M is also oriented, then the volume form $\operatorname{vol}(g)$ from (10.5) agrees with the positively oriented m-form of length ±1. We have $\operatorname{vol}(g) = \sigma^1\wedge\dots\wedge\sigma^m$ if the frame $s=(s_1,\dots,s_m)$ is positively oriented.

We shall use the following notation:
If $I=(i_1<\dots<i_k)$ and $I'=(j_1<\dots<j_{m-k})$ are the ordered tuples with $I\cap I'=\emptyset$ and $I\sqcup I' = \{1,\dots,m\}$, then we put $\sigma^I := \sigma^{i_1}\wedge\dots\wedge\sigma^{i_k}$.

Exercise. The k-forms σ^I for all I as above of length k give an orthonormal basis of g^{-1} on $\Omega^k(U)$. The signature of g^{-1} on $\bigwedge^k T_x^*M$ is

$$(P_+(p,q,k), P_-(p,q,k)) = \left(\textstyle\sum_{j=0, j\text{ even}}^k \binom{p}{k-j}\binom{q}{j}, \sum_{j=0, j\text{ odd}}^k \binom{p}{k-j}\binom{q}{j}\right).$$

On an oriented pseudo-Riemann manifold (M, g) of dimension m and signature (p, q) we have the *Hodge isomorphism* $*$ with its elementary properties:

$$
\begin{aligned}
&* : \bigwedge^k T^*M \to \bigwedge^{m-k} T^*M, \\
(1) \qquad &(*\varphi^k)(X_{k+1}, \dots, X_m)\,\mathrm{vol}(g) = \varphi \wedge g(X_{k+1}) \wedge \dots \wedge g(X_m), \\
&\varphi^k \wedge \psi^{m-k} = g^{-1}(*\varphi^k, \psi^{m-k})\,\mathrm{vol}(g), \\
&g^{-1}(*\varphi^k, *\psi^k) = (-1)^q g^{-1}(\varphi^k, \psi^k), \\
&* * \varphi^k = (-1)^{k(m-k)+q}\varphi^k, \\
&(*\varphi^k) \wedge \psi^k = (*\psi^k) \wedge \varphi^k.
\end{aligned}
$$

In the local orthonormal frame we get

$$
\begin{aligned}
(*\sigma^I)(s_{j_1}, \dots, s_{j_{m-k}})\,\mathrm{vol}(g) &= \sigma^I \wedge g(s_{j_1}) \wedge \dots \wedge g(s_{j_{m-k}}) \\
&= \sigma^I \wedge g(s_{j_1}) \wedge \dots \wedge g(s_{j_{m-k}}) = \sigma^I \wedge \eta_{j_1 j_1}\sigma^{j_1} \wedge \dots \wedge \eta_{j_{m-k} j_{m-k}}\sigma^{j_{m-k}}, \\
*\sigma^I &= \mathrm{sign}\begin{pmatrix} 1 \dots m \\ I \quad I' \end{pmatrix} \eta_{j_1 j_1} \dots \eta_{j_{m-k} j_{m-k}} \sigma^{I'}.
\end{aligned}
$$

To get a geometric interpretation of $*\varphi^k$, we consider

$$
\begin{aligned}
i(X)(*\varphi^k)(X_{k+2}, \dots, X_m)\,\mathrm{vol}(g) &= (*\varphi^k)(X, X_{k+2}, \dots, X_m)\,\mathrm{vol}(g) \\
&= \varphi^k \wedge g(X) \wedge g(X_{k+2}) \wedge \dots \wedge g(X_m) \\
&= *(\varphi^k \wedge g(X))(X_{k+2}, \dots, X_m)\,\mathrm{vol}(g)
\end{aligned}
$$

so that

$$
\begin{aligned}
(2) \qquad & i(X)(*\varphi^k) = *(\varphi^k \wedge g(X)), \\
& \{X : i_X \varphi^k = 0\}^{\perp, g} = \{Y : i_Y(*\varphi^k) = 0\}.
\end{aligned}
$$

25.12. Relations to vector analysis. We consider an oriented pseudo-Riemann manifold (M, g) of signature (p, q). For functions $f \in C^\infty(M, \mathbb{R})$ and vector fields $X \in \mathfrak{X}(M)$ we have the following operations, *gradient* and *divergence*, and their elementary properties:

$$
\begin{aligned}
&\mathrm{grad}^g(f) = g^{-1} \circ df \in \mathfrak{X}(M), \\
&g(X) \in \Omega^1(M), \qquad *g(X) = (-1)^q i_X \mathrm{vol}(g), \\
&* \, df = *g(\mathrm{grad}^g(f)) = (-1)^q i_{\mathrm{grad}^g(f)} \mathrm{vol}(g), \\
&\mathrm{div}^g(X).\,\mathrm{vol}(g) = (-1)^q \, d\, i_X \mathrm{vol}(g) = d * g(X), \\
&\mathrm{grad}^g(f \cdot h) = f \cdot \mathrm{grad}^g(h) + h \cdot \mathrm{grad}^g(f), \\
&\mathrm{div}^g(f \cdot X) = f\,\mathrm{div}^g(X) + (-1)^q df(X), \\
&\mathrm{grad}^g(f)|_U = \textstyle\sum_i \eta_{ii} s_i(f) \cdot s_i, \\
&\mathrm{div}^g(X) = \mathrm{Trace}(\nabla X).
\end{aligned}
$$

Some authors take the negative of our definition of the divergence, so that later the Laplace-Beltrami operator $\Delta f = (-\operatorname{div}^g)\operatorname{grad}^g(f)$ is positive definite on any oriented Riemann manifold.

25.13. In dimension three. On an oriented 3-dimensional pseudo-Riemann manifold we have another operator on vector fields, *curl*, given by

$$\begin{aligned} * g(\operatorname{curl}^g(X)) &= (-1)^q i_{\operatorname{curl}^g(X)} \operatorname{vol}(g) = dg(X), \\ \operatorname{curl}^g(X) &= (-1)^q g^{-1} * dg(X), \end{aligned}$$

and from $d^2 = 0$ we have $\operatorname{curl}^g \operatorname{grad}^g = 0$ and $\operatorname{div}^g \operatorname{curl}^g = 0$.
On the oriented Euclidean space $\mathbb{R}^3$ we have

$$\begin{aligned} \operatorname{grad}(f) &= \tfrac{\partial f}{\partial x^1}\tfrac{\partial}{\partial x^1} + \tfrac{\partial f}{\partial x^2}\tfrac{\partial}{\partial x^2} + \tfrac{\partial f}{\partial x^3}\tfrac{\partial}{\partial x^3}, \\ \operatorname{curl}(X) &= \left(\tfrac{\partial X^3}{\partial x^2} - \tfrac{\partial X^2}{\partial x^3}\right)\tfrac{\partial}{\partial x^1} + \left(\tfrac{\partial X^1}{\partial x^3} - \tfrac{\partial X^3}{\partial x^1}\right)\tfrac{\partial}{\partial x^2} + \left(\tfrac{\partial X^2}{\partial x^1} - \tfrac{\partial X^1}{\partial x^2}\right)\tfrac{\partial}{\partial x^3}, \\ \operatorname{div}(X) &= \tfrac{\partial X^1}{\partial x^1} + \tfrac{\partial X^2}{\partial x^2} + \tfrac{\partial X^3}{\partial x^3}. \end{aligned}$$

Note also that $\operatorname{curl}(f \cdot X) = f \cdot \operatorname{rot}(X) + \operatorname{grad}(f) \times X$ where $\times$ denotes the vector product in $\mathbb{R}^3$.

25.14. The Maxwell equations. Let $U \subset \mathbb{R}^3$ be an open set in the oriented Euclidean 3-space. We will later assume that the first cohomology vanishes: $H^1(U) = 0$. We consider three time dependent vector fields and a function,

$$\begin{aligned} &E : U \times \mathbb{R} \to \mathbb{R}^3, && \text{the electric field,} \\ &B : U \times \mathbb{R} \to \mathbb{R}^3, && \text{the magnetic field,} \\ &J : U \times \mathbb{R} \to \mathbb{R}^3, && \text{the current field,} \\ &\rho : U \times \mathbb{R} \to \mathbb{R}, && \text{the density function of the electric charge.} \end{aligned}$$

Then the *Maxwell equations* are (where c is the speed of light)

$$\begin{aligned} \operatorname{curl}(E) &= -\frac{1}{c}\frac{d}{dt}B, && \operatorname{div}(B) = 0, \\ \operatorname{curl}(B) &= \frac{1}{c}\frac{d}{dt}E + \frac{4\pi}{c}J, && \operatorname{div}(E) = 4\pi\rho. \end{aligned}$$

Now let η be the standard positive definite inner product on $\mathbb{R}^3$. From (25.13) we see that the Maxwell equations can be written as

$$\begin{aligned} *d\,\eta(E) &= -\frac{1}{c}\frac{d}{dt}\eta(B), && d * \eta(B) = 0, \\ *d\,\eta(B) &= \frac{1}{c}\frac{d}{dt}\eta(E) + \frac{4\pi}{c}\eta(J), && d * \eta(E) = 4\pi\rho \cdot \operatorname{vol}(\eta). \end{aligned}$$

Now we assume that $H^1(U) = 0$. Since $d * \eta(B) = 0$, we have

$$*\eta(B) = dA \quad \text{for a function } A, \quad \text{the } \textit{magnetic potential.}$$

Then the first Maxwell equation can be written as

$$d\left(\eta(E) + \frac{1}{c}\frac{d}{dt}A\right) = 0.$$

Using again $H^1(U) = 0$, there exists a function $\Phi : U \times \mathbb{R} \to \mathbb{R}$, called the *electric potential*, such that

$$\eta(E) = -\frac{1}{c}\frac{d}{dt}A - d\Phi.$$

Starting from the magnetic and electric potentials $A, \Phi : U \times \mathbb{R} \to \mathbb{R}$, the electric and magnetic fields are given by

$$\eta(E) = -\frac{1}{c}\frac{d}{dt}A - d\Phi, \qquad \eta(B) = *dA,$$

where all terms are viewed as time dependent functions of forms on $\mathbb{R}^3$. Then the first row of the Maxwell equations is automatically satisfied. The second row then looks like

$$- * d * dA = -\frac{1}{c^2}\frac{d^2}{dt^2}A - \frac{1}{c}\frac{d}{dt}d\Phi + \frac{4\pi}{c}\eta(J), \quad \frac{1}{c}\frac{d}{dt}(*d * A) - \Delta\Phi = 4\pi\rho.$$

26. Riemann Immersions and Submersions

26.1. Riemann submanifolds and isometric immersions. Let $(\bar{M}, \bar{g})$ be a Riemann manifold of dimension $m+p$, and let $M \xrightarrow{i} \bar{M}$ be a manifold of dimension m with an immersion i. Let $g := i^*\bar{g}$ be the induced Riemann metric on M. Let $\bar{\nabla}$ be the Levi-Civita covariant derivative on $\bar{M}$, and let ∇ be the Levi-Civita covariant derivative on M. We denote by $Ti^\perp = TM^\perp := \{X \in T_{i(x)}\bar{M}, x \in M, \bar{g}(X, Ti(T_xM)) = 0\}$ the normal bundle (over M) of the immersion i or the immersed submanifold M.

Let $X, Y \in \mathfrak{X}(M)$. We may regard $Ti.Y$ as vector field with values in $T\bar{M}$ defined along i and thus consider $\bar{\nabla}_X(Ti.Y) : M \to i^*T\bar{M}$.

Lemma. Gauß's formula. *If $X, Y \in \mathfrak{X}(M)$, then*

$$\bar{\nabla}_X(Ti.Y) - Ti \circ \nabla_X Y =: S(X, Y)$$

is normal to M, and $S : TM \times_M TM \to Ti^\perp$ is a symmetric tensor field, which is called the second fundamental form or the shape form of M.

Proof. For $X, Y, Z \in \mathfrak{X}(M)$ and a suitable open set $U \subset M$ we may choose an open subset $\bar{U} \subset \bar{M}$ with $i(U)$ closed in $\bar{U}$ such that $i : U \to \bar{U}$ is an embedding, and then extensions $\bar{X}, \bar{Y}, \bar{Z} \in \mathfrak{X}(\bar{U})$ with $\bar{X} \circ i|_U = Ti.X|_U$, etc. By (22.5.7) we have

$$\begin{aligned}2\bar{g}(\bar{\nabla}_{\bar{X}}\bar{Y}, \bar{Z}) &= \bar{X}(\bar{g}(\bar{Y}, \bar{Z})) + \bar{Y}(\bar{g}(\bar{Z}, \bar{X})) - \bar{Z}(\bar{g}(\bar{X}, \bar{Y})) \\ &\quad + \bar{g}([\bar{X}, \bar{Y}], \bar{Z}) + \bar{g}([\bar{Z}, \bar{X}], \bar{Y}) - \bar{g}([\bar{Y}, \bar{Z}], \bar{X}).\end{aligned}$$

Composing this formula with $i|_U$, we get on U

$$\begin{aligned}2\bar{g}(\bar{\nabla}_X(Ti.Y), Ti.Z) &= X(g(Y, Z)) + Y(g(Z, X)) - Z(g(X, Y)) \\ &\quad + g([X, Y], Z) + g([Z, X], Y) - g([Y, Z], X) \\ &= 2g(\nabla_X Y, Z),\end{aligned}$$

again by (22.5.7). Since this holds for all $Z \in \mathfrak{X}(U)$, the orthonormal projection of $\bar{\nabla}_X Y$ to TM is just $\nabla_X Y$. Thus $S(X, Y) := \bar{\nabla}_X(Ti.Y) - Ti.\nabla_X Y$ is a section of $Ti^\perp$, and it is symmetric in X, Y since

$$\begin{aligned}S(X, Y) &= \bar{\nabla}_X(Ti.Y) - Ti \circ \nabla_X Y = (\bar{\nabla}_{\bar{X}}\bar{Y}) \circ i - Ti \circ \nabla_X Y \\ &= (\bar{\nabla}_{\bar{Y}}\bar{X} + [\bar{X}, \bar{Y}]) \circ i - Ti.(\nabla_Y X + [X, Y]) = S(Y, X).\end{aligned}$$

For $f \in C^\infty(M)$ we have

$$\begin{aligned}S(fX, Y) &= \bar{\nabla}_{fX}(Ti.Y) - Ti \circ \nabla_{fXY} \\ &= f\bar{\nabla}_X(Ti.Y) - fTi \circ \nabla_X Y = fS(X, Y),\end{aligned}$$

and $S(X, fY) = fS(X, Y)$ follows by symmetry. □

26.2. Corollary. *Let $c : [a, b] \to M$ be a smooth curve. Then we have*

$$\bar{\nabla}_{\partial_t}(Ti.c') = \bar{\nabla}_{\partial_t}(i \circ c)' = Ti \circ \nabla_{\partial_t} c' + S(c', c').$$

Consequently c is a geodesic in M if and only if

$$\bar{\nabla}_{\partial_t}(i \circ c)' = S(c', c') \in Ti^\perp,$$

i.e., the acceleration of $i \circ c$ in $\bar{M}$ is orthogonal to M.

Let $i : M \to \bar{M}$ be an isometric immersion. Then the following conditions are equivalent:

(1) *Any geodesic in $\bar{M}$ which starts in $i(M)$ in a direction tangent to $i(M)$ stays in $i(M)$; it is then a geodesic in $i(M)$. We call $i : M \to \bar{M}$ a totally geodesic immersion.*

(2) *The second fundamental form S of $i : M \to \bar{M}$ vanishes.* □

26.3. In the setting of (26.1) we now investigate $\bar\nabla_X\xi$ where $X \in \mathfrak{X}(M)$ and where $\xi \in \Gamma(Ti^\perp)$ is a normal field. We split it into tangential and normal components:

(1) $\bar\nabla_X\xi = -Ti.L_\xi(X) + \nabla^\perp_X\xi \in \mathfrak{X}(M) \oplus \Gamma(Ti^\perp)$ (*Weingarten formula*).

Proposition.
(2) *The mapping $(\xi, X) \mapsto L_\xi(X)$ is $C^\infty(M)$-bilinear; thus $L : Ti^\perp \times_M TM \to TM$ is a tensor field, called the Weingarten mapping or shape operator and we have:*

$$g(L_\xi(X), Y) = \bar g(S(X,Y), \xi), \quad \xi \in \Gamma(Ti^\perp), X, Y \in \mathfrak{X}(M).$$

By the symmetry of S, $L_\xi : TM \to TM$ is a symmetric endomorphism with respect to g, i.e., $g(L_\xi(X), Y) = g(X, L_\xi(Y))$.

(3) *The mapping $(X, \xi) \mapsto \nabla^\perp_X\xi$ is a covariant derivative in the normal bundle $Ti^\perp \to M$ which respects the metric $g^\perp := \bar g|Ti^\perp \times_M Ti^\perp$; i.e.,*

$$\nabla^\perp : \mathfrak{X}(M) \times \Gamma(Ti^\perp) \to \Gamma(Ti^\perp) \quad \textit{is } \mathbb{R}\textit{-bilinear,}$$
$$\nabla^\perp_{f.X}\xi = f.\nabla^\perp_X\xi, \qquad \nabla^\perp_X(f.\xi) = df(X).\xi + \nabla^\perp_X\xi,$$
$$X(g^\perp(\xi,\eta)) = g^\perp(\nabla^\perp_X\xi, \eta) + g^\perp(\xi, \nabla^\perp_X\eta).$$

Note that there does not exist torsion for $\nabla^\perp$.

Proof. The mapping $(\xi, X) \mapsto L_\xi(X)$ is obviously $\mathbb{R}$-bilinear. Moreover,

$$-Ti.L_\xi(f.X) + \nabla^\perp_{f.X}\xi = \bar\nabla_{f.X}\xi = f.\bar\nabla_X\xi = -f.(Ti.L_\xi(X)) + f.\nabla^\perp_X\xi$$

which implies

$$L_\xi(f.X) = f.L_\xi(X), \quad \nabla^\perp_{f.X}\xi = f.\nabla^\perp_X\xi.$$

Furthermore,

$$\begin{aligned} -Ti.L_{f.\xi}(X) + \nabla^\perp_X(f.\xi) &= \bar\nabla_X(f.\xi) = df(X).\xi + f.\bar\nabla_X\xi \\ &= -f.(Ti.L_\xi(X)) + (df(X).\xi + f.\nabla^\perp_X\xi) \end{aligned}$$

implies

$$L_{f.\xi}(X) = f.L_\xi(X), \quad \nabla^\perp_X(f.\xi) = df(X).\xi + f.\nabla^\perp_X\xi.$$

For the rest we enlarge $X, Y \in \mathfrak{X}(M)$ and $\xi, \eta \in \Gamma(Ti^\perp)$ locally to vector fields $\bar X, \bar Y, \bar\xi, \bar\eta$ on $\bar M$. Then we have:

$$\begin{aligned} X(g^\perp(\xi,\eta)) &= \bar X(\bar g(\bar\xi,\bar\eta)) \circ i = \big(\bar g(\bar\nabla_{\bar X}\bar\xi, \bar\eta) + \bar g(\bar\xi, \bar\nabla_{\bar X}\bar\eta)\big) \circ i \\ &= \bar g(\bar\nabla_X\xi, \eta) + \bar g(\xi, \bar\nabla_X\eta) \\ &= \bar g(-Ti.L_\xi(X) + \nabla^\perp_X\xi, \eta) + \bar g(\xi, -Ti.L_\eta(X) + \nabla^\perp_X\eta) \\ &= g^\perp(\nabla^\perp_X\xi, \eta) + g^\perp(\xi, \nabla^\perp_X\eta), \\ \bar X(\bar g(\bar Y, \bar\xi)) &= \bar g(\bar\nabla_{\bar X}\bar Y, \bar\xi) + \bar g(\bar Y, \bar\nabla_{\bar X}\bar\xi). \end{aligned}$$

Pull this back to M:

$$\begin{aligned} 0 = X(\bar g(Y,\xi)) &= \bar g(\bar\nabla_X(Ti.Y),\xi) + \bar g(Ti.Y, \bar\nabla_X\xi) \\ &= \bar g(Ti.\nabla_X Y + S(X,Y),\xi) + \bar g(Ti.Y, -Ti.L_\xi(X) + \nabla^\perp_X\xi) \\ &= g^\perp(S(X,Y),\xi) + g(Y, -L_\xi(X)). \quad \square \end{aligned}$$

26.4. Theorem. *Let* $(M,g) \xrightarrow{i} (\bar M, \bar g)$ *be an isometric immersion of Riemann manifolds with Riemann curvatures* R *and* $\bar R$*, respectively. Then we have:*

(1) *For* $X_i \in \mathfrak{X}(M)$ *or* T_xM *we have (Gauß's equation, theorema egregium)*

$$\begin{aligned} \bar g(\bar R(Ti.X_1,Ti.X_2)(Ti.X_3), Ti.X_4) &= g(R(X_1,X_2)X_3,X_4) \\ &+ g^\perp(S(X_1,X_3),S(X_2,X_4)) - g^\perp(S(X_2,X_3),S(X_1,X_4)). \end{aligned}$$

(2) *The tangential part of* $\bar R(X_1,X_2)X_3$ *is given by:*

$$\begin{aligned} &(\bar R(Ti.X_1,Ti.X_2)(Ti.X_3))^\top \\ &\qquad = R(X_1,X_2)X_3 + L_{S(X_1,X_3)}(X_2) - L_{S(X_2,X_3)}(X_1). \end{aligned}$$

(3) *The normal part of* $\bar R(X_1,X_2)X_3$ *is (Codazzi-Mainardi equation):*

$$\begin{aligned} &(\bar R(Ti.X_1,Ti.X_2)(Ti.X_3))^\perp \\ &\quad = \big(\nabla^{Ti^\perp\otimes T^*M\otimes T^*M}_{X_1} S\big)(X_2,X_3) - \big(\nabla^{Ti^\perp\otimes T^*M\otimes T^*M}_{X_2} S\big)(X_1,X_3). \end{aligned}$$

(4) *The tangential and the normal parts of* $\bar R(Ti.X_1,Ti.X_2)\xi$ *(where* ξ *is a normal field along* i*) are given by:*

$$\begin{aligned} &(\bar R(Ti.X_1,Ti.X_2)\xi)^\top \\ &\quad = Ti.\Big((\nabla^{TM\otimes(Ti^\perp)^*\otimes T^*M}_{X_2} L)_\xi(X_1) - (\nabla^{TM\otimes(Ti^\perp)^*\otimes T^*M}_{X_1} L)_\xi(X_2)\Big), \\ &(\bar R(Ti.X_1,Ti.X_2)\xi)^\perp \\ &\quad = R^{\nabla^\perp}(X_1,X_2)\xi + S(L_\xi(X_1),X_2) - S(L_\xi(X_2),X_1). \end{aligned}$$

Proof. Every $x \in M$ has an open neighborhood U such that $i : U \to \bar M$ is an embedding. Since the assertions are local, we may thus assume that i is an embedding, and we may suppress i in the following proof. For the proof we need vector fields $X_i \in \mathfrak{X}(M)$. We start from the Gauß formula (26.1):

$$\begin{aligned} \bar\nabla_{X_1}(\bar\nabla_{X_2}X_3) &= \bar\nabla_{X_1}(\nabla_{X_2}X_3 + S(X_2,X_3)) \\ &= \nabla_{X_1}\nabla_{X_2}X_3 + S(X_1,\nabla_{X_2}X_3) + \bar\nabla_{X_1}S(X_2,X_3), \\ \bar\nabla_{X_2}(\bar\nabla_{X_1}X_3) &= \nabla_{X_2}\nabla_{X_1}X_3 + S(X_2,\nabla_{X_1}X_3) + \bar\nabla_{X_2}S(X_1,X_3), \\ \bar\nabla_{[X_1,X_2]}X_3 &= \nabla_{[X_1,X_2]}X_3 + S([X_1,X_2],X_3) \\ &= \nabla_{[X_1,X_2]}X_3 + S(\nabla_{X_1}X_2,X_3) - S(\nabla_{X_2}X_1,X_3). \end{aligned}$$

Inserting this, we get for the part which is tangent to M:

$$\begin{aligned}
\bar g(\bar R(X_1,X_2)X_3,X_4) &= \bar g(\bar\nabla_{X_1}\bar\nabla_{X_2}X_3-\bar\nabla_{X_2}\bar\nabla_{X_1}X_3-\bar\nabla_{[X_1,X_2]}X_3,X_4)\\
&= g(\nabla_{X_1}\nabla_{X_2}X_3-\nabla_{X_2}\nabla_{X_1}X_3-\nabla_{[X_1,X_2]}X_3,X_4)\\
&\quad+\bar g\big(S(X_1,\nabla_{X_2}X_3)-S(X_2,\nabla_{X_1}X_3)-S([X_1,X_2],X_3),X_4\big)\quad(\;=0)\\
&\quad+\bar g(\bar\nabla_{X_1}S(X_2,X_3)-\bar\nabla_{X_2}S(X_1,X_3),X_4)\\
&= g(R(X_1,X_2)X_3,X_4)\\
&\quad+g^\perp(S(X_1,X_3),S(X_2,X_4))-g^\perp(S(X_2,X_3),S(X_1,X_4)),
\end{aligned}$$

where we also used (26.3.1) and (26.3.2) in:

$$\begin{aligned}
\bar g(\bar\nabla_{X_1}S(X_2,X_3),X_4) &= \bar g(\nabla^\perp_{X_1}S(X_2,X_3)-L_{S(X_2,X_3)}(X_1),X_4)\\
&= 0-g^\perp(S(X_1,X_4),S(X_2,X_3)).
\end{aligned}$$

So (1) and (2) follow. For equation (3) we have to compute the normal components of the $+--$ sum of the first three equations in this proof:

$$\begin{aligned}
&(\bar R(X_1,X_2)X_3)^\perp\\
&\quad= 0+S(X_1,\nabla_{X_2}X_3)+\big(\bar\nabla_{X_1}S(X_2,X_3)\big)^\perp-0-S(X_2,\nabla_{X_1}X_3)\\
&\qquad-\big(\bar\nabla_{X_2}S(X_1,X_3)\big)^\perp-0-S(\nabla_{X_1}X_2,X_3)+S(\nabla_{X_2}X_1,X_3)\\
&\quad=\big(\nabla^\perp_{X_1}S(X_2,X_3)-S(\nabla_{X_1}X_2,X_3)-S(X_2,\nabla_{X_1}X_3)\big)\\
&\qquad-\big(\nabla^\perp_{X_2}S(X_1,X_3)-S(\nabla_{X_2}X_1,X_3)-S(X_1,\nabla_{X_2}X_3)\big)\\
&\quad=\big(\nabla^{Ti^\perp\otimes T^*M\otimes T^*M}_{X_1}S\big)(X_2,X_3)-\big(\nabla^{Ti^\perp\otimes T^*M\otimes T^*M}_{X_2}S\big)(X_1,X_3).
\end{aligned}$$

For the proof of (4) we start from the Weingarten formula (26.3.1) and use (26.1):

$$\begin{aligned}
\bar\nabla_{X_1}(\bar\nabla_{X_2}\xi) &= \bar\nabla_{X_1}(\nabla^\perp_{X_2}\xi-L_\xi(X_2))\\
&= \nabla^\perp_{X_1}\nabla^\perp_{X_2}\xi-L_{\nabla^\perp_{X_2}\xi}(X_1)-\nabla_{X_1}(L_\xi(X_2))-S(X_1,L_\xi(X_2)),\\
\bar\nabla_{X_2}(\bar\nabla_{X_1}\xi) &= \nabla^\perp_{X_2}\nabla^\perp_{X_1}\xi-L_{\nabla^\perp_{X_1}\xi}(X_2)-\nabla_{X_2}(L_\xi(X_1))-S(X_2,L_\xi(X_1)),\\
\bar\nabla_{[X_1,X_2]}\xi &= \nabla^\perp_{[X_1,X_2]}\xi-L_\xi([X_1,X_2])\\
&= \nabla^\perp_{[X_1,X_2]}\xi-L_\xi(\nabla_{X_1}X_2)+L_\xi(\nabla_{X_2}X_1).
\end{aligned}$$

Inserting this, we get for the tangential part:

$$\begin{aligned}
(\bar R(X_1,X_2)\xi)^\top &= L_{\nabla^\perp_{X_1}\xi}(X_2)-L_{\nabla^\perp_{X_2}\xi}(X_1)\\
&\quad+\nabla_{X_2}(L_\xi(X_1))-L_\xi(\nabla_{X_2}X_1)-\nabla_{X_1}(L_\xi(X_2))+L_\xi(\nabla_{X_1}X_2)\\
&= -(\nabla^{TM\otimes(Ti^\perp)^*\otimes T^*M}_{X_1}L)_\xi(X_2)+(\nabla^{TM\otimes(Ti^\perp)^*\otimes T^*M}_{X_2}L)_\xi(X_1).
\end{aligned}$$

For the normal part we get:

$$(\bar{R}(X_1,X_2)\xi)^\perp = \nabla^\perp_{X_1}\nabla^\perp_{X_2}\xi - \nabla^\perp_{X_2}\nabla^\perp_{X_1}\xi - \nabla^\perp_{[X_1,X_2]}\xi \\ - S(X_1, L_\xi(X_2)) + S(X_2, L_\xi(X_1)). \quad \square$$

26.5. Hypersurfaces. Let $i : (M,g) \to (\bar{M},\bar{g})$ be an isometrically embedded hypersurface, so that $\dim(\bar{M}) = \dim(M)+1$. Let ν be a local unit normal field along M, i.e., $\nu \in \Gamma(Ti^\perp|U)$ with $|\nu|_{\bar{g}} = 1$. There are two choices for ν.

Theorem. *In this situation we have:*

(1) $\bar{\nabla}_X \nu \in TM$ *for all* $X \in TM$.

(2) *For* $X, Y \in \mathfrak{X}(M)$ *we have (Weingarten equation):*

$$\bar{g}(\bar{\nabla}_X\nu, Y) = -\bar{g}(\nu, \bar{\nabla}_X Y) = -g^\perp(\nu, S(X,Y)).$$

(3) $\bar{g}(\bar{\nabla}_X\nu, Y) = \bar{g}(\bar{\nabla}_Y\nu, X)$.

(4) *If we put* $s(X,Y) := g^\perp(\nu, S(X,Y))$, *then* s *is called the classical second fundamental form and the Weingarten equation* (2) *takes the following form:*

$$\bar{g}(\bar{\nabla}_X\nu, Y) = -s(X,Y).$$

(5) *For hypersurfaces the Codazzi-Mainardi equation takes the following form:*

$$\bar{g}(\bar{R}(X_1,X_2)X_3,\nu) = (\nabla_{X_1}s)(X_2,X_3) - (\nabla_{X_2}s)(X_1,X_3).$$

Proof. (1) Since $1 = \bar{g}(\nu,\nu)$, we get $0 = X(\bar{g}(\nu,\nu)) = 2\bar{g}(\bar{\nabla}_X\nu,\nu)$; thus $\bar{\nabla}_X\nu$ is tangent to M.

(2) Since $0 = \bar{g}(\nu, Y)$, we get $0 = X(\bar{g}(\nu,Y)) = \bar{g}(\bar{\nabla}_X\nu, Y) + \bar{g}(\nu, \bar{\nabla}_X Y)$ and thus $\bar{g}(\bar{\nabla}_X\nu, Y) = -\bar{g}(\nu,\bar{\nabla}_X Y) = -\bar{g}(\nu, \nabla_X Y + S(X,Y)) = -\bar{g}(\nu, S(X,Y))$.

(3) follows from (2) and symmetry of $S(X,Y)$. (4) is a reformulation.

(5) We put ourselves back into the proof of (26.4.3) and use $S(X,Y) = s(X,Y).\nu$ and the fact that $s \in \Gamma(S^2T^*M|U)$ is a $\binom{0}{2}$-tensor field so that $\nabla_X s$ makes sense. We have

$$\bar{\nabla}_{X_1}(S(X_2,X_3)) = \bar{\nabla}_{X_1}(s(X_2,X_3).\nu) = X_1(s(X_2,X_3).\nu + s(X_2,X_3).\bar{\nabla}_{X_1}\nu$$

and $\bar{\nabla}_{X_1}\nu$ is tangential to M by (1). Thus the normal part is:

$$\begin{aligned}\big(\bar{\nabla}_{X_1}(S(X_2,X_3))\big)^\perp &= X_1(s(X_2,X_3)).\nu \\ &= (\nabla_{X_1}s)(X_2,X_3).\nu + s(\nabla_{X_1}X_2,X_3).\nu + s(X_2,\nabla_{X_1}X_3).\nu.\end{aligned}$$

Now we put this into the formula of the proof of (26.4.3):

$$(\bar{R}(X_1,X_2)X_3)^\perp = S(X_1,\nabla_{X_2}X_3) + \big(\bar{\nabla}_{X_1}(S(X_2,X_3))\big)^\perp - S(X_2,\nabla_{X_1}X_3)$$

$$
\begin{aligned}
&- \big(\bar\nabla_{X_2}(S(X_1,X_3))\big)^\perp - S(\nabla_{X_1}X_2,X_3) + S(\nabla_{X_2}X_1,X_3)\\
&= \big((\nabla_{X_1}s)(X_2,X_3) - (\nabla_{X_2}s)(X_1,X_3)\big)\nu. \quad \square
\end{aligned}
$$

26.6. Remark (Theorema egregium proper). Let M be a surface in $\mathbb{R}^3$; then $\bar R = 0$ and by (26.4.1) we have for $X, Y \in T_xM$:

$$0 = \langle \bar R(X,Y)X,Y\rangle = \langle R(X,Y)X,Y\rangle + s(X,X).s(Y,Y) - s(Y,X).s(X,Y).$$

Let us now choose a local coordinate system $(U,(x,y))$ on M and put

$$
\begin{aligned}
g &= i^*\langle\ \ ,\ \ \rangle =: E\,dx\otimes dx + F\,dx\otimes dy + F\,dy\otimes dx + G\,dy\otimes dy,\\
s &=: l\,dx\otimes dx + m\,dx\otimes dy + m\,dy\otimes dx + n\,dy\otimes dy,
\end{aligned}
$$

then $K =$ Gauß's curvature $=$ sectional curvature

$$
\begin{aligned}
&= -\frac{\langle R(\partial_x,\partial_y)\partial_x,\partial_y\rangle}{|\partial_x|^2|\partial_y|^2 - \langle\partial_x,\partial_y\rangle^2} = \frac{s(\partial_x,\partial_x).s(\partial_y,\partial_y) - s(\partial_x,\partial_y)^2}{EG-F^2}\\
&= \frac{ln - m^2}{EG-F^2},
\end{aligned}
$$

which is Gauß's formula for his curvature in his notation.

26.7. Adapted frames for isometric embeddings. All the following also hold for immersions. For notational simplicity we stick with embeddings. Let $e : (M,g) \to (\bar M,\bar g)$ be an isometric embedding of Riemann manifolds, and let $\dim(\bar M) = m+p$ and $\dim(M) = m$. An *adapted orthonormal frame* $\bar s = (\bar s_1,\dots,\bar s_{m+p})$ is an orthonormal frame for $\bar M$ over $\bar U \subset \bar M$ such that for $U = \bar U \cap M \subset M$ the fields $s_1 = \bar s_1|_U,\dots,s_m = \bar s_m|_U$ are tangent to M. Thus $s = (s_1,\dots,s_m)$ is an orthonormal frame for M over U. The orthonormal coframe

$$\bar\sigma = \begin{pmatrix}\bar\sigma^1\\ \vdots \\ \bar\sigma^{m+p}\end{pmatrix} = (\bar\sigma^1,\dots,\bar\sigma^{m+p})^\top$$

for $\bar M$ over $\bar U$ dual to $\bar s$ is then given by $\bar\sigma^{\bar\imath}(\bar s_{\bar\jmath}) = \delta^{\bar\imath}_{\bar\jmath}$. We recall from (25.5):

$$
\begin{aligned}
(1)\quad & \bar g = \textstyle\sum_{\bar\imath=1}^{m+p}\bar\sigma^{\bar\imath}\otimes\bar\sigma^{\bar\imath},\\
& \bar\nabla\bar s = \bar s.\bar\omega,\quad \bar\omega^{\bar\imath}_{\bar\jmath} = -\bar\omega^{\bar\jmath}_{\bar\imath},\qquad \text{so } \bar\omega\in\Omega^1(\bar U,\mathfrak{so}(m+p)),\\
& d\bar\sigma + \bar\omega\wedge\bar\sigma = 0,\quad d\bar\sigma^{\bar\imath} + \textstyle\sum_{\bar k=1}^{m+p}\bar\omega^{\bar\imath}_{\bar k}\wedge\bar\sigma^{\bar k} = 0,\\
& \bar R\bar s = \bar s.\bar\Omega,\quad \bar\Omega = d\bar\omega + \bar\omega\wedge\bar\omega \in \Omega^2(\bar U,\mathfrak{so}(m+p)),\\
&\qquad \bar\Omega^{\bar\imath}_{\bar\jmath} = d\bar\omega^{\bar\imath}_{\bar\jmath} + \textstyle\sum_{\bar k=1}^{m+p}\bar\omega^{\bar\imath}_{\bar k}\wedge\bar\omega^{\bar k}_{\bar\jmath},\\
& \bar\Omega\wedge\bar\sigma = 0,\quad \textstyle\sum_{\bar k=1}^{m+p}\bar\Omega^{\bar\imath}_{\bar k}\wedge\bar\sigma^{\bar k} = 0,\quad \text{first Bianchi identity},\\
& d\bar\Omega + \bar\omega\wedge\bar\Omega - \bar\Omega\wedge\bar\omega = d\bar\Omega + [\bar\omega,\bar\Omega]_\wedge = 0,\quad \text{second Bianchi identity}.
\end{aligned}
$$

Likewise, the orthonormal coframe $\sigma = (\sigma^1, \dots, \sigma^m)^\top$ for M over U dual to s is then given by $\sigma^i(s_j) = \delta^i_j$. Recall again from (25.5):

$$\begin{aligned}
(2)\quad & g = \textstyle\sum_{i=1}^m \sigma^i \otimes \sigma^i, \\
& \nabla s = s.\omega, \quad \omega^i_j = -\omega^j_i, \qquad \text{so } \omega \in \Omega^1(U, \mathfrak{so}(m)), \\
& d\sigma + \omega \wedge \sigma = 0, \quad d\sigma^i + \textstyle\sum_{k=1}^m \omega^i_k \wedge \sigma^k = 0, \\
& Rs = s.\Omega, \quad \Omega = d\omega + \omega \wedge \omega \in \Omega^2(U, \mathfrak{so}(m)), \\
& \qquad \Omega^i_j = d\omega^i_j + \textstyle\sum_{k=1}^m \omega^i_k \wedge \omega^k_j, \\
& \Omega \wedge \sigma = 0, \quad \textstyle\sum_{k=1}^m \Omega^i_k \wedge \sigma^k = 0, \quad \text{first Bianchi identity,} \\
& d\Omega + \omega \wedge \Omega - \Omega \wedge \omega = d\Omega + [\omega, \Omega]_\wedge = 0, \quad \text{second Bianchi identity.}
\end{aligned}$$

Obviously we have $\bar\sigma^i|_U = \sigma^i$, more precisely $e^*\bar\sigma^i = \sigma^i$, for $i = 1, \dots, m$, and $e^*\bar\sigma^{\bar\imath} = 0$ for $\bar\imath = m+1, \dots, m+p$. We want to compute $e^*\bar\omega$. From $d\bar\sigma^{\bar\imath} + \sum_{\bar k=1}^{m+p} \bar\omega^{\bar\imath}_{\bar k} \wedge \bar\sigma^{\bar k} = 0$ we get

$$\begin{aligned}
(3)\quad d\sigma^i &= -\textstyle\sum_{\bar k=1}^{m+p} e^*\bar\omega^i_{\bar k} \wedge e^*\bar\sigma^{\bar k} = -\textstyle\sum_{k=1}^m e^*\bar\omega^i_k \wedge \sigma^k \quad \text{for } i = 1, \dots, m, \\
0 &= -\textstyle\sum_{\bar k=1}^{m+p} e^*\bar\omega^{\bar\imath}_{\bar k} \wedge e^*\bar\sigma^{\bar k} = -\textstyle\sum_{k=1}^m e^*\bar\omega^{\bar\imath}_k \wedge \sigma^k \quad \text{for } m+1 \le \bar\imath.
\end{aligned}$$

Since also $e^*\bar\omega^i_j = -e^*\bar\omega^j_i$, the forms $e^*\bar\omega^i_j$ for $1 \le i, j \le m$ satisfy the defining equations for ω^i_j; thus we have:

$$(4)\qquad \omega^i_j = e^*\bar\omega^i_j, \qquad \text{for } 1 \le i, j \le m.$$

Since $\bar g(\bar\nabla_X s_i, s_j) = \bar\omega^j_i(X) = \omega^j_i(X) = g(\nabla_X s_i, s_j)$ for $X \in \mathfrak{X}(M)$, equation (4) also expresses the fact that the tangential part $(\bar\nabla_X s_i)^\top = \nabla_X s_i$.

Next we want to investigate the forms $e^*\bar\omega^i_{\bar\jmath} = -e^*\bar\omega^{\bar\jmath}_i$ for $1 \le i \le m$ and $m+1 \le \bar\jmath \le m+p$. We shall need the following result.

(5) **Lemma (E. Cartan).** *For $\bar U$ open in $\bar M^{m+p}$, let $\lambda^1, \dots, \lambda^m \in \Omega^1(\bar U)$ be everywhere linearly independent, and consider 1-forms $\mu_1, \dots, \mu_m \in \Omega^1(\bar U)$ such that $\sum_{i=1}^m \mu_i \wedge \lambda^i = 0$. Then there exist unique smooth functions $f_{ij} \in C^\infty(\bar U)$ satisfying $\mu_i = \sum_{j=1}^m f_{ij}\lambda^j$ and $f_{ij} = f_{ji}$.*

Proof. Near each point we may find $\lambda^{m+1}, \dots, \lambda^{m+p}$ such that $\lambda^1, \dots, \lambda^{m+p}$ are everywhere linearly independent; thus they form a coframe. Then there exist unique f_{ij} such that $\mu_i = \sum_{\bar k=1}^{m+p} f_{i\bar\jmath}\lambda^{\bar\jmath}$. But we have

$$\begin{aligned}
0 &= \sum_{i=1}^m \mu_i \wedge \lambda^i = \sum_{i=1}^m \sum_{\bar k=1}^{m+p} f_{i\bar k}\, \lambda^{\bar k} \wedge \lambda^i \\
&= \sum_{1 \le k < i \le m} (f_{ik} - f_{ki})\lambda^k \wedge \lambda^i + \sum_{i=1}^m \sum_{\bar k=m+1}^{m+p} f_{i\bar k}\, \lambda^{\bar k} \wedge \lambda^i.
\end{aligned}$$

Since the $\lambda^{\bar{k}} \wedge \lambda^{\bar{\imath}}$ for $\bar{k} < \bar{\imath}$ are linearly independent, we conclude that $f_{ik} = f_{ki}$ for $1 \le i, k \le m$ and $f_{i\bar{k}} = 0$ for $1 \le i \le m < \bar{k} \le m+p$. $\square$

By (3) we have $0 = \sum_{k=1}^{m} e^*\bar{\omega}_k^{\bar{\imath}} \wedge \sigma^k$ for $\bar{\imath} = m+1 \dots m+p$. We use now lemma (5) to see that there exist unique functions $s_{kj}^{\bar{\imath}} \in C^\infty(U)$ for $1 \le j, k \le m$ and $\bar{\imath} = m+1, \dots, m+p$ with:

$$(6) \qquad e^*\bar{\omega}_k^{\bar{\imath}} = \sum_{j=1}^{m} s_{kj}^{\bar{\imath}} \sigma^j, \qquad s_{kj}^{\bar{\imath}} = s_{jk}^{\bar{\imath}}.$$

This is equivalent to the Weingarten formula (26.3.1).

Since $\bar{g}(\bar{\nabla}_{s_k} s_j, \bar{s}_{\bar{\imath}}) = \bar{\omega}_j^{\bar{\imath}}(s_k) = (e^*\bar{\omega}_j^{\bar{\imath}})(s_k) = s_{jk}^{\bar{\imath}}$, we have by (26.1)

$$(7) \qquad S(s_i, s_j) = \sum_{\bar{k}=m+1}^{m+p} (\bar{s}_{\bar{k}}|U)(e^*\omega_j^{\bar{k}})(s_i) = \sum_{\bar{k}=m+1}^{m+p} (\bar{s}_{\bar{k}}|U) s_{ij}^{\bar{k}}.$$

Let us now investigate the second structure equation $\bar{\Omega}_{\bar{j}}^{\bar{\imath}} = d\bar{\omega}_{\bar{j}}^{\bar{\imath}} + \sum_{\bar{k}=1}^{m+p} \bar{\omega}_{\bar{k}}^{\bar{\imath}} \wedge \bar{\omega}_{\bar{j}}^{\bar{k}}$. We look first at indices $1 \le i, j \le m$ and restrict it to M:

$$\begin{aligned} e^*\bar{\Omega}_j^i &= de^*\bar{\omega}_j^i + \sum_{k=1}^{m} e^*\bar{\omega}_k^i \wedge e^*\bar{\omega}_j^k + \sum_{\bar{k}=m+1}^{m+p} e^*\bar{\omega}_{\bar{k}}^i \wedge e^*\bar{\omega}_j^{\bar{k}} \\ &= d\omega_j^i + \sum_{k=1}^{m} \omega_k^i \wedge \omega_j^k + \sum_{\bar{k}=m+1}^{m+p} e^*\bar{\omega}_{\bar{k}}^i \wedge e^*\bar{\omega}_j^{\bar{k}}, \end{aligned}$$

$$(8) \qquad e^*\bar{\Omega}_j^i = \Omega_j^i + \sum_{\bar{k}=m+1}^{m+p} e^*\bar{\omega}_{\bar{k}}^i \wedge e^*\omega_j^{\bar{k}} = \Omega_j^i - \sum_{\bar{k}=m+1}^{m+p} \sum_{l,n=1}^{m} s_{il}^{\bar{k}} s_{jn}^{\bar{k}}\, \sigma^l \wedge \sigma^n.$$

This is equivalent to the Gauß equation (26.4.1).

Then we look at the indices $1 \le j \le m < \bar{\imath} \le m+p$ and restrict the second structure equation to M:

$$\begin{aligned} e^*\bar{\Omega}_j^{\bar{\imath}} &= de^*\bar{\omega}_j^{\bar{\imath}} + \sum_{k=1}^{m} e^*\bar{\omega}_k^{\bar{\imath}} \wedge e^*\bar{\omega}_j^k + \sum_{\bar{k}=m+1}^{m+p} e^*\bar{\omega}_{\bar{k}}^{\bar{\imath}} \wedge e^*\bar{\omega}_j^{\bar{k}} \\ (9) \qquad &= de^*\bar{\omega}_j^{\bar{\imath}} + \sum_{k=1}^{m} e^*\bar{\omega}_k^{\bar{\imath}} \wedge \omega_j^k + \sum_{\bar{k}=m+1}^{m+p} e^*\bar{\omega}_{\bar{k}}^{\bar{\imath}} \wedge e^*\bar{\omega}_j^{\bar{k}}, \end{aligned}$$

which is equivalent to the *Codazzi-Mainardi equation.* In the case of a hypersurface this takes the simpler form:

$$e^*\bar{\Omega}_j^{m+1} = de^*\bar{\omega}_j^{m+1} + \sum_{k=1}^{m} e^*\bar{\omega}_k^{m+1} \wedge \omega_j^k.$$

26.8. Resumé of computing with adapted frames for submanifolds. Let $e : (M, g) \to (\bar{M}, \bar{g})$ be an isometric embedding between Riemann manifolds. Let $\bar{s} = (\bar{s}_1, \dots, \bar{s}_{m+p})$ be an orthonormal local frame on $\bar{M}$ over $\bar{U} \subset \bar{M}$ with connection 1-form $\bar{\omega} = (\bar{\omega}^{\bar{i}}_{\bar{j}}) \in \Omega^1(U, \mathfrak{so}(m+p))$ and curvature 2-form $\bar{\Omega} = (\bar{\Omega}^{\bar{i}}_{\bar{j}}) \in \Omega^2(U, \mathfrak{so}(m+p))$, such that the $s_i := \bar{s}_i|U$ form a local orthonormal frame $s = (s_1, \dots, s_m)$ of TM over $U = \bar{U} \cap M$, with connection 1-form $\omega = (\omega^i_j) \in \Omega^1(U, \mathfrak{so}(m))$ and curvature 2-form $\Omega = (\Omega^i_j) \in \Omega^2(U, \mathfrak{so}(m))$. Let

$$\bar{\sigma} = \begin{pmatrix} \bar{\sigma}^1 \\ \vdots \\ \bar{\sigma}^{m+p} \end{pmatrix}, \qquad \sigma = \begin{pmatrix} \sigma^1 \\ \vdots \\ \sigma^m \end{pmatrix}$$

be the dual coframes. Using the ranges of indices $1 \le i, j, k, l \le m$ and $m+1 \le \bar{\imath}, \bar{\jmath}, \bar{k} \le m+p$, we then have:

$$e^*\bar{\sigma}^i = \sigma^i, \qquad e^*\bar{\sigma}^{\bar{\imath}} = 0,$$

$$e^*\bar{\omega}^i_j = \omega^i_j, \qquad e^*\bar{\omega}^{\bar{\imath}}_j = \textstyle\sum_{k \le m} s^{\bar{\imath}}_{jk}\sigma^k, \qquad s^{\bar{\imath}}_{jk} = s^{\bar{\imath}}_{kj},$$

$$e^*\bar{\Omega}^i_j = \Omega^i_j + \textstyle\sum_{m<\bar{k}} e^*\bar{\omega}^i_{\bar{k}} \wedge e^*\bar{\omega}^{\bar{k}}_j = \Omega^i_j - \sum_{\bar{k}=m+1}^{m+p} \sum_{l,n=1}^{m} s^{\bar{k}}_{il} s^{\bar{k}}_{jn}\, \sigma^l \wedge \sigma^n,$$

$$e^*\bar{\Omega}^{\bar{\imath}}_j = de^*\bar{\omega}^{\bar{\imath}}_j + \sum_{k=1}^{m} e^*\bar{\omega}^{\bar{\imath}}_k \wedge \omega^k_j + \sum_{\bar{k}=m+1}^{m+p} e^*\bar{\omega}^{\bar{\imath}}_{\bar{k}} \wedge e^*\bar{\omega}^{\bar{k}}_j.$$

26.9. Definitions. Let $p : E \to B$ be a submersion of smooth manifolds, that is, $Tp : TE \to TB$ is surjective. Then

$$V = V(p) = V(E) := \ker(Tp)$$

is called the *vertical subbundle* of E. If E is a Riemann manifold with metric g, then we can go on to define the *horizontal subbundle* of E:

$$\operatorname{Hor} = \operatorname{Hor}(p) = \operatorname{Hor}(E) = \operatorname{Hor}(E, g) := V(p)^\perp.$$

If both (E, g_E) and (B, g_B) are Riemann manifolds, then we will call p a *Riemann submersion* if

$$T_x p : \operatorname{Hor}(p)_x \to T_{p(x)} B$$

is an isometric isomorphism for all $x \in E$.

Examples. For any two Riemann manifolds M, N, the projection $pr_1 : M \times N \to M$ is a Riemann submersion. Here the Riemann metric on the product $M \times N$ is given by

$$g_{M \times N}(X_M + X_N, Y_M + Y_N) := g_M(X_M, Y_M) + g_N(X_N, Y_N)$$

using $T(M \times N) \cong TM \oplus TN$. In particular, $\mathbb{R}^{m+n} \to R^m$ with the usual metric, or $pr_2 : S^n \times \mathbb{R}^+ \to R^+$ are Riemann submersions.

26.10. Definition. Let $p : E \to B$ be a Riemann submersion. A vector field $X \in \mathfrak{X}(E)$ is called:

- *vertical* if $X(x) \in V_x(p)$ for all x (i.e., if $Tp.X(x) = 0$),
- *horizontal* if $X(x) \in \operatorname{Hor}_x(p)$ for all x (i.e., if $X(x) \perp V_x(p)$),
- *projectable* if there is an $\eta \in \mathfrak{X}(B)$, such that $Tp.X = \eta \circ p$,
- *basic* if it is horizontal and projectable.

Any vector field $Y \in \mathfrak{X}(E)$ can be uniquely decomposed as

$$Y = Y^{\text{ver}} + Y^{\text{hor}}$$

into its vertical and horizontal components. The orthogonal projection $\Phi : TE \to V(E)$ with respect to the Riemann metric is a (generalized) connection on the bundle (E, p, B) in the sense of (17.3) and defines a local parallel transport over each curve in B (denoted by $\operatorname{Pt}^\Phi(c, .)$) as well as the horizontal lift of tangent vectors:

$$C : TB \times_B E \longrightarrow E, \qquad (X_b, e) \mapsto Y_e,$$

where $Y_e \in \operatorname{Hor}_e(p)$ with $T_e p.Y_e = X_b$. This map also gives us an isomorphism $C_* : \mathfrak{X}(B) \to \mathfrak{X}_{\text{basic}}(E)$ between the vector fields on B and the basic vector fields.

26.11. Lemma. *Consider a Riemann submersion $p : (E, g_E) \to (B, g_B)$ with connection $\Phi : TE \to V(p)$ and $c : [0,1] \to B$, a geodesic. Then we have:*

(1) *The length is preserved by lifting curves horizontally:*

$$L_0^t(c) = L_0^t(\operatorname{Pt}^\Phi(c, ., u)),$$

where $u \in E_{c(0)}$ is the starting point of the parallel transport. Also the energy is preserved, $E_0^t(c) = E_0^t(\operatorname{Pt}^\Phi(c, ., u))$.

(2) $\operatorname{Pt}^\Phi(c, ., u) \perp E_{c(t)}$ *for all t.*

(3) *If c is a geodesic of minimal length in B, then we have*

$$L_0^1(\operatorname{Pt}^\Phi(c, ., u)) = \operatorname{dist}\left(E_{c(0)}, E_{c(1)}\right).$$

(4) *If c is a geodesic in B, then $t \mapsto \operatorname{Pt}^\Phi(c, t, u)$ is a geodesic in E.*

(5) *For vector fields $\xi, \eta \in \mathfrak{X}(B)$ and the corresponding horizontal lifts $C(\xi), C(\eta) \in \mathfrak{X}(E)$, we have*

$$(\nabla^E_{C(\xi)} C(\eta))^{hor} = C(\nabla^B_\xi \eta).$$

Proof. (1) Since $\partial_s \operatorname{Pt}^\Phi(c,s,u)$ is a horizontal vector and by the property of p as Riemann submersion, we have

$$\begin{aligned} L_0^t(\operatorname{Pt}^\Phi(c,.,u)) &= \int_0^t g_E\left(\partial_s \operatorname{Pt}^\Phi(c,s,u), \partial_s \operatorname{Pt}^\Phi(c,s.u)\right)^{\frac{1}{2}} ds \\ &= \int_0^t g_B(c'(s),c'(s))^{\frac{1}{2}} ds = L_0^t(c), \\ E_0^t(\operatorname{Pt}^\Phi(c,.,u)) &= \frac{1}{2}\int_0^t g_E\left(\partial_s \operatorname{Pt}^\Phi(c,s,u), \partial_s \operatorname{Pt}^\Phi(c,s.u)\right) ds = E_0^t(c). \end{aligned}$$

(2) This is due to our choice of Φ as orthogonal projection onto the vertical bundle in terms of the given metric on E. By this choice, the parallel transport is the unique horizontal curve covering c, so it is orthogonal to each fiber $E_{c(t)}$ it meets.

(3) Consider a (piecewise) smooth curve $e : [0,1] \to E$ from $E_{c(0)}$ to $E_{c(1)}$; then $p \circ e$ is a (piecewise) smooth curve from $c(0)$ to $c(1)$. Since c is a minimal geodesic, we have $L_0^1(c) \le L_0^1(p\circ e)$. Furthermore, we can decompose the vectors tangent to e into horizontal and vertical components and use the fact that Tp is an isometry on horizontal vectors to show that $L_0^1(e) \ge L_0^1(p\circ e)$:

$$\begin{aligned} L_0^1(e) &= \int_0^1 |e'(t)^{\text{ver}} + e'(t)^{\text{hor}}|_{g_E} dt \\ &\ge \int_0^1 |e'(t)^{\text{hor}}|_{g_E} dt = \int_0^1 |(p\circ e)'(t)|_{g_M} dt = L_0^1(p\circ e). \end{aligned}$$

Now with (1) we can conclude that for all (piecewise) smooth curves e from $E_{c(0)}$ to $E_{c(1)}$ we have:

$$\begin{aligned} &L_0^1(e) \ge L_0^1(p\circ e) \ge L_0^1(c) = L_0^1(\operatorname{Pt}^\Phi(c,.,u)); \\ \text{thus}\quad &L_0^1(\operatorname{Pt}^\Phi(c,.,u)) = \operatorname{dist}\left(E_{c(0)}, E_{c(1)}\right). \end{aligned}$$

(4) This is a consequence of (3) and the observation from (22.4) that every curve which minimizes length or energy locally is a geodesic.

(5) Since $g_E(C(\xi), C(\eta)) = g_B(\xi,\eta)\circ p$ and since $C(\xi)$ is p-related to ξ and $C(\eta)$ is p-related to η, we get that $[C(\xi), C(\eta)]$ is p-related to $[\xi,\eta]$. We can then apply the implicit equation (22.5.7) for the covariant derivative twice:

$$\begin{aligned} &2g_E((\nabla^E_{C(\xi)} C(\eta))^{\text{hor}}, C(\zeta)) = 2g_E(\nabla^E_{C(\xi)} C(\eta), C(\zeta)) \\ &= C(\xi)(g_E(C(\eta),C(\zeta))) + C(\eta)(g_E(C(\zeta),C(\xi))) - C(\zeta)(g_E(C(\xi),C(\eta))) \\ &- g_E(C(\xi),[C(\eta),C(\zeta)]) + g_E(C(\eta),[C(\zeta),C(\xi)]) + g_E(C(\zeta),[C(\xi),C(\eta)]) \\ &= \big(\xi(g_B(\eta,\zeta)) + \eta(g_B(\zeta,\xi)) - \zeta(g_B(\xi,\eta)) \\ &\qquad - g_B(\xi,[\eta,\zeta]) + g_B(\eta,[\zeta,\xi]) + g_B(\zeta,[\xi,\eta])\big)\circ p = 2g_B(\nabla^B_\xi \eta, \zeta)\circ p. \end{aligned}$$

Since this holds for any $\zeta \in \mathfrak{X}(B)$, we conclude

$$(\nabla^E_{C(\xi)} C(\eta))^{\text{hor}} = C(\nabla^B_\xi \eta). \quad \square$$

26.12. Corollary. *Consider a Riemann submersion $p : E \to B$, and let $c : [0,1] \to E$ be a geodesic in E with the property $c'(t_0) \perp E_{p(c(t_0))}$ for some t_0. Then $c'(t) \perp E_{p(c(t))}$ for all $t \in [0,1]$ and $p \circ c$ is a geodesic in B.*

Proof. Consider the curve $f : t \mapsto \exp^B_{p(c(t_0))}(tT_{c(t_0)}p.c'(t_0))$. It is a geodesic in B and therefore lifts to a geodesic $e(t) = \text{Pt}^\Phi(f, t - t_0, c(t_0))$ in E by (26.11.4). Also $e(t_0) = c(t_0)$ and $e'(t_0) = C(T_{c(t_0)}p.c'(t_0), c(t_0)) = c'(t_0)$ since $c'(t_0) \perp E_{p(c(t_0))}$ is horizontal. But geodesics are uniquely determined by their starting point and starting vector. Therefore $e = c$; thus e is orthogonal to each fiber it meets by (26.11.2) and it projects onto the geodesic f in B. $\square$

26.13. Corollary. *Let $p : E \to B$ be a Riemann submersion. If* $\text{Hor}(E)$ *is integrable, then:*

(1) *Every leaf is totally geodesic in the sense of* (26.2).

(2) *For each leaf L the restriction $p : L \to B$ is a local isometry.*

Proof. (1) follows from corollary (26.12), while (2) is just a direct consequence of the definitions. $\square$

26.14. Remark. If $p : E \to B$ is a Riemann submersion, then $\text{Hor}(E)|_{E_b} = \text{Nor}(E_b)$ for all $b \in B$ and p defines a global parallelism as follows. A section $\tilde{v} \in C^\infty(\text{Nor}(E_b))$ is called p-parallel if $T_e p.\tilde{v}(e) = v \in T_b B$ is the same point for all $e \in E_b$. There is also a second parallelism. It is given by the induced covariant derivative: A section $\tilde{v} \in C^\infty(\text{Nor}(E_b))$ is called parallel if $\nabla^{\text{Nor}}\tilde{v} = 0$. The p-parallelism is always flat and with trivial holonomy which is not generally true for ∇^{Nor}. Yet we will see later on that if $\text{Hor}(E)$ is integrable, then the two parallelisms coincide.

26.15. Definition. *A Riemann submersion $p : E \to B$ is called integrable if* $\text{Hor}(E) = (\ker Tp)^\perp$ *is an integrable distribution.*

26.16. Structure theory of Riemann submersions. Let $p : (E, g^E) \to (B, g^B)$ be a Riemann submersion. We consider first the second fundamental form $S^{E_b} : TE_b \times_{E_b} TE_b \to \text{Hor}(E)$ of the submanifold $E_b := p^{-1}(b)$ in E. By (26.1), S^{E_b} is given as:

$$S^{E_b}(X^{\text{ver}}, Y^{\text{ver}}) = \nabla^E_{X^{\text{ver}}} Y^{\text{ver}} - \nabla^{E_b}_{X^{\text{ver}}} Y^{\text{ver}} = \nabla^E_{X^{\text{ver}}} Y^{\text{ver}} - \left(\nabla^E_{X^{\text{ver}}} Y^{\text{ver}}\right)^{\text{ver}}$$

$$(1) \qquad = (\nabla^E_{X^{\text{ver}}} Y^{\text{ver}})^{\text{hor}} = \left(\nabla^E_{X^{\text{ver}}} Y^{\text{ver}}\right)^{\text{hor}}.$$

The covariant derivative on the normal bundle $\operatorname{Nor}(E_b) = \operatorname{Hor}(E)|_{E_b} \to E_b$ is given by the Weingarten formula (26.3) as the corresponding projection:

$$\nabla^{\mathrm{Nor}} : \mathfrak{X}(E_b) \times \Gamma(\operatorname{Nor}(E_b)) \to \Gamma(\operatorname{Nor}(E_b)),$$

$$\nabla^{\mathrm{Nor}}_{X^{\mathrm{ver}}} Y^{\mathrm{hor}} = (\nabla^E_{X^{\mathrm{ver}}} Y^{\mathrm{hor}})^{\mathrm{hor}}. \tag{2}$$

Yet in the decomposition

$$\nabla^E_X Y = \left(\nabla^E_{X^{\mathrm{ver}}+X^{\mathrm{hor}}}(Y^{\mathrm{ver}} + Y^{\mathrm{hor}})\right)^{\mathrm{ver}+\mathrm{hor}}$$

we can find two more tensor fields (besides S), the so-called O'Neill tensor fields (see [**180**]):

$$X, Y \in \mathfrak{X}(E),$$

$$T(X,Y) := (\nabla^E_{X^{\mathrm{ver}}} Y^{\mathrm{ver}})^{\mathrm{hor}} + \left(\nabla^E_{X^{\mathrm{ver}}} Y^{\mathrm{hor}}\right)^{\mathrm{ver}}, \tag{3}$$

$$A(X,Y) := \left(\nabla^E_{X^{\mathrm{hor}}} Y^{\mathrm{hor}}\right)^{\mathrm{ver}} + (\nabla^E_{X^{\mathrm{hor}}} Y^{\mathrm{ver}})^{\mathrm{hor}}. \tag{4}$$

Each of these four terms making up A and T is a tensor field by itself — the first one restricting to S on E_b. They are combined as two tensors in just this way because of the results below.

Theorem ([**180**]). *For horizontal vectors $X, Y, Z, H \in \operatorname{Hor}(p)_x$ we have*

$$\begin{aligned} g^E_x(R_x(X,Y)Z, H) &= g^B_{p(x)}(R^B_{p(x)}(T_x p.X, T_x p.Y)T_x p.Z, T_p H) \\ &\quad + 2g^E_x(A(X,Y), A(Z,H)) - g^E_x(A(Y,Z), A(X,H)) \\ &\quad - g^E_x(A(Z,X), A(Y,H)). \end{aligned} \tag{5}$$

Proof. Since this is of tensorial character, we can assume that X, Y, Z, U are basic local vector fields which are horizontal lifts of commuting vector fields $\xi, \eta, \zeta, \chi \in \mathfrak{X}(B)$; so $X = C(\xi), Y = C(\eta), Z = C(\zeta), H = C(\chi)$ (see (26.10)) and all Lie brackets $[\xi, \eta]$, etc., on B vanish. Note first that for a vertical field $V = V^{\mathrm{ver}}$ we have

$$\nabla^E_V C(\xi) - \nabla^E_{C(\xi)} V = [V, C(\xi)] = 0$$

since V is projectable to 0. But then

$$\begin{aligned} 0 = \tfrac{1}{2} V g^E(X,X) &= g^E(\nabla^E_V X, X) \\ &= g^E(\nabla^E_X V, X) = 0 - g^E(V, \nabla^E_X X) \\ &= g^E(V, A(X,X)) \end{aligned}$$

and since $A(X,X)$ is vertical for horizontal X, this implies $A(X,X) = 0$. Thus $A(X,Y) = -A(Y,X)$ for basic fields X, Y.

Then $[X,Y] = [C(\xi), C(\eta)]$ is vertical since it projects to $[\xi,\eta] = 0$, and moreover

$$\begin{aligned}
[X,Y] &= \nabla_X Y - \nabla_Y X = (\nabla_X Y)^{\text{ver}} - (\nabla_Y X)^{\text{ver}} \\
&= A(X,Y) - A(Y,X) = 2A(X,Y), \qquad (6)\\
\nabla^E_{[X,Y]} Z &= \nabla_{[X,Y]^{\text{ver}}}(Z^{\text{hor}}) = T([X,Y]^{\text{ver}}, Z^{\text{hor}}) + A([X,Y]^{\text{ver}}, Z^{\text{hor}}) \\
&= 2T(A(X,Y),Z) + 2A(A(X,Y),Z).
\end{aligned}$$

By (26.11.5) we have

$$\begin{aligned}
&\nabla^E_Y Z = A(Y,Z) + (\nabla^E_Y Z)^{\text{hor}} = A(Y,Z) + C(\nabla^B_\eta \zeta),\\
&\nabla^E_X \nabla^E_Y Z = \nabla^E_X(A(Y,Z)) + \nabla^E_{C(\xi)}(C(\nabla^B_\eta\zeta))\\
&\quad = A(X,A(Y,Z)) + (\nabla^E_X(A(Y,Z)))^{\text{ver}} + C(\nabla^B_\xi\nabla^B_\eta\zeta) + A(X, C(\nabla^B_\eta\zeta)).
\end{aligned}$$

Combining, we get

$$\begin{aligned}
&R^E(X,Y)Z = \nabla^E_X\nabla^E_Y Z - \nabla^E_X\nabla^E_Y Z - \nabla^E_{[X,Y]} Z\\
&= A(X,A(Y,Z)) + (\nabla^E_X(A(Y,Z)))^{\text{ver}} + C(\nabla^B_\xi\nabla^B_\eta\zeta) + A(X,C(\nabla^B_\eta\zeta))\\
&\quad - A(Y,A(X,Z)) - (\nabla^E_Y(A(X,Z)))^{\text{ver}} - C(\nabla^B_\eta\nabla^B_\xi\zeta) - A(Y,C(\nabla^B_\xi\zeta))\\
&\quad - 2T(A(X,Y),Z) - 2A(A(X,Y),Z)\\
&= C\big(\nabla^B_\xi\nabla^B_\eta\zeta - \nabla^B_\eta\nabla^B_\xi\zeta - \nabla_{[\xi,\eta]}\zeta\big)\\
&\quad + A(X,A(Y,Z)) - A(Y,A(X,Z)) - 2A(A(X,Y),Z)\\
&\quad + A(X,C(\nabla^B_\eta\zeta)) - A(Y,C(\nabla^B_\xi\zeta)) - 2T(A(X,Y),Z)\\
&\quad + (\nabla^E_X(A(Y,Z)))^{\text{ver}} - (\nabla^E_Y(A(X,Z)))^{\text{ver}}
\end{aligned}$$

using $[\xi,\eta] = 0$, where the first two lines are horizontal and the last two lines are vertical. Take the inner product in E with the horizontal H and use

$$\begin{aligned}
g^E(A(X,A(Y,Z)),H) &= g^E(\nabla_X(A(Y,Z)^{\text{ver}}),H) = g^E(A(Y,Z)^{\text{ver}}, \nabla_X H)\\
&= g^E(A(Y,Z)^{\text{ver}}, (\nabla_X H)^{\text{ver}}) = g^E(A(Y,Z), A(X,H))
\end{aligned}$$

to get the desired formula. □

26.17. Corollary. *Let $p : E \to B$ be a Riemann submersion between manifolds with connections, and consider horizontal vectors $X, Y, Z, H \in \operatorname{Hor}(p)_x$. Then the sectional curvature expression becomes*

$$\begin{aligned}
&g^E_x(R^E(X,Y)X,Y)\\
&\quad = g^B_{p(x)}(R^B(T_xp.X, T_xp.Y)T_xp.X, T_pY) + 3\|A_x(X,Y)\|^2_{g^E}\\
&\quad = g^B_{p(x)}(R^B(T_xp.X, T_xp.Y)T_xp.X, T_pY) + \tfrac{3}{4}\|[\bar X, \bar Y]^{ver}\|^2_{g^E}
\end{aligned}$$

for basic horizontal extensions $\bar X, \bar Y$ of X, Y.

Proof. Again we extend everything to basic horizontal vector fields projecting to fields on B. From a slight generalization of (26.16.6) we have $[X,Y]^{\text{ver}} = 2A(X,Y)$ in this case. By theorem (26.16) we have

$$\begin{aligned} g^E(R(X,Y)X,Y) &= g^B(R^B(\xi,\eta)\xi,\eta) + 3g^E(A(X,Y),A(X,Y)) \\ &= g^B(R^B(\xi,\eta)\xi,\eta) + \tfrac{3}{4}\|[X,Y]^{\text{ver}}\|^2. \end{aligned}$$

Note that the last expression gives another formula in the case where X,Y are horizontal and project to commuting fields. □

26.18. Riemann submersions via local frames. Let $p : (E,g_E) \to (B,g_B)$ be a Riemann submersion. Choose for an open neighborhood U in E an orthonormal frame field $s = (s_1,\dots,s_{m+k}) \in \Gamma(TE|U)^{m+k}$ in such a way that $s_1,\dots,s_m$ are vertical and $s_{m+1},\dots,s_{m+k}$ are basic (horizontal and projectable). That way, if we project $s_{m+1},\dots,s_{m+k}$ onto $TB|p(U)$, we get another orthonormal frame field, $\bar s = (\bar s_{m+1},\dots,\bar s_{m+k}) \in C^\infty(TB|p(U))^k$, since p, as Riemann submersion, is isometric on horizontal vectors. The indices will always run in the domain indicated:

$$1 \le i,j,k \le m, \quad m+1 \le \bar a,\bar b,\bar c \le m+k, \quad 1 \le A,B,C \le m+k.$$

The orthonormal coframe dual to s is given by

$$\sigma^A(s_B) = \delta^A_B, \qquad \sigma = \begin{pmatrix} \sigma^1 \\ \vdots \\ \sigma^{m+k} \end{pmatrix} \in \Omega^1(U)^{m+k}.$$

Analogously, we have the orthonormal coframe $\bar\sigma^{\bar a} \in \Omega^1(p(U))$ on $p(U) \subseteq B$, with $\bar\sigma^{\bar a}(\bar s_{\bar b}) = \delta^{\bar a}_{\bar b}$. It is related to $\sigma^{\bar a}$ by $p^*\bar\sigma^{\bar a} = \sigma^{\bar a}$. By (25.5) we have on $(U \subset E, g_E)$

$$\begin{aligned}
& g_E|_U = \textstyle\sum_A \sigma^A \otimes \sigma^A, \\
& \nabla^E s = s.\omega \quad \text{where} \quad \omega^A_B = -\omega^B_A, \quad \text{so} \quad \omega \in \Omega^1(U,\mathfrak{so}(n+k)), \\
& d\sigma + \omega\wedge\sigma = 0, \quad \text{i.e.,} \quad d\sigma^A + \textstyle\sum_C \omega^A_C \wedge \sigma^C = 0, \\
& Rs = s.\Omega \quad \text{where} \quad \Omega = d\omega + \omega\wedge\omega \in \Omega^2(U,\mathfrak{so}(n+k)), \\
& \qquad\qquad \text{or} \quad \Omega^A_B = d\omega^A_B + \textstyle\sum_C \omega^A_C \wedge \omega^C_B, \\
& \Omega\wedge\sigma = 0 \quad \text{or} \quad \textstyle\sum_C \Omega^A_C \wedge \sigma^C = 0, \quad \text{the first Bianchi identity,} \\
& d\Omega + \omega\wedge\Omega - \Omega\wedge\omega = d\Omega + [\omega,\Omega]_\wedge = 0, \quad \text{the second Bianchi identity,}
\end{aligned}$$

and similarly on $(p(U) \subset B, g^B)$ with bars on all forms.

For the following it will be faster to rederive some results than compiling them from (26.7) and (26.8). We start by pulling back the structure equation

$d\bar\sigma + \bar\omega \wedge \bar\sigma = 0$ from B to E via p^*:

$$\begin{aligned} 0 &= p^*\left(d\bar\sigma^{\bar a} + \sum \bar\omega^{\bar a}_{\bar b} \wedge \bar\sigma^{\bar b}\right) \\ &= dp^*\bar\sigma^{\bar a} + \sum (p^*\bar\omega^{\bar a}_{\bar b}) \wedge (p^*\bar\sigma^{\bar b}) = d\sigma^{\bar a} + \sum (p^*\bar\omega^{\bar a}_{\bar b}) \wedge \sigma^{\bar b}. \end{aligned}$$

The $\bar a$-part of the structure equation on E, $d\sigma^{\bar a} + \sum \omega^{\bar a}_{\bar b} \wedge \sigma^{\bar b} + \sum \omega^{\bar a}_{i} \wedge \sigma^{i} = 0$, combines with this to give

$$(1) \qquad \sum (p^*\bar\omega^{\bar a}_{\bar b}) \wedge \sigma^{\bar b} = \sum \omega^{\bar a}_{\bar b} \wedge \sigma^{\bar b} + \sum \omega^{\bar a}_{i} \wedge \sigma^{i}.$$

The left hand side of this equation contains no $\sigma^i \wedge \sigma^{\bar a}$- or $\sigma^i \wedge \sigma^j$-terms. Let us write out $\omega^{\bar a}_{\bar b}$ and $\omega^{\bar a}_i$ in this basis:

$$\omega^{\bar a}_{\bar b} = -\omega^{\bar b}_{\bar a} =: \sum q^{\bar a}_{\bar b \bar c}\sigma^{\bar c} + \sum b^{\bar a}_{\bar b i}\sigma^i, \qquad \omega^{\bar a}_{i} = -\omega^{i}_{\bar a} =: \sum a^{\bar a}_{i \bar b}\sigma^{\bar b} + \sum r^{\bar a}_{ij}\sigma^j.$$

This gives us for the right hand side of (1):

$$\begin{aligned} &\sum q^{\bar a}_{\bar b \bar c}\sigma^{\bar c} \wedge \sigma^{\bar b} + \sum b^{\bar a}_{\bar b i}\sigma^i \wedge \sigma^{\bar b} + \sum a^{\bar a}_{i \bar b}\sigma^{\bar b} \wedge \sigma^i + \sum r^{\bar a}_{ij}\sigma^j \wedge \sigma^i \\ &\quad = \sum q^{\bar a}_{\bar b \bar c}\sigma^{\bar c} \wedge \sigma^{\bar b} + \sum (b^{\bar a}_{\bar b i} - a^{\bar a}_{i \bar b})\sigma^i \wedge \sigma^{\bar b} + \tfrac12 \sum (r^{\bar a}_{ij} - r^{\bar a}_{ji})\sigma^j \wedge \sigma^i. \end{aligned}$$

So we have found $a^{\bar a}_{i\bar b} = b^{\bar a}_{\bar b i}$ and $r^{\bar a}_{ij} = r^{\bar a}_{ji}$ or, in other words, $\omega^{\bar a}_i(s_{\bar b}) = \omega^{\bar a}_{\bar b}(s_i)$ and $\omega^{\bar a}_i(s_j) = \omega^{\bar a}_j(s_i)$. That is, $\omega^{\bar a}_i(s_A) = \omega^{\bar a}_A(s_i)$, and this just means that the horizontal part of $[s_A, s_i]$ is 0, or $[s_A, s_i]$ is always vertical:

$$(2) \qquad 0 = \sum s_{\bar a}\omega^{\bar a}_i(s_A) - \sum s_{\bar a}\omega^{\bar a}_A(s_i) = (\nabla_{s_A} s_i - \nabla_{s_i} s_A)^{\text{hor}} = [s_A, s_i]^{\text{hor}}.$$

Now we consider again the second fundamental form $S^{E_b} : TE_b \times_{E_b} TE_b \to \operatorname{Hor}(E)$ of the submanifold $E_b := p^{-1}(b)$ in E. By (26.1), S^{E_b} is given as:

$$\begin{aligned} S^{E_b}(X^{\text{ver}}, Y^{\text{ver}}) &= \nabla^E_{X^{\text{ver}}} Y^{\text{ver}} - \nabla^{E_b}_{X^{\text{ver}}} Y^{\text{ver}} = \nabla^E_{X^{\text{ver}}} Y^{\text{ver}} - \left(\nabla^E_{X^{\text{ver}}} Y^{\text{ver}}\right)^{\text{ver}} \\ &= \left(\nabla^E_{X^{\text{ver}}} Y^{\text{ver}}\right)^{\text{hor}} \\ &= \left(\nabla^E_{X^{\text{ver}}} \left(\sum s_i \sigma^i(Y^{\text{ver}})\right)\right)^{\text{hor}} \\ &= \left(\sum (\nabla^E_{X^{\text{ver}}} s_i)\sigma^i(Y^{\text{ver}}) + \sum s_i d(\sigma^i(Y^{\text{ver}})).X^{\text{ver}}\right)^{\text{hor}} \\ &= \left(\sum s_A \omega^A_i(X^{\text{ver}})\sigma^i(Y^{\text{ver}})\right)^{\text{hor}} + 0 = \sum s_{\bar a}\omega^{\bar a}_i(X^{\text{ver}})\sigma^i(Y^{\text{ver}}) \\ &= \sum r^{\bar a}_{ij}\left(s_{\bar a} \otimes \sigma^j \otimes \sigma^i\right)(X^{\text{ver}}, Y^{\text{ver}}). \end{aligned}$$

So

$$\sum s_{\bar a}\sigma^{\bar a}(S^{E_b}) = \sum r^{\bar a}_{ij}\, s_{\bar a} \otimes \sigma^j \otimes \sigma^i.$$

Note that $r^{\bar a}_{ij} = r^{\bar a}_{ji}$ from above corresponds to symmetry of S. The covariant derivative on the normal bundle $\operatorname{Nor}(E_b) = \operatorname{Hor}(E)|_{E_b} \to E_b$ is given by the Weingarten formula (26.3) as the corresponding projection:

$$\begin{aligned} &\nabla^{\text{Nor}} : \mathfrak{X}(E_b) \times \Gamma(\operatorname{Nor}(E_b)) \to \Gamma(\operatorname{Nor}(E_b)), \\ &\nabla^{\text{Nor}}_{X^{\text{ver}}} Y^{\text{hor}} = (\nabla^E_{X^{\text{ver}}} Y^{\text{hor}})^{\text{hor}} = \left(\nabla^E_{X^{\text{ver}}}\left(\sum s_{\bar b}\sigma^{\bar b}(Y^{\text{hor}})\right)\right)^{\text{hor}} \end{aligned}$$

$$\begin{aligned}
&= \Big(\sum(\nabla^E_{X^{\text{ver}}} s_{\bar b})\sigma^{\bar b}(Y^{\text{hor}})\Big)^{\text{hor}} + \sum s_{\bar b} d\sigma^{\bar b}(Y^{\text{hor}}).X^{\text{ver}} \\
&= \sum s_{\bar a}\omega^{\bar a}_{\bar b}(X^{\text{ver}})\sigma^{\bar b}(Y^{\text{hor}}) + \sum s_{\bar b} d\sigma^{\bar b}(Y^{\text{hor}}).X^{\text{ver}} \\
&= \sum b^{\bar a}_{\bar b i} s_{\bar a} \otimes \sigma^i \otimes \sigma^{\bar b}(X^{\text{ver}}, Y^{\text{hor}}) + \sum s_{\bar a} \otimes d\sigma^{\bar a}(Y^{\text{hor}})(X^{\text{ver}}),
\end{aligned}$$

$$\nabla^{\text{Nor}} Y^{\text{hor}} = \sum \Big(b^{\bar a}_{\bar b i}\sigma^{\bar b}(Y^{\text{hor}})\sigma^i + d\sigma^{\bar a}(Y^{\text{hor}})\Big) \otimes s_{\bar a}.$$

We consider now the O'Neill tensor fields from (26.16):

$$\begin{aligned}
&X, Y \in \mathfrak{X}(E), \\
A(X,Y) &= \Big(\nabla^E_{X^{\text{hor}}} Y^{\text{hor}}\Big)^{\text{ver}} + \big(\nabla^E_{X^{\text{hor}}} Y^{\text{ver}}\big)^{\text{hor}} \\
&= \big(\nabla^E_{X^{\text{hor}}} (\textstyle\sum s_{\bar a}\sigma^{\bar a}(Y))\big)^{\text{ver}} + \big(\nabla^E_{X^{\text{hor}}} (\textstyle\sum s_i\sigma^i(Y))\big)^{\text{hor}} \\
&= \sum s_i \omega^i_{\bar a}(X^{\text{hor}})\sigma^{\bar a}(Y) + 0 + \sum s_{\bar a}\omega^{\bar a}_i(X^{\text{hor}})\sigma^i(Y) + 0 \\
&= \sum s_i(-a^{\bar a}_{i\bar b})\sigma^{\bar b}(X)\sigma^{\bar a}(Y) + \sum s_{\bar a} a^{\bar a}_{i\bar b}\sigma^{\bar b}(X)\sigma^i(Y) \\
(3) \qquad &= \sum a^{\bar a}_{i\bar b}(\sigma^{\bar b} \otimes \sigma^i \otimes s_{\bar a} - \sigma^{\bar b} \otimes \sigma^{\bar a} \otimes s_i)(X,Y).
\end{aligned}$$

Analogously:

$$T = \sum r^{\bar a}_{ij}(\sigma^j \otimes \sigma^i \otimes s_{\bar a} - \sigma^j \otimes \sigma^{\bar a} \otimes s_i).$$

If $\operatorname{Hor}(E)$ is integrable, then every leaf L is totally geodesic by (26.13.1), and the $s_{\bar a}|_L$ are a local orthonormal frame field on L. The leaf L is totally geodesic if and only if its second fundamental form which is given by

$$S^L(X^{\text{hor}}, Y^{\text{hor}}) := (\nabla^E_{X^{\text{hor}}} Y^{\text{hor}})^{\text{ver}}$$

vanishes. So it is a necessary condition for the integrability of $\operatorname{Hor}(E)$ that $S^L = 0$, that is,

$$0 = S^L(s_{\bar a}, s_{\bar b}) = (\nabla_{s_{\bar a}} s_{\bar b})^{\text{ver}} = \sum s_i \omega^i_{\bar b}(s_{\bar a}) = \sum s_i(-a^{\bar b}_{i\bar c})\sigma^{\bar c}(s_{\bar a}) = -\textstyle\sum_i s_i a^{\bar b}_{i\bar a}.$$

This is equivalent to the condition $a^{\bar a}_{i\bar b} = 0$ for all ${}^{\bar a}_{i\bar b}$ or to $A = 0$.

Let us now prove the converse: If A vanishes, then the horizontal distribution on E is integrable. In this case, we have $0 = A(s_{\bar a}, s_{\bar b}) = (\nabla^E_{s_{\bar a}} s_{\bar b})^{\text{ver}} + 0$, as well as $0 = A(s_{\bar b}, s_{\bar a}) = (\nabla^E_{s_{\bar b}} s_{\bar a})^{\text{ver}} + 0$. Therefore, $[s_{\bar a}, s_{\bar b}] = \nabla^E_{s_{\bar a}} s_{\bar b} - \nabla^E_{s_{\bar b}} s_{\bar a}$ is horizontal, and the horizontal distribution is integrable.

26.19. Theorem. *Let $p : E \to B$ be a Riemann submersion; then the following conditions are equivalent:*

(1) *p is integrable (that is,* $\operatorname{Hor}(p)$ *is integrable).*

(2) *Every p-parallel normal field along E_b is ∇^{Nor}-parallel.*

(3) *The O'Neill tensor A is zero.*

Proof. We already saw (1) $\iff$ (3) above.

(3) $\Longrightarrow$ (2) Take $s_{\bar{a}}$ for a p-parallel normal field X along E_b. The condition $A = 0$ implies $A(s_{\bar{a}}, s_i) = 0 + (\nabla_{s_{\bar{a}}} s_i)^{\text{hor}} = 0$. Recall that, as we showed in (26.18.2) above, $[s_i, s_{\bar{a}}]$ is vertical. Therefore,

$$\nabla^{\text{Nor}}_{s_i} s_{\bar{a}} = (\nabla^E_{s_i} s_{\bar{a}})^{\text{hor}} = ([s_i, s_{\bar{a}}] + \nabla^E_{s_{\bar{a}}} s_i)^{\text{hor}} = 0.$$

Since for any $e \in E_b$, $T_e p|_{\text{Nor}_b(E_b)}$ is an isometric isomorphism, a p-parallel normal field X along E_b is determined completely by the equation $X(e) = \sum X^{\bar{a}}(e) s_{\bar{a}}(e)$. Therefore it is always a linear combination of the $s_{\bar{a}}$ with constant coefficients and we are done.

(2) $\Longrightarrow$ (3) By (2), $\nabla^{\text{Nor}}_{s_i} s_{\bar{a}} = (\nabla^E_{s_i} s_{\bar{a}})^{\text{hor}} = 0$. Therefore, as above, we have that $([s_i, s_{\bar{a}}] + \nabla^E_{s_{\bar{a}}} s_i)^{\text{hor}} = 0 + (\nabla^E_{s_{\bar{a}}} s_i)^{\text{hor}} = A(s_{\bar{a}}, s_i) = 0$. Thus $\sigma^{\bar{b}} A(s_{\bar{a}}, s_i) = a^{\bar{b}}_{\bar{a}i} = 0$, so A vanishes completely. $\square$

27. Jacobi Fields

27.1. Jacobi fields. Let (M, ∇) be a manifold with covariant derivative ∇, with curvature R and torsion Tor. Let us consider a smooth mapping $\gamma : (-\varepsilon, \varepsilon) \times [0,1] \to M$ such that $t \mapsto \gamma(s,t)$ is a geodesic for each $s \in (-\varepsilon, \varepsilon)$; we call this a 1-*parameter variation through geodesics*. Let us write $\partial_s \gamma =: \gamma'$ and $\partial_t \gamma =: \dot{\gamma}$ in the following. Our aim is to investigate the variation vector field $\partial_s|_0 \gamma(s, \quad) = \gamma'(0, \quad)$.

We first note that by (22.10.4) we have

$$\begin{aligned} \nabla_{\partial_s} \dot{\gamma} &= \nabla_{\partial_s} (T\gamma.\partial_t) = \nabla_{\partial_t} (T\gamma.\partial_s) + T\gamma.[\partial_s, \partial_t] + \text{Tor}(T\gamma.\partial_s, T\gamma.\partial_t) \\ &= \nabla_{\partial_t} \gamma' + \text{Tor}(\gamma', \dot{\gamma}). \end{aligned} \tag{1}$$

We have $\nabla_{\partial_t} \dot{\gamma} = \nabla_{\partial_t} (\partial_t \gamma) = 0$ since $\gamma(s, \quad)$ is a geodesic for each s. Thus by using (24.5), we get

$$\begin{aligned} 0 = \nabla_{\partial_s} \nabla_{\partial_t} \dot{\gamma} &= R(T\gamma.\partial_s, T\gamma.\partial_t)\dot{\gamma} + \nabla_{\partial_t} \nabla_{\partial_s} \dot{\gamma} + \nabla_{[\partial_s, \partial_t]} \dot{\gamma} \\ &= R(\gamma', \dot{\gamma})\dot{\gamma} + \nabla_{\partial_t} \nabla_{\partial_t} \gamma' + \nabla_{\partial_t} \text{Tor}(\gamma', \dot{\gamma}). \end{aligned} \tag{2}$$

Inserting $s = 0$, along the geodesic $c = \gamma(0, \quad)$ we get the *Jacobi differential equation* for the variation vector field $Y = \partial_s|_0 \gamma = \gamma'(0, \quad)$:

$$\boxed{0 = R(Y, \dot{c})\dot{c} + \nabla_{\partial_t} \nabla_{\partial_t} Y + \nabla_{\partial_t} \text{Tor}(Y, \dot{c}).} \tag{3}$$

This is a linear differential equation of second order for vector fields Y along the fixed geodesic $c : [0,1] \to M$. Thus for any $t_0 \in [0,1]$ and any initial values $(Y(t_0), (\nabla_{\partial_t})(t_0)) \in T_{c(t_0)}M \times T_{c(t_0)}M$ there exists a unique global solution Y of (3) along c. These solutions are called *Jacobi fields* along c; they form a $2m$-dimensional vector space.

27.2. The Jacobi flow. Consider a linear connector $K : TTM \to M$ on the tangent bundle with its corresponding horizontal lift mapping $C : TM \times_M TM \to TTM$ (see (22.8)), its spray $S : TM \to TTM$ given by $S(X) := C(X,X)$ (see (22.7)) and its covariant derivative $\nabla_X Y = K \circ TY \circ X$ (see (22.9)).

Theorem ([154]). *Let $S : TM \to TTM$ be a spray on a manifold M. Then $\kappa_{TM} \circ TS : TTM \to TTTM$ is a vector field. Consider a flow line*

$$J(t) = \operatorname{Fl}_t^{\kappa_{TM}\circ TS}(J(0))$$

of this field. Then we have:

- $c := \pi_M \circ \pi_{TM} \circ J$ *is a geodesic on M,*
- $\dot c = \pi_{TM} \circ J$ *is the velocity field of c,*
- $Y := T(\pi_M) \circ J$ *is a Jacobi field along c,*
- $\dot Y = \kappa_M \circ J$ *is the velocity field of Y,*
- $\nabla_{\partial_t} Y = K \circ \kappa_M \circ J$ *is the covariant derivative of Y.*

The Jacobi equation is given by:

$$\begin{aligned} 0 &= \nabla_{\partial_t}\nabla_{\partial_t} Y + R(Y,\dot c)\dot c + \nabla_{\partial_t} \operatorname{Tor}(Y,\dot c)\\ &= K \circ TK \circ TS \circ J. \end{aligned}$$

This implies that in a canonical chart induced from a chart on M the curve $J(t)$ is given by

$$(c(t), \dot c(t); Y(t), \dot Y(t)).$$

Proof. Consider a curve $s \mapsto X(s)$ in TM. Then each $t \mapsto \pi_M(\operatorname{Fl}_t^S(X(s)))$ is a geodesic in M, and in the variable s it is a variation through geodesics. Thus $Y(t) := \partial_s|_0 \pi_M(\operatorname{Fl}_t^S(X(s)))$ is a Jacobi field along the geodesic $c(t) := \pi_M(\operatorname{Fl}_t^S(X(0)))$ by (27.1), and each Jacobi field is of this form, for a suitable curve $X(s)$; see (27.5.4) below. We consider now the curve $J(t) := \partial_s|_0 \operatorname{Fl}_t^S(X(s))$ in TTM. Then by (8.13.6) we have

$$\begin{aligned} \partial_t J(t) &= \partial_t \partial_s|_0 \operatorname{Fl}_t^S(X(s)) = \kappa_{TM}\partial_s|_0 \partial_t \operatorname{Fl}_t^S(X(s)) = \kappa_{TM}\partial_s|_0 S(\operatorname{Fl}_t^S(X(s)))\\ &= (\kappa_{TM}\circ TS)(\partial_s|_0 \operatorname{Fl}_t^S(X(s))) = (\kappa_{TM}\circ TS)(J(t)), \end{aligned}$$

so that $J(t)$ is a flow line of the vector field $\kappa_{TM} \circ TS : TTM \to TTTM$. Moreover using the properties of κ from (8.13) and of S from (22.7), we get

$$\begin{aligned} T\pi_M.J(t) &= T\pi_M.\partial_s|_0 \operatorname{Fl}_t^S(X(s)) = \partial_s|_0 \pi_M(\operatorname{Fl}_t^S(X(s))) = Y(t),\\ \pi_M T\pi_M J(t) &= c(t), \text{ the geodesic},\\ \partial_t Y(t) &= \partial_t T\pi_M.\partial_s|_0 \operatorname{Fl}_t^S(X(s)) = \partial_t\partial_s|_0 \pi_M(\operatorname{Fl}_t^S(X(s)))\\ &= \kappa_M \partial_s|_0 \partial_t \pi_M(\operatorname{Fl}_t^S(X(s))) = \kappa_M \partial_s|_0 \partial_t \pi_M(\operatorname{Fl}_t^S(X(s))) \end{aligned}$$

$$\begin{aligned}
&= \kappa_M \partial_s|_0 T\pi_M.\partial_t \operatorname{Fl}^S_t(X(s)) = \kappa_M \partial_s|_0 (T\pi_M \circ S) \operatorname{Fl}^S_t(X(s)) \\
&= \kappa_M \partial_s|_0 \operatorname{Fl}^S_t(X(s)) = \kappa_M J(t), \\
\nabla_{\partial_t} Y &= K \circ \partial_t Y = K \circ \kappa_M \circ J.
\end{aligned}$$

Finally let us express the Jacobi equation (27.1.3). For the sake of shortness we write $\gamma(s,t) := \pi_M(\operatorname{Fl}^S_t(X(s)))$:

$$\begin{aligned}
&\nabla_{\partial_t}\nabla_{\partial_t} Y + R(Y,\dot c)\dot c + \nabla_{\partial_t}\operatorname{Tor}(Y,\dot c) \\
&\quad = \nabla_{\partial_t}\nabla_{\partial_t}.T\gamma.\partial_s + R(T\gamma.\partial_s, T\gamma.\partial_t)T\gamma.\partial_t + \nabla_{\partial_t}\operatorname{Tor}(T\gamma.\partial_s, T\gamma.\partial_t) \\
&\quad = K.T(K.T(T\gamma.\partial_s).\partial_t).\partial_t \\
&\qquad + (K.TK.\kappa_{TM} - K.TK).TT(T\gamma.\partial_t).T\partial_s.\partial_t \\
&\qquad + K.T((K.\kappa_M - K).TT\gamma.T\partial_s.\partial_t).\partial_t.
\end{aligned}$$

Note that for example for the term in the second summand we have

$$\begin{aligned}
TTT\gamma.TT\partial_t.T\partial_s.\partial_t &= T(T(\partial_t\gamma).\partial_s).\partial_t = \partial_t\partial_s\partial_t\gamma \\
&= \partial_t.\kappa_M.\partial_t.\partial_s\gamma = T\kappa_M.\partial_t.\partial_t.\partial_s\gamma
\end{aligned}$$

which at $s=0$ equals $T\kappa_M\ddot Y$. Using this, we get for the Jacobi equation at $s=0$:

$$\begin{aligned}
&\nabla_{\partial_t}\nabla_{\partial_t} Y + R(Y,\dot c)\dot c + \nabla_{\partial_t}\operatorname{Tor}(Y,\dot c) \\
&= (K.TK + K.TK.\kappa_{TM}.T\kappa_M - K.TK.T\kappa_M + K.TK.T\kappa_M - K.TK).\partial_t\partial_t Y \\
&= K.TK.\kappa_{TM}.T\kappa_M.\partial_t\partial_t Y = K.TK.\kappa_{TM}.\partial_t J = K.TK.TS.J,
\end{aligned}$$

where we used $\partial_t\partial_t Y = \partial_t(\kappa_M.J) = T\kappa_M\partial_t J = T\kappa_M.\kappa_{TM}.TS.J$. Finally the validity of the Jacobi equation $0 = K.TK.TS.J$ follows trivially from $K \circ S = 0_{TM}$. □

Note that the system of Jacobi fields depends only on the geodesic structure, thus on the spray induced by the given covariant derivative. So we may assume that the covariant derivative is torsion-free without loss; we do this from now on.

27.3. Fermi charts. Let (M,g) be a Riemann manifold. Let $c:(-2\varepsilon, 1+2\varepsilon) \to M$ be a geodesic (for $\varepsilon > 0$). We define the *Fermi chart* along c as follows.

Since $c([-\varepsilon, 1+\varepsilon])$ is compact in M, there exists $\rho > 0$ such that

$$B^\perp_{c(0)}(\rho) := \{X \in T^\perp_{c(0)}c := \{Y \in T_{c(0)}M : g(Y, c'(0)) = 0\}, |X|_g < \rho\},$$

$$\exp\circ\operatorname{Pt}(c,\quad): (-\varepsilon, 1+\varepsilon) \times B^\perp_c(0)(\rho) \to M, \tag{1}$$

$$(t,X) \mapsto \exp_{c(t)}(\operatorname{Pt}(c,t)X)$$

is everywhere defined.

Since its tangent mapping along $(-\varepsilon, 1+\varepsilon)\times\{0\}$,

$$T_{t,0}(\exp\circ\operatorname{Pt}(c,\quad)):\mathbb{R}\times T^{\perp}_{c(0)}c\to T_c(t)M=T_{c(t)}(c([0,1]))\times T^{\perp}_{c(t)}c,$$
$$(s,Y)\mapsto s.c'(t)+\operatorname{Pt}(c,t)Y,$$

is a linear isomorphism, we may assume (by choosing ρ smaller if necessary using (22.7.6)) that the mapping $\exp\circ\operatorname{Pt}(c,\quad)$ in (1) is a diffeomorphism onto its image. Its inverse,

$$\text{(2)}\qquad u_{c,\rho}:=(\exp\circ Pt(c,\quad))^{-1}:U_{c.\rho}\to(-\varepsilon,1+\varepsilon)\times B^{\perp}_{c(0)}(\rho),$$
$$U_{c.\rho}:=(\exp\circ\operatorname{Pt}(c,\quad))((-\varepsilon,1+\varepsilon)\times B^{\perp}_{c(0)}(\rho)),$$

is called the *Fermi chart* along c. Its importance is due to the following result.

27.4. Lemma. *Let X be a vector field along the geodesic c. For the Fermi chart along c put $T_{c(t)}(u_{c,\rho})^{-1}.X(t)=:(t,\bar X(t))$. Then we have*

$$T_{c(t)}u_{c,\rho}.(\nabla_{\partial_t}X)(t)=(t,\bar X'(t)).$$

So in the Fermi chart the covariant derivative ∇_{∂_t} along c is just the ordinary derivative. More is true: The Christoffel symbol in the Fermi chart vanishes along $(-\varepsilon,1+\varepsilon)\times\{0\}$.

The last statement is a generalization of the property of Riemann normal coordinates $\exp_x^{-1}$ that the Christoffel symbol vanishes at 0; see (22.7).

Proof. In terms of the Christoffel symbol of the Fermi chart the geodesic equation is given by $\bar c''(t)=\Gamma_{\bar c(t)}(\bar c'(t),\bar c'(t))$; see (22.4). But in the Fermi chart the geodesic c is given by $u_{c,\rho}(c(t))=(t,0)$, so the geodesic equation becomes $0=\Gamma_{\bar c(t)}((1,0),(1,0))=\Gamma_{\bar c(t)}(\bar c'(t),\bar c'(t))$. For $Y_0\in T^{\perp}_{c(0)}c$ the parallel vector field $Y(t)=\operatorname{Pt}(c,t)Y_0$ is represented by $(t,0;0,Y_0)$ in the Fermi chart; thus we get $0=\Gamma_{\bar c(t)}(\bar c'(t),Y_0)$. The geodesic $s\mapsto\exp_{c(t)}(s.\operatorname{Pt}(c,t).Y)$ for $Y\in T^{\perp}_{c(0)}c$ is represented by $s\mapsto(t,s.Y)$ in the Fermi chart. The corresponding geodesic equation is $0=\frac{\partial^2}{\partial s^2}(t,s.Y)=\Gamma_{(t,s.Y)}(Y,Y)$. By symmetry of $\Gamma_{(t,0)}$ these facts imply that $\Gamma_{(t,0)}=0$. Finally,

$$Tu_{c,\rho}.(\nabla_{\partial_t}X)(t)=\bar X'(t)-\Gamma_{(t,0)}(\bar c'(t),\bar X(t))=\bar X'(t).\qquad\square$$

27.5. Let (M^m,g) be a Riemann manifold, and let $c:[0,1]\to M$ be a geodesic which might be constant. Let us denote by $\mathcal{J}_c$ the $2m$-dimensional real vector space of all Jacobi fields along c, i.e., all vector fields Y along c satisfying

$$\nabla_{\partial_t}\nabla_{\partial_t}Y+R(Y,\dot c)\dot c=0.$$

Theorem.
(1) *The vector space $\mathcal{J}_c$ is canonically isomorphic to the vector space $T_{c(t)}M \times T_{c(t)}M$ via $\mathcal{J}_c \ni Y \mapsto (Y(t), (\nabla_{\partial_t} Y)(t))$, for each $t \in [0,1]$.*

(2) *The vector space $\mathcal{J}_c$ carries a canonical symplectic structure (see (20.4)):*

$$\omega_c(Y,Z) = g(Y(t), (\nabla_{\partial_t} Z)(t)) - g(Z(t), (\nabla_{\partial_t} Y)(t)) = \text{constant in } t.$$

(3) *Now let $c' \neq 0$. Then $\mathcal{J}_c$ splits naturally into the direct sum $\mathcal{J}_c = \mathcal{J}_c^\top \oplus \mathcal{J}_c^\perp$. Here $\mathcal{J}_c^\top$ is the 2-dimensional ω_c-nondegenerate subspace of all Jacobi fields which are tangent to c. All these are of the form $t \mapsto (a+tb)c'(t)$ for $(a,b) \in \mathbb{R}^2$. Also, $\mathcal{J}_c^\perp$ is the $(2m-2)$-dimensional ω_c-nondegenerate subspace consisting of all Jacobi fields Y satisfying $g(Y(t), c'(t)) = 0$ for all t. Moreover, $\omega_c(\mathcal{J}_c^\top, \mathcal{J}_c^\perp) = 0$.*

(4) *Each Jacobi field $Y \in \mathcal{J}_c$ is the variation vector field of a 1-parameter variation of c through geodesics, and conversely.*

(5) *Let $\mathcal{J}_c^0$ be the m-dimensional vector space consisting of all Jacobi fields Y with $Y(0) = 0$. Then $\omega_c(\mathcal{J}_c^0, \mathcal{J}_c^0) = 0$, so $\mathcal{J}_c^0$ is a Lagrangian subspace (see (20.4)).*

Proof. First let $c'(t) = 0$ so $c(t) = c(0)$. Then $Y(t) \in T_{c(0)}M$ for all t. The Jacobi equation becomes $\nabla_t \nabla_t Y = Y''$ so $Y(t) = A + tB$ for $A, B \in T_{c(0)}M$. Then (1), (2), and (5) hold.

Let us now assume that $c' \neq 0$. Part (1) follows from (27.1).

(2) For $Y, Z \in \mathcal{J}_c$ consider:

$$\begin{aligned}
\omega_c(Y,Z)(t) &= g(Y(t), (\nabla_{\partial_t} Z)(t)) - g(Z(t), (\nabla_{\partial_t} Y)(t)),\\
\partial_t \omega_c(Y,Z) &= g(\nabla_{\partial_t} Y, \nabla_{\partial_t} Z) + g(Y, \nabla_{\partial_t} \nabla_{\partial_t} Z)\\
&\qquad - g(\nabla_{\partial_t} Z, \nabla_{\partial_t} Y) - g(Z, \nabla_{\partial_t} \nabla_{\partial_t} Y)\\
&= -g(Y, R(Z,c')c') + g(Z, R(Y,c')c')\\
&= -g(R(Z,c')c', Y) + g(R(Y,c')c', Z)\\
&= g(R(Z,c')Y, c') - g(R(Y,c')Z, c') = 0,
\end{aligned}$$

where we used (24.4.4) and (24.4.5). Thus $\omega_c(Y,Z)(t)$ is constant in t. Also it is the standard symplectic structure (see (20.5)) on $T_{c(t)}M \times T_{c(t)}M$ induced by $g_{c(t)}$ via (1).

(3) We have $c' \neq 0$. In the Fermi chart $(U_{c,\rho}, u_{c,\rho})$ along c we have $c' = e_1$, the first unit vector, and the Jacobi equation becomes

(6) $$Y \in \mathcal{J}_c \iff Y''(t) + R(Y, e_1)e_1 = 0.$$

Consider first a Jacobi field $Y(t) = f(t).c'(t)$ which is tangential to c'. From (6) we get

$$0 = Y''(t) + R(Y(t), e_1)e_1 = f''(t).e_1 + f(t).R(e_1, e_1)e_1 = f''(t).e_1$$

so that $f(t) = a + tb$ for $a, b \in \mathbb{R}$. Let $g(t) = a' + tb'$. We use the symplectic structure at $t = 0$ to get $\omega_c(f.c', g.c') = g(a.c', b.c') - g(a'.c', b.c') = (ab' - a'b)|c'|^2$, a multiple of the canonical symplectic structure on $\mathbb{R}^2$.

For an arbitrary $Y \in \mathcal{J}_c$ we can then write $Y = Y_1 + Y_2$ uniquely where $Y_1 \in \mathcal{J}_c^\top$ is tangent to c' and where Y_2 is in the ω_c-orthogonal complement to $\mathcal{J}_c^\top$ in $\mathcal{J}_c$:

$$0 = \omega_c(c', Y_2) = g(c', \nabla_{\partial_t} Y_2) - g(\nabla_{\partial_t} c', Y_2) = g(c', \nabla_{\partial_t} Y_2) \quad \Longrightarrow \nabla_{\partial_t} Y_2 \bot c',$$
$$0 = \omega_c(t.c', Y_2) = g(t.c', \nabla_{\partial_t} Y_2) - g(c', Y_2) = -g(c', Y_2) \quad \Longrightarrow Y_2 \bot c'.$$

Conversely, $Y_2 \bot^g c'$ implies $0 = \partial_t g(c', Y_2) = g(c', \nabla_{\partial_t} Y_2)$ so that $Y_2 \in \mathcal{J}_c^\bot$ and $\mathcal{J}_c^\bot$ equals the ω_c-orthogonal complement of $\mathcal{J}_c^\top$. By symplectic linear algebra the latter space is ω_c-nondegenerate.

(4) for $\dot c \neq 0$ and $\dot c = 0$. Let $Y \in \mathcal{J}_c$ be a Jacobi field. Consider $b(s) := \exp_{c(0)}(s.Y(0))$. We look for a vector field X along b such that $(\nabla_{\partial_s} X)(0) = \nabla_{\partial_t} Y(0)$. We try $X(s) := \operatorname{Pt}(b, s)(\dot c(0) + s.(\nabla_{\partial_t} Y)(0))$, then

$$\begin{aligned} X'(0) &= \partial_s|_0 \left(Pt(b, s)(\dot c(0) + s.(\nabla_{\partial_t} Y)(0))\right) \\ &= \partial_s|_0 \left(Pt(b, s)(\dot c(0))\right) + T(Pt(b, 0))\partial_s|_0 \left(\dot c(0) + s.(\nabla_{\partial_t} Y)(0)\right) \\ &= C(b'(0), \dot c(0)) + \mathrm{vl}_{TM}(\dot c(0), (\nabla_{\partial_t} Y)(0)) \quad \text{using (24.2).} \end{aligned}$$

Now we put $\gamma(s, t) := \exp_{b(s)}(t.X(s))$; then $\gamma(0, t) = \exp_{c(0)}(t.X(0)) = \exp_{c(0)}(t.\dot c(0)) = c(t)$. Obviously, γ is a 1-parameter variation of c through geodesics; thus the variation vector field $Z(t) = \partial_s|_0 \gamma(s, t)$ is a Jacobi vector field. We have

$$\begin{aligned} Z(0) &= \partial_s|_0 \gamma(s, 0) = \partial_s|_0 \exp_{b(s)}(0_{b(s)}) = \partial_s|_0 b(s) = Y(0), \\ (\nabla_{\partial_t} Z)(0) &= \nabla_{\partial_t}(T\gamma.\partial_s)|_{s=0,t=0} \\ &= \nabla_{\partial_s}(T\gamma.\partial_t)|_{s=0,t=0} \quad \text{by (22.10.4) or (27.1.1)} \\ &= \nabla_{\partial_s}\left(\partial_t|_0 \exp_{b(s)}(t.X(s))\right)|_{s=0} = \nabla_{\partial_s} X|_{s=0} \\ &= K(\partial_s|_0 X(s)) = K\big(C(b'(0), \dot c(0)) + \mathrm{vl}(\dot c(0), (\nabla_{\partial_t} Y)(0))\big) \\ &= 0 + (\nabla_{\partial_t} Y)(0). \end{aligned}$$

Thus $Z = Y$ by (1).

(5) follows from (1) and symplectic linear algebra; see (20.5). □

27.6. Lemma. *Let c be a geodesic with $c' \neq 0$ in a Riemann manifold (M, g) and let $Y \in \mathcal{J}_c^0$ be a Jacobi field along c with $Y(0) = 0$. Then we have*

$$Y(t) = T_{t.\dot c(0)}(\exp_{c(0)})\, \mathrm{vl}\big(t.\dot c(0), t.(\nabla_{\partial_t} Y)(0)\big).$$

Proof. Let us step back into the proof of (27.5.4). There we had

$$b(s) = \exp_{c(0)}(s.Y(0)) = c(0),$$

$$\begin{aligned}
X(s) &= \mathrm{Pt}(c,s)(\dot c(0)+s.(\nabla_{\partial_t}Y)(0)) = \dot c(0)+s.(\nabla_{\partial_t}Y)(0),\\
Y(t) &= \partial_s|_0\,\gamma(s,t) = \partial_s|_0\,\exp_{b(s)}(t.X(s)) = T_{t.\dot c(0)}(\exp_{c(0)})\partial_s|_0\, m_tX(s)\\
&= T_{t.\dot c(0)}(\exp_{c(0)}).T(m_t)\partial_s|_0\,(\dot c(0)+s.(\nabla_{\partial_t}Y)(0))\\
&= T_{t.\dot c(0)}(\exp_{c(0)}).T(m_t).\operatorname{vl}\big(\dot c(0),(\nabla_{\partial_t}Y)(0)\big)\\
&= T_{t.\dot c(0)}(\exp_{c(0)}).\operatorname{vl}\big(t.\dot c(0),t.(\nabla_{\partial_t}Y)(0)\big). \quad \square
\end{aligned}$$

27.7. Corollary. *On a Riemann manifold (M,g) consider $\exp_x : T_xM \to M$. Then for $X \in T_xM$ the kernel of $T_X(\exp_x) : T_X(T_xM) \to T_{\exp_x(X)}M$ is isomorphic to the linear space consisting of all Jacobi fields $Y \in \mathcal{J}_c^0$ for $c(t) = \exp|_x(tX)$ which satisfy $Y(0)=0$ and $Y(1)=0$.*

Proof. By (27.6), $Y(t) = T_{tX}(\exp_x).\operatorname{vl}(tX, t(\nabla_{\partial_t}Y)(0))$ is a Jacobi field with $Y(0)=0$. But then $0 = Y(1) = T_X(\exp_x)\operatorname{vl}\big(X,(\nabla_{\partial_t}Y)(0)\big)$ holds if and only if $(\nabla_{\partial_t}Y)(0) \in \ker(T_X(\exp_x))$. $\square$

27.8. Let (M,g) and $(\tilde M,\tilde g)$ be two Riemann manifolds of the same dimension. Let $c:[0,1]\to M$ and $\tilde c:[0,1]\to\tilde M$ be two geodesics of the same length. We choose a linear isometry $I_0 : (T_{c(0)}M, g_{c(0)}) \to (T_{\tilde c(0)}\tilde M, \tilde g_{\tilde c(0)})$ and define the linear isometries:

$$I_t := \tilde{\mathrm{Pt}}(\tilde c,t)\circ I_0\circ \mathrm{Pt}(c,t)^{-1} : T_{c(t)}M \to T_{\tilde c(t)}\tilde M.$$

Lemma. *If Y is a vector field along c, then $t\mapsto (I_*Y)(t) = I_tY(t)$ is a vector field along $\tilde c$ and we have $\tilde\nabla_{\partial_t}(I_*Y) = I_*(\nabla_{\partial_t}Y)$ so that $\tilde\nabla_{\partial_t}\circ I_* = I_*\circ\nabla_{\partial_t}$.*

Proof. We use Fermi charts (with the minimum of the two ρ's)

$$\begin{array}{ccc}
M\supset U_{c,\rho} & \xrightarrow{\;u_{c,\rho}\;} & (-\varepsilon,1+\varepsilon)\times B^\perp_{c(0)}(\rho)\\
 & & \Big\downarrow {\scriptstyle \mathrm{Id}\times I_0}\ \ {\scriptstyle\text{linear}}\\
\tilde M\supset U_{\tilde c,\rho} & \xrightarrow{\;u_{\tilde c,\rho}\;} & (-\varepsilon,1+\varepsilon)\times B^\perp_{\tilde c(0)}(\rho).
\end{array}$$

By construction of the Fermi charts we have $(I_*Y)(t) = T(u_{\tilde c,\rho}^{-1}\circ(\mathrm{Id}\times I_0)\circ u_{c,\rho}).Y(t)$. Thus

$$\begin{aligned}
\tilde\nabla_{\partial_t}(I_*Y)(t) &= \tilde\nabla_{\partial_t}(T(u_{\tilde c,\rho}^{-1}\circ(\mathrm{Id}\times I_0)\circ u_{c,\rho}).Y)(t)\\
&= T(u_{\tilde c,\rho})^{-1}\partial_t\big((\mathrm{Id}\times I_0)\circ T(u_{c,\rho}).Y(t)\big) \quad \text{by (27.4)}\\
&= T(u_{\tilde c,\rho})^{-1}.(\mathrm{Id}\times I_0).\partial_t T(u_{c,\rho}).Y(t)\\
&= T(u_{\tilde c,\rho})^{-1}.(\mathrm{Id}\times I_0).T(u_{c,\rho}).(\nabla_{\partial_t}Y)(t) \quad \text{by (27.4)}\\
&= I_*(\nabla_{\partial_t}Y)(t). \quad \square
\end{aligned}$$

27.9. Jacobi operators. On a Riemann manifold (M, g) with curvature R we consider for each vector field $X \in \mathfrak{X}(M)$ the corresponding *Jacobi operator* $R_X : TM \to TM$ which is given by $R_X(Y) = R(Y, X)X$. It turns out that each R_X is a *self-adjoint* endomorphism, $g(R_X(Y, Z)) = g(Y, R_X(Z))$, since we have $g(R(Y, X)X, Z) = g(R(X, Z)Y, X) = g(R(Z, X)X, Y)$ by (24.4.4) and (24.4.5). One can reconstruct the curvature R from the family of Jacobi operators R_X by polarization and the properties from (24.4).

27.10 Theorem (E. Cartan). *Let (M, g) and $(\tilde{M}, \tilde{g})$ be Riemann manifolds of the same dimension. Let $x \in M$, $\tilde{x} \in \tilde{M}$, and $\varepsilon > 0$ be such that $\exp_x : B_{0_x}(\varepsilon) \to M$ and $\exp_{\tilde{x}} : B_{0_{\tilde{x}}}(\varepsilon) \to \tilde{M}$ are both diffeomorphisms onto their images. Let $I_x : (T_xM, g_x) \to (T_{\tilde{x}}\tilde{M}, \tilde{g}_{\tilde{x}})$ be a linear isometry. Then the following hold:*

The mapping $\Phi := \exp_{\tilde{x}} \circ I_x \circ (\exp_x | B_{0_x}(\varepsilon))^{-1} : B_x(\varepsilon) \to B_{0_x}(\varepsilon) \to B_{0_{\tilde{x}}}(\varepsilon) \to B_{\tilde{x}}(\varepsilon)$ is a diffeomorphism which maps radial geodesics to radial geodesics. The tangent mapping $T\Phi$ maps Jacobi fields Y along radial geodesics with $Y(0) = 0$ to Jacobi fields $\tilde{Y}$ along radial geodesics with $\tilde{Y}(0) = 0$.

Suppose that moreover for all radial geodesics c in $B_x(\varepsilon)$ and their images $\tilde{c} = \Phi \circ c$ the property

$$I_t \circ R_{\dot{c}(t)} = \tilde{R}_{\dot{\tilde{c}}(t)} \circ I_t \tag{1}$$

holds where $I_t : T_{c(t)}M \to T_{\tilde{c}(t)}\tilde{M}$ is defined in (27.8). Then Φ is an isometry. Conversely, if Φ is an isometry, then (1) holds.

Proof. It is clear that Φ maps radial geodesics in $B_x(\varepsilon) \subset M$ to radial geodesics in $B_{\tilde{x}}(\varepsilon) \subset \tilde{M}$. Any Jacobi field Y along a radial geodesic c can be written as variation vector field $Y(t) = \partial_s|_0 \, \gamma(s, t)$ where $\gamma(s, \quad)$ is a radial geodesic for all s and $\gamma(0, t) = c(t)$. Then $T\Phi.Y(t) = T\Phi.\partial_s|_0 \, \gamma(s, t) = \partial_s|_0 \, (\Phi\gamma(s, t))$, and any $\Phi\gamma(s, \quad)$ is a radial geodesic in $B_{\tilde{x}}(\varepsilon)$. Thus $T\Phi.Y$ is a Jacobi field along the radial geodesic $\Phi \circ c$ with $T\Phi.Y(0) = 0$. This proves the first assertion.

Now let Y be a Jacobi field along the radial geodesic c with $Y(0) = 0$. Then the Jacobi equation $0 = \nabla_{\partial_t}\nabla_{\partial_t}Y + R_{\dot{c}}(Y)$ holds. Consider $(I_*Y)(t) = I_tY(t)$. By (27.8) and (1) we then have

$$\tilde{\nabla}_{\partial_t}\tilde{\nabla}_{\partial_t}(I_*Y) + \tilde{R}_{\dot{\tilde{c}}}(I_*Y) = I_*(\nabla_{\partial_t}\nabla_{\partial_t}Y + R_{\dot{c}}Y) = 0.$$

Thus I_*Y is again a Jacobi field along the radial geodesic $\tilde{c}$ with $(I_*Y)(0) = 0$. Since also $\tilde{\nabla}_{\partial_t}(I_*Y)(0) = I_*(\nabla_{\partial_t}Y)(0) = I_0 \, (\nabla_{\partial_t}Y)(0) = T\Phi.(\nabla_{\partial_t}Y)(0)$, we get $I_*Y = T\Phi.Y$. Since the vectors $Y(t)$ for Jacobi fields Y along c with $Y(0) = 0$ span $T_{c(t)}M$ by (27.6), we may conclude that $T_{c(t)}\Phi = I_t : T_{c(t)}M \to T_{\tilde{c}(t)}\tilde{M}$ is an isometry. The converse statement is obvious since an isometry intertwines the curvatures. □

27.11. Conjugate points. Let $c : [0, a] \to M$ be a geodesic on a Riemann manifold (M, g) with $c(0) = x$. A parameter $t_0 \in [0, a]$ or its image $c(t_0) \in c([0, a])$ is called a *conjugate point* for $x = c(0)$ on $c([0, a])$ if the tangent mapping

$$T_{t_0 \dot{c}(0)}(\exp_x) : T_{t_0 \dot{c}(0)}(T_x M) \to T_{c(t_0)} M$$

is **not** an isomorphism. Then $t_0 > 0$. The *multiplicity* of the conjugate point is the dimension of the kernel of $T_{t_0 \dot{c}(0)}(\exp_x)$ which equals the dimension of the subspace of all Jacobi fields Y along c with $Y(0) = 0$ and $Y(t_0) = 0$, by (27.7).

27.12. Example. Let $M = \rho \cdot S^m \subset \mathbb{R}^{M+1}$, the sphere of radius $\rho > 0$. Then any geodesic c with $|\dot{c}| = 1$ satisfies $c(\rho\pi) = -c(0)$, so $-c(0)$ is conjugate to $c(0)$ along c with multiplicity $m - 1$.

27.13. Lemma. *Let $c : [0, a] \to M$ be a geodesic in a Riemann manifold (M, g). Then the vector $\partial_t(t.\dot{c}(0))|_{t=t_0} = \operatorname{vl}(t_0.\dot{c}(0), \dot{c}(0)) \in T_{t_0.\dot{c}(0)}(T_{c(0)} M)$ is orthogonal to the kernel $\ker(T_{t_0 \dot{c}(0)}(\exp_{c(0)}))$, for any $t_0 \in [0, a]$.*

Proof. If $c(t_0)$ is not a conjugate point to $x = c(0)$ of c, this is clearly true. If it is, let Y be the Jacobi field along c with $Y(0) = 0$ and $(\nabla_{\partial_t} Y)(0) = X \neq 0$ where $\operatorname{vl}(t_0.\dot{c}(0), X) \in \ker(T_{t_0 \dot{c}(0)}(\exp_x))$. Then we have

$$T_{t_0 \dot{c}(0)}(\exp_x) \operatorname{vl}(t_0.\dot{c}(0), X) = Y(t_0) = 0.$$

Let $\hat{c}(t) = (t - t_0)\dot{c}(0) \in \mathcal{J}_c^{\top}$, a tangential Jacobi field along c. By (27.5.2) applied for $t = 0$ and for $t - t_0$ we get

$$\begin{aligned}
\omega_c(\hat{c}, Y) &= g(\hat{c}(0), (\nabla_{\partial_t})Y(0)) - g(Y(0), (\nabla_{\partial_t} Y)(0)) = g(t_0.\dot{c}(0), X) - 0 \\
&= g(\hat{c}(t_0), (\nabla_{\partial_t})Y(t_0)) - g(Y(t_0), (\nabla_{\partial_t} Y)(t_0)) = 0.
\end{aligned}$$

Thus $t_0.g(\dot{c}(0), X) = 0$ and since $t_0 > 0$, we get $X \perp \dot{c}(0)$. □

We can extract more information about the Jacobi field Y from this proof. We showed that then $(\nabla_{\partial_t} Y)(0) \perp^g \dot{c}(0)$. We use this in the following application of (27.5.2) for $t = 0$: Now

$$\omega_c(\dot{c}, Y) = g(\dot{c}(0), (\nabla_{\partial_t} Y)(0)) - g(Y(0, (\nabla_{\partial_t} \dot{c})(0))) = 0.$$

Together with $\omega_c(\hat{c}, Y) = 0$ from the proof this says that $Y \in \mathcal{J}_c^{\perp}$, so by (27.5.3), $Y(t) \perp^g \dot{c}(t)$ for all t.

Let us denote by $\mathcal{J}_c^{\perp,0} = \mathcal{J}_c^{\perp} \cap \mathcal{J}_c^0$ the space of all Jacobi fields Y with $Y(0) = 0$ and $Y(t) \perp^g \dot{c}(t)$ for all t. Then the dimension of the kernel of $T_{t_0 \dot{c}(0)}(\exp_x)$ equals the dimension of the space of all $Y \in \mathcal{J}_c^{\perp,0}$ which satisfy $Y(t_0) = 0$.

Thus, if $c(0)$ and $c(t_0)$ are conjugate, then there are 1-parameter variations of c through geodesics which all start at $c(0)$ and end at $c(t_0)$, at least infinitesimally in the variation parameter. For this reason conjugate points are also called *focal points.* We will strengthen this later on.

27.14. The Hessian of the energy alias second variation formulas. Let (M,g) be a Riemann manifold. Let $c:[0,a]\to M$ be a geodesic with $c(0)=x$ and $c(a)=y$. A smooth *variation* of c with fixed ends is a smooth mapping $F:(-\varepsilon,\varepsilon)\times[0,a]\to M$ with $F(0,t)=c(t)$, $F(s,0)=x$, and $F(s,a)=y$. The variation vector field for F is the vector field $X=\partial_s|_0 F(s,\quad)$ along c, with $X(0)=0$ and $X(a)=0$.

The space $C^\infty(([0,a],0,a),(M,x,y))$ of all smooth curves $\gamma:[0,a]\to M$ with $c(0)=x$ and $c(a)=y$ is an infinite-dimensional smooth manifold modeled on Fréchet spaces. See [**113**] for a thorough account of this. The tangent space $T_c(C^\infty(([0,a],0,a),(M,x,y)))$ at the geodesic c of this infinite-dimensional manifold consists of all variation vector fields along c as above. We consider again the energy as a smooth function

$$E:C^\infty(([0,a],0,a),(M,x,y))\to\mathbb{R},\qquad E(\gamma)=\tfrac12\int_0^a|\dot\gamma(t)|_g^2\,dt.$$

Let now F be a variation with fixed ends of the geodesic c. Then we have:

$$\begin{aligned}\partial_s E(F(s,\quad))&=\tfrac12\int_0^a\partial_s g(\partial_t F,\partial_t F)\,dt=\int_0^a g(\nabla_{\partial_s}\partial_t F,\partial_t F)\,dt\\&=\int_0^a g(\nabla_{\partial_t}\partial_s F,\partial_t F)\,dt,\qquad\text{by (22.10.4) or (27.1.1).}\end{aligned}$$

Therefore,

$$\begin{aligned}\partial_s^2|_0E(F(s,\quad))&=\int_0^a\Big(g(\nabla_{\partial_s}\nabla_{\partial_t}\partial_sF,\partial_tF)+g(\nabla_{\partial_t}\partial_sF,\nabla_{\partial_s}\partial_tF)\Big)\Big|_{s=0}dt\\&=\int_0^a\Big(g(\nabla_{\partial_t}\nabla_{\partial_s}\partial_sF,\partial_tF)+g(R(\partial_sF,\partial_tF)\partial_sF,\partial_tF)\\&\qquad+g(\nabla_{\partial_t}\partial_sF,\nabla_{\partial_t}\partial_sF)\Big)\Big|_{s=0}dt\quad\text{by (24.5) and (22.10.4)}\\&=\int_0^a\Big(g(\nabla_{\partial_t}\partial_sF,\nabla_{\partial_t}\partial_sF)+g(R(\partial_sF,\partial_tF)\partial_sF,\partial_tF)\Big)\Big|_{s=0}dt\\&\quad+\int_0^a\Big(g(\nabla_{\partial_t}\nabla_{\partial_s}\partial_sF,\partial_tF)|_{s=0}+g(\nabla_{\partial_s}\partial_sF|_{s=0},\underbrace{\nabla_{\partial_t}\partial_tF|_{s=0}}_{\nabla_{\partial_t}\dot c=0})\Big)\,dt.\end{aligned}$$

The last summand equals $\int_0^a\partial_t g(\nabla_{\partial_s}\partial_sF,\partial_tF)|_{s=0}\,dt$ which vanishes since we have a variation with fixed ends and thus $(\nabla_{\partial_s}\partial_sF)(s,0)=0$ and also $(\nabla_{\partial_s}\partial_sF)(s,a)=0$. Recall $X=\partial_s|_0 F$, a vector field along c with $X(0)=0$ and $X(a)=0$. Thus

$$d^2E(c)(X,X)=\partial_s^2|_0E(F(s,\quad))=\int_0^a\big(g(\nabla_{\partial_t}X,\nabla_{\partial_t}X)+g(R(X,\dot c)X,\dot c)\big)\,dt.$$

If we polarize this, we get the Hessian of the energy at a geodesic c as follows (the boundary terms vanish since X, Y vanish at the ends 0 and a):

$$dE(c)(X) = \int_0^a g(\nabla_{\partial_t} X, \dot c)\, dt = -\int_0^a g(X, \nabla_{\partial_t} \dot c)\, dt = 0,$$

$$d^2E(c)(X,Y) = \int_0^a \big(g(\nabla_{\partial_t} X, \nabla_{\partial_t} Y) - g(R_{\dot c}(X), Y)\big)\, dt, \tag{1}$$

$$d^2E(c)(X,Y) = -\int_0^a g\big(\nabla_{\partial_t}\nabla_{\partial_t} X + R_{\dot c}(X), Y\big)\, dt. \tag{2}$$

We see that among all vector fields X along c with $X(0) = 0$ and $X(a) = 0$ those which satisfy $d^2E(c)(X,Y) = 0$ for all Y are exactly the Jacobi fields.

We shall need a slight generalization. Let X, Y be continuous vector fields along c which are smooth on $[t_i, t_{i+1}]$ for $0 = t_0 < t_1 < \cdots < t_k = a$, and which vanish at 0 and a. These are tangent vectors at c to the smooth manifold of all curves from x to y which are piecewise smooth in the same manner. Then we take the following formula as a definition, which can be motivated by the computations above (with considerable care). We will just need that $d^2E(c)$, to be defined below, is continuous in the natural uniform C^2-topology on the space of piecewise smooth vector fields so that later we can approximate a broken vector field by a smooth one.

$$\begin{aligned}
d^2E(c)(X,Y) &= \int_0^a \big(g(\nabla_{\partial_t} X, \nabla_{\partial_t} Y) + g(R(X,\dot c)Y, \dot c)\big)\, dt \\
&= \sum_{i=0}^{k-1} \int_{t_i}^{t_{i+1}} \big(g(\nabla_{\partial_t} X, \nabla_{\partial_t} Y) + g(R(X,\dot c)Y, \dot c)\big)\, dt \\
&= \sum_{i=0}^{k-1} \int_{t_i}^{t_{i+1}} \Big(\partial_t g(\nabla_{\partial_t} X, Y) - g(\nabla_{\partial_t}\nabla_{\partial_t} X, Y) - g(R(X,\dot c)\dot c, Y)\Big)\, dt \\
&= -\int_0^a g\big(\nabla_{\partial_t}\nabla_{\partial_t} X + R_{\dot c}(X), Y\big)\, dt \qquad (3) \\
&\quad + \sum_{i=0}^{k-1} \Big(g\big((\nabla_{\partial_t} X)(t_{i+1}-), Y(t_{i+1})\big) - g\big((\nabla_{\partial_t} X)(t_i+), Y(t_{i+1})\big)\Big).
\end{aligned}$$

27.15. Theorem. *Let (M,g) be a Riemann manifold and let $c : [0,a] \to M$ be a geodesic with $c(0) = x$ and $c(a) = y$.*

(1) If $T_{t\dot c(0)}(\exp_x) : T_{t\dot c(0)}(T_x M) \to T_{c(t)}M$ is an isomorphism for all $t \in [0,a]$, then for any smooth curve e from x to y which is near enough to c the length $L(e) \geq L(c)$ with equality if and only if e is a reparameterization of c. Moreover, $d^2E(c)(X,X) \geq 0$ for each smooth vector field X along c which vanishes at the ends.

(2) *If there are conjugate points $c(0)$, $c(t_1)$ along c with $0 < t_1 < a$, then there exists a smooth vector field X along c with $X(0) = 0$ and $X(a) = 0$ such that $d^2E(c)(X,X) < 0$. Thus for any smooth variation F of c with $\partial_s|_0 F(s,\ \ ) = X$ the curve $F(s,\ \ )$ from x to y is shorter than c for all $0 < |s| < \varepsilon$.*

Proof. (1) Since $T_{t\dot c(0)}(\exp_x) : T_{t\dot c(0)}(T_xM) \to T_{c(t)}M$ is an isomorphism, for each $t \in [0,a]$ there exist an open neighborhood $U(t.\dot c(0)) \subset T_xM$ of $t\dot c(0)$ such that $\exp_x|U(t.\dot c(0))$ is a diffeomorphism onto its image. Since $[0,a].\dot c(0)$ is compact in T_xM, there exists an $\varepsilon > 0$ such that $U(t.\dot c(0)) \supset B_{t\dot c(0)}(\varepsilon)$ for all t.

Now let $e : [0,a] \to M$ be a smooth curve with $e(0) = x$ and $e(a) = y$ which is near c in the sense that there exists a subdivision $0 = t_0 < t_1 < \dots < t_k = a$ with $e([t_i, t_{i+1}]) \subset \exp_x(B_{t_i\dot c(0)}(\varepsilon))$. We put:

$$\tilde e : [0,a] \to T_xM,$$
$$\tilde e(t) := (\exp_x|B_{t_i\dot c(0)}(\varepsilon))^{-1}(e(t)), \quad t \in [t_i, t_{i+1}].$$

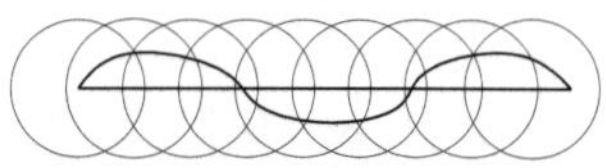

Then $\tilde e$ is smooth, $\tilde e(0) = 0_x$, $\tilde e(a) = a.\dot c(0)$, and $\exp_x(\tilde e(t)) = e(t)$. We consider the polar representation $\tilde e(t) = r(t).\varphi(t)$ in T_xM where $\varphi(t) = \frac{\tilde e(t)}{|\tilde e(t)|}$ and $r(t) = |\tilde e(t)|$. Let $r = |\tilde e(a)| = a|\dot c(0)|$. Then we put:

$$\gamma(s,t) = \exp_x(r.t.\varphi(s))$$

which implies

$$e(t) = \gamma(t, \tfrac{r(t)}{r}) = \exp_x(r(t).\varphi(t)), \quad \dot e(t) = \partial_s\gamma(t, \tfrac{r(t)}{r}) + \partial_t\gamma(t, \frac{r(t)}{r})\frac{\dot r(t)}{r}.$$

Note that $\nabla_{\partial_t}\partial_t\gamma = 0$ since $\gamma(s,\ \ )$ is a geodesic. From

$$\begin{aligned}\partial_t g(\partial_s\gamma,\partial_t\gamma) &= g(\nabla_{\partial_t}\partial_s\gamma, \partial_t\gamma) + g(\partial_s\gamma, \nabla_{\partial_t}\partial_t\gamma)\\ &= g(\nabla_{\partial_s}\partial_t\gamma, \partial_t\gamma) + 0 \qquad \text{by (22.10.1)}\\ &= \tfrac12\partial_s g(\partial_t\gamma,\partial_t\gamma) = \tfrac12\partial_s|\partial_t\gamma(s,\ \ )|^2 = \tfrac12\partial_s r^2|\varphi(s)|^2 = \tfrac12\partial_s r^2 = 0\end{aligned}$$

we get that $g(\partial_s\gamma(s,t), \partial_t\gamma(s,t)) = g(\partial_s\gamma(s,0), \partial_t\gamma(s,0)) = g(0, r.\varphi(s)) = 0$. Thus

$$(3) \qquad g_{\gamma(s,t)}(\partial_s\gamma(s,t), \partial_t\gamma(s,t)) = 0 \quad \text{for all } s,t.$$

By Pythagoras

$$\begin{aligned}|\dot e(t)|_g^2 &= |\partial_s\gamma(t, \tfrac{r(t)}{r})|_g^2 + |\partial_t\gamma(t, \tfrac{r(t)}{r})|_g^2\frac{|\dot r(t)|^2}{r^2}\\ &= |\partial_s\gamma(t, \tfrac{r(t)}{r})|_g^2 + r^2|\varphi(t)|_g^2\frac{|\dot r(t)|^2}{r^2} \geq |\dot r(t)|^2\end{aligned}$$

with equality if and only if $\partial_s\gamma(t,\frac{r(t)}{r})=0$, i.e., $\varphi(t)$ is constant in t. So

(4)
$$L(e)=\int_0^a |\dot e(t)|_g\,dt \ge \int_0^a |\dot r(t)|\,dt \ge \int_0^a \dot r(t)\,dt = r(a)-r(0)=r=L(c)$$

with equality if and only if $\dot r(t)\ge 0$ and $\varphi(t)$ is constant, i.e., e is a reparameterization of c.

Note that (3) and (4) generalize Gauß's lemma (23.2) and its corollary (23.3) to more general assumptions.

Now consider a vector field X along c with $X(0)=0$ and $X(a)=0$ and let $F:(-\varepsilon,\varepsilon)\times[0,a]\to M$ be a smooth variation of c with $F(s,0)=x$, $F(s,a)=y$, and $\partial_s|_0 F = X$. We have

$$\begin{aligned}
2E(F(s,\quad)).a &= \int_0^a |\partial_t F|_g^2\,dt\cdot\int_0^a 1^2\,dt \ge \Big(\int_0^a |\partial_t F|_g.1\,dt\Big)^2\\
&= L(F(s,\quad))^2 \ge L(c)^2 \quad \text{by (4)} \qquad (5)\\
&= \Big(\int_0^a |\dot c(0)|_g\,dt\Big)^2 = |\dot c(0)|^2.a^2 = \int_0^a |\dot c(0)|^2\,dt\cdot a = 2E(c).a.
\end{aligned}$$

Since c is a geodesic, we have $\partial_s|_0 E(F(s,\quad))=0$ and therefore we also get $d^2E(c)(X,X)=\partial_s^2|_0 E(F(s,\quad))\ge 0$.

(2) Let $c(0)$, $c(t_1)$ be conjugate points along c with $0<t_1<a$. By (27.11) there exists a Jacobi field $Y\ne 0$ along c with $Y(0)=0$ and $Y(t_1)=0$. Choose $0<t_0<t_1<t_2<a$ and a vector field Z along c with $Z|[0,t_0]=0$, $Z|[t_2,a]=0$, and $Z(t_1)=-(\nabla_{\partial_t}Y)(t_1)\ne 0$ (since $Y\ne 0$). Let $\tilde Y$ be the continuous piecewise smooth vector field along c which is given by $\tilde Y|[0,t_1]=Y|[0,t_1]$ and $\tilde Y|[t_1,a]=0$. Then $\tilde Y+\eta Z$ is a continuous piecewise smooth vector field along c which is broken at t_1 and vanishes at 0 and at a. Then we have

$$d^2E(c)(\tilde Y+\eta Z,\tilde Y+\eta Z)=d^2E(c)(\tilde Y,\tilde Y)+\eta^2\,d^2E(c)(Z,Z)+2\eta\,d^2E(c)(\tilde Y,Z)$$

and by (22.12.3)

$$\begin{aligned}
d^2E(c)(\tilde Y,\tilde Y) &= -\int_0^{t_1} g\big(\nabla_{\partial_t}\nabla_{\partial_t}Y+R_{\dot c}(Y),Y\big)\\
&\quad -\int_{t_1}^a g\big(\nabla_{\partial_t}\nabla_{\partial_t}0+R_{\dot c}(0),0\big)\\
&\quad + g((\nabla_{\partial_t}Y)(t_1-),0)-g((\nabla_{\partial_t}Y)(0+),0)\\
&\quad + g((\nabla_{\partial_t}\tilde Y)(a-),0)-g((\nabla_{\partial_t}\tilde Y)(t_1+),0)=0,\\
d^2E(c)(\tilde Y,\tilde Z) &= -\int_0^{t_1} g\big(\nabla_{\partial_t}\nabla_{\partial_t}Y+R_{\dot c}(Y),Z\big)
\end{aligned}$$

$$
\begin{aligned}
&- \int_{t_1}^{a} g(\nabla_{\partial_t}\nabla_{\partial_t}0 + R_{\dot c}(0), Z) \\
&\quad + g((\nabla_{\partial_t}Y)(t_1-), Z(t_1)) - g((\nabla_{\partial_t}Y)(0+), 0) \\
&\quad + g((\nabla_{\partial_t}\tilde{Y})(a-), 0) - g((\nabla_{\partial_t}0)(t_1+), Z(t_1)) \\
&= g((\nabla_{\partial_t}Y)(t_1), Z(t_1)) = -g((\nabla_{\partial_t}Y)(t_1), (\nabla_{\partial_t}Y)(t_1)) \\
&= -|(\nabla_{\partial_t}Y)(t_1)|_g^2 < 0, \\
d^2E(c)(\tilde{Y} + \eta Z, \tilde{Y} + \eta Z) &= \eta^2\, d^2E(c)(Z,Z) - 2\eta\, |(\nabla_{\partial_t}Y)(t_1)|_g^2.
\end{aligned}
$$

The last expression will be negative for η small enough. Since $d^2E(c)$ is continuous in the C^2-topology for continuous piecewise smooth vector fields along c, we can approximate $\tilde{Y} + \eta Z$ by a smooth vector field X vanishing at the ends such that we still have $d^2E(c)(X,X) < 0$.

Finally, let $F : (-\varepsilon, \varepsilon) \times [0,a] \to M$ be any smooth variation of c with fixed ends and $\partial_s|_0 F = X$. Consider the Taylor expansion

$$E(F(s,\quad)) = E(c) + s\, dE(c)(X) + \tfrac{s^2}{2} d^2E(c)(X,X) + s^3 h(s)$$

where $h(s) = \int_0^1 \frac{(1-u)^2}{2} \partial_v^3 E(F(v,\quad))|_{v=us}\, du$. Since $dE(c)(X) = 0$, this implies $E(F(s,\quad)) < E(c)$ for $s \neq 0$ small enough. Using both halves of (5), this implies $L(F(s,\quad))^2 \le 2E(F(s,\quad))\, a < 2E(c)\, a = L(c)^2$. □

27.16. Theorem. *Let (M,g) be a Riemann manifold with sectional curvature $k \ge k_0 > 0$. Then for any geodesic c in M the distance between two conjugate points along c is $\le \frac{\pi}{\sqrt{k_0}}$.*

Proof. Let $c : [0,a] \to M$ be a geodesic with $|\dot c| = 1$ such that $c(a)$ is the first point which is conjugate to $c(0)$ along c. We choose a parallel unit vector field Z along c, $Z(t) = \mathrm{Pt}(c,t).Z(0)$, $|Z(0)|_g = 1$, $Z(t) \perp^g \dot c(t)$, so that $\nabla_{\partial_t} Z = 0$. Consider $f \in C^\infty([0,a], \mathbb{R})$ with $f(0) = 0$ and $f(a) = 0$, and let $0 < b < a$. By (27.15.1) we have $d^2E_0^b(c)(fZ, fZ) \ge 0$. By (27.14.1) we have

$$
\begin{aligned}
d^2E_0^b(c)(fZ,fZ) &= \int_0^b \big(g(\nabla_{\partial_t}(fZ), \nabla_{\partial_t}(fZ)) - g(R(fZ,\dot c)\dot c, fZ)\big)\, dt \\
&= \int_0^b (f'^2 - f^2 k(Y \wedge \dot c))\, dt \le \int_0^b (f'^2 - f^2 k_0)\, dt
\end{aligned}
$$

since $Y, \dot c$ form an orthonormal basis. Now we choose $f(t) = \sin(\pi t b)$ so that $\int_0^b f^2\, dt = \frac{b}{2}$ and $\int_0^b f'^2\, dt = \frac{\pi^2}{2b}$. Thus

$$0 \le \int_0^b (f'^2 - f^2 k_0)\, dt = \frac{\pi^2}{2b} - \frac{b}{2} k_0$$

which implies $b \le \frac{\pi}{\sqrt{k_0}}$. Since this holds for all $b < a$, we get $a \le \frac{\pi}{\sqrt{k_0}}$. □

27.17. Corollary (Myers, 1935). *Let M be a complete connected Riemann manifold with sectional curvature $k \geq k_0 > 0$. Then the diameter of M is bounded:*

$$\operatorname{diam}(M) := \sup\{\operatorname{dist}(x, y) : x, y \in M\} \leq \frac{\pi}{\sqrt{k_0}}.$$

Thus M is compact and each covering space of M is also compact, so the the fundamental group $\pi_1(M)$ is finite.

Proof. By (23.6.6) any two points $x, y \in M$ can be connected by a geodesic c of minimal length. Assume for contradiction that $\operatorname{dist}(x, y) > \frac{\pi}{\sqrt{k_0}}$; then by (27.16) there exists an interior point z on the geodesic c which is conjugate to x. By (27.15.2) there exist smooth curves in M from x to y which are shorter than c, contrary to the minimality of c □

27.18. Theorem. *Let M be a connected complete Riemann manifold with sectional curvature $k \leq 0$. Then $\exp_x : T_xM \to M$ is a covering mapping for each $x \in M$. If M is also simply connected, then $\exp_x : T_xM \to M$ is a diffeomorphism.*

This result is due to [**81**] for surfaces and to [**33**] in the general case.

Proof. Let $c : [0, \infty) \to M$ be a geodesic with $c(0) = x$. If $c(a)$ is a point conjugate to $c(0)$ along c, then by (27.11) and (27.7) there exists a Jacobi field $Y \neq 0$ along c with $Y(0) = 0$ and $Y(a) = 0$. By (27.13) we have $Y(t) \perp^g \dot{c}(t)$ for all t. Now use (27.14.2) and (27.14.1) to get

$$\begin{aligned}
d^2E(c)(Y, Y) &= -\int_0^a g\big(\nabla_{\partial_t}\nabla_{\partial_t}Y + R_{\dot{c}}(Y), Y\big)\,dt = 0,\\
d^2E_0^a(c)(Y, Y) &= \int_0^a \big(g(\nabla_{\partial_t}Y, \nabla_{\partial_t}Y) - g(R(Y, \dot{c})\dot{c}, Y)\big)\,dt\\
&= \int_0^a \big(|\nabla_{\partial_t}Y|_g^2 - k(Y \wedge \dot{c})(|Y|^2|\dot{c}|^2 - g(Y, \dot{c}))\big)\,dt > 0,
\end{aligned}$$

a contradiction. Thus there are no conjugate points. Thus the surjective (by (23.6)) mapping $\exp_x : T_xM \to M$ is a local diffeomorphism by (27.11). Lemma (27.20) below then finishes the proof. □

27.19. A smooth mapping $f : (M, g) \to (\bar{M}, \bar{g})$ between Riemann manifolds is called *distance increasing* if $f^*\bar{g} \geq g$; in detail, $\bar{g}_{f(x)}(T_xf.X, T_xf.X) \geq g_x(X, X)$ for all $X \in T_xM$, all $x \in M$.

Lemma. *Let (M, g) be a connected complete Riemann manifold. If a smooth mapping $f : (M, g) \to (\bar{M}, \bar{g})$ is surjective and distance increasing, then f is a covering mapping.*

Proof. Obviously, f is locally injective; thus $T_x f$ is injective for all x and $\dim(M) \le \dim(\bar{M})$. Since f is surjective, $\dim(M) \ge \dim(\bar{M})$ by the theorem of Sard (1.18).

For each curve $c : [0,1] \to M$ we have

$$L_g(c) = \int_0^1 |c'|_g \, dt \le \int_0^1 |c'|_{f^*\bar{g}} \, dt = L_{f^*\bar{g}}(c);$$

thus $\operatorname{dist}_g(x,y) \le \operatorname{dist}_{f^*\bar{g}}(x,y)$ for $x, y \in M$. So $(M, \operatorname{dist}_{f^*\bar{g}})$ is a complete metric space and $(M, f^*\bar{g})$ is a complete Riemann manifold also. Without loss we may thus assume that $g = f^*\bar{g}$, so that f is a local isometry. Then $(\bar{M} = f(M), \bar{g})$ is also complete.

For fixed $\bar{x} \in \bar{M}$ let $r > 0$ such that $\exp_{\bar{x}} : B_{0_{\bar{x}}}(2r) \to B_{\bar{x}}(2r) \subset \bar{M}$ is a diffeomorphism. Let $f^{-1}(\bar{x}) = \{x_1, x_2, \dots\}$. For each i the following diagram commutes:

$$\begin{array}{ccccccc}
T_{x_i}M & \hookleftarrow & B_{0_{x_i}}(2r) & \xrightarrow{\exp_{x_i}} & B_{x_i}(2r) & \hookrightarrow & M \\
\downarrow{\scriptstyle T_{x_i}f} & & \downarrow{\scriptstyle T_{x_i}f} & & \downarrow{\scriptstyle f} & & \downarrow{\scriptstyle f} \\
T_{\bar{x}}\bar{M} & \hookleftarrow & B_{0_{\bar{x}}}(2r) & \xrightarrow{\exp_{\bar{x}}} & B_{\bar{x}}(2r) & \hookrightarrow & \bar{M}.
\end{array}$$

We claim (which finishes the proof):

(1) $f : B_{x_i}(2r) \to B_{\bar{x}}(2r)$ is a diffeomorphism for each i,

(2) $f^{-1}(B_{\bar{x}}(r)) = \bigcup_i B_{x_i}(r)$,

(3) $B_{x_i}(r) \cup B_{x_j}(r) = \emptyset$ for $i \ne j$.

(1) From the diagram we conclude that $\exp_{x_i}$ is injective and f is surjective. Since $\exp_{x_i} : B_{0_{x_i}}(r) \to B_{x_i}(r)$ is also surjective (by completeness), $f : B_{x_i}(r) \to B_{\bar{x}}(r)$ is injective too and thus a diffeomorphism.

(2) From the diagram (with $2r$ replaced by r) we see that $f^{-1}(B_{\bar{x}}(r)) \supseteq B_{x_i}(r)$ for all i. If conversely $y \in f^{-1}(B_{\bar{x}}(r))$, let $\bar{c} : [0,s] \to B_{\bar{x}}(r)$ be the minimal geodesic from $f(y)$ to $\bar{x}$ in $\bar{M}$ where $s = \operatorname{dist}_{\bar{g}}(f(y), \bar{x})$. Let c be the geodesic in M which starts at y and satisfies $T_y f.c'(0) = \bar{c}'(0)$. Since f is an infinitesimal isometry, $f \circ c = \bar{c}$ and thus $f(c(s)) = \bar{x}$. So $c(s) = x_i$ for some i. Since $\operatorname{dist}_g(y, x_i) \le s < r$, we have $y \in B_{0_{x_i}}(r)$. Thus $f^{-1}(B_{\bar{x}}(r)) \subseteq \bigcup_i B_{x_i}(r)$.

(3) If $y \in B_{x_i}(r) \cup B_{x_j}(r)$, then $x_j \in B_{x_i}(2r)$ and by (1) we get $x_j = x_i$. □

27.20. Lemma ([106]). *If M is a connected complete Riemann manifold without conjugate points, then $\exp_x : T_x M \to M$ is a covering mapping.*

Proof. Since (M, g) is complete and connected, $\exp_x : T_x M \to M$ is surjective; and it is also a local diffeomorphism by (27.11) since M has no conjugate points. We will construct a complete Riemann metric $\tilde g$ on T_xM such that $\exp_x : (T_xM, \tilde g) \to (M, g)$ is distance increasing. By (27.19) this finishes the proof.

Define the continuous function $h : T_xM \to \mathbb{R}_{>0}$ by

$$\begin{aligned} h(X) &= \sup\{r : |T_X(\exp_x).\xi|^2_{g_{\exp_x(X)}} \ge r|\xi|^2_{g_x} \text{ for all } \xi \in T_xM\} \\ &= \min\{|T_X(\exp_x).\xi|^2_{g_{\exp_x(X)}} : |\xi|_{g_x} = 1\} \\ &= 1\Big/\sqrt{\text{operator norm}(T_X(\exp_x)^{-1} : T_{\exp_x(X)}M \to T_xM)}. \end{aligned}$$

We use polar coordinates $\varphi : \mathbb{R}_{>0} \times S^{m-1} \to T_xM \setminus \{0_x\}$ given by $\varphi(r, \vartheta) = r.\vartheta$ and express the metric by $\varphi^*(g_x) = dr^2 + r^2 g^S$ where g^S is the metric on the sphere. Now we choose an even smooth function $f : \mathbb{R} \to \mathbb{R}$ which satisfies $0 < f(r(X)) \le h(X)$. Consider the Riemann metric $\tilde g = dr^2 + r^2 f(r)$ on T_xM.

For every $R > 0$ we have

$$\overline{B}^{\tilde g}_{0_x}(R) = \{X \in T_xM : \operatorname{dist}_{\tilde g}(X, 0_x) \le R\} \subseteq \{X \in T_xM : r(X) \le R\}$$

which is compact; thus $(T_xM, \tilde g)$ is complete.

It remains to check that $\exp_x : (T_xM, \tilde g) \to (M, g)$ is distance increasing. Let $\xi \in T_X(T_xM)$. If $X = 0_x$, then $T_{0_x}(\exp_x).\xi = \xi$, so $\exp_x$ is distance increasing at 0_x since $f(0) \le 1$.

So let $X \ne 0_x$. Then $\xi = \xi_1 + \xi_2$ where $dr(\xi_2) = 0$; thus ξ_2 tangent to the sphere through X, and $\xi_1 \perp \xi_2$ (with respect to both g_x and $\tilde g_X$). Then

$$|\xi|^2_{g_x} = |\xi_1|^2_{g_x} + |\xi_2|^2_{g_x}, \quad |\xi|^2_{\tilde g} = |\xi_1|^2_{\tilde g} + |\xi_2|^2_{\tilde g}, \quad |\xi|_{g_x} = |\xi|_{\tilde g} = |dr(\xi_1)| = |dr(\xi)|.$$

By variant (27.15.3) of Gauß's lemma the vector $T_X(\exp_x).\xi_1 \in T_{\exp_x(X)}M$ is tangent to the geodesic $t \mapsto \exp_x(t.X)$ in (M, g) and $T_X(\exp_x).\xi_2$ is normal to it. Thus $|T_X(\exp_x).\xi_1|_g = |\xi_1|_g = |\xi_1|_{\tilde g}$ and

$$|T_X(\exp_x).\xi|^2_g = |T_X(\exp_x).\xi_1|^2_g + |T_X(\exp_x).\xi_2|^2_g = |\xi_1|_{\tilde g} + |T_X(\exp_x).\xi_2|^2_g,$$
$$|T_X(\exp_x).\xi|^2_g - |\xi|^2_{\tilde g} = |T_X(\exp_x).\xi_2|^2_g - |\xi_2|^2_{\tilde g}.$$

In order to show that $|T_X(\exp_x).\xi|_g \ge |\xi|_{\tilde g}$, we can thus assume that $\xi = \xi_2$ is normal to the ray $t \mapsto t.X$. But for these ξ we have $|\xi|^2_{\tilde g} = f(r(X))|\xi|^2_{g_x}$ by construction of $\tilde g$ and

$$|T_X(\exp_x).\xi|^2_g \ge h(X)\,|\xi|^2_{g_x} \ge f(r(X))\,|\xi|^2_{g_x} = |\xi|^2_{\tilde g}.$$

So $\exp_x : (T_xM, \tilde g) \to (M, g)$ is distance increasing. □

CHAPTER VI. Isometric Group Actions or Riemann G-Manifolds

In this chapter, a Riemann or pseudo-Riemann metric will usually be called γ since g is usually a group element.

28. Isometries, Homogeneous Manifolds, and Symmetric Spaces

28.1. The group of isometries. Let (M, γ) be a connected pseudo-Riemann manifold. Recall that a diffeomorphism $\varphi : M \to M$ is an *isometry* if

$$\varphi^* \gamma = \gamma.$$

A vector field $\xi \in \mathfrak{X}(M)$ is called a *Killing vector field* if its flow Fl^ξ_t consists of local isometries. By (8.16.3) this is the case if and only if the Lie derivative satisfies $\mathcal{L}_\xi \gamma = 0$. By (8.20) the space of all Killing vector fields is a Lie algebra.

Theorem. *The space $\mathfrak{X}(M, \gamma) = \{\xi \in \mathfrak{X}(M) : \mathcal{L}_\xi \gamma = 0\}$ of all Killing vector fields on a pseudo-Riemann manifold (M, γ) is a finite-dimensional Lie algebra of dimension at most $m^2 + m$ where $m = \dim(M)$.*

The subspace of all complete Killing fields ξ (see (3.8)) is also a finite-dimensional Lie algebra.

The group $\operatorname{Isom}(M,\gamma)$ *of all isometries of* (M,γ) *is a Lie group with Lie algebra the algebra of all complete Killing fields. It acts smoothly on* M. *The Lie group topology equals the pointwise open topology and also the compact open toplogy in the Riemann case.*

Proof. Any Killing vector field ξ is a Jacobi field along each geodesic since its flow Fl^ξ_t deforms geodesics through geodesics; see (27.1). Let us fix a point $x_0 \in M$. By (27.5.1) the field ξ is then uniquely determined by its value $\xi(x_0)$ and by its covariant derivative $T_{x_0}M \ni X \mapsto \nabla_X\xi \in T_{x_0}M$: First we know ξ on $\exp^\gamma_{x_0}(U_{x_0})$ for a 0-neighborhood U_{x_0}, and each point $x \in M$ can be connected to x_0 by a curve consisting of geodesic arcs. Thus the dimension of the Lie algebra $\mathfrak{X}(M,\gamma) = \{\xi \in \mathfrak{X}(M) : \mathcal{L}_\xi\gamma = 0\}$ of all Killing vector fields is at most $m^2 + m$ if $\dim(M) = m$.

We now use theorem (6.5): Let G be a simply connected Lie group with Lie algebra $\mathfrak{X}(M,\gamma)$. Then there is a smooth local action $G \times M \supseteq U \xrightarrow{\ell} M$.

Let H be the subgroup of G consisting of all $g \in G$ with ℓ_g defined on the whole of M. Let $\mathfrak{g} \subset \mathfrak{X}(M,\gamma)$ be the vector space of all ξ such that $\exp(t\xi) \in H$ for all t. By (4.30.3) the space $\mathfrak{g}$ is a Lie subalgebra of $\mathfrak{X}(M,\gamma)$, and $\mathfrak{g}$ consists of complete vector fields. By theorem (6.5) again, there is a Lie group $\operatorname{Isom}_0(M,\gamma)$ consisting of diffeomorphisms of M which implements the infinitesimal action of the Lie algebra $\mathfrak{g}$ on M. This is the connected component of the isometry group $\operatorname{Isom}(M,\gamma)$. Therefore the isometry group is a Lie group.

Obviously, the pointwise open topology on $\operatorname{Isom}(M,\gamma)$ equals the topology as a Lie group. In the Riemann case, $\operatorname{Isom}(M,\gamma)$ consists of equicontinuous mappings (namely isometries) for the Riemann distance function on M; thus the pointwise open topology equals the compact open topology. □

28.2. Invariant covariant derivatives on homogeneous spaces. Let G be a Lie group and H a closed subgroup with Lie algebras $\mathfrak{g}$ and $\mathfrak{h}$, respectively. Let $\ell : G \times G/H$ be the left action of G on the homogeneous space G/H with notation $\ell_g(x) = \ell^x(g) = \ell(g,x) = g.x$ for $g \in G, x \in G/H$. Let $\zeta : \mathfrak{g} \to \mathfrak{X}(G/H)$ be the corresponding infinitesimal action, called the fundamental vector field mapping. It is a Lie algebra antihomomorphism. Let $p : G \to G/H$ be the projection, $p(e) = o \in G/H$. Then $T_ep : \mathfrak{g} \to T_o(G/H)$ factors to a linear isomorphism $\overline{p} : \mathfrak{g}/\mathfrak{h} \to T_o(G/H)$ which is equivariant under $\underline{\mathrm{Ad}} : H \to \operatorname{Aut}(\mathfrak{g}/\mathfrak{h})$ (induced from the adjoint action) and $h \mapsto T_o\ell_h \in GL(T_o(G/H))$. We shall also use $T_e(\underline{\mathrm{Ad}}) =: \underline{\mathrm{ad}} : \mathfrak{h} \to L(\mathfrak{g}/\mathfrak{h}, \mathfrak{g}/\mathfrak{h})$.

Let ∇ be a G-invariant linear connection for $T(G/H)$, and let $\xi, \eta \in \mathfrak{X}(G/H)$. Recall its properties: $\nabla : \mathfrak{X}(G/H) \times \mathfrak{X}(G/H) \to \mathfrak{X}(G/H)$ is $\mathbb{R}$-bilinear,

$\nabla_{f\xi}\eta = f.\nabla_\xi\eta$, and $\nabla_\xi(f.\eta) = df(\xi).\eta + f.\nabla_\xi\eta$. The Nomizu operator $N_\xi : \mathfrak{X}(G/H) \to \mathfrak{X}(G/H)$ is defined by

$$N_\xi\eta := \nabla_\xi\eta - [\xi,\eta]. \tag{1}$$

Then $N_\xi(f.\eta) = f.N_\xi\eta$, so N_ξ is tensorial, $N_\xi : TG/H \to TG/H$. Moreover, $N_{f.\xi}\eta = f.N_\xi\eta + df(\eta).\xi$. Now G-invariance for ∇ means $\ell_{g^{-1}}^*(\nabla_{\ell_g^*\xi}\ell_g^*\eta) = \nabla_\xi\eta$ so that $\nabla_{\ell_g^*\xi}\ell_g^*\eta = \ell_g^*\nabla_\xi\eta$ implies

$$N_{\ell_g^*\xi}\ell_g^*\eta = \nabla_{\ell_g^*\xi}\ell_g^*\eta - [\ell_g^*\xi,\ell_g^*\eta] = \ell_g^*\nabla_\xi\eta - \ell_g^*[\xi,\eta] = \ell_g^*N_\xi\eta.$$

Let us apply this for a fundamental vector field $\xi = \zeta_X$:

$$T(\ell_{g^{-1}})\circ(N_{\zeta_X}\eta)\circ\ell_g = \ell_g^*(N_{\zeta_X}\eta) = N_{\ell_g^*\zeta_X}(\ell_g^*\eta) = N_{\zeta_{\mathrm{Ad}(g^{-1})X}}(T(\ell_{g^{-1}})\circ\eta\circ\ell_g).$$

We evaluate this at the origin $o \in G/H$:

$$T(\ell_{g^{-1}}).(N_{\zeta_X}\eta)|_{gH} = N_{\zeta_{\mathrm{Ad}(g^{-1})X}}.(T_{gH}(\ell_{g^{-1}}).\eta|_{gH}),$$

$$N_{\zeta_X}|_{gH} = T_o(\ell_g).N_{\zeta_{\mathrm{Ad}(g^{-1})X}}|_o.T_{gH}(\ell_{g^{-1}}) : T_{gH}G/H \to T_{gH}G/H. \tag{2}$$

Let us now define

$$\Phi : \mathfrak{g} \to L(\mathfrak{g}/\mathfrak{h},\mathfrak{g}/\mathfrak{h}) \quad \text{by} \quad \Phi_X(Y+\mathfrak{h}) = \bar{p}^{-1}.N_{\zeta_X}|_o.\bar{p}.(Y+\mathfrak{h}). \tag{3}$$

We have $\zeta_Z(gH) = T_o(\ell_g).\zeta_{\mathrm{Ad}(g^{-1})Z}|_o$, and for $h \in H$ we have $T_o(\ell_h).\bar{p} = \underline{\mathrm{Ad}}(h).\bar{p}$. Using (2) we obtain:

$$\begin{aligned}\Phi_X &= \bar{p}^{-1}.N_{\zeta_X}|_o.\bar{p} = \bar{p}^{-1}.T_o(\ell_h).N_{\zeta_{\mathrm{Ad}(h^{-1})X}}|_o.T_o(\ell_{h^{-1}}).\bar{p}\\ &= \underline{\mathrm{Ad}}(h).\bar{p}^{-1}.N_{\zeta_{\mathrm{Ad}(h^{-1})X}}|_o.\bar{p}.\underline{\mathrm{Ad}}(h^{-1}) = \underline{\mathrm{Ad}}(h).\Phi_{\mathrm{Ad}(h^{-1})X}.\underline{\mathrm{Ad}}(h^{-1}).\end{aligned}$$

$$\boxed{\Phi_{\mathrm{Ad}(h)X} = \underline{\mathrm{Ad}}(h).\Phi_X.\underline{\mathrm{Ad}}(h^{-1}), \quad h \in H, X \in \mathfrak{g}.} \tag{4}$$

If $X \in \mathfrak{h}$, then $\zeta_X(o) = 0$ so $(N_{\zeta_X}\eta)|_o = \nabla_{\zeta_X|_o}\eta - [\zeta_X,\eta]|_o = 0 - [\zeta_X,\eta]|_o$ which depends only on $\eta(o)$ since $\zeta_X(o) = 0$ so $[\zeta_X, f.\eta]|_o = f(o).[\zeta_X,\eta]|_o + df(\zeta_X|_o).\eta(o)$. Thus for $X \in \mathfrak{h}$:

$$\begin{aligned}\Phi_X(Y+\mathfrak{h}) &= \bar{p}^{-1}.N_{\zeta_X}|_o.\bar{p}.(Y+\mathfrak{h}) = \bar{p}^{-1}.N_{\zeta_X}|_o.T_ep.Y = \bar{p}^{-1}.N_{\zeta_X}|_o(\zeta_Y|_o)\\ &= -\bar{p}^{-1}.[\zeta_X,\zeta_Y]|_o = \bar{p}^{-1}.\zeta_{[X,Y]}|_o = \underline{\mathrm{ad}}(X)(Y+\mathfrak{h}).\end{aligned} \tag{5}$$

Theorem. *G-invariant linear connections ∇ on a homogeneous space G/H correspond bijectively to H-homomorphisms $\Phi : \mathfrak{g} \to L(\mathfrak{g}/\mathfrak{h},\mathfrak{g}/\mathfrak{h})$ such that $\Phi_X = \underline{\mathrm{ad}}(X)$ for $X \in \mathfrak{h}$. If one such homomorphism exists, then the space of all G-invariant linear connections on G/H is an affine space modeled on* $\mathrm{Hom}_H(\otimes^2(\mathfrak{g}/\mathfrak{h}),\mathfrak{g}/\mathfrak{h})$.

The torsion of ∇ corresponds to the linear mapping $\bigwedge^2 \mathfrak{g}/\mathfrak{h} \to \mathfrak{g}/\mathfrak{h}$ which is induced by

$$(X,Y) \mapsto \Phi_X(Y+\mathfrak{h}) - \Phi_Y(X+\mathfrak{h}) - ([X,Y]+\mathfrak{h}).$$

The curvature of ∇ *corresponds to the mapping* $\bigwedge^2 \mathfrak{g}/\mathfrak{h} \to L(\mathfrak{g}/\mathfrak{h}, \mathfrak{g}/\mathfrak{h})$ *which is induced by*

$$(X, Y) \mapsto \Phi_X \circ \Phi_Y - \Phi_Y \circ \Phi_X - \Phi_{[X,Y]}.$$

Proof. Just unravel all computations from before backwards. Note that

$$\begin{aligned}
(\nabla_{\zeta_X}\zeta_Y)|_{gH} &= (N_{\zeta_X}\zeta_Y)|_{gH} + [\zeta_X, \zeta_Y]|_{gH} \\
&= T_o(\ell_g).N_{\zeta_{\mathrm{Ad}(g^{-1})X}}|_o.T_{gH}(\ell_{g^{-1}}).\zeta_Y|_{gH} - \zeta_{[X,Y]}|_{gH} \\
&= T_o(\ell_g).N_{\zeta_{\mathrm{Ad}(g^{-1})X}}|_o.\zeta_{\mathrm{Ad}\, g^{-1}Y}|_o - T_o(\ell_g).T_{gH}(\ell_{g^{-1}}).\zeta_{[X,Y]}|_{gH} \\
&= T_o(\ell_g).\overline{p}.\Phi_{\mathrm{Ad}(g^{-1})X}(\mathrm{Ad}\, g^{-1}Y + \mathfrak{h}) - T_o(\ell_g).\zeta_{\mathrm{Ad}(g^{-1})[X,Y]}|_o \\
&= T_o(\ell_g).\overline{p}.\big(\Phi_{\mathrm{Ad}(g^{-1})X}(\mathrm{Ad}(g^{-1})Y + \mathfrak{h}) \qquad (6)\\
&\qquad - ([\mathrm{Ad}(g^{-1})X, \mathrm{Ad}(g^{-1})Y] + \mathfrak{h})\big). \quad \square
\end{aligned}$$

28.3. Invariant pseudo-Riemann metrics on homogenous spaces. Let G be a Lie group and let H be a closed subgroup. A *G-invariant pseudo-Riemann metric* γ on G/H (if it exists) is uniquely determined by the H-invariant nondegenerate bilinear form γ_o on T_oG/H, and this in turn is determined by the $\underline{\mathrm{ad}}(H)$-invariant bilinear form $B = \overline{p}^*\gamma_o$ on $\mathfrak{g}/\mathfrak{h}$, if it exists. Suppose that such a nondegenerate bilinear form exists. Then

$$\gamma_{gH}(\xi, \eta) = (\ell^*_{g^{-1}}\gamma)_{gH}(\xi, \eta) = \gamma_o(T(\ell_{g^{-1}}).\xi, T(\ell_{g^{-1}}).\eta).$$

For fundamental vector fields we get:

$$\begin{aligned}
&\gamma_{gH}(\zeta_X(gH), \zeta_Y(gH)) = \gamma_o(T(\ell_{g^{-1}}).\zeta_X(gH), T(\ell_{g^{-1}}).\zeta_Y(gH)) \\
&\quad = B(\mathrm{Ad}(g^{-1})X + \mathfrak{h}, \mathrm{Ad}(g^{-1})Y + \mathfrak{h}), \\
&d(\mathrm{Ad}\circ\nu)|_g(T_e(\mu_g)X) = \mathrm{Ad}(\nu(g)).(\mathrm{ad}\circ\kappa^l)(T_g\nu.T_e(\mu_g).X) \\
&\quad = \mathrm{Ad}(g^{-1}).\,\mathrm{ad}(\mathrm{Ad}(g).X), \\
&dB((\mathrm{Ad}\circ\nu).X + \mathfrak{h}, (\mathrm{Ad}\circ\nu).X + \mathfrak{h})|_g(T_e(\mu_g).Z) \\
&\quad = -2B(\mathrm{Ad}(g^{-1}).\,\mathrm{ad}(\mathrm{Ad}(g).Z).X + \mathfrak{h}, \mathrm{Ad}(g^{-1})X + \mathfrak{h}) \\
&\quad = -2B(\mathrm{Ad}(g^{-1}).[\mathrm{Ad}(g).Z, X] + \mathfrak{h}, \mathrm{Ad}(g^{-1})X + \mathfrak{h}) \\
&\quad = -2B([Z, \mathrm{Ad}(g^{-1}).X] + \mathfrak{h}, \mathrm{Ad}(g^{-1})X + \mathfrak{h}), \\
&d\gamma(\zeta_X, \zeta_X)(\zeta_Z(gH)) = d\gamma(\zeta_X, \zeta_X)|_{gH}(T_o(\ell_g).\zeta_{\mathrm{Ad}(g^{-1})Z}(o)) \\
&\quad = d\gamma(\zeta_X, \zeta_X)|_{gH}(Tp.T_e(\mu_g).\,\mathrm{Ad}(g^{-1}).Z) \\
&\quad = dB((\mathrm{Ad}\circ\nu).X + \mathfrak{h}, (\mathrm{Ad}\circ\nu).X + \mathfrak{h})|_g(T_e(\mu_g).\,\mathrm{Ad}(g^{-1})Z) \\
&\quad = -2B([\mathrm{Ad}(g^{-1})Z, \mathrm{Ad}(g^{-1}).X] + \mathfrak{h}, \mathrm{Ad}(g^{-1})X + \mathfrak{h}) \\
&\quad = -B(\mathrm{Ad}(g^{-1})[Z, X] + \mathfrak{h}, \mathrm{Ad}(g^{-1})X + \mathfrak{h}).
\end{aligned}$$

On the other hand we have for a G-invariant linear connection on G/H corresponding to $\Phi : \mathfrak{g} \to L(\mathfrak{g}/\mathfrak{h}, \mathfrak{g}/\mathfrak{h})$ as in (28.2.3):

$$\begin{aligned}2\gamma(\nabla_{\zeta_Z}\zeta_X, \zeta_X)|_{gH} = 2B(&\Phi_{\operatorname{Ad}(g^{-1})Z}(\operatorname{Ad} g^{-1}X + \mathfrak{h}) \\ &- ([\operatorname{Ad}(g^{-1})Z, \operatorname{Ad}(g^{-1})X] + \mathfrak{h}), \operatorname{Ad}(g^{-1})X + \mathfrak{h}).\end{aligned}$$

Collecting, we get

$$\begin{aligned}d\gamma(\zeta_X, \zeta_X)(\zeta_Z(gH)) &- 2\gamma(\nabla_{\zeta_Z}\zeta_X, \zeta_X)|_{gH} \\ &= 2B\big(\Phi_{\operatorname{Ad}(g^{-1})Z}(\operatorname{Ad} g^{-1}X + \mathfrak{h}), \operatorname{Ad}(g^{-1})X + \mathfrak{h}\big)\end{aligned}$$

so that the connection

$$(1)\quad \nabla \text{ respects } \gamma \iff B(\Phi_Z(X + \mathfrak{h}), Y + \mathfrak{h}) + B(X + \mathfrak{h}, \Phi_Z(Y + \mathfrak{h})) = 0.$$

The Levi-Civita connection ∇ is uniquely determined by the pseudo-metric γ. We now derive a formula for Φ corresponding to the connection ∇ directly from B. ∇ is torsion-free iff

$$0 = \Phi_X(Y + \mathfrak{h}) - \Phi_Y(X + \mathfrak{h}) - ([X, Y] + \mathfrak{h}) \text{ for all } X, Y \in \mathfrak{g}.$$

Consider the symmetric bilinear form $\tilde{B} : \mathfrak{g} \times \mathfrak{g} \to \mathfrak{g}/\mathfrak{h} \times \mathfrak{g}/\mathfrak{h} \xrightarrow{B} \mathbb{R}$ on $\mathfrak{g}$. Then torsion-freeness corresponds to

$$\tilde{B}([X, Y], Z) = B(\Phi_X(Y + \mathfrak{h}), Z + \mathfrak{h}) - B(\Phi_Y(X + \mathfrak{h}), Z + \mathfrak{h}).$$

We permute this cyclically:

$$\begin{aligned}+\tilde{B}([X, Y], Z) &= +B(\Phi_X(Y + \mathfrak{h}), Z + \mathfrak{h}) - B(\Phi_Y(X + \mathfrak{h}), Z + \mathfrak{h}),\\ +\tilde{B}([Y, Z], X) &= +B(\Phi_Y(Z + \mathfrak{h}), X + \mathfrak{h}) - B(\Phi_Z(Y + \mathfrak{h}), X + \mathfrak{h}),\\ -\tilde{B}([Z, X], Y) &= -B(\Phi_Z(X + \mathfrak{h}), Y + \mathfrak{h}) + B(\Phi_X(Z + \mathfrak{h}), Y + \mathfrak{h}).\end{aligned}$$

We add, using (1):

$$(2)\quad \tilde{B}([X, Y], Z) + \tilde{B}([Y, Z], X) - \tilde{B}([Z, X], Y) = 2B(\Phi_Y(Z + \mathfrak{h}), X + \mathfrak{h}).$$

It remains to check that the trilinear expression

$$(X, Y, Z) \mapsto \tilde{B}([X, Y], Z) + \tilde{B}([Y, Z], X) - \tilde{B}([Z, X], Y)$$

factors to $\mathfrak{g}/\mathfrak{h} \times \mathfrak{g}/\mathfrak{h} \times \mathfrak{g}/\mathfrak{h} \to \mathbb{R}$. If one of X, Y, Z is in $\mathfrak{h}$ (say, X), then the second term vanishes and the first term cancels with the third one since B is $\underline{\operatorname{Ad}}(\mathfrak{h})$-invariant. So (2) defines a mapping Φ which in turn gives rise to the Levi-Civita connection ∇.

28.4. Pseudo-Riemann locally symmetric spaces. Let (M, γ) be a connected pseudo-Riemann manifold. For $x \in M$ let U_x be an open neighborhood of x in M and let B_{0_x} be an open absolutely convex 0-neighborhood in T_xM such that $\exp_x^\gamma : B_{0_x} \to U_x$ is a diffeomorphism. We consider the exponential mapping

$$T_xM \supset B_{0_x} \xrightarrow{\exp_x^\gamma} U_x \subset M$$

and the local *geodesic symmetry*

$$s_x : U_x \to U_x, \quad s_x(\exp_x^\gamma(X)) = \exp_x^\gamma(-X).$$

Note that $T_x(s_x) = -\operatorname{Id}$ on T_xM. The pseudo-Riemann manifold (M, γ) is called a *pseudo-Riemann locally symmetric space* if each local symmetry s_x is a local isometry, i.e., $s_x^*\gamma = \gamma$.

Proposition. *A pseudo-Riemann manifold (M, γ) is locally symmetric if and only if its curvature is parallel: $\nabla R^\nabla = 0$.*

Proof. If (M, γ) is locally symmetric, we have $s_x^*(\nabla R) = (s^*\nabla)(s^*R) = \nabla R$, but $(s_x^*(\nabla R))_x = (-1)^5(\nabla R)_x$ so that $(\nabla R)_x = 0$ for all $x \in M$. If conversely $\nabla R = 0$, then R is constant under parallel transport. Thus by theorem (27.10) each local symmetry is an isometry. □

28.5. Symmetric spaces. A connected (pseudo-)Riemann manifold (M, γ) is called a *(pseudo-)Riemann symmetric space* if for each $x \in M$ the local symmetry extends to a globally defined isometry $s_x : M \to M$. Let us choose a point $o \in M$ which we call the origin.

(1) *An isometry f on (M, γ) which is involutive ($f^2 = \operatorname{Id}$) and has x as isolated fixed point equals s_x*, by considering the linear involution T_xf: Among the possible eigenvalues ± 1 only -1 is admissible since x is a locally isolated fixed point.

(2) *(M, γ) is a geodesically complete Riemann manifold.*

Namely, let $c : (a, b) \to M$ be a geodesic. Then $s_{c(t)}$ maps the geodesic to itself (suitably reparameterized) and thus prolongs c (if t is not the midpoint). So any geodesic is extendable to $\mathbb{R}$ and by (23.6), (M, γ) is a complete Riemann manifold.

(3) *The group* $\operatorname{Isom}(M, \gamma)$ *of all isometries of M acts transitively on M.* In the Riemann case, by (23.6.6) for any point $x \in M$ there exists a geodesic $c : [0, 1] \to M$ with $c(0) = o$ and $c(1) = x$. But then $s_{c(1/2)}(o) = x$. So every point of M lies in the orbit through o. In the pseudo-Riemann case where the Hopf-Rinov theorem (23.6) does not hold, we can choose a piecewise smooth curve c from o to x which consists of geodesic segments. Then we

can apply the reflections s_y with respect to the midpoint of each geodesic segment, iteratively, to map o to x.

(4) *The group* $G := \operatorname{Isom}(M,\gamma)$ *of isometries of* M *is a Lie group and the action* $\ell : G \times M \to M$ *is smooth.* This is an immediate consequence of theorem (28.1).

(5) *The mapping* $s : M \times M \to M$ *given by* $(x,y) \mapsto s_x(y)$ *is smooth.* This is obvious.

(6) Let $\sigma : G \to G$ be given by $\sigma(\varphi) = s_o \circ \varphi \circ s_o$. Let $H = G_o$ be the isotropy group of the origin o. *Then* σ *is an involutive automorphism of* G *and we have* $G^\sigma = \{g \in G : \sigma(g) = g\} \supseteq H \subseteq G^\sigma_0$. Namely, for $g \in H$ we have

$$\begin{aligned}(s_o \circ g \circ s_o)(\exp^\gamma_o(t.X)) &= (s_o \circ g)(\exp^\gamma_o(-t.X)) = s_o(\exp^\gamma_o(-t.T_og.X))\\ &= \exp^\gamma_o(t.T_og.X) = g(\exp^\gamma_o(t.X))\end{aligned}$$

so that $\sigma(g) = g$ near o and thus everywhere, since g is an isometry. The Lie algebra of G^σ is $\mathfrak{g}^{\sigma'}$, the space of all $\xi \in \mathfrak{g} = \mathfrak{X}(M,\gamma)$ such that $\sigma'(\xi) = \xi$. But if $\xi = \sigma'(\xi) = Ts_o \circ \xi \circ s_o$, then at o we have $\xi(o) = T_o(s_o).\xi(o) = -\xi(o)$ so that ξ vanishes at o. But then $\exp^G(t\xi) = \operatorname{Fl}^\xi_t$ has o as fixed point.

(7) *For* $x, y \in M$ *we have* $s_x \circ s_y = \ell_g$ *for some* $g \in G$. Choose $g_x, g_y \in G$ with $g_x.o = x$ and $g_y.o = y$. Then

$$\begin{aligned}s_x &= \ell_{g_x} \circ s_o \circ \ell_{g_x^{-1}}, \qquad s_y = \ell_{g_y} \circ s_o \circ \ell_{g_y^{-1}},\\ s_x \circ s_y &= \ell_{g_x} \circ s_o \circ \ell_{g_x^{-1}} \circ \ell_{g_y} \circ s_o \circ \ell_{g_y^{-1}}\\ &= \ell_{g_x} \circ (s_o \circ \ell_{g_x^{-1}.g_y} \circ s_o) \circ \ell_{g_y^{-1}}\\ &= \ell_{g_x} \circ \ell_{\sigma(g_x^{-1}.g_y)} \circ \ell_{g_y^{-1}} = \ell_{g_x.\sigma(g_x^{-1}.g_y).g_y^{-1}}.\end{aligned}$$

(8) Since $\sigma' : \mathfrak{g} \to \mathfrak{g}$ is an involutive automorphism of $\mathfrak{g} = \mathfrak{X}(M,\gamma)$, we can decompose $\mathfrak{g}$ into the ± 1 eigenspaces of σ' and obtain

$$\mathfrak{g} = \mathfrak{X}(M,\gamma) = \mathfrak{h} \oplus \mathfrak{m}, \quad [\mathfrak{h},\mathfrak{h}] \subset \mathfrak{h}, \quad [\mathfrak{h},\mathfrak{m}] \subset \mathfrak{m}, \quad [\mathfrak{m},\mathfrak{m}] \subset \mathfrak{h},$$

which is called a *reductive decomposition* of $\mathfrak{g}$. Note that this decomposition is invariant under $\operatorname{ad}(H)$ acting on $\mathfrak{g}$.

(9) Let $p : G \to M \cong G/H$ be the submersion $p(g) = g.o = \ell(g,o) = \ell^o(g)$. Then $T_ep : \mathfrak{g} \to T_oM$ induces a linear isomorphism $p' := T_ep|_\mathfrak{m} : \mathfrak{m} \to T_oM$ which is equivariant for the action of $H = G_o$ on $\mathfrak{m}$ via ad and on T_oM via $h \mapsto T_o\ell_h$. The bilinear form $B := (p')\gamma_o$ on $\mathfrak{m}$ is nondegenerate and $\operatorname{ad}(H)$-invariant. We identify $\mathfrak{m} \cong \mathfrak{g}/\mathfrak{h} \cong T_oM$ and make use of (28.3). From now on let $X, X_i \in \mathfrak{h}$, $Y, Y_i \in \mathfrak{m}$. If B is an H-invariant inner product on $\mathfrak{m}$, we have $B(\operatorname{ad}_X Y_1, Y_2) + B(Y_1, \operatorname{ad}_X Y_2) = 0$. We extend B to a bilinear

form $\tilde{B}$ on $\mathfrak{g}$ which has $\mathfrak{h}$ as its kernel. Equation (28.3.1) then becomes

$$\begin{aligned}
B(\Phi_X Y_1, Y_2) &= \tilde{B}([Y_2, X], Y_1) + \tilde{B}([X, Y_1], Y_2) - \tilde{B}([Y_1, Y_2], X) \\
&= -2B(\operatorname{ad}_X Y_1, Y_2), \\
B(\Phi_Y Y_1, Y_2) &= \tilde{B}([Y_2, Y], Y_1) + \tilde{B}([Y, Y_1], Y_2) - \tilde{B}([Y_1, Y_2], Y) = 0, \\
\Phi_{X+Y} Y_1 &= -\operatorname{ad}_X Y_1.
\end{aligned}$$

Note that the G-invariant connection on the symmetric space $M = G/H$ is prescribed by Φ uniquely and is independent of the choice of the metric γ. From theorem (28.2) we conclude that the curvature operator is given by

$$R(Y_1, Y_2)Y_3 = \Phi_{Y_1}.\Phi_{Y_2}.Y_3 - \Phi_{Y_2}.\Phi_{Y_1}.Y_3 - \Phi[Y_1, Y_2]Y_3 = [[Y_1, Y_2], Y_3].$$

(10) *Geodesics emanating from o are given by*

$$\exp_o^M(t.p'.Y) = p(\exp^G(tY)), \quad Y \in \mathfrak{m}.$$

The fundamental vector field $\zeta_Y \in \mathfrak{X}(M)$ is

$$\begin{aligned}
\zeta_Y(\ell_g(o)) &= T_o(\ell_g).\zeta_{\operatorname{ad}(g^{-1})Y}(o) \text{ by (6.2.2)} \\
&= T_o(\ell_g).T_e(\ell^o)\operatorname{ad}(g^{-1})Y = T_o(\ell_g).p'.\operatorname{ad}(g^{-1})Y.
\end{aligned}$$

By (28.2.6), for $g = \exp(tY)$ we have

$$\begin{aligned}
\nabla_{\zeta_Y}\zeta_Y|_{\exp(tY).o} &= T_o(\ell_{\exp(tY)}).\bar{p}.\big(\Phi_{\operatorname{Ad}(\exp(-tY))Y}(\operatorname{Ad}(\exp(-tY))Y + \mathfrak{h}) \\
&\qquad - ([\operatorname{Ad}(\exp(-tY))Y, \operatorname{Ad}(\exp(-tY))Y] + \mathfrak{h})\big) \\
&= T_o(\ell_{\exp(tY)}).\bar{p}.\big(\Phi_Y(Y) - ([Y, Y] + \mathfrak{h})\big) = 0.
\end{aligned}$$

So ζ_Y is parallel along the flowline $\operatorname{Fl}_t^{\zeta_Y}(o) = \exp(tY).o$ and thus $\exp(tY).o$ is a geodesic in M.

(11) We consider now a geodesically complete connected submanifold N of M. Without loss we assume that $o \in N$. *If N is totally geodesic, then N is itself a symmetric space* with group of isometries $G' = N_G(N) = \{g \in G : \ell_g(N) \subset N\}$ and isotropy group $G'_o = H \cap G'$. For $x \in N$ the submanifolds N and $s_x(N)$ are both totally geodesic with the same tangent space at x; thus $s_x(N) = N$. For $g \in G'$ we have $\ell_{\sigma(g)} = s_o \circ \ell_g \circ s_o : N \to N$, so $\sigma(g) \in G'$. Finally, G' acts transitively on N: For $x \in N$ choose a piecewise geodesic in N from o to x. For each geodesic piece $c : [0, 1] \to N$ the mapping $s_{c(3/4)} \circ s_{c(1/4)}$ is in G' by (7) and maps $c(0)$ to $c(1)$.

(12) *There is a bijective correspondence between*

- *totally geodesic connected geodesically complete submanfolds N of M containing o and*
- *linear subspaces $\mathfrak{n} \subseteq \mathfrak{m}$ with $[[\mathfrak{n}, \mathfrak{n}], \mathfrak{n}] \subseteq \mathfrak{n}$.*

The correspondence is given by $\mathfrak{n} = (T_e p|_{\mathfrak{m}})^{-1}(T_o N)$. *The submanifold* N *is flat if and only if* $[[\mathfrak{n},\mathfrak{n}],\mathfrak{n}] = 0$. Given N, then by (11) we have $N \cong G'/H'$ where $H' = G \cap H$, and $\sigma \in \operatorname{Aut}(\mathfrak{g})$ respects the Lie subalgebra $\mathfrak{g}' = \operatorname{Lie}(G')$. Thus $\mathfrak{g}' = \mathfrak{h}' \oplus \mathfrak{n} \subseteq \mathfrak{h} \oplus \mathfrak{m}$ are compatible reductive decompositions and thus $[\mathfrak{n},\mathfrak{n}] \subset \mathfrak{h}'$ and $[\mathfrak{h}',\mathfrak{n}] \subset \mathfrak{n}$.

Conversely, given $\mathfrak{n} \subseteq \mathfrak{m}$ with $[[\mathfrak{n},\mathfrak{n}],\mathfrak{n}] \subseteq \mathfrak{n}$, we put $\mathfrak{h}' := [\mathfrak{n},\mathfrak{n}] \subset \mathfrak{h}'$ and $\mathfrak{g}' = \mathfrak{h}' \oplus \mathfrak{n}$. This is a Lie subalgebra of $\mathfrak{g}$. Let G' be the connected Lie subgroup of G with Lie algebra $\mathfrak{h}'$. Then $G'.o = \ell(G', o)$ is a connected geodesically complete submanifold of M which is a symmetric space and thus totally geodesic in M by (10).

This ends our very short treatment of symmetric spaces. From here on it becomes quite algebraic, and there are may good books on this subject; [**82**] is the standard reference. The theory of homogeneous manifolds however is best understood by using Cartan connections; for this see the book [**32**].

29. Riemann G-Manifolds

29.1. Preliminaries. Let (M,γ) be a Riemann G-manifold. If $\varphi : M \to M$ is an isometric diffeomorphism, then

(1) $\varphi(\exp^M_x(tX)) = \exp^M_{\varphi(x)}(tT_x\varphi.X)$. This is due to the fact that isometries map geodesics to geodesics, and the starting vector of the geodesic $t \mapsto \varphi(\exp^M_x(t.X))$ is $T_x\varphi.X$.

(2) If $\varphi(x) = x$, then, in the chart $(U_x, (\exp^M_x)^{-1})$, φ is a linear isometry (where U_x is a neighborhood of x so small that $(\exp^M_x)^{-1} : U_x \to T_xM$ is a diffeomorphism onto a neighborhood of 0 in T_xM):

$$\begin{aligned}\bar\varphi(X) :&= (\exp^M_x)^{-1} \circ \varphi \circ \exp^M_x(X)\\ &= (\exp^M_x)^{-1} \exp^M_x(T_x\varphi.X) = T_x\varphi.X.\end{aligned}$$

(3) $M^\varphi := \{x \in M : \varphi(x) = x\}$ *is a totally geodesic submanifold of* M*:* If we choose $X \in T_x(M^\varphi)$, then, since $T_x\varphi.X = X$ and by (1), we have

$$\varphi(\exp^M_x(tX)) = \exp^M_x(T_x\varphi.tX) = \exp^M_x(tX).$$

So the geodesic through x with starting vector X stays in M^φ.

(4) If H is a set of isometries, then $M^H = \{x \in M : \varphi(x) = x$ for all $\varphi \in H\}$ is also a totally geodesic submanifold in M.

29.2. Definition. Let M be a proper Riemann G-manifold, $x \in M$. The *normal bundle* to the orbit $G.x$ is defined as

$$\operatorname{Nor}(G.x) := T(G.x)^\perp.$$

Let $\operatorname{Nor}_\varepsilon(G.x) = \{X \in \operatorname{Nor}(G.x) : \|X\|_\gamma < \varepsilon\}$, and choose $r > 0$ small enough for $\exp_x : T_xM \supseteq B_r(0_x) \to M$ to be a diffeomorphism onto its image and for $\exp_x(B_r(0_x)) \cap G.x$ to have only one component. Then, since the action of G is isometric, exp defines a diffeomorphism from $\mathrm{Nor}_{r/2}(G.x)$ onto an open neighborhood of $G.x$, so $\exp(\mathrm{Nor}_{r/2}(G.x)) =: U_{r/2}(G.x)$ is a *tubular neighborhood* of $G.x$. We define the *normal slice* at x by

$$S_x := \exp_x(\mathrm{Nor}_{r/2}(G.x)_x).$$

29.3. Lemma. *Under these conditions we have that*
(1) $S_{g.x} = g.S_x$,
(2) S_x *is a slice at* x.

Proof. (1) Since G acts isometrically and by (29.1.1),

$$S_{g.x} = \exp_{g.x}\big(T_x\ell_g(\mathrm{Nor}_{r/2}(G.x))_x\big) = \ell_g \exp_x\big(\mathrm{Nor}_{r/2}(G.x)_x\big) = g.S_x.$$

(2) The mapping $r : G.S_x \to G.x$ given by $\exp_{g.x} X \mapsto g.x$ defines a smooth equivariant retraction (note that S_x and S_y are disjoint if $x \neq y$). □

29.4. Isotropy representation. Let M be a G-manifold and $x \in M$; then the representation of the isotropy group

$$G_x \longrightarrow GL(T_xM), \quad g \mapsto T_x\ell_g,$$

is called the *isotropy representation.* If M is a Riemann G-manifold, then the isotropy representation is orthogonal and $T_x(G.x)$ is an invariant subspace under G_x. So $T_x(G.x)^\perp$ is also invariant, and

$$G_x \longrightarrow (\mathrm{Nor}_x(G.x)), \quad g \mapsto T_x\ell_g,$$

is called the *slice representation.*

29.5. Example. Let $M = G$ be a compact Lie group with a bi-invariant metric. Then $G\times G$ acts on G by $(g_1, g_2).g := g_1 g g_2^{-1}$, making G a Riemann $(G \times G)$-space. The isotropy group of e is $(G \times G)_e = \{(g, g) : g \in G\}$, and the isotropy representation coincides with the adjoint representation of $G \cong (G \times G)_e$ on $\mathfrak{g} = T_e(G)$.

29.6. Example. Let G/K be a semisimple symmetric space (K compact) and let $\mathfrak{g} = \mathfrak{k} + \mathfrak{m}$ be the corresponding reductive decomposition of the Lie algebra $\mathfrak{g}$; see (28.5). Then $T_e(G/K) \cong \mathfrak{g}/\mathfrak{k} \cong \mathfrak{m}$, and the isotropy subgroup of G at e is K. The isotropy representation is $\mathrm{Ad}^\perp_{K,G} : K \to (\mathfrak{m})$. The slices are points since the action is transitive.

29.7. Lemma. *Let M be a proper Riemann G-manifold, $x \in M$. Then the following three statements are equivalent:*
(1) *x is a regular point.*
(2) *The slice representation at x is trivial.*
(3) *$G_y = G_x$ for all $y \in S_x$ for a sufficiently small slice S_x.*

Proof. Clearly, (2) $\iff$ (3). To see (3) $\implies$ (1), let S_x be a small slice at x. Then $U := G.S$ is an open neighborhood of $G.x$ in M, and for all $g.s \in U$ we have $G_{g.s} = gG_sg^{-1} = gG_xg^{-1}$. Therefore $G.x$ is a principal orbit. The converse is true by (6.16.3), since G_x is compact. □

29.8. Definition. Let M be a Riemann G-manifold and $G.x$ some orbit; then a smooth section u of the normal bundle $\operatorname{Nor}(G.x)$ is called an *equivariant normal field* if

$$T_y(\ell_g).u(y) = u(g.y) \qquad \text{for all } y \in G.x, g \in G.$$

29.9. Corollary. *Let M be a proper Riemann G-manifold and x a regular point. If $X \in \operatorname{Nor}_x(G.x)$, then $\hat{X}(g.x) := T_x(\ell_g).X$ is a well defined equivariant normal field along $G.x$ in M.*

Proof. If $g.x = h.x$, then $h^{-1}g \in G_x \Rightarrow T_x(\ell_{h^{-1}g}).X = X$, since the slice representation is trivial by (29.7.2). Now by the chain rule, $T_x(\ell_g).X = T_x(\ell_h).X$. Therefore $\hat{X}$ is a well defined, smooth section of $\operatorname{Nor}(G.x)$. It is equivariant by definition. □

29.10. Corollary. *Let M be a Riemann G-manifold, $G.x$ a principal orbit, and $(u_1, \dots, u_n)$ an orthonormal basis of* $\operatorname{Nor}_x(G.x)$. *By corollary* (29.9), *each u_i defines an equivariant normal field $\hat{u}_i$. So $(\hat{u}_1, \dots, \hat{u}_n)$ is a global equivariant orthonormal frame field for* $\operatorname{Nor}(G.x)$, *and* $\operatorname{Nor}(G.x)$ *is a trivial bundle.* □

This follows also from the tubular neighborhood description $G.S_x \cong G \times_{G_x} S_x$, where S_x is a normal slice at x with trivial G_x-action; see (29.7).

29.11. Orbits as Riemann submanifolds. Let (M, γ) be a Riemann G-manifold and u an equivariant normal field along an orbit $G.x_0$. Considering this orbit as a Riemann submanifold, we recall from (26.1) the second fundamental form $S \in \Gamma(S^2T^*(G.x_0) \otimes \operatorname{Nor}(G.x_0))$ and from (26.3.2) the Weingarten mapping or shape operator $L_u : T(G.x_0) \to T(G.x_0)$ along the normal field, which are related by

$$\gamma|_{TP}(L_{u(x)}(X_x), Y_x) = \gamma(S(X_x, Y_x), u(x)), \quad X_x, Y_x \in T_x(G.x_0), x \in G.x_0.$$

Its eigenvalues are called the main curvatures of $G.x_0$ along u. Since γ and the submanifold $G > x_0$ are G-invariant, the second fundamental form S is G-equivariant. Since u is an equivariant normal form, the shape operator L_u along u is also G-equivariant.

29.12. Lemma. *Let u be an equivariant normal field along an orbit $G.x$; then*

(1) $L_{u(g.x)} = T_x(\ell_g).L_{u(x)}.T_{g.x}(\ell_{g^{-1}})$,

(2) *the main curvatures of $G.x$ along u are all constant,*

(3) $\{\exp^M(u(y)) : y \in G.x\}$ *is another G-orbit.*

Proof. (1) was already proved in (29.11) above. This implies (2) since the eigenvalues are invariant under conjugation.
(3) $\{\exp^M(u(y)) : y \in G.x\} = G.\exp^M(u(x))$, since

$$g.\exp^M(u(x)) = \exp^M(T\ell_g.u(x)) = \exp^M(u(g.x)). \quad \square$$

29.13. Example. Let $N^n(c)$ be the simply connected space form with constant sectional curvature c, that is,

$$N^n(c) = \begin{cases} S^n, \text{ sphere with radius } \frac{1}{c}, & \text{if } c > 0, \\ \mathbb{R}^n, & \text{if } c = 0, \\ H^n, \text{ hyperbolic sphere with radius } \frac{1}{|c|}, & \text{if } c < 0. \end{cases}$$

Let G be a closed subgroup of $\operatorname{Isom}(N^n(c))$. If P is a G-orbit, then so is the subset $\{\exp(u(x)) : x \in P\}$ for any equivariant normal field u along P. For instance:

(1) If $G = SO(n) \subset \operatorname{Isom}(\mathbb{R}^n)$, then the G-orbits are the spheres with center 0. A radial vector field with constant length on each sphere, $u(x) := f(|x|).x$, defines an equivariant normal field on each orbit. Clearly its flow carries orbits to orbits.

(2) Another example is the subgroup of $\operatorname{Isom}(\mathbb{R}^n)$ consisting only of affine translations in directions corresponding to a linear subspace $V \subset \mathbb{R}^n$. Here the orbits of G are then affine planes parallel to V. An equivariant normal field on an orbit is a constant vector field orthogonal to V.

29.14. Theorem. *Let M be a proper G-manifold; then the set of all regular points M_{reg} is open and dense in M. In particular, there is always a principal orbit type.*

Proof. Suppose $x \in M_{\text{reg}}$. By (6.27) there is a slice S at x, and by (6.16.3) S can be chosen small enough for all orbits through S to be principal as well. Therefore $G.S$ is an open neighborhood of x in M_{reg} which itself is open by (6.15.3).

To see that M_{reg} is dense, let $U \subseteq M$ be open, $x \in U$, and S be a slice at x. Now choose a $y \in G.S \cap U$ for which the isotropy group G_y has minimal dimension and the smallest number of connected components for this dimension in all of $G.S \cap U$. Let S_y be a slice at y; then $G.S_y \cap G.S \cap U$ is open, and for any $z \in G.S_y \cap G.S \cap U$ we have $z \in g.S_y = S_{g.y}$, so $G_z \subseteq G_{g.y} = gG_yg^{-1}$. By choice of y, this implies $G_z = gG_yg^{-1}$ for all $z \in G.S_y \cap G.S \cap U$, and $G.y$ is a principal orbit. □

29.15. Theorem. *Let M be a proper G-manifold and $x \in M$. Then there is a G-invariant neighborhood U of x in which only finitely many orbit types occur.*

Proof. By theorem (6.30) there is a G-invariant Riemann metric on M. Let S be the normal slice at x. Then S is again a Riemann manifold, and the compact group G_x acts isometrically on S. In (6.16.4) we saw that if $G_x.s_1$ and $G_x.s_2$ have the same orbit type in S, then $G.s_1$ and $G.s_2$ have the same orbit type in M. So the number of G-orbit types in $G.S$ can be no more than the number of G_x-orbit types in S. Therefore it is sufficient to consider the case where G is a compact Lie group. Let us now prove the assertion under this added assumption. We carry out induction on the dimension of M.

For $n = 0$ there is nothing to prove. Suppose the assertion is proved for $\dim M < n$. Again, it will do to find a slice S at x with only a finite number of G_x-orbit types. If $\dim S < \dim M$, this follows from the induction hypothesis. Now suppose $\dim S = n$. The slice S is equivariantly diffeomorphic (by $\exp^\gamma$) to an open ball in T_xM under the slice representation. Since the slice representation is orthogonal, it restricts to a G_x-action on each sphere $r.S^{n-1}$ in this ball. By the induction hypothesis, locally, S^{n-1} has only finitely many G_x-orbit types. Since S^{n-1} is compact, it has only finitely many orbit types globally. The orbit types are the same on all spheres $r.S^{n-1}$ $(r > 0)$, since $x \mapsto \frac{1}{r}x$ is G-equivariant. Therefore, S has only finitely many orbit types: those of S^{n-1} and the 0-orbit. □

29.16. Theorem. *If M is a proper G-manifold, then the set M_{sing}/G of all singular G-orbits does not locally disconnect the orbit space M/G (that is, to every point in M/G the connected neighborhoods remain connected even after removal of all singular orbits).*

Proof. As in the previous theorem, we will reduce the statement to an assertion about the slice representation. By theorem (6.30), there is a G-invariant Riemann metric on M. Let S be the normal slice at x. Then S is again a Riemann manifold, and the compact group G_x acts isometrically on S. A principal G_x-orbit is the restriction of a principal G-orbit, since $G_x.s$ is principal means that all orbits in a sufficiently small neighborhood of $G_x.s$ have the same orbit type as the orbit $G_x.s$; see (29.7). Therefore, by (6.16.4), the corresponding orbits in $G.U$ are also of the same type, and

$G.s$ is principal as well. So there are 'fewer' singular G-orbits in $G.S$ than there are singular G_x-orbits in S. Now cover M with tubular neighborhoods like $G.S_x$, and recall that $G.S_x/G \cong S_x/G_x$ by (6.16.5). This together with the above argument shows us that it will suffice to prove the statement for the slice action. Furthermore, as in the proof of theorem (29.15), we can restrict our considerations to the slice representation. So we have reduced the statement to the following:

If V is a real, n-dimensional vector space and G a compact Lie group acting on V, then the set V_{sing}/G of all singular G-orbits does not locally disconnect the orbit space V/G (that is, to every point in V/G the connected neighborhoods remain connected even after removal of all singular orbits).

We will prove this by induction on the dimension n of V. For $n = 1$, i.e., $V = \mathbb{R}$, the only nontrivial choice for G is $O(1) \cong \mathbb{Z}_2$. In this case, $\mathbb{R}/G = [0,\infty)$ and $\mathbb{R}_{\text{sing}}/G = \{0\}$. Clearly, $\{0\}$ does not locally disconnect $[0,\infty)$, and we can proceed to the general case.

Suppose the assertion is proved for all dimensions smaller than n. Now for $G \subseteq O(n)$ we consider the induced action on the invariant submanifold S^{n-1}. For any $x \in S^{n-1}$ we can apply the induction hypothesis to the slice representation $G_x \to (\operatorname{Nor}_x(G.x))$. This implies for the G_x-action on S_x that $S_x/G_x \cong G.S_x/G$ is not locally disconnected after removing all its singular points. As above, we can again cover S^{n-1} with tubular neighborhoods like $G.S_x$, and we see that all of S^{n-1}/G is not locally disconnected by its singular orbits. Now we need to verify that the orbit space of the unit ball D^n is not locally disconnected by its singular orbits. Since scalar multiplication is a G-equivariant diffeomorphism, the singular orbits in V (not including $\{0\}$) project radially onto singular orbits in S^{n-1}. So if we view the ball D^n as cone over S^{n-1} and denote the cone construction by $\operatorname{cone}(S^{n-1})$, then $D^n_{\text{sing}} = \operatorname{cone}(S^{n-1}_{\text{sing}})$. Furthermore, we have a homeomorphism

$$\operatorname{cone}(S^{n-1})/G \longrightarrow \operatorname{cone}(S^{n-1}/G), \quad G.[x,t] \mapsto [G.x,t],$$

since G preserves the 'radius' t. Therefore

$$\begin{aligned} D^n/G &= (\operatorname{cone}(S^{n-1}))/G \cong \operatorname{cone}(S^{n-1}/G), \\ D^n_{\text{sing}}/G &= \operatorname{cone}(S^{n-1}_{\text{sing}})/G \cong \operatorname{cone}(S^{n-1}_{\text{sing}}/G). \end{aligned}$$

Since S^{n-1}_{sing}/G does not locally disconnect S^{n-1}/G, we also see that

$$\operatorname{cone}(S^{n-1}_{\text{sing}}/G) \cong D^n_{\text{sing}}/G$$

does not locally disconnect $\operatorname{cone}(S^{n-1}/G) \cong D^n/G$. □

29.17. Corollary. *Let M be a connected proper G-manifold; then:*

(1) *M/G is connected.*

(2) *M has precisely one principal orbit type.*

Proof. (1) Since M is connected and the quotient map $\pi : M \to M/G$ is continuous, its image M/G is connected as well.

(2) By the theorem (29.16) we have that $M/G \setminus M_{\text{sing}}/G = M_{\text{reg}}/G$ is connected. On the other hand by (29.7), the orbits of a certain principal orbit type form an open subset of M/G, in particular of M_{reg}/G. Therefore if there were more than one principal orbit type, these orbit types would partition M_{reg}/G into disjoint nonempty open subsets contradicting the fact that M_{reg}/G is connected. □

29.18. Corollary. *Let M be a connected, proper G-manifold of dimension n and let k be the least number of connected components of all isotropy groups of dimension $m := \inf\{\dim G_x | x \in M\}$. Then the following two assertions are equivalent:*

(1) *$G.x_0$ is a principal orbit.*

(2) *The isotropy group G_{x_0} has dimension m and k connected components.*

If furthermore G is connected and simply connected, these conditions are again equivalent to:

(3) *The orbit $G.x_0$ has dimension $n - m$ and for the order of the fundamental group we have $|\pi_1(G.x_0)| = k$.*

Proof. Recall that we proved the existence of a principal orbit in (29.14) just by taking a G_{x_0} as described above. The other direction of the proof follows from corollary (29.17). Since there is only one principal orbit type, this must be it.

If moreover G is connected and simply connected, we look at the fibration $G_{x_0} \to G \to G/G_{x_0} = G.x_0$ and at the following portion of its long exact homotopy sequence:

$$0 = \pi_1(G) \to \pi_1(G.x_0) \to \pi_0(G_{x_0}) \to \pi_0(G) = 0$$

from which we see that $|\pi_1(G.x_0)| = k$ if and only if the isotropy group G_{x_0} has k connected components. □

29.19. Theorem ([197]). *Let $\pi : G \to O(V)$ be an orthogonal, real, finite-dimensional representation of a compact Lie group G. Let $\rho_1, \dots, \rho_k \in \mathbb{R}[V]^G$ be homogeneous generators for the algebra $\mathbb{R}[V]^G$ of invariant polynomials on V. For $v \in V$, let $\operatorname{Nor}_v(G.v) := T_v(G.v)^\perp$ be the normal space*

to the orbit at v, *and let* $\operatorname{Nor}_v(G.v)^{G_v}$ *be the subspace of those vectors which are invariant under the isotropy group* G_v.
Then $\operatorname{grad}\rho_1(v), \dots, \operatorname{grad}\rho_k(v)$ *span* $\operatorname{Nor}_v(G.v)^{G_v}$ *as a real vector space.*

Proof. Clearly each $\operatorname{grad}\rho_i(v) \in \operatorname{Nor}_v(G.v)^{G_v}$. In the following we will identify G with its image $\pi(G) \subseteq O(V)$. Its Lie algebra is then a subalgebra of $\mathfrak{o}(V)$ and can be realized as a Lie algebra consisting of skew-symmetric matrices. Let $v \in V$, and let S_v be the normal slice at v which is chosen so small that the projection of the tubular neighborhood (see (6.18)) $p_{G.v} : G.S_v \to G.v$ from the diagram

$$\begin{array}{ccccc}
G \times S_v & \xrightarrow{q} & G \times_{G_v} S_v & \xrightarrow{\pi} & G.S_v \\
 & & \Big\downarrow{\scriptstyle p} & & \Big\downarrow{\scriptstyle p_{G.v}} \\
 & & G/G_v & \xrightarrow[\cong]{\pi} & G.v
\end{array}$$

has the property that for any $w \in G.S_v$ the point $p_{G.v}(w) \in G.v$ is the unique point in the orbit $G.v$ which minimizes the distance between w and the orbit $G.v$.
Choose $n \in \operatorname{Nor}_v(G.v)^{G_v}$ so small that $x := v + n \in S_v$. So $p_{G.v}(x) = v$. For the isotropy groups we have $G_x \subseteq G_v$ by (6.16.2). But we have also $G_v \subseteq G_v \cap G_n \subseteq G_x$, so that $G_v = G_x$. Let S_x be the normal slice at x which we also choose so small that $p_{G.x} : G.S_x \to G.x$ has the same minimizing property as $p_{G.v}$ above, but so large that $v \in G.S_x$ (choose n smaller if necessary). We also have $p_{G.x}(v) = x$ since for the Euclidean distance in V we have

$$\begin{aligned}
|v - x| &= \min_{g\in G} |g.v - x| && \text{since } v = p_{G.v}(x) \\
&= \min_{g\in G} |h.g.v - h.x| && \text{for all } h \in G \\
&= \min_{g\in G} |v - g^{-1}.x| && \text{by choosing } h = g^{-1}.
\end{aligned}$$

For $w \in G.S_x$ we consider the local, smooth, G-invariant function

$$\begin{aligned}
\operatorname{dist}(w, G.x)^2 &= \operatorname{dist}(w, p_{G.x}(w))^2 = \langle w - p_{G.x}(w), w - p_{G.x}(w)\rangle \\
&= \langle w, w\rangle + \langle p_{G.x}(w), p_{G.x}(w)\rangle - 2\langle w, p_{G.x}(w)\rangle \\
&= \langle w, w\rangle + \langle x, x\rangle - 2\langle w, p_{G.x}(w)\rangle.
\end{aligned}$$

Its derivative with respect to w is

$$(1) \qquad d(\operatorname{dist}(\ \ , G.x)^2)(w)y = 2\langle w, y\rangle - 2\langle y, p_{G.x}(w)\rangle - 2\langle w, dp_{G.x}(w)y\rangle.$$

We shall show below that

$$(2) \qquad \langle v, dp_{G.x}(v)y\rangle = 0 \quad \text{for all } y \in V,$$

so that the derivative at v is given by

$$(3)\ d(\operatorname{dist}(\ \ ,G.x)^2)(v)y = 2\langle v,y\rangle - 2\langle y, p_{G.x}(v)\rangle = 2\langle v-x, y\rangle = -2\langle n, y\rangle.$$

Now choose a smooth G_x-invariant function $f : S_x \to \mathbb{R}$ with compact support which equals 1 in an open ball around x and extend it smoothly (see the diagram above, but for S_x) to $G.S_x$ and then to the whole of V. We assume that f is still equal to 1 in a neighborhood of v. Then $g = f.\operatorname{dist}(\ \ ,G.x)^2$ is a smooth G-invariant function on V which coincides with $\operatorname{dist}(\ \ ,G.x)^2$ near v. By the theorem of Schwarz (7.13) there is a smooth function $h \in C^\infty(\mathbb{R}^k,\mathbb{R})$ such that $g = h\circ\rho$, where $\rho = (\rho_1,\dots,\rho_k) : V \to \mathbb{R}^k$. Then we have finally by (3)

$$\begin{aligned} -2n &= \operatorname{grad}(\operatorname{dist}(\ \ ,G.x)^2)(v) = \operatorname{grad} g(v) \\ &= \operatorname{grad}(h\circ\rho)(v) = \sum_{i=1}^k \frac{\partial h}{\partial y_i}(\rho(v))\operatorname{grad}\rho_i(v), \end{aligned}$$

which proves the result.

It remains to check equation (2). Since $T_vV = T_v(G.v)\oplus \operatorname{Nor}_v(G.v)$, the normal space $\operatorname{Nor}_x(G.x) = \ker dp_{G.x}(v)$ is still transversal to $T_v(G.v)$ if n is small enough; so it remains to show that $\langle v, dp_{G.x}(v).X.v\rangle = 0$ for each $X \in \mathfrak{g}$. Since $x = p_{G.x}(v)$, we have $|v-x|^2 = \min_{g\in G}|v-g.x|^2$, and thus the derivative of $g\mapsto \langle v-g.x, v-g.x\rangle$ at e vanishes: For all $X\in\mathfrak{g}$ we have

$$(4) \qquad 0 = 2\langle -X.x, v-x\rangle = 2\langle X.x, x\rangle - 2\langle X.x, v\rangle = 0 - 2\langle X.x, v\rangle,$$

since the action of X on V is skew-symmetric. Now we consider the equation $p_{G.x}(g.v) = g.p_{G.x}(v)$ and differentiate it with respect to g at $e\in G$ in the direction $X\in\mathfrak{g}$ to obtain in turn

$$\begin{aligned} dp_{G.x}(v).X.v &= X.p_{G.x}(v) = X.x, \\ \langle v, dp_{G.x}(v).X.v\rangle &= \langle v, X.x\rangle = 0, \quad \text{by (4).} \quad \square \end{aligned}$$

29.20. Lemma. *Let $\pi : G \to O(V)$ be an orthogonal representation. Let $\omega \in \Omega^p_{hor}(V)^G$ be an invariant differential form on V which is horizontal in the sense that $i_w\omega_x = 0$ if w is tangent to the orbit $G.x$. Let $v\in V$ and let $w\in T_vV$ be orthogonal to the space $\operatorname{Nor}_v(G.v)^{G_v^0}$ of those orthogonal vectors which are invariant under the connected component G_v^0 of the isotropy group G_v. Then $i_w\omega_v = 0$.*

Proof. We consider the orthogonal decomposition

$$T_vV = T_v(G.v)\oplus W\oplus \operatorname{Nor}_v(G.v)^{G_v^0}.$$

We may assume that $w \in W$ since $i_u\omega_v = 0$ for $u \in T_v(G.v)$.

We claim that each $w \in W$ is a linear combination of elements of the form $X.u$ for $u \in W$ and $X \in \mathfrak{g}_v := \ker(d\pi(\quad)v)$. Since G_v^0 is compact, the representation space W has no fixed point other than zero and is completely reducible under G_v^0 and thus also under its Lie algebra $\mathfrak{g}_v$, and we may treat each irreducible component separately or assume that W is irreducible. Then $\mathfrak{g}_v(W)$ is an invariant subspace which is not 0. So it agrees with W, and the claim follows.

So we may assume that $w = X.u$ for $u \in W$. But then

$$(v + \tfrac{1}{n}u, X.u = nX.(v + \tfrac{1}{n}u)) \in T_{v+\frac{1}{n}u}(G.(v + \tfrac{1}{n}u))$$

satisfies $i_{X.u}\omega_{v+u/n} = 0$ by horizontality and thus we have

$$i_w\omega_v = i_{X.u}\omega_v = \lim_n i_{X.u}\omega_{v+u/n} = 0. \quad \square$$

29.21. ***G*-manifold with a single orbit type as fiber bundle.** Let (M, γ) be a proper Riemann G-manifold and suppose that M has only one orbit type (see 6.11), (H). We then want to study the quotient map $\pi : M \to M/G$. Let us first consider the orbit space M/G. Choose $x \in M$ and let S_x denote the normal slice at x. Then by (6.16.2) we have $G_y \subseteq G_x$ for all $y \in S_x$. Since G_y must additionally be conjugate to G_x and both are compact, they must be the same by (6.12). So $G_x = G_y$ and therefore G_x acts trivially on S_x (this can also be seen as a special case of (29.7)). From (6.16.5) it follows that $\pi(S_x) \cong S_x/G_x = S_x$, and with (6.18) we have that $G.S_x$ is isomorphic to $G/G_x \times S_x$. Therefore, for any $x \in M$, $(\pi(S_x), \exp_x^{-1}|_{S_x})$ can serve as a chart for M/G:

$$\begin{array}{ccccc} M & \hookleftarrow & S_x & & \\ \downarrow{\scriptstyle\pi} & & \downarrow{\scriptstyle\pi} & & \\ M/G & \hookleftarrow & \pi(S_x) & = & S_x/G_x. \end{array}$$

To make an atlas out of these charts, we have to check whether they are compatible — which is obvious. By (6.29), M/G is Hausdorff, and therefore it is a smooth manifold.

Now let us study the smooth submersion $\pi : M \to M/G$. We want to find a Riemann metric on M/G which will make π a Riemann submersion.

Claim. *For $X_x, Y_x \in \operatorname{Hor}_x(\pi) = \operatorname{Nor}_x(G.x)$, the following inner product is well defined:*

$$\bar{\gamma}_{\pi(x)}(T\pi X_x, T\pi Y_x) := \gamma_x(X_x, Y_x).$$

Proof. Choose $X'_{gx}, Y'_{gx} \in \mathrm{Hor}_{gx}(\pi)$ such that $T\pi.X'_{gx} = T\pi.X_x$ and $T\pi.Y'_{gx} = T\pi.Y_x$. Then we see that $X'_{gx} = T(\ell_g)X_x$ by the following argumentation: Clearly $T\pi(X'_{gx} - T(\ell_g).X_x) = 0$, so the difference $X'_{gx} - T(\ell_g).X_x$ is vertical. The map ℓ_g leaves $G.x$ invariant; consequently, $T\ell_g$ maps vertical vectors to vertical vectors and since it is an isometry, it also maps horizontal vectors to horizontal vectors. Therefore $X'_{gx} - T(\ell_g).X_x$ is horizontal as well as vertical and must be zero.

Now we can conclude, that

$$\gamma_{gx}(X'_{gx}, Y'_{gx}) = \gamma_{gx}(T(\ell_g)X_x, T(\ell_g)Y_x) = \gamma_x(X_x, Y_x). \qquad \square$$

So we have found a Riemann metric $\bar{\gamma}$ on M/G which makes π a Riemann submersion.

Let us finally try to understand in which sense $\pi : M \to M/G$ is an associated bundle. Let $x \in M$ be such that $G_x = H$. By (29.1.4) the set $M^H = \{x \in M : g.x = x \text{ for all } g \in H\}$ is a geodesically complete submanifold of M. It is $N_G(H)$-invariant, and the restriction $\pi : M^H \to M/G$ is a smooth submersion since for each $y \in M^H$ the slice S_y is also contained in M^H. The fiber of $\pi : M^H \to M/G$ is a free $N_G(H)/H$-orbit: If $\pi(x) = \pi(y)$ and $G_x = H = G_y$, then $g \in N_G(H)$. So $\pi : M^H \to M/G$ is a principal $N_G(H)/H$-bundle, and M is the associated bundle with fiber G/H as follows:

$$\begin{array}{ccc}
M^H \times G/H & & \\
\downarrow & \searrow^{(x,[g]) \mapsto g.x} & \\
M^H \times_{N_G(H)/H} G/H & \xrightarrow{\cong} & M \\
\downarrow & & \downarrow \\
M/G & = & M/G.
\end{array}$$

29.22. Another fiber bundle construction. Let M again be a proper Riemann G-manifold with only one orbit type. Then we can 'partition' M into the totally geodesic submanifolds

$$M^{gHg^{-1}} := \{x \in M : ghg^{-1}.x = x \text{ for all } h \in H\}$$

where $H = G_{x_0}$ ($x_0 \in M$ arbitrary) is fixed and g varies. This is not a proper partitioning in the sense that if $g \neq e$ commutes with H, for instance, then $M^{gHg^{-1}} = M^{eHe^{-1}}$. We want to find out just which g give the same sets $M^{gHg^{-1}}$.

Claim.

$$M^{gHg^{-1}} = M^{g'Hg'^{-1}} \quad \Longleftrightarrow \quad gN(H) = g'N(H)$$

where $N(H)$ denotes the normalizer of H in G.

Proof. First let us show the following identity:

$$N(H) = \{g \in G : g(M^H) \subseteq M^H\}.$$

($\subseteq$) Let $n \in N(H)$ and $y \in M^H$. Then $n.y$ is H-invariant:

$$hn.y = nn^{-1}hn.y = n(n^{-1}hn).y = n.y.$$

($\supseteq$) $gM^H \subseteq M^H$ implies that $hg.y = g.y$, or equivalently $g^{-1}hg.y = y$, for any $y \in M^H$ and $h \in H$. Recall at this point that $H = G_{x_0}$ for some $x_0 \in M$. Therefore, we have $g^{-1}hg.x_0 = x_0$ and consequently $g^{-1}hg \in G_{x_0} = H$.

Using this characterization for $N(H)$ and the identity

$$g'\{g \in G : gM^H \subseteq M^H\} = \{g \in G : gM^H \subseteq g'M^H\},$$

we can convert the right hand side of our equality, $gN(H) = g'N(H)$, to the following:

$$\{a \in G : aM^H \subseteq g.M^H\} = \{a \in G : aM^H \subseteq g'.M^H\}.$$

In particular, this is the case if

$$g.M^H = g'.M^H.$$

In fact, let us show that the two equations are equivalent. Suppose indirectly that $g.y \notin g'.M^H$ for some $y \in M^H$. Then $a = g$ has the property $a.M^H \not\subseteq g'.M^H$, so $\{a \in G : aM^H \subseteq g.M^H\} \neq \{a \in G : aM^H \subseteq g'.M^H\}$.
So far we have shown that $gN(H) = g'N(H) \Leftrightarrow g.M^H = g'.M^H$. To complete the proof, it only remains to check whether

$$M^{gHg^{-1}} = g.M^H.$$

This is easily done (as well as plausible, since it strongly resembles the 'dual' notion $G_{gx} = gG_xg^{-1}$):

$$\begin{aligned} y \in M^{gHg^{-1}} &\iff ghg^{-1}.y = y \text{ for all } h \in H \\ &\iff hg^{-1}.y = g^{-1}y \text{ for all } h \in H \\ &\iff g^{-1}.y \in M^H \\ &\iff y \in gM^H. \quad \square \end{aligned}$$

Claim. The map $\bar{\pi} : M \to G/N(H)$ defined by $M^{gHg^{-1}} \ni x \mapsto g.N(H)$ is a fiber bundle with typical fiber M^H.

Proof. To prove this, let us consider the following diagram:

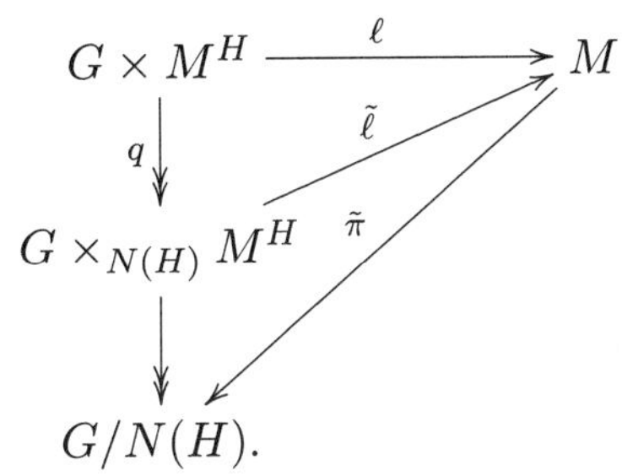

Here we use the restricted action $\ell : N(H) \times M^H \to M^H$ to associate to the principal bundle $G \to G/N(H)$ the bundle $G[M^H, \ell] = G \times_{N(H)} M^H$. It remains to show that $\tilde{\ell}$ is a diffeomorphism, since then $\bar{\pi}$ has the desired fiber bundle structure. The map $\tilde{\ell}$ is smooth, since $\tilde{\ell} \circ q = \ell$ is smooth and q is a submersion. Now let us show that $\tilde{\ell}$ is bijective.
(1) $\tilde{\ell}$ is surjective: Since H is the only orbit type, for every $x \in M$ there is a $g \in G$, such that $G_x = gHg^{-1}$, which implies $x \in M^{gHg^{-1}} = gM^H \subseteq \ell(G \times M^H)$. So ℓ is surjective and, by the commutativity of the diagram, so is $\tilde{\ell}$.
(2) $\tilde{\ell}$ is injective: Suppose $\ell(a, x) = a.x = b.y = \ell(b, y)$, for some $a, b \in G$, $x, y \in M^H$. Then $b^{-1}a.x = y \in M^H$ implies $hb^{-1}a.x = y = b^{-1}a.x$ which implies again $(b^{-1}a)^{-1}hb^{-1}a.x = x$. Since there is only one orbit type and all isotropy groups are compact, we know that $x \in M^H \Rightarrow H = G_x$ (by (6.12)). So $(b^{-1}a)^{-1}hb^{-1}a$ is again in H, and $b^{-1}a \in N(H)$. In this case, $q(a, x) = [a, x] = [bb^{-1}a, x] = [b, b^{-1}a.x] = [b, y] = q(b, y)$.
The inverse map $\tilde{\ell}^{-1}$ is smooth, since ℓ is a submersion. So $\tilde{\ell}$ is a diffeomorphism and $\bar{\pi}$ a fiber bundle with typical fiber M^H. □

29.23. Construction for more than one orbit type. Let (H) be one particular orbit type $(H = G_x)$. To reduce the case at hand to the previous one, we must partition the points in M into sets with common orbit type:

$$M_{(H)} := \{x \in M : (G_x) = (H)\}.$$

Claim. *For a proper Riemann G-manifold, the space $M_{(H)}$ as defined above is a smooth G-invariant submanifold.*

Proof. The set $M_{(H)}$ is of course G-invariant as a collection of orbits of a certain type. We only have to prove that it is a smooth submanifold. Take any x in $M_{(H)}$ (then, without loss, $H = G_x$), and let S_x be a slice at x. Consider the tubular neighborhood $G.S \cong G \times_H S_x$ (see (6.18)). Then the orbits of type (H) in $G.S$ are just those orbits that meet S_x in S_x^H

(where S_x^H shall denote the fixed point set of H in S_x). Or, equivalently, $(G \times_H S_x)_{(H)} = G \times_H S_x^H$:

($\subseteq$) $[g, s] \in (G \times_H S_x)_{(H)} \Rightarrow g.s \in G.S_{(H)} \Rightarrow gHg^{-1} = G_s \subseteq H \Rightarrow G_s = H \Rightarrow s \in S_x^H \Rightarrow [g, s] \in G \times_H S_x^H$.

($\supseteq$) $[g, s] \in G \times_H S_x^H \Rightarrow s \in S_x^H \Rightarrow H \subseteq G_s$, but since $s \in S_x$, we have $G_s \subseteq G_x = H$ by (6.16.2); therefore $G_s = H$ and $[g, s] \in (G \times_H S_x)_{(H)}$.

Now, let $S_x = \exp_x(\mathrm{Nor}_r(G.x))$ be the normal slice at x. That is, r is chosen so small that $\exp_x$ is a diffeomorphism on $\mathrm{Nor}_r(G.x) =: V$. Notice that V is not only diffeomorphic to S_x, but G-equivariantly so, if we let G act on $\mathrm{Nor}_x(G.x)$ via the slice representation. Since the slice action is orthogonal, in particular linear, the set of points fixed by the action of H is a linear subspace of $\mathrm{Nor}_x(G.x)$ and its intersection with V, a "linear" submanifold. Therefore S_x^H is also a submanifold of S_x. Now consider the diagram

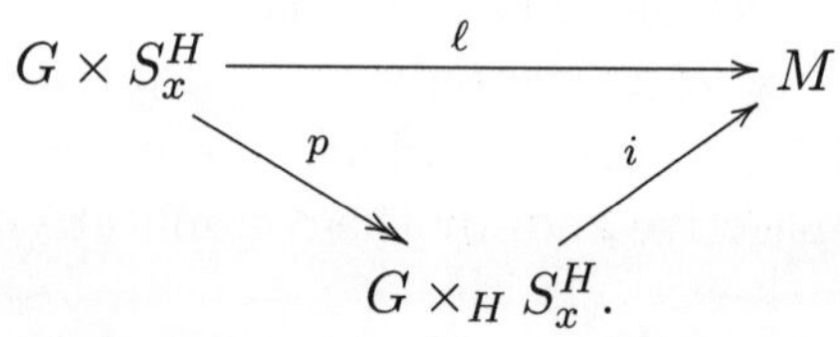

The map i is well defined, injective and smooth, since p is a submersion and ℓ is smooth. Furthermore, p is open, and so is ℓ. Just consider any open set of the form $U \times W$ in $G \times S_x^H$. Then $\ell(U \times W)$ is the union of all sets $\ell_u(W)$ for $u \in U$. Since ℓ_u is a diffeomorphism, each one of these is open, so $\ell(U \times W)$ is open as well. Therefore, i must be open, and so i is an embedding and $G.S^H \cong G \times_H S_x^H$ is an embedded submanifold of M. $\square$

Let (H) be one particular orbit type ($H = G_x$); then M^H is again a closed, totally geodesic submanifold of M; see (29.1.3).

Claim. *$\{x \in M : G_x = H\}$ is an open submanifold of M^H.*

For one orbit type, $x \in M^H$ implied $H = G_x$, and thus $\{x \in M : G_x = H\} = M^H$. For more than one orbit type, M^H is not necessarily contained in $M_{(H)}$. Therefore, it is necessary to study $\{x \in M : G_x = H\} = M^H \cap M_{(H)}$.

Proof. In (29.22) we saw that $N(H)$ is the largest subgroup of G acting on M^H. It induces a proper $N(H)/H$-action on M^H. Now, $\{x \in M : G_x = H\}$ is the set of all points in M^H with trivial isotropy group with respect to this action. So by (29.18) it is simply the set of all regular points. Therefore, by (29.14), $\{x \in M : G_x = H\}$ is an open, dense submanifold of M^H. $\square$

Now, $M_{(H)}$ can be turned into a fiber bundle over $G/N(H)$ with typical fiber $\{x \in M : G_x = H\}$ just as before. On the other hand, $M_{(H)}$ is a fiber bundle over $M_{(H)}/G$ with typical fiber G/H. The partition of M into submanifolds

$M_{(H)}$ and that of M/G into the different orbit types is locally finite by (29.15). So M and M/G are in a sense stratified, and $\pi : M \to M/G$ is a stratified Riemann submersion (see also [**40**]).

29.24. Remark. Let M be a connected Riemann G-manifold and (H) the principal orbit type, then we saw in (29.23) that $\pi : M_{(H)} \to M_{(H)}/G$ is a Riemann submersion. Now we can prove:

Claim. *For $x \in M_{reg} = M_{(H)}$ a vector field $\xi \in \Gamma(\mathrm{Nor}(G.x))$ is π-parallel if and only if ξ is G-equivariant.*

Proof. ($\Longleftarrow$) If $\xi(g.x) = T_x\ell_g.\xi(x)$, then $T_{g.x}\pi.\xi(g.x) = T_{g.x}\pi \circ T_x\ell_g.\xi(x) = T_x\pi.\xi(x)$ for all $g \in G$. Therefore ξ is π-parallel.

($\Longrightarrow$) The tangent vectors $\xi(g.x)$ and $T_x\ell_g\xi(x)$ are both in $\mathrm{Nor}_{g.x}(G.x)$, and since ξ is π-parallel, we have $T_{g.x}\pi.\xi(g.x) = T_x\pi.\xi(x) = T_{g.x}\pi \circ T_x\ell_g.\xi(x)$. So $\xi(g.x)$ and $T_x\ell_g.\xi(x)$ both have the same image under $T_{g.x}\pi$. Because $T_{g.x}\pi$ restricted to $\mathrm{Nor}_{g.x}(G.x)$ is an isomorphism, $\xi(g.x) = T_x\ell_g.\xi(x)$. $\square$

30. Polar Actions

In this chapter, let (M, γ) always denote a connected, complete Riemann G-manifold, and assume that the action of G on M is effective and isometric.

30.1. Lemma. *Consider $X \in \mathfrak{g}$, the Lie algebra of G, ζ_X, the associated fundamental vector field to X, and c, a geodesic in M. Then $\gamma(c'(t), \zeta_X(c(t)))$ is constant in t.*

This is an example of a momentum mapping if we lift the whole situation to the symplectic manifold T^*M and identify this with TM via γ. See section (34).

Proof. Let ∇ be the Levi-Civita covariant derivative on M. Then

$$\partial_t.\gamma(c'(t), \zeta_X(c(t))) = \gamma\big(\nabla_{\partial_t} c'(t), \zeta_X(c(t))\big) + \gamma\big(c'(t), \nabla_{\partial_t}(\zeta_X \circ c)\big).$$

Since c is a geodesic, $\nabla_{\partial_t} c'(t) = 0$, and so is the entire first summand. So it remains to show that $\gamma\big(c'(t), \nabla_{\partial_t}(\zeta_X \circ c)\big)$ vanishes as well.

Let $s_1, \dots, s_n$ be a local orthonormal frame field on an open neighborhood U of $c(t)$, and let $\sigma^1, \dots, \sigma^n$ be the orthonormal coframe. Then $\gamma = \sum \sigma^i \otimes \sigma^i$. Let us use the notation

$$\zeta_X|_U =: \sum s_i X^i,$$
$$\nabla \zeta_X|_U =: \sum X_i^j s_j \otimes \sigma^i.$$

Then we have

$$\nabla_{\partial_t}(\zeta_X \circ c) = \sum X_i^j(c(t)) s_j(c(t)) \sigma^i(c'(t)).$$

So

$$\begin{aligned}\gamma(c'(t), \nabla_{\partial_t}(\zeta_X \circ c)) &= \sum \sigma^j(c'(t))\sigma^j(\nabla_{\partial_t}(\zeta_X \circ c))\\ &= \sum X_i^j(c(t))\sigma^j(c'(t))\sigma^i(c'(t)).\end{aligned}$$

If we now show that $X_i^j + X_j^i = 0$, then $\gamma(c'(t), \nabla_{\partial_t}(\zeta_X \circ c))$ will be zero, and the proof will be complete. Since the action of G is isometric, ζ_X is a Killing vector field; that is, $\mathcal{L}_{\zeta_X}\gamma = 0$. So we have

$$\sum \mathcal{L}_{\zeta_X}\sigma^i \otimes \sigma^i + \sum \sigma^i \otimes \mathcal{L}_{\zeta_X}\sigma^i = 0.$$

Now we must try to express $\mathcal{L}_{\zeta_X}\sigma^i$ in terms of X_i^j. For this, recall the structure equation: $d\sigma^k + \sum \omega_j^k \wedge \sigma^j = 0$. We have

$$\begin{aligned}\mathcal{L}_{\zeta_X}\sigma^i &= i_{\zeta_X} d\sigma^i + d(i_{\zeta_X}\sigma^i) = -i_{\zeta_X}(\sum \omega_j^i \wedge \sigma^j) + d(\sigma^i(\zeta_X))\\ &= -i_{\zeta_X}\sum \omega_j^i \wedge \sigma^j + dX^i = \sum \omega_j^i . X^j - \sum \omega_j^i(\zeta_X)\sigma^j + dX^i.\end{aligned}$$

Since

$$\nabla\zeta_X|_U = \nabla(\sum s_j X^j) = \sum s_i . \omega_j^i . X^j + \sum s_i \otimes dX^i = \sum X_j^i s_i \otimes \sigma^j,$$

we can replace $\sum \omega_j^i . X^j$ by $\sum X_j^i \sigma^j - dX^i$. Therefore,

$$\mathcal{L}_{\zeta_X}\sigma^i = \sum (X_j^i \sigma^j - \omega_j^i(\zeta_X)\sigma^j) = \sum (X_j^i - \omega_j^i(\zeta_X))\sigma^j.$$

Now, let us insert this into $0 = \mathcal{L}_{\zeta_X}\gamma$:

$$\begin{aligned}0 &= \sum \mathcal{L}_{\zeta_X}\sigma^i \otimes \sigma^i + \sum \sigma^i \otimes \mathcal{L}_{\zeta_X}\sigma^i\\ &= \sum (X_j^i - \omega_j^i(\zeta_X))\sigma^j \otimes \sigma^i + \sum (X_j^i - \omega_j^i(\zeta_X))\sigma^i \otimes \sigma^j\\ &= \sum (X_j^i + X_i^j)\sigma^j \otimes \sigma^i - \sum (\omega_j^i(\zeta_X) + \omega_i^j(\zeta_X))\sigma^j \otimes \sigma^i\\ &= \sum (X_j^i + X_i^j)\sigma^j \otimes \sigma^i - 0\end{aligned}$$

since $\omega(Y)$ is skew-symmetric. This implies $X_j^i + X_i^j = 0$, and we are done. □

30.2. Definition. *For any x in M_{reg} we define:*

$$\begin{aligned}E(x) &:= \exp_x^\gamma(\mathrm{Nor}_x(G.x)) \subseteq M,\\ E_{reg}(x) &:= E(x) \cap M_{reg}.\end{aligned}$$

In a neighborhood of x, $E(x)$ is a manifold; globally, it can intersect itself.

30.3. Lemma. *Let $x \in M_{reg}$, then:*

(1) $g.E(x) = E(g.x)$, $g.E_{reg}(x) = E_{reg}(g.x)$.

(2) *For* $X_x \in \operatorname{Nor}(G.x)$ *the geodesic* $c : t \mapsto \exp(t.X_x)$ *is orthogonal to every orbit it meets.*

(3) *If* G *is compact, then* $E(x)$ *meets every orbit in* M.

Proof. (1) This is a direct consequence of (29.1.1):

$$g.\exp_x(t.X) = \exp_{g.x}(t.T_x\ell_g.X).$$

(2) By choice of starting vector X_x, the geodesic c is orthogonal to the orbit $G.x$, which it meets at $t = 0$. Therefore it intersects every orbit it meets orthogonally, by lemma (30.1).

(3) For arbitrary $x, y \in M$, we will prove that $E(x)$ intersects $G.y$. Since G is compact, by continuity of $\ell^y : G \to M$, the orbit $G.y$ is compact as well. Therefore we can choose $g \in G$ in such a way that $\operatorname{dist}(x, G.y) = \operatorname{dist}(x, g.y)$. Let $c(t) := \exp_x(t.X_x)$ be a minimal geodesic connecting $x = c(0)$ with $g.y = c(1)$. We now have to show that $X_x \in \operatorname{Nor}_x(G.x)$: Take a point $p = c(t)$ on the geodesic very close to $g.y$ — close enough so that $\exp_p$ is a diffeomorphism into a neighborhood U_p of p containing $g.y$ (it shall have domain $V \subseteq T_pM$). In this situation Gauß's lemma (23.2) states that all geodesics through p are orthogonal to the geodesic spheres: $\exp_p(k.S^{m-1})$ (where $S^{m-1} := \{X_p \in T_pM : \gamma(X_p, X_p) = 1\}$, and $k > 0$ is small enough for $k.S^{m-1} \subseteq V$ to hold). From this it can be concluded that c is orthogonal to $G.y$: Take the smallest geodesic sphere around p touching $G.y$. By the minimality of c, c must leave the geodesic sphere at a touching point, and by Gauß's lemma, it must leave at a right angle to the geodesic sphere. Clearly, the touching point is just $g.y = c(1)$, and there c also meets $G.y$ at a right angle. By (2), c encloses a right angle with every other orbit it meets as well. In particular, c starts orthogonally to $G.x$. Therefore, X_x is in $\operatorname{Nor}_x(G.x)$, and $g.y = c(1) \in E(x)$. □

30.4. Remark. Let $x \in M$ be a regular point and S_x the normal slice at x. If S_x is orthogonal to every orbit it meets, then so are all $g.S_x$ ($g \in G$ arbitrary). So the submanifolds $g.S_x$ can be considered as leaves of the horizontal foliation (local solutions of the horizontal distribution — which has constant rank in a neighborhood of a regular point), and the Riemann submersion $\pi : M_{\text{reg}} \to M_{\text{reg}}/G$ is integrable. Since this is not always the case (the horizontal distribution is not generally integrable), it must also be false, in general, that the normal slice is orthogonal to every orbit it meets. But it does always meet orbits transversally.

Example. Consider the isometric action of the circle group S^1 on $\mathbb{C}\times\mathbb{C}$ (as real vector spaces) defined by $e^{it}.(z_1, z_2) := (e^{it}.z_1, e^{it}.z_2)$. Then $p = (0, 1)$ is a regular point: $G_p = \{1\}$. The subspace $\operatorname{Nor}_p(S^1.p)$ of $T_p\mathbb{C}\times\mathbb{C}$ takes on the following form: $\operatorname{Nor}_p(S^1.p) = \langle(1,0), (i,0), (0,1)\rangle_{\mathbb{R}} = \mathbb{C}\times\mathbb{R}$. Therefore, we get $E(0,1) = \{(u, 1+r) : u \in \mathbb{C}, r \in \mathbb{R}\}$. In particular, $y = (1,1) \in E(0,1)$, but $S^1.y = \{(e^{it}, e^{it}) : t \in \mathbb{R}\}$ is not orthogonal to $E(0,1)$. Its tangent space, $T_y(S^1.y) = \mathbb{R}.(i,i)$, is not orthogonal to $\mathbb{C} \times \mathbb{R}$.

30.5. Definition. A connected closed complete submanifold $\Sigma \subset M$ is called a *section* for the G-action, and the action is called a *polar action* if:

(1) Σ meets every orbit, or equivalently, $G.\Sigma = M$.
(2) Where Σ meets an orbit, it meets it orthogonally: For $x \in \Sigma$ we have $T_x\Sigma \subseteq \operatorname{Nor}_x(G.x)$; equivalently, for $x \in \Sigma, X \in \mathfrak{g}$ we have $\zeta_X(x) \perp T_x\Sigma$.

If Σ is a section, then so is $g.\Sigma$ for all g in G. Since $G.\Sigma = M$, there is a section through every point in M. We say that M *admits sections.*

The notion of a section was introduced in [**215, 216**] and in slightly different form in [**188, 189**]. The case of linear representations was considered in [**23**], [**36**], and then in [**38**] where representations admitting sections were called polar representations (see (30.16)) and where all polar representations of connected Lie groups were completely classified. Also, [**35**] considered Riemann manifolds admitting flat sections. We follow here the notion of [**188**].

30.6. Examples. For the standard action of $O(n)$ on $\mathbb{R}^n$ the orbits are spheres, and every line through 0 is a section.

If G is a compact, connected Lie group with bi-invariant metric, then conj : $G \times G \to G$, $\operatorname{conj}_g(h) = ghg^{-1}$ is an isometric action on G. The orbits are just the conjugacy classes of elements.

Proposition. *Every maximal torus H of a compact connected Lie group G is a section.*

A torus is a product of circle groups, or equivalently, a compact connected abelian Lie group; a maximal torus of a compact Lie group is a toral subgroup which is not properly contained in any larger toral subgroup.

Proof. We check (30.5.1), $\operatorname{conj}(G).H = G$: This states that any $g \in G$ can be found in some subgroup which is conjugate to H, $g \in aHa^{-1}$. This is equivalent to $ga \in aH$ or $gaH = aH$. So the claim now presents itself as a fixed point problem: Does the map $\ell_g : G/H \to G/H : aH \mapsto gaH$ have a fixed point? It is solved in the following way:

The **fixed point theorem of Lefschetz** [**214**, 11.6.2, p. 297] says that *a smooth mapping $f : M \to M$ from a connected compact manifold to itself has no fixed point if and only if*

$$\sum_{i=0}^{\dim M} (-1)^i \operatorname{Trace}(H^i(f) : H^i(M) \to H^i(M)) = 0.$$

Since G is connected, ℓ_g is homotopic to the identity, so

$$\sum_{i=0}^{\dim G/H} (-1)^i \operatorname{Trace}(H^i(\ell_g) : H^i(G/H) \to H^i(G/H))$$
$$= \sum_{i=0}^{\dim G/H} (-1)^i \operatorname{Trace}(H^i(\mathrm{Id})) = \sum_{i=0}^{\dim G/H} (-1)^i \dim H^i(G/H) = \chi(G/H),$$

the Euler characteristic of G/H. This is given by the following theorem [**181**, Sec. 13, Theorem 2, p. 217]: *If G is a connected compact Lie group and H is a connected compact subgroup, then the Euler characteristic $\chi(G/H) \geq 0$. Moreover $\chi(G/H) > 0$ if and only if the rank of G equals the rank of H. In the case when $\chi(G/H) > 0$, then $\chi(G/H) = |W_G|/|W_H|$, the quotient of the respective Weyl groups.*

Since the Weyl group of a torus is trivial, in our case we have $\chi(G/H) = |W_G| > 0$, and thus there exists a fixed point.

Now we show that (30.5.2) holds, $h \in H, X \in \mathfrak{g} \Rightarrow \zeta_X(h) \perp T_h H$: We have

$$\zeta_X(h) = \partial_t|_0 \exp(tX) h \exp(-tX) = T_e\mu^h.X - T_e\mu_h.X.$$

Now choose $Y \in \mathfrak{h}$. Then we have $T_e\mu_h.Y \in T_h H$, and

$$\begin{aligned}\gamma_h(T_e\mu_h.Y, T_e\mu^h.X - T_e\mu_h.X) &= \gamma_e(Y, \mathrm{Ad}(h).X - X)\\ &= \gamma_e(Y, \mathrm{Ad}(h).X) - \gamma_e(Y, X)\\ &= \gamma_e(\mathrm{Ad}(h).Y, \mathrm{Ad}(h).X) - \gamma_e(Y, X) = 0\end{aligned}$$

by the right, left, and therefore Ad-invariance of γ and by the commutativity of H. □

30.7. Example. Let G be a compact semisimple Lie group acting on its Lie algebra by the adjoint action $\mathrm{Ad} : G \times \mathfrak{g} \to \mathfrak{g}$. Then every Cartan subalgebra $\mathfrak{h}$ of $\mathfrak{g}$ is a section.

Proof. Every element of a semisimple Lie algebra $\mathfrak{g}$ is contained in a Cartan subalgebra, and any two Cartan subalgebras are conjugated by an element $g \in G$, since G is compact. This is a consequence of (30.6) above, since the subgroup in G corresponding to a Cartan subalgebra is a maximal torus. Thus every Ad_G-orbit meets the Cartan subalgebra $\mathfrak{h}$. It meets orthogonally with respect to the Cartan-Killing form B: Let $H_1, H_2 \in \mathfrak{h}$ and $X \in \mathfrak{g}$. Then $\partial_t|_0 \mathrm{Ad}(\exp(tX)).H_1 = \mathrm{ad}(X)H_1$ is a typical vector tangent to the orbit through $H_1 \in \mathfrak{h}$, and H_2 is tangent to $\mathfrak{h}$. Then

$$B(\mathrm{ad}(X)H_1, H_2) = B([X, H_1], H_2) = B(X, [H_1, H_2]) = 0$$

since $\mathfrak{h}$ is commutative. □

30.8. Example. In (7.1) we showed that for the $O(n)$-action on $S(n)$ by conjugation the space Σ of all diagonal matrices is a section.

Similarly, when the $SU(n)$ act on the Hermitian matrices by conjugation, the (real) diagonal matrices turn out to be a section.

30.9. Definition. The **principal horizontal distribution** on a Riemann G-manifold M is the horizontal distribution on $\pi : M_{\text{reg}} \to M_{\text{reg}}/G$.

Theorem. *If a connected, complete Riemann G-manifold M has a section Σ, then:*

(1) *The principal horizontal distribution is integrable.*
(2) *Every connected component of Σ_{reg} is a leaf for the principal horizontal distribution.*
(3) *If L is the leaf of* $\operatorname{Hor}(M_{reg})$ *through $x \in M_{reg}$, then $\pi|_L : L \to M_{reg}/G$ is an isometric covering map.*
(4) *Σ is totally geodesic.*
(5) *Every regular point $x \in M$ is contained in a unique section $E(x) =$* $\exp_x^\gamma(\operatorname{Nor}_x(G.x))$.
(6) *A G-equivariant normal field along a principal orbit is parallel in terms of the induced covariant derivative ∇^{Nor}.*

Proof. (1) The submanifolds $g.\Sigma_{\text{reg}}$ of M_{reg} are integral manifolds to the horizontal distribution, since they are orthogonal to each orbit and by an argument of dimension.

(2) is clear. (3) is (26.13.2). (4) follows from (26.13.1).

(5) For $x \in M$ choose $g \in G$ such that $g.x \in \Sigma \cap G.x$; then $g^{-1}.\Sigma$ is a section through x. By (2) and (4) we have $E(x) \subseteq g^{-1}.\Sigma$. The converse can be seen as follows: Let $y \in g^{-1}.\Sigma$ and choose a minimal geodesic from x to y. By the argument given in the proof of (30.3.2) this geodesic is orthogonal to the orbit through x and thus lies in $E(x)$. So $y \in E(x)$.

(6) See (26.19) and recall that by (29.24) a normal field is G-equivariant if and only if it is π-parallel, where $\pi : M \to M/G$ is the orbit map. □

30.10. Remark. The converse of (30.9.1) is not true. Namely, an integral manifold of $\operatorname{Hor}(M_{\text{reg}})$ is not, in general, a section.

Example. Consider the Lie group $G = S^1 \times \{1\}$, and let it act on $M := S^1 \times S^1$ by translation. Let $\xi = (1,0)$ denote the fundamental vector field of the action, and choose any $\eta \in \operatorname{Lie}(S^1 \times S^1) = \mathbb{R} \times \mathbb{R}$ which generates a 1-parameter subgroup c which is dense in $S^1 \times S^1$ (irrational ascent). Now, endow $S^1 \times S^1$ with a Riemann metric making ξ and η an orthonormal frame field. Any section of M would then have to be a coset of c, and therefore dense. This contradicts the assumption that a section is a closed embedded submanifold.

30.11. Example. If $(G/H, \sigma)$ is a symmetric space, then the totally geodesic connected submanifolds N of G/H through $e \in G/H$ correspond exactly to the linear subspaces

$$T_eN = \mathfrak{n} \subseteq \mathfrak{m} := T_eG/H \cong \{X \in \mathfrak{g} : \sigma'(X) = -X\}$$

which fulfill $[[\mathfrak{n}, \mathfrak{n}], \mathfrak{n}] \subseteq \mathfrak{n}$; see (28.5.12).

This implies that a locally totally geodesic submanifold of a simply connected symmetric space can be extended uniquely to a complete, totally geodesic submanifold. Here we mean by locally geodesic submanifold that a geodesic can leave the submanifold only at its "boundary". In other words, the second fundamental form must be zero.

30.12. Corollary. *Let $M = G/H$ be a simply connected, complete symmetric space, and let $K \subseteq G$ be a Lie subgroup. Then the action of K on G/H admits sections if and only if* $\operatorname{Hor}(M_{reg})$ *is integrable. In particular, if the principal K-orbits have codimension* 1, *there exist sections.*

30.13. Theorem. *Consider any Riemann G-manifold M. Then the following statements are equivalent.*

(1) $\operatorname{Hor}(M_{reg})$ *is integrable.*
(2) *Each G-equivariant normal field along a principal orbit is ∇^{Nor}-parallel.*
(3) *For $x \in M_{reg}$, S the normal slice at x and $X \in \mathfrak{g}$ and $s \in S$ arbitrary, $\zeta_X(s) \perp T_s(S)$.*

Proof. The equivalence of (1) and (2) is a direct consequence of (26.19) and remark (29.24). Furthermore, suppose (1); then there is an integral submanifold H of the horizontal distribution going through x. The submanifold H is totally geodesic by (26.13.1), and so $S = \exp_x(\operatorname{Nor}_r(G.x))$ is contained in H. Therefore, (3) holds: The fundamental vector field ζ_X is tangent to the orbit $G.s$ and so it is perpendicular to the horizontal distribution and to $T_s(S)$. Now if we suppose (3), then S is an integral submanifold of $\operatorname{Hor}(M_{\mathrm{reg}})$, and (1) holds. □

30.14. Remark. We already saw in (29.10) that $\operatorname{Nor} G.x$ is a trivial bundle. Now we even have a parallel global frame field. So the normal bundle to a regular orbit is flat.

30.15. Corollary. *Consider an orthogonal representation $\rho : G \to O(V)$ Let $x \in V$ be any regular point and Σ the linear subspace of V that is orthogonal to the orbit through x. Then the following statements are equivalent:*

(1) *V admits sections.*
(2) *Σ is a section.*
(3) *For all $y \in \Sigma$ and $X \in \mathfrak{g}$ we have $\zeta_X(y) \perp \Sigma$.*

Proof. (3) implies that the horizontal bundle is integrable (see (30.13)). In this case (30.11) implies (1). Also, (1) implies (2) by (30.9.5), and (2) obviously implies (3). □

30.16. Definition. An orthogonal representation of G is called a **polar representation** if it admits sections.

Corollary. *Let $G \subset O(V)$ be a polar representation of a compact Lie group, and let $v \in V$ be a regular point. Then*

$$\Sigma := \{w \in V : \zeta_{\mathfrak{g}}(w) \subseteq \zeta_{\mathfrak{g}}(v)\}$$

is the section through v, where $\zeta_{\mathfrak{g}}(w) := \{\zeta_X(w) : X \in \mathfrak{g}\} \subseteq V$.

Proof. Since $\zeta_{\mathfrak{g}}(v) = T_v(G.v)$ and by (30.15), a section through v is given by $\Sigma' := \zeta_{\mathfrak{g}}(v)^\perp$. If $z \in \Sigma'$, then $\zeta_{\mathfrak{g}}(z) \subseteq (\Sigma')^\perp$, which in our case implies that $\zeta_{\mathfrak{g}}(z) \subseteq \zeta_{\mathfrak{g}}(v)$. So $z \in \Sigma$.

Conversely, suppose z is a regular point in Σ. Consider the section $\Sigma'' = \zeta_{\mathfrak{g}}(z)^\perp$ through z. Then, since $\zeta_{\mathfrak{g}}(z) \subseteq \zeta_{\mathfrak{g}}(v)$, we also have that $\Sigma' = \zeta_{\mathfrak{g}}(v)^\perp \subseteq \zeta_{\mathfrak{g}}(z)^\perp = \Sigma''$. Therefore $\Sigma' = \Sigma''$ and, in particular, $z \in \Sigma'$. □

30.17. Lemma. *Let $G \subset O(V)$ be an orthogonal representation of a compact Lie group. Then for every $v \in V$ the normal space to the orbit*

$$\operatorname{Nor}_v := \operatorname{Nor}_v(G.v) = T_v(G.v)^\perp$$

meets every orbit.

Proof. Let $w \in V$ and consider $f : G \to \mathbb{R}$, $f(g) = \langle g.w, v\rangle$. Let g_0 be a critical point, e.g., a minimum on the compact group G; then $0 = df(g_0).(X.g_0) = \langle X.g_0.w, v\rangle = -\langle g_0.w, X.v\rangle$ for all $X \in \mathfrak{g}$. Thus $g_0.w \in \operatorname{Nor}_v(G.v)$. □

30.18. Lemma. *Let $G \subset O(V)$ be an orthogonal representation of a compact Lie group. For any regular $v_0 \in V$ the following assertions are equivalent:*

(1) *For any $v \in V_{reg}$ there exists $g \in G$ with $g.T_v(G.v) = T_{v_0}(G.v_0)$.*
(2) *$\operatorname{Nor}_{v_0}(G.v_0) = T_{v_0}(G.v_0)^\perp$ is a section.*

Proof. (1) $\Rightarrow$ (2) Let $\mathfrak{g} \subset \mathfrak{o}(V)$ be the Lie algebra of G. Consider the linear subspace

$$A := \{v \in \operatorname{Nor}_{v_0}(G.v_0) : \langle \mathfrak{g}.v, \operatorname{Nor}_{v_0}(G.v_0)\rangle = 0\}$$

of $\operatorname{Nor}_{v_0}(G.v_0) \subset V$. If (2) does not hold, then $A \subsetneq \operatorname{Nor}_{v_0}(G.v_0)$, and then $\dim(G.A) < \dim(V)$. So there exists $w \in V_{\text{reg}} \setminus G.A$, and by lemma (30.17) we may assume that $w \in \operatorname{Nor}_{v_0}(G.v_0)$. By (1) there exists $g \in G$ with $g.\operatorname{Nor}_w(G.w) = \operatorname{Nor}_{v_0}(G.v_0)$. This means $\operatorname{Nor}_{g.w}(G.w) = \operatorname{Nor}_{v_0}(G.v_0)$, a contradiction to $g.w \notin A$.

(2) $\Rightarrow$ (1) For any $w \in V_{\text{reg}}$ there exists $g \in G$ with $g.w \in \operatorname{Nor}_{v_0}(G.v_0)$, by (3.18). But then $g.\operatorname{Nor}_w(G.w) = \operatorname{Nor}_{g.w}(G.w) = \operatorname{Nor}_{v_0}(G.v_0)$, so (1) holds. $\square$

30.19. Theorem. *If $G \subset O(V)$ is a polar representation, then for any $v \in V$ with a section $\Sigma \subset \operatorname{Nor}_v(G.v)$, the isotropy representation $G_v \subset O(\operatorname{Nor}_v(G.v))$ is also polar with the same section $\Sigma \subset \operatorname{Nor}_v(G.v)$.*

Conversely, if there exists some $v \in V$ such that the isotropy representation $G_v \subset O(\operatorname{Nor}_v(G.v))$ is polar with section $\Sigma \subset \operatorname{Nor}_v(G.v)$, then also $G \subset O(V)$ is polar with the same section $\Sigma \subset V$.

Proof. Let $G \subset O(V)$ be polar with section Σ, and let $v \in \Sigma$ and $w \in \Sigma_{\text{reg}} = \Sigma \cap V_{\text{reg}}$.

Claim. Then $V = \Sigma \oplus \mathfrak{g}_v.w \oplus \mathfrak{g}.v$ is an orthogonal direct sum decomposition. Namely, we have $\langle \mathfrak{g}.\Sigma, \Sigma\rangle = 0$ so that

$$\langle \mathfrak{g}_v.w, \mathfrak{g}.v\rangle = \langle w, \mathfrak{g}.\underbrace{\mathfrak{g}_v.v}_{0}\rangle - \langle w, \underbrace{[\mathfrak{g}_v, \mathfrak{g}]}_{\subset \mathfrak{g}}.v\rangle = 0.$$

Since w is in V_{reg}, we have the orthogonal direct sum $V = \Sigma \oplus \mathfrak{g}.w$, so that $\dim(V) = \dim(\Sigma) + \dim(\mathfrak{g}) - \dim(\mathfrak{g}_w)$; and also we have $(\mathfrak{g}_v)_w = \mathfrak{g}_w$. Thus we get

$$\begin{aligned}\dim(\Sigma \oplus \mathfrak{g}_v.w \oplus \mathfrak{g}.v) &= \dim(\Sigma) + \dim(\mathfrak{g}_v) - \dim((\mathfrak{g}_v)_w) + \dim(\mathfrak{g}) - \dim(\mathfrak{g}_v)\\ &= \dim(\Sigma) + \dim(\mathfrak{g}_v) - \dim(\mathfrak{g}_w) + \dim(\mathfrak{g}) - \dim(\mathfrak{g}_v)\\ &= \dim(V)\end{aligned}$$

and the claim follows.

But then we see from the claim that $\operatorname{Nor}_v = \Sigma \oplus \mathfrak{g}_v.w$ is an orthogonal decomposition and that (30.18.1) holds, so that $G_v \subset \operatorname{Nor}_v$ is polar with section Σ.

Conversely, if $G_v \subset \operatorname{Nor}_v$ is polar with section Σ, we get the orthogonal decomposition $\operatorname{Nor}_v(G.v) = \Sigma \oplus \mathfrak{g}_v.\Sigma$ of $\operatorname{Nor}_v(G.v)$. This implies $\langle \Sigma, \mathfrak{g}.\Sigma\rangle = 0$. By lemma (30.17) we have $G.\operatorname{Nor}_v = V$. By polarity we have $G_v.\Sigma = \operatorname{Nor}_v$; thus finally $G.\Sigma = V$. So $G \subset O(V)$ is polar. $\square$

30.20. Theorem. *Let G be connected and $G \subset O(V = V_1 \oplus V_2)$ be a polar reducible representation, which is decomposed as $V = V_1 \oplus V_2$ as G-module. Then we have:*

(1) *Both G-modules V_1 and V_2 are polar, and any section Σ of V is of the form $\Sigma = \Sigma_1 \oplus \Sigma_2$ for sections Σ_i in V_i.*

(2) *Consider the connected subgroups*

$$G_1 := \{g \in G : g|\Sigma_2 = 0\}^o, \qquad G_2 := \{g \in G : g|\Sigma_1 = 0\}^o.$$

Then $G = G_1.G_2$, and $G_1 \times G_2$ acts on $V = V_1 \oplus V_2$ componentwise by $(g_1, g_2)(v_1 + v_2) = g_1.v_1 + g_2.v_2$, with the same orbits as G: $G.v = (G_1 \times G_2).v$ for any v.

Proof. Let $v = v_1 + v_2 \in \Sigma \cap V_{\text{reg}} \subset V = V_1 \oplus V_2$. Then $V = \Sigma \oplus \mathfrak{g}.v$; thus $v_i = s_i + X_i.v$ for $s_i \in \Sigma_i$ and $X_i \in \mathfrak{g}$. But then $s_i \in \Sigma_i \cap V_i =: \Sigma_i$ and $V_i = (\Sigma \cap V_i) \oplus \mathfrak{g}.v_i$ and the assertion (1) follows.

Moreover $\text{Nor}_{v_1} = (\mathfrak{g}.v_1)^\perp = \Sigma_1 \oplus V_2$, and by theorem (30.19) the action of G_{v_1} on this space is polar with section $\Sigma_1 \oplus \Sigma_2$. Thus we have $\mathfrak{g}_{v_1} = \mathfrak{g}_2 := \mathfrak{g}_{\Sigma_1}$ and $\mathfrak{g}_{v_1}$ acts only on V_2 and vanishes on V_1 and we get $V_2 = \Sigma_2 \oplus \mathfrak{g}_{v_1} v_2 = \Sigma_2 \oplus \mathfrak{g}.v_2$. Similarly $\mathfrak{g}_{v_2} = \mathfrak{g}_1 := \mathfrak{g}_{\Sigma_2}$ and $\mathfrak{g}_{v_2}$ acts only on V_1 and vanishes on V_2, and $V_1 = \Sigma_1 \oplus \mathfrak{g}_{v_2} v_1 = \Sigma_1 \oplus \mathfrak{g}.v_1$. Thus $\mathfrak{g} = \mathfrak{g}_1 + \mathfrak{g}_2$ and consequently $G = G_1.G_2 = G_2.G_1$ by compactness of G_i. For any $g \in G$ we have $g = g_1.g_2 = g_2'.g_1'$ for $g_i, g_i' \in G_i$. For $u = u_1 + u_2 \in V_1 \oplus V_2 = V$ we then have $g.(u_1 + u_2) = g_1.g_2.u_1 + g_2'.g_1'.u_2 = g_1.u_1 + g_2'.u_2$; thus $G.u \subseteq (G_1 \times G_2).u$. Since both orbits have the same dimension, $G.u$ is open in $(G_1 \times G_2).u$; since all groups are compact and connected, the orbits coincide. □

30.21. The generalized Weyl group of a section. Consider a complete Riemann G-manifold M which admits sections. For any closed subset S of M we define the largest subgroup of G which induces an action on S:

$$N(S) := \{g \in G : \ell_g(S) = S\}$$

and the subgroup consisting of all $g \in G$ which act trivially on S:

$$Z(S) := \{g \in G : \ell_g(s) = s, \text{ for all } s \in S\}.$$

Then, since S is closed, $N(S)$ is closed, hence a Lie subgroup of G. The centralizer $Z(S) = \bigcap_{s \in S} G_s$ is closed as well and is a normal subgroup of $N(S)$. Therefore, $N(S)/Z(S)$ is a Lie group, and it acts on S effectively.

If we take for S a section Σ, then the above constructed group is called the *generalized Weyl group* of Σ and is denoted by

$$W(\Sigma) = N(\Sigma)/Z(\Sigma).$$

For any regular point $x \in \Sigma$, G_x acts trivially on the normal slice S_x at x (by (29.7)). Since $\Sigma = \exp_x \operatorname{Nor}_x(G.x)$ by (30.9.5), S_x is an open subset of Σ, and we see that G_x acts trivially on all of Σ. So we have $G_x \subseteq Z(\Sigma)$. On the other hand, $Z(\Sigma) \subseteq G_x$ is obvious; therefore

$$Z(\Sigma) = G_x \qquad \text{for } x \in \Sigma \cap M_{\text{reg}}.$$

Now, since $Z(\Sigma)$ is a normal subgroup of $N(\Sigma)$, we have $N(\Sigma) \subseteq N(G_x)$ where the second N stands for the normalizer in G. So we have

$$W(\Sigma) \subseteq N(G_x)/G_x \qquad \text{for } x \in \Sigma \cap M_{\text{reg}}.$$

30.22. Proposition. *Let M be a proper Riemann G-manifold and let Σ be a section; then the associated Weyl group $W(\Sigma)$ is discrete. If Σ' is a different section, then there is an isomorphism $W(\Sigma) \to W(\Sigma')$ induced by an inner automorphism of G. It is uniquely determined up to an inner automorphism of $W(\Sigma)$.*

Proof. Take a regular point $x \in \Sigma$ and consider the normal slice S_x. Then $S_x \subseteq \Sigma$ is open. Therefore, any g in $N(\Sigma)$ close to the identity element maps x back into S_x. By (6.15.2), the element g then lies in $G_x = Z(\Sigma)$. So $Z(\Sigma)$ is an open subset of $N(\Sigma)$, and the quotient $W(\Sigma)$ is discrete.

If Σ' is another section, then $\Sigma' = g.\Sigma$ where $g \in G$ is uniquely determined up to $N(\Sigma)$. Clearly, $\operatorname{conj}_g : G \to G$ induces isomorphisms

$$\operatorname{conj}_g : N(\Sigma) \xrightarrow{\cong} N(\Sigma'),$$
$$Z(\Sigma) \xrightarrow{\cong} Z(\Sigma'),$$

and therefore it factors to an isomorphism $W(\Sigma) \xrightarrow{\cong} W(\Sigma')$. □

30.23. Example. Any finite group is a generalized Weyl group in the appropriate setting. That is, to an arbitrary finite group W we will now construct a setting in which it occurs as a Weyl group. Let G be a compact Lie group and H a closed subgroup such that $W \subseteq N(H)/H$ (this is always possible since any finite group can be regarded as a subgroup of $O(V) =: G$ so we need only choose $H = \{e\}$). Next, take a smooth manifold Σ on which W acts effectively. Consider the inverse image of W under the quotient map $\pi : N(H) \to N(H)/H$, $K := \pi^{-1}(W)$. Then the action of W induces a K-action on Σ as well. The smooth manifold $M := G \times_K \Sigma$ has a left G-action. Let $-B$ denote the G-invariant Riemann metric on G induced by the Cartan-Killing form on the semisimple part and any inner product on the center, and let γ_Σ be a W-invariant Riemann metric on Σ. Then the Riemann metric $-B \times \gamma_\Sigma$ on $G \times \Sigma$ induces a G-invariant Riemann metric on the quotient space $G \times_K \Sigma$. With this, $G \times_K \Sigma$ is a Riemann G-manifold, and if $q : G \times \Sigma \to G \times_K \Sigma$ is the quotient map, then $q(\{e\} \times \Sigma) \cong \Sigma$ meets every G-orbit orthogonally. So it is a section. The largest subgroup of G acting on Σ is K and the largest one acting trivially on Σ is H. Therefore, $W(\Sigma) = K/H = W$ is the Weyl group associated to the section Σ.

30.24. Theorem. *Let M be a proper Riemann G-manifold with sections. Then, for any $x \in M$, the slice representation $G_x \to O(\mathrm{Nor}_x(G.x))$ is a polar representation. If Σ is a section through x in M, then $T_x\Sigma$ is a section in $\mathrm{Nor}_x(G.x)$ for the slice representation. Furthermore,*

$$W(T_x\Sigma) = W(\Sigma)_x.$$

Proof. Clearly $T_x\Sigma \subseteq \mathrm{Nor}_x(G.x)$. We begin by showing that it has the right codimension. Take a $\xi \in \mathrm{Nor}_x(G.x)$ close to 0_x; then $(G_x)_\xi = G_y$ for $y = \exp_x^\gamma \xi$, since $\exp_x$ is a G_x-equivariant diffeomorphism in a neighborhood of 0_x. So $G_x.\xi \cong G_x/(G_x)_\xi = G_x/G_y$. Let us now calculate the codimension of $G_x.\xi$ in $\mathrm{Nor}_x(G.x)$:

$$\begin{aligned}\dim(\mathrm{Nor}_x(G.x)) - \dim(G_x.\xi) &= \dim(\mathrm{Nor}_x(G.x)) - \dim(G_x) + \dim(G_y)\\ &= \underbrace{\dim(\mathrm{Nor}_x(G.x)) + \dim(G/G_x)}_{=\dim M} - \underbrace{(\dim G - \dim(G_y))}_{=\dim G/G_y} = \mathrm{codim}_M(G.y).\end{aligned}$$

Since regular points form a dense subset, we can choose $\xi \in T_x\Sigma$ regular by assuming that $y = \exp_x^\gamma(X)$ is regular in Σ. Then y is regular as well and we get

$$\mathrm{codim}_{\mathrm{Nor}_x(G.x)}(G_x.\xi) = \mathrm{codim}_M(G.y) = \dim\Sigma = \dim(T_x\Sigma).$$

So $T_x\Sigma$ is a linear subspace of $\mathrm{Nor}_x G.x$ with the right codimension for a section. Therefore, if we show that $T_x\Sigma$ is orthogonal to each orbit it meets, then it is already the entire orthogonal complement of a regular orbit, and by corollary (30.15), we know that it meets every orbit.

Denote the G-action on M by $\ell : G \to \mathrm{Isom}(M)$. If $\xi \in T_x\Sigma$ is arbitrary, then it remains to prove that for all $\eta \in T_x\Sigma$ and $X \in \mathfrak{g}_x$:

$$\gamma_x(\eta, \zeta_X^{T\ell|G_x}(\xi)) = 0.$$

To do this, choose a smooth 1-parameter family $\eta(t) \in T_{\exp(t\xi)}\Sigma$ such that $\eta(0) = \eta$ and $\nabla_{\partial_t}\eta = 0$. Since Σ is a section in M, we know for each single t that

$$\gamma_{\exp(t\xi)}(\zeta_X^\ell(\exp^\gamma(t\xi)), \eta(t)) = 0.$$

If we differentiate this equation, we get

$$0 = \partial_s|_0\gamma(\zeta_X^\ell(\exp^\gamma(s\xi)), \eta(s)) = \gamma\big(\nabla_{\partial_s}\zeta_X^\ell(\exp^\gamma(s\xi)), \eta(0)\big).$$

So it remains to show that $\nabla_{\partial_s}\zeta^\ell_X(\exp^\gamma(s\xi))$ is the fundamental vector field of X at ξ for the slice representation:

$$\begin{aligned}\nabla_{\partial_s}\zeta^\ell_X(\exp^\gamma(s\xi)) &= \nabla_\xi\zeta^\ell_X = K\circ T\zeta^\ell_X.\xi\\ &= K\circ T(\partial_t|_0\ell_{\exp^G(tX)}).\partial_s|_0\exp^\gamma_x(s\xi)\\ &= K.\partial_s|_0.\partial_t|_0\ell_{\exp^G(tX)}(\exp^\gamma_x(s\xi))\\ &= K.\kappa_M.\partial_t|_0.\partial_s|_0\ell_{\exp^G(tX)}(\exp^\gamma_x(s\xi))\\ &= K.\kappa_M.\partial_t|_0.T(\ell_{\exp^G(tX)})(\xi).\end{aligned}$$

Here, K denotes the connector and κ_M the canonical flip between the two structures of TTM, and we use the identity $K\circ\kappa = K$, which is a consequence of the symmetry of the Levi-Civita connection. The argument of K in the last expression is vertical already since $X\in\mathfrak{g}_x$. Therefore we can replace K by the vertical projection and get

$$\nabla_{\partial_s}\zeta^\ell_X(\exp^\gamma(s\xi)) = \operatorname{vpr}\partial_t|_0T_x(\ell_{\exp^G(tX)}).\xi = \zeta^{T_2\ell|G_x}_X(\xi).$$

So $\zeta^{T_2\ell|G_x}_X(\xi)$ intersects $T_x\Sigma$ orthogonally, and therefore $T_x\Sigma$ is a section.

Now consider $N_{G_x}(T_x(\Sigma)) = \{g\in G_x : T_x(\ell_g).T_x\Sigma = T_x\Sigma\}$. Clearly, $N_G(\Sigma)\cap G_x\subseteq N_{G_x}(T_x(\Sigma))$. On the other hand, any $g\in N_{G_x}(T_x(\Sigma))$ leaves Σ invariant as the following argument shows.

For any regular $y\in\Sigma$ we have $\Sigma=\exp_y\operatorname{Nor}(G.y)$. Therefore $x=\exp_y\eta$ for a suitable $\eta\in T_y\Sigma$, and conversely, y can be written as $y=\exp_x\xi$ for $\xi=-\partial_t|_1\exp_y(t\eta)\in T_x\Sigma$. Now $g.y=g.\exp_x\xi=\exp_x T_x\ell_g.\xi$ lies in Σ, since $T_x\ell_g.\xi$ lies in $T_x\Sigma$. So g maps all regular points in Σ back into Σ. Since these form a dense subset and since ℓ_g is continuous, we get $g\in N_G(\Sigma)$.

We have now shown that

$$N_{G_x}(T_x\Sigma) = N_G(\Sigma)\cap G_x.$$

Analogous arguments used on $Z_{G_x}(T_x\Sigma)$ give

$$Z_{G_x}(T_x\Sigma) = Z_G(\Sigma),$$

and we see that

$$W_{G_x}(T_x\Sigma) = (N(\Sigma)\cap G_x)/Z(\Sigma) = W(\Sigma)_x. \qquad \square$$

30.25. Corollary. *Let M be a Riemann G-manifold admitting sections and let $x\in M$. Then for any section Σ through x we have*

$$\operatorname{Nor}_x(G.x)^{G^0_x}\subseteq T_x\Sigma,$$

where G^0_x is the connected component of the isotropy group G_x at x.

Proof. By theorem (30.24) the tangent space $T_x\Sigma$ is a section for the slice representation $G_x\to O(\operatorname{Nor}_x(G.x))$. Let $\xi\in T_x\Sigma$ be a regular vector for the slice representation. By corollary (30.16) we have $T_x\Sigma = \{\eta \in \operatorname{Nor}_x(G.x) : \zeta_{\mathfrak{g}_x}(\eta) \subset \zeta_{\mathfrak{g}_x}(\xi)\}$. Since $\operatorname{Nor}_x(G.x)^{G_x^0}$ consists of all η in $\operatorname{Nor}_x(G.x)$ with $\zeta_{\mathfrak{g}_x}(\eta) = 0$, the result follows. □

30.26. Corollary. *Let M be a proper Riemann G-manifold admitting sections and let $x \in M$. Then G_x acts transitively on the set of all sections through x.*

Proof. Consider two arbitrary sections Σ_1 and Σ_2 through x and a normal slice S_x at x. By theorem (30.24), $T_x\Sigma_2$ is a section for the slice representation. Since $\exp_x$ can be restricted to a G_x-equivariant diffeomorphism onto S_x, $\Sigma_2 \cap S_x$ is a section for the G_x-action on S_x. Next, choose a regular point $y \in \Sigma_1 \cap S_x$. Its G_x-orbit meets the section $\Sigma_2 \cap S_x$, that is, we can find a $g \in G_x$ such that $g.y \in \Sigma_2$. Now Σ_2 and $g.\Sigma_1$ are both sections containing the regular point $g.y$. Therefore they are equal. □

30.27. Corollary. *Let M be a proper G-manifold with sections, let Σ be a section of M and let $x \in \Sigma$. Then*

$$G.x \cap \Sigma = W(\Sigma).x.$$

Proof. The inclusion ($\supseteq$) is clear. Now we have

$$y \in G.x \cap \Sigma \quad \iff \quad y = g.x \in \Sigma \text{ for some } g \in G.$$

Take this g and consider the section $\Sigma' := g.\Sigma$. Then Σ and Σ' are both sections through y, and by (30.26) there is a $g' \in G_y$ which carries Σ' back into Σ. Now $g'g.\Sigma = \Sigma$, that is, $g'g \in N(\Sigma)$, and $g'g.x = g'.y = y$. So $y \in N(\Sigma).x = W(\Sigma).x$. □

30.28. Corollary. *If M is a proper G-manifold with section Σ, then the inclusion of Σ into M induces a homeomorphism j between the orbit spaces:*

$$\begin{array}{ccc} \Sigma & \xrightarrow{\ i\ } & M \\ {\scriptstyle \pi_\Sigma}\downarrow & & \downarrow{\scriptstyle \pi_M} \\ \Sigma/W(\Sigma) & \xrightarrow{\ j\ } & M/G \end{array}$$

(but it does not necessarily preserve orbit types; see remark (6.17)).

Proof. By the preceding corollary there is a one to one correspondence between the G-orbits in M and the $W(G)$-orbits in Σ, so j is well defined and bijective. Since $j \circ \pi_\Sigma = \pi_M \circ i$ and π_Σ is open, j is continuous.

Consider any open set $U \subseteq \Sigma/W(\Sigma)$. We now have to show that

$$\pi_M^{-1} j(U) = G.\pi_\Sigma^{-1}(U)$$

is an open subset of M (since then $j(U)$ is open and j^{-1} is continuous). Take any $x \in \pi_M^{-1} j(U)$. We assume $x \in \Sigma$ (otherwise it can be replaced by a suitable $g.x \in \Sigma$). So $x \in \pi_\Sigma^{-1}(U)$. Let S_x be a normal slice at x; then $\Sigma \cap S_x$ is a submanifold of S_x of dimension $\dim \Sigma$. In S_x, x has arbitrarily small G_x-invariant neighborhoods, since the slice action is orthogonal and S_x is G-equivariantly diffeomorphic to an open ball in $\operatorname{Nor}_x(G.x)$. Let V_x be such an open neighborhood of x, small enough for $V_x \cap \Sigma$ to be contained in $\pi_\Sigma^{-1}(U)$. Then V_x is again a slice; therefore $G.V_x$ is open in M by (6.15.3). Now we have to check whether $G.V_x$ is really a subset of $\pi_M^{-1} j(U)$. Using corollary (30.26), we get

$$G.(V_x \cap \Sigma) = G.G_x(V_x \cap \Sigma) = G.(V_x \cap G_x.\Sigma) = G.V_x.$$

Therefore, $G.V_x \subseteq G.\pi_\Sigma^{-1}(U) = \pi_M^{-1} j(U)$ where it is an open neighborhood of x. So $\pi_M^{-1} j(U)$ is an open subset of M, $j(U)$ is open in M/G, and j^{-1} is continuous. □

30.29. Corollary. *Let M be a proper Riemann G-manifold and $\Sigma \subseteq M$ a section with Weyl group W. Then the inclusion $i : \Sigma \hookrightarrow M$ induces an isomorphism*

$$C^0(M)^G \xrightarrow{i^*} C^0(\Sigma)^W.$$

Proof. By corollary (30.27) we see that every $f \in C^0(\Sigma)^W$ has a unique G-equivariant extension $\tilde{f}$ onto V. If we consider once more the diagram

$$\begin{array}{ccc} \Sigma & \xrightarrow{i} & M \\ {\scriptstyle \pi_\Sigma}\downarrow & & \downarrow{\scriptstyle \pi_M} \\ \Sigma/W(\Sigma) & \xrightarrow{j} & M/G, \end{array}$$

we see that f factors over π_Σ to a map $f' \in C^0(\Sigma/W(\Sigma))$, and since j is a homeomorphism by (30.28), we get for the G-invariant extension $\tilde{f}$ of f:

$$\tilde{f} = f' \circ j^{-1} \circ \pi_M \in C^0(M)^G. \qquad \square$$

30.30. Theorem ([188, 4.12] or [219, theorem D]). *Let $G \to GL(V)$ be a polar representation of a compact Lie group G, with section Σ and generalized Weyl group $W = W(\Sigma)$.*

Then the algebra $\mathbb{R}[V]^G$ of G-invariant polynomials on V is isomorphic to the algebra $\mathbb{R}[\Sigma]^W$ of W-invariant polynomials on the section Σ, via the restriction mapping $f \mapsto f|_\Sigma$.

Remark. This seemingly very algebraic theorem is actually a consequence of the geometry of the orbits. This already becomes evident in the case of a degree 1 homogeneous polynomial. To see that the G-invariant extension of $p \in \mathbb{B}[\Sigma]_1^W$ to V is again a polynomial (and again of degree 1), we we must assume the following convexity result of Terng.

Under the conditions of the theorem, for every regular orbit $G.x$ the orthogonal projection onto Σ, $\mathrm{pr}(G.x)$, is contained in the convex hull of $G.x \cap \Sigma$ (this is a finite subset of Σ by (30.27) since G is compact and $W(\Sigma)$ is discrete).

Let us make this assumption. Denote by $\tilde{p}$ the unique G-invariant extension of p; then clearly $\tilde{p}$ is homogeneous. Now, notice that for any orbit $G.x$, p is constant on the convex hull of $G.x \cap \Sigma =: \{g_1.x, g_2.x, \dots, g_k.x\}$. Just take any $s = \sum \lambda_i g_i.x$ with $\sum \lambda_i = 1$; then

$$p(s) = \sum \lambda_i p(g_i.x) = p(g_1.x) \sum \lambda_i = p(g_1.x).$$

With this and with our assumption we can show that for regular points $u, v \in M$, $\tilde{p}(u+v) = \tilde{p}(u) + \tilde{p}(v)$. Suppose without loss of generality that $u + v \in \Sigma$; then

$$p(u+v) = p(\mathrm{pr}(u) + \mathrm{pr}(v)) = p(\mathrm{pr}(u)) + p(\mathrm{pr}(v)).$$

At this point, the convexity theorem asserts that $\mathrm{pr}(u)$ and $\mathrm{pr}(v)$ can be written as convex combinations of elements of $G.u \cap \Sigma$ and $G.v \cap \Sigma$, respectively. If we fix an arbitrary g_u (resp. g_v) in G such that $g_u.u$ (resp. $g_v.v$) lie in Σ, then by the above argument we get

$$p(\mathrm{pr}(u)) = p(g_u.u) \quad \text{and} \quad p(\mathrm{pr}(v)) = p(g_v.v).$$

So we have

$$p(u+v) = p(g_u.u) + p(g_v.v) = \tilde{p}(u) + \tilde{p}(v),$$

and $\tilde{p}$ is linear on V_{reg}. Since the regular points are a dense subset of V and since $\tilde{p}$ is continuous by (30.29), $\tilde{p}$ is linear altogether.

A proof of the convexity theorem can be found in [**218**] or again in [**189**, pp. 168–170]. For a proof of theorem (30.30) we refer to [**219**]. In both sources the assertions are shown for the more general case where the principal orbits are replaced by *isoparametric submanifolds* (i.e., submanifolds of a space form with flat normal bundle and whose principal curvatures along any parallel normal field are constant; compare (29.12) and (30.14)). To any isoparametric submanifold there is a singular foliation which generalizes the orbit foliation of a polar action but retains many of its fascinating properties.

In connection with the example we studied in (7.1), the convexity theorem from above yields the following classical result of [**202**]:

Let $M \subseteq S(n)$ be the subset of all symmetric matrices with fixed distinct eigenvalues $a_1, \dots, a_n$ and $\mathrm{pr} : S(n) \to R^n$ defined by

$$\mathrm{pr}(x_{ij}) := (x_{11}, x_{22}, \dots, x_{nn});$$

then $\mathrm{pr}(M)$ is contained in the convex hull of the permutation group orbit $\mathcal{S}_n.a$ through $a = (a_1, \dots, a_n)$.

30.31. Theorem. *Let M be a proper Riemann G-manifold with section Σ and Weyl group W. Then the inclusion $i : \Sigma \hookrightarrow M$ induces an isomorphism*

$$C^\infty(M)^G \xrightarrow{i^*} C^\infty(\Sigma)^{W(\Sigma)}.$$

Proof. Clearly $f \in C^\infty(M)^G$ implies $i^* f \in C^\infty(\Sigma)^W$. By (30.29) we know that every $f \in C^\infty(\Sigma)^W$ has a unique continuous G-invariant extension $\tilde{f}$. We now have to show that $\tilde{f} \in C^\infty(M)^G$.

Let us take an $x \in M$ and show that $\tilde{f}$ is smooth at x. Actually, we can assume $x \in \Sigma$, because if $\tilde{f}$ is smooth at x, then $\tilde{f} \circ \ell_{g^{-1}}$ is smooth at $g.x$, so $\tilde{f}$ is smooth at $g.x$ as well. Now let S_x denote a normal slice at x. Then we have

$$\begin{array}{ccccc} G.S_x & \xleftarrow{\quad I \quad} & G \times_{G_x} S_x & \xleftarrow{\quad q \quad} & G \times S_x \\ & \searrow \qquad \swarrow & & {\scriptstyle \tilde{f}\circ I}\searrow \qquad \swarrow{\scriptstyle \tilde{f}|_{S_x}\circ \mathrm{pr}_2} & \\ & G/G_x & & \mathbb{R}. & \end{array}$$

Since in the above diagram I is an isomorphism and q a submersion, it is sufficient to show that $\tilde{f}|_{S_x} \circ \mathrm{pr}_2$, or equivalently, that $\tilde{f}|_{S_x}$ is smooth at x. Let $B \subseteq T_x S_x$ be a ball around 0_x such that $B \cong S_x$ and $T_x \Sigma \cap B \cong \Sigma \cap S_x$. Then, by theorem (30.24), the G_x-action on S_x is basically a polar representation (up to diffeomorphism). So it remains to show the following:

Claim. If Σ is a section of a polar representation $G_x \to O(V)$ with Weyl group W_x and f is a smooth W_x-invariant function on Σ, then f extends to a smooth G_x-invariant function $\tilde{f}$ on V.

In order to show this, let $\rho_1, \dots, \rho_k$ be a system of homogeneous Hilbert generators for $\mathbb{R}[\Sigma]^{W_x}$. Then, by Schwarz's theorem (7.13), there is an $f' \in C^\infty(\mathbb{R}^k)$ such that $f = f' \circ (\rho_1, \dots, \rho_k)$. By theorem (30.30), each ρ_i extends to a polynomial $\tilde{\rho}_i \in \mathbb{R}[V]^{G_x}$. Therefore we get that

$$\tilde{f} := f' \circ (\tilde{\rho}_1, \dots, \tilde{\rho}_k) : V \to \mathbb{R}$$

is a smooth G_x-invariant extension of f. $\square$

30.32. Basic differential forms. Our next aim is to show that pull-back along the embedding $\Sigma \to M$ induces an isomorphism $\Omega^p_{\text{hor}}(M)^G \cong \Omega^p(\Sigma)^{W(\Sigma)}$ for each p, where a differential form ω on M is called *horizontal* if it kills each vector tangent to some orbit. For each point x in M, the slice representation of the isotropy group G_x on the normal space $T_x(G.x)^\perp$ to the tangent space to the orbit through x is a polar representation. The first step is to show that the result holds for polar representations. This is done in theorem (30.40). The method used there is inspired by [**211**]. Then the general result is proven, following [**153, 155**].

As usual, for a Lie group G we denote by $\mathfrak{g}$ its Lie algebra, the multiplication by $\mu : G \times G \to G$; for $g \in G$ let $\mu_g, \mu^g : G \to G$ denote the left and right translation. Let $\ell : G \times M \to M$ be a left action of the Lie group G on a smooth manifold M. We consider the partial mappings $\ell_g : M \to M$ for $g \in G$ and $\ell^x : G \to M$ for $x \in M$ and the fundamental vector field mapping $\zeta : \mathfrak{g} \to \mathfrak{X}(M)$ given by $\zeta_X(x) = T_e(\ell^x)X$. Since ℓ is a left action, the negative $-\zeta$ is a Lie algebra homomorphism.

A differential form $\varphi \in \Omega^p(M)$ is called *G-invariant* if $(\ell_g)^*\varphi = \varphi$ for all $g \in G$ and *horizontal* if φ kills each vector tangent to a G-orbit: $i_{\zeta_X}\varphi = 0$ for all $X \in \mathfrak{g}$. We denote by $\Omega^p_{\text{hor}}(M)^G$ the space of all horizontal G-invariant p-forms on M. They are also called *basic forms.*

30.33. Lemma. *Under the exterior differential, the space $\Omega_{hor}(M)^G$ of basic forms is a subcomplex of $\Omega(M)$.*

The cohomology of the complex $(\Omega_{\text{hor}}(M)^G, d)$ is called the *basic cohomology* of the G-manifold M.

Proof. If $\varphi \in \Omega_{\text{hor}}(M)^G$, then the exterior derivative $d\varphi$ is clearly G-invariant. For $X \in \mathfrak{g}$ we have

$$i_{\zeta_X} d\varphi = i_{\zeta_X} d\varphi + d i_{\zeta_X}\varphi = \mathcal{L}_{\zeta_X}\varphi = 0,$$

so $d\varphi$ is also horizontal. □

30.34. Lemma. *Let f, $g : M \to N$ be smooth G-equivariant mappings between G-manifolds which are G-equivariantly C^∞-homotopic: $\mathbb{R} \times M$ is again a G-manifold (with the action on M only), and there exists a G-equivariant $h \in C^\infty(\mathbb{R} \times M, N)$ with $h(0,x) = f(x)$ and $h(1,x) = g(x)$.*

Then f and g induce the same mapping in basic cohomology:

$$f^* = g^* : H_{basic}(N) \to H_{basic}(M).$$

Proof. We recall the proof of (11.4) where we showed this without G. For $\omega \in \Omega^k_{\text{hor}}(N)^G$ we have $h^*\omega \in \Omega^k_{\text{hor}}(\mathbb{R} \times M)^G$ since h is equivariant. The insertion operator $\text{ins}_t : M \to \mathbb{R} \times M$, given by $\text{ins}_t(x) = (t, x)$, is also equivariant. The integral operator $I_0^1(\varphi) := \int_0^1 \mathrm{ins}_t^* \varphi\, dt$ commutes with the insertion of fundamental vector fields and with the G-action, so it induces an operator $I_0^1 : \Omega^k_{\mathrm{hor}}(\mathbb{R}\times M)^G \to \Omega^k_{\mathrm{hor}}(M)^G$. Let $T := \frac{\partial}{\partial t} \in \mathfrak{X}(\mathbb{R}\times M)$ be the unit vector field in direction $\mathbb{R}$. Thus the *homotopy operator* $\bar h := I_0^1 \circ i_T \circ h^* : \Omega^k_{\mathrm{hor}}(N)^G \to \Omega^{k-1}_{\mathrm{hor}}(M)^G$ is well defined, and from the proof of (11.4) we conclude that it still satisfies

$$\begin{aligned} g^* - f^* &= (h\circ \mathrm{ins}_1)^* - (h\circ \mathrm{ins}_0)^* \\ &= (\mathrm{ins}_1^* - \mathrm{ins}_0^*)\circ h^* \\ &= (d\circ I_0^1 \circ i_T + I_0^1\circ i_T \circ d)\circ h^* \\ &= d\circ \bar h - \bar h \circ d, \end{aligned}$$

which implies the desired result since for $\omega \in \Omega^k_{\mathrm{hor}}(M)^G$ with $d\omega = 0$ we have $g^*\omega - f^*\omega = \bar h d\omega + d\bar h\omega = d\bar h\omega$. □

30.35. Basic lemma of Poincaré ([111]). *Let $\ell : G\times M \to M$ be a proper G-manifold. For $k > 0$ let $\omega \in \Omega^k_{hor}(M)^G$ be a basic k-form on M with $d\omega = 0$ in a G-invariant neighborhood of an orbit $G.x$. Then there exists a basic form $\varphi \in \Omega^{k-1}_{hor}(M)^G$ such that $d\omega = \varphi$ in a G-invariant neighborhood of $G.x$.*

Proof. By (6.30) we may assume that M is a Riemann G-manifold. Let S_x be a slice with center x which is diffeomorphic to a small ball in $T_x(G.x)^\perp \subset T_xM$ and thus contractible.

We denote again by $\ell : G_x \times S_x \to S_x$ the induced action of the isotropy group G_x on the slice. Then $d\omega = 0$ on the G-invariant neighborhood $G.S_x$ which is G-equivariantly diffeomorphic to the associated bundle $G\times_{G_x} S_x$ by (6.18). The quotient mapping

$$q : G\times S_x \to G\times_{G_x} S_x$$

is the projection of a principal G_x bundle by (18.7.3) for the right action $R^h(g,y) = (gh, h^{-1}.y)$, and it is equivariant for the left G-action (acting on G alone). Thus $q^*\omega$ is still G-horizontal and G-invariant on $G\times S_x$ and thus $q^*\omega$ is of the form $\mathrm{pr}_{S_x}\,\alpha$ for a unique form α on S_x. Moreover $q^*\omega$ is also horizontal and invariant for the right G_x-action by (19.14). So α is a G_x-basic form on S_x.

Now S_x is G_x-equivariantly diffeomorphic to a ball in a vector space where G_x acts linearly and isometrically. This ball is G_x-equivariantly contractible to 0 via $v \to r.v$, $r\in[0,1]$. Thus the basic cohomology of $H^k_{\mathrm{basic}}(S_x)$ vanishes for $k>0$, and there exists a G_x-basic form $\beta \in \Omega^{k-1}_{\mathrm{hor}}(S_x)^{G_x}$ with $d\beta = \alpha$. Then $\mathrm{pr}^*_{S_x}\beta$ is G_x-basic and G-basic on $G\times S_x$, so it induces a form $\varphi \in \Omega^{k-1}_{\mathrm{hor}}(G\times_{G_x} S_x)^G$ which satisfies $d\psi = \omega$. □

30.36. Theorem. *For a proper G-manifold M the basic cohomology $H^*_{basic}(M)$ coincides with the real cohomology of the Hausdorff orbit space M/G in the sense of Čech or in the sense of singular cohomology.*

Sketch of proof. On the category of proper G-manifolds and smooth G-equivariant mappings the basic cohomology satisfies the axioms for cohomology listed in (11.11). We proved all but the Mayer-Vietoris property, for which the proof (11.10) applies without any change. Pushing these properties down to the orbit spaces, they suffice to prove that basic cohomology equals singular or Čech cohomology of the orbit space with real coefficients, via the abstract theorem of de Rham in sheaf theory. □

30.37. Theorem ([153, 155]). *Let $M \times G \to M$ be a proper isometric right action of a Lie group G on a smooth Riemann manifold M, which admits a section Σ.*

Then the restriction of differential forms induces an isomorphism

$$\Omega^p_{hor}(M)^G \xrightarrow{\cong} \Omega^p(\Sigma)^{W(\Sigma)}$$

between the space of horizontal G-invariant differential forms on M and the space of all differential forms on Σ which are invariant under the action of the generalized Weyl group $W(\Sigma)$ of the section Σ.

Proof of injectivity in (30.37). Let $i : \Sigma \to M$ be the embedding of the section. It clearly induces a linear mapping $i^* : \Omega^p_{\text{hor}}(M)^G \to \Omega^p(\Sigma)^{W(\Sigma)}$ which is injective by the following argument: Let $\omega \in \Omega^p_{\text{hor}}(M)^G$ with $i^*\omega = 0$. For $x \in \Sigma$ we have $i_X\omega_x = 0$ for $X \in T_x\Sigma$ since $i^*\omega = 0$ and also for $X \in T_x(G.x)$ since ω is horizontal. Let $x \in \Sigma \cap M_{\text{reg}}$ be a regular point; then $T_x\Sigma = (T_x(G.x))^\perp$ and so $\omega_x = 0$. This holds along the whole orbit through x since ω is G-invariant. Thus $\omega|M_{\text{reg}} = 0$, and since M_{reg} is dense in M, $\omega = 0$. □

So it remains to show that i^* is surjective. This will be done in (30.44) below.

30.38. Lemma. *Let $\ell \in V^*$ be a linear functional on a finite-dimensional vector space V, and let $f \in C^\infty(V, \mathbb{R})$ be a smooth function which vanishes on the kernel of ℓ, so that $f|\ell^{-1}(0) = 0$. Then there is a unique smooth function g such that $f = \ell.g$.*

Proof. Choose coordinates $x^1, \dots, x^n$ on V such that $\ell = x^1$. Then we have $f(0, x^2, \dots, x^n) = 0$ and therefore

$$f(x^1, \dots, x^n) = \int_0^1 \partial_1 f(tx^1, x^2, \dots, x^n)dt.x^1 = g(x^1, \dots, x^n).x^1. \quad \square$$

30.39. Question. *Let $G \to GL(V)$ be a representation of a compact Lie group in a finite-dimensional vector space V. Let*

$$\rho = (\rho_1, \dots, \rho_m) : V \to \mathbb{R}^m$$

be the polynomial mapping whose components ρ_i are a minimal set of homogeneous generators for the algebra $\mathbb{R}[V]^G$ of invariant polynomials.

We consider the pullback homomorphism $\rho^* : \Omega^p(\mathbb{R}^m) \to \Omega^p(V)$. Is it surjective onto the space $\Omega^p_{hor}(V)^G$ of G-invariant horizontal smooth p-forms on V?

See remark (30.41) for a class of representations where the answer is yes.

In general the answer is no. A counterexample is the following: Let the cyclic group $\mathbb{Z}_n = \mathbb{Z}/n\mathbb{Z}$ of order n, viewed as the group of n-th roots of unity, act on $\mathbb{C} = \mathbb{R}^2$ by complex multiplication. A generating system of polynomials consists of $\rho_1 = |z|^2$, $\rho_2 = \operatorname{Re}(z^n)$, $\rho_3 = \operatorname{Im}(z^n)$. But then each $d\rho_i$ vanishes at 0 and there is no chance of having the horizontal invariant volume form $dx \wedge dy$ in $\rho^*\Omega(\mathbb{R}^3)$.

30.40. Theorem ([153, 155]). *Let $G \to GL(V)$ be a polar representation of a compact Lie group G, with section Σ and generalized Weyl group $W = W(\Sigma)$. Then the pullback to Σ of differential forms induces an isomorphism*

$$\Omega^p_{hor}(V)^G \xrightarrow{\cong} \Omega^p(\Sigma)^{W(\Sigma)}.$$

According to [**38**, remark after proposition 6], for any polar representation of a connected Lie group the generalized Weyl group $W(\Sigma)$ is a reflection group. This theorem is true for polynomial differential forms and also for real analytic differential forms, by essentially the same proof.

Proof. Let $i : \Sigma \to V$ be the embedding. It is proved in (30.37) that the restriction $i^* : \Omega^p_{\mathrm{hor}}(V)^G \to \Omega^p(\Sigma)^{W(G)}$ is injective, so it remains to prove surjectivity.

Let us first suppose that $W = W(\Sigma)$ is generated by reflections (a reflection group or Coxeter group). Let $\rho_1, \dots, \rho_n$ be a minimal set of homogeneous generators of the algebra $\mathbb{R}[\Sigma]^W$ of W-invariant polynomials on Σ. Then this is a set of algebraically independent polynomials, $n = \dim \Sigma$, and their degrees $d_1, \dots, d_n$ are uniquely determined up to order. We even have (see [**95**])

(1) $d_1 \dots d_n = |W|$, the order of W,

(2) $d_1 + \dots + d_n = n + N$, where N is the number of reflections in W,

(3) $\prod_{i=1}^n (1 + (d_i - 1)t) = a_0 + a_1 t + \dots + a_n t^n$, where a_i is the number of elements in W whose fixed point set has dimension $n - i$.

Let us consider the mapping $\rho = (\rho_1, \dots, \rho_n) : \Sigma \to \mathbb{R}^n$ and its Jacobian $J(x) = \det(d\rho(x))$. Let $x^1, \dots, x^n$ be coordinate functions in Σ. Then for each $\sigma \in W$ we have

$$\begin{aligned} J.dx^1 \wedge \dots \wedge dx^n &= d\rho_1 \wedge \dots \wedge d\rho_n = \sigma^*(d\rho_1 \wedge \dots \wedge d\rho_n) \\ &= (J \circ \sigma)\sigma^*(dx^1 \wedge \dots \wedge dx^n) \\ &= (J \circ \sigma)\det(\sigma)(dx^1 \wedge \dots \wedge dx^n), \end{aligned}$$

$$J \circ \sigma = \det(\sigma^{-1})J. \tag{4}$$

If $J(x) \neq 0$, then in a neighborhood of x the mapping ρ is a diffeomorphism by the inverse function theorem, so that the 1-forms $d\rho_1, \dots, d\rho_n$ are a local coframe there. Since the generators $\rho_1, \dots, \rho_n$ are algebraically independent over $\mathbb{R}$, $J \neq 0$. Since J is a polynomial of degree $(d_1 - 1) + \dots + (d_n - 1) = N$ (see (2)), the set $U = \Sigma \setminus J^{-1}(0)$ is open and dense in Σ, and $d\rho_1, \dots, d\rho_n$ form a coframe on U.

Now let $(\sigma_\alpha)_{\alpha=1,\dots,N}$ be the set of reflections in W, with reflection hyperplanes H_α. Let $\ell_\alpha \in \Sigma^*$ be a linear functional with $H_\alpha = \ell^{-1}(0)$. If $x \in H_\alpha$, we have $J(x) = \det(\sigma_\alpha)J(\sigma_\alpha.x) = -J(x)$, so that $J|H_\alpha = 0$ for each α, and by lemma (30.38) we have

$$J = c.\ell_1 \dots \ell_N. \tag{5}$$

Since J is a polynomial of degree N, c must be a constant. Repeating the last argument for an arbitrary function g and using (5), we get:

(6) If $g \in C^\infty(\Sigma, \mathbb{R})$ satisfies $g \circ \sigma = \det(\sigma^{-1})g$ for each $\sigma \in W$, we have $g = J.h$ for $h \in C^\infty(\Sigma, \mathbb{R})^W$.

(7) **Claim.** Let $\omega \in \Omega^p(\Sigma)^W$. Then we have

$$\omega = \sum_{j_1 < \dots < j_p} \omega_{j_1 \dots j_p} d\rho_{j_1} \wedge \dots \wedge d\rho_{j_p},$$

where $\omega_{j_1 \dots j_p} \in C^\infty(\Sigma, \mathbb{R})^W$.

Since $d\rho_1, \dots, d\rho_n$ form a coframe on the W-invariant dense open set $U = \{x : J(x) \neq 0\}$, we have

$$\omega|U = \sum_{j_1 < \dots < j_p} g_{j_1 \dots j_p} d\rho_{j_1}|U \wedge \dots \wedge d\rho_{j_p}|U$$

for $g_{j_1 \dots j_p} \in C^\infty(U, \mathbb{R})$. Since ω and all $d\rho_i$ are W-invariant, we may replace $g_{j_1 \dots j_p}$ by

$$\tfrac{1}{|W|} \sum_{\sigma \in W} g_{j_1 \dots j_p} \circ \sigma \in C^\infty(U, \mathbb{R})^W,$$

or assume without loss that $g_{j_1 \dots j_p} \in C^\infty(U, \mathbb{R})^W$.

Let us choose now a form index $i_1 < \cdots < i_p$ with

$$\{i_{p+1} < \cdots < i_n\} = \{1, \ldots, n\} \setminus \{i_1 < \cdots < i_p\}.$$

Then for some sign $\varepsilon = \pm 1$ we have

$$\begin{aligned} \omega|U \wedge d\rho_{i_{p+1}} \wedge \cdots \wedge d\rho_{i_n} &= \varepsilon.g_{i_1 \ldots i_p}.d\rho_1 \wedge \cdots \wedge d\rho_n \\ &= \varepsilon.g_{i_1 \ldots i_p}.J.dx^1 \wedge \cdots \wedge dx^n, \end{aligned}$$

$$\omega \wedge d\rho_{i_{p+1}} \wedge \cdots \wedge d\rho_{i_n} = \varepsilon.k_{i_1 \ldots i_p} dx^1 \wedge \cdots \wedge dx^n \tag{8}$$

for a function $k_{i_1 \ldots i_p} \in C^\infty(\Sigma, \mathbb{R})$. Thus

$$k_{i_1 \ldots i_p}|U = g_{i_1 \ldots i_p}.J|U. \tag{9}$$

Since ω and all $d\rho_i$ are W-invariant, by (8) we get $k_{i_1 \ldots i_p} \circ \sigma = \det(\sigma^{-1}) k_{i_1 \ldots i_p}$ for each $\sigma \in W$. But then by (6) we have $k_{i_1 \ldots i_p} = \omega_{i_1 \ldots i_p}.J$ for unique $\omega_{i_1 \ldots i_p} \in C^\infty(\Sigma, \mathbb{R})^W$, and (9) then implies $\omega_{i_1 \ldots i_p}|U = g_{i_1 \ldots i_p}$, so that the claim (7) follows since U is dense.

Now we may finish the proof of the theorem in the case where $W = W(\Sigma)$ is a reflection group. Let $i : \Sigma \to V$ be the embedding. By theorem (30.30) the algebra $\mathbb{R}[V]^G$ of G-invariant polynomials on V is isomorphic to the algebra $\mathbb{R}[\Sigma]^W$ of W-invariant polynomials on the section Σ, via the restriction mapping i^*. Choose polynomials $\tilde{\rho}_1, \ldots, \tilde{\rho}_n \in \mathbb{R}[V]^G$ with $\tilde{\rho}_i \circ i = \rho_i$ for all i. Put $\tilde{\rho} = (\tilde{\rho}_1, \ldots, \tilde{\rho}_n) : V \to \mathbb{R}^n$. In the setting of claim (7), use Schwarz's theorem (7.13) to find $h_{i_1, \ldots, i_p} \in C^\infty(\mathbb{R}^n, \mathbb{R})$ with $h_{i_1, \ldots, i_p} \circ \rho = \omega_{i_1, \ldots, i_p}$ and consider

$$\tilde{\omega} = \sum_{j_1 < \cdots < j_p} (h_{j_1 \ldots j_p} \circ \tilde{\rho}) d\tilde{\rho}_{j_1} \wedge \cdots \wedge d\tilde{\rho}_{j_p},$$

which is in $\Omega^p_{\text{hor}}(V)^G$ and satisfies $i^* \tilde{\omega} = \omega$.

Thus the mapping $i^* : \Omega^p_{\text{hor}}(V)^G \to \Omega^p_{\text{hor}}(\Sigma)^W$ is surjective in the case where $W = W(\Sigma)$ is a reflection group.

Now we treat the general case. Let G_0 be the connected component of G. From (30.15.3) one concludes: *A subspace Σ of V is a section for G if and only if it is a section for G_0. Thus ρ is a polar representation for G if and only if it is a polar representation for G_0.*

The generalized Weyl groups of Σ with respect to G and to G_0 are related by

$$W(G_0) = N_{G_0}(\Sigma)/Z_{G_0}(\Sigma) \subset W(G) = N_G(\Sigma)/Z_G(\Sigma),$$

since $Z_G(\Sigma) \cap N_{G_0}(\Sigma) = Z_{G_0}(\Sigma)$.

Let $\omega \in \Omega^p(\Sigma)^{W(G)} \subset \Omega^p(\Sigma)^{W(G_0)}$. Since G_0 is connected, the generalized Weyl group $W(G_0)$ is generated by reflections (a Coxeter group) by [**38**, remark after proposition 6]. Thus by the first part of the proof

$$i^* : \Omega^p_{\text{hor}}(V)^{G_0} \xrightarrow{\cong} \Omega^p(\Sigma)^{W(G_0)}$$

is an isomorphism, and we get $\varphi \in \Omega^p_{\text{hor}}(M)^{G_0}$ with $i^*\varphi = \omega$. Let us consider

$$\psi := \int_G g^*\varphi \, dg \in \Omega^p_{\text{hor}}(V)^G,$$

where dg denotes Haar measure on G. In order to show that $i^*\psi = \omega$, it suffices to check that $i^*g^*\varphi = \omega$ for each $g \in G$. Now $g(\Sigma)$ is again a section of G, thus also of G_0. Since any two sections are related by an element of the group, there exists $h \in G_0$ such that $hg(\Sigma) = \Sigma$. Then $hg \in N_G(\Sigma)$ and we denote by $[hg]$ the coset in $W(G)$, and we may compute as follows:

$$\begin{aligned}
(i^*g^*\varphi)_x &= (g^*\varphi)_x.\bigwedge^p Ti = \varphi_{g(x)}.\bigwedge^p Tg.\bigwedge^p Ti \\
&= (h^*\varphi)_{g(x)}.\bigwedge^p Tg.\bigwedge^p Ti, \qquad \text{since } \varphi \in \Omega^p_{\text{hor}}(M)^{G_0} \\
&= \varphi_{hg(x)}.\bigwedge^p T(hg).\bigwedge^p Ti = \varphi_{i[hg](x)}.\bigwedge^p Ti.\bigwedge^p T([hg]) \\
&= \varphi_{i[hg](x)}.\bigwedge^p Ti.\bigwedge^p T([hg]) = (i^*\varphi)_{[hg](x)}.\bigwedge^p T([hg]) \\
&= \omega_{[hg](x)}.\bigwedge^p T([hg]) = [hg]^*\omega = \omega. \qquad \square
\end{aligned}$$

30.41. Remark. The proof of theorem (30.40) shows that the answer to question (30.39) is yes for the representations treated in (30.40).

30.42. Corollary. *Let $\rho : G \to O(V, \langle\ ,\ \rangle)$ be an orthogonal polar representation of a compact Lie group G, with section Σ and generalized Weyl group $W = W(\Sigma)$. Let $B \subset V$ be an open ball centered at 0.*

Then the restriction of differential forms induces an isomorphism

$$\Omega^p_{hor}(B)^G \xrightarrow{\cong} \Omega^p(\Sigma \cap B)^{W(\Sigma)}.$$

Proof. Check the proof of (30.40) or use the following argument. Suppose that $B = \{v \in V : |v| < 1\}$ and consider a smooth diffeomorphism $f : [0,1) \to [0,\infty)$ with $f(t) = t$ near 0. Then $g(v) := \frac{f(|v|)}{|v|}v$ is a G-equivariant diffeomorphism $B \to V$ and by (30.40) we get

$$\Omega^p_{\text{hor}}(B)^G \xrightarrow{(g^{-1})^*} \Omega^p_{\text{hor}}(V)^G \xrightarrow{\cong} \Omega^p(\Sigma)^{W(\Sigma)} \xrightarrow{g^*} \Omega^p(\Sigma \cap B)^{W(\Sigma)}. \qquad \square$$

30.43. Let us assume that we are in the situation of the main theorem (30.37) for the rest of this section. For $x \in M$ let S_x be a (normal) slice and G_x the isotropy group, which acts on the slice. Then $G.S_x$ is open in M and G-equivariantly diffeomorphic to the associated bundle $G \to G/G_x$ via

$$\begin{array}{ccccc} G \times S_x & \xrightarrow{q} & G \times_{G_x} S_x & \xrightarrow{\cong} & G.S_x \\ & & \big\downarrow & & \big\downarrow r \\ & & G/G_x & \xrightarrow{\cong} & G.x, \end{array}$$

where r is the projection of a tubular neighborhood. Since $q : G \times S_x \to G \times_{G_x} S_x$ is a principal G_x-bundle with principal right action $(g, s).h = (gh, h^{-1}.s)$, we have an isomorphism $q^* : \Omega(G\times_{G_x} S_x) \to \Omega_{G_x-\text{hor}}(G\times S_x)^{G_x}$. Since q is also G-equivariant for the left G-actions, the isomorphism q^* maps the subalgebra $\Omega^p_{\text{hor}}(G.S_x)^G \cong \Omega^p_{\text{hor}}(G \times_{G_x} S_x)^G$ of $\Omega(G \times_{G_x} S_x)$ to the subalgebra $\Omega^p_{G_x-\text{hor}}(S_x)^{G_x}$ of $\Omega_{G_x-\text{hor}}(G \times S_x)^{G_x}$. So we have proved:

Lemma. *In this situation there is a canonical isomorphism*

$$\Omega^p_{hor}(G.S_x)^G \xrightarrow{\cong} \Omega^p_{G_x-hor}(S_x)^{G_x}$$

which is given by pullback along the embedding $S_x \to G.S_x$.

30.44. Rest of the proof of theorem (30.37). Let us consider $\omega \in \Omega^p(\Sigma)^{W(\Sigma)}$. We want to construct a form $\tilde{\omega} \in \Omega^p_{\text{hor}}(M)^G$ with $i^*\tilde{\omega} = \omega$. This will finish the proof of theorem (30.40).

Choose $x \in \Sigma$ and an open ball B_x with center 0 in T_xM such that the Riemann exponential mapping $\exp_x : T_xM \to M$ is a diffeomorphism on B_x. We consider now the compact isotropy group G_x and the slice representation $\rho_x : G_x \to O(V_x)$, where $V_x = \text{Nor}_x(G.x) = (T_x(G.x))^\perp \subset T_xM$ is the normal space to the orbit. This is a polar representation with section $T_x\Sigma$, and its generalized Weyl group is given by $W(T_x\Sigma) \cong N_G(\Sigma) \cap G_x/Z_G(\Sigma) = W(\Sigma)_x$ (see (30.24)). Then $\exp_x : B_x \cap V_x \to S_x$ is a diffeomorphism onto a slice and $\exp_x : B_x \cap T_x\Sigma \to \Sigma_x \subset \Sigma$ is a diffeomorphism onto an open neighborhood Σ_x of x in the section Σ.

Let us now consider the pullback $(\exp|B_x \cap T_x\Sigma)^*\omega \in \Omega^p(B_x \cap T_x\Sigma)^{W(T_x\Sigma)}$. By corollary (30.42) there exists a unique form $\varphi^x \in \Omega^p_{G_x-\text{hor}}(B_x \cap V_x)^{G_x}$ such that $i^*\varphi^x = (\exp|B_x \cap T_x\Sigma)^*\omega$, where i_x is the embedding. Then we have

$$((\exp|B_x \cap V_x)^{-1}) * \varphi^x \in \Omega^p_{G_x-\text{hor}}(S_x)^{G_x}$$

and by lemma (30.43) this form corresponds uniquely to a differential form $\omega^x \in \Omega^p_{\text{hor}}(G.S_x)^G$ which satisfies $(i|\Sigma_x)^*\omega^x = \omega|\Sigma_x$, since the exponential mapping commutes with the respective restriction mappings. Now the intersection $G.S_x \cap \Sigma$ is the disjoint union of all the open sets $w_j(\Sigma_x)$ where we pick one w_j in each left coset of the subgroup $W(\Sigma)_x$ in $W(\Sigma)$. If we choose $g_j \in N_G(\Sigma)$ projecting on w_j for all j, then

$$\begin{aligned}(i|w_j(\Sigma_x))^*\omega^x &= (\ell_{g_j} \circ i|\Sigma_x \circ w_j^{-1})^*\omega^x \\ &= (w_j^{-1})^*(i|\Sigma_x)^*\ell_{g_j}^*\omega^x \\ &= (w_j^{-1})^*(i|\Sigma_x)^*\omega^x = (w_j^{-1})^*(\omega|\Sigma_x) = \omega|w_j(\Sigma_x),\end{aligned}$$

so that $(i|G.S_x \cap \Sigma)^*\omega^x = \omega|G.S_x \cap \Sigma$. We can do this for each point $x \in \Sigma$. Using the method of (6.28) and (6.30), we may find a sequence of points $(x_n)_{n\in\mathbb{N}}$ in Σ such that the $\pi(\Sigma_{x_n})$ form a locally finite open cover of the orbit space $M/G \cong \Sigma/W(\Sigma)$ and a smooth partition of unity f_n consisting of G-invariant functions with $\operatorname{supp}(f_n) \subset G.S_{x_n}$. Then $\tilde\omega := \sum_n f_n\omega^{x_n} \in \Omega^p_{\text{hor}}(M)^G$ has the required property $i^*\tilde\omega = \omega$. □

CHAPTER VII. Symplectic and Poisson Geometry

31. Symplectic Geometry and Classical Mechanics

31.1. Motivation. A particle with mass $m > 0$ moves in a potential $V(q)$ along a curve $q(t)$ in $\mathbb{R}^3$ in such a way that Newton's second law is satisfied: $m\ddot{q}(t) = -\operatorname{grad} V(q(t))$. Let us consider the the quantity $p_i := m \cdot \dot{q}^i$ as an independent variable. It is called the i-th *momentum*. Let us define the energy function (as the sum of the kinetic and potential energy) by

$$E(q,p) := \frac{1}{2m}|p|^2 + V(q) = \frac{m|\dot{q}|^2}{2} + V(q).$$

Then $m\ddot{q}(t) = -\operatorname{grad} V(q(t))$ is equivalent to

$$\begin{cases} \dot{q}^i = \frac{\partial E}{\partial p_i}, \\ \dot{p}_i = -\frac{\partial E}{\partial q^i}, \qquad i = 1,2,3, \end{cases}$$

which are *Hamilton's equations* of motion. In order to study this equation for a general energy function $E(q,p)$, we consider the matrix

$$J = \begin{pmatrix} 0 & \mathbb{I}_{\mathbb{R}^3} \\ -\mathbb{I}_{\mathbb{R}^3} & 0 \end{pmatrix}.$$

Then the equation is equivalent to $\dot{u}(t) = J \cdot \operatorname{grad} E(u(t))$, where $u = (q,p) \in \mathbb{R}^6$. In complex notation, where $z^i = q^i + \sqrt{-1}\,p_i$, this is equivalent to $\dot{z}^i = -2\sqrt{-1}\frac{\partial E}{\partial \bar{z}^i}$.

Consider the *Hamiltonian vector field* $H_E := J \cdot \operatorname{grad} E$ associated to the energy function E; then we have $\dot{u}(t) = H_E(u(t))$, so the orbit of the particle with initial position and momentum $(q_0, p_0) = u_0$ is given by $u(t) = \operatorname{Fl}_t^{H_E}(u_0)$.

Let us now consider the *symplectic structure*

$$\omega(x,y) = \sum_{i=1}^{3}(x^i y^{3+i} - x^{3+i}y^i) = (x|Jy) \quad \text{for } x, y \in \mathbb{R}^6.$$

Then the Hamiltonian vector field H_E is given by

$$\begin{aligned}\omega(H_E(u), v) &= (H_E|Jv) = (J \operatorname{grad} E(u)|Jv)\\ &= (J^\top J \operatorname{grad} E(u)|v) = (\operatorname{grad} E(u)|v) = dE(u)v.\end{aligned}$$

The Hamiltonian vector field is therefore the 'gradient' of E with respect to the symplectic structure ω; we write $H_E = \operatorname{grad}^\omega E$.

How does this equation react to coordinate transformations? So let $f : \mathbb{R}^3 \times \mathbb{R}^3 \to \mathbb{R}^3 \times \mathbb{R}^3$ be a (local) diffeomorphism. We consider the energy $E \circ f$ and put $u = f(w)$. Then

$$\begin{aligned}\omega(\operatorname{grad}^\omega(E\circ f)(w), v) &= d(E\circ f)(w)v = dE(f(w)).df(w)v\\ &= \omega(\operatorname{grad}^\omega E(f(w)), df(w)v)\\ &= \omega(df(w)\, df(w)^{-1} \operatorname{grad}^\omega E(f(w)), df(w)v)\\ &= \omega(df(w)\,(f^* \operatorname{grad}^\omega E)(w), df(w)v)\\ &= (f^*\omega)((f^* \operatorname{grad}^\omega E)(w), v).\end{aligned}$$

So we see that $f^* \operatorname{grad}^\omega E = \operatorname{grad}^\omega(E\circ f)$ if and only if $f^*\omega = \omega$, i.e., $df(w) \in Sp(3,\mathbb{R})$ for all w. Such diffeomorphisms are called *symplectomorphisms.* By (3.14) we have

$$\operatorname{Fl}_t^{f^* \operatorname{grad}^\omega E} = f^{-1} \circ \operatorname{Fl}_t^{\operatorname{grad}^\omega E} \circ f$$

in general.

31.2. Lemma (E. Cartan). *Let V be a real finite-dimensional vector space, and let $\omega \in \bigwedge^2 V^*$ be a 2-form on V. Consider the linear mapping $\check{\omega} : V \to V^*$ given by $\langle \check{\omega}(v), w\rangle = \omega(v,w)$.*

If $\omega \neq 0$, then the rank of the linear mapping $\check{\omega} : V \to V^*$ is $2p$, and there exist linearly independent $l^1, \dots, l^{2p} \in V^*$ which form a basis of $\check{\omega}(V) \subset V^*$ such that $\omega = \sum_{k=1}^{p} l^{2k-1} \wedge l^{2k}$. Furthermore, l^2 can be chosen arbitrarily in $\check{\omega}(V) \setminus 0$.

Proof. Let $v_1, \dots, v_n$ be a basis of V and let $v^1, \dots, v^n$ be the dual basis of V^*. Then $\omega = \sum_{i<j} \omega(v_i, v_j) v^i \wedge v^j =: \sum_{i<j} a_{ij}\, v^i \wedge v^j$. Since $\omega \neq 0$, not all

$a_{ij} = 0$. Suppose that $a_{12} \neq 0$. Put

$$l^1 = \frac{1}{a_{12}}\check{\omega}(v_1) = \frac{1}{a_{12}}i(v_1)\omega = \frac{1}{a_{12}}i(v_1)\Big(\sum_{i<j} a_{ij}\, v^i \wedge v^j\Big)$$

$$= v^2 + \frac{1}{a_{12}}\sum_{j=3}^{n} a_{1j}\, v^j,$$

$$l^2 = \check{\omega}(v_2) = i(v_2)\omega = i(v_2)\Big(\sum_{i<j} a_{ij}\, v^i \wedge v^j\Big) = -a_{12}v^1 + \sum_{j=3}^{n} a_{2j}\, v^j.$$

So, $l^1, l^2, v^3, \dots, v^n$ is still a basis of V^*. Put $\omega_1 := \omega - l^1 \wedge l^2$. Then

$$i_{v_1}\omega_1 = i_{v_1}\omega - i_{v_1}l^1 \wedge l^2 + l^1 \wedge i_{v_1}l^2 = a_{12}l^1 - 0 - a_{12}l^1 = 0,$$
$$i_{v_2}\omega_1 = i_{v_2}\omega - i_{v_2}l^1 \wedge l^2 + l^1 \wedge i_{v_2}l^2 = l^2 - l^2 + 0 = 0.$$

So the 2-form ω_1 is in the subalgebra of $\bigwedge V^*$ generated by $v^3, v^4, \dots, v^n$. If $\omega_1 = 0$, then $\omega = l^1 \wedge l^2$. If $\omega_1 \neq 0$, we can repeat the procedure and get the form of ω.

If $l = \check{\omega}(v) \in \check{\omega}(V) \subset V^*$ is arbitrary but $\neq 0$, there is some $w \in V$ with $\langle l, w\rangle = \omega(v, w) \neq 0$. Choose a basis $v_1, \dots, v_n$ of V with $v_1 = w$ and $v_2 = v$. Then $l^2 = i(v_2)\omega = i(v)\omega = l$. □

31.3. Corollary. *Let $\omega \in \bigwedge^2 V^*$ and let $2p = \operatorname{rank}(\check{\omega} : V \to V^*)$. Then p is the maximal number k such that $\omega^{\wedge k} = \omega \wedge \dots \wedge \omega \neq 0$.*

Proof. By (31.2) we have $\omega^{\wedge p} = p!\, l^1 \wedge l^2 \wedge \dots \wedge l^{2p}$ and $\omega^{\wedge(2p+1)} = 0$. □

31.4. Symplectic vector spaces. A symplectic form on a vector space V is a 2-form $\omega \in \bigwedge^2 V^*$ such that $\check{\omega} : V \to V^*$ is an isomorphism. Then $\dim(V) = 2n$ and there is a basis $l^1, \dots, l^{2n}$ of V^* such that $\omega = \sum_{i=1}^n l^i \wedge l^{n+i}$, by (31.2).

For a linear subspace $W \subset V$ we define the *symplectic orthogonal* by $W^{\omega\perp} = W^\perp := \{v \in V : \omega(w, v) = 0 \text{ for all } w \in W\}$; which coincides with the annihilator (or polar) $\check{\omega}(W)^\circ = \{v \in V : \langle\check{\omega}(w), v\rangle = 0 \text{ for all } w \in W\}$ in V.

Lemma. *For linear subspaces $W, W_1, W_2 \subset V$ we have:*

1. $W^{\perp\perp} = W$.
2. $\dim(W) + \dim(W^\perp) = \dim(V) = 2n$.
3. $\check{\omega}(W^\perp) = W^\circ$ *and* $\check{\omega}(W) = (W^\perp)^\circ$ *in* V^*.
4. *For two linear subspace $W_1, W_2 \subset V$ we have $W_1 \subseteq W_2 \Leftrightarrow W_1^\perp \supseteq W_2^\perp$, $(W_1 \cap W_2)^\perp = W_1^\perp + W_2^\perp$, and $(W_1 + W_2)^\perp = W_1^\perp \cap W_2^\perp$.*
5. $\dim(W_1 \cap W_2) - \dim(W_1^\perp \cap W_2^\perp) = \dim W_1 + \dim W_2 - 2n$.

Proof. (1) – (4) are obvious, using duality and the annihilator. (5) can be seen as follows. By (4) we have

$$\begin{aligned}\dim(W_1\cap W_2)^\perp &= \dim(W_1^\perp + W_2^\perp)\\ &= \dim(W_1^\perp)+\dim(W_2^\perp)-\dim(W_1^\perp\cap W_2^\perp),\\ \dim(W_1\cap W_2) &= 2n-\dim(W_1\cap W_2)^\perp \qquad \text{by (2)}\\ &= 2n-\dim(W_1^\perp)-\dim(W_2^\perp)+\dim(W_1^\perp\cap W_2^\perp)\\ &= \dim(W_1)+\dim(W_2)-2n+\dim(W_1^\perp\cap W_2^\perp). \quad\square\end{aligned}$$

A linear subspace $W \subseteq V$ is called:

isotropic	if	$W\subseteq W^\perp$	$\Rightarrow \dim(W)\le n$,
coisotropic	if	$W\supseteq W^\perp$	$\Rightarrow \dim(W)\ge n$,
Lagrangian	if	$W=W^\perp$	$\Rightarrow \dim(W)=n$,
symplectic	if	$W\cap W^\perp=0$	$\Rightarrow \dim(W)=$ even.

31.5. Example. Let W be a vector space with dual W^*. Then $(W\times W^*,\omega)$ is a symplectic vector space where $\omega((v,v^*),(w,w^*)) = \langle w^*,v\rangle - \langle v^*,w\rangle$. Choose a basis $w_1,\dots,w_n$ of $W = W^{**}$ and let $w^1,\dots,w^n$ be the dual basis. Then $\omega = \sum_i w^i\wedge w_i$. The two subspaces $W\times 0$ and $0\times W^*$ are Lagrangian.

Consider now a symplectic vector space (V,ω) and suppose that $W_1, W_2\subseteq V$ are two Lagrangian subspaces such that $W_1\cap W_2=0$. Then $\omega: W_1\times W_2\to\mathbb{R}$ is a duality pairing, so we may identify W_2 with W_1^* via ω. Then (V,ω) is isomorphic to $W_1\times W_1^*$ with the symplectic structure described above.

31.6. Let $\mathbb{R}^{2n} = \mathbb{R}^n\times(\mathbb{R}^n)^*$ with the standard symplectic structure ω from (31.5). Recall from (4.7) the Lie group $Sp(n,\mathbb{R})$ of symplectic automorphisms of $(\mathbb{R}^{2n},\omega)$,

$$Sp(n,\mathbb{R}) = \{A\in L(\mathbb{R}^{2n},\mathbb{R}^{2n}) : A^\top JA = J\},\qquad \text{where } J = \begin{pmatrix}0 & \mathbb{I}_{\mathbb{R}^n}\\ -\mathbb{I}_{\mathbb{R}^n} & 0\end{pmatrix}.$$

Let $(\ \mid\ )$ be the standard inner product on $\mathbb{R}^{2n}$ and let $\mathbb{R}^{2n}\cong\sqrt{-1}\mathbb{R}^n\oplus\mathbb{R}^n=\mathbb{C}^n$, where the scalar multiplication by $\sqrt{-1}$ is given by $J\binom{x}{y}=\binom{y}{-x}$. Then we have:

$$\begin{aligned}\omega\left(\tbinom{x}{y},\tbinom{x'}{y'}\right) &= \langle y',x\rangle-\langle y,x'\rangle = \left(\tbinom{x}{y}\Big|\tbinom{y'}{-x'}\right)=\left(\tbinom{x}{y}\Big|\,J\tbinom{x'}{y'}\right)\\ &= (x^T,y^T)J\begin{pmatrix}x'\\ y'\end{pmatrix}.\end{aligned}$$

Since $J^2 = -\mathbb{I}_{R^{2n}}$ we have $J \in Sp(n,\mathbb{R})$, and since $J^\top = -J = J^{-1}$ we also have $J \in O(2n,\mathbb{R})$. We consider now the Hermitian inner product $h : \mathbb{C}^n \times \mathbb{C}^n \to \mathbb{C}$ given by

$$\begin{aligned}
h(u,v) :&= (u|v) + \sqrt{-1}\omega(u,v) = (u|v) + \sqrt{-1}(u|Jv), \\
h(v,u) &= (v|u) + \sqrt{-1}(v|Ju) = (u|v) + \sqrt{-1}(J^\top v|u) \\
&= (u|v) - \sqrt{-1}(u|Jv) = \overline{h(u,v)}, \\
h(Ju,v) &= (Ju|v) + \sqrt{-1}(Ju|Jv) = \sqrt{-1}((u|J^\top Jv) - \sqrt{-1}(u|J^\top v)) \\
&= \sqrt{-1}((u|v) + \sqrt{-1}\omega(u,v)) = \sqrt{-1}h(u,v).
\end{aligned}$$

Lemma. *The subgroups $Sp(n,\mathbb{R})$, $O(2n,\mathbb{R})$, and $U(n)$ of $GL(2n,\mathbb{R})$ acting on $\mathbb{R}^{2n} \cong \mathbb{C}^n$ are related by*

$$O(2n,\mathbb{R}) \cap GL(n,\mathbb{C}) = Sp(n,\mathbb{R}) \cap GL(n,\mathbb{C}) = Sp(n,\mathbb{R}) \cap O(2n,\mathbb{R}) = U(n).$$

Proof. For $A \in GL(2n,\mathbb{R})$ (and all $u, v \in \mathbb{R}^{2n}$) we have in turn

$$\begin{aligned}
&h(Au,Av) = h(u,v) && \Leftrightarrow A \in U(n), \\
&\left\{\begin{matrix} (Au|Av) = (u|v) & \text{(real part)} \\ \omega(Au,Av) = \omega(u,v) & \text{(imagin. part)} \end{matrix}\right\} && \Leftrightarrow A \in O(2n,\mathbb{R}) \cap Sp(n,\mathbb{R}), \\
&\left\{\begin{matrix} (Au|Av) = (u|v) \\ JA = AJ \end{matrix}\right\} && \Leftrightarrow A \in O(2n,\mathbb{R}) \cap GL(n,\mathbb{C}), \\
&\left\{\begin{matrix} JA = AJ \\ (Au|JAv) = (Au|AJv) = (u|Jv) \end{matrix}\right\} && \Leftrightarrow A \in Sp(n,\mathbb{R}) \cap GL(n,\mathbb{C}). \quad \square
\end{aligned}$$

31.7. The Lagrange-Grassmann manifold. Let $L(\mathbb{R}^{2n},\omega) = L(2n)$ denote the space of all Lagrangian linear subspaces of $\mathbb{R}^{2n}$; we call it the *Lagrange-Grassmann manifold.* It is a subset of the Grassmann manifold $G(n,2n;\mathbb{R})$; see (18.5).

In the situation of (31.6) we consider a linear subspace $W \subset (\mathbb{R}^{2n},\omega)$ of dimension n. Then we have:

$$\begin{aligned}
&W \text{ is a Lagrangian subspace} \\
\Leftrightarrow\ &\omega|W = 0 \quad \Leftrightarrow \quad (\ \ |J(\ \ ))|W = 0 \\
\Leftrightarrow\ &J(W) \text{ is orthogonal to } W \text{ with respect to } (\ \ |\ \ ) = \operatorname{Re}(h).
\end{aligned}$$

Thus the group $O(2n,\mathbb{R}) \cap GL(n,\mathbb{C}) = U(n)$ acts transitively on the Lagrange-Grassmann manifold $L(2n)$. The isotropy group of the Lagrangian subspace $\mathbb{R}^n \times 0$ is the subgroup $O(n,\mathbb{R}) \subset U(n)$ consisting of all unitary matrices with all entries real. So $L(2n) = U(n)/O(n,\mathbb{R})$ by (5.11), which is a compact homogenous space and a smooth manifold. The dimension of $L(2n)$ is given by

$$\dim L(2n) = \dim U(n) - \dim O(n,\mathbb{R}) = (n + 2\tfrac{n(n-1)}{2}) - \tfrac{n(n-1)}{2} = \tfrac{n(n+1)}{2}.$$

Which choices did we make in this construction? Starting with a general symplectic vector space (V,ω), we first choose a basic Lagrangian subspace L $(= \mathbb{R}^n \times 0)$, and then we identify V/L with L^* via ω. Next we chose a positive inner product on L, transport it to L^* via ω and extend it to $L \times L^*$ by putting L and L^* orthogonal to each other. All these possible choices are homotopic to each other.

Finally we consider $\det_{\mathbb{C}} = \det : U(n) \to S^1 \subset \mathbb{C}$. Then $\det(O(n)) = \{\pm 1\}$. So $\det^2 : U(n) \to S^1$ and $\det^2(O(n)) = \{1\}$. For $U \in U(n)$ and $A \in O(n,\mathbb{R})$ we have $\det^2(UA) = \det^2(U)\det^2(A) = \det^2(U)$, so this factors to a well defined smooth mapping $\det^2 : U(n)/O(n) = L(2n) \to S^1$.

Claim. The group $SU(n)$ acts (from the left) transitively on each fiber of $\det^2 : L(2n) = U(n)/O(n) \to S^1$.

Namely, for $U_1, U_2 \in U(n)$ with $\det^2(U_1) = \det^2(U_2)$ we get $\det(U_1) = \pm\det(U_2)$. There exists $A \in O(n)$ such that $\det(U_1) = \det(U_2.A)$; thus $U_1(U_2A)^{-1} \in SU(n)$ and $U_1(U_2A)^{-1}U_2A\,O(n) = U_1\,O(n)$. The claim is proved.

Now $SU(n)$ is simply connected and each fiber of $\det^2 : U(n)/O(n) \to S^1$ is diffeomorphic to $SU(n)/SO(n)$ which again is simply connected by the exact homotopy sequence of the fibration $SO(n) \to SU(n) \to SU(n)/SO(N)$,

$$\dots \to (0 = \pi_1(SU(n))) \to \pi_1(SU(n)/SO(n)) \to (\pi_0(SO(n)) = 0) \to \cdots.$$

Now we consider the fibration $SU(n)/SO(n) \to L(2n) \to S^1$; from its exact homotopy sequence

$$\to 0 = \pi_1(SU(n)/SO(n)) \to \pi_1(L(2n)) \to \pi_1(S^1) \to \pi_0(SU(n)/SO(n)) = 0$$

we conclude that $\pi_1(L(2n)) = \pi_1(S^1) = \mathbb{Z}$. Also (by the Hurewicz homomorphism) we have $H^1(L(2n),\mathbb{Z}) = \mathbb{Z}$ and thus $H^1(L(2n),\mathbb{R}) = \mathbb{R}$.

Let $\frac{dz}{2\pi\sqrt{-1}z}|_{S^1} = \frac{xdy-ydx}{2\pi\sqrt{-1}}|_{S^1} \in \Omega^1(S^1)$ be a generator of $H^1(S^1,\mathbb{Z})$. Then the pullback $(\det^2)^*\frac{dz}{2\pi\sqrt{-1}z} = (\det^2)^*\frac{xdy-ydx}{2\pi\sqrt{-1}} \in \Omega^1(L(2n))$ is a generator of $H^1(L(2n))$. Its cohomology class is called the *Maslov class*.

31.8. Symplectic manifolds and their submanifolds. A *symplectic manifold* (M,ω) is a manifold M together with a 2-form $\omega \in \Omega^2(M)$ such that $d\omega = 0$ and $\omega_x \in \bigwedge^2 T_x^*M$ is a symplectic structure on T_xM for each $x \in M$. So $\dim(M)$ is even; $\dim(M) = 2n$, say. Moreover, $\omega^{\wedge n} = \omega \wedge \dots \wedge \omega$ is a volume form on M (nowhere zero) called the *Liouville volume*, which fixes also an orientation of M.

Among the submanifolds N of M we can single out those whose tangent spaces T_xN have special relations to the symplectic structure ω_x on T_xM as listed in (31.4): Thus a submanifold N of M is called:

isotropic	if	$T_xN \subseteq T_xN^{\omega\perp}$ for each $x \in N$	$\Rightarrow \dim(N) \leq n$,
coisotropic	if	$T_xN \supseteq T_xN^{\omega\perp}$ for each $x \in N$	$\Rightarrow \dim(N) \geq n$,
Lagrangian	if	$T_xN = T_xN^{\omega\perp}$ for each $x \in N$	$\Rightarrow \dim(N) = n$,
symplectic	if	$T_xN \cap T_xN^{\omega\perp} = 0$ for each $x \in N$	$\Rightarrow \dim(N) =$ even,

where for a linear subspace $W \subset T_xN$ its symplectic orthogonal is given by $W^{\omega\perp} = \{X \in T_xM : \omega_x(X,Y) = 0 \text{ for all } Y \in W\}$, as in (31.4).

31.9. The cotangent bundle. Consider the manifold $M = T^*Q$, where Q is a manifold. Recall that for any smooth $f : Q \to P$ which is locally a diffeomorphism we get a homomorphism of vector bundles $T^*f : T^*Q \to T^*P$ covering f by $T_x^*f = ((T_xf)^{-1})^* : T_x^*Q \to T_{f(x)}^*P$.

There is a canonical 1-form $\vartheta = \vartheta_Q \in \Omega^1(T^*Q)$, called the *Liouville form*, which is given by

$$\vartheta(X) = \langle \pi_{T^*Q}(X), T(\pi_Q)(X) \rangle, \qquad X \in T(T^*Q),$$

where we used the projections (and their local forms):

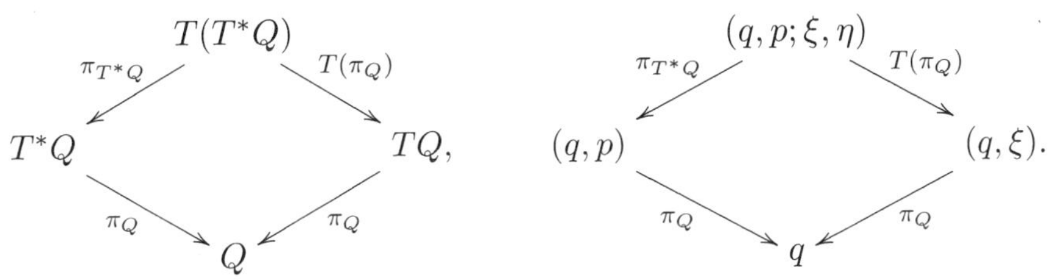

For a chart $q = (q^1, \dots, q^n) : U \to \mathbb{R}^n$ on Q and its induced (cotangent) chart $T^*q : T^*U \to \mathbb{R}^n \times (\mathbb{R}^n)^*$, where $T_x^*q = (T_xq^{-1})^*$, we consider the 'momentum coordinates' $p_i := \langle T^*q(\ \ ), e_i \rangle : T^*U \to \mathbb{R}$. Then $(q^1, \dots, q^n, p_1, \dots, p_n) : T^*U \to \mathbb{R}^n \times (\mathbb{R}^n)^*$ are the canonically induced coordinates on the cotangent bundle. In these coordinates we have

$$\vartheta_Q = \sum_{i=1}^n \Big(\vartheta_Q(\tfrac{\partial}{\partial q^i}) dq^i + \vartheta_Q(\tfrac{\partial}{\partial p_i}) dp_i \Big) = \sum_{i=1}^n p_i \, dq^i + 0,$$

since $\vartheta_Q(\frac{\partial}{\partial q^i}) = \vartheta_{\mathbb{R}^n}(q, p; e_i, 0) = \langle p, e_i \rangle = p_i$.

Now we define the *canonical symplectic structure* $\omega_Q = \omega \in \Omega^2(T^*Q)$ by

$$\omega_Q := -d\vartheta_Q \overset{\text{locally}}{=} \sum_{i=1}^n dq^i \wedge dp_i.$$

Note that $\check{\omega}(\frac{\partial}{\partial q^i}) = dp_i$ and $\check{\omega}(\frac{\partial}{\partial p_i}) = -dq^i$.

Lemma. *The 1-form $\vartheta_Q \in \Omega^1(T^*Q)$ has the following universal property and is uniquely determined by it:*

Any 1-form $\varphi \in \Omega^1(Q)$ is a smooth section $\varphi : Q \to T^*Q$ and for the pullback we have $\varphi^*\vartheta_Q = \varphi \in \Omega^1(Q)$. Moreover, $\varphi^*\omega_Q = -d\varphi \in \Omega^2(Q)$.

The 1-form ϑ_Q is natural in $Q \in \mathcal{M}f_n$: For every local diffeomorphism $f : Q \to P$ the local diffeomorphism $T^*f : T^*Q \to T^*P$ satisfies $(T^*f)^*\vartheta_P = \vartheta_Q$, and a fortiori $(T^*f)^*\omega_P = \omega_Q$.

In this sense ϑ_Q is a *universal* 1-*form*, or a *universal connection*, and ω_Q is the *universal curvature*, for $\mathbb{R}^1$-principal bundles over Q. Compare with section (19).

Proof. For a 1-form $\varphi \in \Omega^1(Q)$ we have

$$\begin{aligned}(\varphi^*\vartheta_Q)(X_x) &= (\vartheta_Q)_{\varphi_x}(T_x\varphi.X_x) = \varphi_x(T_{\varphi_x}\pi_Q.T_x\varphi.X_x)\\ &= \varphi_x(T_x(\pi_Q \circ \varphi).X_x) = \varphi_x(X_x).\end{aligned}$$

Thus $\varphi^*\vartheta_Q = \varphi$. Clearly this equation describes ϑ_Q uniquely. For ω we have $\varphi^*\omega_Q = -\varphi^*d\vartheta_Q = -d\varphi^*\vartheta_Q = -d\varphi$.

For a local diffeomorphism $f : Q \to P$, for $\alpha \in T^*_xQ$, and for $X_\alpha \in T_\alpha(T^*Q)$ we compute as follows:

$$\begin{aligned}((T^*f)^*\vartheta_P)_\alpha(X_\alpha) &= (\vartheta_P)_{T^*f.\alpha}(T_\alpha(T^*f).X_\alpha) = (T^*f.\alpha)(T(\pi_P).T(T^*f).X_\alpha)\\ &= (\alpha \circ T_xf^{-1})(T(\pi_P \circ T^*f).X_\alpha) = \alpha.T_xf^{-1}.T(f \circ \pi_Q).X_\alpha\\ &= \alpha(T(\pi_Q).X_\alpha) = \vartheta_Q(X_\alpha). \quad \square\end{aligned}$$

31.10. Lemma. *Let $\varphi : T^*Q \to T^*P$ be a (globally defined) local diffeomorphism such that $\varphi^*\vartheta_P = \vartheta_Q$. Then there exists a local diffeomorphism $f : Q \to P$ such that $\varphi = T^*f$.*

Proof. Let $\xi_Q = -\check{\omega}^{-1} \circ \vartheta_Q \in \mathfrak{X}(T^*Q)$ be the so-called *Liouville vector field:*

$$\begin{array}{ccc} T(T^*Q) & \xrightarrow{\quad -\check{\omega}_Q \quad} & T^*(T^*Q). \\ {\scriptstyle \xi_Q} \nwarrow & & \nearrow {\scriptstyle \theta_Q} \\ & T^*Q & \end{array}$$

Then locally $\xi_Q = \sum_{i=1}^n p_i \frac{\partial}{\partial p_i}$. Its flow is given by $\mathrm{Fl}_t^{\xi_Q}(\alpha) = e^t.\alpha$. Since $\varphi^*\vartheta_P = \vartheta_Q$, we also have that the Liouville vector fields ξ_Q and ξ_P are φ-dependent. Since $\xi_Q = 0$ exactly at the zero section, we have $\varphi(0_Q) \subseteq 0_P$, so there is a smooth mapping $f : Q \to P$ with $0_P \circ f = \varphi \circ 0_Q : Q \to T^*P$. By (3.14) we have $\varphi \circ \mathrm{Fl}_t^{\xi_Q} = \mathrm{Fl}_t^{\xi_P} \circ \varphi$, so the image of φ of the closure of a flow line of ξ_Q is contained in the closure of a flow line of ξ_P. For $\alpha_x \in T^*_x Q$ the closure of the flow line is $[0,\infty).\alpha_x$ and $\varphi(0_x) = 0_{f(x)}$; thus $\varphi([0,\infty).\alpha_x) \subset T^*_{f(x)}P$, and φ is fiber respecting: $\pi_P \circ \varphi = f \circ \pi_Q : T^*Q \to P$. Finally, for $X_\alpha \in T_\alpha(T^*Q)$ we have

$$\begin{aligned}\alpha(T_\alpha \pi_Q.X_\alpha) &= \vartheta_Q(X_\alpha) = (\varphi^*\vartheta_P)(X_\alpha) = (\vartheta_P)_{\varphi(\alpha)}(T_\alpha\varphi.X_\alpha)\\ &= (\varphi(\alpha))(T_{\varphi(\alpha)}\pi_P.T_\alpha\varphi.X_\alpha) = (\varphi(\alpha))(T_\alpha(\pi_P\circ\varphi).X_\alpha)\\ &= (\varphi(\alpha))(T_\alpha(f\circ\pi_Q).X_\alpha) = (\varphi(\alpha))(Tf.T_\alpha\pi_Q.X_\alpha),\\ \text{thus}\qquad \alpha &= \varphi(\alpha)\circ T_{\pi_Q(\alpha)}f,\\ \varphi(\alpha) &= \alpha\circ T_{\pi_Q(\alpha)}f^{-1} = (T_{\pi_Q(\alpha)}f^{-1})^*(\alpha) = T^*f(\alpha). \qquad \square\end{aligned}$$

31.11. Time dependent vector fields. Let f_t be a smooth curve of diffeomorphisms on a manifold M locally defined for each t, with $f_0 = \mathrm{Id}_M$, as in (3.6). We define two time dependent vector fields

$$\xi_t(x) := (T_x f_t)^{-1}\tfrac{\partial}{\partial t}f_t(x), \qquad \eta_t(x) := (\tfrac{\partial}{\partial t}f_t)(f_t^{-1}(x)).$$

Then $T(f_t).\xi_t = \frac{\partial}{\partial t}f_t = \eta_t \circ f_t$, so ξ_t and η_t are f_t-related.

Lemma. *In this situation, for $\omega \in \Omega^k(M)$ we have:*

(1) $i_{\xi_t} f_t^*\,\omega = f_t^*\, i_{\eta_t}\omega$.

(2) $\frac{\partial}{\partial t} f_t^*\omega = f_t^*\mathcal{L}_{\eta_t}\omega = \mathcal{L}_{\xi_t} f_t^*\,\omega$.

Proof. (1) is by computation:

$$\begin{aligned}(i_{\xi_t} f_t^*\,\omega)_x(X_2,\dots,X_k) &= (f_t^*\,\omega)_x(\xi_t(x),X_2,\dots,X_k)\\ &= \omega_{f_t(x)}(T_xf_t.\xi_t(x),T_xf_t.X_2,\dots,T_xf_t.X_k)\\ &= \omega_{f_t(x)}(\eta_t(f_t(x)),T_xf_t.X_2,\dots,T_xf_t.X_k)\\ &= (f_t^*\, i_{\eta_t}\omega)_x(X_2,\dots,X_k).\end{aligned}$$

(2) We put $\bar\eta \in \mathfrak{X}(\mathbb{R}\times M)$, $\bar\eta(t,x) = (\partial_t, \eta_t(x))$. We recall from (3.30) the evolution operator for η_t:

$$\Phi^\eta : \mathbb{R}\times\mathbb{R}\times M \to M, \quad \tfrac{\partial}{\partial t}\Phi^\eta_{t,s}(x) = \eta_t(\Phi^\eta_{t,s}(x)), \quad \Phi^\eta_{s,s}(x) = x,$$

which satisfies

$$(t, \Phi^\eta_{t,s}(x)) = \mathrm{Fl}^{\bar\eta}_{t-s}(s,x), \quad \Phi^\eta_{t,s} = \Phi^\eta_{t,r}\circ\Phi^\eta_{r,s}(x).$$

Since f_t satisfies $\frac{\partial}{\partial t}f_t = \eta_t\circ f_t$ and $f_0 = \mathrm{Id}_M$, we may conclude that $f_t = \Phi^\eta_{t,0}$, or $(t, f_t(x)) = \mathrm{Fl}^{\bar\eta}_t(0,x)$, so $f_t = \mathrm{pr}_2\circ\mathrm{Fl}^{\bar\eta}_t\circ\mathrm{ins}_0$. Thus

$$\begin{aligned}\tfrac{\partial}{\partial t}f_t^*\omega &= \tfrac{\partial}{\partial t}(\mathrm{pr}_2\circ\mathrm{Fl}^{\bar\eta}_t\circ\mathrm{ins}_0)^*\omega = \mathrm{ins}_0^*\,\tfrac{\partial}{\partial t}(\mathrm{Fl}^{\bar\eta}_t)^*\,\mathrm{pr}_2^*\,\omega\\ &= \mathrm{ins}_0^*(\mathrm{Fl}^{\bar\eta}_t)^*\mathcal{L}_{\bar\eta}\,\mathrm{pr}_2^*\,\omega.\end{aligned}$$

For time independent vector fields X_i on M we have, using (9.6):

$$\begin{aligned}
(\mathcal{L}_{\bar\eta}\operatorname{pr}_2^*\omega)(0\times X_1,\dots,0\times X_k)|_{(t,x)} &= \bar\eta((\operatorname{pr}_2^*\omega)(0\times X_1,\dots))|_{(t,x)}\\
&\quad-\textstyle\sum_i(\operatorname{pr}_2^*\omega)(0\times X_1,\dots,[\bar\eta,0\times X_i],\dots,0\times X_k)|_{(t,x)}\\
&= (\partial_t,\eta_t(x))(\omega(X_1,\dots,X_k))-\textstyle\sum_i\omega(X_1,\dots,[\eta_t,X_i],\dots,X_k)|_x\\
&= (\mathcal{L}_{\eta_t}\omega)_x(X_1,\dots,X_k).
\end{aligned}$$

This implies for $X_i\in T_xM$

$$\begin{aligned}
(\tfrac{\partial}{\partial t}f_t^*\omega)_x(X_1,\dots,X_k) &= (\operatorname{ins}_0^*(\operatorname{Fl}_t^{\bar\eta})^*\mathcal{L}_{\bar\eta}\operatorname{pr}_2^*\omega)_x(X_1,\dots,X_k)\\
&= ((\operatorname{Fl}_t^{\bar\eta})^*\mathcal{L}_{\bar\eta}\operatorname{pr}_2^*\omega)_{(0,x)}(0\times X_1,\dots,0\times X_k)\\
&= (\mathcal{L}_{\bar\eta}\operatorname{pr}_2^*\omega)_{(t,f_t(x))}(0_t\times T_xf_t.X_1,\dots,0_t\times T_xt_x.X_k)\\
&= (\mathcal{L}_{\eta_t}\omega)_{f_t(x)}(T_xf_t.X_1,\dots,T_xt_x.X_k)\\
&= (f_t^*\mathcal{L}_{\eta_t}\omega)_x(X_1,\dots,X_k),
\end{aligned}$$

which proves the first part of (2). The second part now follows by using (1):

$$\begin{aligned}
\tfrac{\partial}{\partial t}f_t^*\omega &= f_t^*\mathcal{L}_{\eta_t}\omega = f_t^*(di_{\eta_t}+i_{\eta_t}d)\omega = d\,f_t^*\,i_{\eta_t}\omega + f_t^*\,i_{\eta_t}d\omega\\
&= d\,i_{\xi_t}\,f_t^*\,\omega + i_{\xi_t}\,f_t^*\,d\omega = d\,i_{\xi_t}\,f_t^*\,\omega + i_{\xi_t}\,d\,f_t^*\,\omega = \mathcal{L}_{\xi_t}\,f_t^*\,\omega. \qquad\square
\end{aligned}$$

31.12. Surfaces. Let M be an orientable 2-dimensional manifold. Let $\omega\in\Omega^2(M)$ be a volume form on M. Then $d\omega=0$, so (M,ω) is a symplectic manifold. There are not many different symplectic structures on M if M is compact, since we have:

31.13. Theorem (J. Moser). *Let M be a connected compact oriented manifold. Let $\omega_0,\omega_1\in\Omega^{\dim M}(M)$ be two volume forms (both >0).*

If $\int_M\omega_0=\int_M\omega_1$, then there is a diffeomorphism $f:M\to M$ such that $f^*\omega_1=\omega_0$.

Proof. Put $\omega_t:=\omega_0+t(\omega_1-\omega_0)$ for $t\in[0,1]$; then each ω_t is a volume form on M since these form a convex set.

We look for a curve of diffeomorphisms $t\mapsto f_t$ with $f_t^*\omega_t=\omega_0$; then $\frac{\partial}{\partial t}(f_t^*\omega_t)=0$. Since $\int_M(\omega_1-\omega_0)=0$, we have $[\omega_1-\omega_0]=0\in H^m(M)$, so $\omega_1-\omega_0=d\psi$ for some $\psi\in\Omega^{m-1}(M)$. Put $\eta_t:=(\frac{\partial}{\partial t}f_t)\circ f_t^{-1}$; then by (31.11) we have:

$$0\overset{\text{wish}}{=}\tfrac{\partial}{\partial t}(f_t^*\omega_t)=f_t^*\mathcal{L}_{\eta_t}\omega_t+f_t^*\tfrac{\partial}{\partial t}\omega_t=f_t^*(\mathcal{L}_{\eta_t}\omega_t+\omega_1-\omega_0),$$

$$0\overset{\text{wish}}{=}\mathcal{L}_{\eta_t}\omega_t+\omega_1-\omega_0=di_{\eta_t}\omega_t+i_{\eta_t}d\omega_t+d\psi=di_{\eta_t}\omega_t+d\psi.$$

We can choose η_t uniquely by $i_{\eta_t}\omega_t=-\psi$, since ω_t is nondegenerate for all t. Then the evolution operator $f_t=\Phi^\eta_{t,0}$ exists for $t\in[0,1]$ since M is compact, by (3.30). We have, using (31.11.2),

$$\tfrac{\partial}{\partial t}(f_t^*\omega_t) = f_t^*(\mathcal{L}_{\eta_t}\omega_t + d\psi) = f_t^*(di_{\eta_t}\omega_t + d\psi) = 0,$$

so $f_t^*\omega_t = \text{ constant } = \omega_0$. □

31.14. Coadjoint orbits of a Lie group. Let G be a Lie group with Lie algebra $\mathfrak{g}$ and dual space $\mathfrak{g}^*$, and consider the adjoint representation $\mathrm{Ad} : G \to GL(\mathfrak{g})$. The *coadjoint representation* $\mathrm{Ad}^* : G \to GL(\mathfrak{g}^*)$ is then given by $\mathrm{Ad}^*(g)\alpha := \alpha\circ\mathrm{Ad}(g^{-1}) = \mathrm{Ad}(g^{-1})^*(\alpha)$. For $\alpha \in \mathfrak{g}^*$ we consider the coadjoint orbit $G.\alpha \subset \mathfrak{g}^*$ which is diffeomorphic to the homogenous space G/G_α, where G_α is the isotropy group $\{g \in G : \mathrm{Ad}^*(g)\alpha = \alpha\}$ at α.

As in (6.2), for $X \in \mathfrak{g}$ we consider the fundamental vector field $\zeta_X = -\mathrm{ad}(X)^* \in \mathfrak{X}(\mathfrak{g}^*)$ of the coadjoint action. For any $Y \in \mathfrak{g}$ we consider the linear function $\mathrm{ev}_Y : \mathfrak{g}^* \to \mathbb{R}$. The Lie derivative of the fundamental vector field ζ_X on the function ev_Y is then given by

(1) $\mathcal{L}_{\zeta_X}\mathrm{ev}_Y = -d\,\mathrm{ev}_Y \circ \mathrm{ad}(X)^* = -\mathrm{ev}_Y \circ \mathrm{ad}(X)^* = \mathrm{ev}_{[Y,X]}, \quad X, Y \in \mathfrak{g}.$

Note that the tangent space to the orbit is $T_\beta(G.\alpha) = \{\zeta_X(\beta) : X \in \mathfrak{g}\}$. Now we define the symplectic structure on the orbit $O = G.\alpha$ by

(2)
$$\begin{aligned}(\omega_O)_\alpha(\zeta_X, \zeta_Y) &= \alpha([X,Y]) = \langle\alpha, [X,Y]\rangle, \qquad \alpha \in \mathfrak{g}^*, \quad X, Y \in \mathfrak{g},\\ \omega_O(\zeta_X, \zeta_Y) &= \mathrm{ev}_{[X,Y]}\,.\end{aligned}$$

Theorem (Kirillov, Kostant, Souriau). *If G is a Lie group, then any coadjoint orbit $O \subset \mathfrak{g}^*$ carries a canonical symplectic structure ω_O which is invariant under the coadjoint action of G.*

Proof. First we claim that for $X \in \mathfrak{g}$ we have $X \in \mathfrak{g}_\alpha = \{Z \in \mathfrak{g} : \zeta_Z(\alpha) = 0\}$ if and only if $\alpha([X, \quad]) = (\omega_O)_\alpha(\zeta_X, \quad) = 0$. Indeed, for $Y \in \mathfrak{g}$ we have

$$\begin{aligned}\langle\alpha, [X,Y]\rangle &= \langle\alpha, \partial|_0 \mathrm{Ad}(\exp(tX))Y\rangle = \partial|_0\langle\alpha, \mathrm{Ad}(\exp(tX))Y\rangle\\ &= \partial|_0\langle\mathrm{Ad}^*(\exp(-tX))\alpha, Y\rangle = -\langle\zeta_X(\alpha), Y\rangle = 0.\end{aligned}$$

This shows that ω_O as defined in (2) is well defined and also nondegenerate along each orbit.

Now we show that $d\omega_O = 0$, using (2):

$$\begin{aligned}(d\omega_O)(\zeta_X, \zeta_Y, \zeta_Z) &= \sum_{\text{cyclic}} \zeta_X\,\omega_O(\zeta_Y, \zeta_Z) - \sum_{\text{cyclic}} \omega_O([\zeta_X, \zeta_Y], \zeta_Z)\\ &= \sum_{\text{cyclic}} \zeta_X\,\mathrm{ev}_{[Y,Z]} - \sum_{\text{cyclic}} \omega_O(\zeta_{-[X,Y]}, \zeta_Z) \quad \text{(now use (1))}\\ &= \sum_{\text{cyclic}} \mathrm{ev}_{[[Y,Z],X]} + \sum_{\text{cyclic}} \mathrm{ev}_{[[X,Y],Z]} = 0 \quad \text{by Jacobi.}\end{aligned}$$

Finally we show that ω_O is G-invariant: For $g \in G$ we have

$$\begin{aligned}
&((\operatorname{Ad}^*(g))^*\omega_O)_\alpha(\zeta_X(\alpha), \zeta_Y(\alpha)) \\
&= (\omega_O)_{\operatorname{Ad}^*(g)\alpha}(T(\operatorname{Ad}^*(g)).\zeta_X(\alpha), T(\operatorname{Ad}^*(g)).\zeta_Y(\alpha)) \\
&= (\omega_O)_{\operatorname{Ad}^*(g)\alpha}(\zeta_{\operatorname{Ad}(g)X}(\operatorname{Ad}^*(g)\alpha), \zeta_{\operatorname{Ad}(g)Y}(\operatorname{Ad}^*(g)\alpha)), \quad \text{by (6.2.2)}, \\
&= (\operatorname{Ad}^*(g)\alpha)([\operatorname{Ad}(g)X, \operatorname{Ad}(g)Y]) \\
&= (\alpha \circ \operatorname{Ad}(g^{-1}))(\operatorname{Ad}(g)[X,Y]) = \alpha([X,Y]) = (\omega_O)_\alpha(\zeta_X, \zeta_Y). \quad \square
\end{aligned}$$

31.15. Theorem (Darboux). *Let (M, ω) be a symplectic manifold of dimension $2n$. Then for each $x \in M$ there exists a chart (U, u) of M centered at x such that $\omega|U = \sum_{i=1}^n du^i \wedge du^{n+i}$. So each symplectic manifold is locally symplectomorphic to a cotangent bundle.*

Proof. Take any chart $(U, u : U \to u(U) \subset \mathbb{R}^{2n})$ centered at x. Choose linear coordinates on $\mathbb{R}^{2n}$ in such a way that $\omega_x = \sum_{i=1}^n du^i \wedge du^{n+i}|_x$ at x only. Then $\omega_0 = \omega|U$ and $\omega_1 = \sum_{i=1}^n du^i \wedge du^{n+i}$ are two symplectic structures on the open set $U \subset M$ which agree at x. Put $\omega_t := \omega_0 + t(\omega_1 - \omega_0)$. By making U smaller if necessary, we may assume that ω_t is a symplectic structure for all $t \in [0,1]$.

We want to find a curve of diffeomorphisms f_t near x with $f_0 = \mathrm{Id}$ such that $f_t(x) = x$ and $f_t^*\omega_t = \omega_0$. Then $\frac{\partial}{\partial t} f_t^*\omega_t = \frac{\partial}{\partial t}\omega_0 = 0$. We may assume that U is contractible; thus $H^2(U) = 0$, so $d(\omega_1 - \omega_0) = 0$ implies that $\omega_1 - \omega_0 = d\psi$ for some $\psi \in \Omega^1(U)$. By adding a constant form (in the chart on U), we may assume that $\psi_x = 0$. So we get for the time dependent vector field $\eta_t = \frac{\partial}{\partial t} f_t \circ f_t^{-1}$, using (31.11.2),

$$\begin{aligned}
0 &= \tfrac{\partial}{\partial t} f_t^*\omega_t = f_t^*(\mathcal{L}_{\eta_t}\omega_t + \tfrac{\partial}{\partial t}\omega_t) = f_t^*(d\, i_{\eta_t}\omega_t + i_{\eta_t} d\omega_t + \omega_1 - \omega_0) \\
&= f_t^*\, d(i_{\eta_t}\omega_t + \psi).
\end{aligned}$$

We can now prescribe η_t uniquely by $i_{\eta_t}\omega_t = -\psi$, since ω_t is nondegenerate on x. Moreover $\eta_t(x) = 0$ since $\psi_x = 0$. On a small neighborhood of x the left evolution operator f_t of η_t exists for all $t \in [0,1]$, and then clearly $\frac{\partial}{\partial t}(f_t^*\omega_t) = 0$, so $f_t^*\omega_t = \omega_0$ for all $t \in [0,1]$. $\square$

31.16. Relative Poincaré lemma. *Let M be a smooth manifold, let $N \subset M$ be a submanifold, and let $k \geq 0$. Let ω be a closed $(k+1)$-form on M which vanishes when pulled back to N. Then there exists a k-form φ on an open neighborhood U of N in M such that $d\varphi = \omega|U$ and $\varphi = 0$ along N. If moreover $\omega = 0$ along N (on $\bigwedge^k TM|N$), then we may choose φ such that the first derivatives of φ vanish on N.*

Proof. By restricting to a tubular neighborhood of N in M, we may assume that $p : M =: E \to N$ is a smooth vector bundle and that $i : N \to E$ is the zero section of the bundle. We consider $\mu : \mathbb{R} \times E \to E$, given by $\mu(t,x) = \mu_t(x) = tx$; then $\mu_1 = \mathrm{Id}_E$ and $\mu_0 = i \circ p : E \to N \to E$. Let $\xi \in \mathfrak{X}(E)$ be the vertical vector field $\xi(x) = \mathrm{vl}(x,x) = \partial|_0(x+tx)$; then $\mathrm{Fl}^\xi_t = \mu_{e^t}$. So locally for t near $(0,1]$ we have

$$\begin{aligned}\tfrac{d}{dt}\mu_t^*\omega &= \tfrac{d}{dt}(\mathrm{Fl}^\xi_{\log t})^*\omega = \tfrac{1}{t}(\mathrm{Fl}^\xi_{\log t})^*\mathcal{L}_\xi\omega \text{ by (31.11) or (8.16)}\\ &= \tfrac{1}{t}\mu_t^*(i_\xi d\omega + di_\xi\omega) = \tfrac{1}{t}d\mu_t^* i_\xi\omega.\end{aligned}$$

For $x \in E$ and $X_1,\dots,X_k \in T_xE$ we may compute

$$\begin{aligned}(\tfrac{1}{t}\mu_t^* i_\xi\omega)_x(X_1,\dots,X_k) &= \tfrac{1}{t}(i_\xi\omega)_{tx}(T_x\mu_t.X_1,\dots,T_x\mu_t.X_k)\\ &= \tfrac{1}{t}\omega_{tx}(\xi(tx),T_x\mu_t.X_1,\dots,T_x\mu_t.X_k)\\ &= \omega_{tx}(\mathrm{vl}(tx,x),T_x\mu_t.X_1,\dots,T_x\mu_t.X_k).\end{aligned}$$

So if $k \geq 0$, the k-form $\frac{1}{t}\mu_t^* i_\xi\omega$ is defined and smooth in (t,x) for all t near $[0,1]$ and describes a smooth curve in $\Omega^k(E)$. Note that for $x \in N = 0_E$ we have $(\frac{1}{t}\mu_t^* i_\xi\omega)_x = 0$, and if $\omega = 0$ along N, then $\frac{1}{t}\mu_t^* i_\xi\omega$ vanishes of second order along N. Since $\mu_0^*\omega = p^*i^*\omega = 0$ and $\mu_1^*\omega = \omega$, we have

$$\begin{aligned}\omega = \mu_1^*\omega - \mu_0^*\omega &= \int_0^1 \tfrac{d}{dt}\mu_t^*\omega dt\\ &= \int_0^1 d(\tfrac{1}{t}\mu_t^* i_\xi\omega)dt = d\left(\int_0^1 \tfrac{1}{t}\mu_t^* i_\xi\omega dt\right) =: d\varphi.\end{aligned}$$

If $x \in N$, we have $\varphi_x = 0$, and also the last claim is obvious from the explicit form of φ. □

31.17. Lemma. *Let M be a smooth finite-dimensional manifold, let $N \subset M$ be a submanifold, and let ω_0 and ω_1 be symplectic forms on M which are equal along N (on $\bigwedge^2 TM|N$).*

Then there exists a diffeomorphism $f : U \to V$ between two open neighborhoods U and V of N in M which satisfies $f|N = \mathrm{Id}_N$, $Tf|(TM|N) = \mathrm{Id}_{TM|N}$, and $f^*\omega_1 = \omega_0$.

Proof. Let $\omega_t = \omega_0 + t(\omega_1 - \omega_0)$ for $t \in [0,1]$. Since the restrictions of ω_0 and ω_1 to $\bigwedge^2 TM|N$ are equal, there is an open neighborhood U_1 of N in M such that ω_t is a symplectic form on U_1, for all $t \in [0,1]$. If $i : N \to M$ is the inclusion, we also have $i^*(\omega_1 - \omega_0) = 0$, and by assumption $d(\omega_1 - \omega_0) = 0$. Thus by lemma (31.16) there is a smaller open neighborhood U_2 of N such that $\omega_1|U_2 - \omega_0|U_2 = d\varphi$ for some $\varphi \in \Omega^1(U_2)$ with $\varphi_x = 0$ for $x \in N$, such that also all first derivatives of φ vanish along N.

Let us now consider the time dependent vector field $X_t := -(\check{\omega_t})^{-1} \circ \varphi$ given by $i_{X_t}\omega_t = \varphi$, which vanishes together with all first derivatives along N. Let f_t be the curve of local diffeomorphisms with $\frac{\partial}{\partial t} f_t = X_t \circ f_t$; then $f_t|N = \mathrm{Id}_N$

and $Tf_t|(TM|N) = \mathrm{Id}$. There is a smaller open neighborhood U of N such that f_t is defined on U for all $t \in [0,1]$. Then by (31.11) we have

$$\begin{aligned}\tfrac{\partial}{\partial t}(f_t^*\omega_t) &= f_t^*\mathcal{L}_{X_t}\omega_t + f_t^*\tfrac{\partial}{\partial t}\omega_t = f_t^*(di_{X_t}\omega_t + \omega_1 - \omega_0)\\ &= f_t^*(-d\varphi + \omega_1 - \omega_0) = 0,\end{aligned}$$

so $f_t^*\omega_t$ is constant in t, equals $f_0^*\omega_0 = \omega_0$, and finally $f_1^*\omega_1 = \omega_0$ as required. □

31.18. Lemma (Ehresmann). *Let (V,ω) be a symplectic vector space of real dimension $2n$, and let g be a nondegenerate symmetric bilinear form on V. Let $K := \check{g}^{-1}\circ\check{\omega} : V \to V^* \to V$ so that $g(Kv,w) = \omega(v,w)$.*
Then $K \in GL(V)$ and the following properties are equivalent:

(1) $K^2 = -\mathrm{Id}_V$, *so K is a complex structure.*
(2) $\omega(Kv,Kw) = \omega(v,w)$, *so $K \in Sp(V,\omega)$.*
(3) $g(Kv,Kw) = g(v,w)$, *so $K \in O(V,g)$.*

If these conditions are satisfied, any subpair of the triple ω, g, J is said to be *compatible*.

Proof. Starting from the definition, we have in turn:

$$\begin{aligned}g(Kv,w) &= \langle \check{g}K(v), w\rangle = \langle \check{g}\check{g}^{-1}\check{\omega}(v), w\rangle = \langle\check{\omega}(v), w\rangle = \omega(v,w),\\ \omega(Kv,Kw) &= g(K^2v,Kw) = g(Kw,K^2v) = \omega(w,K^2v) = -\omega(K^2v,w),\\ g(K^2v,w) &= \omega(Kv,w) = -\omega(w,Kv) = -g(Kw,Kv) = -g(Kv,Kw).\end{aligned}$$

The second line shows that (1) $\Leftrightarrow$ (2), and the third line shows that (1) $\Leftrightarrow$ (3).

31.19. Lemma (Polar decomposition of ω). *Let (V,g) be a Euclidean real vector space (positive definite). Let ω be a symplectic structure on V, let $A = \check{g}^{-1}\circ\check{\omega} \in GL(V)$, and let $A = BJ$ be the polar decomposition from* (4.38). *Then A is g-skew-symmetric, J is a complex structure, and the nondegenerate symmetric inner product $g_1(v,w) = \omega(v,Jw)$ is positive definite.*

Proof. We have $g(Av,w) = \omega(v,w) = -\omega(w,v) = -g(Aw,v) = -g(v,Aw)$; thus $A^\top = -A$. This has the consequence (see the proof of (4.38)) that $B = \exp(\frac12\log(AA^\top)) = \exp(\frac12\log(-A^2))$ commutes with A; thus also $J = B^{-1}A$ commutes with A and thus with B. Since $B^\top = B$, we get $J^{-1} = J^\top = (B^{-1}A)^\top = A^\top(B^{-1})^\top = -AB^{-1} = -B^{-1}A = -J$; thus J is a complex structure. Moreover, we have

$$\omega(Jv,Jw) = g(AJv,Jw) = g(JAv,Jw) = g(Av,w) = \omega(v,w);$$

thus by (31.18) the symplectic form ω and the complex structure J are compatible, and the symmetric (by (31.18)) bilinear form g_1 defined by $g_1(v,w) = \omega(v,Jw)$ is positive definite: $g_1(v,v) = \omega(v,Jv) = g(Av,Jv) = g(BJv,Jv) > 0$ since B is positive definite. □

31.20. Relative Darboux theorem (Weinstein). *Let (M,ω) be a symplectic manifold, and let $L \subset M$ be a Lagrangian submanifold.*

Then there exist a tubular neighborhood U of L in M, an open neigborhood V of the zero section 0_L in T^*L and a symplectomorphism

$$(T^*L,\omega_L) \supset (V,\omega_L) \xrightarrow{\varphi} (U,\omega|U) \subset (M,\omega)$$

such that $\varphi \circ 0_L : L \to V \to M$ is the embedding $L \subset M$.

Moreover, suppose that for the Lagrangian subbundle TL in the symplectic vector bundle $TM|L \to L$ we are given a complementary Lagrangian subbundle $E \to L$; then the symplectomorphism φ may be chosen in such a way that $T_{0_x}\varphi.V_{0_x}(T^*L) = E_{\varphi(0_x)}$ for $x \in L$.

Here $V(T^*L)$ denotes the vertical bundle in the tangent bundle of T^*L.

Proof. The tangent bundle $TL \to L$ is a Lagrangian subbundle of the symplectic vector bundle $TM|L \to L$.

Claim. *There exists a Lagrangian complementary vector bundle $E \to L$ in the symplectic vector bundle $TM|L$.* Namely, we choose a fiberwise Riemann metric g in the vector bundle $TM|L \to L$ and consider the vector bundle homomorphism $A = \check{g}^{-1}\check{\omega} : TM|L \to T^*M|L \to TM|L$ and its polar decomposition $A = BJ$ with respect to g as explained in (31.19). Then J is a fiberwise complex structure, and $g_1(u,v) := \omega(u,Jv)$ defines again a positive definite fiberwise Riemann metric. Since $g_1(J\ \ ,\ \ ) = \omega(\ \ ,\ \ )$ vanishes on TL, the Lagrangian subbundle $E = JTL \subset TM|L$ is g_1-orthogonal to TL, thus a complement.

We may use either the constructed or the given Lagrangian complement to TL in what follows.

The symplectic structure ω induces a duality pairing between the vector bundles E and TL; thus we may identify $(TM|L)/TL \cong E \to L$ with the cotangent bundle T^*L by $\langle X_x, \check{\omega}(Y_x)\rangle = \omega(X_x,Y_x)$ for $x \in L$, $X_x \in T_xL$ and $Y_x \in E_x$.

Let $\psi := \exp^g \circ \check{\omega}^{-1} : T^*L \to M$ where $\exp^g$ is any geodesic exponential mapping on TM restricted to E. Then ψ is a diffeomorphism from a neighborhood V of the zero section in T^*L to a tubular neighborhood U of L in M, which equals the embedding of L along the zero section.

Let us consider the pullback $\psi^*\omega$ and compare it with ω_L on V. For $0_x \in 0_L$ we have $T_{0_x}V = T_xL \oplus T_x^*L \cong T_xL \oplus E_x$. The linear subspace T_xL is

Lagrangian for both ω_L and $\psi^*\omega$ since L is a Lagrange submanifold. The linear subspace T_x^*L is Lagrangian for ω_L, and it is also Lagrangian for $\psi^*\omega$ since E was a Lagrangian bundle. Both $(\omega_L)_{0_x}$ and $(\psi^*\omega)_{0_x}$ induce the same duality between T_xL and T_x^*L since the identification $E_x \cong T_x^*L$ was via ω_x. Thus ω_L equals $\psi^*\omega$ along the zero section.

Finally, by lemma (31.17) the identity of the zero section extends to a diffeomorphism ρ on a neighborhood with $\rho^*\psi^*\omega = \omega_L$. The diffeomorphism $\varphi = \psi \circ \rho$ then satisfies the theorem. □

31.21. The Poisson bracket. Let (M, ω) be a symplectic manifold. For $f \in C^\infty(M)$ the *Hamiltonian vector field* or *symplectic gradient* $H_f = \operatorname{grad}^\omega(f) \in \mathfrak{X}(M)$ is defined by any of the following equivalent prescriptions:

(1) $\quad i(H_f)\omega = df, \quad H_f = \check{\omega}^{-1}df, \quad \omega(H_f, X) = X(f)$ for $X \in TM$.

For two functions $f, g \in C^\infty(M)$ we define their *Poisson bracket* $\{f, g\}$ by

(2)
$$\begin{aligned}\{f,g\} :&= i(H_f)i(H_g)\omega = \omega(H_g, H_f)\\ &= H_f(g) = \mathcal{L}_{H_f}g = dg(H_f) \in C^\infty(M).\end{aligned}$$

Let us furthermore put

(3)
$$\mathfrak{X}(M,\omega) := \{X \in \mathfrak{X}(M) : \mathcal{L}_X\omega = 0\}$$

and call this the space of *locally Hamiltonian vector fields* or *ω-respecting vector fields.*

Theorem. *Let (M, ω) be a symplectic manifold.*

Then $(C^\infty(M), \{\ ,\ \})$ is a Lie algebra which also satisfies $\{f, gh\} = \{f, g\}h + g\{f, h\}$, i.e., $\operatorname{ad}_f = \{f,\ \}$ is a derivation of $(C^\infty(M), \cdot)$.

Moreover, there is an exact sequence of Lie algebras and Lie algebra homomorphisms

$$\begin{array}{ccccccccc} 0 \longrightarrow & H^0(M) & \xrightarrow{\alpha} & C^\infty(M) & \xrightarrow{H=\operatorname{grad}^\omega} & \mathfrak{X}(M,\omega) & \xrightarrow{\gamma} & H^1(M) & \longrightarrow 0 \\ & 0 & & \{\ ,\ \} & & [\ ,\ ] & & 0 & \end{array}$$

where the brackets are written under the spaces, where α is the embedding of the space of all locally constant functions, and where $\gamma(X) := [i_X\omega] \in H^1(M)$.

The whole situation behaves invariantly (resp. equivariantly) under pullback by symplectomorphisms $\varphi : M \to M$: For example $\varphi^*\{f, g\} = \{\varphi^*f, \varphi^*g\}$, $\varphi^*(H_f) = H_{\varphi^*f}$, and $\varphi^*\gamma(X) = \gamma(\varphi^*X)$. Consequently for $X \in \mathfrak{X}(M, \omega)$ we have $\mathcal{L}_X\{f, g\} = \{\mathcal{L}_Xf, g\} + \{f, \mathcal{L}_Xg\}$ and $\gamma(\mathcal{L}_XY) = 0$.

Proof. The operator H takes values in $\mathfrak{X}(M,\omega)$ since

$$\mathcal{L}_{H_f}\omega = i_{H_f}\,d\omega + d\,i_{H_f}\,\omega = 0 + ddf = 0.$$

The mapping H is a Lie algebra homomorphism, i.e., $H(\{f,g\}) = [H_f, H_g]$, since by (9.9) and (9.7) we have

$$\begin{aligned} i_{H(\{f,g\})}\omega &= d\{f,g\} = d\mathcal{L}_{H_f}g = \mathcal{L}_{H_f}dg - 0 = \mathcal{L}_{H_f}i_{H_g}\omega - i_{H_g}\mathcal{L}_{H_f}\omega \\ &= [\mathcal{L}_{H_f}, i_{H_g}]\omega = i_{[H_f,H_g]}\omega. \end{aligned}$$

The sequence is exact at $H^0(M)$ since the embedding α of the locally constant functions is injective.

The sequence is exact at $C^\infty(M)$: For a locally constant function c we have $H_c = \check{\omega}^{-1}dc = \check{\omega}^{-1}0 = 0$. If $H_f = \check{\omega}^{-1}df = 0$ for $f \in C^\infty(M)$, then $df = 0$, so f is locally constant.

The sequence is exact at $\mathfrak{X}(M,\omega)$: For $X \in \mathfrak{X}(M,\omega)$ we have $di_X\omega = di_X\omega + i_Xd\omega = \mathcal{L}_X\omega = 0$; thus $\gamma(X) = [i_X\omega] \in H^1(M)$ is well defined. For $f \in C^\infty(M)$ we have $\gamma(H_f) = [i_{H_f}\omega] = [df] = 0 \in H^1(M)$. If $X \in \mathfrak{X}(M,\omega)$ with $\gamma(X) = [i_X\omega] = 0 \in H^1(M)$, then $i_X\omega = df$ for some $f \in \Omega^0(M) = C^\infty(M)$, but then $X = H_f$.

The sequence is exact at $H^1(M)$: The mapping γ is surjective since for $\varphi \in \Omega^1(M)$ with $d\varphi = 0$ we may consider $X := \check{\omega}^{-1}\varphi \in \mathfrak{X}(M)$ which satisfies $\mathcal{L}_X\omega = i_Xd\omega + di_X\omega = 0 + d\varphi = 0$ and $\gamma(X) = [i_X\omega] = [\varphi] \in H^1(M)$.

The Poisson bracket $\{\ \ ,\ \ \}$ is a Lie bracket and $\{f,gh\} = \{f,g\}h + g\{f,h\}$:

$$\begin{aligned} \{f,g\} &= \omega(H_g, H_f) = -\omega(H_f, H_g) \\ &= \{g,f\}, \\ \{f,\{g,h\}\} &= \mathcal{L}_{H_f}\mathcal{L}_{H_g}h = [\mathcal{L}_{H_f}, \mathcal{L}_{H_g}]h + \mathcal{L}_{H_g}\mathcal{L}_{H_f}h \\ &= \mathcal{L}_{[H_f,H_g]}h + \{g,\{f,h\}\} \\ &= \mathcal{L}_{H_{\{f,g\}}}h + \{g,\{f,h\}\} \\ &= \{\{f,g\},h\} + \{g,\{f,h\}\}, \\ \{f,gh\} &= \mathcal{L}_{H_f}(gh) = \mathcal{L}_{H_f}(g)h + g\mathcal{L}_{H_f}(h) \\ &= \{f,g\}h + g\{f,h\}. \end{aligned}$$

All mappings in the sequence are Lie algebra homomorphisms: For local constants $\{c_1, c_2\} = H_{c_1}c_2 = 0$. We already checked for H. For $X, Y \in \mathfrak{X}(M,\omega)$ we have

$$i_{[X,Y]}\omega = [\mathcal{L}_X, i_Y]\omega = \mathcal{L}_Xi_Y\omega - i_Y\mathcal{L}_X\omega = di_Xi_Y\omega + i_Xdi_Y\omega - 0 = di_Xi_Y\omega;$$

thus $\gamma([X,Y]) = [i_{[X,Y]}\omega] = 0 \in H^1(M)$.

Let us now transform the situation by a symplectomorphism $\varphi : M \to M$ via pullback. Then

$$\begin{aligned}
\varphi^*\omega = \omega \quad &\Leftrightarrow \quad (T\varphi)^* \circ \check{\omega} \circ T\varphi = \check{\omega} \\
\Rightarrow H_{\varphi^* f} &= \check{\omega}^{-1} d\,\varphi^* f = \check{\omega}^{-1}(\varphi^* df) \\
&= (T\varphi^{-1} \circ \check{\omega}^{-1} \circ (T\varphi^{-1})^*) \circ ((T\varphi)^* \circ df \circ \varphi) \\
&= (T\varphi^{-1} \circ \check{\omega}^{-1} \circ df \circ \varphi) = \varphi^*(H_f), \\
\varphi^*\{f,g\} = \varphi^*(dg(H_f)) &= (\varphi^* dg)(\varphi^* H_f) = d(\varphi^* g)(H_{\varphi^* f}) = \{\varphi^* f, \varphi^* g\}.
\end{aligned}$$

For the assertions about the Lie derivative apply $\mathcal{L}_X = \partial|_0 (\mathrm{Fl}^X_t)^*$. □

31.22. Basic example. In the situation of (31.1), where $M = T^*\mathbb{R}^n$ with $\omega = \omega_{\mathbb{R}^n} = -d\vartheta_{\mathbb{R}^n} = \sum_{i=1}^n dq^i \wedge dp_i$, we have

$$\begin{aligned}
\check{\omega} : T(T^*\mathbb{R}^n) \to T^*(T^*\mathbb{R}^n), \quad \check{\omega}(\partial_{q^i}) = dp_i, \quad \check{\omega}(\partial_{p^i}) = -dq_i, \\
H_f = \check{\omega}^{-1}.df = \check{\omega}^{-1}\Big(\textstyle\sum_i \Big(\frac{\partial f}{\partial q^i} dq^i + \frac{\partial f}{\partial p_i} dp_i\Big)\Big) = \textstyle\sum_i \Big(\frac{\partial f}{\partial p_i}\frac{\partial}{\partial q^i} - \frac{\partial f}{\partial q^i}\frac{\partial}{\partial p_i}\Big), \\
\{f,g\} = H_f g = \textstyle\sum_i \Big(\frac{\partial f}{\partial p_i}\frac{\partial g}{\partial q^i} - \frac{\partial f}{\partial q^i}\frac{\partial g}{\partial p_i}\Big), \\
\{p_i, p_j\} = 0, \quad \{q^i, q^j\} = 0, \quad \{q^i, p_j\} = -\delta^i_j.
\end{aligned}$$

31.23. Kepler's laws: Elementary approach. Here we give an elementary approach to the derivation of Kepler's laws.

Let us choose the orthonormal coordinate system in the oriented Euclidean space $\mathbb{R}^3$ with standard inner product $(\ \ |\ \ )$ and vector product $q \times q'$ in such a way that the sun with mass M is at $0 \in \mathbb{R}^3$. A planet now moves in a force field F on an orbit $q(t)$ according to Newton's law:

(1) $$F(q(t)) = m\ddot{q}(t).$$

(2) *If the force field is centripetal,* $F(q) = -f(q)\,q$ *for* $f \geq 0$, *then the angular momentum* $q(t) \times \dot{q}(t) = J$ *is a constant vector,* since

$$\partial_t (q \times \dot{q}) = \dot{q} \times \dot{q} + q \times \ddot{q} = 0 + \tfrac{1}{m} f(q)\, q \times q = 0.$$

Thus the planet moves in the plane orthogonal to the angular momentum vector J and we may choose coordinates such that this is the plane $q^3 = 0$. Let $z = q^1 + iq^2 = re^{i\varphi}$; then

$$\begin{aligned}
J = \begin{pmatrix} 0 \\ 0 \\ j \end{pmatrix} = z \times \dot{z} = \begin{pmatrix} q^1 \\ q^2 \\ 0 \end{pmatrix} \times \begin{pmatrix} \dot{q}^1 \\ \dot{q}^2 \\ 0 \end{pmatrix} = \begin{pmatrix} 0 \\ 0 \\ q^1\dot{q}^2 - q^2\dot{q}^1 \end{pmatrix}, \\
j = q^1\dot{q}^2 - q^2\dot{q}^1 = \mathrm{Im}(\bar{z}.\dot{z}) = \mathrm{Im}(re^{-i\varphi}(\dot{r}e^{i\varphi} + ir\dot{\varphi}e^{i\varphi})) \\
= \mathrm{Im}(r\dot{r} + ir^2\dot{\varphi}) = r^2\dot{\varphi}.
\end{aligned}$$

(3) *Thus in a centripetal force field area is swept out at a constant rate* $j = r^2\dot\varphi$ *(2nd law of Kepler, 1602, published 1606)*, since

$$\begin{aligned}\mathrm{Area}(t_1,t_2) &= \int_{\varphi(t_1)}^{\varphi(t_2)}\int_0^{r(\varphi)} r\,dr\,d\varphi = \int_{\varphi(t_1)}^{\varphi(t_2)} \tfrac12 r(\varphi)^2 d\varphi\\ &= \int_{t_1}^{t_2} \tfrac12 r(\varphi(t))^2\dot\varphi(t)\,dt = \tfrac{j}{2}(t_2-t_1).\end{aligned}$$

Now we specify the force field. According to Newton's law of gravity the sun acts on a planet of mass m at the point $0 \neq q \in \mathbb{R}^3$ by the force

$$\begin{aligned}F(q) &= -G\frac{Mm}{|q|^3}q = -\operatorname{grad} U(q),\\ U(q) &= -G\frac{Mm}{|q|},\end{aligned} \tag{4}$$

where $G = (6.67428 \pm 0.00067)\cdot 10^{-11}\, m^3\, kg^{-1}\, s^{-2}$ is the gravitational constant and U is the gravitational potential. We consider now the energy function (compare with (31.1)) along the orbit as the sum of the kinetic and the potential energies

$$E(t) := \frac{m}{2}|\dot q(t)|^2 + U(q(t)) = \frac{m}{2}|\dot q(t)|^2 - G\frac{Mm}{|q(t)|} \tag{5}$$

which is constant along the orbit, since

$$\partial_t E(t) = m(\ddot q(t)|\dot q(t)) + (\operatorname{grad} U(q(t))|\dot q(t)) = 0.$$

We have in the coordinates specified above for the velocity $v = |\dot q|$

$$v^2 = |\dot q|^2 = \mathrm{Re}(\bar{\dot z}\dot z) = \mathrm{Re}((\dot r e^{-i\varphi} - ir\dot\varphi e^{-i\varphi})(\dot r e^{i\varphi} + ir\dot\varphi e^{i\varphi})) = \dot r^2 + r^2\dot\varphi^2.$$

We look now for a solution in the form $r = r(\varphi)$. From (3) we have $\dot\varphi = j/r^2$ so that

$$v^2 = \dot r^2 + r^2\dot\varphi^2 = \left(\frac{dr}{d\varphi}\right)^2\dot\varphi^2 + r^2\dot\varphi^2 = \left(\frac{dr}{d\varphi}\right)^2\frac{j^2}{r^4} + \frac{j^2}{r^2}.$$

Plugging into the conservation of energy (5), we get

$$\left(\frac{dr}{d\varphi}\right)^2\frac{j^2}{r^4} + \frac{j^2}{r^2} - 2GM\frac{1}{r(t)} = \gamma = \text{ constant},$$

$$\frac{1}{r^4}\left(\frac{dr}{d\varphi}\right)^2 = \frac{\gamma}{j^2} + \frac{2GM}{j^2}\frac{1}{r(t)} - \frac{1}{r^2}. \tag{6}$$

Excluding the catastrophe of the planet falling into the sun, we may assume that r is never 0 and substitute

$$u(\varphi) = \frac{1}{r(\varphi)},\quad \frac{du}{d\varphi} = -\frac{1}{r^2}\frac{dr}{d\varphi}$$

into (6) to obtain

$$\left(\frac{du}{d\varphi}\right)^2 = \frac{\gamma}{j^2} + \frac{2GM}{j^2}u - u^2 = \frac{G^2M^2}{j^4}\left(1 + \frac{\gamma j^2}{G^2M^2}\right) - \left(u - \frac{GM}{j^2}\right)^2,$$

$$(7)\quad \left(\frac{du}{d\varphi}\right)^2 = \frac{\varepsilon^2}{p^2} - \left(u - \frac{1}{p}\right)^2, \quad \text{where } p := \frac{j^2}{GM}, \quad \varepsilon := \sqrt{1 + \frac{\gamma j^2}{G^2M^2}}$$

are parameters suitable to describe conic sections.

If $\varepsilon = 0$, then $(\frac{du}{d\varphi})^2 = -(u - \frac{1}{p})^2$ so that both sides have to be zero: $u = 1/p$ or $r = p =$ constant and the planet moves on a circle.

If $\varepsilon > 0$, then (7) becomes

$$\frac{du}{d\varphi} = \sqrt{\frac{\varepsilon^2}{p^2} - \left(u - \frac{1}{p}\right)^2} \quad \text{or} \quad d\varphi = \frac{du}{\sqrt{\frac{\varepsilon^2}{p^2} - \left(u - \frac{1}{p}\right)^2}},$$

$$\varphi + C = \int d\varphi = \int \frac{du}{\sqrt{\frac{\varepsilon^2}{p^2} - \left(u - \frac{1}{p}\right)^2}} \qquad \text{now use } w = u - \frac{1}{p}$$

$$= \int \frac{dw}{\sqrt{\frac{\varepsilon^2}{p^2} - w^2}} = \frac{p}{\varepsilon}\int \frac{dw}{\sqrt{1 - (\frac{pw}{\varepsilon})^2}} \qquad \text{now use } z = \frac{pw}{\varepsilon}$$

$$= \int \frac{dz}{\sqrt{1 - z^2}} = \arcsin(z) = \arcsin\left(\frac{pw}{\varepsilon}\right) = \arcsin\left(\frac{pu - 1}{\varepsilon}\right).$$

This implies

$$\sin(\varphi + C) = \frac{pu - 1}{\varepsilon}, \qquad u = \frac{1 + \varepsilon\sin(\varphi + C)}{p},$$

$$r = \frac{1}{u} = \frac{p}{1 + \varepsilon\sin(\varphi + C)}.$$

We choose the parameter C such that the minimal distance $\frac{p}{1+\varepsilon}$ of the planet from the sun (its *perihel*) is attained at $\varphi = 0$ so that $\sin(C) = 1$ or $C = \pi/2$; then $\sin(\varphi + \pi/2) = \cos(\varphi)$ and the *planetary orbit* is described by the equation

$$(8)\qquad r = \frac{p}{1 + \varepsilon\cos\varphi}, \qquad p > 0, \quad \varepsilon \geq 0.$$

Equation (8) *describes a conic section in polar coordinates with one focal point at* 0. *We have:*

- *a circle for* $\varepsilon = 0$,
- *an ellipse for* $0 \leq \varepsilon < 1$,
- *a parabola for* $\varepsilon = 1$,
- *the left branch of a hyperbola for* $\varepsilon > 1$.

For the ellipse with the right hand focal point at 0:

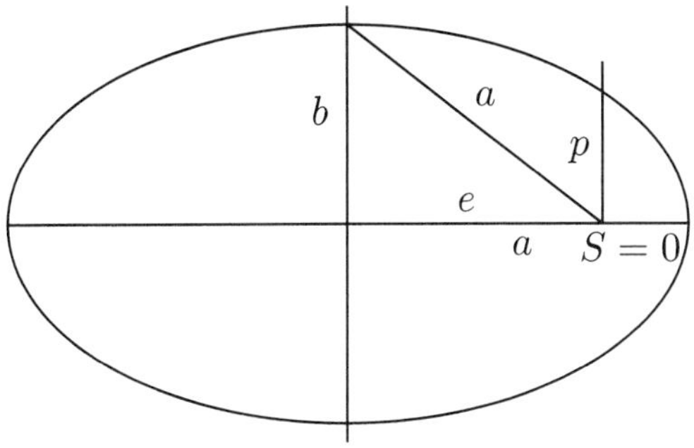

$$\frac{(q_1+e)^2}{a^2}+\frac{{q_2}^2}{b^2}=1,$$
$$e=\sqrt{a^2-b^2},$$
$$\frac{(r\cos\varphi+e)^2}{a^2}+\frac{r^2\sin^2\varphi}{b^2}=1,$$
$$(b^2-a^2)r^2\cos^2\varphi + 2b^2r\sqrt{a^2-b^2}\cos\varphi + a^2r^2-b^4=0.$$

Solving for $\cos\varphi$, we get

$$\begin{aligned}\cos\varphi &= \frac{-2b^2r\sqrt{a^2-b^2}\pm\sqrt{4b^4r^2(a^2-b^2)+4(a^2-b^2)r^2(a^2r^2-b^4)}}{-2(a^2-b^2)r^2}\\ &= \frac{-2b^2re\pm 2r^2ea}{-2r^2e^2}=\frac{b^2}{re}\pm\frac{a}{e},\\ \frac{b^2}{re}&=\cos\varphi\pm\frac{a}{e},\\ r&=\frac{b^2}{e(\cos\varphi\pm\frac{a}{e})}=\frac{b^2}{e\frac{a}{e}(\pm1+\frac{e}{a}\cos\varphi)}=\frac{\frac{b^2}{a}}{\pm1+\frac{e}{a}\cos\varphi}.\end{aligned}$$

Put $p=b^2/a$ and $0\le\varepsilon=\sqrt{1-b^2/a^2}=e/a\le1$ and note that $r>0$ to obtain the desired equation (8), i.e.,

$$r=\frac{p}{1+\varepsilon\cos\varphi}.$$

For the parabola with focal point at 0:

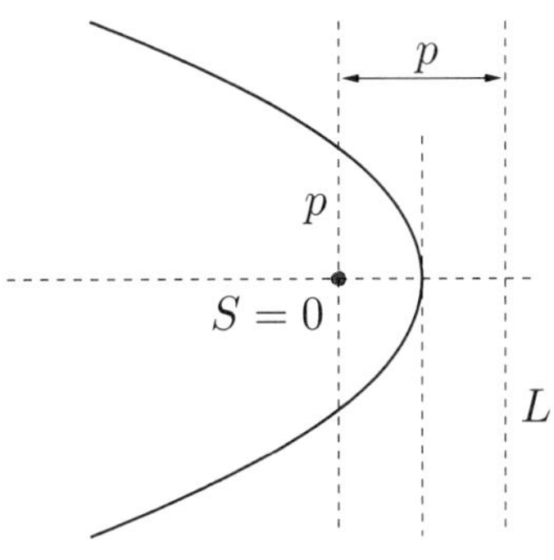

$$\begin{aligned}{q_2}^2 &= -2p(q_1-\frac{p}{2})=-2pq_1+p^2,\\ r^2(1-\cos^2\varphi)&=-2pr\cos\varphi+p^2,\\ r^2\cos^2\varphi-2pr\cos\varphi+p^2-r^2&=0,\\ \cos\varphi&=\frac{2pr\pm\sqrt{4p^2r^2-4r^2(p^2-r^2)}}{2r^2}\\ &=\frac{p\pm r}{r}=\frac{p}{r}\pm1,\\ r&=\frac{p}{1+\cos\varphi}\quad>0.\end{aligned}$$

For the hyperbola with left hand focal point at 0:

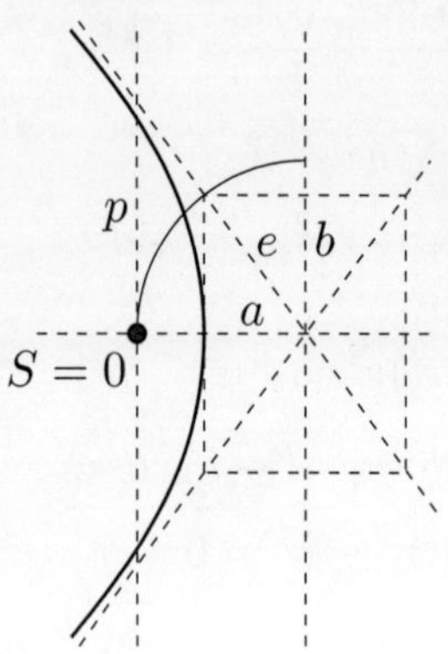

$$\begin{aligned}
&\frac{(q_1 - e)^2}{a^2} - \frac{{q_2}^2}{b^2} = 1,\\
&e = \sqrt{a^2 + b^2},\\
&\frac{(r\cos\varphi - e)^2}{a^2} - \frac{r^2\sin^2\varphi}{b^2} = 1,\\
&b^2r^2\cos^2\varphi - 2b^2r\sqrt{a^2+b^2}\cos\varphi\\
&\qquad + a^2b^2 + b^4 - a^2r^2(1-\cos^2\varphi) = a^2b^2,\\
&(b^2+a^2)r^2\cos^2\varphi\\
&\qquad - 2b^2r\sqrt{a^2+b^2}\cos\varphi + b^4 - a^2r^2 = 0.
\end{aligned}$$

Solving again for $\cos\varphi$, we get

$$\begin{aligned}
\cos\varphi &= \frac{2b^2r\sqrt{a^2+b^2} \pm \sqrt{4b^4r^2(a^2+b^2) - 4(a^2+b^2)r^2(b^4-a^2r^2)}}{2(a^2+b^2)r^2}\\
&= \frac{2b^2re \pm 2r^2ea}{2r^2e^2}.
\end{aligned}$$

Put $p = b^2/a$ and $\varepsilon = \sqrt{1 + b^2/a^2} = e/a > 1$ and note that $r > 0$ to obtain the desired equation (8), i.e., $r = \frac{p}{1+\varepsilon\cos\varphi}$.

(Kepler's 3rd law) If T is the orbital period of a planet on an elliptic orbit with major half-axis a, then:

$$\frac{T^2}{a^3} = \frac{(2\pi)^2}{GM}$$

is a constant depending only on the mass of the sun and not on the planet.

Let a and b be the major and minor half-axes of an elliptic planetary orbit with period T. The area of this ellipse is $ab\pi$. But by (3) this area equals $ab\pi = jT/2$. In (7) we had $p = j^2/(GM)$, and for an ellipse we have $p = b^2/a$; thus we get

$$\frac{j}{2}T = ab\pi = a^{3/2}p^{1/2}\pi = a^{3/2}\frac{j}{\sqrt{GM}}\pi, \qquad T = \frac{2\pi a^{3/2}}{\sqrt{GM}}, \qquad \frac{T^2}{a^3} = \frac{(2\pi)^2}{GM}.$$

31.24. Kepler's laws II: The 2-body system. Here we start to treat the 2-body system with methods like Poisson brackets, etc., as explained in (31.22). So the symplectic manifold (the *phase space*) is $T^*(\mathbb{R}^3 \setminus \{0\})$ with symplectic form $\omega = \omega_{\mathbb{R}^3} = -d\vartheta_{\mathbb{R}^3} = \sum_{i=1}^3 dq^i \wedge dp_i$. As in (31.1) we use the canonical coordinates q^i on $\mathbb{R}^3$ and $p_i := m \cdot \dot q^i$ on the cotangent fiber. The Hamiltonian function of the system is the energy from (31.23.5) written in these coordinates:

$$(1)\ E(q,p) := \frac{1}{2m}|p|^2+U(q) = \frac{1}{2m}|p|^2-G\frac{Mm}{|q|} = \frac{1}{2m}\sum p_i^2-G\frac{Mm}{\sqrt{\sum(q^i)^2}}.$$

The Hamiltonian vector field is then given by

$$H_E = \sum_{i=1}^{3}\Big(\frac{\partial E}{\partial p_i}\frac{\partial}{\partial q^i} - \frac{\partial E}{\partial q^i}\frac{\partial}{\partial p_i}\Big) = \sum_{i=1}^{3}\Big(\frac{1}{m}p_i\frac{\partial}{\partial q^i} - \frac{GMm}{|q|^3}q^i\frac{\partial}{\partial p_i}\Big).$$

The flow lines of this vector field can be expressed in terms of elliptic functions. Briefed by (31.23.2), we consider the 3 components of the vector product $J(q,p) = q \times p$ and we may compute that

$$J^1 = q^2p_3 - q^3p_2, \quad J^2 = -q^1p_3 + q^3p_1, \quad J^3 = q^1p_2 - q^2p_1,$$
$$\{E, J^i\} = 0, \quad \{J^1, J^2\} = -J^3$$

We shall later interpret (J^1, J^2, J^3) as a momentum mapping.

32. Completely Integrable Hamiltonian Systems

32.1. Introduction. The pioneers of analytical mechanics, Euler, Lagrange, Jacobi, Kowalewska, ..., were deeply interested in completely integrable systems, of which they discovered many examples: the motion of a rigid body with a fixed point in the three classical cases (the Euler-Lagrange, Euler-Poisot, and Kowalewska cases), Kepler's system, the motion of a massive point in the gravitational field created by fixed attracting points, geodesics on an ellipsoid, etc. To analyze such systems, Jacobi developed a method which now bears his name, based on a search for a complete integral of the first order partial differential equation associated with the Hamitonian system under consideration, called the Hamilton-Jacobi equation. Later it turned out, with many contributions by Poincaré, that complete integrability is very exceptional: A small perturbation of the Hamiltonian function can destroy it. Thus this topic fell into disrespect.

Later Kolmogorov, Arnold, and Moser showed that certain qualitative properties of completely integrable systems persist after perturbation: certain invariant tori on which the quasi-periodic motion of the nonperturbed, completely integrable system takes place survive the perturbation.

More recently it has been shown that certain nonlinear partial differential equations such as the Korteweg-de Vries equation $u_t + 3u_xu + au_{xxx} = 0$ or the Camassa-Holm equation $u_t - u_{txx} = u_{xxx}.u + 2u_{xx}.u_x - 3u_x.u$ may be regarded as infinit-dimensional ordinary differential equations which have many properties of completely integrable Hamiltonian systems. This started new, very active research in completely integrable systems. See [**68, 69**] for an overview.

32.2. Completely integrable systems. Let (M,ω) be a symplectic manifold with $\dim(M) = 2n$ with a Hamiltonian function $h \in C^\infty(M)$.

(1) The *Hamiltonian system* (M,ω,h) is called *completely integrable* if there are n functions $f_1,\dots,f_n \in C^\infty(M)$ which

- are pairwise in involution: $\{f_i,f_j\} = 0$ for all i,j,
- are first integrals of the system: $\{h,f_i\} = 0$ for all i,
- are nondegenerate: their differentials are linearly independent on a dense open subset of M.

We shall keep this notation throughout this section.

(2) The $n+1$ functions $h, f_1,\dots,f_n \in C^\infty(M)$ are pairwise in involution. At each point $x \in M$ the Hamiltonian fields $H_h(x), H_{f_1}(x),\dots,H_{f_n}(x)$ span an isotropic subset of T_xM which has dimension $\le n$; thus they are linearly dependent. On the dense open subset $U \subseteq M$ where the differentials df_i are linearly independent, $dh(x)$ is a linear combination of $df_1(x),\dots,df_n(x)$. Thus *each $x \in U$ has an open neighborhood $V \subset U$ such that $h|V = \tilde h \circ (f_1,\dots,f_n)|V$ for a smooth local function on $\mathbb{R}^n$.* To see this, note that the H_{f_i} span an integrable distribution of constant rank in U whose leaves are given by the connected components of the sets described by the equations $f_i = c_i$, c_i constant, for $i = 1,\dots,n$ of maximal rank. Since $\{h,f_i\} = 0$, the function h is constant along each leaf and thus factors locally over the mapping $f := (f_1,\dots,f_n) : U \to f(U) \subset \mathbb{R}^n$. The Hamiltonian vector field H_h is then a linear combination of the Hamiltonian fields H_{f_i},

$$H_h = \check\omega^{-1}(dh) = \check\omega^{-1}\Big(\sum_{i=1}^n \frac{\partial \tilde h}{\partial f_i}(f_1,\dots,f_n)\, df_i\Big) = \sum_{i=1}^n \frac{\partial \tilde h}{\partial f_i}(f_1,\dots,f_n)\, H_{f_i}.$$

whose coefficients $\frac{\partial \tilde h}{\partial f_i}(f_1,\dots,f_n)$ depend only on the first integrals $f_1,\dots,f_n$. The f_i are constant along the flow lines of H_h since $\{h,f_i\} = 0$ implies $(\mathrm{Fl}_t^{H_h})^* f_i = f_i$ and $(\mathrm{Fl}_t^{H_h})^* H_{f_i} = H_{f_i}$. This last argument also shows that a trajectory of H_h intersecting U is completely contained in U. Therefore these coefficients $\frac{\partial \tilde h}{\partial f_i}(f_1,\dots,f_n)$ are constant along each trajectory of H_h which is contained in U.

(3) *The Hamiltonian vector fields $H_{f_1},\dots,H_{f_n}$ span a smooth integrable distribution of nonconstant rank on M according to* (3.28), since $[H_{f_i},H_{f_j}] = H_{\{f_i,f_j\}} = 0$ and $(\mathrm{Fl}_t^{H_{f_i}})^* H_{f_j} = H_{f_j}$, so the dimension of the span is constant along each flow. Thus we have a foliation of jumping dimension on M: Each point of M lies in an initial submanifold which is an integral manifold for the distribution spanned by the H_{f_i}. Each trajectory of H_h or of any H_{f_i} is completely contained in one of these leaves. The restriction of this foliation to the open set U is a foliation of U by Lagrangian submanifolds, whose

leaves are defined by the equations $f_i = c_i$, $i = 1, \dots, n$, where the c_i are constants.

32.3. Lemma ([10]). *Let $\mathbb{R}^{2n} = \mathbb{R}^n \times \mathbb{R}^n$ be the standard symplectic vector space with standard basis e_i such that $\omega = \sum_{i=1}^n e^i \wedge e^{n+i}$. Let $W \subset \mathbb{R}^{2n}$ be a Lagrangian subspace.*

Then there is a partition $\{1, \dots, n\} = I \sqcup J$ such that the Lagrangian subspace U of $\mathbb{R}^{2n}$ spanned by the e_i for $i \in I$ and the e_{n+j} for $j \in J$ is a complement to W in $\mathbb{R}^{2n}$.

Proof. Let $k = \dim(W \cap (\mathbb{R}^n \times 0))$. If $k = n$, we may take $I = \emptyset$. If $k < n$, there exist $n-k$ elements $e_{i_1}, \dots, e_{i_{n-k}}$ of the basis $e_1, \dots, e_n$ of $\mathbb{R}^n \times 0$ which span a complement U' of $W \cap (\mathbb{R}^n \times 0)$ in $\mathbb{R}^n \times 0$. Put $I = \{i_1, \dots, i_{n-k}\}$ and let J be the complement. Let U'' be the span of the e_{n+j} for $j \in J$, and let $U = U' \oplus U''$. Then U is a Lagrangian subspace. We have

$$\mathbb{R}^n \times 0 = (W \cap (\mathbb{R}^n \times 0)) \oplus U', \quad W \cap (\mathbb{R}^n \times 0) \subset W, \quad U' = U \cap (\mathbb{R}^n \times 0) \subset U.$$

Thus $\mathbb{R}^n \times 0 \subset W + U$. Since $\mathbb{R}^n \times 0$, W, U are Lagrangian, by (31.4.4) we have $W \cap U = W^\perp \cap U^\perp = (W+U)^\perp \subset (\mathbb{R}^n \times 0)^\perp = \mathbb{R}^n \times 0$; thus $W \cap U = (W \cap (\mathbb{R}^n \times 0)) \cap (U \cap (\mathbb{R}^n \times 0)) = W \cap (\mathbb{R}^n \times 0) \cap U' = 0$, and U is a complement of W. $\square$

32.4. Lemma. *Let (M, ω) be a symplectic manifold of dimension $2n$, and let $x \in M$. Suppose that $2n$ smooth functions $u^1, \dots, u^n, f_1, \dots, f_n$ are given near x, that their differentials are linearly independent, and that they satisfy the following properties:*

- *The submanifold defined by the equations $u^i = u^i(x)$ for $i = 1, \dots, n$ is Lagrangian.*
- *The functions $f_1, \dots, f_n$ are pairwise in involution: $\{f_i, f_j\} = 0$ for all i, j.*

Then on an open neighborhood U of x in M we may determine n other smooth functions $g_1, \dots, g_n$ such that

$$\omega | U = \sum_{i=1}^n df_i \wedge dg_i.$$

The determination of g_i uses exclusively the operations of integration, elimination (solving linear equations), and partial differentiation.

Proof. Without loss we may assume that $u^i(x) = 0$ for all i. There exists a contractible open neighborhood U of x in M such that $(u, f) := (u^1, \dots, u^n, f_1, \dots, f_n)$ is a chart defined on U and such that each diffeomorphism $\psi_t(u, f) := (t\,u, f)$ is defined on the whole of U for t near $[0, 1]$ and maps U into itself. Since ψ_0 maps U onto the Lagrange submanifold

$N := \{y \in U : u^i(y) = 0 \text{ for } i = 1, \dots, n\}$, we have $\psi_0^* \omega = 0$. Using the homotopy invariance (11.4), we have

$$\omega|U = \psi_1^*\omega = \psi_0^*\omega + d\,\bar{h}(\omega) - \bar{h}(d\omega) = 0 + d\,\bar{h}(\omega) + 0,$$

where $\bar{h}(\omega) = \int_0^1 \operatorname{ins}_t^* i_{\partial_t} \psi^* \omega \, dt$ is from the proof of (11.4).

Since $f_1, \dots, f_n$ are pairwise in involution and have linearly independent differentials, $\omega|U$ belongs to the ideal in $\Omega^*(U)$ generated by $df_1, \dots, df_n$. This is a pointwise property. At $y \in U$ the tangent vectors $H_{f_1}(y), \dots, H_{f_n}(y)$ span a Lagrangian vector subspace L of T_yM with annihilator $L^o \subset T_y^*M$ spanned by $df_1(y), \dots, df_n(y)$. Choose a complementary Lagrangian subspace $W \subset T_yM$; see the proof of (31.20). Let $\alpha_1, \dots, \alpha_n \in T_y^*M$ be a basis of the annihilator W^o. Then $\omega_y = \sum_{i,j=1}^n \omega_{ij} \alpha_i \wedge df_j(y)$ since ω vanishes on L, on W, and induces a duality between L and W.

From the form of $\bar{h}(\omega)$ above we then see that $\bar{h}(\omega)$ also belongs to this ideal, since $\psi_t^* f_i = f_i$ for all i. Namely,

$$\bar{h}(\omega) = \sum_{i,j=1}^n \int_0^1 \Big(\operatorname{ins}_t^* \, i_{\partial_t} \, \psi^*(\omega_{ij}.\alpha_i) \Big) df_j \; dt =: -\sum_{j=1}^n g_j \, df_j$$

for smooth functions g_i. Finally we remark that the determination of the components of ω in the chart (u, f) uses partial differentiations and eliminations, whereas the calculation of the components of $\bar{h}(\omega)$ uses integration. □

32.5. Lemma. *Let (M, ω) be a symplectic manifold of dimension $2n$. We assume that the following data are known on an open subset U of M:*

- *a canonical system of local coordinates $(q^1, \dots, q^n, p_1, \dots, p_n)$ on U such that the symplectic form is given by $\omega|U = \sum_{i=1}^n dq^i \wedge dp_i$,*
- *smooth functions $f_1, \dots, f_n$ which are pairwise in involution, $\{f_i, f_j\} = 0$ for all i, j, and whose differentials are linearly independent.*

Then each $x \in U$ admits an open neighborhood $V \subseteq U$ on which we can determine other smooth functions $g_1, \dots, g_n$ such that

$$\omega|_V = \sum_{i=1}^n df_i \wedge dg_i.$$

The determination of g_i uses exclusively the operations of integration, elimination (use of the implicit function theorem), and partial differentiation.

Proof. If the functions $q^1, \dots, q^n, f_1, \dots, f_n$ have linearly independent differentials at a point $x \in U$, the result follows from (32.4). In the general case consider the Lagrangian subspace $L \subset T_xM$ spanned by $H_{f_1}(x), \dots, H_{f_n}(x)$. By lemma (32.3) there exists a partition $\{1, \dots, n\} = I \sqcup J$ such that the

Langrangian subspace $W \subset T_xM$ spanned by $H_{q^i}(x)$ for $i \in I$ and $H_{p_j}(x)$ for $j \in J$ is complementary to L. Now the result follows from lemma (32.4) by calling u^k, $k = 1, \dots, n$, the functions q^i for $i \in I$ and p_j for $j \in J$. □

32.6. Proposition. *Let (M, ω, h) be a Hamiltonian system on a symplectic manifold of dimension $2n$. We assume that the following data are known on an open subset U of M:*

- *a canonical system of local coordinates $(q^1, \dots, q^n, p_1, \dots, p_n)$ on U such that the symplectic form is given by $\omega|U = \sum_{i=1}^n dq^i \wedge dp_i$,*
- *a family $f = (f_1, \dots, f_n)$ of smooth first integrals for the Hamiltonian function h which are pairwise in involution, i.e., $\{h, f_i\} = 0$ and $\{f_i, f_j\} = 0$ for all i, j, and whose differentials are linearly independent.*

Then for each $x \in U$ the integral curve of H_h passing through x can be determined locally by using exclusively the operations of integration, elimination, and partial differentiation.

Proof. By lemma (32.5) there exists an open neighborhood V of x in U and functions $g_1, \dots, g_n \in C^\infty(V)$ such that $\omega|V = \sum_{i=1}^n df_i \wedge dg_i$. The determination uses only integration, partial differentiation, and elimination. We may choose V so small that $(f, g) := (f_1, \dots, f_n, g_1, \dots, g_n)$ is a chart on V with values in a cube in $\mathbb{R}^{2n}$.

We have already seen in (32.2.2) that $h|V = \tilde{h} \circ (f, g)$ where $\tilde{h} = h \circ (f, g)^{-1}$ is a smooth function on the cube which does not depend on the g_i. In fact $\tilde{h}$ may be determined by elimination since h is constant on the leaves of the foliation given by $f_i = c_i$, c_i constant.

The differential equation for the trajectories of H_h in V is given by

$$\dot{f}_k = \frac{\partial \tilde{h}}{\partial g_k} = 0, \qquad \dot{g}_k = -\frac{\partial \tilde{h}}{\partial f_k}, \qquad k = 1, \dots, n;$$

thus the integral curve $\operatorname{Fl}_t^{H_h}(x)$ is given by

$$\begin{aligned} f_k(\operatorname{Fl}_t^{H_h}(x)) &= f_k(x), \\ g_k(\operatorname{Fl}_t^{H_h}(x)) &= g_k(x) - t\frac{\partial \tilde{h}}{\partial f_k}(f(x)), \end{aligned} \qquad k = 1, \dots, n. \quad \square$$

32.7. Proposition. *Let (M, ω, h) be a Hamiltonian system with* $\dim(M) =$ *$2n$ and let $f = (f_1, \dots, f_n)$ be a family first integrals of h which are pairwise in involution, $\{h, f_i\} = 0$ and $\{f_i, f_j\} = 0$ for all i, j. Suppose that all Hamiltonian vector fields H_{f_i} are complete. Then we have:*

(1) *The vector fields H_{f_i} are the infinitesimal generators of a smooth action $\ell : \mathbb{R}^n \times M \to M$ whose orbits are the isotropic leaves of the*

foliation with jumping dimension described in (32.2.3) *and which can be described by*

$$\ell_{(t_1,\dots,t_n)}(x) = (\mathrm{Fl}^{H_{f_1}}_{t_1} \circ \dots \circ \mathrm{Fl}^{H_{f_n}}_{t_n})(x).$$

Each orbit is invariant under the flow of H_h.

(2) *(Liouville's theorem) If* $a \in f(M) \subset \mathbb{R}^n$ *is a regular value of* f *and if* $N \subseteq f^{-1}(a)$ *is a connected component, then* N *is a Lagrangian submanifold and is an orbit of the action of* $\mathbb{R}^n$ *which acts transitively and locally freely on* N*: For any point* $x \in N$ *the isotopy subgroup* $(\mathbb{R}^n)_x := \{t \in \mathbb{R}^n : \ell_t(x) = x\}$ *is a discrete subgroup of* $\mathbb{R}^n$*. Thus it is a lattice* $\sum_{i=1}^k 2\pi\mathbb{Z}\, v_i$ *generated by* $k = \operatorname{rank}(\mathbb{R}^n)_x$ *linearly independent vectors* $2\pi v_i \in \mathbb{R}^n$*. The orbit* N *is diffeomorphic to the quotient group* $\mathbb{R}^n/(\mathbb{R}^n)_x \cong \mathbb{T}^k \times \mathbb{R}^{n-k}$*, a product of the* k*-dimensional torus by an* $(n-k)$*-dimensional vector space.*
Moreover, there exist constants $(w_1, \dots, w_n) \in \mathbb{R}^n$ *such that the flow of the Hamiltonian* h *on* N *is given by* $\mathrm{Fl}^{H_h}_t = \ell_{(tw_1,\dots,tw_n)}$*. If we use coordinates* $(b_1 \bmod 2\pi, \dots, b_k \bmod 2\pi, b_{k+1}, \dots, b_n)$ *corresponding to the diffeomorphic description* $N \cong \mathbb{T}^k \times \mathbb{R}^{n-k}$*, the flow of* h *is given by*

$$\begin{aligned}\mathrm{Fl}^{H_h}_t&(b_1 \bmod 2\pi, \dots, b_k \bmod 2\pi, b_{k+1}, \dots, b_n)\\ &= (b_1 + tc_1 \bmod 2\pi, \dots, b_k + tc_k \bmod 2\pi, b_{k+1} + tc_{k+1}, \dots, b_n + tc_n)\end{aligned}$$

for constant $c_i = w_i/|v_i|$ *for* $i \le k$ *and* $c_j = w_j$ *for* $j > k$*. If* N *is compact so that* $k = n$*, this is called a quasi-periodic flow.*

Proof. The action ℓ is well defined since the complete vector fields H_{f_i} commute; see the proof of (3.17). Or we conclude the action directly from theorem (6.5). The rest of this theorem follows already from (32.2), or it is obvious. The form of discrete subgroups of $\mathbb{R}^n$ is proved in the next lemma. □

32.8. Lemma. *Let* G *be a discrete subgroup of* $\mathbb{R}^n$*. Then* G *is the lattice* $\sum_{i=1}^k \mathbb{Z}\, v_i$ *generated by* $0 \le k = \operatorname{rank}(G) \le n$ *linearly independent vectors* $v_i \in \mathbb{R}^n$*.*

Proof. We use the standard Euclidean structure of $\mathbb{R}^n$. If $G \neq 0$, there is $0 \neq v \in G$. Let v_1 be the point in $\mathbb{R}v \cap G$ which is nearest to 0 but nonzero. Then $G \cap \mathbb{R}v = \mathbb{Z}v_1$: If there were $w \in G$ in one of the intervals $(m, m+1)v_1$, then $w - mv_1 \in \mathbb{R}v_1$ would be nonzero and closer to 0 than v_1.

If $G \neq \mathbb{Z}v_1$, there exists $v \in G \setminus \mathbb{R}v_1$. We will show that there exists a point v_2 in G with minimal distance to the line $\mathbb{R}v_1$ but not in the line. Suppose that the orthogonal projection $\mathrm{pr}_{\mathbb{R}v_1}(v)$ of v onto $\mathbb{R}v_1$ lies in the interval

$P = [m, m+1]v_1$ for $m \in \mathbb{Z}$, consider the cylinder

$$C = \{z \in \mathrm{pr}_{\mathbb{R}v_1}^{-1}(P) : \mathrm{dist}(z, P) \le \mathrm{dist}(v, P)\}$$

and choose a point $v_2 \in G \setminus \mathbb{R}v_1$ in this cylinder nearest to P. Then v_2 has minimal distance to $\mathbb{R}v_1$ in $G \setminus (\mathbb{R}v_1)$ since any other point in G with smaller distance can be shifted into the cylinder C by adding some suitable mv_1.

Then $\mathbb{Z}v_1 + \mathbb{Z}v_2$ forms a lattice in the plane $\mathbb{R}v_1 + \mathbb{R}v_2$ which is partitioned into parallelograms $Q = \{a_1v_1 + a_2v_2 : m_i \le a_i < m_i + 1\}$ for $m_i \in \mathbb{Z}$. If there is a point $w \in G$ in one of these parallelograms Q, then a suitable translate $w - n_1v_1 - n_2v_2$ would be nearer to $\mathbb{R}v_1$ than v_2. Thus $G \cap (\mathbb{R}v_1 + \mathbb{R}v_2) = \mathbb{Z}v_1 + \mathbb{Z}v_2$.

If there is a point of G outside this plane, we may find as above a point v_3 of G with minimal distance to the plane, and by covering the 3-space $\mathbb{R}v_1 + \mathbb{R}v_2 + \mathbb{R}v_3$ with parallelepipeds, we may show as above that $G \cap (\mathbb{R}v_1 + \mathbb{R}v_2 + \mathbb{R}v_3) = \mathbb{Z}v_1 + \mathbb{Z}v_2 + \mathbb{Z}v_3$, and so on. □

33. Poisson Manifolds

33.1. Poisson manifolds. A *Poisson structure* on a smooth manifold M is a Lie bracket $\{\ ,\ \}$ on the vector space of smooth functions $C^\infty(M)$ also satisfying

$$\{f, gh\} = \{f, g\}h + g\{f, h\}. \tag{1}$$

This means that for each $f \in C^\infty(M)$ the mapping $\mathrm{ad}_f = \{f,\ \}$ is a derivation of $(C^\infty(M), \cdot)$, so by (3.3) there exists a unique vector field $H(f) = H_f \in \mathfrak{X}(M)$ such that $\{f, h\} = H_f(h) = dh(H_f)$ holds for each $h \in C^\infty(M)$. We also have $H(fg) = f\,H_g + g\,H_f$ since

$$H_{fg}(h) = \{fg, h\} = f\{g, h\} + g\{f, h\} = (f\,H_g + g\,H_f)(h).$$

Thus there exists a unique tensor field $P \in \Gamma(\bigwedge^2 TM)$ such that

$$\{f, g\} = H_f(g) = P(df, dg) = \langle df \wedge dg, P\rangle. \tag{2}$$

The choice of sign is motivated by the following. If ω is a symplectic form on M, we consider, using (31.21):

$$\begin{aligned}
\check\omega : TM \to T^*M, &\qquad \langle \check\omega(X), Y\rangle = \omega(X, Y),\\
\check P = \check\omega^{-1} : T^*M \to TM, &\qquad \langle \psi, \check P(\varphi)\rangle = P(\varphi, \psi),\\
H_f = \check\omega^{-1}(df) = \check P(df), &\qquad i_{H_f}\omega = df,\\
\{f, g\} = H_f(g) = i_{H_f}\,dg &= i_{H_f} i_{H_g}\omega = \omega(H_g, H_f)\\
&= \langle dg, H_f\rangle = \langle dg, \check P(df)\rangle = P(df, dg) = \tfrac12\langle df \wedge dg, P\rangle.
\end{aligned}$$

33.2. Proposition. Schouten-Nijenhuis bracket. *Let M be a smooth manifold. We consider the space $\Gamma(\bigwedge TM)$ of multivector fields on M. This space carries a graded Lie bracket for the grading $\Gamma(\bigwedge^{*+1} TM), * = -1, 0, 1, 2, \dots$, called the Schouten-Nijenhuis bracket, which is given by*

(1) $[X_1 \wedge \dots \wedge X_p, Y_1 \wedge \dots \wedge Y_q]$
$$= \sum_{i,j} (-1)^{i+j} [X_i, Y_j] \wedge X_1 \wedge \dots \widehat{X_i} \dots \wedge X_p \wedge Y_1 \wedge \dots \widehat{Y_j} \dots \wedge Y_q,$$

(2) $[f, U] = -\bar{\imath}(df)U,$

where $\bar{\imath}(df)$ is the insertion operator $\bigwedge^k TM \to \bigwedge^{k-1} TM$, the adjoint of $df \wedge (\quad) : \bigwedge^l T^*M \to \bigwedge^{l+1} T^*M$.

Let $U \in \Gamma(\bigwedge^u TM)$, $V \in \Gamma(\bigwedge^v TM)$, $W \in \Gamma(\bigwedge^w TM)$, and $f \in C^\infty(M, \mathbb{R})$. Then we have:

$$(3) \qquad [U, V] = -(-1)^{(u-1)(v-1)} [V, U].$$

$$(4) \qquad [U, [V, W]] = [[U, V], W] + (-1)^{(u-1)(v-1)} [V, [U, W]].$$

$$(5) \qquad [U, V \wedge W] = [U, V] \wedge W + (-1)^{(u-1)v} V \wedge [U, W].$$

$$(6) \qquad [X, U] = \mathcal{L}_X U.$$

(7) Let $P \in \Gamma(\bigwedge^2 TM)$. Then the product $\{f, g\} := \frac{1}{2}\langle df \wedge dg, P \rangle$ on $C^\infty(M)$ satisfies the Jacobi identity if and only if $[P, P] = 0$.

Proof. The bilinear mapping $\bigwedge^k \Gamma(TM) \times \bigwedge^l \Gamma(TM) \to \bigwedge^{k+l-1} \Gamma(TM)$ given by (1) factors over $\bigwedge^k \Gamma(TM) \to \bigwedge^k_{C^\infty(M)} \Gamma(TM) = \Gamma(\bigwedge^k TM)$ since we may easily compute that

$$\begin{aligned}[X_1 \wedge \dots \wedge X_p, Y_1 \wedge \dots \wedge fY_j \wedge \dots \wedge Y_q] &= f[X_1 \wedge \dots \wedge X_p, Y_1 \wedge \dots \wedge Y_q] \\ &\quad + (-1)^p \bar{\imath}(df)(X_1 \wedge \dots \wedge X_p) \wedge Y_1 \wedge \dots \wedge Y_q.\end{aligned}$$

So the bracket $[\quad,\quad] : \Gamma(\bigwedge^{k-1} TM) \times \Gamma(\bigwedge^{l-1} TM) \to \Gamma(\bigwedge^{k+l-1} TM)$ is a well defined operation. Properties (3)–(6) have to be checked by direct computations.

Property (7) can be seen as follows: We have

$$(8) \qquad 2\{f, g\} = \langle df \wedge dg, P \rangle = \langle dg, \bar{\imath}(df) P \rangle = -\langle dg, [f, P] \rangle = [g, [f, P]].$$

Now a straightforward computation involving the graded Jacobi identity and the graded skew-symmetry of the Schouten-Nijenhuis bracket gives

$$[h, [g, [f, [P, P]]]] = -8(\{f, \{g, h\}\} + \{g, \{h, f\}\} + \{h, \{f, g\}\}).$$

Since $[h, [g, [f, [P, P]]]] = \langle df \wedge dg \wedge dh, [P, P] \rangle$, the result follows. □

In [**199**] there is an expression for $(-1)^{u-1}[U, V]$ in terms of covariant derivatives which does not depend on the covariant derivative, and in [**175**] it is found that it satisfied the graded Jacobi identity. In [**123**] the relation of the Schouten-Nijenhuis bracket to Poisson manifolds was spelled out. See also [**221**], [**146**] for the version presented here and [**222**] for more information.

Let us point out here that the skew-symmetric Schouten-Nijenhuis bracket has a symmetric counterpart. It is an ordinary (non-graded) Lie bracket extending the Lie bracket from the space of vector fields to the space $\Gamma(\bigvee TM)$ of symmetric multivector fields. It satisfies $[X,f] = \mathcal{L}_X f$ for $X \in \mathfrak{X}(M)$ and

$$\begin{aligned}&[X_1 \vee \dots \vee X_p, Y_1 \vee \dots \vee Y_q]\\&\qquad = \sum_{i,j} [X_i, Y_j] \vee X_1 \vee \dots \widehat{X_i} \dots \vee X_p \vee Y_1 \vee \dots \widehat{Y_j} \dots \vee Y_q.\end{aligned}$$

A symmetric multivector field on M can be viewed as a smooth function on T^*M which is a homogeneous polynomial on each fiber. The symmetric Schouten-Nijenhuis bracket is then just the restriction of the canonical Poisson bracket on $C^\infty(T^*M)$ to the subalgebra of these fiberwise polynomial functions.

33.3. Hamiltonian vector fields for Poisson structures. Let (M,P) be a Poisson manifold. As usual we denote by $\check{P} : T^*M \to TM$ the associated skew-symmetric homomorphism of vector bundles. Let $\mathfrak{X}(M,P) := \{X \in \mathfrak{X}(M) : \mathcal{L}_X P = 0\}$ be the *Lie algebra of infinitesimal automorphisms* of the Poisson structure. For $f \in C^\infty(M)$ we define the *Hamiltonian vector field* by

$$\text{(1)} \qquad \operatorname{grad}^P(f) = H_f = \check{P}(df) = -[f,P] = -[P,f] \in \mathfrak{X}(M),$$

and we recall the relation between Poisson structure and Poisson bracket, (33.1.2) and (33.2.8),

$$\{f,g\} = H_f(g) = P(df,dg) = \tfrac{1}{2}\langle df \wedge dg, P\rangle = [g,[f,P]].$$

Lemma. *The Hamiltonian vector field mapping takes values in $\mathfrak{X}(M,P)$ and is a Lie algebra homomorphism*

$$(C^\infty(M), \{\ \ ,\ \ \}_P) \xrightarrow{H=\operatorname{grad}^P} \mathfrak{X}(M,P).$$

Proof. For $f \in C^\infty(M)$ we have:

$$\begin{aligned}0 = [f,[P,P]] &= [[f,P],P] - [P,[f,P]] = 2[[f,P],P],\\ \mathcal{L}_{H_f} P &= [H_f, P] = -[[f,P],P] = 0.\end{aligned}$$

For $f,g \in C^\infty(M)$ we get

$$\begin{aligned}[H_f, H_g] &= [[f,P],[g,P]] = [g,[[f,P],P]] - [[g,[f,P]],P]\\ &= [g, -\mathcal{L}_{H_f}P] - [\{f,g\},P] = 0 + H(\{f,g\}). \qquad \square\end{aligned}$$

33.4. Theorem. *Let (M, P) be a Poisson manifold. Then $\check{P}(T^*M) \subseteq TM$ is an integrable smooth distribution (with jumping dimension) in the sense of (3.23). On each leaf L (which is an initial submanifold of M by (3.25)) the Poisson structure P induces the inverse of a symplectic structure on L.*

One says that the Poisson manifold M is stratified into symplectic leaves.

Proof. We use theorem (3.28). Consider the set

$$\mathcal{V} := \{\check{P}(df) = H_f = -[f, P] : f \in C^\infty(M)\} \subset \mathfrak{X}(\check{P}(T^*M))$$

of sections of the distribution. The set $\mathcal{V}$ spans the distribution since through each point in T^*M we may find a form df. The set $\mathcal{V}$ is involutive since $[H_f, H_g] = H_{\{f,g\}}$. Finally we have to check that the dimension of $\check{P}(T^*M)$ is constant along flow lines of vector fields in $\mathcal{V}$, i.e., of vector fields H_f:

$$\begin{aligned}
&\check{P} = (\mathrm{Fl}_t^{H_f})^* \check{P} = T(\mathrm{Fl}_{-t}^{H_f}) \circ \check{P} \circ (T\,\mathrm{Fl}_{-t}^{H_f})^* \quad \text{since } \mathcal{L}_{H_f} P = 0 \\
&\qquad \Longrightarrow \dim \check{P}(T^*_{\mathrm{Fl}_t^{H_f}(x)} M) = \text{constant in } t.
\end{aligned}$$

So all assumptions of theorem (3.28) are satisfied and thus the distribution $P(T^*M)$ is integrable.

Now let L be a leaf of the distribution $P(T^*M)$, a maximal integral manifold. The 2-vector field $P|L$ is tangent to L, since a local smooth function f on M is constant along each leaf if and only if $\check{P}(df) = -df \circ \check{P} : T^*M \to \mathbb{R}$ vanishes. Therefore, $\check{P}|L : T^*L \to TL$ is an injective homomorphism of vector bundles of the same fiber dimension and is thus an isomorphism. Then $\check{\omega}_L := (\check{P}|L)^{-1} : TL \to T^*L$ defines a 2-form $\omega_L \in \Omega^2(L)$ which is nondegenerate. It remains to check that ω_L is closed. For each $x \in L$ there exists an open neighborhood $U \subset M$ and functions $f, g, h \in C^\infty(U)$ such that the vector fields $H_f = \check{P}(df)|L$, H_g, and H_h on L take arbitrary prescribed values in T_xL at $x \in L$. Thus $d\omega_L = 0 \in \Omega^3(L)$ results from the following computation:

$$\begin{aligned}
\omega_L(H_f, H_g) &= (i_{H_f}\omega_L)(H_g) = \check{\omega}_L(H_f)(H_g) \\
&= df(H_g) = \{g, f\}, \\
d\omega_L(H_f, H_g, H_h) &= H_f(\omega_L(H_g, H_h)) + H_g(\omega_L(H_h, H_f)) \\
&\quad + H_h(\omega_L(H_f, H_g)) - \omega_L([H_f, H_g], H_h) \\
&\quad - \omega_L([H_g, H_h], H_f) - \omega_L([H_h, H_f], H_g) \\
&= \{\{h, g\}, f\} + \{\{f, h\}, g\} + \{\{g, f\}, h\} \\
&\quad - \{h, \{f, g\}\} - \{f, \{g, h\}\} - \{g, \{h, f\}\} = 0. \qquad \square
\end{aligned}$$

33.5. Proposition. Poisson morphisms. *Let (M_1, P_1) and (M_2, P_2) be two Poisson manifolds. A smooth mapping $\varphi : M_1 \to M_2$ is called a Poisson morphism if any of the following equivalent conditions is satisfied:*

(1) *For all $f, g \in C^\infty(M_2)$ we have $\varphi^*\{f, g\}_2 = \{\varphi^* f, \varphi^* g\}_1$.*

(2) *For all $f \in C^\infty(M_2)$ the Hamiltonian vector fields $H^1_{\varphi^* f} \in \mathfrak{X}(M_1, P_1)$ and $H^2_f \in \mathfrak{X}(M_2, P_2)$ are φ-related.*

(3) *We have $\bigwedge^2 T\varphi \circ P_1 = P_2 \circ \varphi : M_1 \to \bigwedge^2 TM_2$.*

(4) *For each $x \in M_1$ we have*
$$T_x\varphi \circ (\check{P}_1)_x \circ (T_x\varphi)^* = (\check{P}_2)_{\varphi(x)} : T^*_{\varphi(x)} M_2 \to T_{\varphi(x)} M_2.$$

Proof. For $x \in M_1$ we have
$$\begin{aligned}
\{\varphi^* f, \varphi^* g\}_1(x) &= (P_1)_x(d(f\circ\varphi)|_x, d(g\circ\varphi)|_x) \\
&= (P_1)_x(df|_{\varphi(x)}.T_x\varphi, dg|_{\varphi(x)}.T_x\varphi) \\
&= \tfrac{1}{2}(P_1)_x.\overset{2}{\bigwedge}(T_x\varphi)^*.(df|_{\varphi(x)} \wedge dg|_{\varphi(x)}) \\
&= \big(\overset{2}{\bigwedge} T_x\varphi.(P_1)_x\big)(df|_{\varphi(x)}, dg|_{\varphi(x)}), \\
\varphi^*\{f, g\}_2(x) &= \{f, g\}_2(\varphi(x)) = (P_2)_{\varphi(x)}(df|_{\varphi(x)}, dg|_{\varphi(x)}).
\end{aligned}$$
This shows that (1) and (3) are equivalent since $df(y)$ meets each point of T^*M_2. (3) and (4) are obviously equivalent.

(2) and (4) are equivalent since we have
$$\begin{aligned}
T_x\varphi.H^1_{\varphi^* f}(x) &= T_x\varphi.(\check{P}_1)_x.d(f\circ\varphi)|_x = T_x\varphi.(\check{P}_1)_x.(T_x\varphi)^*.df|_{\varphi(x)}, \\
H^2_f(\varphi(x)) &= (\check{P}_2)_{\varphi(x)}.df|_{\varphi(x)}. \quad \square
\end{aligned}$$

33.6. Proposition. *Let (M_1, P_1), (M_2, P_2), and (M_3, P_3) be Poisson manifolds and let $\varphi : M_1 \to M_2$ and $\psi : M_2 \to M_3$ be smooth mappings.*

(1) *If φ and ψ are Poisson morphisms, then also $\psi \circ \varphi$ is a Poisson morphism.*

(2) *If φ and $\psi \circ \varphi$ are Poisson morphisms and if φ is surjective, then also ψ is a Poisson morphism. In particular, if φ is Poisson and a diffeomorphism, then also φ^{-1} is Poisson.*

Proof. Part (1) follows from (33.5.1), say. For (2) we use (33.5.3) as follows:
$$\overset{2}{\bigwedge} T\varphi \circ P_1 = P_2 \circ \varphi \quad \text{and} \quad \overset{2}{\bigwedge} T(\psi\circ\varphi) \circ P_1 = P_3 \circ \psi \circ \varphi$$
imply $\overset{2}{\bigwedge} T\psi \circ P_2 \circ \varphi = \overset{2}{\bigwedge} T\psi \circ \overset{T}{\bigwedge} \varphi \circ P_1 = \overset{2}{\bigwedge} T(\psi\circ\varphi) \circ P_1 = P_3 \circ \psi \circ \varphi$,

which implies the result since φ is surjective. □

33.7. Example and theorem. *For a Lie algebra $\mathfrak{g}$ there is a canonical Poisson structure P on the dual $\mathfrak{g}^*$, given by the dual of the Lie bracket:*

$$[\quad,\quad]:\bigwedge^2\mathfrak{g}\to\mathfrak{g},\qquad P=-[\quad,\quad]^*:\mathfrak{g}^*\to\bigwedge^2\mathfrak{g}^*,$$
$$\{f,g\}(\alpha)=\langle\alpha,[dg(\alpha),df(\alpha)]\rangle\quad \text{for } f,g\in C^\infty(\mathfrak{g}^*),\alpha\in\mathfrak{g}^*.$$

The symplectic leaves are exactly the connected components of coadjoint orbits with their symplectic structures from (31.14).

Proof. We check directly the properties (33.1) of a Poisson structure. Skew symmetry is clear. The derivation property (33.1.1) is:

$$\begin{aligned}\{f,gh\}(\alpha)&=\langle\alpha,[h(\alpha)dg(\alpha)+g(\alpha)dh(\alpha),df(\alpha)]\rangle\\&=\langle\alpha,[dg(\alpha),df(\alpha)]\rangle h(\alpha)+g(\alpha)\langle\alpha,[dh(\alpha),df(\alpha)]\rangle\\&=(\{f,g\}h+g\{f,h\})(\alpha).\end{aligned}$$

For the Jacobi identity (33.1.1) we compute

$$\begin{aligned}\langle\beta,d\{g,h\}|_\alpha\rangle&=\langle\beta,[dh(\alpha),dg(\alpha)]\rangle+\langle\alpha,[d^2h(\alpha)(\beta,\quad),dg(\alpha)]\rangle\\&\qquad+\langle\alpha,[dh(\alpha),d^2g(\alpha)(\beta,\quad)]\rangle\\&=\langle\beta,[dh(\alpha),dg(\alpha)]\rangle-\langle(\mathrm{ad}_{dg(\alpha)})^*\alpha,d^2h(\alpha)(\beta,\quad)\rangle\\&\qquad+\langle(\mathrm{ad}_{dh(\alpha)})^*\alpha,d^2g(\alpha)(\beta,\quad)\rangle\\&=\langle\beta,[dh(\alpha),dg(\alpha)]\rangle-d^2h(\alpha)(\beta,(\mathrm{ad}_{dg(\alpha)})^*\alpha)\\&\qquad+d^2g(\alpha)(\beta,(\mathrm{ad}_{dh(\alpha)})^*\alpha)\end{aligned}$$

and we use this to obtain

$$\begin{aligned}\{f,&\{g,h\}\}(\alpha)=\langle\alpha,[d\{g,h\}(\alpha),df(\alpha)]\rangle\\&=\langle\alpha,[[dh(\alpha),dg(\alpha)],df(\alpha)]\rangle-\langle\alpha,[d^2h(\alpha)(\quad,(\mathrm{ad}_{dg(\alpha)})^*\alpha),df(\alpha)]\rangle\\&\qquad+\langle\alpha,[d^2g(\alpha)(\quad,(\mathrm{ad}_{dh(\alpha)})^*\alpha),df(\alpha)]\rangle\\&=\langle\alpha,[[dh(\alpha),dg(\alpha)],df(\alpha)]\rangle-d^2h(\alpha)((\mathrm{ad}_{df(\alpha)})^*\alpha,(\mathrm{ad}_{dg(\alpha)})^*\alpha)\\&\qquad+d^2g(\alpha)((\mathrm{ad}_{df(\alpha)})^*\alpha,(\mathrm{ad}_{dh(\alpha)})^*\alpha).\end{aligned}$$

The cyclic sum over the last expression vanishes. Comparing with (31.14) and (31.21.2), we see that the symplectic leaves are exactly the coadjoint orbits, since

$$\begin{aligned}\langle H_f(\alpha),dg(\alpha)\rangle=H_f(g)|_\alpha&=\{f,g\}(\alpha)=\langle\alpha,[dg(\alpha),df(\alpha)]\rangle\\&=-\langle(\mathrm{ad}_{df(\alpha)})^*\alpha,dg(\alpha)\rangle,\\H_f(\alpha)&=-(\mathrm{ad}_{df(\alpha)})^*\alpha.\end{aligned}$$

The symplectic structure on an orbit $O = \operatorname{Ad}(G)^*\alpha$ is the same as in (31.14) which was given by $\omega_O(\zeta_X, \zeta_Y) = \operatorname{ev}_{[X,Y]}$ where $\zeta_X = -\operatorname{ad}(X)^*$ is the fundamental vector field of the (left) adjoint action. But then $d\operatorname{ev}_Y(\zeta_X(\alpha)) = -\langle \operatorname{ad}(X)^*\alpha, Y\rangle = \langle \alpha, [Y,X]\rangle = \omega_O(\zeta_Y, \zeta_X)$ so that on the orbit the Hamiltonian vector field is given by $H_{\operatorname{ev}_Y} = \zeta_Y = -\operatorname{ad}(Y)^* = -\operatorname{ad}(d\operatorname{ev}_Y(\alpha))^*$, as for the Poisson structure above. □

33.8. Theorem. Poisson reduction. *Let (M,P) be a Poisson manifold and let $r : M \times G \to M$ be the right action of a Lie group on M such that each $r^g : M \to M$ is a Poisson morphism. Let us suppose that the orbit space M/G is a smooth manifold such that the projection $p : M \to M/G$ is a submersion.*

Then there exists a unique Poisson structure $\bar P$ on M/G such that $p : (M,P) \to (M/G, \bar P)$ is a Poisson morphism.

The quotient M/G is a smooth manifold if the action is proper and all orbits of G are of the same type: All isotropy groups G_x are conjugate in G. See (29.21)

Proof. We work with Poisson brackets. A function $f \in C^\infty(M)$ is of the form $f = \bar f \circ p$ for $\bar f \in C^\infty(M/G)$ if and only if f is G-invariant. Thus $p^* : C^\infty(M/G) \to C^\infty(M)$ is an algebra isomorphism onto the subalgebra $C^\infty(M)^G$ of G-invariant functions. If $f, h \in C^\infty(M)$ are G-invariant, then so is $\{f,h\}$ since $(r^g)^*\{f,h\} = \{(r^g)^f, (r^g)^*h\} = \{f,h\}$ by (33.5), for all $g \in G$. So $C^\infty(M)^G$ is a subalgebra for the Poisson bracket which we may regard as a Poisson bracket on $C^\infty(M/G)$. □

33.9. Poisson cohomology. Let (M,P) be a Poisson manifold. We consider the mapping

$$\delta_P := [P,\quad] : \Gamma(\bigwedge^{k-1} TM) \to \Gamma(\bigwedge^{k} TM)$$

which satisfies $\delta_P \circ \delta_P = 0$ since $[P,[P,U]] = [[P,P],U] + (-1)^{1.1}[P,[P,U]]$ by the graded Jacobi identity. Thus we define the *Poisson cohomology* by

$$(1)\qquad H^k_{\text{Poisson}}(M) := \frac{\ker(\delta_P : \Gamma(\bigwedge^k TM) \to \Gamma(\bigwedge^{k+1} TM))}{\operatorname{im}(\delta_P : \Gamma(\bigwedge^{k-1} TM) \to \Gamma(\bigwedge^k TM))}.$$

The direct sum

$$H^*_{\text{Poisson}}(M) = \bigoplus_{k=0}^{\dim(M)} H^k_{\text{Poisson}}(M)$$

is a graded commutative algebra via $U \wedge V$ since $\operatorname{im}(\delta_P)$ is an ideal in $\ker(\delta_P)$ by (33.2.5). The degree 0 part of Poisson cohomology is given by

$$(2)\qquad H^0_{\text{Poisson}}(M) = \{f \in C^\infty(M) : H_f = \{f,\quad\} = 0\},$$

i.e., the vector space of all functions which are constant along each symplectic leaf of the Poisson structure, since $[P,f]=[f,P]=-\bar{\imath}(df)P=-\check{P}(df)=-H_f=-\{f,\quad\}$ by (33.2.2), (33.2.8), and (33.1.2). The degree 1 part of Poisson cohomology is given by

$$H^1_{\text{Poisson}}(M)=\frac{\{X\in\mathfrak{X}(M):[P,X]=-\mathcal{L}_XP=0\}}{\{[P,f]:f\in C^\infty(M)\}}$$

(3) $$=\frac{\mathfrak{X}(M,P)}{\{H_f:f\in C^\infty(M)\}}.$$

Thus we get the following refinement of lemma (33.3). *There exists an exact sequence of homomorphisms of Lie algebras:*

$$0\longrightarrow H^0_{\text{Poisson}}(M)\xrightarrow{\alpha}C^\infty(M)\xrightarrow[\text{grad}^P]{H=}\mathfrak{X}(M,P)\xrightarrow{\gamma}H^1_{\text{Poisson}}(M)\longrightarrow 0,$$

$$0\qquad\qquad\{\ ,\ \}\qquad\qquad[\ ,\ ]\qquad\qquad[\ ,\ ]$$

where the brackets are written under the spaces, where α is the embedding of the space of all functions which are constant on all symplectic leaves, and where γ is the quotient mapping from (3). The bracket on $H^1_{\text{Poisson}}(M)$ is induced by the Lie bracket on $\mathfrak{X}(M,P)$ since $\{H_f:f\in C^\infty(M)\}$ is an ideal: $[H_f,X]=[-[f,P],X]=-[f,[P,X]]-[P,[f,X]]=0+[X(f),P]=-H_{X(f)}$.

33.10. Lemma ([67], [129]). *Let (M,P) be a Poisson manifold.*

Then there exists a Lie bracket $\{\quad,\quad\}^1:\Omega^1(M)\times\Omega^1(M)\to\Omega^1(M)$ which is given by

(1) $$\begin{aligned}\{\varphi,\psi\}^1&=\mathcal{L}_{\check{P}(\varphi)}\psi-\mathcal{L}_{\check{P}(\psi)}\varphi-d(P(\varphi,\psi))\\&=\mathcal{L}_{\check{P}(\varphi)}\psi-\mathcal{L}_{\check{P}(\psi)}\varphi-di_{\check{P}(\varphi)}\psi.\end{aligned}$$

It is the unique $\mathbb{R}$-bilinear skew-symmetric bracket satisfying

(2) $$\{df,dg\}^1=d\{f,g\}\quad\textit{for } f,g\in C^\infty(M),$$

(3) $$\{\varphi,f\psi\}^1=f\{\varphi,\psi\}^1+\mathcal{L}_{\check{P}(\varphi)}(f)\,\psi\quad\textit{for } \varphi,\psi\in\Omega^1(M).$$

Furthermore $\check{P}_*:\Omega^1(M)\to\mathfrak{X}(M)$ is a homomorphism of Lie algebras:

(4) $$\check{P}(\{\varphi,\psi\}^1)=[\check{P}(\varphi),\check{P}(\psi)]\ \textit{for } \varphi,\psi\in\Omega^1(M).$$

The coboundary operator of Poisson cohomology has a similar form in terms of the bracket $\{\quad,\quad\}^1$ as the exterior derivative has in terms of the usual Lie bracket. Namely, for $U\in\Gamma(\bigwedge^kTM)$ and $\varphi_0,\dots,\varphi_k\in\Omega^1(M)$ we have

(5) $$\begin{aligned}(-1)^k(\delta_PU)(\varphi_0,\dots,\varphi_k):=&\sum_{i=0}^k(-1)^i\mathcal{L}_{P(\varphi_i)}(U(\varphi_0,\dots,\widehat{\varphi_i},\dots,\varphi_k))\\&+\sum_{i<j}(-1)^{i+j}U(\{\varphi_i,\varphi_j\}^1,\varphi_0,\dots,\widehat{\varphi_i},\dots,\widehat{\varphi_j},\dots,\varphi_k).\end{aligned}$$

Proof. (1) is skew-symmetric $\mathbb{R}$-bilinear and satisfies (2) and (3) since by (33.3) we have

$$\begin{aligned}\{df, dg\}^1 &= \mathcal{L}_{\check{P}(df)}dg - \mathcal{L}_{\check{P}(dg)}df - d(P(df, dg)) = d\mathcal{L}_{H_f}g - d\mathcal{L}_{H_g}f - d\{f,g\}\\ &= d\{f,g\},\\ \{\varphi, f\psi\}^1 &= \mathcal{L}_{\check{P}(\varphi)}(f\psi) - \mathcal{L}_{f\check{P}(\psi)}\varphi - d(fP(\varphi,\psi))\\ &= \mathcal{L}_{\check{P}(\varphi)}(f)\psi + f\mathcal{L}_{\check{P}(\varphi)}(\psi) - f\mathcal{L}_{\check{P}(\psi)}\varphi - \varphi(\check{P}(\psi))\, df\\ &\quad - P(\varphi,\psi)\, df - f\, d(P(\varphi,\psi))\\ &= f\{\varphi,\psi\}^1 + \mathcal{L}_{\check{P}(\varphi)}(f)\,\psi.\end{aligned}$$

So an $\mathbb{R}$-bilinear and skew-symmetric operation satisfying (2) and (3) exists. It is uniquely determined since from (3) we see that is local in ψ, i.e., if $\psi|U = 0$ for some open U, then also $\{\varphi,\psi\}^1|U = 0$ by using appropriate bump functions. By skew-symmetry it is also local in φ. But locally each 1-form is a linear combination of expressions $f\, df'$. Thus (2) and (3) determine the bracket $\{\ \ ,\ \ \}^1$ uniquely. By locality it suffices to check condition (4) for 1-forms $f\, df'$ only:

$$\begin{aligned}\check{P}(\{f\, df', g\, dg'\}^1) &= \check{P}\big(fg\,\{df', dg'\}^1 + f\, H_{f'}(g)\, dg' - g\, H_{g'}(f)\, df'\big)\\ &= fg\,\check{P}(d\{f',g'\}) + f\, H_{f'}(g)\,\check{P}(dg') - g\, H_{g'}(f)\,\check{P}(df')\\ &= fg\, H_{\{f',g'\}} + f\, H_{f'}(g)\,\check{P}(dg') - g\, H_{g'}(f)\,\check{P}(df')\\ &= fg\,[H_{f'}, H_{g'}] + f\, H_{f'}(g)\, H_{g'} - g\, H_{g'}(f)\, H_{f'}\\ &= [f\, H_{f'}, g\, H_{g'}] = [\check{P}(f\, df'), \check{P}(g\, dg')].\end{aligned}$$

Now we can check the Jacobi identity. Again it suffices to do this for 1-forms $f\, df'$. We shall use:

$$\begin{aligned}\{f\, df', g\, dg'\}^1 &= fg\,\{df', dg'\}^1 + f\, H_{f'}(g)\, dg' - g\, H_{g'}(f)\, df'\\ &= fg\, d\{f',g'\} + f\,\{f',g\}\, dg' - g\,\{g',f\}\, df'\end{aligned}$$

in order to compute

$$\begin{aligned}&\{\{f\, df', g\, dg'\}^1, h\, dh'\}^1 = \big\{fg\, d\{f',g'\} + f\{f',g\}\, dg' - g\{g',f\}\, df', h\, dh'\big\}^1\\ &= \{fg\, d\{f',g'\}, h\, dh'\}^1 + \{f\{f',g\}\, dg', h\, dh'\}^1 - \{g\{g',f\}\, df', h\, dh'\}^1\\ &= fgh\, d\{\{f',g'\},h'\} + fg\{\{f',g'\},h\}\, dh' - h\{h',fg\}\, d\{f',g'\}\\ &\quad + f\{f',g\}h\, d\{g',h'\} + f\{f',g\}\{g',h\}\, dh' - h\{h', f\{f',g\}\}\, dg'\\ &\quad - g\{g',f\}h\, d\{f',h'\} - g\{g',f\}\{f',h\}\, dh' + h\{h', g\{g',f\}\}\, df'\\ &= fgh\, d\{\{f',g'\},h'\} + (fg\{f',\{g',h\}\}\, dh' - fg\{g'\{f',h\}\}\, dh')\\ &\qquad + (-gh\{h',f\}\, d\{f',g'\} - fh\{h',g\}\, d\{f',g'\})\\ &\quad + hf\{f',g\}\, d\{g',h'\} + f\{f',g\}\{g',h\}\, dh'\end{aligned}$$

$$
\begin{aligned}
&+(-h\{h',f\}\{f',g\}\,dg' - hf\{h',\{f',g\}\}\,dg')\\
&- hg\{g',f\}\,d\{f',h'\} - g\{g',f\}\{f',h\}\,dh'\\
&+(h\{h',g\}\{g',f\}\,df' + gh\{h',\{g',f\}\}\,df').
\end{aligned}
$$

The cyclic sum over these expressions vanishes by using once the Jacobi identity for the Poisson bracket and many pairwise cancellations.

It remains to check formula (5) for the coboundary operator of Poisson cohomology. We use induction on k. For $k = 0$ we have

$$(\delta_P f)(dg) = \mathcal{L}_{H_g} f = \{g,f\} = -\mathcal{L}_{H_f} g = -H_f(dg) = [P,f](dg).$$

For $k = 1$ we have

$$
\begin{aligned}
(\delta_P X)(df,dg) &= \mathcal{L}_{H_f}(X(dg)) - \mathcal{L}_{H_g}(X(df)) - X(\{df,dg\}^1)\\
&= \mathcal{L}_{H_f}(X(dg)) - \mathcal{L}_{H_g}(X(df)) - X(d\{f,g\}),\\
[P,X](df,dg) &= -(\mathcal{L}_X P)(df,dg)\\
&= -\mathcal{L}_X(P(df,dg)) + P(\mathcal{L}_X df, dg) + P(df, \mathcal{L}_X dg)\\
&= -X(d\{g,f\}) + \{g, X(df)\} + \{X(dg), f\}\\
&= -(X(d\{f,g\}) - \mathcal{L}_{H_g}(X(df)) - \mathcal{L}_{H_f}(X(dg)))\\
&= -(\delta_P X)(df,dg).
\end{aligned}
$$

Finally we note that the algebraic consequences of the definition of δ_P are the same as for the exterior derivative d; in particular, we have $\delta_P(U \wedge V) = (\delta_P U) \wedge V + (-1)^u U \wedge (\delta_P V)$. So formula (5) now follows since both sides are graded derivations and agree on the generators of $\Gamma(\bigwedge^* TM)$, namely on $C^\infty(M)$ and on $\mathfrak{X}(M)$. □

33.11. Remark: The Koszul bracket. Lemma (33.10) has the following generalization which we present without proof. For a Poisson field $P \in \Gamma(\bigwedge^2 TM)$, the insertion operator $i_P : \Omega^k(M) \to \Omega^{k-2}(M)$ is the adjoint of multiplication by P:

$$\langle i_P \varphi, U\rangle = \langle \varphi . P \wedge U\rangle \quad \text{for } \varphi \in \Omega^p(M) \text{ and } U \in \Gamma(\textstyle\bigwedge^{p-2} TM).$$

Then $\partial_P := [i_P, d] = i_P \circ d - d \circ i_P$ is the Poisson homology operator of Koszul and satisfies $\partial_P \circ \partial_P = 0$.

Result ([112]). *Let (M,P) be a Poisson manifold. On the exterior algebra $\Omega^{*+1}(M)$ of differential forms,*

$$\{\varphi,\psi\}^1 := (-1)^p\big(\partial_P(\varphi\wedge\psi) - \partial_P(\varphi)\wedge\psi - (-1)^p\varphi\wedge\partial_P(\psi)\big)$$

defines a graded Lie bracket, called the Koszul bracket. It satisfies the Leibniz rule

$$\{\varphi,\psi\wedge\tau\}^1 = \{\varphi,\psi\}^1\wedge\tau + (-1)^{(p-1)q}\psi\wedge\{\varphi,\tau\}^1$$

where $\varphi \in \Omega^p(M)$, $\psi \in \Omega^q(M)$, *and* $\tau \in \Omega(M)$. *The exterior derivative is a derivation of the bracket*

$$d\{\varphi, \psi\}^1 = \{d\varphi, \psi\}^1 + (-1)^{p-1}\{\varphi, d\psi\}^1.$$

On the space $\Omega^1(M)$ *of* 1*-forms this bracket coincides with the Lie bracket from lemma* (33.10). *Moreover, the algebra homomorphism (for the wedge product)*

$$\bigwedge \check{P} : \Omega(M) \to \Gamma(\bigwedge TM)$$

is a homomorphism of graded Lie algebras from the Koszul bracket into the Schouten-Nijenhuis bracket.

33.12. The graded Poisson bracket for differential forms. We consider a Poisson manifold (M, P). Recall $\Omega(M; TM) = \Gamma(\Lambda^* T^*M \otimes TM)$, the space of tangent bundle valued differential forms on M, equipped with the Frölicher-Nijenhuis bracket $[\ \ ,\ \ ]$; see (16.5). Recall for $K \in \Omega^k(M; TM)$ the graded Lie derivative $\mathcal{L}_K \in \operatorname{Der}_k \Omega(M)$ from (16.3) and the graded algebraic derivative $i_K \in \operatorname{Der}_{k-1} \Omega(M)$ from (16.2).

We first extend $\check{P} : T^*M \to TM$ to a linear mapping

$$\check{P} : \Omega(M) \to \Omega(M; TM)$$

of degree -1 by $\check{P}|\Omega^0(M) = 0$, and for $\varphi_i \in \Omega^1(M)$ by

$$\check{P}(\varphi_1 \wedge \cdots \wedge \varphi_k) = \sum_{i=1}^{k} (-1)^{i-1} \varphi_1 \wedge \ldots \widehat{\varphi_i} \cdots \wedge \varphi_k \otimes \check{P}(\varphi_i).$$

This extension is an $\Omega(M)$-bimodule valued graded derivation of degree -1, i.e., for $\varphi \in \Omega^p(M)$ and $\psi \in \Omega^q(M)$ we have:

$$\check{P}(\varphi \wedge \psi) = \check{P}(\varphi) \wedge \psi + (-1)^p \varphi \wedge \check{P}(\psi).$$

Then we have the *Hamiltonian mapping*:

$$H = \operatorname{grad}^{\check{P}} : \Omega(M) \to \Omega(M; TM), \quad H(\psi) := \check{P}(d\psi).$$

Result ([79]). *The Poisson bracket on* $C^\infty(M) = \Omega^0(M)$ *extends to a graded Lie bracket* $\{\ \ ,\ \ \}$ *on the space* $\Omega(M)$ *of all differential forms which is given by*

$$\begin{aligned} \{\varphi, \psi\} : &= \mathcal{L}_{H_\varphi} \psi + d\mathcal{L}_{\check{P}(\varphi)} \psi \\ &= i_{\check{P}(d\varphi)} d\psi + d i_{\check{P}(\varphi)} d\psi - (-1)^{pq} d\, i_{\check{P}(\psi)} d\varphi, \end{aligned}$$

such that the Hamiltonian mapping

$$H : (\Omega(M), \{\ \ ,\ \ \}) \to (\Omega(M; TM), [\ \ ,\ \ ])$$

is a Lie algebra homomorphism into the Frölicher-Nijenhuis bracket. Moreover we have

$$\{f,\psi\} = \mathcal{L}_{H_f}\psi \quad \textit{for } f \in C^\infty(M),$$
$$d\{\varphi,\psi\} = \{d\varphi,\psi\} = (-1)^p\{\varphi, d\psi\}.$$

Thus $Z(M) = \ker(d : \Omega(M) \to \Omega(M))$ *is a commutative Lie ideal. Exterior derivative* $d : (\Omega^*(M), \{\ ,\ \}) \to (\Omega^{*+1}(M), \{\ ,\ \}^1)$ *is a Lie algebra homomorphism into the Koszul bracket from* (33.11). *But this bracket does not act as a derivation for the exterior product; there is no extension of the Poisson bracket doing this and mapping to the Koszul bracket via* d.

In [**144**], for the case of symplectic manifolds, it was shown that the Poisson bracket on $C^\infty(M) = \Omega^0(M)$ extends to a graded Lie bracket on the space $\Omega(M)/B(M)$ of differential forms modulo exact forms such that the Hamiltonian mapping H is a homomorphism of Lie algebras. This bracket was given by the quotient modulo $B(M)$ of either $i(H_\varphi)d\psi$ or $\mathcal{L}_{H(\varphi)}\psi$. The first bracket is graded anticommutative, the second satisfies one form of the graded Jacobi identity, and the two differ by something exact. See also [**160**] and [**30**]. Later Grabowski in [**79**] found the correct expression for the bracket on $\Omega(M)$. See also [**109**] for a still more general view on this.

33.13. Dirac structures — a common generalization of symplectic and Poisson structures ([37], [29], [28]). Let M be a smooth manifold of dimension m. By a *Dirac structure* on M we mean a vector subbundle $D \subset TM \times_M T^*M$ with the following two properties:

(1) Each fiber D_x is maximally isotropic with respect to the metric of signature (m,m) on $TM \times_M T^*M$ given by

$$\langle (X,\alpha),(X',\alpha')\rangle_+ = \alpha(X') + \alpha'(X).$$

So D is of fiber dimension m.

(2) The space of sections of D is closed under the non-skew-symmetric version of the Courant bracket

$$[(X,\alpha),(X',\alpha')] = ([X,X'], \mathcal{L}_X\alpha' - i_{X'}d\alpha).$$

If (X,α) and (X',α') are sections of D, then $i_X\alpha' = -i_{X'}\alpha$ by isotropy; thus $\mathcal{L}_X\alpha' - i_{X'}d\alpha = i_Xd\alpha' + \frac{1}{2}d(i_X\alpha' - i_{X'}\alpha) - i_{X'}d\alpha$ so the Courant bracket is skew-symmetric on $\Gamma(D)$.

Natural examples of Dirac structures are the following:

(3) Symplectic structures ω on M, where $D = D^\omega = \{(X, \check\omega(X)) : X \in TM\}$ is just the graph of $\check\omega : TM \to T^*M$.

More generally, for a 2-form ω on M the graph D^ω of $\check\omega : TM \to T^*M$ is a Dirac structure if and only if $d\omega = 0$ (a presymplectic structure); these are precisely the Dirac structures D with $TM \cap D = \{0\}$. Namely,

$$\begin{aligned}\langle (X, \check{\omega}(X)), (Y, \check{\omega}(Y))\rangle_+ &= \omega(X,Y) + \omega(Y,X) = 0,\\ [(X, i_X\omega), (Y, i_Y\omega)] &= ([X,Y], \mathcal{L}_X i_Y\omega - i_Y d i_X\omega)\\ &= ([X,Y], i_{[X,Y]}\omega).\end{aligned}$$

(4) Poisson structures P on M where $D = D^P = \{(\check{P}(\alpha), \alpha) : \alpha \in T^*M\}$ is the graph of $P : T^*M \to TM$; these are precisely the Dirac structures D which are transversal to T^*M. Namely,

$$\begin{aligned}\langle (\check{P}(\alpha), \alpha), (\check{P}(\beta), \beta)\rangle_+ &= P(\alpha,\beta) + P(\beta,\alpha) = 0,\\ [(\check{P}(\alpha), \alpha), (\check{P}(\beta), \beta)] &= ([\check{P}(\alpha), \check{P}(\beta)], \mathcal{L}_{\check{P}(\alpha)}\beta - i_{\check{P}(\beta)} d\alpha)\\ &= (\check{P}(\{\alpha,\beta\}^1), \{\alpha,\beta\}^1), \quad \text{using (33.10)}.\end{aligned}$$

Given a Dirac structure D on M, we consider its range $R(D) = \mathrm{pr}_{TM}(D) = \{X \in TM : (X,\alpha) \in D \text{ for some } \alpha \in T^*M\}$. There is a skew-symmetric 2-form Θ_D on $R(D)$ which is given by $\Theta_D(X, X') = \alpha(X')$ where $\alpha \in T^*M$ is such that $(X, \alpha) \in D$. The range $R(D)$ is an integrable distribution of nonconstant rank in the sense of (3.28), so M is foliated into maximal integral submanifolds L of $R(D)$ of varying dimensions, which are all initial submanifolds. The form Θ_D induces a closed 2-form on each leaf L and (L, Θ_D) is thus a presymplectic manifold (Θ_D might be degenerate on some L). If the Dirac structure corresponds to a Poisson structure, then the (L, Θ_D) are exactly the symplectic leaves of the Poisson structure.

The main advantage of Dirac structures is that one can apply arbitrary pushforwards and pullbacks to them. So if $f : N \to M$ is a smooth mapping and D_M is a Dirac structure on M, then the pullback is defined by $f^*D_M = \{(X, f^*\alpha) \in TN \times_N T^*N : (Tf.X, \alpha) \in D_M\}$. Likewise the pushforward of a Dirac structure D_N on N is given by $f_*D_N = \{(Tf.X, \alpha) \in TM \times_M T^*M : (X, f^*\alpha) \in D_N\}$. If $D = D^\omega$ for a closed 2-form ω on M, then $f^*(D^\omega) = D^{f^*\omega}$. If P_N and P_M are Poisson structures on N and M, respectively, which are f-related, then $f_*D^{P_N} = D^{f_*P_N} = D^{P_M}$.

34. Hamiltonian Group Actions and Momentum Mappings

34.1. Symplectic and Hamiltonian group actions. Let us suppose that a Lie group G acts from the right on a symplectic manifold (M, ω) by $r : M \times G \to M$ in a way which respects ω, so that each transformation r^g is a symplectomorphism. This is called a *symplectic group action.* Let us list some immediate consequences:

(1) *The space $C^\infty(M)^G$ of G-invariant smooth functions is a Lie subalgebra for the Poisson bracket,* since $(r^g)^*\{f,h\} = \{(r^g)^*f, (r^g)^*h\} = \{f,h\}$ holds for each $g \in G$ and $f, h \in C^\infty(M)^G$.

(2) *For $x \in M$ the pullback of ω to the orbit $x.G$ is a closed 2-form of constant rank and is invariant under the action of G on the orbit.* Note first that the orbit is an initial submanifold by (6.4). If $i : x.G \to M$ is the embedding of the orbit, then $r^g \circ i = i \circ r^g$, so that $i^*\omega = i^*(r^g)^*\omega = (r^g)^* i^*\omega$ holds for each $g \in G$ and thus $i^*\omega$ is invariant. Since G acts transitively on the orbit, $i^*\omega$ has constant rank (as a mapping $T(x.G) \to T^*(x.G)$).

(3) By (6.3) the fundamental vector field mapping $\zeta : \mathfrak{g} \to \mathfrak{X}(M, \omega)$, given by $\zeta_X(x) = T_e(r(x, \quad))X$ for $X \in \mathfrak{g}$ and $x \in M$, is a homomorphism of Lie algebras, where $\mathfrak{g}$ is the Lie algebra of G. (For a left action we get an antihomomorphism of Lie algebras; see (6.2)). Moreover, ζ takes values in $\mathfrak{X}(M, \omega)$. Let us consider again the exact sequence of Lie algebra homomorphisms from (31.21):

$$\begin{array}{ccccccccccc} 0 & \longrightarrow & H^0(M) & \xrightarrow{\alpha} & C^\infty(M) & \xrightarrow{H} & \mathfrak{X}(M,\omega) & \xrightarrow{\gamma} & H^1(M) & \longrightarrow & 0. \\ & & & & & \nwarrow^{j} & \uparrow{\scriptstyle \zeta} & & & & \\ & & & & & & \mathfrak{g} & & & & \end{array}$$

One can lift ζ to a linear mapping $j : \mathfrak{g} \to C^\infty(M)$ if and only if $\gamma \circ \zeta = 0$. In this case the action of G is called a *Hamiltonian group action*, and the linear mapping $j : \mathfrak{g} \to C^\infty(M)$ is called a *generalized Hamiltonian function* for the group action. It is unique up to addition of a mapping $\alpha \circ \tau$ for $\tau : \mathfrak{g} \to H^0(M)$.

(4) *If $H^1(M) = 0$, then any symplectic action on (M, ω) is a Hamiltonian action. If not and $\gamma \circ \zeta \neq 0$, we may either* (a) *lift ω to the universal cover of M for which the first cohomology then vanishes, and try to lift the group action also (where we might have to enlarge the group by the discrete group of deck transformations), or* (b) *replace $\mathfrak{g}$ by its Lie subalgebra* $\ker(\gamma \circ \zeta) \subset \mathfrak{g}$ *and consider the corresponding Lie subgroup G; in both cases we get a Hamiltonian action.*

(5) *If the Lie algebra $\mathfrak{g}$ is equal to its commutator subalgebra $[\mathfrak{g}, \mathfrak{g}]$ (i.e., to the linear span of all $[X, Y]$ for $X, Y \in \mathfrak{g}$), then any infinitesimal symplectic action $\zeta : \mathfrak{g} \to \mathfrak{X}(M, \omega)$ is a Hamiltonian action,* since then any $Z \in \mathfrak{g}$ can be written as $Z = \sum_i [X_i, Y_i]$ so that $\zeta_Z = \sum [\zeta_{X_i}, \zeta_{Y_i}] \in \operatorname{im}(\operatorname{grad}^\omega)$ since $\gamma : \mathfrak{X}(M, \omega) \to H^1(M)$ is a homomorphism into the zero Lie bracket.

34.2. Lemma. Momentum mappings. *For an infinitesimal symplectic action, i.e., a homomorphism $\zeta : \mathfrak{g} \to \mathfrak{X}(M, \omega)$ of Lie algebras, we can find a linear lift $j : \mathfrak{g} \to C^\infty(M)$ if and only if there exists a mapping $J : M \to \mathfrak{g}^*$ such that*

$$H_{\langle J, X \rangle} = \zeta_X \quad \textit{for all } X \in \mathfrak{g}.$$

Proof. Namely, for $y \in M$ we have

$$J : M \to \mathfrak{g}^*, \quad \langle J(y), X\rangle = j(X)(y) \in \mathbb{R}, \quad j : \mathfrak{g} \to C^\infty(M). \quad \square$$

The mapping $J : M \to \mathfrak{g}^*$ is called the *momentum mapping* for the infinitesimal action $\zeta : \mathfrak{g} \to \mathfrak{X}(M, \omega)$. This holds even for a Poisson manifold (M, P) (see section (33)) and an infinitesimal action of a Lie algebra $\zeta : \mathfrak{g} \to \mathfrak{X}(M, P)$ by Poisson morphisms. Let us note again the relations between the generalized Hamiltonian j and the momentum mapping J:

$$\begin{aligned} &J : M \to \mathfrak{g}^*, \quad j : \mathfrak{g} \to C^\infty(M), \quad \zeta : \mathfrak{g} \to \mathfrak{X}(M, P), \\ (1) \qquad &\langle J, X\rangle = j(X) \in C^\infty(M), \quad H_{j(X)} = \zeta(X), \quad X \in \mathfrak{g}, \end{aligned}$$

where $\langle\ ,\ \rangle$ is the duality pairing.

34.3. Basic properties of the momentum mapping. Consider a Hamiltonian right action $r : M \times G \to M$ of a Lie group G on a symplectic manifold M, let $j : \mathfrak{g} \to C^\infty(M)$ be a generalized Hamiltonian and let $J : M \to \mathfrak{g}^*$ be the associated momentum mapping.

(1) *For $x \in M$, the transposed mapping of $dJ(x) : T_xM \to \mathfrak{g}^*$ is*

$$dJ(x)^\top : \mathfrak{g} \to T_x^*M, \qquad dJ(x)^\top = \check{\omega}_x \circ \zeta,$$

since for $\xi \in T_xM$ and $X \in \mathfrak{g}$ we have

$$\langle dJ(\xi), X\rangle = \langle i_\xi dJ, X\rangle = i_\xi d\langle J, X\rangle = i_\xi i_{\zeta_X}\omega = \langle \check{\omega}_x(\zeta_X(x)), \xi\rangle.$$

(2) *The image $dJ(T_xM)$ of $dJ(x) : T_xM \to \mathfrak{g}^*$ is the annihilator $\mathfrak{g}_x^\circ$ of the isotropy Lie algeba $\mathfrak{g}_x := \{X \in \mathfrak{g} : \zeta_X(x) = 0\}$ in $\mathfrak{g}^*$*, since the annihilator of the image is the kernel of the transposed mapping,

$$\operatorname{im}(dJ(x))^\circ = \ker(dJ(x)^\top) = \ker(\check{\omega}_x \circ \zeta) = \ker(\operatorname{ev}_x \circ \zeta) = \mathfrak{g}_x.$$

(3) *The kernel of $dJ(x)$ is the symplectic orthogonal $(T_x(x.G))^\perp \subseteq T_xM$*, since for the annihilator of the kernel we have

$$\begin{aligned} \ker(dJ(x))^\circ &= \operatorname{im}(dJ(x)^\top) = \operatorname{im}(\check{\omega}_x \circ \zeta) = \{\check{\omega}_x(\zeta_X(x)) : X \in \mathfrak{g}\} \\ &= \check{\omega}_x(T_x(x.G)). \end{aligned}$$

(4) *For each $x \in M$ the rank of $dJ(x) : T_xM \to \mathfrak{g}^*$ equals the dimension of the orbit $x.G$, i.e., the codimension in $\mathfrak{g}$ of the isotropy Lie algebra $\mathfrak{g}_x$.* This follows from (3) since

$$\operatorname{rank}(dJ(x)) = \operatorname{codim}_{T_xM}(\ker dJ(x)) = \dim(\ker(dJ(x))^\circ) = \dim(T_x(x.G)).$$

(5) *The momentum mapping $J : M \to \mathfrak{g}^*$ is a submersion at $x \in M$ if and only if the isotropy group G_x is discrete.*

(6) *If G is connected, $x \in M$ is a fixed point for the G-action if and only if x is a critical point of J, i.e., $dJ(x) = 0$.*

(7) *Suppose that all orbits of the G-action on M have the same dimension. Then $J : M \to \mathfrak{g}^*$ is of constant rank. Moreover, the distribution $\mathcal{F}$ of all symplectic orthogonals to the tangent spaces to all orbits is then an integrable distribution of constant rank and its leaves are exactly the connected components of the fibers of J.* Namely, the rank of J is constant by (3). For each $x \in M$ the subset $J^{-1}(J(x))$ is then a submanifold by (1.13), and by using (3) we see that $J^{-1}(J(x))$ is a maximal integral submanifold of $\mathcal{F}$ through x.

A direct proof that the distribution $\mathcal{F}$ is integrable is as follows: It has constant rank and is involutive, since for $\xi \in \mathfrak{X}(M)$ we have $\xi \in \mathfrak{X}(\mathcal{F})$ if and only if $i_\xi i_{\zeta_X}\omega = -\omega(\xi, \zeta_X) = 0$ for all $X \in \mathfrak{g}$. For $\xi, \eta \in \mathfrak{X}(\mathcal{F})$ and $X \in \mathfrak{g}$ we have

$$\begin{aligned} i_{[\xi,\eta]} i_{\zeta_X}\omega &= [\mathcal{L}_\xi, i_\eta] i_{\zeta_X}\omega = \mathcal{L}_\xi i_\eta i_{\zeta_X}\omega - i_\eta \mathcal{L}_\xi i_{\zeta_X}\omega \\ &= 0 - i_\eta i_\xi d i_{\zeta_X}\omega - i_\eta d i_\xi i_{\zeta_X}\omega = 0. \end{aligned}$$

(8) *(E. Noether's theorem) Let $h \in C^\infty(M)$ be a Hamiltonian function which is invariant under the Hamiltonian G action. Then the momentum mapping $J : M \to \mathfrak{g}^*$ is constant on each trajectory of the Hamiltonian vector field H_h.* Namely,

$$\begin{aligned} \tfrac{d}{dt}\langle J \circ \operatorname{Fl}^{H_h}_t, X\rangle &= \langle dJ \circ \tfrac{d}{dt}\operatorname{Fl}^{H_h}_t, X\rangle = \langle dJ(H_h) \circ \operatorname{Fl}^{H_h}_t, X\rangle \\ &= (i_{H_h} d\langle J, X\rangle) \circ \operatorname{Fl}^{H_h}_t = \{h, \langle J, X\rangle\} \circ \operatorname{Fl}^{H_h}_t \\ &= -\{\langle J, X\rangle, h\} \circ \operatorname{Fl}^{H_h}_t = -(\mathcal{L}_{\zeta_X} h) \circ \operatorname{Fl}^{H_h}_t = 0. \end{aligned}$$

E. Noether's theorem admits the following generalization.

34.4. Theorem (Marsden and Weinstein). *Let G_1 and G_2 be two Lie groups which act by Hamiltonian actions r_1 and r_2 on the symplectic manifold (M, ω), with momentum mappings J_1 and J_2, respectively. We assume that J_2 is G_1-invariant, i.e., J_2 is constant along all G_1-orbits and that G_2 is connected.*

Then J_1 is constant on the G_2-orbits and the two actions commute.

Proof. Let $\zeta^i : \mathfrak{g}_i \to \mathfrak{X}(M, \omega)$ be the two infinitesimal actions. Then for $X_1 \in \mathfrak{g}_1$ and $X_2 \in \mathfrak{g}_2$ we have

$$\begin{aligned} \mathcal{L}_{\zeta^2_{X_2}} \langle J_1, X_1\rangle &= i_{\zeta^2_{X_2}} d\langle J_1, X_1\rangle = i_{\zeta^2_{X_2}} i_{\zeta^1_{X_1}} \omega \\ &= \{\langle J_2, X_2\rangle, \langle J_1, X_1\rangle\} = -\{\langle J_1, X_1\rangle, \langle J_2, X_2\rangle\} \\ &= -i_{\zeta^1_{X_1}} d\langle J_2, X_2\rangle = -\mathcal{L}_{\zeta^1_{X_1}} \langle J_2, X_2\rangle = 0 \end{aligned}$$

since J_2 is constant along each G_1-orbit. Since G_2 is assumed to be connected, J_1 is also constant along each G_2-orbit. We also saw that each Poisson bracket $\{\langle J_2, X_2\rangle, \langle J_1, X_1\rangle\}$ vanishes; by $H_{\langle J_i, X_i\rangle} = \zeta^i_{X_i}$ we conclude that $[\zeta^1_{X_1}, \zeta^2_{X_2}] = 0$ for all $X_i \in \mathfrak{g}_i$ which implies the result if also G_1 is connected. In the general case we can argue as follows:

$$\begin{aligned}(r_1^{g_1})^*\zeta^2_{X_2} &= (r_1^{g_1})^* H_{\langle J_2, X_2\rangle} = (r_1^{g_1})^*(\check{\omega}^{-1} d\langle J_2, X_2\rangle) \\ &= (((r_1^{g_1})^*\omega)\check{})^{-1} d\langle (r_1^{g_1})^* J_2, X_2\rangle = \check{\omega}^{-1} d\langle J_2, X_2\rangle = H_{\langle J_2, X_2\rangle} = \zeta^2_{X_2}.\end{aligned}$$

Thus $r_1^{g_1}$ commutes with each $r_2^{\exp(tX_2)}$ and thus with each $r_2^{g_2}$, since G_2 is connected. □

34.5. Remark. The classical first integrals of mechanical systems can be derived by Noether's theorem, where the group G is the group of isometries of Euclidean 3-space $\mathbb{R}^3$, the semidirect product $\mathbb{R}^3 \rtimes SO(3)$. Let (M, ω, h) be a Hamiltonian mechanical system consisting of several rigid bodies moving in physical 3-space. Then the Hamiltonian function is the sum of the kinetic energy and the potential energy. This system is said to be free if the Hamiltonian function h describing the movement of the system is invariant under the group of isometries acting on $\mathbb{R}^3$ and its induced action on phase space $M \subseteq T^*(\mathbb{R}^{3k})$. This action is Hamiltonian since for the motion group G we have $[\mathfrak{g}, \mathfrak{g}] = \mathfrak{g}$, by (34.1.5). Equivalently, the action is free if there is no potential. Then there exists a momentum mapping $J = (J_l, J_a) : M \to (\mathbb{R}^3 \rtimes \mathfrak{so}(3))^* = (\mathbb{R}^3)^* \times \mathfrak{so}(3)^*$. Its component J_l is the momentum mapping for the action of the translation group and is called the *linear momentum*; the component J_a is the momentum mapping for the action of the rotation group and is called the *angular momentum*.

The momentum map is essentially due to Lie [**124**, pp. 300–343]. The modern notion is due to [**110**], [**212**], and [**105**]. Also, [**132**], [**122**] and [**134**] are convenient references, and [**134**] has a large and updated bibliography. The momentum map has a strong tendency to have *convex image* and is important for *representation theory*; see [**105**] and [**171**]. Recently, there has also been a proposal for a group valued momentum mapping; see [**3**].

34.6. Strongly Hamiltonian group actions. Suppose that we have a Hamiltonian action $M \times G \to M$ on the symplectic manifold (M, ω), and consider a generalized Hamiltonian $j : \mathfrak{g} \to C^\infty(M)$, which is unique up to addition of $\alpha \circ \tau$ for some $\tau : \mathfrak{g} \to H^0(M)$:

$$\begin{array}{ccccccccc} 0 & \longrightarrow & H^0(M) & \xrightarrow{\alpha} & C^\infty(M) & \xrightarrow{H} & \mathfrak{X}(M,\omega) & \xrightarrow{\gamma} & H^1(M) \longrightarrow 0. \\ & & & & & \nwarrow{\scriptstyle j} & \uparrow{\scriptstyle \zeta} & & \\ & & & & & & \mathfrak{g} & & \end{array}$$

We want to investigate whether we can change j into a homomorphism of Lie algebras.

(1) The map $\mathfrak{g} \ni X, Y \mapsto \{jX, jY\} - j([X,Y]) =: \bar{j}(X,Y)$ takes values in $\ker(H) = \operatorname{im}(\alpha)$ since

$$H(\{jX, jY\}) - H(j([X,Y])) = [H_{jX}, H_{jY}] - \zeta_{[X,Y]} = [\zeta_X, \zeta_Y] - \zeta_{[X,Y]} = 0.$$

Moreover, $\bar{j} : \bigwedge^2 \mathfrak{g} \to H^0(M)$ is a cocycle for the Chevalley cohomology of the Lie algebra $\mathfrak{g}$, as explained in (14.14):

$$\begin{aligned}
d\bar{j}(X,Y,Z) &= -\sum_{\text{cyclic}} \bar{j}([X,Y],Z) = -\sum_{\text{cyclic}} (\{j([X,Y]), jZ\} - j([[X,Y],Z])) \\
&= -\sum_{\text{cyclic}} \{\{jX, jY\} - \bar{j}(X,Y), jZ\} - 0 \\
&= -\sum_{\text{cyclic}} (\{\{jX, jY\}, jZ\} - \{\bar{j}(X,Y), jZ\}) = 0,
\end{aligned}$$

by the Jacobi identity and since $\bar{j}(X,Y) \in H^0(M)$ which equals the center of the Poisson algebra. Recall that the linear mapping $j : \mathfrak{g} \to C^\infty(M)$ was unique up to addition of a mapping $\alpha \circ \tau$ for $\tau : \mathfrak{g} \to H^0(M)$. But

$$\begin{aligned}
\overline{j+\tau}(X,Y) &= \{(j+\tau)X, (j+\tau)Y\} - (j+\tau)([X,Y]) \\
&= \{jX, jY\} + 0 - j([X,Y]) - \tau([X,Y]) = (\bar{j} + d\tau)(X,Y).
\end{aligned}$$

Thus, if $\gamma \circ \zeta = 0$, there is a unique Chevalley cohomology class $\tilde{\zeta} := [\bar{j}] \in H^2(\mathfrak{g}, H^0(M))$.

(2) The cohomology class $\tilde{\zeta} = [\bar{j}]$ is automatically zero if $H^2(\mathfrak{g}, H^0(M)) = H^2(\mathfrak{g}) \otimes H^0(M) = 0$. This is the case for semisimple $\mathfrak{g}$, by the Whitehead lemmas; see [**83**, p. 249].

(3) The cohomology class $\tilde{\zeta} = [\bar{j}]$ is automatically zero if the symplectic structure ω on M is exact, $\omega = -d\vartheta$ for $\vartheta \in \Omega^1(M)$, and $\mathcal{L}_{\zeta_X}\vartheta = 0$ for each $X \in \mathfrak{g}$: Then we may use $j(X) = i_{\zeta_X}\vartheta = \vartheta(\zeta_X)$, since $i(H_{jX})\omega = d(jX) = di_{\zeta_X}\vartheta = \mathcal{L}_{\zeta_X}\vartheta - i_{\zeta_X}d\vartheta = 0 + i_{\zeta_X}\omega$ implies $H_{jX} = \zeta_X$. For this choice of j we have

$$\begin{aligned}
\bar{j}(X,Y) &= \{jX, jY\} - j([X,Y]) = \mathcal{L}_{H_{jX}}(jY) - i_{\zeta([X,Y])}\vartheta \\
&= \mathcal{L}_{\zeta_X} i_{\zeta_Y}\vartheta - i_{[\zeta_X,\zeta_Y]}\vartheta = \mathcal{L}_{\zeta_X} i_{\zeta_Y}\vartheta - [\mathcal{L}_{\zeta_X}, i_{\zeta_Y}]\vartheta = -i_{\zeta_Y}\mathcal{L}_{\zeta_X}\vartheta = 0.
\end{aligned}$$

(4) The condition of (3) holds if $M = T^*Q$ is a cotantent bundle and if $\zeta : \mathfrak{g} \to \mathfrak{X}(T^*Q, \omega_Q)$ is induced by $\sigma : \mathfrak{g} \to \mathfrak{X}(Q)$ in the sense that its flow is given by $\mathrm{Fl}_t^{\zeta_X} = T^*(\mathrm{Fl}_t^{\sigma_X}) = T(\mathrm{Fl}_{-t}^{\sigma_X})^*$. Namely, by (31.9) we have:

$$\mathcal{L}_{\zeta_X}\vartheta_Q = \partial|_0(\mathrm{Fl}_t^{\zeta_X})^*\vartheta_Q = \partial|_0(T^*(\mathrm{Fl}_t^{\sigma_X}))^*\vartheta_Q = 0.$$

Let us note here for further use that the j is given by the following formula: For $p_q \in T_q^*Q$ we have:

$$j(X)(p_q) = \vartheta(\zeta_X(p_q)) = \langle \pi_{T^*Q}(\zeta_X(p_q)), T(\pi_Q)(\zeta_X(p_q))\rangle = \langle p_q, \sigma_X(q)\rangle.$$

(5) An example where the cohomology class $\tilde{\zeta} = [\bar{\jmath}] \in H^2(\mathfrak{g}, H^0(M))$ does not vanish: Let $\mathfrak{g} = (\mathbb{R}^2, [\quad,\quad] = 0)$ with coordinates a, b. Let $M = T^*\mathbb{R}$ with coordinates q, p, and let $\omega = dq \wedge dp$. Let $\zeta_{(a,b)} = a\partial_q + b\partial_p$. A lift is given by $j(a,b)(q,p) = ap - bq$. Then

$$\begin{aligned}\bar{\jmath}((a_1,b_1),(a_2,b_2)) &= \{j(a_1,b_1), j(a_2,b_2)\} - j(0) = \{a_1p - b_1q, a_2p - b_2q\}\\ &= -a_1b_2 + a_2b_1.\end{aligned}$$

(6) For a symplectic group action $r : M \times G \to M$ of a Lie group G on a symplectic manifold M, let us suppose that the cohomology class $\tilde{\zeta} = [\bar{\jmath}] \in H^2(\mathfrak{g}, H^0(M))$ from (34.1.1) vanishes. Then there exists $\tau \in L(\mathfrak{g}, H^0(M))$ with $d\tau = \bar{\jmath}$, i.e.,

$$\begin{aligned}d\tau(X,Y) &= -\tau([X,Y]) = \bar{\jmath}(X,Y) = \{jX, jY\} - j([X,Y]),\\ \overline{j-\tau}(X,Y) &= \{(j-\tau)X, (j-\tau)Y\} - (j-\tau)([X,Y])\\ &= \{jX, jY\} + 0 - j([X,Y]) + \tau([X,Y]) = 0,\end{aligned}$$

so that $j - \tau : \mathfrak{g} \to C^\infty(M)$ is a homomorphism of Lie algebras. Then the action of G is called a *strongly Hamiltonian group action* and the homomorphism $j - \tau : \mathfrak{g} \to C^\infty(M)$ is called the associated *infinitesimal strongly Hamiltonian action.*

34.7. Theorem. *The momentum mapping $J : M \to \mathfrak{g}^*$ for an infinitesimal strongly Hamiltonian action $j : \mathfrak{g} \to C^\infty(M)$ on a Poisson manifold (M, P^M) has the following properties:*

(1) *J is infinitesimally equivariant: For each $X \in \mathfrak{g}$ the Hamiltonian vector fields $H_{j(X)} = \zeta_X \in \mathfrak{X}(M,P)$ and $\operatorname{ad}(X)^* : \mathfrak{g}^* \to \mathfrak{g}^*$ are J-related.*

(2) *J is a Poisson morphism $J : (M, P^M) \to (\mathfrak{g}^*, P^{\mathfrak{g}^*})$ into the canonical Poisson structure on $\mathfrak{g}^*$ from* (33.7).

(3) *The momentum mapping for a strongly Hamiltonian action of a connected Lie group G on a Poisson manifold is G-equivariant: $J(x.g) = \operatorname{Ad}(g)^*.J(x)$.*

Proof. (1) By definition (34.2.1) we have $\langle J(x), X\rangle = j(X)(x)$; differentiating this, we get $\langle dJ(x)(\xi_x), X\rangle = d(j(X))(\xi_x)$ or $d\langle J, X\rangle = dj(X) \in \Omega^1(M)$. Then we have

$$\begin{aligned}\langle dJ(\zeta_X), Y\rangle &= dj(Y)(\zeta_X) = H_{j(X)}(j(Y))\\ &= \{j(X), j(Y)\} = j[X,Y],\\ \langle \operatorname{ad}(X)^* \circ J, Y\rangle &= \langle J, \operatorname{ad}(X)Y\rangle = \langle J, [X,Y]\rangle,\\ dJ.\zeta_X &= \operatorname{ad}(X)^* \circ J.\end{aligned}$$

(2) We have to show that $\bigwedge^2 dJ(x).P^M = P^{\mathfrak{g}^*}(J(x))$, according to (33.5.3):

$$\begin{aligned}\langle P^{\mathfrak{g}^*} \circ J, X \wedge Y\rangle &= 2\langle J, [X,Y]\rangle \qquad \text{by (33.7)}\\ &= j[X,Y] = \{j(X), j(Y)\},\end{aligned}$$

$$\begin{aligned}\langle \bigwedge^2 dJ(x).P^M, X \wedge Y\rangle &= \langle \bigwedge^2 dJ(x)^*.(X\wedge Y), P^M\rangle\\ &= \langle dJ(x)^* X \wedge dJ(x)^* Y, P^M\rangle\\ &= \langle P^M, d\langle J, X\rangle \wedge d\langle J,Y\rangle\rangle(x)\\ &= \langle P^M, dj(X)\wedge dj(Y)\rangle(x) = 2\{j(X), j(Y)\}(x).\end{aligned}$$

(3) is an immediate consequence of (1). □

34.8. Equivariance of momentum mappings. Let $J : M \to \mathfrak{g}^*$ be a momentum mapping for a Hamiltonian right group action $r : M \times G \to M$ on a symplectic manifold (M, ω). We do not assume here that the lift $j : \mathfrak{g} \to C^\infty(M)$ given by $j(X) = \langle J, X\rangle$ is a Lie algebra homomorphism. Recall that for the fundamental vector field mapping $\zeta : \mathfrak{g} \to \mathfrak{X}(M, \omega)$ we have $\zeta_X = H_{j(X)} = H_{\langle J,X\rangle}$. We also assume that M is connected; otherwise one has to treat each connected component separately.

For $X \in \mathfrak{g}$ and $g \in G$ we have (compare with the proof of (34.4))

$$\begin{aligned}(r^g)^*\zeta_X &= (r^g)^* H_{\langle J,X\rangle} = (r^g)^*(\check{\omega}^{-1} d\langle J, X\rangle)\\ &= (((r^g)^*\omega)\check{})^{-1} d\langle (r^g)^* J, X\rangle = \check{\omega}^{-1} d\langle J\circ r^g, X\rangle = H_{\langle J\circ r^g, X\rangle},\\ (r^g)^*\zeta_X &= T(r^{g^{-1}})\circ \zeta_X \circ r^g = \zeta_{\mathrm{Ad}(g)X} \qquad \text{by (6.3.2)}\\ &= H_{\langle J, \mathrm{Ad}(g)X\rangle} = H_{\langle \mathrm{Ad}(g)^* J, X\rangle}.\end{aligned}$$

So we conclude that $\langle J\circ r^g - \mathrm{Ad}(g)^* J, X\rangle \in H^0(M)$ is a constant function on M (which we assumed to be connected) for every $X \in \mathfrak{g}$ and we get a smooth mapping

(1) $$\bar{J} : G \to \mathfrak{g}^*,$$

$$\bar{J}(g) := J\circ r^g - \mathrm{Ad}(g)^* \circ J = J(x.g) - \mathrm{Ad}(g)^* J(x) \in \mathfrak{g}^* \quad \text{for each } x \in M,$$

which satisfies for $g_1, g_2 \in G$ and each $x \in M$

(2) $$\begin{aligned}\bar{J}(g_0 g_1) &= J(x.g_0 g_1) - \mathrm{Ad}(g_0 g_1)^* J(x)\\ &= J((x.g_0).g_1) - \mathrm{Ad}(g_1)^* \mathrm{Ad}(g_0)^* J(x)\\ &= J((x.g_0).g_1) - \mathrm{Ad}(g_1)^* J(x.g_0) + \mathrm{Ad}(g_1)^*(J(x.g_0) - \mathrm{Ad}(g_0)^* J(x))\\ &= \bar{J}(g_1) + \mathrm{Ad}(g_1)^*\bar{J}(g_0) = \bar{J}(g_1) + \bar{J}(g_0).\mathrm{Ad}(g_1).\end{aligned}$$

This equation says that $\bar{J} : G \to \mathfrak{g}^*$ is a smooth 1-cocycle with values in the right G-module $\mathfrak{g}^*$ for the smooth group cohomology which is given by the following coboundary operator, which for completeness sake we write for a

G-bimodule V, i.e., a vector space V with a linear left action $\lambda : G\times V \to V$ and a linear right action $\rho : V \times G \to V$ which commute:

(3) $C^k(G,V) := C^\infty(G^k = G\times\ldots\times G, V), \quad C^0(G,V) = V, \quad k\geq 0,$

$$\delta : C^k(G,V) \to C^{k+1}(G,V),$$

$$\delta\Phi(g_0,\ldots,g_k) = g_0.\Phi(g_1,\ldots,g_k) + \sum_{i=1}^{k}(-1)^i\Phi(g_0,\ldots,g_{i-2},g_{i-1}g_i,\ldots,g_k) + (-1)^{k+1}\Phi(g_0,\ldots,g_{k-1}).g_k.$$

It is easy to check that $\delta\circ\delta = 0$. As noted in (15.16), the *group cohomology* is defined by

$$H^k(G;V) := \frac{\ker(\delta : C^k(G,V)\to C^{k+1}(G,V))}{\operatorname{im}(\delta : C^{k-1}(G,V)\to C^k(G,V))}.$$

Since for $v\in V = C^0(G,V)$ we have $\delta v(g_0) = g_0.v - v.g_0$, it follows that $H^0(G,V) = \{v\in V : g.v = v.g\} = Z_V(G)$. A smooth mapping $\Phi : G\to V$ is a cocycle $\delta\Phi = 0$ if and only if $\Phi(g_0g_1) = g_0.\Phi(g_1) + \Phi(g_0).g_1$, i.e., Φ is a 'derivation'.

In our case $V = \mathfrak{g}^*$ with trivial left G-action (each $g\in G$ acts by the identity) and right action $\operatorname{Ad}(\quad)^*$. Any other moment mapping $J' : M\to\mathfrak{g}^*$ is of the form $J' = J+\alpha$ for constant $\alpha\in\mathfrak{g}^*$ since M is connected. The associated group cocycle is then

$$\begin{aligned}\overline{J+\alpha}(g) &= J(x.g) + \alpha - \operatorname{Ad}(g)^*(J(x)+\alpha) = \bar J(g) + \alpha - \alpha.\operatorname{Ad}(g)\\ &= (\bar J + \delta\alpha)(g),\end{aligned} \tag{4}$$

so that the group cohomology class $\tilde r = [\bar J]\in H^1(G,\mathfrak{g}^*)$ of the Hamiltonian G-action does not depend on the choice of the momentum mapping.

(5) *The differential $d\bar J(e) : \mathfrak{g}\to\mathfrak{g}^*$ at $e\in G$ of the group cocycle $\bar J : G\to\mathfrak{g}*$ satisfies*

$$\langle d\bar J(e)X, Y\rangle = \bar\jmath(X,Y),$$

where $\bar\jmath$, given by $\bar\jmath(X,Y) = \{\jmath(X),\jmath(Y)\} - \jmath([X,Y])$, is the Lie algebra cocycle from (34.6.1), since

$$\begin{aligned}\{\jmath(X),\jmath(Y)\}(x) &= H_{\jmath(X)}(\jmath(Y))(x) = i(H_{\langle J,X\rangle}(x))d\langle J,Y\rangle\\ &= \langle dJ(\zeta_X(x)),Y\rangle = \partial|_0\langle J(x.\exp(tX)),Y\rangle\\ &= \partial|_0\langle \operatorname{Ad}(\exp(tX))^*J(x) + \bar J(\exp(tX)),Y\rangle\\ &= \langle \operatorname{ad}(X)^*J(x) + d\bar J(e)(X),Y\rangle = \langle J(x),\operatorname{ad}(X)Y\rangle + \langle d\bar J(e)(X),Y\rangle\\ &= \jmath[X,Y](x) + \langle d\bar J(e)(X),Y\rangle.\end{aligned}$$

(6) *If the group cohomology class $\tilde{r}$ of the Hamiltonian group action vanishes, then there exists a G-equivariant momentum mapping $J : M \to \mathfrak{g}^*$, i.e.,*

$$J(x.g) = \operatorname{Ad}(g)^* J(x).$$

Namely, let the group cohomology class be given by $\tilde{r} = [\bar{J}] \in H^1(G, \mathfrak{g}^*)$. Then $\bar{J} = \delta\alpha$ for some constant $\alpha \in \mathfrak{g}^*$. Then $J_1 = J - \alpha$ is a G-equivariant momentum mapping since $J_1(x.g) = J(x.g) - \alpha = \operatorname{Ad}(g)^* J(x) + \bar{J}(g) - \alpha = \operatorname{Ad}(g)^* J(x) + \delta\alpha(g) - \alpha = \operatorname{Ad}(g)^* J(x) - \operatorname{Ad}(g)^*\alpha = \operatorname{Ad}(g)^* J_1(x)$.

For $X, Y \in \mathfrak{g}$ and $g \in G$ we have

$$\langle \bar{J}(g), [X, Y] \rangle = -\bar{\jmath}(X, Y) + \bar{\jmath}(\operatorname{Ad}(g)X, \operatorname{Ad}(g)Y). \tag{7}$$

To see this, we use the cocycle property $\bar{J}(g_0 g_1) = \bar{J}(g_1) + \operatorname{Ad}(g_1)^* \bar{J}(g_0)$ from part (2) to get

$$\begin{aligned}
d\bar{J}(g)(T(\mu^g)X) &= \partial|_0 \bar{J}(\exp(tX)g) = \partial|_0 \big(\bar{J}(g) + \operatorname{Ad}(g)^* \bar{J}(\exp(tX))\big) \\
&= \operatorname{Ad}(g)^* d\bar{J}(e)X, \\
\langle \bar{J}(g), [X, Y] \rangle &= \partial|_0 \langle \bar{J}(g), \operatorname{Ad}(\exp(tX))Y \rangle = \partial|_0 \langle \operatorname{Ad}(\exp(tX))^* \bar{J}(g), Y \rangle \\
&= \partial|_0 \langle \bar{J}(g \exp(tX)) - \bar{J}(\exp(tX)), Y \rangle \\
&= \langle \partial|_0 \bar{J}(g \exp(tX) g^{-1} g) - \partial|_0 \bar{J}(\exp(tX)), Y \rangle \\
&= \langle \partial|_0 \bar{J}(\exp(t \operatorname{Ad}(g)X)g) - d\bar{J}(e)X, Y \rangle \\
&= \langle \operatorname{Ad}(g)^* d\bar{J}(e) \operatorname{Ad}(g)X - d\bar{J}(e)X, Y \rangle \\
&= \bar{\jmath}(\operatorname{Ad}(g)X, \operatorname{Ad}(g)Y) - \bar{\jmath}(X, Y).
\end{aligned}$$

34.9. Theorem. *Let $J : M \to \mathfrak{g}^*$ be a momentum mapping for a Hamiltonian right group action $r : M \times G \to M$ on a connected symplectic manifold (M, ω) with group 1-cocycle $\bar{J} : G \to \mathfrak{g}^*$ and Lie algebra 2-cocycle $\bar{\jmath} : \bigwedge^2 \mathfrak{g} \to \mathbb{R}$. Then we have:*

(1) *There is a unique affine right action of G on $\mathfrak{g}^*$ with linear part the coadjoint action,*

$$a^g = a^g_{\bar{J}} : \alpha \mapsto \operatorname{Ad}(g)^*\alpha + \bar{J}(g),$$

such that $J : M \to \mathfrak{g}^*$ is G-equivariant.

(2) *There is a Poisson structure on $\mathfrak{g}^*$, given by*

$$\{f, h\}_{\bar{\jmath}}(\alpha) = \langle \alpha, [df(\alpha), dh(\alpha)]_{\mathfrak{g}} \rangle + \bar{\jmath}(df(\alpha), dh(\alpha)),$$

which is invariant under the affine G-action $a_{\bar{J}}$ from (1) and has the property that the momentum mapping $J : (M, \omega) \to (\mathfrak{g}^*, \{\ ,\ \}_{\bar{\jmath}})$ is a Poisson morphism. The symplectic leaves of this Poisson structure are exactly the orbits under the connected component G_0 of e for the affine action in (1).

Proof. (1) By (34.8.1), J is G-equivariant. It remains to check that we have a right action:

$$\begin{aligned}
a^{g_2}a^{g_1}(\alpha) &= a^{g_2}(\operatorname{Ad}(g_1)^*\alpha + \bar J(g_1))\\
&= \operatorname{Ad}(g_2)^*\operatorname{Ad}(g_1)^*\alpha + \operatorname{Ad}(g_2)^*\bar J(g_1) + \bar J(g_2)\\
&= \operatorname{Ad}(g_1g_2)^*\alpha + \bar J(g_1g_2)\\
&= a^{g_1g_2}\alpha, \qquad \text{by (34.8.2)}.
\end{aligned}$$

(2) Let $X_1,\dots,X_n$ be a basis of $\mathfrak g$ with dual basis $\xi^1,\dots,\xi^n$ of $\mathfrak g^*$. Then we have in terms of the structure constants of the Lie algebra $\mathfrak g$

$$\begin{aligned}
[X_i,X_j] &= \sum_k c_{ij}^k X_k,\\
[\ \ ,\ \ ] &= \tfrac12\sum_{ijk} c_{ij}^k X_k\otimes(\xi^i\wedge\xi^j),\\
P^{\mathfrak g^*} &= -[\ \ ,\ \ ]^* = -\tfrac12\sum_{ijk} c_{ij}^k(\xi^i\otimes X_k)\wedge\xi^j,\\
\bar\jmath &= \tfrac12\sum_{ij}\bar\jmath_{ij}\xi^i\wedge\xi^j,\\
P_{\bar\jmath}^{\mathfrak g^*} &= -\tfrac12\sum_{ijk} c_{ij}^k(\xi^i\otimes X_k)\wedge\xi^j + \tfrac12\sum_{ij}\bar\jmath_{ij}\xi^i\wedge\xi^j : \mathfrak g^*\to\bigwedge^2\mathfrak g^*.
\end{aligned}$$

Let us now compute the Schouten bracket. We note that $[P^{\mathfrak g^*},P^{\mathfrak g^*}]=0$ since this is a Poisson structure, and $[\bar\jmath,\bar\jmath]=0$ since it is a constant 2-vector field on the vector space $\mathfrak g^*$. Then

$$\begin{aligned}
[P_{\bar\jmath}^{\mathfrak g^*},P_{\bar\jmath}^{\mathfrak g^*}] &= [P^{\mathfrak g^*}+\bar\jmath, P^{\mathfrak g^*}+\bar\jmath]\\
&= [P^{\mathfrak g^*},P^{\mathfrak g^*}] + 2[P^{\mathfrak g^*},\bar\jmath] + [\bar\jmath,\bar\jmath]\\
&= 0 + 2[P^{\mathfrak g^*},\bar\jmath] + 0\\
&= -\tfrac12\sum_{ijklm} c_{ij}^k\,\bar\jmath_{lm}\Big([\xi^i\otimes X_k,\xi^l]\wedge\xi^j\wedge\xi^m - [\xi^i\otimes X_k,\xi^m]\wedge\xi^j\wedge\xi^l\\
&\qquad - [\xi^j,\xi^l]\wedge(\xi^i\otimes X_k)\wedge\xi^m + [\xi^j,\xi^m]\wedge(\xi^i\otimes X_k)\wedge\xi^l\Big)\\
&= -\tfrac12\sum_{ijklm} c_{ij}^k\,\bar\jmath_{lm}\Big(-\delta_k^l\,\xi^i\wedge\xi^j\wedge\xi^m + \delta_k^m\,\xi^i\wedge\xi^j\wedge\xi^l - 0 + 0\Big)\\
&= \sum_{ijkm} c_{ij}^k\,\bar\jmath_{km}\,\xi^i\wedge\xi^j\wedge\xi^m = -2d\bar\jmath = 0.
\end{aligned}$$

which is zero since $\bar\jmath$ is a Lie algebra cocycle. Thus $P_{\bar\jmath}^{\mathfrak g^*}$ is a Poisson structure.

The Poisson structure $P^{\mathfrak{g}^*}_{\bar{\jmath}}$ is invariant under the affine action since

$$\begin{aligned}
\{f\circ a^g, h\circ a^g\}_{\bar{\jmath}}(\alpha) &= \langle \alpha, [df(a^g(\alpha)).T(a^g), dh(a^g(\alpha)).T(a^g)]\rangle \\
&\quad + \bar{\jmath}(df(a^g(\alpha)).T(a^g), dh(a^g(\alpha)).T(a^g)) \\
&= \langle \alpha, [df(a^g(\alpha)).\operatorname{Ad}(g)^*, dh(a^g(\alpha)).\operatorname{Ad}(g)^*]\rangle \\
&\quad + \bar{\jmath}(df(a^g(\alpha)).\operatorname{Ad}(g)^*, dh(a^g(\alpha)).\operatorname{Ad}(g)^*) \\
&= \langle \alpha, \operatorname{Ad}(g)[df(a^g(\alpha)), dh(a^g(\alpha))]\rangle + \bar{\jmath}(\operatorname{Ad}(g)df(a^g(\alpha)), \operatorname{Ad}(g)dh(a^g(\alpha))) \\
&= \langle \operatorname{Ad}(g)^*\alpha, [df(a^g(\alpha)), dh(a^g(\alpha))]\rangle + \langle \bar{J}(g), [df(a^g(\alpha)), dh(a^g(\alpha))]\rangle \\
&\quad + \bar{\jmath}(df(a^g(\alpha)), dh(a^g(\alpha))), \qquad \text{by (34.8.7)} \\
&= \langle a^g(\alpha), [df(a^g(\alpha)), dh(a^g(\alpha))]\rangle + \bar{\jmath}(df(a^g(\alpha)), dh(a^g(\alpha))) \\
&= \{f, g\}_{\bar{\jmath}}(a^g(\alpha)).
\end{aligned}$$

To see that the momentum mapping $J : (M,\omega) \to (\mathfrak{g}^*, P^{\mathfrak{g}^*}_{\bar{\jmath}})$ is a Poisson morphism, we have to show that $\bigwedge^2 dJ(x).P^\omega(x) = P^{\mathfrak{g}^*}_{\bar{\jmath}}(J(x)) \in \bigwedge^2 \mathfrak{g}^*$ for $x \in M$, according to (33.5.3). Recall from the definition (34.2.1) that $\langle J, X\rangle = j(X)$; thus also $\langle dJ(x), X\rangle = dj(X)(x) : T_xM \to \mathbb{R}$.

$$\begin{aligned}
\langle \bigwedge^2 dJ(x).P^\omega(x), X\wedge Y\rangle &= \langle P^\omega(x), \bigwedge^2 dJ(x)^*(X\wedge Y)\rangle \\
&= \langle P^\omega(x), dJ(x)^*X \wedge dJ(x)^*Y\rangle = \langle P^\omega(x), d\langle J, X\rangle \wedge d\langle J, Y\rangle\rangle \\
&= \langle P^\omega(x), dj(X)\wedge dj(Y)\rangle = 2\{j(X), j(Y)\}_\omega \\
&= 2\bar{\jmath}(X,Y) + 2j([X,Y])(x) \qquad\qquad \text{by (34.6.1)} \\
&= 2\langle J(x), [X,Y]\rangle + 2\bar{\jmath}(X,Y) \\
&= \langle P^{\mathfrak{g}^*}_{\bar{\jmath}}(J(x)), X\wedge Y\rangle.
\end{aligned}$$

It remains to investigate the symplectic leaves of the Poisson structure $P^{\mathfrak{g}^*}_{\bar{\jmath}}$. The fundamental vector fields for the twisted right action $a_{\bar{J}}$ is given by

$$\zeta^{a_{\bar{J}}}_X(\alpha) = \partial|_0(\operatorname{Ad}(\exp(tX))^*\alpha + \bar{J}(\exp(tX))) = \operatorname{ad}(X)^*\alpha + d\bar{J}(e)X.$$

This fundamental vector field is also the Hamiltonian vector field for the function $\operatorname{ev}_X : \mathfrak{g}^* \to \mathbb{R}$ since

$$\begin{aligned}
\text{(3)}\qquad H^{\bar{\jmath}}_{\operatorname{ev}_X}(f)(\alpha) &= \{\operatorname{ev}_X, f\}_{\bar{\jmath}}(\alpha) = \langle \alpha, [X, df(\alpha)]\rangle + \bar{\jmath}(X, df(\alpha)) \\
&= \langle \operatorname{ad}(X)^*\alpha, df(\alpha)\rangle + \langle d\bar{J}(e)X, df(\alpha)\rangle \\
&= \zeta^{a_{\bar{J}}}_X(f)(\alpha).
\end{aligned}$$

Hamiltonian vector fields of linear functions suffice to span the integrable distribution with jumping dimension which generates the symplectic leaves. Thus the symplectic leaves are exactly the orbits of the G_0-action $a_{\bar{J}}$. □

34.10. Corollary (Kostant, Souriau). *Let $J : M \to \mathfrak{g}^*$ be a momentum mapping for a transitive Hamiltonian right group action $r : M \times G \to M$ on a connected symplectic manifold (M, ω) with group 1-cocycle $\bar{J} : G \to \mathfrak{g}^*$ and Lie algebra 2-cocycle $\bar{\jmath} : \bigwedge^2 \mathfrak{g} \to \mathbb{R}$.*

Then the image $J(M)$ of the momentum mapping is an orbit O of the affine action $a_{\bar{J}}$ of G on $\mathfrak{g}^*$ for which J is equivariant, and the map $J : M \to O$ is locally a symplectomorphism and a covering mapping of O.

Proof. Since G acts transitively on M and J is G-equivariant, $J(M) = O$ is an orbit for the twisted action a_J of G on $\mathfrak{g}^*$. Since M is connected, O is connected and is thus a symplectic leaf of the twisted Poisson structure $P^{\mathfrak{g}^*}_{\bar{\jmath}}$ for which $J : M \to \mathfrak{g}^*$ is a Poisson mapping. Along O the Poisson structure is symplectic and its pullback via J equals ω; thus $T_xJ : T_xM \to T_{J(x)}O$ is invertible for each $x \in M$ and J is a local diffeomorphism. Since J is equivariant, it is diffeomorphic to a mapping $M \cong G/G_x \to G/G_{J(x)}$ and is thus a covering mapping. $\square$

34.11. Let us suppose that for some symplectic infinitesimal action of a Lie algebra $\zeta : \mathfrak{g} \to \mathfrak{X}(M, \omega)$ the cohomology class $\tilde{\zeta} = [\bar{\jmath}] \in H^2(\mathfrak{g}, H^0(M))$ does not vanish. Then we replace the Lie algebra $\mathfrak{g}$ by the *central extension*, see section (15),

$$0 \to H^0(M) \to \tilde{\mathfrak{g}} \to \mathfrak{g} \to 0$$

which is defined by $\tilde{\zeta} = [\bar{\jmath}]$ in the following way: $\tilde{\mathfrak{g}} = H^0(M) \times \mathfrak{g}$ with bracket $[(a, X), (b, Y)] := (\bar{\jmath}(X, Y), [X, Y])$. This satisfies the Jacobi identity since

$$\begin{aligned} [[(a, X), (b, Y)], (c, Z)] &= [(\bar{\jmath}(X, Y), [X, Y]), (c, Z)] \\ &= (\bar{\jmath}([X, Y], Z), [[X, Y], Z]) \end{aligned}$$

and the cyclic sum of this expression vanishes. The map $j_1 : \tilde{g} \to C^\infty(M)$, given by $j_1(a, X) = j(X) + a$, fits into the diagram

$$\begin{array}{ccccccccc} 0 \longrightarrow & H^0(M) & \xrightarrow{\alpha} & C^\infty(M) & \xrightarrow{H} & \mathfrak{X}(M, \omega) & \xrightarrow{\gamma} & H^1(M) & \longrightarrow 0 \\ & \| & & \uparrow{\scriptstyle j_1} & \nwarrow{\scriptstyle j} & \uparrow{\scriptstyle \zeta} & & & \\ 0 \longrightarrow & H^0(M) & \longrightarrow & \tilde{\mathfrak{g}} & \longrightarrow & \mathfrak{g} & \longrightarrow & 0 & \end{array}$$

and is a homomorphism of Lie algebras since

$$\begin{aligned} j_1([(a, X), (b, Y)]) &= j_1(\bar{\jmath}(X, Y), [X, Y]) = j([X, Y]) + \bar{\jmath}(X, Y) \\ &= j([X, Y]) + \{jX, jY\} - j([X, Y]) \\ &= \{jX, jY\} = \{jX + a, jY + b\} \\ &= \{j_1(a, X), j_1(b, Y)\}. \end{aligned}$$

In this case we can consider the momentum mapping

$$\begin{aligned}
J_1 : M \to \tilde{\mathfrak{g}}^* &= (H^0(M) \times \mathfrak{g})^*, \\
\langle J_1(x), (a, X)\rangle &= j_1(a, X)(x) \\
&= j(X)(x) + a, \\
H_{j_1(a,X)} &= \zeta_X, \quad x \in M, \quad X \in \mathfrak{g}, \quad a \in H^0(M),
\end{aligned}$$

which has all the properties of theorem (34.7).

Let us describe this in more detail. Property (34.7.1) says that for all $(a, X) \in H^0(M)\times\mathfrak{g}$ the vector fields $H_{j(X)+a} = \zeta_X \in \mathfrak{X}(M)$ and $\operatorname{ad}(a, X)^* \in \mathfrak{X}(\tilde{g}^*)$ are J_1-related. We have

$$\begin{aligned}
\langle \operatorname{ad}(a, X)^*(\alpha, \xi),(b, Y)\rangle &= \langle(\alpha, \xi), [(a, X)(b, Y)]\rangle \\
&= \langle(\alpha, \xi), (\bar{\jmath}(X, Y), [X, Y])\rangle \\
&= \alpha\bar{\jmath}(X, Y) + \langle\xi, [X, Y]\rangle \\
&= \alpha\bar{\jmath}(X, Y) + \langle\operatorname{ad}(X)^*\xi, Y\rangle \\
&= \langle(0, \alpha\bar{\jmath}(X,\quad) + \operatorname{ad}(X)^*\xi), (b, Y)\rangle, \\
\operatorname{ad}(a, X)^*(\alpha, \xi) &= (0, \alpha\bar{\jmath}(X,\quad) + \operatorname{ad}(X)^*\xi).
\end{aligned}$$

This is related to formula (34.9.3) which describes the infinitesimal twisted right action corresponding to the twisted group action of (34.9.1).

The Poisson bracket on $\tilde{\mathfrak{g}}^* = (H^0(M) \times \mathfrak{g})^* = H^0(M)^* \times \mathfrak{g}^*$ is given by

$$\begin{aligned}
\{f, h\}^{\tilde{\mathfrak{g}}^*}(\alpha, \xi) &= \langle(\alpha, \xi), [(d_1f(\alpha, \xi), d_2f(\alpha, \xi)), (d_1h(\alpha, \xi), d_2h(\alpha, \xi))]\rangle \\
&= \langle(\alpha, \xi), (\bar{\jmath}(d_2f(\alpha, \xi), d_2h(\alpha, \xi)), [d_2f(\alpha, \xi), d_2h(\alpha, \xi)])\rangle \\
&= \alpha\bar{\jmath}(d_2f(\alpha, \xi), d_2h(\alpha, \xi)) + \langle\xi, [d_2f(\alpha, \xi), d_2h(\alpha, \xi)]\rangle
\end{aligned}$$

which for $\alpha = 1$ and connected M is the twisted Poisson bracket in (34.9.2). We may continue and derive all properties of (34.9) for a connected Lie group from here, with some interpretation.

34.12. Symplectic reduction. Let $J : M \to \mathfrak{g}^*$ be a momentum mapping for a Hamiltonian right group action $r : M \times G \to M$ on a connected symplectic manifold (M, ω) with group 1-cocycle $\bar{J} : G \to \mathfrak{g}^*$ and Lie algebra 2-cocycle $\bar{\jmath} : \bigwedge^2 \mathfrak{g} \to \mathbb{R}$.

(1) ([**22**]) *A point $\alpha \in J(M) \subset \mathfrak{g}^*$ is called a weakly regular value for J if $J^{-1}(\alpha) \subset M$ is a submanifold such that for each $x \in J^{-1}(\alpha)$ we have $T_xJ^{-1}(\alpha) = \ker(T_xJ)$.*

This is the case if α is a regular value for J, or if J is of constant rank in a neighborhood of $J^{-1}(\alpha)$, by (1.13). Let us fix a weakly regular value $\alpha \in \mathfrak{g}^*$ of J for the following. The submanifold $J^{-1}(\alpha) \subset M$ then has the following properties:

(2) *For a weakly regular value α of J, the submanifold $J^{-1}(\alpha)$ is invariant under the action of the isotropy group $G_\alpha = \{g \in G : a_{\bar{J}}^g(\alpha) = \alpha\}$. The dimension of the the isotropy group G_x of $x \in J^{-1}(\alpha)$ does not depend on $x \in J^{-1}(\alpha)$ and is given by*

$$\dim(G_x) = \dim(G) - \dim(M) + \dim(J^{-1}(\alpha)).$$

Namely, $J : M \to \mathfrak{g}^*$ is equivariant for these actions by (34.9.1). Thus $J^{-1}(\alpha)$ is invariant under G_α and $G_x \subseteq G_\alpha$. For each $x \in J^{-1}(\alpha)$, by (34.3.4) we have $\operatorname{im}(dJ(x)) = \mathfrak{g}_x^\circ \subset \mathfrak{g}^*$. Since $T_x(J^{-1}(\alpha)) = \ker(dJ(x))$, we get

$$\begin{aligned}
\dim(T_xM) &= \dim(T_xJ^{-1}(\alpha)) + \operatorname{rank}(dJ(x)),\\
\dim(G_x) &= \dim(G) - \dim(x.G)\\
&= \dim(G) - \dim(\mathfrak{g}_x^\circ)\\
&= \dim(G) - \operatorname{rank}(dJ(x))\\
&= \dim(G) - \dim(M) + \dim(J^{-1}(\alpha)).
\end{aligned}$$

(3) *At any $x \in J^{-1}(\alpha)$ the kernel of the pullback $\omega^{J^{-1}(\alpha)}$ of the symplectic form ω equals $T_x(x.G_\alpha)$ and its rank is constant and is given by*

$$\operatorname{rank}(\omega^{J^{-1}(\alpha)}) = 2\dim(J^{-1}(\alpha)) + \dim(a_{\bar{J}}^G(\alpha)) - \dim(M).$$

Namely, $T_xJ^{-1}(\alpha) = \ker(dJ(x))$ implies

$$\begin{aligned}
\ker(\omega^{J^{-1}(\alpha)}) &= T_x(J^{-1}(\alpha)) \cap T_x(J^{-1}(\alpha))^\perp\\
&= T_x(J^{-1}(\alpha)) \cap \ker(dJ(x))^\perp\\
&= T_x(J^{-1}(\alpha)) \cap T_x(x.G), \qquad \text{by (34.3.3)}\\
&= T_x(x.G_\alpha),
\end{aligned}$$

$$\begin{aligned}
\operatorname{rank}(\omega_x^{J^{-1}(\alpha)}) &= \dim(J^{-1}(\alpha)) - \dim(x.G_\alpha)\\
&= \dim(J^{-1}(\alpha)) - \dim(G_\alpha) + \dim(G_x)\\
&= \dim(J^{-1}(\alpha)) - \dim(G_\alpha) + \dim(G) - \dim(M) + \dim(J^{-1}(\alpha)) \text{ by (2)}\\
&= 2\dim(J^{-1}(\alpha)) + \dim(a_{\bar{J}}^G(\alpha)) - \dim(M).
\end{aligned}$$

(4) *If α is a regular value of $J : M \to \mathfrak{g}^*$, the action of G on M is locally free in a neighborhood of every point $x \in J^{-1}(\alpha)$, by (34.3.5), i.e., the isotropy group G_x is discrete.*

This follows from $\operatorname{codim}_M(J^{-1}(\alpha)) = \dim(\mathfrak{g}) - \dim(G)$.

34.13. Theorem. Weakly regular symplectic reduction. *Consider a momentum mapping $J : M \to \mathfrak{g}^*$ for a Hamiltonian right group action $r : M \times G \to M$ on a connected symplectic manifold (M, ω) with group 1-cocycle $\bar{J} : G \to \mathfrak{g}^*$ and Lie algebra 2-cocycle $\bar{\jmath} : \bigwedge^2 \mathfrak{g} \to \mathbb{R}$. Let $\alpha \in J(M) \subset \mathfrak{g}^*$ be a weakly regular value of J.*

Then the pullback 2-form $\omega^{J^{-1}(\alpha)} \in \Omega^2(J^{-1}(\alpha))$ of ω is of constant rank, invariant under the action of G_α, and the leaves of the foliation described by its kernel are the orbits of the action of the connected component G_α^0 of the isotropy group $G_\alpha := \{g \in G : a^g_{\bar{J}}(\alpha) = \alpha\}$ in $J^{-1}(\alpha)$.

If moreover the orbit space $M_\alpha := J^{-1}(\alpha)/G_\alpha^0$ is a smooth manifold, then there exists a unique symplectic form ω^α on it such that for the canonical projection $\pi : J^{-1}(\alpha) \to M_\alpha$ we have $\pi^*\omega^\alpha = \omega^{J^{-1}(\alpha)}$.

Let $h \in C^\infty(M)^G$ be a Hamiltonian function on M which is G-invariant; then $h|J^{-1}(\alpha)$ factors to $\bar{h} \in C^\infty(M_\alpha)$ with $\bar{h} \circ \pi = h|J^{-1}(\alpha)$. The Hamiltonian vector field $\operatorname{grad}^\omega(h) = H_h$ *is tangent to $J^{-1}(\alpha)$ and the vector fields $H_h|J^{-1}(\alpha)$ and $H_{\bar{h}}$ are π-related. Thus their trajectories are mapped onto each other:*

$$\pi(\mathrm{Fl}_t^{H_h}(x)) = \mathrm{Fl}_t^{H_{\bar{h}}}(\pi(x)).$$

In this case we call $(M_\alpha = J^{-1}(\alpha)/G_\alpha^0, \omega^\alpha)$ the *reduced symplectic manifold.*

Proof. By (34.12.3) the 2-form $\omega^{J^{-1}(\alpha)} \in \Omega^2(J^{-1}(\alpha))$ is of constant rank and the foliation corresponding to its kernel is given by the orbits of the unit component G_α^0 of the isotropy group G_α. Let us now suppose that the orbit space $M_\alpha = J^{-1}(\alpha)/G_\alpha^0$ is a smooth manifold. Since the 2-form $\omega^{J^{-1}(\alpha)}$ is G_α^0-invariant and horizontal for the projection

$$\pi : J^{-1}(\alpha) \to J^{-1}(\alpha)/G_\alpha^0 = M_\alpha,$$

it factors to a smooth 2-form $\omega^\alpha \in \Omega^2(M_\alpha)$ which is closed and nondegenerate since we just factored out its kernel. Thus $(M_\alpha, \omega^\alpha)$ is a symplectic manifold and $\pi^*\omega^\alpha = \omega^{J^{-1}(\alpha)}$ by construction.

Now let $h \in C^\infty(M)$ be a Hamiltonian function which is invariant under G. By E. Noether's theorem (34.3.8) the momentum mapping J is constant along each trajectory of the Hamiltonian vector field H_h; thus H_h is tangent to $J^{-1}(\alpha)$ and G_α-invariant on $J^{-1}(\alpha)$. Let $\bar{h} \in C^\infty(M_\alpha)$ be the factored function with $\bar{h} \circ \pi = h$, and consider $H_{\bar{h}} \in \mathfrak{X}(M_\alpha, \omega^\alpha)$. Then for $x \in J^{-1}(\alpha)$ we have

$$(T_x\pi)^*(i_{T_x\pi.H_h(x)}\omega^\alpha) = i_{H_h(x)}\pi^*\omega^\alpha = dh(x) = (T_x\pi)^*(d\bar{h}(\pi(x))).$$

Since $(T_x\pi)^* : T^*_{\pi(x)}M_\alpha \to T_x(J^{-1}(\alpha))$ is injective, we see that $i_{T_x\pi.H_h(x)}\omega^\alpha = d\bar{h}(\pi(x))$ and hence $T_x\pi.H_h(x) = H_{\bar{h}}(\pi(x))$. Thus $H_h|J^{-1}(\alpha)$ and $H_{\bar{h}}$ are π-related and the remaining assertions follow from (3.14). □

34.14. Proposition. Constant rank symplectic reduction. *Consider a momentum mapping $J : M \to \mathfrak{g}^*$ for a Hamiltonian right group action $r : M \times G \to M$ on a connected symplectic manifold (M, ω) with group 1-cocycle $\bar{J} : G \to \mathfrak{g}^*$ and Lie algebra 2-cocycle $\bar{\jmath} : \bigwedge^2 \mathfrak{g} \to \mathbb{R}$. Let G be connected. Let $\alpha \in J(M) \subset \mathfrak{g}^*$ be such that J has constant rank in a neighborhood of $J^{-1}(\alpha)$. We consider the orbit $\alpha.G = a^G_{\bar{J}}(\alpha) \subset \mathfrak{g}^*$.*

(1) *$J^{-1}(\alpha.G) \subset M$ is an initial G-invariant submanifold.*

(2) *The smooth map $J^{-1}(\alpha) \times G \to J^{-1}(\alpha.G)$, $(x, g) \mapsto x.g$ factors to a diffeomorphism $J^{-1}(\alpha) \times_{G_\alpha} G \cong J^{-1}(\alpha.G)$.*

(3) *Let $\iota : J^{-1}(\alpha.G) \to M$ be the inclusion. Then $\iota^*\omega - \iota^* J^* \omega^{\overline{J}}$ is closed, of constant rank and G-invariant. The leaves of the foliation described by its kernel are the orbits of the G-action restricted to $J^{-1}(\alpha.G)$. Here $\omega^{\overline{J}}$ is the symplectic structure on the affine orbit $\alpha.G$ from theorem (34.9.2).*

(4) *If $M_{\alpha.G} := J^{-1}(\alpha.G)/G$ is a manifold, then $\iota^*\omega - \iota^* J^* \omega^{\overline{J}}$ factors to a symplectic form $\omega^{M_{\alpha.G}}$ on $M_{\alpha.G}$ which is thus characterized by*

$$\iota^*\omega = \pi^*\omega^{M_{\alpha.G}} + (J \circ \iota)^* \omega^{\overline{J}}$$

where $\pi : J^{-1}(\alpha.G) \to M_{\alpha.G}$ is the projection.

(5) *The orbit spaces $J^{-1}(\alpha)/G_\alpha$ and $M_{\alpha.G}$ are homeomorphic, and they are symplectomorphic if one of the orbit spaces is a manifold.*

(6) *Let $h \in C^\infty(M)^G$ be a G-invariant Hamiltonian function on M. Then $h|J^{-1}(\alpha.G)$ factors to $\bar{h} \in C^\infty(M_\alpha)$ with $\bar{h} \circ \pi = h|J^{-1}(\alpha.G)$. The Hamiltonian vector field $\operatorname{grad}^\omega(f) = H_h$ is tangent to $J^{-1}(\alpha.G)$ and the vector fields $H_h|J^{-1}(\alpha.G)$ and $H_{\bar{h}}$ are π-related. Thus their trajectories are mapped onto each other:*

$$\pi(\mathrm{Fl}^{H_h}_t(x)) = \mathrm{Fl}^{H_{\bar{h}}}_t(\pi(x)).$$

Proof. (1) Let $\alpha \in J(M) \subset \mathfrak{g}^*$ be such that J is of constant rank on a neighborhood of $J^{-1}(\alpha)$. Let $\alpha.G = a^G_{\bar{J}}(\alpha)$ be the orbit though α under the twisted coadjoint action. Then $J^{-1}(\alpha.G) = J^{-1}(\alpha).G$ by the G-equivariance of J. Thus the dimension of the isotropy group G_x of a point $x \in J^{-1}(\alpha.G)$ does not depend on x and is given by (34.12.2). It remains to show that the inverse image $J^{-1}(\alpha.G)$ is an initial submanifold which is invariant under G.

If α is a regular value for J, then J is a submersion on an open neighborhood of $J^{-1}(\alpha.G)$ and $J^{-1}(\alpha.G)$ is an initial submanifold by lemma (2.16).

Under the weaker assumption that J is of constant rank on a neighborhood of $J^{-1}(\alpha)$, we will construct an initial submanifold chart as in (2.13.1) centered at each $x \in J^{-1}(\alpha.G)$. Using a suitable transformation in G, we may assume without loss that $x \in J^{-1}(\alpha)$. We shall use the method of the proof of theorem (3.25).

Let $m = \dim(M)$, $n = \dim(\mathfrak{g})$, $r = \operatorname{rank}(dJ(x))$, $p = m - r = \dim(J^{-1}(\alpha))$ and $k = \dim(\alpha.G) \leq l = \dim(x.G)$. Using that $\mathfrak{g}_x \subseteq \mathfrak{g}_\alpha$, we choose a basis $X_1, \dots, X_n$ of $\mathfrak{g}$ such that

- $\zeta^{\mathfrak{g}^*}_{X_1}(\alpha), \dots, \zeta^{\mathfrak{g}^*}_{X_k}(\alpha)$ is a basis of $T_\alpha(\alpha.G)$ and $X_{k+1}, \dots, X_n$ is a basis of the isotropy algebra $\mathfrak{g}_\alpha$,
- $\zeta^M_{X_1}(x), \dots, \zeta^M_{X_l}(x)$ is a basis of $T_x(x.G)$ and $X_{l+1}, \dots, X_n$ is a basis of the isotropy algebra $\mathfrak{g}_x$.

By the constant rank theorem (1.13) there exists a chart (U, u) on M centered at x and a chart (V, v) on $\mathfrak{g}^*$ centered at α such that

$$v \circ J \circ u^{-1} : u(U) \to v(V)$$

has the following form:

$$(x^1, \dots, x^m) \mapsto (x^1, \dots, x^k, x^{k+1}, \dots, x^r, 0, \dots, 0),$$

and we may also assume that

- $\zeta^{\mathfrak{g}^*}_{X_1}(\alpha), \dots, \zeta^{\mathfrak{g}^*}_{X_k}(\alpha), \frac{\partial}{\partial v^{k+1}}|_\alpha, \dots, \frac{\partial}{\partial v^n}|_\alpha$ is a basis of $T_\alpha(\mathfrak{g}^*)$,
- $\zeta^M_{X_1}(x), \dots, \zeta^M_{X_l}(x), \frac{\partial}{\partial u^{l+1}}|_x, \dots, \frac{\partial}{\partial u^m}|_x$ is a basis of $T_x(M)$.

Then the mapping

$$f(y^1, \dots, y^n) = (\mathrm{Fl}^{\zeta^{\mathfrak{g}^*}_{X_1}}_{y^1} \circ \dots \circ \mathrm{Fl}^{\zeta^{\mathfrak{g}^*}_{X_k}}_{y^k} \circ v^{-1})(0, \dots, 0, y^{k+1}, \dots, y^n)$$

is a diffeomorphism from a neighborhood of 0 in $\mathbb{R}^n$ onto a neighborhood of α in $\mathfrak{g}^*$. Let $(\tilde{V}, \tilde{v})$ be the chart f^{-1}, suitably restricted. We have

$$\beta \in \alpha.G \iff (\mathrm{Fl}^{\zeta^{\mathfrak{g}^*}_{X_1}}_{y^1} \circ \dots \circ \mathrm{Fl}^{\zeta^{\mathfrak{g}^*}_{X_k}}_{y^k})(\beta) \in \alpha.G$$

for all β and all $y^1, \dots, y^k$ for which both expressions make sense. So we have

$$f(y^1, \dots, y^n) \in \alpha.G \iff f(0, \dots, 0, y^{k+1}, \dots, y^n) \in \alpha.G,$$

and consequently $\alpha.G \cap \tilde{V}$ is the disjoint union of countably many connected sets of the form $\{\beta \in \tilde{V} : (\tilde{v}^{k+1}(\beta), \dots, \tilde{v}^n(\beta)) = \text{ constant}\}$, since $\alpha.G$ is second countable.

Now let us consider the situation on M. Since $J^{-1}(\alpha)$ is G_α-invariant, exactly the vectors $\zeta^M_{X_{k+1}}(x), \dots, \zeta^M_{X_l}(x)$ are tangent to $x.G_\alpha \subseteq J^{-1}(\alpha)$. The mapping

$$g(x^1, \dots, x^m) = (\mathrm{Fl}^{\zeta^M_{X_1}}_{x^1} \circ \dots \circ \mathrm{Fl}^{\zeta^M_{X_k}}_{x^k} \circ u^{-1})(0, \dots, 0, x^{k+1}, \dots, x^m)$$

is a diffeomorphisms from a neighborhood of 0 in $\mathbb{R}^m$ onto a neighborhood of x in M. Let $(\tilde{U}, \tilde{u})$ be the chart g^{-1}, suitably restricted. By G-invariance of J we have

$$\begin{aligned}(J \circ g)(x^1, \dots, x^m) &= (J \circ \mathrm{Fl}^{\zeta^M_{X_1}}_{x^1} \circ \dots \circ \mathrm{Fl}^{\zeta^M_{X_k}}_{x^k} \circ u^{-1})(0, \dots, 0, x^{k+1}, \dots, x^m) \\ &= (\mathrm{Fl}^{\zeta^{\mathfrak{g}^*}_{X_1}}_{x^1} \circ \dots \circ \mathrm{Fl}^{\zeta^{\mathfrak{g}^*}_{X_k}}_{x^k} \circ v^{-1} \circ v \circ J \circ u^{-1})(0, \dots, 0, x^{k+1}, \dots, x^m) \\ &= (\mathrm{Fl}^{\zeta^{\mathfrak{g}^*}_{X_1}}_{x^1} \circ \dots \circ \mathrm{Fl}^{\zeta^{\mathfrak{g}^*}_{X_k}}_{x^k} \circ v^{-1})(0, \dots, 0, x^{k+1}, \dots, x^r, 0, \dots, 0) \\ &= f(x^1, \dots, x^k, x^{k+1}, \dots, x^r, 0, \dots, 0)\end{aligned}$$

and thus

$$\begin{aligned}&g(x^1, \dots, x^m) \in J^{-1}(\alpha.G) \\ &\iff (J \circ g)(x^1, \dots, x^m) = f(x^1, \dots, x^k, x^{k+1}, \dots, x^r, 0, \dots, 0) \in \alpha.G \\ &\iff f(0_{\mathbb{R}^k}, x^{k+1}, \dots, x^r, 0_{\mathbb{R}^{n-r}}) \in \alpha.G.\end{aligned}$$

Consequently, $(J^{-1}(\alpha.G)) \cap \tilde{U}$ is the disjoint union of countably many connected sets of the form $\{x \in \tilde{U} : (\tilde{u}^{k+1}(x), \dots, \tilde{u}^r(x)) = \text{constant}\}$, since $\alpha.G$ is second countable. We have proved now that $J^{-1}(\alpha.G)$ is an initial submanifold or M.

(2) The induced map $J^{-1}(\alpha) \times_{G_\alpha} G \to J^{-1}(\alpha.G)$, $[(x, g)] \mapsto x.g$ is a bijective submersion, and thus a diffeomorphism.

(3) Let $x \in J^{-1}(\alpha)$ and $X, Y \in \mathfrak{g}$. Then

$$\begin{aligned}(\iota^*\omega)_x(\zeta_X(x), \zeta_Y(x)) &= \omega_x(H_{jX}(x), H_{jY}(x)) \\ &= -\{jX, jY\}(x) && \text{by (31.21)} \\ &= -\{\mathrm{ev}_X, \mathrm{ev}_Y\}^{\overline{J}}(\alpha) && \text{by (34.9.2)} \\ &= \omega^{\overline{J}}_\alpha(\zeta^{a_{\overline{J}}}_X(\alpha), \zeta^{a_{\overline{J}}}_Y(\alpha)) && \text{by (34.9.3)}\end{aligned} \tag{7}$$

where $\omega^{\overline{J}}$ is the symplectic structure from (34.9.2) on the affine orbit $\alpha.G$. Let $\xi_1, \xi_2 \in T_x J^{-1}(\alpha.G)$. By (2) we may (nonuniquely) decompose ξ_i as $\xi_i = \eta_i + \zeta_{X_i}(x) \in T_x J^{-1}(\alpha) + T_x(x.G)$, where $i = 1, 2$. By (34.3.3) and (7) we have

$$(\iota^*\omega)_x(\xi_1, \xi_2) = \omega^{J^{-1}(\alpha)}_x(\eta_1, \eta_2) + (\iota^* J^* \omega^{\overline{J}})_x(\zeta_{X_1}(x), \zeta_{X_2}(x))$$

where we also use the notation from theorem (34.13). Thus

$$\ker(\iota^*\omega - \iota^* J^* \omega^{\overline{J}})_x = T_x(x.G) + \ker \omega^{J^{-1}(\alpha)}_x = T_x(x.G)$$

by (34.12.3). Therefore, $\iota^*\omega - \iota^* J^* \omega^{\overline{J}}$ is closed, G-invariant and the leaves of the foliation described by its kernel coincide with the orbits of the G-action. By (34.12.2) this form is also of constant rank.

(4) follows immediately from (3).

(5) By (2) the orbit spaces in question are homeomorphic and diffeomorphic if one of them is a manifold. In the latter case they are also symplectomorphic because of the formula in (4).

(6) Hamiltonian reduction follows similarly as in theorem (34.13). □

34.15. Example: Coadjoint orbits. Let G be a Lie group acting upon itself by inversion of left multiplication, i.e., $x.g = g^{-1}x$. Consider T^*G with its canonical symplectic structure ω_G from (31.9). The cotangent lifted action by G on $T^*G = G \times \mathfrak{g}^*$ (trivialized via left multiplication) is given by $(x, \alpha).g = (g^{-1}x, \alpha)$. According to (34.6.3) this action is strongly Hamiltonian with momentum mapping given by

$$\langle J(x, \alpha), X\rangle = \langle \alpha, \zeta_X(x)\rangle = \langle -\operatorname{Ad}^*(x^{-1}).\alpha, X\rangle$$

where $X \in \mathfrak{g}$. The G action is free whence all points of $\mathfrak{g}^*$ are regular values for J. Let $O \subset \mathfrak{g}^*$ be a coadjoint orbit. Then $J^{-1}(O) = G \times (-O)$ and $\iota^*\omega_G - (J \circ \iota)^*\omega_O$ is basic with respect to the projection $J^{-1}(O) \to J^{-1}(O)/G = -O$. (Here $\iota : J^{-1}(O) \to G \times \mathfrak{g}^*$ is the inclusion and ω_O is the coadjoint orbit symplectic form from (31.14).) The reduced symplectic space is thus given by $(-O, -\omega_O) \cong (O, \omega_O)$.

If we consider the action by G on itself given by right multiplication, we see that $(O, -\omega_O)$ is the symplectic reduction of (T^*G, ω_G).

34.16. Example of a symplectic reduction: The space of Hermitian matrices. Let $G = SU(n)$ act on the space $H(n)$ of complex Hermitian $(n \times n)$-matrices by conjugation, where the inner product is given by the (always real) trace $\operatorname{Tr}(AB)$. We also consider the linear subspace $\Sigma \subset H(n)$ of all diagonal matrices; they have real entries. For each Hermitian matrix A there exists a unitary matrix g such that gAg^{-1} is diagonal with eigenvalues decreasing in size. Thus a fundamental domain (we will call it a chamber) for the group action is here given by the quadrant $C \subset \Sigma$ consisting of all real diagonal matrices with eigenvalues $\lambda_1 \geq \lambda_2 \geq \dots \geq \lambda_n$. There are no further identifications in the chamber; thus $H(n)/SU(n) \cong C$.

We are interested in the following problem: Consider a straight line $t \mapsto A + tV$ of Hermitian matrices. We want to describe the corresponding curve of eigenvalues $t \mapsto \lambda(t) = (\lambda_1(t) \geq \dots \geq \lambda_n(t))$ of the Hermitian matrix $A + tV$ as precisely as possible. In particular, we want to find an ordinary differential equation describing the evolution of eigenvalues. We follow here the development in [**4**] which was inspired by [**103**].

(1) *Hamiltonian description.* Let us describe the curves of eigenvalues as trajectories of a Hamiltonian system on a reduced phase space. Let $T^*H(n) = H(n) \times H(n)$ be the cotangent bundle where we identified $H(n)$ with its

dual by the inner product, so the duality is given by $\langle \alpha, A\rangle = \operatorname{Tr}(A\alpha)$. Then the canonical 1-form is given by $\vartheta(A,\alpha,A',\alpha') = \operatorname{Tr}(\alpha A')$, the symplectic form is $\omega_{(A,\alpha)}((A',\alpha'),(A'',\alpha'')) = \operatorname{Tr}(A'\alpha'' - A''\alpha')$, and the Hamiltonian function for the straight lines $(A+t\alpha,\alpha)$ on $H(n)$ is $h(A,\alpha) = \frac{1}{2}\operatorname{Tr}(\alpha^2)$. The action $SU(n) \ni g \mapsto (A \mapsto gAg^{-1})$ lifts to the action $SU(n) \ni g \mapsto ((A,\alpha) \mapsto (gAg^{-1}, g\alpha g^{-1}))$ on $T^*H(n)$ with fundamental vector fields $\zeta_X(A,\alpha) = (A,\alpha,[X,A],[X,\alpha])$ for $X \in \mathfrak{su}(n)$, and with generating functions $j_X(A,\alpha) = \vartheta(\zeta_X(A,\alpha)) = \operatorname{Tr}(\alpha[X,A]) = \operatorname{Tr}([A,\alpha]X)$. Thus the momentum mapping $J : T^*H(n) \to \mathfrak{su}(n)^*$ is given by $\langle X, J(A,\alpha)\rangle = j_X(A,\alpha) = \operatorname{Tr}([A,\alpha]X)$. If we identify $\mathfrak{su}(n)$ with its dual via the inner product $\operatorname{Tr}(XY)$, the momentum mapping is $J(A,\alpha) = [A,\alpha]$. Along the line $t \mapsto A+t\alpha$ the momentum mapping is constant: $J(A+t\alpha,\alpha) = [A,\alpha] = Y \in \mathfrak{su}(n)$. Note that for $X \in \mathfrak{su}(n)$ the evaluation on X of $J(A+t\alpha,\alpha) \in \mathfrak{su}(n)^*$ equals the inner product:

$$\langle X, J(A+t\alpha,\alpha)\rangle = \operatorname{Tr}(\tfrac{d}{dt}(A+t\alpha), \zeta_X(A+t\alpha)),$$

which is obviously constant in t; compare with the general result of Riemann transformation groups (30.1).

According to principles of symplectic reduction (34.12) we have to consider for a regular value Y (and later for an arbitrary value) of the momentum mapping J the submanifold $J^{-1}(Y) \subset T^*H(n)$. The null distribution of $\omega|J^{-1}(Y)$ is integrable (with constant dimensions since Y is a regular value) and its leaves are exactly the orbits in $J^{-1}(Y)$ of the isotropy group $SU(n)_Y$ for the coadjoint action, by (34.13). So we have to consider the orbit space $J^{-1}(Y)/SU(n)_Y$. If Y is not a regular value of J, the inverse image $J^{-1}(Y)$ is a subset which is described by polynomial equations since J is polynomial (in fact quadratic), so $J^{-1}(Y)$ is stratified into submanifolds; symplectic reduction works also for this case; see [**209**], [**16**], or [**184**].

(2) *The case of momentum $Y = 0$ gives billiard of straight lines in C, reflected at the walls.* If $Y = 0$, then $SU(n)_Y = SU(n)$ and $J^{-1}(0) = \{(A,\alpha) : [A,\alpha] = 0\}$, so A and α commute. If A is regular (i.e., all eigenvalues are distinct), using a uniquely determined transformation $g \in SU(n)$, we move the point A into the open chamber $C^o \subset H(n)$, so $A = \operatorname{diag}(a_1 > a_2 > \cdots > a_n)$ and since α commutes with A, it is also in diagonal form. The symplectic form ω restricts to the canonical symplectic form on $C^o \times \Sigma = C^o \times \Sigma^* = T^*(C^o)$. Thus symplectic reduction gives $(J^{-1}(0) \cap (T^*H(n))_{\text{reg}})/SU(n) = T^*(C^o) \subset T^*H(n)$. By [**209**] we also use symplectic reduction for nonregular A and we get (see in particular [**121**, 3.4]) $J^{-1}(0)/SU(n) = T^*C$, the stratified cotangent cone bundle of the chamber C considered as stratified space. Namely, if one root $\varepsilon_i(A) = a_i - a_{i+1}$ vanishes on the diagonal matrix A, then the isotropy group $SU(n)_A$ contains a subgroup $SU(2)$ corresponding to these coordinates. Any matrix α with $[A,\alpha] = 0$ contains an arbitrary

Hermitian submatrix corresponding to the coordinates i and $i+1$, which may be brought into diagonal form with the help of this $SU(2)$ so that $\varepsilon_i(\alpha) = \alpha_i - \alpha_{i+1} \geq 0$. Thus the tangent vector α with foot point in a wall is either tangent to the wall (if $\alpha_i = \alpha_{i+1}$) or points into the interior of the chamber C. The Hamiltonian h restricts to $C^o \times \Sigma \ni (A, \alpha) \mapsto \frac{1}{2}\sum_i \alpha_i^2$, so the trajectories of the Hamiltonian system here are again straight lines which are reflected at the walls.

(3) *The case of general momentum* Y. If $Y \neq 0 \in \mathfrak{su}(n)$ and if $SU(n)_Y$ is the isotropy group of Y for the adjoint representation, then by the references at the end of (1) (concerning the singular version of (34.14) with stratified orbit space) we may pass from Y to the coadjoint orbit $\mathcal{O}(Y) = \mathrm{Ad}^*(SU(n))(Y)$ and get

$$J^{-1}(Y)/SU(n)_Y = J^{-1}(\mathcal{O}(Y))/SU(n),$$

where the (stratified) diffeomorphism is symplectic.

(4) *The Calogero-Moser system.* As the simplest case we assume that $Y' \in \mathfrak{su}(n)$ is not zero but has maximal isotropy group, and we follow [**103**]. So we assume that Y' has complex rank 1 plus an imaginary multiple of the identity, $Y' = \sqrt{-1}(c\mathbb{I}_n + v \otimes v^*)$ for $0 \neq v = (v^i)$ a column vector in $\mathbb{C}^n$. The coadjoint orbit is then $\mathcal{O}(Y') = \{\sqrt{-1}(c\mathbb{I}_n + w \otimes w^*) : w \in \mathbb{C}^n, |w| = |v|\}$, isomorphic to $S^{2n-1}/S^1 = \mathbb{C}P^n$, of real dimension $2n-2$. Consider (A', α') with $J(A', \alpha') = Y'$, choose $g \in SU(n)$ such that $A = gA'g^{-1} = \mathrm{diag}(a_1 \geq a_2 \geq \cdots \geq a_n)$, and let $\alpha = g\alpha'g^{-1}$. Then the entry of the commutator is $[A, \alpha]_{ij} = \alpha_{ij}(a_i - a_j)$. So $[A, \alpha] = gY'g^{-1} =: Y = \sqrt{-1}(c\mathbb{I}_n + gv \otimes (gv)^*) = \sqrt{-1}(c\mathbb{I}_n + w \otimes w^*)$ has zero diagonal entries; thus $0 < w^i\bar{w}^i = -c$ and $w^i = \exp(\sqrt{-1}\vartheta_i)\sqrt{-c}$ for some ϑ_i. But then all off-diagonal entries $Y_{ij} = \sqrt{-1}w^i\bar{w}^j = -\sqrt{-1}\,c\exp(\sqrt{-1}(\vartheta_i - \vartheta_j)) \neq 0$, and A has to be regular. We may use the remaining gauge freedom in the isotropy group $SU(n)_A = S(U(1)^n)$ to put $w^i = \exp(\sqrt{-1}\vartheta)\sqrt{-c}$ where $\vartheta = \sum \vartheta_i$. Then $Y_{ij} = -c\sqrt{-1}$ for $i \neq j$.

So the reduced space $(T^*H(n))_Y$ is diffeomorphic to the submanifold of $T^*H(n)$ consisting of all $(A, \alpha) \in H(n) \times H(n)$ where $A = \mathrm{diag}(a_1 > a_2 > \cdots > a_n)$ and where α has arbitrary diagonal entries $\alpha_i := \alpha_{ii}$ and off-diagonal entries $\alpha_{ij} = Y_{ij}/(a_i - a_j) = -c\sqrt{-1}/(a_i - a_j)$. We can thus use $a_1, \ldots, a_n, \alpha_1, \ldots, \alpha_n$ as coordinates. The invariant symplectic form pulls back to $\omega_{(A,\alpha)}((A'\alpha'), (A'', \alpha'')) = \mathrm{Tr}(A'\alpha'' - A''\alpha') = \sum(a_i'\alpha_i'' - a_i''\alpha_i')$. The invariant Hamiltonian h restricts to the Hamiltonian

$$h(A, \alpha) = \tfrac{1}{2}\mathrm{Tr}(\alpha^2) = \frac{1}{2}\sum_i \alpha_i^2 + \frac{1}{2}\sum_{i \neq j} \frac{c^2}{(a_i - a_j)^2}.$$

This is the famous Hamiltonian function of the Calogero-Moser completely integrable system; see [**167**], [**179**], [**103**], and [**192**, 3.1 and 3.3]. The corresponding Hamiltonian vector field and the differential equation for the eigenvalue curve are then

$$H_h = \sum_i \alpha_i \frac{\partial}{\partial a_i} + 2\sum_i \sum_{j:j\neq i} \frac{c^2}{(a_i - a_j)^3} \frac{\partial}{\partial \alpha_i},$$

$$\ddot{a}_i = 2\sum_{j\neq i} \frac{c^2}{(a_i - a_j)^3},$$

$$(a_i - a_j)^{\cdot\cdot} = 2\sum_{k:k\neq i} \frac{c^2}{(a_i - a_k)^3} - 2\sum_{k:k\neq j} \frac{c^2}{(a_j - a_k)^3}.$$

Note that the curve of eigenvalues avoids the walls of the Weyl chamber C.

(5) *Degenerate cases of nonzero momenta of minimal rank.* Let us discuss now the case of nonregular diagonal A. Namely, if one root, say $\varepsilon_{12}(A) = a_1 - a_2$, vanishes on the diagonal matrix A, then the isotropy group $SU(n)_A$ contains a subgroup $SU(2)$ corresponding to these coordinates. Consider α with $[A, \alpha] = Y$; then $0 = \alpha_{12}(a_1 - a_2) = Y_{12}$. Thus α contains an arbitrary Hermitian submatrix corresponding to the first two coordinates, which may be brought into diagonal form with the help of this $SU(2) \subset SU(n)_A$ so that $\varepsilon_{12}(\alpha) = \alpha_1 - \alpha_2 \geq 0$. Thus the tangent vector α with foot point A in a wall is either tangent to the wall (if $\alpha_1 = \alpha_2$) or points into the interior of the chamber C (if $\alpha_1 > \alpha_2$). Note that then $Y_{11} = Y_{22} = Y_{12} = 0$.

Let us now assume that the momentum Y is of the form $Y = \sqrt{-1}(c\mathbb{I}_{n-2} + v \otimes v^*)$ for some vector $0 \neq v \in \mathbb{C}^{n-2}$. We can repeat the analysis of (4) in the subspace $\mathbb{C}^{n-2}$ and get for the Hamiltonian function (where $I_{1,2} = \{(i,j) : i \neq j\} \setminus \{(1,2),(2,1)\}$)

$$h(A, \alpha) = \tfrac{1}{2}\operatorname{Tr}(\alpha^2) = \frac{1}{2}\sum_{i=1}^{n} \alpha_i^2 + \frac{1}{2}\sum_{(i,j)\in I_{1,2}} \frac{c^2}{(a_i - a_j)^2},$$

$$H_h = \sum_{i=1}^{n} \alpha_i \frac{\partial}{\partial a_i} + 2\sum_{(i,j)\in I_{1,2}} \frac{c^2}{(a_i - a_j)^3} \frac{\partial}{\partial \alpha_i},$$

$$\ddot{a}_i = 2\sum_{\{j:(i,j)\in I_{1,2}\}} \frac{c^2}{(a_i - a_j)^3}.$$

(6) *The case of general momentum Y and regular A.* Starting again with some regular A', consider (A', α') with $J(A', \alpha') = Y'$, choose $g \in SU(n)$ such that $A = gA'g^{-1} = \operatorname{diag}(a_1 > a_2 > \cdots > a_n)$, and let $\alpha = g\alpha' g^{-1}$ and $Y = gY'g^{-1} = [A, \alpha]$. Then the entry of the commutator is $Y_{ij} = [A, \alpha]_{ij} = \alpha_{ij}(a_i - a_j)$; thus $Y_{ii} = 0$. We may pass to the coordinates a_i and $\alpha_i := \alpha_{ii}$ for $1 \leq i \leq n$ on the one hand, and Y_{ij} for $i \neq j$ on the other hand, with the linear relation $Y_{ji} = -\overline{Y_{ij}}$ and with $n - 1$ nonzero entries

$Y_{ij} > 0$ with $i > j$ (chosen in lexicographic order) by applying the remaining isotropy group $SU(n)_A = S(U(1)^n) = \{\operatorname{diag}(e^{\sqrt{-1}\vartheta_1}, \dots, e^{\sqrt{-1}\vartheta_n}) : \sum \vartheta_i \in 2\pi\mathbb{Z}\}$. This choice of coordinates (a_i, α_i, Y_{ij}) shows that the reduced phase space $J^{-1}(\mathcal{O}(Y))/SU(n)$ is stratified symplectomorphic to $T^*C^o \times ((\mathcal{O}(Y) \cap \mathfrak{su}(n)_A^{\perp})/SU(n)_A)$; see [**86**], [**85**] and [**87**]. In these coordinates, the Hamiltonian function is as follows:

$$\begin{aligned}
h(A,\alpha) &= \tfrac{1}{2}\operatorname{Tr}(\alpha^2)\\
&= \frac{1}{2}\sum_i \alpha_i^2 - \frac{1}{2}\sum_{i\neq j}\frac{Y_{ij}Y_{ji}}{(a_i-a_j)^2},\\
dh &= \sum_i \alpha_i\, d\alpha_i + \sum_{i\neq j}\frac{Y_{ij}Y_{ji}}{(a_i-a_j)^3}(da_i - da_j) - \frac{1}{2}\sum_{i\neq j}\frac{dY_{ij}.Y_{ji} + Y_{ij}.dY_{ji}}{(a_i-a_j)^2}\\
&= \sum_i \alpha_i\, d\alpha_i + 2\sum_{i\neq j}\frac{Y_{ij}Y_{ji}}{(a_i-a_j)^3}da_i - \sum_{i\neq j}\frac{Y_{ji}}{(a_i-a_j)^2}dY_{ij}. \qquad (7)
\end{aligned}$$

The invariant symplectic form on $TH(n)$ pulls back, in these coordinates, to the symplectic form which is the product of the following two structures. The first one is $\omega_{(A,\alpha)}((A'\alpha'), (A'', \alpha'')) = \operatorname{Tr}(A'\alpha'' - A''\alpha') = \sum(a_i'\alpha_i'' - a_i''\alpha_i')$ which equals $\sum_i da_i \wedge d\alpha_i$. The second one comes by reduction from the Poisson structure on $\mathfrak{su}(n)$ which is given by

$$\begin{aligned}
P_Y(U,V) &= \operatorname{Tr}(Y[U,V]) = \sum_{m,n,p}(Y_{mn}U_{np}V_{pm} - Y_{mn}V_{np}U_{pm}),\\
P_Y &= \sum_{i\neq j, k\neq l} P_Y(dY_{ij}, dY_{kl})\partial_{Y_{ij}} \otimes \partial_{Y_{kl}}\\
&= \sum_{i\neq j, k\neq l}\sum_{m,n}(Y_{mn}\delta_{ni}\delta_{jk}\delta_{lm} - Y_{mn}\delta_{nk}\delta_{li}\delta_{jm})\partial_{Y_{ij}} \otimes \partial_{Y_{kl}}\\
&= \sum_{i\neq j, k\neq l}(Y_{li}\delta_{jk} - Y_{jk}\delta_{li})\partial_{Y_{ij}} \otimes \partial_{Y_{kl}}.
\end{aligned}$$

Since this Poisson 2-vector field is tangent to the orbit $\mathcal{O}(Y)$ and is $SU(n)$-invariant, we can push it down to the stratified orbit space

$$(\mathcal{O}(Y) \cap \mathfrak{su}(n)_A^{\perp})/SU(n)_A.$$

The latter space is the singular reduction of $\mathcal{O}(Y)$ with respect to the $SU(n)_A$-action. There it maps dY_{ij} to (remember that $Y_{ii} = 0$)

$$\check{P}_Y(dY_{ij}) = \sum_{k\neq l}(Y_{li}\delta_{jk} - Y_{jk}\delta_{li})\partial_{Y_{kl}} = \sum_k (Y_{ki}\partial_{Y_{jk}} - Y_{jk}\partial_{Y_{ki}}).$$

So the Hamiltonian vector field is

$$
\begin{aligned}
H_h &= \sum_i \alpha_i\, \partial_{a_i} - 2\sum_{i\neq j} \frac{Y_{ij}Y_{ji}}{(a_i-a_j)^3}\,\partial_{\alpha_i} \\
&\quad - \sum_{i\neq j} \frac{Y_{ji}}{(a_i-a_j)^2} \sum_k (Y_{ki}\,\partial_{Y_{jk}} - Y_{jk}\,\partial_{Y_{ki}}) \\
&= \sum_i \alpha_i\, \partial_{a_i} - 2\sum_{i\neq j} \frac{Y_{ij}Y_{ji}}{(a_i-a_j)^3}\,\partial_{\alpha_i} + \sum_{i,j,k} \left(\frac{Y_{ji}Y_{jk}}{(a_i-a_j)^2} - \frac{Y_{ij}Y_{kj}}{(a_j-a_k)^2} \right) \partial_{Y_{ki}}.
\end{aligned}
$$

The differential equation thus becomes (remember that $Y_{jj}=0$):

$$
\begin{aligned}
\dot a_i &= \alpha_i, \\
\dot\alpha_i &= -2\sum_j \frac{Y_{ij}Y_{ji}}{(a_i-a_j)^3} \\
&= 2\sum_j \frac{|Y_{ij}|^2}{(a_i-a_j)^3}, \\
\dot Y_{ki} &= \sum_j \left(\frac{Y_{ji}Y_{jk}}{(a_i-a_j)^2} - \frac{Y_{ij}Y_{kj}}{(a_j-a_k)^2} \right).
\end{aligned}
$$

Consider the matrix Z with $Z_{ii}=0$ and $Z_{ij}=Y_{ij}/(a_i-a_j)^2$. Then the differential equations become:

$$
\begin{aligned}
\ddot a_i &= 2\sum_j \frac{|Y_{ij}|^2}{(a_i-a_j)^3}, \\
\dot Y &= [Z, Y^*].
\end{aligned}
$$

This is the Calogero-Moser integrable system with spin; see [**13**], [**14**], and [**86, 85**].

(8) *The case of general momentum Y and singular A.* Let us consider the situation of (6), when A is not regular. Let us assume again that one root, say $\varepsilon_{12}(A)=a_1-a_2$, vanishes on the diagonal matrix A. Consider α with $[A,\alpha]=Y$. From $Y_{ij}=[A,\alpha]_{ij}=\alpha_{ij}(a_i-a_j)$ we conclude that $Y_{ii}=0$ for all i and also $Y_{12}=0$. The isotropy group $SU(n)_A$ contains a subgroup $SU(2)$ corresponding to the first two coordinates and we may use this to move α into the form that $\alpha_{12}=0$ and $\varepsilon_{12}(\alpha)\geq 0$. Thus the tangent vector α with foot point A in the wall $\{\varepsilon_{12}=0\}$ is either tangent to the wall when $\alpha_1=\alpha_2$ or points into the interior of the chamber C when $\alpha_1>\alpha_2$. We can then use the same analysis as in (6) where we use now that $Y_{12}=0$.

In the general case, when some roots vanish, we get for the Hamiltonian function, vector field, and differential equation:

$$h(A,\alpha) = \tfrac{1}{2}\operatorname{Tr}(\alpha^2) = \frac{1}{2}\sum_i \alpha_i^2 + \frac{1}{2}\sum_{\{(i,j):a_i(0)\neq a_j(0)\}} \frac{|Y_{ij}|^2}{(a_i-a_j)^2},$$

$$H_h = \sum_i \alpha_i\partial_{a_i} + 2\sum_{(i,j):a_j(0)\neq a_i(0)} \frac{|Y_{ij}|^2}{(a_i-a_j)^3}\partial_{\alpha_i}$$

$$+ \sum_{(i,j):a_j(0)\neq a_i(0)}\sum_k \frac{Y_{ji}Y_{jk}}{(a_i-a_j)^2}\partial_{Y_{ki}} - \sum_{(j,k):a_j(0)\neq a_k(0)}\sum_i \frac{Y_{ij}Y_{kj}}{(a_j-a_k)^2}\partial_{Y_{ki}},$$

$$\ddot{a}_i = 2\sum_{j:a_j(0)\neq a_i(0)} \frac{|Y_{ij}|^2}{(a_i-a_j)^3}, \qquad \dot{Y} = [Z, Y^*],$$

where we use the same notation as above. It would be very interesting to investigate the reflection behavior of this curve at the walls.

34.17. Example: Symmetric matrices. We finally treat the action of $SO(n) = SO(n,\mathbb{R})$ on the space $S(n)$ of real symmetric matrices by conjugation. Following the method of (34.16.6) and (34.16.7), we get the following result. Let $t \mapsto A' + t\alpha'$ be a straight line in $S(n)$. Then the ordered set of eigenvalues $a_1(t), \dots, a_n(t)$ of $A' + t\alpha'$ is part of the integral curve of the following vector field:

$$H_h = \sum_i \alpha_i\partial_{a_i} + 2\sum_{(i,j):a_j(0)\neq a_i(0)} \frac{Y_{ij}^2}{(a_i-a_j)^3}\partial_{\alpha_i}$$

$$- \sum_{(i,j):a_i(0)\neq a_j(0)}\sum_k \frac{Y_{ij}Y_{jk}}{(a_i-a_j)^2}\partial_{Y_{ki}} + \sum_{(j,k):a_j(0)\neq a_k(0)}\sum_i \frac{Y_{ij}Y_{jk}}{(a_j-a_k)^2}\partial_{Y_{ki}},$$

$$\ddot{a}_i = 2\sum_{(i,j):a_j(0)\neq a_i(0)} \frac{Y_{ij}^2}{(a_i-a_j)^3},$$

$$\dot{Y} = [Z, Y], \qquad \text{where } Z_{ij} = -\frac{Y_{ij}}{(a_i-a_j)^2},$$

where we also note that $Y_{ij} = Z_{ij} = 0$ whenever $a_i(0) = a_j(0)$.

List of Symbols

(a, b) open interval or pair

$[a, b]$ closed interval

$[X, Y]$ Lie bracket, commutator, Frölicher-Nijehuis bracket

$\langle \alpha, X \rangle$ usually a duality $V^* \times V \to \mathbb{R}$

$\mathrm{Ad}(g)$ adjoint action of a Lie group on its Lie algebra

$\mathrm{ad}(X) = [X, \quad]$ adjoint derivative of a Lie algebra

$\alpha : J^r(M, N) \to M$ the source mapping of jets

$\beta : J^r(M, N) \to N$ the target mapping of jets

$B_x(r)$ open ball with center x and radius $r > 0$

$\mathrm{conj}_g(h) = ghg^{-1}$ conjugation in a Lie group

$\Gamma(E)$, also $\Gamma(E \to M)$ the space of smooth sections of a fiber bundle

$\mathbb{C}$ field of complex numbers

$C : TM \times_M TM \to TTM$ connection or horizontal lift

$C^\infty(M, \mathbb{R})$ or $C^\infty(M)$ the space of smooth functions on a manifold M

d usually the exterior derivative

(E, p, M, S), also simply E usually a fiber bundle with total space E, base M, and standard fiber S

$\exp$ exponential mapping from a Lie algebra to its Lie group

$\exp^g_x$ geodesic exponential mapping centered at x

Fl^X_t, also $\mathrm{Fl}(t, X)$ the flow of a vector field X

$K : TTM \to M$ the connector of a covariant derivative

G usually a general Lie group with multiplication $\mu : G \times G \to G$; we use
$$gh = \mu(g,h) = \mu_g(h) = \mu^h(g)$$

$\mathfrak{g} = \text{Lie}(G)$ usually a Lie algebra for a Lie group G

$\mathbb{H}$ skew field of quaternions

$\mathbb{I}_k$ or $\mathbb{I}_{\mathbb{R}^k}$ short for the $k \times k$-identity matrix $\text{Id}_{\mathbb{R}^k}$.

i_X insertion operator of a vector field in a form

$J^r(E)$ the bundle of r-jets of sections of a fiber bundle $E \to M$

$J^r(M,N)$ the bundle of r-jets of smooth functions from M to N

$j^r f(x)$, also $j^r_x f$ the r-jet of a mapping or function f

$\kappa_M : TTM \to TTM$ the canonical flip mapping

$\ell : G \times S \to S$ usually a left action

$\mathcal{L}_X$ Lie derivative along a vector field X

M usually a manifold

$\mu : G \times G \to G$ usually the multiplication on a Lie group, $\mu(g,h) = g.h = \mu_g(h) = \mu^h(g)$, so μ_g is left translation by g and μ^h is right translation by h

$\mathbb{N}$ natural numbers > 0

$\mathbb{N}_0$ nonnegative integers

∇_X, pronounced 'Nabla', covariant derivative along X

$\nu : G \to G, \nu(g) = g^{-1}$ usually the inversion on a Lie group

$\text{Pt}(c,t)$ parallel transport along a curve c from time 0 to time t

$p : P \to M$ or (P,p,M,G) a principal bundle with structure group G

$\pi^r_l : J^r(M,N) \to J^l(M,N)$ projections of jets

$\mathbb{R}$ field of real numbers

$r : P \times G \to P$ usually a right action, in particular the principal right action of a principal bundle

TM the tangent bundle of a manifold M with projection $\pi_M : TM \to M$

$Tf : TM \to TN$ tangent mapping of $f : M \to N$

$\mathbb{Z}$ integers

Bibliography

1. R. Abraham, *Lectures of Smale on differential topology*, Columbia University, New York, 1962, Lecture Notes.
2. R. Abraham and J. Marsden, *Foundations of mechanics*, Addison-Wesley, 1978.
3. A. Alekseev, A. Malkin, and E. Meinrenken, *Lie group valued moment maps*, J. Differ. Geom **48** (1998), 445–495.
4. D. Alekseevsky, A. Kriegl, M. Losik, and P.W. Michor, *The Riemannian geometry of orbit spaces. The metric, geodesics, and integrable systems*, Publicationes Mathematicae **62** (2003), 247–276.
5. D. Alekseevsky and P. Michor, *Differential geometry of $\mathfrak{g}$-manifolds*, Diff. Geom. Appl. **5** (1995), 371–403.
6. D. Alekseevsky, P.W. Michor, and W.A.F. Ruppert, *Extensions of Lie algebras*, math.DG/0005042, 2000.
7. ———, *Extensions of super Lie algebras*, J. Lie Theory **15** (2005), 125–134, math.QA/0101190.
8. R. Almeida and P. Molino, *Suites d'atiyah et feuilletages transversalement complets*, C. R. Acad. Sci. Paris **300, Ser. 1** (1985), 13–15.
9. W. Ambrose and I.M. Singer, *A theorem on holonomy*, Trans. Amer. Math. Soc. **75** (1953), 428–443.
10. V.I. Arnold, *Mathematical methods of classical mechanics*, Springer-Verlag, New York, 1978.
11. M. Atiyah, R. Bott, and V.K. Patodi, *On the heat equation and the index theorem*, Inventiones Math. **19** (1973), 279–330.
12. J.A. Azcárraga and J.M. Izquierdo, *Lie groups, Lie algebras, cohomology and some applications in physics*, Cambridge University Press, Cambridge, UK, 1995, Cambridge Monographs on Mathematical Physics.
13. O. Babelon and M. Talon, *The symplectic structure of the spin Calogero model*, Phys. Lett. A **236, 5-6** (1997), 462–468, arXiv:q-alg/9707011.
14. ———, *The symplectic structure of rational Lax pair*, Phys. Lett. A **257, 3-4** (1999), 139–144.

15. R. Baer, *Erweiterungen von Gruppen und ihren Isomorphismen*, Math. Zeit. **38** (1934), 375–416.

16. L. Bates and E. Lerman, *Proper group actions and symplectic stratified spaces*, Pacific J. Math. **181** (1997), 201–229.

17. A.L. Besse, *Einstein manifolds*, Springer-Verlag, Berlin, 1987, Ergebnisse der Mathematik, 3. Folge, Band 10.

18. E. Bierstone and P.D. Milman, *Composite differentiable functions*, Ann. of Math. **116** (1982), 541–558.

19. E. Bierstone, P.D. Milman, and W. Pawłucki, *Composite differentiable functions*, Duke Math. J. **83** (1996), 607–620.

20. J. Boman, *Differentiability of a function and of its composition with a function of one variable*, Math. Scand. **20** (1967), 249–268.

21. B. Boos and D.D. Bleeker, *Topology and analysis*, Springer-Verlag, New York, 1985 (1977).

22. R. Bott, *Non-degenerate critical manifolds*, Ann. Math. **60** (1954), 248–261.

23. R. Bott and H. Samelson, *Applications of the theory of Morse to symmetric spaces*, Amer. J. Math. **80** (1958), 964–1029.

24. J.-P. Bourguignon, *Une stratification de l'espace des structures Riemanniens*, Compositio Mathematica **30** (1975), 1–41.

25. G.E. Bredon, *Introduction to compact transformation groups*, Academic Press, New York, 1972.

26. T. Bröcker and K. Jänich, *Einfürung in die Differentialtopologie*, Springer-Verlag, Heidelberg, 1973.

27. K.S. Brown, *Cohomology of groups*, Springer-Verlag, New-York, 1982, Graduate Texts 87.

28. H. Bursztyn, M. Crainic, A. Weinstein, and C. Zhu, *Integration of twisted Dirac brackets*, Duke Math. J. **123** (2004), 549–607.

29. H. Bursztyn and O. Radko, *Gauge equivalence of Dirac structures*, Ann. Inst. Fourier **53** (2003), 309–337.

30. A. Cabras and A. M. Vinogradov, *Extensions of the Poisson bracket to differential forms and multi-vector fields*, J. Geom. Physics **9** (1992), 75–100.

31. E. Calabi, *Sur les extensions des groupes topologiques*, Brioschi Annali di Mat. Pura e Appl. Ser 4 **32** (1951), 295–370.

32. A. Cap and J. Slovak, *Parabolic geometries I: Background and general theory*, AMS, Providence, 2009, Surveys and Monographs.

33. É. Cartan, *Leçons sur la géométrie des espaces de riemann*, Gauthiers-Villars, Paris, 1928.

34. R. Cohen, *Immersions of manifolds*, Proc. Nat. Acad. Sci. USA **79** (1982), 3390–3392.

35. L. Conlon, *Variational completeness and K-transversal domains*, J. Differential Geom. **5** (1971), 135–147.

36. ———, *A class of variationally complete representations*, J. Differential Geom. **7** (1972), 149–160.

37. T. Courant, *Dirac manifolds*, Trans. AMS **319** (1990), 631–661.

38. J. Dadok, *Polar coordinates induced by actions of compact Lie groups*, TAMS **288** (1985), 125–137.

39. J. Dadok and V. Kac, *Polar representations*, J. of Algebra **92** (1985), 504–524.

40. M. Davis, *Smooth g-manifolds as collections of fiber bundles*, Pac. J. Math. **77** (1978), 315–363.

41. J. Dieudonne, *Foundations of modern analysis I*, Academic Press, 1960.

42. S. Donaldson, *An application of gauge theory to the topology of 4-manifolds*, J. Diff. Geo. **18** (1983), 269–316.

43. J.J. Duistermaat and J.A.C. Kolk, *Lie groups*, Springer-Verlag, 1999.

44. J.L. Dupont, *Curvature and characteristic classes*, Lecture Notes in Mathematics, vol. 640, Springer-Verlag, Berlin, 1978.

45. D. Ebin, *The manifold of Riemannian metrics*, Proc. Symp. Pure Math. AMS **15** (1970), 11–40.

46. D.G. Ebin and J.E. Marsden, *Groups of diffeomorphisms and the motion of an incompressible fluid*, Ann. Math. **92** (1970), 102–163.

47. D.J. Eck, *Gauge-natural bundles and generalized gauge theories*, Mem. Amer. Math. Soc. **247** (1981), vi + 48.

48. ______, *Product preserving functors on smooth manifolds*, J. Pure and Applied Algebra **42** (1986), 133–140.

49. J. Eells, *A setting for global analysis*, Bull AMS **72** (1966), 571–807.

50. J. Eells and L. Lemaire, *A report on harmonic maps*, Bull. London Math. Soc. **10** (1978), 1–68.

51. J Eells and L. Lemaire, *Selected topics in harmonic mappings*, vol. 50, Am. Math. Soc., Providence, 1983.

52. ______, *Another report on harmonic maps*, Bull. London Math. Soc. **20** (1988), 385–524.

53. C. Ehresmann, *Les prolongements d'une variété différentiable I, II, III*, C. R. Acad. Sci. Paris **233** (1951), I: 598–600, II: 777–779, III: 1081–1083.

54. ______, *Gattungen von Lokalen Strukturen*, Jahresberichte der Deutschen Math. Ver. **60-2** (1957), 49–77.

55. S. Eilenberg, *Foundations of fiber bundles*, University of Chicago, Summer 1957.

56. S. Eilenberg and S. MacLane, *Cohomology theory in abstract groups, II. Group extensions with non-abelian kernel*, Ann. Math. (2) **48** (1947), 326–341.

57. T. Eisenkölbl, Personal communication, 2002.

58. C. Emmrich and H. Römer, *Orbifolds as configuration spaces of systems with gauge-symmetries*, Commun. Math. Phys. **129** (1990), 69–94.

59. D.B.A. Epstein, *Natural tensors on Riemannian manifolds*, J. Diff. Geom. **10**, 631–645.

60. D.B.A. Epstein and W.P. Thurston, *Transformation groups and natural bundles*, Proc. London Math. Soc. **38** (1979), 219–236.

61. D.S. Freed and D. Groisser, *The basic geometry of the manifold of Riemannian metrics and of its quotient by the diffeomorphism group*, Michigan Math. J. **36** (1989), 323–344.

62. M.H. Freedman, *The topology of four dimensional manifolds*, J. Diff. Geo. **17** (1982), 357–454.

63. M.H. Freedman and F. Luo, *Selected applications of geometry to low-dimensional topology*, University Lecture Series, vol. 1, Amer. Math. Soc., 1989.

64. A. Frölicher and A. Kriegl, *Linear spaces and differentiation theory*, J. Wiley, Chichester, 1988, Pure and Applied Mathematics.
65. A. Frölicher and A. Nijenhuis, *Theory of vector valued differential forms. Part I*, Indagationes Math **18** (1956), 338–359.
66. P. Garcia Pérez, *Gauge algebras, curvature, and symplectic structure*, J. Diff. Geom. **12** (1977), 209–227.
67. I.M. Gelfand and I.Ya. Dorfman, *Hamiltonian operators and the classical Yang-Baxter equation*, Funct. Anal. Appl. **16** (1982), 241–248.
68. F. Gesztesy and H. Holden, *Soliton equations and their algebro-geometric solutions, volume I:* $(1+1)$*-dimensional continuous models*, Cambridge Studies in Advanced Mathematics, vol. 79, Cambridge University Press, Cambridge, 2003.
69. F. Gesztesy, H. Holden, J. Michor, and G. Teschl, *Soliton equations and their algebro-geometric solutions, volume II:* $(1+1)$*-dimensional discrete models*, Cambridge Studies in Advanced Mathematics, Cambridge University Press, Cambridge, 2008.
70. O. Gil-Medrano and P.W. Michor, *The Riemannian manifold of all Riemannian metrics*, Quarterly J. Math. Oxford **42** (1991), 183–202.
71. O. Gil-Medrano, P.W. Michor, and M. Neuwirther, *Pseudoriemannian metrics on spaces of bilinear structures*, Quarterly J. Math. Oxford **43** (1992), 201–221.
72. P.B. Gilkey, *The spectral geometry of a Riemannian manifold*, J. Diff. Geom. **10** (1975), 601–618.
73. ———, *Invariance theory, the heat equation and the index theorem*, Publish or Perish, Wilmington, Del. (USA), 1984.
74. J. Giraud, *Cohomologie non abélienne*, Springer-Verlag, Berlin, 1971, Grundlehren 179.
75. G. Glaeser, *Fonctions composées différentiables*, Ann. of Math. **77** (1963), 193–209.
76. Yu. Gliklikh, *Global analysis in mathematical physics. Geometric and stochastic methods*, Springer-Verlag, New York, 1997, Aplied Mathematical Sciences.
77. R. Godement, *Topologie algébrique et théorie des fausceaux*, Hermann, Paris, 1958.
78. R. Gompf, *Three exotic* $\mathbb{R}^4$*'s and other anomalies*, J. Diff. Geo. **18** (1983), 317–328.
79. J. Grabowski, $\mathbf{Z}$*-graded extensions of Poisson brackets*, Rev. Math. Phys. **9** (1997), no. 1, 1–27.
80. W. Greub, S. Halperin, and R. Vanstone, *Connections, curvature, and cohomology I, II, III*, Academic Press, New York and London, 1972, 1973, 1976.
81. J. Hadamard, *Les surfaces à courbures opposées et leurs lignes géodésiques*, J. Math. Pures Appl. **4** (1898), 27–73.
82. S. Helgason, *Differential geometry, lie groups, and symmetric spaces*, Graduate Studies in Mathematics, vol. 34, AMS, Providence, 1978, 2001.
83. P.J. Hilton and U. Stammbach, *A course in homological algebra*, Graduate Texts in Mathematics, vol. 4, Springer-Verlag, New York, 1971, 1997.
84. M.W. Hirsch, *Differential topology*, Springer-Verlag, New York, 1976, GTM 33.
85. S. Hochgerner, *Spinning particles in a Yang-Mills field*, math.SG/0602062.
86. ———, *Singular cotangent bundle reduction & spin Calogero-Moser systems*, Diff. Geom. Appl. (2008), math.SG/0411068.
87. S. Hochgerner and A. Rainer, *Singular Poisson reduction of cotangent bundles*, Rev. Mat. Complut. **19** (2006), 431–466, math.SG/0508455.

88. G. Hochschild, *Group extensions of Lie groups I, II*, Ann Math. **54** (1951), 96–109; 537–551.
89. ———, *Cohomology classes of finite type and finite dimensional kernels for Lie algebras*, Am. J. Math. **76** (1954), 763–778.
90. ———, *The structure of Lie groups*, Holden Day, San Francisco, 1965.
91. G.P. Hochschild and J.-P. Serre, *Cohomology of group extensions*, Trans. AMS **74** (1953), 110–134.
92. ———, *Cohomology of Lie algebras*, Ann. Math. **57** (1953), 591–603.
93. O. Hölder, *Die Gruppen der Ordnungen p^3, pq^2, pqr, p^4*, Mathematische Annalen **43** (1893), 301–412.
94. L. Hörmander, *The analysis of linear partial differential operators I*, Springer-Verlag, Berlin, 1983, Grundlehren 256.
95. J.E. Humphreys, *Reflection groups and Coxeter groups*, Cambridge Studies in Advanced Mathematics, vol. 29, Cambridge University Press, Cambridge, 1990, 1992.
96. N. Jacobson, *Lie algebras*, J. Wiley-Interscience, 1962.
97. H Joris, *Une c^∞-application non-immersive qui possède la propriété universelle des immersions*, Archiv Math. **39** (1982), 269–277.
98. H. Joris and E. Preissmann, *Germes pseudo-immersifs de* $(\mathbb{R}^n, 0)$ *dans* $(\mathbb{R}^n, 0)$, C. R. Acad. Sci., Paris, Ser. I **305** (1987), 261–263.
99. ———, *Pseudo-immersions*, Ann. Inst. Fourier **37** (1987), 195–221.
100. G. Kainz, A. Kriegl, and P.W. Michor, *c^∞-algebras from the functional analytic viewpoint*, J. Pure Appl. Algebra **46** (1987), 89–107.
101. G. Kainz and P.W. Michor, *Natural transformations in differential geometry*, Czechoslovak Math. J. **37** (1987), 584–607.
102. F. Kamber and P.W. Michor, *Completing Lie algebra actions to Lie group actions*, Electron. Res. Announc. Amer. Math. Soc. **10** (2004), 1–10.
103. D. Kazhdan, B. Kostant, and S. Sternberg, *Hamiltonian group actions and dynamical systems of Calogero type*, Comm. Pure Appl. Math. **31** (1978), 481–501.
104. R. Kerner, *Generalization of the Kaluza-Klein theory for an arbitrary non-Abelian group*, Ann. Inst. Henri Poincare **9** (1968), 143–152.
105. A.A. Kirillov, *Elements of the theory of representations*, Springer-Verlag, Berlin, 1976.
106. S. Kobayashi, *Riemannian manifolds without conjugate points*, Ann. Mat. Pura Appl. **53** (1961), 149–155.
107. S. Kobayashi and K. Nomizu, *Foundations of differential geometry. I, II*, J. Wiley – Interscience, I: 1963, II: 1969.
108. I. Kolář, P.W. Michor, and J. Slovák, *Natural operations in differential geometry*, Springer-Verlag, Berlin, Heidelberg, New York, 1993.
109. Y. Kosmann-Schwarzbach, *Derived brackets*, Lett. Math. Phys. **69** (2004), 61–87.
110. B. Kostant, *Orbits, symplectic structures, and representation theory*, p. 71, Nippon Hyoronsha, Tokyo, 1966.
111. J.L. Koszul, *Sur certains groupes de transformations de Lie*, pp. 137–141, Centre National de la Recherche Scientifique, Paris, 1953.
112. ———, *Crochet de Schouten-Nijenhuis et cohomologie*, Astérisque (1985), no. Numero Hors Serie, 257–271, The mathematical heritage of Élie Cartan (Lyon, 1984).

113. A. Kriegl and P.W. Michor, *The convenient setting for global analysis*, Surveys and Monographs, vol. 53, AMS, Providence, 1997.

114. J. Kubarski, *About Stefan's definition of a foliation with singularities: a reduction of the axioms*, Bull. Soc. Math. France **118** (1990), 391–394.

115. M. Kummer, *On the construction of the reduced phase space of a Hamiltonian system with symmetry*, Univ. Math. J. **30** (1981), pp. 281–291.

116. H. Lawson and M.L. Michelsohn, *Spin geometry*, Princeton University Press, Princeton, 1989.

117. P. Lecomte, *Sur la suite exacte canonique associée à un fibré principal*, Bul. Soc. Math. France **13** (1985), 259–271.

118. ——, *On some sequence of graded Lie algebras asociated to manifolds*, Ann. Global Analysis Geom. **12** (1994), 183–192.

119. P. Lecomte, P.W. Michor, and H. Schicketanz, *The multigraded Nijenhuis-Richardson algebra, its universal property and application*, J. Pure Appl. Algebra **77** (1992), 87–102.

120. P. Lecomte and C. Roger, *Sur les déformations des algèbres de courants de type réductif*, C. R. Acad. Sci. Paris, I **303** (1986), 807–810.

121. E. Lerman, R. Montgomery, and R. Sjamaar, *Examples of singular reduction*, London Math. Soc. Lecture Note Ser., vol. 192, pp. 127–155, Cambridge Univ. Press, Cambridge, 1993.

122. P. Libermann and C.M. Marle, *Symplectic geometry and analytical mechanics*, D. Reidel, Dordrecht, 1987.

123. A. Lichnerowicz, *Les variétés de Poisson et leur algébres de Lie associées*, J. Diff. Geom. **12** (1977), 253–300.

124. S. Lie, *Theorie der Transformationsgruppen. Zweiter Abschnitt*, Teubner, Leipzig, 1890.

125. D. Luna, *Fonctions différentiables invariantes sous l'operation d'un groupe réductif*, Ann. Inst. Fourier **26** (1976), 33–49.

126. K. Mackenzie, *Lie groupoids and Lie algebroids in differential geometry*, London Math. Soc. Lecture Notes Ser., vol. 124, Cambridge Univ. Press, Cambridge, 1987.

127. G.W. Mackey, *Induced representations of locally compact groups I*, Ann. Math. **55** (1952), 101–139.

128. ——, *Les ensembles boréliens et les extensions des groupes*, J. Math. **36** (1957), 171–178.

129. F. Magri and C. Morosi, *A geometrical characterization of integrable Hamiltonian systems through the theory of Poisson-Nijenhuis manifolds*, Quaderno, Univ. Milano **19** (1984).

130. B. Malgrange, *Ideals of differentiable functions*, Oxford Univ. Press, 1966.

131. L. Mangiarotti and M. Modugno, *Graded Lie algebras and connections on a fibred space*, Journ. Math. Pures et Appl. **83** (1984), 111–120.

132. G. Marmo, E. Saletan, A. Simoni, and B. Vitale, *Dynamical systems. A differential geometric approach to symmetry and reduction*, Wiley-Interscience, Chichester, 1985.

133. J. Marsden and M. Perlmutter, *The orbit bundle picture of cotangent bundle reduction*, C. R. Math. Acad. Sci. Soc. R. Can. **22** (2000), 35–54.

134. J. Marsden and T. Ratiu, *Introduction to mechanics and symmetry*, Springer-Verlag, New York, 1999.

135. J. Marsden and A. Weinstein, *Some comments on the history, theory, and applications of symplectic reduction*, Quantization of singular symplectic quotients, Birkhäuser, 2001, Progress in Math. 198.

136. J.N. Mather, *Notes on topological stability*, Harvard, 1970, mimeographed notes.

137. ______, *Differentiable invariants*, Topology **16** (1977), 145–155.

138. M. Mauhart, *Iterierte Lie Ableitungen und Integrabilität*, Master's thesis, Univ. Wien, 1990.

139. M. Mauhart and P.W. Michor, *Commutators of flows and fields*, Archivum Mathematicum (Brno) **28** (1992), 228–236.

140. F. Michor, *Die Mathematik der Planetenbewegung*, BRG Klosterneuburg, 2000, Fachbereichsarbeit aus Mathematik.

141. P.W. Michor, *Manifolds of differentiable mappings*, Shiva, Orpington, 1980.

142. ______, *Manifolds of smooth mappings IV: Theorem of de Rham*, Cahiers Top. Geo. Diff. **24** (1983), 57–86.

143. ______, *A convenient setting for differential geometry and global analysis I, II*, Cahiers Topol. Geo. Diff. **25** (1984), 63–109, 113–178.

144. ______, *A generalisation of Hamiltonian mechanics*, J. Geometry and Physics **2** (1985), 67–82.

145. ______, *Remarks on the Frölicher-Nijenhuis bracket*, Proceedings of the Conference on Differential Geometry and its Applications, Brno 1986, D. Reidel, 1987, pp. 197–220.

146. ______, *Remarks on the Schouten-Nijenhuis bracket*, Suppl. Rendiconti del Circolo Matematico di Palermo, Serie II, **16** (1987), 208–215.

147. ______, *Gauge theory for diffeomorphism groups*, Proceedings of the Conference on Differential Geometric Methods in Theoretical Physics (Como, 1987) (K Bleuler and M Werner, eds.), Kluwer, 1988, pp. 345–371.

148. ______, *Graded derivations of the algebra of differential forms associated with a connection*, Differential Geometry, Proceedings (Peniscola, 1988), Lecture Notes in Math., vol. 1410, Springer-Verlag, 1989, pp. 249–261.

149. ______, *Knit products of graded Lie algebras and groups*, Suppl. Rendiconti Circolo Matematico di Palermo, Ser. II **22** (1989), 171–175.

150. ______, *The moment mapping for a unitary representation*, J. Global Analysis and Geometry **8** (1990), 299–313.

151. ______, *Gauge theory for fiber bundles*, Monographs and Textbooks in Physical Sciences, Lecture Notes, vol. 19, Bibliopolis, Napoli, 1991.

152. ______, *The relation between systems and associated bundles*, Ann. Mat. Pura Appl. **163** (1993), 385–399.

153. ______, *Basic differential forms for actions of Lie groups*, Proc. AMS **124** (1996), 1633–1642.

154. ______, *The Jacobi flow*, Rend. Sem. Mat. Univ. Pol. Torino **54** (1996), 365–372.

155. ______, *Basic differential forms for actions of Lie groups II*, Proc. AMS **125** (1997), 2175–2177.

156. J. Milnor, *On manifolds homeomorphic to the 7-sphere*, Annals of Math. **64** (1956), 399–405.

157. J. Milnor and J. Stasheff, *Characteristic classes*, Annals of Math. Studies, vol. 76, Princeton University Press, Princeton, 1974.

158. I. Moerdijk and Reyes G.E., *Rings of smooth funcions and their localizations I*, J. Algebra **99** (1986), 324–336.

159. ———, *Rings of smooth funcions and their localizations II*, Mathematical logic and theoretical computer science (College Park, Md., 1984–1985), Lecture Notes in Pure and Appl. Math., vol. 106, Dekker, New York, 1987, pp. 277–300.

160. J. Monterde, *Generalized symplectomorphisms*, Lecture Notes in Math, vol. 1410, pp. 262–271, Springer-Verlag, Berlin, Peñiscola, 1988, 1989.

161. J. Monterde and A. Montesinos, *Integral curves of derivations*, J. Global Analysis and Geometry **6** (1988), 177–189.

162. D. Montgomery and L. Zippin, *Transformation groups*, J. Wiley-Interscience, New York, 1955.

163. M. Mori, *On the three-dimensional cohomology group of Lie algebras*, J. Math. Soc. Japan **5** (1953), 171–183.

164. A. Morimoto, *Prolongations of geometric structures*, Math. Inst. Nagoya University, Chiknsa-Ku, Nagoya, Japan, 1969.

165. J. Morrow, *The denseness of complete Riemannian metrics*, J. Diff. Geo. **4** (1970), 225–226.

166. M. Morse, *The behavior of a function on its critical set*, Ann. Math. **40** (1939), 345–396.

167. J. Moser, *Three integrable Hamiltonian systems connected with isospectral deformations*, Adv. Math. **16** (1975), 197–220.

168. J. Nagata, *Modern dimension theory*, North Holland, 1965.

169. M. Nagata, *On the 14-th problem of Hilbert*, Amer. J. Math. **81** (1959), 766–772.

170. K.-H. Neeb, *A note on central extensions of Lie groups*, J. Lie Theory **6** (1996), 207–213.

171. ———, *Holomorphy and convexity in Lie theory*, de Gruyter, Berlin, 1999.

172. A. Newlander and L. Nirenberg, *Complex analytic coordinates in almost complex manifolds*, Ann. of Math. **65** (1957), 391–404.

173. A. Nijenhuis, *On the holonomy groups of linear connections. IA. IB*, Indagationes Math. **15** (1953), 233–240, 241–249.

174. ———, *On the holonomy groups of linear connections. II*, Indagationes Math. **16** (1954), 17–25.

175. ———, *Jacobi-type identities for bilinear differential concomitants of certain tensor fields I*, Indagationes Math. **17** (1955), 390–403.

176. ———, *Natural bundles and their general properties*, Differential Geometry in Honor of K. Yano, Kinokuniya, Tokio, 1972, pp. 317–334.

177. A. Nijenhuis and R. Richardson, *Deformation of Lie algebra structures*, J. Math. Mech. **17** (1967), 89–105.

178. K. Nomizu and H. Ozeki, *The existence of complete Riemannian metrics*, Proc. AMS **12** (1961), 889–891.

179. M. Olshanetskii and A. Perelomov, *Geodesic flows on symmetric spaces of zero curvature and explicit solution of the generalized Calogero model for the classical case*, Funct. Anal. Appl. **10, 3** (1977), 237–239.

180. B. O'Neill, *The fundamental equations of a submersion*, Mich. Math. J. **13** (1966), 459–469.

181. A.L. Onishchik, *Topology of transitive transformation groups*, Johann Ambrosius Barth, Leipzig, Berlin, Heidelberg, 1994.

182. A.L. Onishchik and E. Vinberg, *Lie groups and Lie algebras I*, Encyclopedia of Math. Sciences, vol. 20, Springer-Verlag, Heidelberg, 1993.

183. J.-P. Ortega and T Ratiu, *Singular reduction of Poisson manifolds*, Lett. Math. Phys. **46** (1998), 359–372.

184. J.-P. Ortega and T. Ratiu, *Momentum maps and Hamiltonian reduction*, Progress in Mathematics, vol. 222, Birkhäuser, Boston, 2004.

185. R.S. Palais, *A global formulation of the Lie theory of transformation groups*, Mem. AMS **22** (1957).

186. ______, *On the existence of slices for the actions of non-compact Lie groups*, Ann. of Math. **73** (1961), 295–323.

187. R.S. Palais and C.L. Terng, *Natural bundles have finite order*, Topology **16** (1977), 271–277.

188. ______, *A general theory of canonical forms*, Trans. AMS **300** (1987), 771–789.

189. ______, *Critical point theory and submanifold geometry*, Springer-Verlag, Berlin, 1988, Lecture Notes in Mathematics 1353.

190. S. Parameswaran, *Skew-symmetric determinants*, Amer. Math. Monthly **61** (1954), 116.

191. R. Penrose and W. Rindler, *Spinors and space-time, vol. 1., vol. 2*, Cambridge University Press, 1987, 1988.

192. A. Perelomov, *Integrable systems of classical mechanics and Lie algebras*, Birkhäuser, Basel, 1990.

193. H.V. Pitie, *Characteristic classes of foliations*, Research Notes in Mathematics, vol. 10, Pitman, London, 1976.

194. A. Pressley and G. Segal, *Loop groups*, Oxford Mathematical Monographs, Oxford University Press, 1986.

195. F. Quinn, *Ends III*, J. Diff. Geo. **17** (1982), 503–521.

196. A. Sard, *The measure of the critical points of differentiable maps*, Bull. AMS **48** (1942), 883–890.

197. G. Sartori, *A theorem on orbit structures (strata) of compact linear Lie groups*, J. Math. Phys. **24** (1983), 765–768.

198. H. Schicketanz, *On derivations and cohomology of the Lie algebra of vector valued forms related to a smooth manifold*, Bull. Soc. Roy. Sci. Liége **57** (1988), 599–617.

199. J.A. Schouten, *Über Differentialkonkomitanten zweier kontravarianten Grössen*, Indagationes Math. **2** (1940), 449–452.

200. O. Schreier, *Über die Erweiterungen von Gruppen I*, Monatsh. Math. Physik **34** (1926), 164–180.

201. ______, *Über die Erweiterungen von Gruppen II*, Hamburger Abh. **4** (1926), 321–346.

202. I. Schur, *Über eine Klasse von Mittelbildungen mit Anwendungen in der Determinantentheorie*, Sitzungsberichte der Berliner Mathematischen Gesellschaft **22** (1923), 9–20.

203. G.W. Schwarz, *Smooth functions invariant under the action of a compact Lie group*, Topology **14** (1975), 63–68.

204. R.T. Seeley, *Extension of c^∞-functions defined in a half space*, Proc. AMS **15** (1964), 625–626.

205. J.-P. Serre, *Cohomologie des groupes discrets*, Séminaire Bourbaki **399** (1970/71).

206. ———, *Cohomologie des groupes discrets*, Ann. of Math. Studies **70** (1971), 77–169, Princeton University Press.

207. P. Shanahan, *The Atiyah-Singer index theorem*, Lecture Notes in Mathematics, vol. 638, Springer-Verlag, Berlin, 1978.

208. U. Shukla, *A cohomology for Lie algebras*, J. Math. Soc. Japan **18** (1966), 275–289.

209. R. Sjamaar and E. Lerman, *Stratified symplectic spaces and reduction*, Ann. Math. **134** (1991), 375–422.

210. J. Slovák, *Peetre theorem for nonlinear operators*, Annals of Global Analysis and Geometry **6** (1988), 273–283.

211. L. Solomon, *Invariants of finite reflection groups*, Nagoya Math. J. **22** (1963), 57–64.

212. J.M. Souriau, *Quantification géométrique*, Comm. Math. Phys. **1** (1966), 374–398.

213. P. Stefan, *Accessible sets, orbits, and foliations with singularities*, Proc. London Math. Soc. **29** (1974), 699–713.

214. R. Stöcker and H. Zieschang, *Algebraische Topologie*, Teubner, Stuttgart, 1988.

215. J. Szenthe, *A generalization of the Weyl group*, Acta Math. Hungarica **41** (1983), 347–357.

216. ———, *Orthogonally transversal submanifolds and the generalizations of the Weyl group*, Period. Math. Hungarica **15** (1984), 281–299.

217. C.L. Terng, *Natural vector bundles and natural differential operators*, American J. of Math. **100** (1978), 775–828.

218. ———, *A convexity theorem for isoparametric submanifolds*, Invent. Math. **85** (1986), 487–492.

219. ———, *Isoparametric submanifolds and their Coxeter groups*, J. Diff. Geom. **1985** (21), 79–107.

220. J.C. Tougeron, *Idéaux de fonctions différentiables*, Springer-Verlag, 1972, Ergebnisse d. Math. 71.

221. W.M. Tulczyjew, *The graded Lie algebra of multivector fields and the generalized Lie derivative of forms*, Bull. Acad. Polon. Sci. **22, 9** (1974), 937–942.

222. I. Vaisman, *Lectures on the geometry of Poisson manifolds*, Progress in Mathematics, vol. 118, Birkhäuser Verlag, Basel, 1994.

223. V.S. Varadarajan, *Lie groups, Lie algebras, and their representations*, Prentice-Hall, Springer-Verlag, Englewood Cliffs, N.J., New York, 1974 1984, 2nd edition.

224. A. Weil, *Théorie des points proches sur les variétés differentielles*, Colloque de topologie et géométrie différentielle (Strasbourg), 1953, pp. 111–117.

225. A. Weinstein, *A universal phase space for particles in a Yang-Mills field*, Lett. Math. Phys. **2** (1978), 417–420.

226. H. Whitney, *Analytic extensions of differentiable functions defined in closed sets*, Trans. AMS **36** (1934), 63–89.

227. ———, *Differentiable even functions*, Duke Math. J. **10** (1943), 159–166.

228. ———, *The selfintersections of a smooth n-manifold in $2n$-space*, Annals of Math. **45** (1944), 220–293.

229. S. Wong, *Field and particle equations for the classical Yang-Mills field and particles with isotopic spin*, Il Nuovo Cimento **65A** (1970), 689–694.

230. H. Yamabe, *On an arcwise connected subgroup of a Lie group*, Osaka Math. J. **2** (1950), 13–14.

Index

Titles in This Series

93 **Peter W. Michor,** Topics in differential geometry, 2008
92 **I. Martin Isaacs,** Finite group theory, 2008
91 **Louis Halle Rowen,** Graduate algebra: Noncommutative view, 2008
90 **Larry J. Gerstein,** Basic quadratic forms, 2008
89 **Anthony Bonato,** A course on the web graph, 2008
88 **Nathanial P. Brown and Narutaka Ozawa,** C*-algebras and finite-dimensional approximations, 2008
87 **Srikanth B. Iyengar, Graham J. Leuschke, Anton Leykin, Claudia Miller, Ezra Miller, Anurag K. Singh, and Uli Walther,** Twenty-four hours of local cohomology, 2007
86 **Yulij Ilyashenko and Sergei Yakovenko,** Lectures on analytic differential equations, 2007
85 **John M. Alongi and Gail S. Nelson,** Recurrence and topology, 2007
84 **Charalambos D. Aliprantis and Rabee Tourky,** Cones and duality, 2007
83 **Wolfgang Ebeling,** Functions of several complex variables and their singularities (translated by Philip G. Spain), 2007
82 **Serge Alinhac and Patrick Gérard,** Pseudo-differential operators and the Nash–Moser theorem (translated by Stephen S. Wilson), 2007
81 **V. V. Prasolov,** Elements of homology theory, 2007
80 **Davar Khoshnevisan,** Probability, 2007
79 **William Stein,** Modular forms, a computational approach (with an appendix by Paul E. Gunnells), 2007
78 **Harry Dym,** Linear algebra in action, 2007
77 **Bennett Chow, Peng Lu, and Lei Ni,** Hamilton's Ricci flow, 2006
76 **Michael E. Taylor,** Measure theory and integration, 2006
75 **Peter D. Miller,** Applied asymptotic analysis, 2006
74 **V. V. Prasolov,** Elements of combinatorial and differential topology, 2006
73 **Louis Halle Rowen,** Graduate algebra: Commutative view, 2006
72 **R. J. Williams,** Introduction the the mathematics of finance, 2006
71 **S. P. Novikov and I. A. Taimanov,** Modern geometric structures and fields, 2006
70 **Seán Dineen,** Probability theory in finance, 2005
69 **Sebastián Montiel and Antonio Ros,** Curves and surfaces, 2005
68 **Luis Caffarelli and Sandro Salsa,** A geometric approach to free boundary problems, 2005
67 **T.Y. Lam,** Introduction to quadratic forms over fields, 2004
66 **Yuli Eidelman, Vitali Milman, and Antonis Tsolomitis,** Functional analysis, An introduction, 2004
65 **S. Ramanan,** Global calculus, 2004
64 **A. A. Kirillov,** Lectures on the orbit method, 2004
63 **Steven Dale Cutkosky,** Resolution of singularities, 2004
62 **T. W. Körner,** A companion to analysis: A second first and first second course in analysis, 2004
61 **Thomas A. Ivey and J. M. Landsberg,** Cartan for beginners: Differential geometry via moving frames and exterior differential systems, 2003
60 **Alberto Candel and Lawrence Conlon,** Foliations II, 2003
59 **Steven H. Weintraub,** Representation theory of finite groups: algebra and arithmetic, 2003
58 **Cédric Villani,** Topics in optimal transportation, 2003

TITLES IN THIS SERIES

57 **Robert Plato,** Concise numerical mathematics, 2003

56 **E. B. Vinberg,** A course in algebra, 2003

55 **C. Herbert Clemens,** A scrapbook of complex curve theory, second edition, 2003

54 **Alexander Barvinok,** A course in convexity, 2002

53 **Henryk Iwaniec,** Spectral methods of automorphic forms, 2002

52 **Ilka Agricola and Thomas Friedrich,** Global analysis: Differential forms in analysis, geometry and physics, 2002

51 **Y. A. Abramovich and C. D. Aliprantis,** Problems in operator theory, 2002

50 **Y. A. Abramovich and C. D. Aliprantis,** An invitation to operator theory, 2002

49 **John R. Harper,** Secondary cohomology operations, 2002

48 **Y. Eliashberg and N. Mishachev,** Introduction to the h-principle, 2002

47 **A. Yu. Kitaev, A. H. Shen, and M. N. Vyalyi,** Classical and quantum computation, 2002

46 **Joseph L. Taylor,** Several complex variables with connections to algebraic geometry and Lie groups, 2002

45 **Inder K. Rana,** An introduction to measure and integration, second edition, 2002

44 **Jim Agler and John E. McCarthy,** Pick interpolation and Hilbert function spaces, 2002

43 **N. V. Krylov,** Introduction to the theory of random processes, 2002

42 **Jin Hong and Seok-Jin Kang,** Introduction to quantum groups and crystal bases, 2002

41 **Georgi V. Smirnov,** Introduction to the theory of differential inclusions, 2002

40 **Robert E. Greene and Steven G. Krantz,** Function theory of one complex variable, third edition, 2006

39 **Larry C. Grove,** Classical groups and geometric algebra, 2002

38 **Elton P. Hsu,** Stochastic analysis on manifolds, 2002

37 **Hershel M. Farkas and Irwin Kra,** Theta constants, Riemann surfaces and the modular group, 2001

36 **Martin Schechter,** Principles of functional analysis, second edition, 2002

35 **James F. Davis and Paul Kirk,** Lecture notes in algebraic topology, 2001

34 **Sigurdur Helgason,** Differential geometry, Lie groups, and symmetric spaces, 2001

33 **Dmitri Burago, Yuri Burago, and Sergei Ivanov,** A course in metric geometry, 2001

32 **Robert G. Bartle,** A modern theory of integration, 2001

31 **Ralf Korn and Elke Korn,** Option pricing and portfolio optimization: Modern methods of financial mathematics, 2001

30 **J. C. McConnell and J. C. Robson,** Noncommutative Noetherian rings, 2001

29 **Javier Duoandikoetxea,** Fourier analysis, 2001

28 **Liviu I. Nicolaescu,** Notes on Seiberg-Witten theory, 2000

27 **Thierry Aubin,** A course in differential geometry, 2001

26 **Rolf Berndt,** An introduction to symplectic geometry, 2001

25 **Thomas Friedrich,** Dirac operators in Riemannian geometry, 2000

24 **Helmut Koch,** Number theory: Algebraic numbers and functions, 2000

23 **Alberto Candel and Lawrence Conlon,** Foliations I, 2000

22 **Günter R. Krause and Thomas H. Lenagan,** Growth of algebras and Gelfand-Kirillov dimension, 2000

21 **John B. Conway,** A course in operator theory, 2000

For a complete list of titles in this series, visit the
AMS Bookstore at **www.ams.org/bookstore/**.